Computer Vision
A Modern Approach

David A. Forsyth
University of California at Berkeley

Jean Ponce
University of Illinois at Urbana-Champaign

An Alan R. Apt Book

Prentice Hall

Prentice Hall
Upper Saddle River, New Jersey 07458

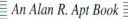

Library of Congress Cataloging-in-Publication Data

CIP data on file.

Vice President and Editorial Director, ECS: *Marcia J. Horton*
Publisher: *Alan Apt*
Associate Editor: *Toni D. Holm*
Editorial Assistant: *Patrick Lindner*
Vice President and Director of Production and Manufacturing, ESM: *David W. Riccardi*
Executive Managing Editor: *Vince O'Brien*
Assistant Managing Editor: *Camille Trentacoste*
Production Editor: *Leslie Galen*
Director of Creative Services: *Paul Belfanti*
Creative Director: *Carole Anson*
Art Director: *Wanda Espana*
Art Editor: *Greg Dulles*
Manufacturing Manager: *Trudy Pisciotti*
Manufacturing Buyer: *Lynda Castillo*
Marketing Manager: *Pamela Shaffer*
Marketing Assistant: *Barrie Reinhold*

About the cover: Image courtesy of Gamma/Superstock

Prentice Hall

Pearson Education, Inc.
Upper Saddle River, NJ 07458

Printed in the United States of America

10 9 8 7 6 5 4 3 2 1

ISBN 0-13-085198-1

Pearson Education Ltd., *London*
Pearson Education Australia Pty. Ltd., *Sydney*
Pearson Education Singapore, Pte. Ltd.
Pearson Education North Asia Ltd., *Hong Kong*
Pearson Education Canada, Inc., *Toronto*
Pearson Educacíon de Mexico, S.A. de C.V.
Pearson Education–Japan, *Tokyo*
Pearson Education Malaysia, Pte. Ltd.
Pearson Education, *Upper Saddle River, New Jersey*

To my family—DAF

To Camille and Oscar—JP

Contents

Preface

Computer vision as a field is an intellectual frontier. Like any frontier, it is exciting and disorganised; there is often no reliable authority to appeal to—many useful ideas have no theoretical grounding, and some theories are useless in practice; developed areas are widely scattered, and often one looks completely inaccessible from the other. Nevertheless, we have attempted in this book to present a fairly orderly picture of the field.

We see computer vision—or just "vision"; apologies to those who study human or animal vision—as an enterprise that uses statistical methods to disentangle data using models constructed with the aid of geometry, physics and learning theory. Thus, in our view, vision relies on a solid understanding of cameras and of the physical process of image formation (part I of this book) to obtain simple inferences from individual pixel values (part II), combine the information available in multiple images into a coherent whole (part III), impose some order on groups of pixels to separate them from each other or infer shape information (part IV), and recognize objects using geometric information (part V) or probabilistic techniques (part VI). Computer vision has a wide variety of applications, old (e.g., mobile robot navigation, industrial inspection, and military intelligence) and new (e.g., human computer interaction, image retrieval in digital libraries, medical image analysis, and the realistic rendering of synthetic scenes in computer graphics). We discuss some of these applications in part VII.

WHY STUDY VISION?

Computer vision's great trick is extracting descriptions of the world from pictures or sequences of pictures. This is unequivocally useful. Taking pictures is usually non-destructive and sometimes discreet. It is also easy and (now) cheap. The descriptions that users seek can differ widely between applications. For example, a technique known as structure from motion makes it possible to extract a representation of what is depicted and how the camera moved from a series of pictures. People in the entertainment industry use these techniques to build three-dimensional (3D) computer models of buildings, typically keeping the structure and throwing away the motion. These models are used where real buildings cannot be; they are set fire to, blown up, etc. Good, simple, accurate and convincing models can be built from quite small sets of photographs. People who wish to control mobile robots usually keep the motion and throw away the structure. This is because they generally know something about the area where the robot is working, but don't usually know the precise robot location in that area. They can determine it from information about how a camera bolted to the robot is moving.

There are a number of other, important applications of computer vision. One is in medical imaging: One builds software systems that can enhance imagery, or identify important phenomena or events, or visualize information obtained by imaging. Another is in inspection: One takes pictures of objects to determine whether they are within specification. A third is in interpreting satellite images, both for military purposes—a program might be required to determine what militarily interesting phenomena have occurred in a given region recently; or what damage was caused by a bombing—and for civilian purposes—what will this year's maize crop be? How much rainforest is left? A fourth is in organizing and structuring collections of pictures. We know how to search and browse text libraries (though this is a subject that still has difficult open questions) but don't really know what to do with image or video libraries.

Computer vision is at an extraordinary point in its development. The subject itself has been around since the 1960s, but it is only recently that it has been possible to build useful computer systems using ideas from computer vision. This flourishing has been driven by several trends: Computers and imaging systems have become very cheap. Not all that long ago, it took tens of thousands of dollars to get good digital color images; now it takes a few hundred, at most. Not all that long ago, a color printer was something one found in few, if any, research labs; now they are in many homes. This means it is easier to do research. It also means that there are many people with problems to which the methods of computer vision apply. For example, people would like to organize their collection of photographs, make 3D models of the world around them, and manage and edit collections of videos. Our understanding of the basic geometry and physics underlying vision and, what is more important, what to do about it, has improved significantly. We are beginning to be able to solve problems that lots of people care about, but none of the hard problems have been solved and there are plenty of easy ones that have not been solved either (to keep one intellectually fit while trying to solve hard problems). It is a great time to be studying this subject.

What Is in This Book?

This book covers what we feel a computer vision professional ought to know. However, it is addressed to a wider audience. We hope that those engaged in computational geometry, computer graphics, image processing, imaging in general, and robotics will find it an informative reference. We have tried to make the book accessible to senior undergraduates or graduate students with a passing interest in vision. Each chapter covers a different part of the subject, and, as a glance at Table 1 will confirm, chapters are relatively independent. This means that one can dip into the book as well as read it from cover to cover. Generally, we have tried to make chapters run from easy material at the start to more arcane matters at the end. Each chapter has brief notes at the end, containing historical material and assorted opinions. We have tried to produce a book that describes ideas that are useful, or likely to be so in the future. We have put emphasis on understanding the basic geometry and physics of imaging, but have tried to link this with actual applications. In general, the book reflects the enormous recent influence of geometry and various forms of applied statistics on computer vision.

A reader who goes from cover to cover will hopefully be well informed, if exhausted; there is too much in this book to cover in a one-semester class. Of course, prospective (or active) computer vision professionals should read every word, do all the exercises, and report any bugs found for the second edition (of which it is probably a good idea to plan buying a copy!). While the study of computer vision does not require deep mathematics, it does require facility with a lot of different mathematical ideas. We have tried to make the book self contained, in the sense that readers with the level of mathematical sophistication of an engineering senior should be comfortable with the material of the book, and should not need to refer to other texts. We have also tried to keep the mathematics to the necessary minimum—after all, this book is about computer vision, not applied mathematics—and have chosen to insert what mathematics we have kept in the main chapter bodies instead of a separate appendix.

Generally, we have tried to reduce the interdependence between chapters, so that readers interested in particular topics can avoid wading through the whole book. It is not possible to make each chapter entirely self contained, and Table 1 indicates the dependencies between chapters.

TABLE 1 Dependencies between chapters: It will be difficult to read a chapter if you don't have a good grasp of the material in the chapters it "requires." If you have not read the chapters labeled "helpful," you may need to look one or two things up.

Part	Chapter	Requires	Helpful
I	1: Cameras		
	2: Geometric camera models	1	
	3: Geometric camera calibration	2	
	4: Radiometry—measuring light		
	5: Sources, shadows and shading		4, 1
	6: Color		5
II	7: Linear filters		
	8: Edge detection	7	
	9: Texture	7	8
III	10: The geometry of multiple views	3	
	11: Stereopsis	10	
	12: Affine structure from motion	10	
	13: Projective structure from motion	12	
IV	14: Segmentation by clustering		9, 6, 5
	15: Segmentation by fitting a model		14
	16: Segmentation and fitting using probabilistic methods		15,10
	17: Tracking with linear dynamic models		
V	18: Model-based vision	3	
	19: Smooth surfaces and their outlines	2	
	20: Aspect graphs	19	
	21: Range data		20, 19, 3
VI	22: Finding templates using classifiers		9, 8, 7, 6, 5
	23: Recognition by relations between templates		9, 8, 7, 6, 5
	24: Geometric templates from spatial relations	2, 1	16, 15, 14
VII	25: Application: Finding in digital libraries		16, 15, 14, 6
	26: Application: Image-based rendering	10	13, 12, 11, 6, 5, 3, 2, 1

What Is Not in This Book

The computer vision literature is vast, and it was not easy to produce a book about computer vision that can be lifted by ordinary mortals. To do so, we had to cut material, ignore topics, and so on. We cut two entire chapters close to the last moment: One is an introduction to probability and inference, the other an account of methods for tracking objects with non-linear dynamics. These chapters appear on the book's web page `http://www.cs.berkeley.edu/~daf/book.html`.

We left out some topics because of personal taste, or because we became exhausted and stopped writing about a particular area, or because we learned about them too late to put them in, or because we had to shorten some chapter, or any of hundreds of other reasons. We have tended to omit detailed discussions of material that is mainly of historical interest, and offer instead some historical remarks at the end of each chapter. Neither of us claims to be a fluent intellectual archaeologist, meaning that ideas may have deeper histories than we have indicated. We just didn't get around to writing up deformable templates and mosaics, two topics of considerable practical importance; we will try to put them into the second edition.

ACKNOWLEDGMENTS

In preparing this book, we have accumulated a significant set of debts. A number of anonymous reviewers have read several drafts of the book and have made extremely helpful contributions. We are grateful to them for their time and efforts. Our editor, Alan Apt, organized these reviews with the help of Jake Warde. We thank them both. Leslie Galen, Joe Albrecht, and Dianne Parish, of Integre Technical Publishing, helped us over numerous issues with proofreading and illustrations. Some images used herein were obtained from IMSI's Master Photos Collection, 1895 Francisco Blvd. East, San Rafael, CA 94901-5506, USA. In preparing the bibliography, we have made extensive use of Keith Price's excellent computer vision bibliography, which can be found at `http://iris.usc.edu/Vision-Notes/bibliography/contents.html`.

Both the overall coverage of topics and several chapters were reviewed by various colleagues, who made valuable and detailed suggestions for their revision. We thank Kobus Barnard, Margaret Fleck, David Kriegman, Jitendra Malik and Andrew Zisserman. A number of our students contributed suggestions, ideas for figures, proofreading comments, and other valuable material. We thank Okan Arikan, Sébastien Blind, Martha Cepeda, Stephen Chenney, Frank Cho, Yakup Genc, John Haddon, Sergey Ioffe, Svetlana Lazebnik, Cathy Lee, Sung-il Pae, David Parks, Fred Rothganger, Attawith Sudsang, and the students in several offerings of our vision classes at U.C. Berkeley and UIUC. We have been very lucky to have colleagues at various universities use (often rough) drafts of our book in their vision classes. Institutions whose students suffered through these drafts include, besides ours, Carnegie-Mellon University, Stanford University, the University of Wisconsin at Madison, the University of California at Santa Barbara and the University of Southern California; there may be others we are not aware of. We are grateful for all the helpful comments from adopters, in particular Chris Bregler, Chuck Dyer, Martial Hebert, David Kriegman, B.S. Manjunath, and Ram Nevatia, who sent us many detailed and helpful comments and corrections. The book has also benefitted from comments and corrections from Aydin Alaylioglu, Srinivas Akella, Marie Banich, Serge Belongie, Ajit M. Chaudhari, Navneet Dalal, Richard Hartley, Glen Healey, Mike Heath, Hayley Iben, Stéphanie Jonquières, Tony Lewis, Benson Limketkai, Simon Maskell, Brian Milch, Tamara Miller, Cordelia Schmid, Brigitte and Gerry Serlin, Ilan Shimshoni, Eric de Sturler, Camillo J. Taylor, Jeff Thompson, Claire Vallat, Daniel S. Wilkerson, Jinghan Yu, Hao Zhang, and Zhengyou Zhang. If you find an apparent typographic error, please email DAF (daf@cs.berkeley.edu) with the details, using the phrase "book typo" in your email; we will try to credit the first finder of each typo in the second edition.

We also thank P. Besl, B. Boufama, J. Costeira, P. Debevec, O. Faugeras, Y. Genc, M. Hebert, D. Huber, K. Ikeuchi, A.E. Johnson, T. Kanade, K. Kutulakos, M. Levoy, S. Mahamud, R. Mohr, H. Moravec, H. Murase, Y. Ohta, M. Okutami, M. Pollefeys, H. Saito, C. Schmid, S. Sullivan, C. Tomasi, and M. Turk for providing the originals of some of the figures shown in this book.

DAF acknowledges a wide range of intellectual debts, starting at kindergarten. Important figures in the very long list of his creditors include Gerald Alanthwaite, Mike Brady, Tom Fair, Margaret Fleck, Jitendra Malik, Joe Mundy, Mike Rodd, Charlie Rothwell and Andrew Zisserman. JP cannot even remember kindergarten, but acknowledges his debts to Olivier Faugeras, Mike Brady, and Tom Binford. He also wishes to thank Sharon Collins for her help. Without her, this book, like most of his work, probably would have never been finished. Both authors would also like to acknowledge the profound influence of Jan Koenderink's writings on their work at large and on this book in particular.

SAMPLE SYLLABI

The whole book can be covered in two (rather intense) semesters, by starting at the first page and plunging on. Ideally, one would cover one application chapter—probably the chapter on image-based rendering—in the first semester, and the other one in the second. Few departments will experience heavy demand for so detailed a sequence of courses. We have tried to structure this book so that instructors can choose areas according to taste. Sample syllabi for busy 15-week semesters appear in Tables 2 to 6, structured according to needs that can reasonably be expected. We would encourage (and expect!) instructors to rearrange these according to taste.

Table 2 contains a suggested syllabus for a one-semester introductory class in computer vision for seniors or first-year graduate students in computer science, electrical engineering, or other engineering or science disciplines. The students receive a broad presentation of the field, including application areas such as digital libraries and image-based rendering. Although the

TABLE 2 A one-semester introductory class in computer vision for seniors or first-year graduate students in computer science, electrical engineering, or other engineering or science disciplines.

Week	Chapter	Sections	Key topics
1	1, 4	1.1, 4 (summary only)	pinhole cameras, radiometric terminology
2	5	5.1–5.5	local shading models; point, line and area sources; photometric stereo
3	6	all	color
4	7, 8	7.1–7.5, 8.1–8.3	linear filters; smoothing to suppress noise; edge detection
5	9	all	texture: as statistics of filter outputs; synthesis; shape from
6	10, 11	10.1, 11	basic multi-view geometry; stereo
7	14	all	segmentation as clustering
8	15	15.1–15.4	fitting lines, curves; fitting as maximum likelihood; robustness
9	16	16.1,16.2	hidden variables and EM
10	17	all	tracking with a Kalman filter; data association
11	2, 3	2.1, 2.2, all of 3	camera calibration
12	18	all	model-based vision using correspondence and camera calibration
13	22	all	template matching using classifiers
14	23	all	matching on relations
15	25, 26	all	finding images in digital libraries; image based rendering

TABLE 3 A syllabus for students of computer graphics who want to know the elements of vision that are relevant to their topic.

Week	Chapter	Sections	Key topics
1	1, 4	1.1, 4 (summary only)	pinhole cameras, radiometric terminology
2	5	5.1–5.5	local shading models; point, line and area sources; photometric stereo
3	6.1–6.4	all	color
4	7, 8	7.1–7.5, 8.1–8.3	linear filters; smoothing to suppress noise; edge detection
5	9	9.1–9.3	texture: as statistics of filter outputs; synthesis
6	2, 3	2.1, 2.2, all of 3	camera calibration
7	10, 11	10.1, 11	basic multi-view geometry; stereo
8	12	all	affine structure from motion
9	13	all	projective structure from motion
10	26	all	image-based rendering
11	15	all	fitting; robustness; RANSAC
12	16	all	hidden variables and EM
13	19	all	surfaces and outlines
14	21	all	range data
15	17	all	tracking, the Kalman filter and data association

TABLE 4 A syllabus for students who are primarily interested in the applications of computer vision.

Week	Chapter	Sections	Key topics
1	1, 4	1.1, 4 (summary only)	pinhole cameras, radiometric terminology
2	5, 6	5.1,5.3, 5.4, 5.5, 6.1–6.4	local shading models; point, line and area sources; photometric stereo; color—physics, human perception, color spaces
3	2, 3	all	camera models and their calibration
4	7, 9	all of 7; 9.1–9.3	linear filters; texture as statistics of filter outputs; texture synthesis
5	10, 11	all	multiview geometry, stereo as an example
6	12,13	all	affine structure from motion; projective structure from motion
7	13, 26	all	projective structure from motion; image-based rendering
8	14	all	segmentation as clustering, particular emphasis on shot boundary detection and background subtraction
9	15	all	fitting lines, curves; robustness; RANSAC
10	16	all	hidden variables and EM
11	25	all	finding images in digital libraries
12	17	all	tracking, the Kalman filter and data association
13	18	all	model-based vision
14	22	all	finding templates using classifiers
15	20	all	range data

hardest theoretical material is omitted, there is a thorough treatment of the basic geometry and physics of image formation. We assume that students will have a wide range of backgrounds, and can be assigned background readings in probability (we suggest the chapter on the book's web page) around week 2 or 3. We have put off the application chapters to the end, but many may prefer to do chapter 20 around week 10 and chapter 21 around week 6.

Table 3 contains a syllabus for students of computer graphics who want to know the elements of vision that are relevant to their topic. We have emphasized methods that make it possible to recover object models from image information; understanding these topics needs a working knowledge of cameras and filters. Tracking is becoming useful in the graphics world, where it is particularly important for motion capture. We assume that students will have a wide range of backgrounds, and have some exposure to probability.

Table 4 shows a syllabus for students who are primarily interested in the applications of computer vision. We cover material of most immediate practical interest. We assume that students will have a wide range of backgrounds, and can be assigned background reading on probability around week 2 or 3.

Table 5 is a suggested syllabus for students of cognitive science or artificial intelligence who want a basic outline of the important notions of computer vision. This syllabus is less aggressively paced, and assumes less mathematical experience. Students will need to read some material on probability (e.g., the chapter on the book's web page) around week 2 or 3.

Table 6 shows a sample syllabus for students who have a strong interest in applied mathematics, electrical engineering or physics. This syllabus makes for a very busy semester; we move fast, assuming that students can cope with a lot of mathematical material. We assume that students will have a wide range of backgrounds, and can be assigned some reading on probability around week 2 or 3. As a break in a pretty abstract and demanding syllabus, we have inserted a brief review of digital libraries; the chapter on image-based rendering or that on range data could be used instead.

TABLE 5 For students of cognitive science or artificial intelligence who want a basic outline of the important notions of computer vision.

Week	Chapter	Sections	Key topics
1	1, 4	1, 4 (summary only)	pinhole cameras; lenses; cameras and the eye; radiometric terminology
2	5	all	local shading models; point, line and area sources; photometric stereo; interreflections; lightness computations
3	6	all	color: physics, human perception, spaces; image models; color constancy
4	7	7.1–7.5, 7.7	linear filters; sampling; scale
5	8	all	edge detection
6	9	all	texture; representation, synthesis, shape from
7	10.1,10.2	all	basic multiple view geometry
8	11	all	stereopsis
9	14	all	segmentation by clustering
10	15	all	fitting lines, curves; robustness; RANSAC
11	16	all	hidden variables and EM
12	18	all	model-based vision
13	22	all	finding templates using classifiers
14	23	all	recognition by relations between templates
15	24	all	geometric templates from spatial relations

TABLE 6 A syllabus for students who have a strong interest in applied mathematics, electrical engineering or physics.

Week	Chapter	Sections	Key topics
1	1, 4	all	cameras, radiometry
2	5	all	shading models; point, line and area sources; photometric stereo; interreflections and shading primitives
3	6	all	color:—physics, human perception, spaces, color constancy
4	2, 3	all	camera parameters and calibration
5	7, 8	all	linear filters and edge detection
6	8, 9	all	finish edge detection; texture: representation, synthesis, shape from
7	10, 11	all	multiple view geometry, stereopsis as an example
8	12, 13	all	structure from motion
9	14, 15	all	segmentation as clustering; fitting lines, curves; robustness; RANSAC
10	15, 16	all	finish fitting; hidden variables and EM
11	17, 25	all	tracking: Kalman filters, data association; finding images in digital libraries
12	18	all	model-based vision
13	19	all	surfaces and their outlines
14	20	all	aspect graphs
15	22	all	template matching

NOTATION

We use the following notation throughout the book: points, lines, and planes are denoted by Roman or Greek letters in italic font (e.g., P, Δ, or Π). Vectors are usually denoted by Roman or Greek bold-italic letters (e.g., $\boldsymbol{v}$, $\boldsymbol{P}$, or $\boldsymbol{\xi}$), but the vector joining two points P and Q is often denoted by $\overrightarrow{PQ}$. Lower-case letters are normally used to denote geometric figures in the image plane (e.g., p, $\boldsymbol{p}$, δ), and upper-case letters are used for scene objects (e.g., P, Π). Matrices are denoted by Roman letters in calligraphic font (e.g., $\mathcal{U}$).

The familiar three-dimensional Euclidean space is denoted by $\mathbb{E}^3$, and the vector space formed by n-tuples of real numbers with the usual laws of addition and multiplication by a scalar is denoted by $\mathbb{R}^n$, with $\mathbf{0}$ being used to denote the zero vector. Likewise, the vector space formed by $m \times n$ matrices with real entries is denoted by $\mathbb{R}^{m\times n}$. When $m = n$, Id is used to denote the identity matrix—that is, the $n \times n$ matrix whose diagonal entries are equal to 1 and nondiagonal entries are equal to 0. The transpose of the $m\times n$ matrix $\mathcal{U}$ with coefficients u_{ij} is the $n\times m$ matrix denoted by $\mathcal{U}^T$ with coefficients u_{ji}. Elements of $\mathbb{R}^n$ are often identified with column vectors or $n \times 1$ matrices, e.g., $\boldsymbol{a} = (a_1, a_2, a_3)^T$ is the transpose of a 1×3 matrix (or *row vector*), i.e., an 3×1 matrix (or *column vector*), or equivalently an element of $\mathbb{R}^3$.

The *dot product* (or *inner product*) of two vectors $\boldsymbol{a} = (a_1, \dots, a_n)^T$ and $\boldsymbol{b} = (b_1, \dots, b_n)^T$ in $\mathbb{R}^n$ is defined by

$$\boldsymbol{a} \cdot \boldsymbol{b} = a_1 b_1 + \cdots + a_n b_n,$$

and it can also be written as a matrix product, i.e., $\boldsymbol{a} \cdot \boldsymbol{b} = \boldsymbol{a}^T \boldsymbol{b} = \boldsymbol{b}^T \boldsymbol{a}$. We denote by $|\boldsymbol{a}|^2 = \boldsymbol{a} \cdot \boldsymbol{a}$ the square of the Euclidean norm of the vector $\boldsymbol{a}$ and denote by d the distance function induced by the Euclidean norm in $\mathbb{E}^n$, i.e., $d(P, Q) = |\overrightarrow{PQ}|$. Given a matrix $\mathcal{U}$ in $\mathbb{R}^{m \times n}$, we generally use $|\mathcal{U}|$ to denote its *Frobenius norm*, i.e., the square root of the sum of its squared entries.

When the vector $\boldsymbol{a}$ has unit norm, the dot product $\boldsymbol{a} \cdot \boldsymbol{b}$ is equal to the (signed) length of the projection of $\boldsymbol{b}$ onto $\boldsymbol{a}$. More generally,

$$\boldsymbol{a} \cdot \boldsymbol{b} = |\boldsymbol{a}|\,|\boldsymbol{b}|\,\cos\theta,$$

where θ is the angle between the two vectors, which shows that a necessary and sufficient condition for two vectors to be orthogonal is that their dot product be zero.

The *cross product* (or *outer product*) of two vectors $\boldsymbol{a} = (a_1, a_2, a_3)^T$ and $\boldsymbol{b} = (b_1, b_2, b_3)^T$ in $\mathbb{R}^3$ is the vector

$$\boldsymbol{a} \times \boldsymbol{b} \stackrel{\text{def}}{=} \begin{pmatrix} a_2 b_3 - a_3 b_2 \\ a_3 b_1 - a_1 b_3 \\ a_1 b_2 - a_2 b_1 \end{pmatrix}.$$

Note that $\boldsymbol{a} \times \boldsymbol{b} = [\boldsymbol{a}_\times]\boldsymbol{b}$, where

$$[\boldsymbol{a}_\times] \stackrel{\text{def}}{=} \begin{pmatrix} 0 & -a_3 & a_2 \\ a_3 & 0 & -a_1 \\ -a_2 & a_1 & 0 \end{pmatrix}.$$

The cross product of two vectors $\boldsymbol{a}$ and $\boldsymbol{b}$ in $\mathbb{R}^3$ is orthogonal to these two vectors, and a necessary and sufficient condition for $\boldsymbol{a}$ and $\boldsymbol{b}$ to have the same direction is that $\boldsymbol{a} \times \boldsymbol{b} = \boldsymbol{0}$. If θ denotes as before the angle between the vectors $\boldsymbol{a}$ and $\boldsymbol{b}$, it can be shown that

$$|\boldsymbol{a} \times \boldsymbol{b}| = |\boldsymbol{a}|\,|\boldsymbol{b}|\,|\sin\theta|.$$

PROGRAMMING ASSIGNMENTS AND RESOURCES

The programming assignments given throughout the book sometimes require routines for numerical linear algebra, singular value decomposition, and linear and nonlinear least squares. An extensive set of such routines is available in MATLAB as well as in public-domain libraries such as LINPACK, LAPACK, and MINPACK, which can be downloaded from the Netlib repository (`http://www.netlib.org/`). We offer some pointers to other software on the book's web page `http://www.cs.berkeley.edu/~daf/book.html`. Datasets—or pointers to datasets—for the programming assignment are also available there.

PART I
Image Formation and Image Models

1

Cameras

There are many types of imaging devices, from animal eyes to video cameras and radio telescopes. They may or may not be equipped with lenses. For example, the first models of the *camera obscura* (literally, dark chamber) invented in the 16th century did not have lenses, but instead used a *pinhole* to focus light rays onto a wall or translucent plate and demonstrate the laws of perspective discovered a century earlier by Brunelleschi. Pinholes were replaced by more and more sophisticated lenses as early as 1550, and the modern photographic or digital camera is essentially a camera obscura capable of recording the amount of light striking every small area of its backplane (Figure 1.1).

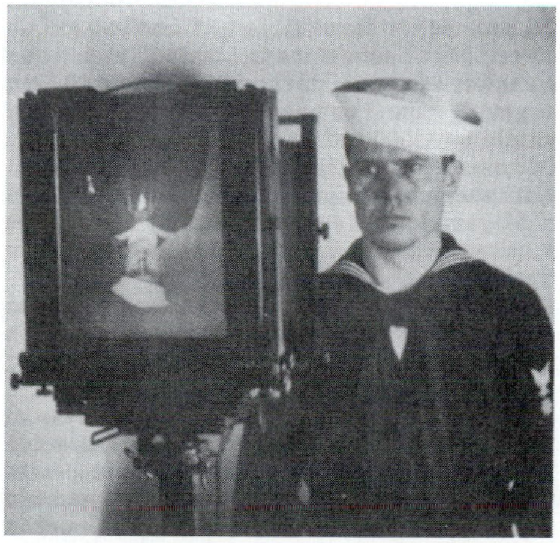

Figure 1.1 Image formation on the backplate of a photographic camera. *Figure from US NAVY MANUAL OF BASIC OPTICS AND OPTICAL INSTRUMENTS, prepared by the Bureau of Naval Personnel, reprinted by Dover Publications, Inc., (1969).*

The imaging surface of a camera is generally a rectangle, but the shape of the human retina is much closer to a spherical surface, and panoramic cameras may be equipped with cylindrical retinas. Imaging sensors have other characteristics. They may record a spatially discrete picture (like our eyes with their rods and cones, 35 mm cameras with their grain, and digital cameras with their rectangular picture elements or pixels) or a continuous one (in the case of old-fashioned TV tubes, for example). The signal that an imaging sensor records at a point on its retina may be discrete or continuous, and it may consist of a single number (black-and-white camera), a few values (e.g., the R G B intensities for a color camera or the responses of the three types of cones for the human eye), many numbers (e.g., the responses of hyperspectral sensors), or even a continuous function of wavelength (which is essentially the case for spectrometers). Examining these characteristics is the subject of this chapter.

1.1 PINHOLE CAMERAS

1.1.1 Perspective Projection

Imagine taking a box, pricking a small hole in one of its sides with a pin, and then replacing the opposite side with a translucent plate. If you hold that box in front of you in a dimly lit room, with the pinhole facing some light source (say a candle), you see an inverted image of the candle appearing on the translucent plate (Figure 1.2). This image is formed by light rays issued from the scene facing the box. If the pinhole were really reduced to a point (which is of course physically impossible), exactly one light ray would pass through each point in the plane of the plate (or *image plane*), the pinhole, and some scene point.

In reality, the pinhole has a finite (albeit small) size, and each point in the image plane collects light from a cone of rays subtending a finite solid angle, so this idealized and extremely simple model of the imaging geometry does not strictly apply. In addition, real cameras are normally equipped with lenses, which further complicates things. Still, the *pinhole perspective* (also called *central perspective*) projection model, first proposed by Brunelleschi at the beginning of the 15th century, is mathematically convenient. Despite its simplicity, it often provides an acceptable approximation of the imaging process. Perspective projection creates inverted images, and it is sometimes convenient to consider instead a *virtual image* associated with a plane lying *in front* of the pinhole at the same distance from it as the actual image plane (Figure 1.2). This virtual image is not inverted, but is otherwise strictly equivalent to the actual one. Depending on the context, it may be more convenient to think about one or the other. Figure 1.3(a) illustrates an obvious effect of perspective projection: The apparent size of objects depends on their distance. For example, the images B' and C' of the posts B and C have the same height, but A and C

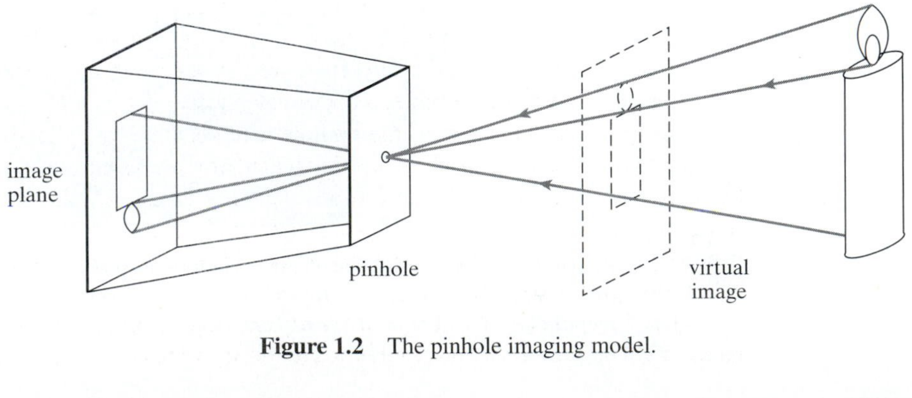

Figure 1.2 The pinhole imaging model.

(a)

(b)

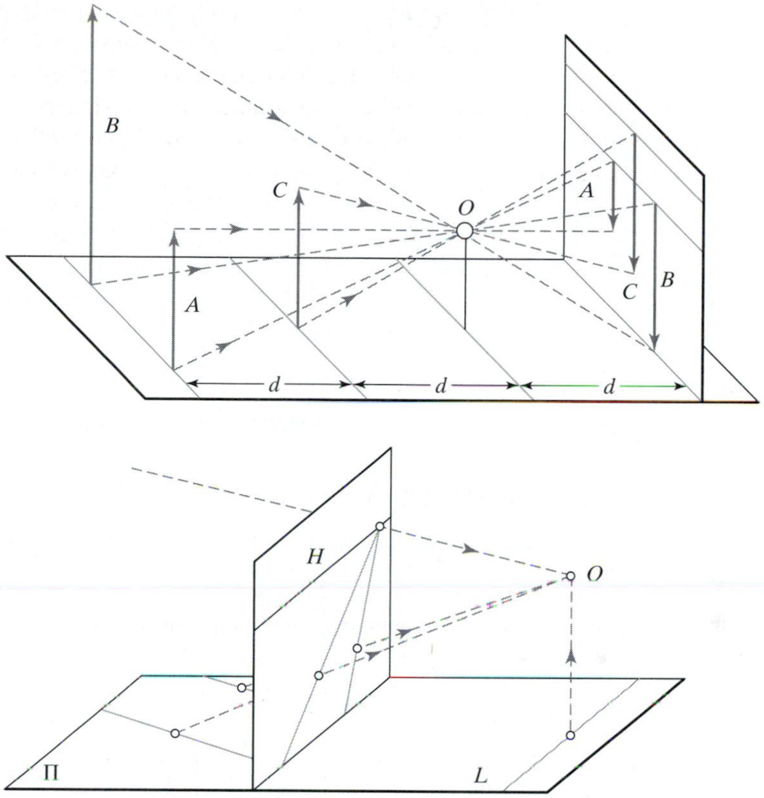

Figure 1.3 Perspective effects: (a) far objects appear smaller than close ones: the distance d from the pinhole O to the plane containing C is half the distance from O to the plane containing A and B; (b) the images of parallel lines intersect at the horizon (after Hilbert and Cohn-Vossen, 1952, Figure 127). Note that the image plane is *behind* the pinhole in (a) (physical retina), and *in front* of it in (b) (virtual image plane). Most of the diagrams in this chapter and the rest of this book feature the physical image plane, but a virtual one is also used when appropriate, as in (b).

are really half the size of B. Figure 1.3(b) illustrates another well-known effect: The projections of two parallel lines lying in some plane Π appear to converge on a horizon line H formed by the intersection of the image plane with the plane parallel to Π and passing through the pinhole. Note that the line L in Π that is parallel to the image plane has no image at all.

These properties are easy to prove in a purely geometric fashion. However, it is often convenient (if not quite as elegant) to reason in terms of reference frames, coordinates, and equations. Consider, for example, a coordinate system $(O, \boldsymbol{i}, \boldsymbol{j}, \boldsymbol{k})$ attached to a pinhole camera, whose origin O coincides with the pinhole, and vectors $\boldsymbol{i}$ and $\boldsymbol{j}$ form a basis for a vector plane parallel to the image plane Π', which is located at a positive distance f' from the pinhole along the vector $\boldsymbol{k}$ (Figure 1.4). The line perpendicular to Π' and passing through the pinhole is called the *optical axis*, and the point C' where it pierces Π' is called the *image center*. This point can be used as the origin of an image plane coordinate frame, and it plays an important role in camera calibration procedures.

Let P denote a scene point with coordinates (x, y, z) and P' denote its image with coordinates (x', y', z'). Since P' lies in the image plane, we have $z' = f'$. Since the three points P, O,

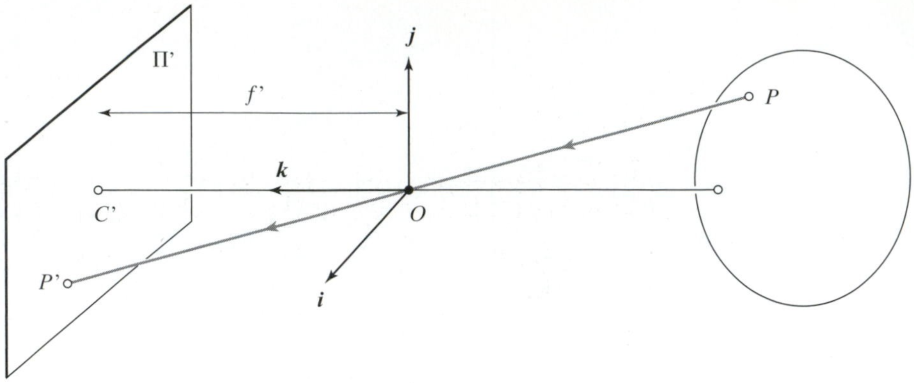

Figure 1.4 The perspective projection equations are derived in this section from the collinearity of the point P, its image P', and the pinhole O.

and P' are collinear, we have $\overrightarrow{OP'} = \lambda \overrightarrow{OP}$ for some number λ, so

$$\begin{cases} x' = \lambda x \\ y' = \lambda y \\ f' = \lambda z \end{cases} \iff \lambda = \frac{x'}{x} = \frac{y'}{y} = \frac{f'}{z},$$

and therefore

$$\begin{cases} x' = f' \dfrac{x}{z}, \\ y' = f' \dfrac{y}{z}. \end{cases} \tag{1.1}$$

1.1.2 Affine Projection

As noted in the previous section, pinhole perspective is only an approximation of the geometry of the imaging process. This section discusses a class of coarser approximations, called *affine projection models*, that are also useful on occasion. We focus on two specific affine models—namely, *weak-perspective* and *orthographic* projections. A third one, the *paraperspective* model, is introduced in Chapter 12, where the name affine projection is also justified.

Consider the *fronto-parallel plane* Π_0 defined by $z = z_0$ (Figure 1.5). For any point P in Π_0 we can rewrite the perspective projection Eq. (1.1) as

$$\begin{cases} x' = -mx \\ y' = -my \end{cases} \quad \text{where} \quad m = -\frac{f'}{z_0}. \tag{1.2}$$

Physical constraints impose that z_0 be negative (the plane must be in front of the pinhole), so the *magnification* m associated with the plane Π_0 is positive. This name is justified by the following remark: Consider two points P and Q in Π_0 and their images P' and Q' (Figure 1.5); obviously the vectors $\overrightarrow{PQ}$ and $\overrightarrow{P'Q'}$ are parallel, and we have $|\overrightarrow{P'Q'}| = m|\overrightarrow{PQ}|$. This is the dependence of image size on object distance noted earlier.

When the scene depth is small relative to the average distance from the camera, the magnification can be taken to be constant. This projection model is called *weak perspective* or *scaled*

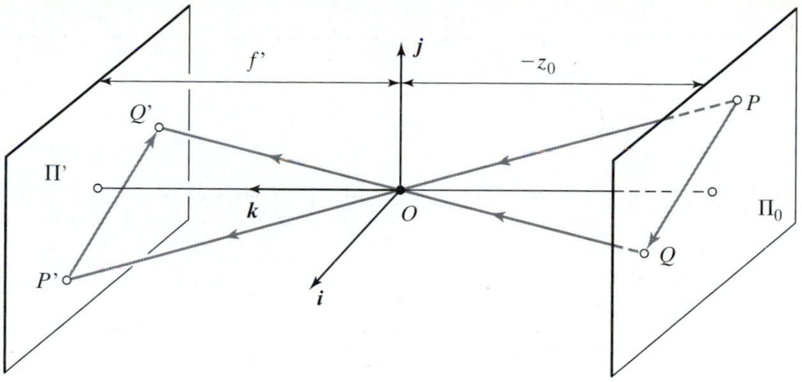

Figure 1.5 Weak-perspective projection: All line segments in the plane Π_0 are projected with the same magnification.

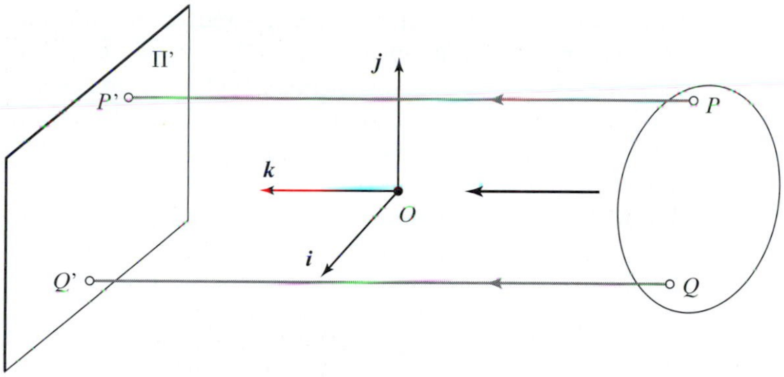

Figure 1.6 Orthographic projection. Unlike other geometric models of the image-formation process, orthographic projection does not involve a reversal of image features. Accordingly, the magnification is taken to be negative, which is a bit unnatural, but simplifies the projection equations.

orthography. When it is a priori known that the camera always remains at a roughly constant distance from the scene, we can go further and normalize the image coordinates so that $m = -1$. This is *orthographic projection* defined by

$$\begin{cases} x' = x, \\ y' = y, \end{cases} \tag{1.3}$$

with all light rays parallel to the $\boldsymbol{k}$ axis and orthogonal to the image plane Π' (Figure 1.6). Although weak-perspective projection is an acceptable model for many imaging conditions, assuming pure orthographic projection is usually unrealistic.

1.2 CAMERAS WITH LENSES

Most cameras are equipped with lenses. There are two main reasons for this: The first one is to gather light since a single ray of light would otherwise reach each point in the image plane under ideal pinhole projection. Real pinholes have a finite size of course, so each point in the image

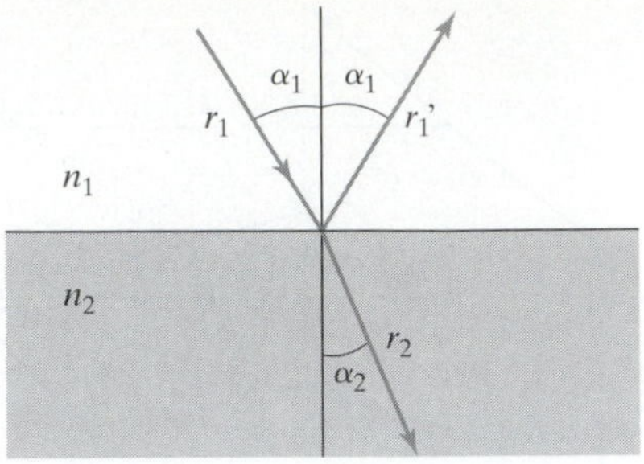

Figure 1.7 Reflection and refraction at the interface between two homogeneous media with indexes of refraction n_1 and n_2.

plane is illuminated by a cone of light rays subtending a finite solid angle. The larger the hole, the wider the cone and the brighter the image, but a large pinhole gives blurry pictures. Shrinking the pinhole produces sharper images, but reduces the amount of light reaching the image plane, and may introduce *diffraction* effects. The second main reason for using a lens is to keep the picture in sharp focus while gathering light from a large area.

Ignoring diffraction, interferences, and other physical optics phenomena, the behavior of lenses is dictated by the laws of geometric optics (Figure 1.7): (1) light travels in straight lines (*light rays*) in homogeneous media; (2) when a ray is reflected from a surface, this ray, its reflection, and the surface normal are coplanar, and the angles between the normal and the two rays are complementary; and (3) when a ray passes from one medium to another, it is *refracted* (i.e., its direction changes). According to Snell's law, if r_1 is the ray incident to the interface between two transparent materials with indexes of refraction n_1 and n_2, and r_2 is the refracted ray, then r_1, r_2 and the normal to the interface are coplanar, and the angles α_1 and α_2 between the normal and the two rays are related by

$$n_1 \sin \alpha_1 = n_2 \sin \alpha_2. \tag{1.4}$$

In this chapter, we only consider the effects of refraction and ignore those of reflection. In other words, we concentrate on lenses as opposed to *catadioptric optical systems* (e.g., telescopes) that may include both reflective (mirrors) and refractive elements. Tracing light rays as they travel through a lens is simpler when the angles between these rays and the refracting surfaces of the lens are assumed to be small. The next section discusses this case.

1.2.1 Paraxial Geometric Optics

In this section, we consider *paraxial* (or *first-order*) geometric optics, where the angles between all light rays going through a lens and the normal to the refractive surfaces of the lens are small. In addition, we assume that the lens is rotationally symmetric about a straight line, called its *optical axis*, and that all refractive surfaces are spherical. The symmetry of this setup allows us to determine the projection geometry by considering lenses with circular boundaries lying in a plane that contains the optical axis.

Let us consider an incident light ray passing through a point P_1 on the optical axis and refracted at the point P of the circular interface of radius R separating two transparent media with indexes of refraction n_1 and n_2 (Figure 1.8). Let us also denote by P_2 the point where the refracted ray intersects the optical axis a second time (the roles of P_1 and P_2 are completely symmetric) and by C the center of the circular interface.

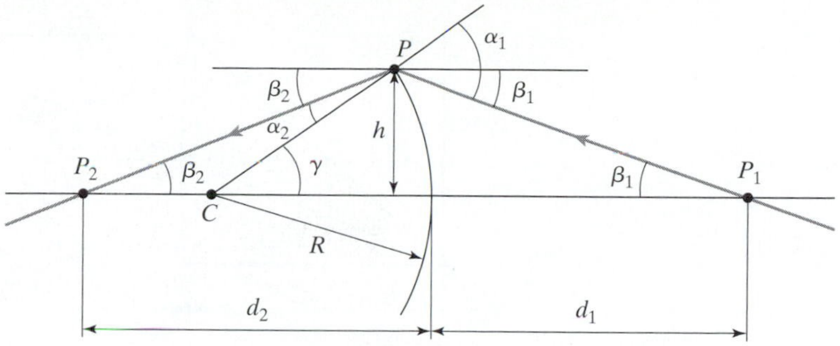

Figure 1.8 Paraxial refraction: A light ray passing through the point P_1 is refracted at the point P where it intersects a circular interface. The refracted ray intersects the optical axis in P_2. The center of the interface is at the point C of the optical axis, and its radius is R. The angles α_1, β_1, α_2, and β_2 are all assumed to be small.

Let α_1 and α_2, respectively, denote the angles between the two rays and the chord joining C to P. If β_1 (resp. β_2) is the angle between the optical axis and the line joining P_1 (resp. P_2) to P, the angle between the optical axis and the line joining C to P is, as shown by Figure 1.8, $\gamma = \alpha_1 - \beta_1 = \alpha_2 + \beta_2$. Now let h denote the distance between P and the optical axis and R the radius of the circular interface. If we assume all angles are small and thus, to first order, equal to their sines and tangents, we have

$$\alpha_1 = \gamma + \beta_1 \approx h\left(\frac{1}{R} + \frac{1}{d_1}\right) \quad \text{and} \quad \alpha_2 = \gamma - \beta_2 \approx h\left(\frac{1}{R} - \frac{1}{d_2}\right).$$

Writing Snell's law for small angles yields the *paraxial refraction equation*:

$$n_1\alpha_1 \approx n_2\alpha_2 \Longleftrightarrow \frac{n_1}{d_1} + \frac{n_2}{d_2} = \frac{n_2 - n_1}{R}. \tag{1.5}$$

Note that the relationship between d_1 and d_2 depends on R, n_1, and n_2, but not on β_1 or β_2. This is the main simplification introduced by the paraxial assumption. It is easy to see that Eq. (1.5) remains valid when some (or all) of the values of d_1, d_2, and R become negative, corresponding to the points P_1, P_2, or C switching sides.

Of course, real lenses are bounded by at least two refractive surfaces. The corresponding ray paths can be constructed iteratively using the paraxial refraction equation. The next section illustrates this idea in the case of thin lenses.

1.2.2 Thin Lenses

Let us now consider a lens with two spherical surfaces of radius R and index of refraction n. We assume that this lens is surrounded by vacuum (or, to an excellent approximation, by air), with an index of refraction equal to 1, and that it is *thin* (i.e., that a ray entering the lens and refracted at its right boundary is immediately refracted again at the left boundary).

Consider a point P located at (negative) depth z off the optical axis and denote by (PO) the ray passing through this point and the center O of the lens (Figure 1.9). As shown in the exercises, it follows from Snell's law and Eq. (1.5) that the ray (PO) is not refracted and that all other rays passing through P are focused by the thin lens on the point P' with depth z' along

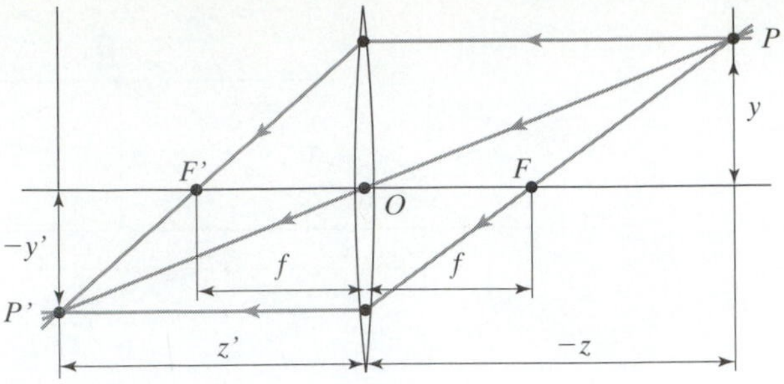

Figure 1.9 A thin lens. Rays passing through the point O are not refracted. Rays parallel to the optical axis are focused on the focal point F'.

(PO) such that

$$\frac{1}{z'} - \frac{1}{z} = \frac{1}{f}, \tag{1.6}$$

where $f = \frac{R}{2(n-1)}$ is the *focal length* of the lens.

Note that the equations relating the positions of P and P' are exactly the same as under pinhole perspective projection if we take $z' = f'$, since P and P' lie on a ray passing through the center of the lens, but that points located at a distance $-z$ from O are only in sharp focus when the image plane is located at a distance z' from O on the other side of the lens that satisfies Eq. (1.6) (i.e., the *thin lens equation*). Letting $z \to -\infty$ shows that f is the distance between the center of the lens and the plane where objects such as stars, which are effectively located at $z = -\infty$, focus. The two points F and F' located at distance f from the lens center on the optical axis are called the *focal points* of the lens.

In practice, objects within some range of distances (called *depth of field* or *depth of focus*) are in acceptable focus. As shown in the exercises, the depth of field increases with the *f number* of the lens (i.e., the ratio between the focal length of the lens and its diameter). The *field of view* of a camera is the portion of scene space that actually projects onto the retina of the camera. It is not defined by the focal length alone, but also depends on the effective area of the retina (e.g., the area of film that can be exposed in a photographic camera, or the area of the CCD sensor in a digital camera; Figure 1.10).

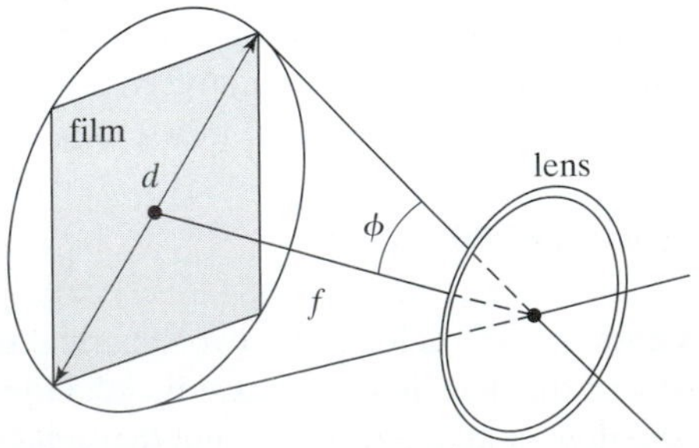

Figure 1.10 The field of view of a camera is 2ϕ, where $\phi \stackrel{\text{def}}{=} \arctan \frac{d}{2f}$, d is the diameter of the sensor (film or CCD chip) and f is the focal length of the camera. When f is (much) shorter than d, we have a wide-angle lens with rays that can be off the optical axis by more than 45°. Telephoto lenses have a small field of view and produce pictures closer to affine ones.

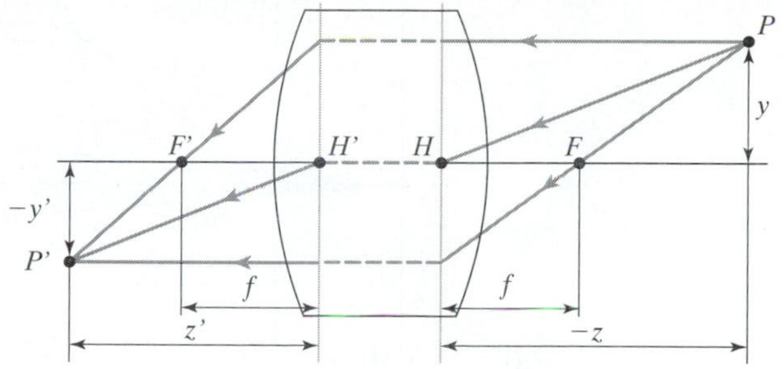

Figure 1.11 A simple thick lens with two spherical surfaces.

1.2.3 Real Lenses

A more realistic model of simple optical systems is the *thick lens*. The equations describing its behavior are easily derived from the paraxial refraction equation, and they are the same as the pinhole perspective and thin lens projection equations except for an offset (Figure 1.11): If H and H' denote the *principal points* of the lens, then Eq. (1.6) holds when $-z$ (resp. z') is the distance between P (resp. P') and the plane perpendicular to the optical axis and passing through H (resp. H'). In this case, the only undeflected ray is along the optical axis.

Simple lenses suffer from a number of *aberrations*. To understand why, let us remember first that the paraxial refraction Eq. (1.5) is only an approximation—valid when the angle α between each ray along the optical path and the optical axis of the length is small and $\sin\alpha \approx \alpha$. For larger angles, a third-order Taylor expansion of the sine function yields the following refinement of the paraxial equation:

$$\frac{n_1}{d_1} + \frac{n_2}{d_2} = \frac{n_2 - n_1}{R} + h^2 \left[\frac{n_1}{2d_1} \left(\frac{1}{R} + \frac{1}{d_1} \right)^2 + \frac{n_2}{2d_2} \left(\frac{1}{R} - \frac{1}{d_2} \right)^2 \right].$$

Here, h denotes, as in Figure 1.8, the distance between the optical axis and the point where the incident ray intersects the interface. In particular, rays striking the interface farther from the optical axis are focused closer to the interface.

The same phenomenon occurs for a lens and it is the source of two types of *spherical aberrations* (Figure 1.12[a]): Consider a point P on the optical axis and its paraxial image P'. The distance between P' and the intersection of the optical axis with a ray issued from P and refracted by the lens is called the *longitudinal spherical aberration* of that ray. Note that if an image plane Π' were erected in P, the ray would intersect this plane at some distance from the axis, called the *transverse spherical aberration* of that ray. Together, all rays passing through P and refracted by the lens form a circle of confusion centered in P as they intersect Π'. The size of that circle changes when we move Π' along the optical axis. The circle with minimum diameter is called the *circle of least confusion*, and it is not (in general) located in P'.

Besides spherical aberration, there are four other types of *primary aberrations* caused by the differences between first- and third-order optics—namely, *coma, astigmatism, field curvature*, and *distortion*. A precise definition of these aberrations is beyond the scope of this book. Suffice to say that, like spherical aberration, they degrade the image by blurring the picture of every object point. Distortion plays a different role and changes the shape of the image as a whole (Figure 1.12[b]). This effect is due to the fact that different areas of a lens have slightly different focal lengths. The aberrations mentioned so far are monochromatic (i.e., they are independent of the response of the lens to various wavelengths). However, the index of refraction of a transparent

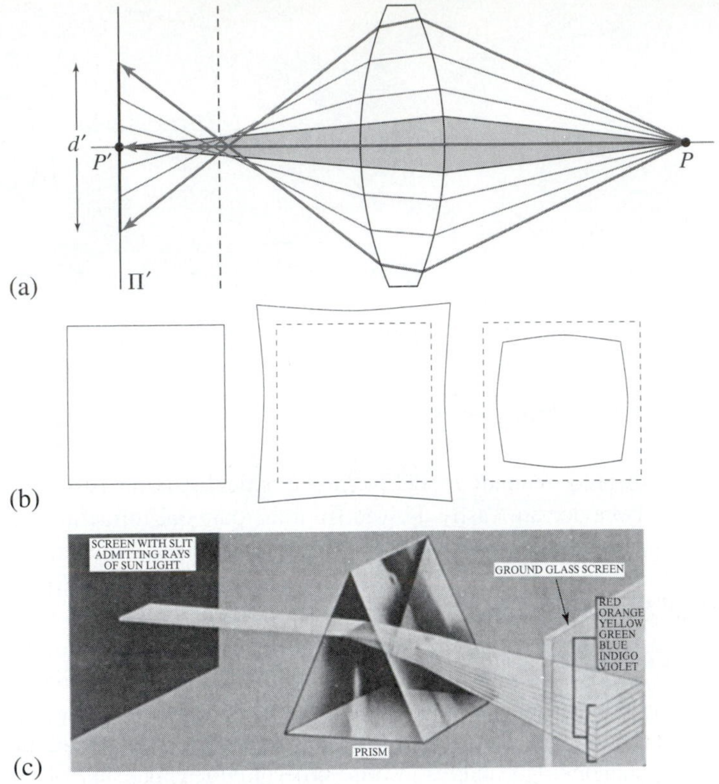

Figure 1.12 Aberrations. (a) Spherical aberration: The grey region is the paraxial zone where the rays issued from P intersect at its paraxial image P'. If an image plane Π' is erected in P', the image of P' in that plane forms a circle of confusion of diameter d'. The focus plane yielding the circle of least confusion is indicated by a dashed line. (b) Distortion: From left to right, the nominal image of a fronto-parallel square, pincushion distortion, and barrel distortion. (c) Chromatic aberration: The index of refraction of a transparent medium depends on the wavelength (or color) of the incident light rays. Here, a prism decomposes white light into a palette of colors. *Figure from US NAVY MANUAL OF BASIC OPTICS AND OPTICAL INSTRUMENTS, prepared by the Bureau of Naval Personnel, reprinted by Dover Publications, Inc., (1969).*

medium depends on wavelength (Figure 1.12[c]), and it follows from the thin lens Eq. (1.6) that the focal length depends on wavelength as well. This causes the phenomenon of *chromatic aberration*: Refracted rays corresponding to different wavelengths intersect the optical axis at different points (*longitudinal chromatic aberration*) and form different circles of confusion in the same image plane (*transverse chromatic aberration*).

Aberrations can be minimized by aligning several simple lenses with well-chosen shapes and refraction indexes, separated by appropriate stops. These *compound lenses* can still be modeled by the thick lens equations. They suffer from one more defect relevant to machine vision: Light beams emanating from object points located off-axis are partially blocked by the various apertures (including the individual lens components) positioned inside the lens to limit aberrations (Figure 1.13). This phenomenon, called *vignetting*, causes the brightness to drop in the image periphery. Vignetting may pose problems to automated image analysis programs, but it is not as important in photography thanks to the human eye's remarkable insensitivity to smooth

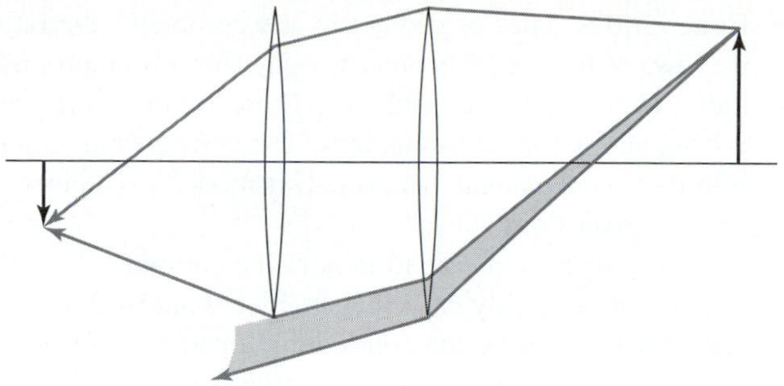

Figure 1.13 Vignetting effect in a two-lens system. The shaded part of the beam never reaches the second lens. Additional apertures and stops in a lens further contribute to vignetting.

brightness gradients. Speaking of which, it is time to look at this extraordinary organ in a bit more detail.

1.3 THE HUMAN EYE

Here we give a (brief) overview of the anatomical structure of the eye. It is largely based on the presentation in Wandell (1995), and the interested reader is invited to read this excellent book for more details. Figure 1.14 (left) is a sketch of the section of an eyeball through its vertical plane of symmetry, showing the main elements of the eye: the *iris* and the *pupil*, which control the amount of light penetrating the eyeball; the *cornea* and the crystalline *lens*, which together refract the light to create the retinal image; and finally the *retina*, where the image is formed.

Despite its globular shape, the human eyeball is functionally similar to a camera with a field of view covering a 160° (width) × 135° (height) area. Like any other optical system, it suffers

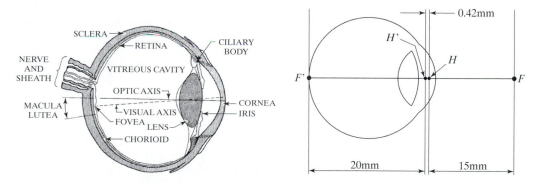

Figure 1.14 Left: the main components of the human eye. *Reproduced with permission, the American Society for Photogrammetry and Remote Sensing. A.L. Nowicki, "Stereoscopy." MANUAL OF PHOTOGRAMMETRY, edited by M.M. Thompson, R.C. Eller, W.A. Radlinski, and J.L. Speert, third edition, pp. 515–536. Bethesda: American Society of Photogrammetry, (1966).* Right: Helmoltz's schematic eye as modified by Laurance (after Driscoll and Vaughan, 1978). The distance between the pole of the cornea and the anterior principal plane is 1.96 mm, and the radii of the cornea, anterior, and posterior surfaces of the lens are respectively 8 mm, 10 mm, and 6 mm.

from various types of geometric and chromatic aberrations. Several models of the eye obeying the laws of first-order geometric optics have been proposed, and Figure 1.14 (right) shows one of them, *Helmoltz's schematic eye*. There are only three refractive surfaces, with an infinitely thin cornea and a homogeneous lens. The constants given in Figure 1.14 are for the eye focusing at infinity (*unaccommodated eye*). This model is of course only an approximation of the real optical characteristics of the eye.

Let us have a second look at the components of the eye one layer at a time: the cornea is a transparent, highly curved, refractive window through which light enters the eye before being partially blocked by the colored and opaque surface of the iris. The pupil is an opening at the center of the iris whose diameter varies from about 1 to 8 mm in response to illumination changes, dilating in low light to increase the amount of energy that reaches the retina and contracting in normal lighting conditions to limit the amount of image blurring due to spherical aberration in the eye. The refracting power (reciprocal of the focal length) of the eye is, in large part, an effect of refraction at the the air–cornea interface, and it is fine tuned by deformations of the crystalline lens that accommodates to bring objects into sharp focus. In healthy adults, it varies between 60 (unaccommodated case) and 68 diopters (1 diopter $= 1\ \text{m}^{-1}$), corresponding to a range of focal lengths between 15 and 17 mm. The retina itself is a thin, layered membrane populated by two types of photoreceptors—*rods* and *cones*—that respond to light in the 330 to 730 nm wavelength range (violet to red). As mentioned in Chapter 6, there are three types of cones with different spectral sensitivities, and these play a key role in the perception of color. There are about 100 million rods and 5 million cones in a human eye. Their spatial distribution varies across the retina: The *macula lutea* is a region in the center of the retina where the concentration of cones is particularly high and images are sharply focused whenever the eye fixes its attention on an object (Figure 1.14). The highest concentration of cones occurs in the *fovea*, a depression in the middle of the macula lutea where it peaks at $1.6 \times 10^5/\text{mm}^2$, with the centers of two neighboring cones separated by only half a minute of visual angle (Figure 1.15). Conversely, there are no rods in the center of the fovea, but the rod density increases toward the periphery of the visual field. There is also a *blind spot* on the retina, where the ganglion cell axons exit the retina and form the optic nerve.

The rods are extremely sensitive photoreceptors; they are capable of responding to a single photon, but they yield relatively poor spatial detail despite their high number because many rods converge to the same neuron within the retina. In contrast, cones become active at higher light

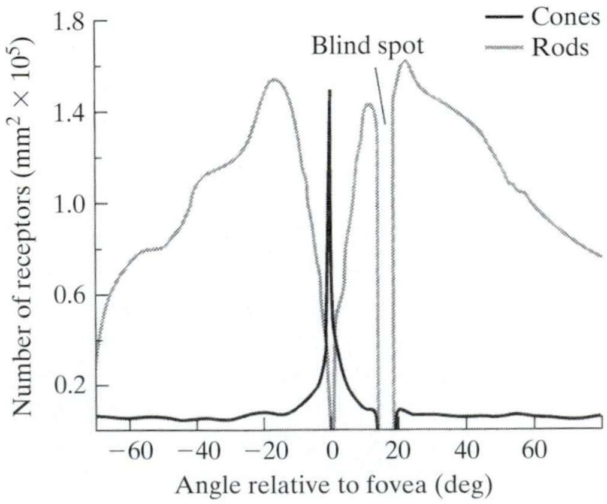

Figure 1.15 The distribution of rods and cones across the retina. *Reprinted from FOUNDATIONS OF VISION, by B. Wandell, Sinauer Associates, Inc., (1995). © 1995 Sinauer Associates, Inc.*

levels, but the signal output by each cone in the fovea is encoded by several neurons, yielding a high resolution in that area. More generally, the area of the retina influencing a neuron's response is traditionally called its *receptive field*, although this term now also characterizes the actual electrical response of neurons to light patterns.

Of course, much more could (and should) be said about the human eye—for example how our two eyes verge and fixate on targets, cooperate in stereo vision, and so on. Besides, vision only starts with this camera of our mind, which leads to the fascinating (and still largely unsolved) problem of deciphering the role of the various portions of our brain in human vision. We come back to various aspects of this endeavor later in this book.

1.4 SENSING

What differentiates a camera (in the modern sense of the world) from the portable camera obscura of the 17th century is its ability to record the pictures that form on its backplane. Although it had been known since at least the Middle Ages that certain silver salts rapidly darken under the action of sunlight, it was only in 1816 that Niepce obtained the first true photographs by exposing paper treated with silver chloride to the light rays striking the image plane of a camera obscura, then fixing the picture with nitric acid. These first images were negatives, and Niepce soon switched to other photosensitive chemicals to obtain positive pictures. The earliest photographs have been lost, and the first one to have been preserved is *la table servie* (the set table) reproduced in Figure 1.16.

Niepce invented photography, but Daguerre would be the one to popularize it. After the two became associates in 1826, Daguerre went on to develop his own photographic process using mercury fumes to amplify and reveal the latent image formed on an iodized plating of silver on copper. *Daguerréotypes* were an instant success when Arago presented Daguerre's process at the French Academy of Sciences in 1839, three years after Niepce's death. Other milestones in the long history of photography include the introduction of the wet-plate negative/positive process by Legray and Archer in 1850, which required the pictures to be developed on the spot but produced excellent negatives; the invention of the gelatin process by Maddox in 1870, which eliminated the need for immediate development; the introduction in 1889 of the photographic film (that has replaced glass plates in most modern applications) by Eastman; and the invention by the Lumière brothers of cinema in 1895 and color photography in 1908.

Figure 1.16 The first photograph on record, *la table servie*, obtained by Nicéphore Niepce in 1822. *Collection Harlinge–Viollet.*

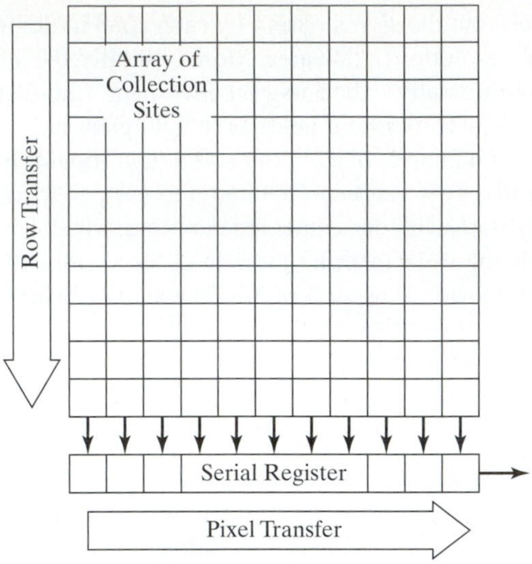

Figure 1.17 A CCD Device.

The invention of television in the 1920s by people like Baird, Farnsworth, and Zworykin was of course a major impetus for the development of electronic sensors. The *vidicon* is a common type of TV vacuum tube. It is a glass envelope with an electron gun at one end and a faceplate at the other. The back of the faceplate is coated with a thin layer of photoconductor material laid over a transparent film of positively charged metal. This double coating forms the *target*. The tube is surrounded by focusing and deflecting coils that are used to repeatedly scan the target with the electron beam generated by the gun. This beam deposits a layer of electrons on the target to balance its positive charge. When a small area of the faceplate is struck by light, electrons flow through, locally depleting the charge of the target. As the electron beam scans this area, it replaces the lost electrons, creating a current proportional to the incident light intensity. The current variations are then transformed into a video signal by the vidicon circuitry.

1.4.1 CCD Cameras

Let us now turn to *charge-coupled-device (CCD)* cameras that were proposed in 1970 and have replaced vidicon cameras in most modern applications, from consumer camcorders to special-purpose cameras geared toward microscopy or astronomy applications. A CCD sensor uses a rectangular grid of electron-collection sites laid over a thin silicon wafer to record a measure of the amount of light energy reaching each of them (Figure 1.17). Each site is formed by growing a layer of silicon dioxide on the wafer and then depositing a conductive gate structure over the dioxide. When photons strike the silicon, electron-hole pairs are generated (*photo-conversion*), and the electron are captured by the *potential well* formed by applying a positive electrical potential to the corresponding gate. The electrons generated at each site are collected over a fixed period of time T.

At this point, the charges stored at the individual sites are moved using *charge coupling*: Charge packets are transfered from site to site by manipulating the gate potentials, preserving the separation of the packets. The image is read out of the CCD one row at a time, each row being transfered in parallel to a serial output register with one element in each column. Between two row reads, the register transfers its charges one at a time to an output amplifier that generates a signal proportional to the charge it receives. This process continues until the entire image has been read out. It can be repeated 30 times per second (TV rate) for video applications or at a much slower pace, leaving ample time (seconds, minutes, even hours) for electron collection in low-light-level applications such as astronomy. It should be noted that the digital output of most CCD cameras is transformed internally into an analog video signal before being passed to a *frame grabber* that constructs the final digital image.

Consumer-grade color CCD cameras essentially use the same chips as black-and-white cameras, except that successive rows or columns of sensors are made sensitive to red, green or blue light often using a filter coating that blocks the complementary light. Other filter patterns are possible, including mosaics of 2×2 blocks formed by two green, one red, and one blue receptors (*Bayer patterns*). The spatial resolution of single-CCD cameras is of course limited, and higher-quality cameras use a beam splitter to ship the image to three different CCDs via color filters. The individual color channels are then either digitized separately (*RGB* output) or combined into a composite color video signal (*NTSC* output in the United States, *SECAM* or *PAL* in Europe and Japan) or into a *component video* format separating color and brightness information.

1.4.2 Sensor Models

For simplicity, we restrict our attention in this section to black-and-white CCD cameras: Color cameras can be treated in a similar fashion by considering each color channel separately and taking the effect of the associated filter response explicitly into account.

The number I of electrons recorded at the cell located at row r and column c of a CCD array can be modeled as

$$I(r, c) = T \int_\lambda \int_{p \in S(r, c)} E(p, \lambda) R(p) q(\lambda) dp \, d\lambda,$$

where T is the electron-collection time and the integral is calculated over the spatial domain $S(r, c)$ of the cell and the range of wavelengths to which the CCD has a nonzero response. In this integral, E is is the power per unit area and unit wavelength (i.e., the *irradiance*, see chapter 4 for a formal definition) arriving at the point p, R is the spatial response of the site, and q is the *quantum efficiency* of the device (i.e., the number of electrons generated per unit of incident light energy). In general, both E and q depend on the light wavelength λ, and E and R depend on the point location p within $S(r, c)$.

The output amplifier of the CCD transforms the charge collected at each site into a measurable voltage. In most cameras, this voltage is then transformed into a low-pass-filtered[1] video signal by the camera electronics with a magnitude proportional to I. The analog image can be once again transformed into a digital one using a frame grabber that spatially samples the video signal and quantizes the brightness value at each image point or *pixel* (from *picture element*).

There are several physical phenomena that alter the ideal camera model presented earlier: *Blooming* occurs when the light source illuminating a collection site is so bright that the charge stored at that site overflows into adjacent ones. It can be avoided by controlling the illumination, but other factors such as fabrication defects, thermal and quantum effects, and quantization noise are inherent to the imaging process. As shown next, these factors are appropriately captured by simple statistical models.

Quantum physics effects introduce an inherent uncertainty in the photoconversion process at each site (*shot noise*). More precisely, the number of electrons generated by this process can be modeled by a random integer variable $N_I(r, c)$ obeying a Poisson distribution with mean $\beta(r, c) I(r, c)$, where $\beta(r, c)$ is a number between 0 and 1 that reflects the variation of the spatial response and quantum efficiency across the image and also accounts for bad pixels. Electrons

[1]That is, roughly speaking, spatially or temporally averaged; more on this later.

freed from the silicon by thermal energy add to the charge of each collection site. Their contribution is called *dark current* and it can be modeled by a random integer variable $N_{DC}(r, c)$ whose mean $\mu_{DC}(r, c)$ increases with temperature. The effect of dark current can be controlled by cooling down the camera. Additional electrons are introduced by the CCD electronics (*bias*), and their number can also be modeled by a Poisson-distributed random variable $N_B(r, c)$ with mean $\mu_B(r, c)$. The output amplifier adds read-out noise that can be modeled by a real-valued random variable R obeying a Gaussian distribution with mean μ_R and standard deviation σ_R.

There are other sources of uncertainty (e.g., charge transfer efficiency), but they can often be neglected. Finally, the discretization of the analog voltage by the frame grabber introduces both geometric effects (*line jitter*), which can be corrected via calibration, and a quantization noise, which can be modeled as a zero-mean random variable $Q(r, c)$ with a uniform distribution in the $[-\frac{1}{2}\delta, \frac{1}{2}\delta]$ interval and a variance of $\frac{1}{12}\delta^2$, where δ is the quantization step. This yields the following composite model for the digital signal $D(r, c)$:

$$D(r, c) = \gamma(N_I(r, c) + N_{DC}(r, c) + N_B(r, c) + R(r, c)) + Q(r, c).$$

In this equation, γ is the combined gain of the amplifier and camera circuitry. The statistical properties of this model can be estimated via radiometric camera calibration: For example, dark current can be estimated by taking a number of sample pictures in a dark environment ($I = 0$).

1.5 NOTES

The classical textbook by Hecht (1987) is an excellent introduction to geometric optics. It includes a detailed discussion of paraxial optics as well as the various aberrations briefly mentioned in this chapter (see also Driscoll and Vaughan, 1978). Vignetting is discussed in Horn (1986) and Russ (1995). Wandell (1995) gives an excellent treatment of image formation in

TABLE 1.1 Reference card: Camera models.

Perspective projection	$\begin{cases} x' = f'\frac{x}{z} \\ y' = f'\frac{y}{z} \end{cases}$	x, y: world coordinates ($z < 0$) x', y': image coordinates f': pinhole-to-retina distance
Weak-perspective projection	$\begin{cases} x' = -mx \\ y' = -my \\ m = -\frac{f'}{z_0} \end{cases}$	x, y: world coordinates x', y': image coordinates f': pinhole-to-retina distance z_0: reference-point depth (< 0) m: magnification (> 0)
Orthographic projection	$\begin{cases} x' = x \\ y' = y \end{cases}$	x, y: world coordinates x', y': image coordinates
Snell's law	$n_1 \sin\alpha_1 = n_2 \sin\alpha_2$	n_1, n_2: refraction indexes α_1, α_2: normal-to-ray angles
Paraxial refraction	$\frac{n_1}{d_1} + \frac{n_2}{d_2} = \frac{n_2 - n_1}{R}$	n_1, n_2: refraction indexes d_1, d_2: point-to-interface distances R: interface radius
Thin lens equation	$\frac{1}{z'} - \frac{1}{z} = \frac{1}{f}$	z: object-point depth (< 0) z': image-point depth (> 0) f: focal length

the human visual system. The Helmoltz schematic model of the eye is detailed in Driscoll and Vaughan (1978).

CCD devices were introduced in Boyle and Smith (1970) and Amelio *et al.* (1970). Scientific applications of CCD cameras to microscopy and astronomy are discussed in Aiken *et al.* (1989), Janesick *et al.* (1987), Snyder *et al.* (1993), and Tyson (1990). The statistical sensor model presented in this chapter is based on Snyder *et al.* (1993), with an additional term for the quantization noise taken from Healey and Kondepudy (1994). These two articles contain interesting applications of sensor modeling to image restoration in astronomy and radiometric camera calibration in machine vision.

Given the fundamental importance of the notions introduced in this chapter, the main equations derived in its course have been collected in Table 1.1 for reference.

PROBLEMS

1.1. Derive the perspective equation projections for a virtual image located at a distance f' *in front* of the pinhole.

1.2. Prove geometrically that the projections of two parallel lines lying in some plane Π appear to converge on a horizon line H formed by the intersection of the image plane with the plane parallel to Π and passing through the pinhole.

1.3. Prove the same result algebraically using the perspective projection Eq. (1.1). You can assume for simplicity that the plane Π is orthogonal to the image plane.

1.4. Use Snell's law to show that rays passing through the optical center of a thin lens are not refracted, and derive the thin lens equation.

Hint: consider a ray r_0 passing through the point P and construct the rays r_1 and r_2 obtained respectively by the refraction of r_0 by the right boundary of the lens and the refraction of r_1 by its left boundary.

1.5. Consider a camera equipped with a thin lens, with its image plane at position z' and the plane of scene points in focus at position z. Now suppose that the image plane is moved to $\hat{z}'$. Show that the diameter of the corresponding blur circle is

$$d\frac{|z' - \hat{z}'|}{z'},$$

where d is the lens diameter. Use this result to show that the depth of field (i.e., the distance between the near and far planes that will keep the diameter of the blur circles below some threshold ε) is given by

$$D = 2\varepsilon f z(z + f)\frac{d}{f^2 d^2 - \varepsilon^2 z^2},$$

and conclude that, for a *fixed* focal length, the depth of field increases as the lens diameter decreases, and thus the f number increases.

Hint: Solve for the depth $\hat{z}$ of a point whose image is focused on the image plane at position $\hat{z}'$, considering both the case where $\hat{z}'$ is larger than z' and the case where it is smaller.

1.6. Give a geometric construction of the image P' of a point P given the two focal points F and F' of a thin lens.

1.7. Derive the thick lens equations in the case where both spherical boundaries of the lens have the same radius.

2

Geometric Camera Models

The fundamental laws of perspective projection were introduced in Chapter 1 in a camera-centered coordinate system. This chapter introduces the analytical machinery necessary to establish quantitative constraints between image measurements and the position and orientation of geometric figures measured in some arbitrary *external* coordinate system. We start by briefly recalling elementary notions of analytical Euclidean geometry, including homogeneous coordinates and matrix representations of geometric transformations. We then introduce the various physical parameters (the so-called *intrinsic* and *extrinsic* parameters) that relate the world and the camera coordinate frames and derive the general form of the perspective projection equation in this setting. We conclude with a brief presentation of *affine* models of the imaging process, that approximate pinhole perspective projection for distant objects, and include the orthographic and weak-perspective models briefly discussed in Chapter 1.

2.1 ELEMENTS OF ANALYTICAL EUCLIDEAN GEOMETRY

We assume that the reader has some familiarity with elementary Euclidean geometry and linear algebra. This section discusses useful analytical concepts such as coordinate systems, homogeneous coordinates, rotation matrices, and the like.

2.1.1 Coordinate Systems and Homogeneous Coordinates

We already used three-dimensional coordinate systems in chapter 1. This section introduces them a bit more formally. We assume throughout a fixed system of units, say meters or inches, so unit length is well defined.

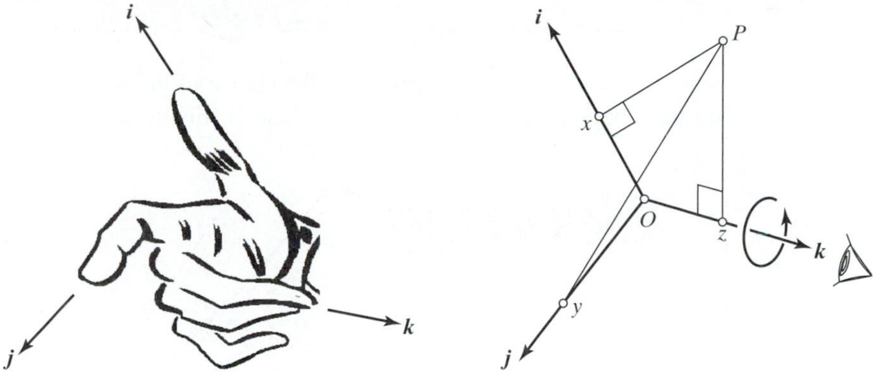

Figure 2.1 A right-handed coordinate system and the coordinates x, y, and z of a point P.

Picking a point O in the physical three-dimensional Euclidean space $\mathbb{E}^3$ and three unit vectors $\boldsymbol{i}$, $\boldsymbol{j}$, and $\boldsymbol{k}$ orthogonal to each other defines an *orthonormal coordinate frame* (F) as the quadruple $(O, \boldsymbol{i}, \boldsymbol{j}, \boldsymbol{k})$. The point O is the *origin* of the coordinate system (F), and $\boldsymbol{i}$, $\boldsymbol{j}$, and $\boldsymbol{k}$ are its *basis vectors*. We restrict our attention to *right-handed* coordinate systems, such that the vectors $\boldsymbol{i}$, $\boldsymbol{j}$ and $\boldsymbol{k}$ can be thought of as being attached to fingers of your right hand, with the thumb pointing up, index pointing straight, and middle finger pointing left as shown in Figure 2.1.[1]

The coordinates x, y, and z of a point P in this coordinate frame are defined as the (signed) lengths of the orthogonal projections of the vector $\overrightarrow{OP}$ onto the vectors $\boldsymbol{i}$, $\boldsymbol{j}$, and $\boldsymbol{k}$ (right side of Figure 2.1), with

$$\begin{cases} x = \overrightarrow{OP} \cdot \boldsymbol{i} \\ y = \overrightarrow{OP} \cdot \boldsymbol{j} \\ z = \overrightarrow{OP} \cdot \boldsymbol{k} \end{cases} \iff \overrightarrow{OP} = x\boldsymbol{i} + y\boldsymbol{j} + z\boldsymbol{k}.$$

The column vector

$$\boldsymbol{P} = \begin{pmatrix} x \\ y \\ z \end{pmatrix} \in \mathbb{R}^3$$

is called the *coordinate vector* of the point P in (F). We can also define the coordinate vector associated with any free vector $\boldsymbol{v}$ by the lengths of its projections onto the basis vectors of (F), and these coordinates are of course independent of the choice of the origin O. Let us now consider a plane Π, an arbitrary point A in Π, and a unit vector $\boldsymbol{n}$ perpendicular to the plane. The points lying in Π are characterized by

$$\overrightarrow{AP} \cdot \boldsymbol{n} = 0.$$

In a coordinate system (F), where the coordinates of the point P are x, y, and z and the coordinates of $\boldsymbol{n}$ are a, b, and c, this can be rewritten as $\overrightarrow{OP} \cdot \boldsymbol{n} - \overrightarrow{OA} \cdot \boldsymbol{n} = 0$, or

[1] This is the traditional way of defining right-handed coordinate systems. One of the authors, who is left-handed, has always found it a bit confusing and prefers to identify these coordinate systems using the fact that when one looks down the $\boldsymbol{k}$ axis at the $(\boldsymbol{i}, \boldsymbol{j})$ plane, the vector $\boldsymbol{i}$ is mapped onto the vector $\boldsymbol{j}$ by a *counterclockwise* 90° rotation (Figure 2.1). Left-handed coordinate systems correspond to *clockwise* rotations. Left- and right-handed readers alike may find this characterization useful as well.

$$ax + by + cz - d = 0, \tag{2.1}$$

where $d \stackrel{\text{def}}{=} \overrightarrow{OA} \cdot \boldsymbol{n}$ is independent of the choice of the point A in Π and is simply the (signed) distance between the origin O and the plane Π (Figure 2.2).

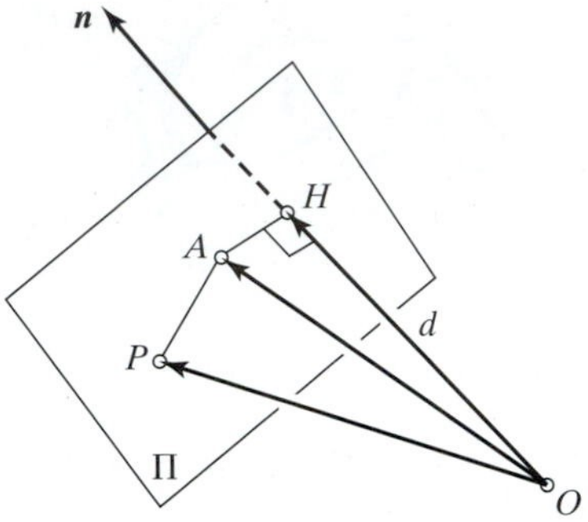

Figure 2.2 The geometric definition of the equation of a plane. The distance d between the origin and plane is reached at the point H where the normal vector passing through the origin pierces the plane.

At times, it is useful to use *homogeneous coordinates* to represent points, vectors, and planes. We formally justify their definition later in this book when we introduce affine and projective geometry in chapters 12 and 13, but for the time being let us note that Eq. (2.1) can be rewritten as

$$(a, b, c, -d)\begin{pmatrix} x \\ y \\ z \\ 1 \end{pmatrix} = 0,$$

or, more concisely, as

$$\boldsymbol{\Pi} \cdot \boldsymbol{P} = 0, \quad \text{where} \quad \boldsymbol{\Pi} \stackrel{\text{def}}{=} \begin{pmatrix} a \\ b \\ c \\ -d \end{pmatrix} \quad \text{and} \quad \boldsymbol{P} \stackrel{\text{def}}{=} \begin{pmatrix} x \\ y \\ z \\ 1 \end{pmatrix}. \tag{2.2}$$

The vector $\boldsymbol{P}$ is called the *homogeneous coordinate vector* of the point P in the coordinate system (F), and it is simply obtained by adding a fourth coordinate equal to 1 to the ordinary coordinate vector of P. Likewise, the vector $\boldsymbol{\Pi}$ is the vector of homogeneous coordinates of the plane Π in the coordinate frame (F), and Eq. (2.2) is called the equation of Π in that coordinate system. Note that $\boldsymbol{\Pi}$ is only defined up to scale since multiplying this vector by any nonzero constant does not change the solutions of Eq. (2.2). We use the convention that homogeneous coordinates are only defined up to scale, whether they represent points or planes (this may appear a bit counterintuitive for points, but it is fully justified in chapter 13). To go back to the ordinary nonhomogeneous coordinates of points, one just divides all coordinates by the fourth one.

Before proceeding, let us point out that, although our presentation focuses on three-dimensional Euclidean geometry in this chapter, the concepts discussed throughout also apply to planar geometry: A coordinate frame (F) is defined in the plane by its origin o and a right-handed orthonormal basis $(\boldsymbol{i}, \boldsymbol{j})$; the coordinates of the point p in this frame are $x = \overrightarrow{op} \cdot \boldsymbol{i}$ and $y = \overrightarrow{op} \cdot \boldsymbol{j}$, and homogeneous coordinates can be defined as well; in particular, the equation of a line δ in the plane is

$$ax + by - d = 0 \iff \boldsymbol{\delta} \cdot \boldsymbol{p} = 0, \quad \text{where} \quad \boldsymbol{\delta} = \begin{pmatrix} a \\ b \\ -d \end{pmatrix} \text{ and } \boldsymbol{p} = \begin{pmatrix} x \\ y \\ 1 \end{pmatrix},$$

and a, b, and d denote, respectively, the coordinates to the unit normal to δ in (F) and the signed distance from o to δ.

Let us go back to three-dimensional geometry and show that homogeneous coordinates can be used to describe more complex geometric figures than points and planes.[2] Consider, for example, a sphere S of radius R centered at the origin. A necessary and sufficient condition for the point P with coordinates x, y, and z to belong to S is of course that

$$x^2 + y^2 + z^2 = R^2,$$

which is equivalent to

$$(x, y, z, 1)^T \begin{pmatrix} 1 & 0 & 0 & 0 \\ 0 & 1 & 0 & 0 \\ 0 & 0 & 1 & 0 \\ 0 & 0 & 0 & -R^2 \end{pmatrix} \begin{pmatrix} x \\ y \\ z \\ 1 \end{pmatrix} = 0.$$

More generally, a *quadric surface* is the locus of the points P whose coordinates satisfy the equation

$$a_{200}x^2 + a_{110}xy + a_{020}y^2 + a_{011}yz + a_{002}z^2 + a_{101}xz + a_{100}x + a_{010}y + a_{001}z + a_{000} = 0,$$

and it is straightforward to check that this condition is equivalent to

$$\boldsymbol{P}^T \mathcal{Q} \boldsymbol{P} = 0, \quad \text{where} \quad \mathcal{Q} = \begin{pmatrix} a_{200} & \frac{1}{2}a_{110} & \frac{1}{2}a_{101} & \frac{1}{2}a_{100} \\ \frac{1}{2}a_{110} & a_{020} & \frac{1}{2}a_{011} & \frac{1}{2}a_{010} \\ \frac{1}{2}a_{101} & \frac{1}{2}a_{011} & a_{002} & \frac{1}{2}a_{001} \\ \frac{1}{2}a_{100} & \frac{1}{2}a_{010} & \frac{1}{2}a_{001} & a_{000} \end{pmatrix}. \tag{2.3}$$

In this equation, $\boldsymbol{P}$ denotes the homogeneous coordinate vector of P. Note that $\mathcal{Q}$ is a 4×4 symmetric matrix and, like the parameters a_{ijk}, it is only defined up to scale.

2.1.2 Coordinate System Changes and Rigid Transformations

When several different coordinate systems are considered at the same time, it is convenient to follow Craig (1989) and denote by ${}^F P$ (resp. ${}^F \boldsymbol{v}$) the coordinate vector of the point P (resp. vector $\boldsymbol{v}$) in the frame (F)—that is,

$${}^F P = {}^F \overrightarrow{OP} = \begin{pmatrix} x \\ y \\ z \end{pmatrix} \iff \overrightarrow{OP} = x\boldsymbol{i} + y\boldsymbol{j} + z\boldsymbol{k}.$$

Although the superscripts and subscripts preceding points, vectors, and matrices in Craig's notation may be awkward at first, the rest of this section clearly demonstrates their convenience. Let us now consider two coordinate systems: $(A) = (O_A, \boldsymbol{i}_A, \boldsymbol{j}_A, \boldsymbol{k}_A)$ and $(B) = (O_B, \boldsymbol{i}_B, \boldsymbol{j}_B, \boldsymbol{k}_B)$. The rest of this section shows how to express ${}^B P$ as a function of ${}^A P$. Let us

[2]The inquisitive reader may be wondering about lines in $\mathbb{E}^3$. A line can of course be defined as the intersection of two planes. More generally, lines in $\mathbb{E}^3$ can be defined in terms of *Plücker coordinates*, see Exercises.

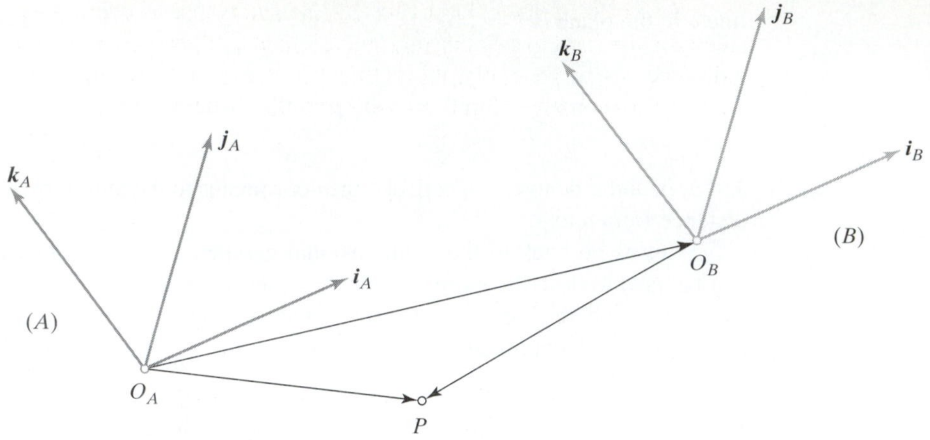

Figure 2.3 Change of coordinates between two frames: pure translation.

suppose first that the basis vectors of both coordinate systems are parallel to each other (i.e., $\boldsymbol{i}_A = \boldsymbol{i}_B$, $\boldsymbol{j}_A = \boldsymbol{j}_B$ and $\boldsymbol{k}_A = \boldsymbol{k}_B$), but the origins O_A and O_B are distinct (Figure 2.3).

We say in this case that the two coordinate systems are separated by a *pure translation*, and we have $\overrightarrow{O_B P} = \overrightarrow{O_B O_A} + \overrightarrow{O_A P}$, thus

$$^B P = {}^A P + {}^B O_A.$$

When the origins of the two frames coincide (i.e., $O_A = O_B = O$), we say that the frames are separated by a *pure rotation* (Figure 2.4). Let us define the *rotation matrix* ${}^B_A\mathcal{R}$ as the 3×3 array of numbers

$${}^B_A\mathcal{R} \stackrel{\text{def}}{=} \begin{pmatrix} \boldsymbol{i}_A \cdot \boldsymbol{i}_B & \boldsymbol{j}_A \cdot \boldsymbol{i}_B & \boldsymbol{k}_A \cdot \boldsymbol{i}_B \\ \boldsymbol{i}_A \cdot \boldsymbol{j}_B & \boldsymbol{j}_A \cdot \boldsymbol{j}_B & \boldsymbol{k}_A \cdot \boldsymbol{j}_B \\ \boldsymbol{i}_A \cdot \boldsymbol{k}_B & \boldsymbol{j}_A \cdot \boldsymbol{k}_B & \boldsymbol{k}_A \cdot \boldsymbol{k}_B \end{pmatrix}.$$

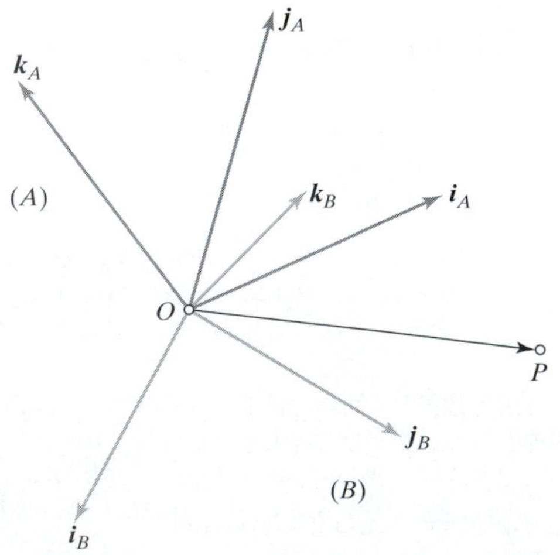

Figure 2.4 Change of coordinates between two frames: pure rotation.

Note that the first column of ${}^B_A\mathcal{R}$ is formed by the coordinates of $\boldsymbol{i}_A$ in the basis $(\boldsymbol{i}_B, \boldsymbol{j}_B, \boldsymbol{k}_B)$. Likewise, the third row of this matrix is formed by the coordinates of $\boldsymbol{k}_B$ in the basis $(\boldsymbol{i}_A, \boldsymbol{j}_A, \boldsymbol{k}_A)$, and so on. More generally, the matrix ${}^B_A\mathcal{R}$ can be written in a more compact fashion using a combination of three column vectors or three row vectors:

$$ {}^B_A\mathcal{R} = \begin{pmatrix} {}^B\boldsymbol{i}_A & {}^B\boldsymbol{j}_A & {}^B\boldsymbol{k}_A \end{pmatrix} = \begin{pmatrix} {}^A\boldsymbol{i}_B{}^T \\ {}^A\boldsymbol{j}_B{}^T \\ {}^A\boldsymbol{k}_B{}^T \end{pmatrix}. $$

It follows that ${}^A_B\mathcal{R} = {}^B_A\mathcal{R}^T$.

As noted earlier, all these subscripts and superscripts may be somewhat confusing at first. To keep everything straight, it is useful to remember that, in a change of coordinates, subscripts refer to the object being described, whereas superscripts refer to the coordinate system in which the object is described. For example, ${}^AP$ refers to the coordinate vector of the point P in the frame (A), ${}^B\boldsymbol{j}_A$ is the coordinate vector of the vector $\boldsymbol{j}_A$ in the frame (B), and ${}^B_A\mathcal{R}$ is the rotation matrix describing the frame (A) in the coordinate system (B).

Let us give an example of pure rotation: Suppose that $\boldsymbol{k}_A = \boldsymbol{k}_B = \boldsymbol{k}$, and denote by θ the angle such that the vector $\boldsymbol{i}_B$ is obtained by applying to the vector $\boldsymbol{i}_A$ a counterclockwise rotation of angle θ about $\boldsymbol{k}$ (Figure 2.5). The angle between the vectors $\boldsymbol{j}_A$ and $\boldsymbol{j}_B$ is also θ in this case, and we have

$$ {}^B_A\mathcal{R} = \begin{pmatrix} \cos\theta & \sin\theta & 0 \\ -\sin\theta & \cos\theta & 0 \\ 0 & 0 & 1 \end{pmatrix}. \tag{2.4} $$

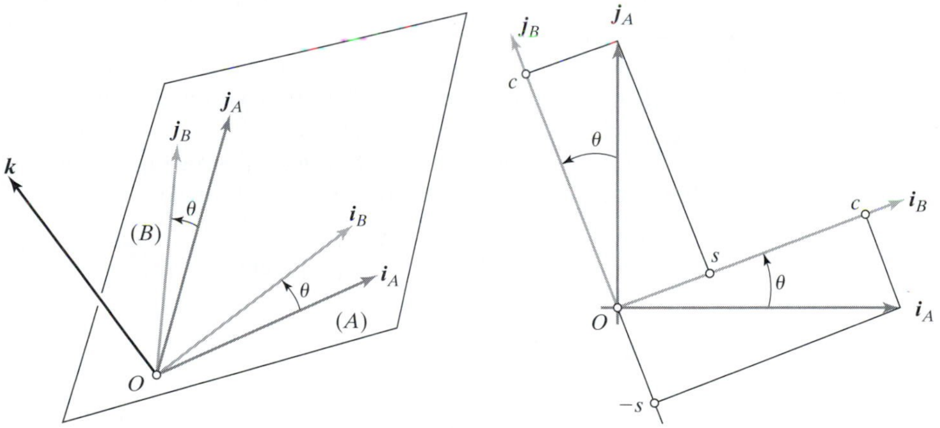

Figure 2.5 Two coordinate frames separated by a rotation of angle θ about their common $\boldsymbol{k}$ basis vector. As shown in the right of the figure, $\boldsymbol{i}_A = c\boldsymbol{i}_B - s\boldsymbol{j}_B$ and $\boldsymbol{j}_A = s\boldsymbol{i}_B + c\boldsymbol{j}_B$, where $c = \cos\theta$ and $s = \sin\theta$.

Similar formulas can be written when the two coordinate systems are deduced from each other via rotations about the $\boldsymbol{i}_A$ or $\boldsymbol{j}_A$ axes (see Exercises). In general, it can be shown that any rotation matrix can be written as the product of three elementary rotations about the $\boldsymbol{i}$, $\boldsymbol{j}$, and $\boldsymbol{k}$ vectors of some coordinate system.

Let us go back to characterizing the change of coordinates associated with an arbitrary rotation matrix. Writing

$$\overrightarrow{OP} = (\boldsymbol{i}_A \quad \boldsymbol{j}_A \quad \boldsymbol{k}_A)\begin{pmatrix} {}^A x \\ {}^A y \\ {}^A z \end{pmatrix} = (\boldsymbol{i}_B \quad \boldsymbol{j}_B \quad \boldsymbol{k}_B)\begin{pmatrix} {}^B x \\ {}^B y \\ {}^B z \end{pmatrix}$$

in the frame (B) yields immediately

$$^B P = {}^B_A\mathcal{R}\, {}^A P,$$

since the rotation matrix ${}^B_B\mathcal{R}$ is obviously the identity. Note how the subscript matches the following superscript. This property remains true for more general coordinate changes, and it can be used after some practice to reconstruct the corresponding formulas without calculations.

It is easy to show (see Exercises) that rotation matrices are characterized by the following properties: (1) the inverse of a rotation matrix is equal to its transpose, and (2) its determinant is equal to 1. By definition, the columns of a rotation matrix form a right-handed orthonormal coordinate system. It follows from Properties (1) and (2) that their rows also form such a coordinate system.

It should be noted that the set of rotation matrices, equipped with the matrix product, forms a *group*, that is, (a) the product of two rotation matrices is also a rotation matrix (this is intuitively obvious and easily verified analytically); (b) the matrix product is associative—that is, $(\mathcal{R}\mathcal{R}')\mathcal{R}'' = \mathcal{R}(\mathcal{R}'\mathcal{R}'')$ for any rotation matrices $\mathcal{R}$, $\mathcal{R}'$ and $\mathcal{R}''$; (c) there is a unit element, the 3×3 identity matrix Id, that is indeed a rotation matrix and verifies $\mathcal{R}\,\text{Id} = \text{Id}\,\mathcal{R} = \mathcal{R}$ for any rotation matrix $\mathcal{R}$; and (d) every rotation matrix $\mathcal{R}$ admits an inverse $\mathcal{R}^{-1} = \mathcal{R}^T$ such that $\mathcal{R}\mathcal{R}^{-1} = \mathcal{R}^{-1}\mathcal{R} = \text{Id}$. This group is not, however, commutative (i.e., given two rotation matrices $\mathcal{R}$ and $\mathcal{R}'$, the two products $\mathcal{R}\mathcal{R}'$ and $\mathcal{R}'\mathcal{R}$ are in general different).

When the origins and basis vectors of the two coordinate systems are different, we say that the frames are separated by a general *rigid transformation* (Figure 2.6), and we have

$$^B P = {}^B_A\mathcal{R}\, {}^A P + {}^B O_A, \tag{2.5}$$

where ${}^B_A\mathcal{R}$ and ${}^B O_A$ are defined as before. It should be clear that related formulas express coordinate changes for the homogeneous coordinate vectors of planes and the symmetric matrices associated with quadric surfaces (see Exercises).

Homogeneous coordinates can be used to rewrite Eq. (2.5) as a matrix product: Let us first note that matrices can be multiplied in blocks—that is, if

$$\mathcal{A} = \begin{pmatrix} \mathcal{A}_{11} & \mathcal{A}_{12} \\ \mathcal{A}_{21} & \mathcal{A}_{22} \end{pmatrix} \quad \text{and} \quad \mathcal{B} = \begin{pmatrix} \mathcal{B}_{11} & \mathcal{B}_{12} \\ \mathcal{B}_{21} & \mathcal{B}_{22} \end{pmatrix}, \tag{2.6}$$

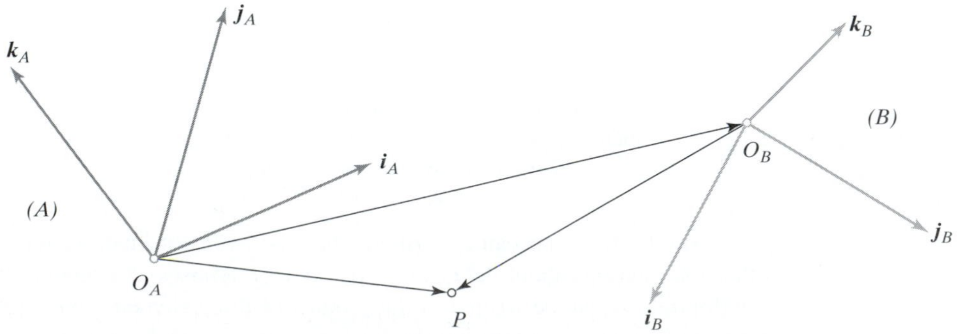

Figure 2.6 Change of coordinates between two frames: general rigid transformation.

where the number of columns of the submatrices $\mathcal{A}_{11}$ and $\mathcal{A}_{21}$ (resp. $\mathcal{A}_{12}$ and $\mathcal{A}_{22}$) is equal to the number of rows of $\mathcal{B}_{11}$ and $\mathcal{B}_{12}$ (resp. $\mathcal{B}_{21}$ and $\mathcal{B}_{22}$), then

$$\mathcal{A}\mathcal{B} = \begin{pmatrix} \mathcal{A}_{11}\mathcal{B}_{11} + \mathcal{A}_{12}\mathcal{B}_{21} & \mathcal{A}_{11}\mathcal{B}_{12} + \mathcal{A}_{12}\mathcal{B}_{22} \\ \mathcal{A}_{21}\mathcal{B}_{11} + \mathcal{A}_{22}\mathcal{B}_{21} & \mathcal{A}_{21}\mathcal{B}_{12} + \mathcal{A}_{22}\mathcal{B}_{22} \end{pmatrix}.$$

In particular, Eq. (2.6) allows us to rewrite the change of coordinates given by Eq. (2.5) as

$$\begin{pmatrix} {}^B P \\ 1 \end{pmatrix} = {}^B_A\mathcal{T} \begin{pmatrix} {}^A P \\ 1 \end{pmatrix}, \quad \text{where} \quad {}^B_A\mathcal{T} \stackrel{\text{def}}{=} \begin{pmatrix} {}^B_A\mathcal{R} & {}^B O_A \\ \mathbf{0}^T & 1 \end{pmatrix} \tag{2.7}$$

and $\mathbf{0} = (0, 0, 0)^T$. In other words, using homogeneous coordinates allows us to write a general change of coordinates as the product of a 4×4 matrix and a 4 vector. It is easy to show that the set of rigid transformations defined by Eq. (2.7), equipped with the matrix product operation, is also a group.

A rigid transformation maps a coordinate system onto another one. In a given coordinate frame (F), it can also be considered as a mapping between points—that is, a point P is mapped onto the point P' such that

$${}^F P' = \mathcal{R}\, {}^F P + \boldsymbol{t} \iff \begin{pmatrix} {}^F P' \\ 1 \end{pmatrix} = \begin{pmatrix} \mathcal{R} & \boldsymbol{t} \\ \mathbf{0}^T & 1 \end{pmatrix} \begin{pmatrix} {}^F P \\ 1 \end{pmatrix}, \tag{2.8}$$

where $\mathcal{R}$ is a rotation matrix and $\boldsymbol{t}$ is an element of $\mathbb{R}^3$ (Figure 2.7). The set of rigid transformations considered as mappings of $\mathbb{E}^3$ onto itself and equipped with the law of composition is once again easily shown to form a group. It is also easy to show that rigid transformations preserve the distance between two points and the angle between two vectors. However, the 4×4 matrix associated with a rigid transformation depends on the choice of (F).

For example, let us consider the rotation of angle θ about the $\boldsymbol{k}$ axis of the frame (F). As shown in the exercises, this mapping can be represented by

$${}^F P' = \mathcal{R}^F P, \quad \text{where} \quad \mathcal{R} = \begin{pmatrix} \cos\theta & -\sin\theta & 0 \\ \sin\theta & \cos\theta & 0 \\ 0 & 0 & 1 \end{pmatrix}.$$

In particular, if (F') is the coordinate system obtained by applying this rotation to (F), we have, according to Eq. (2.4), ${}^{F'} P = {}^{F'}_F\mathcal{R}^F P$ and $\mathcal{R} = {}^{F'}_F\mathcal{R}^{-1}$. More generally, the matrix

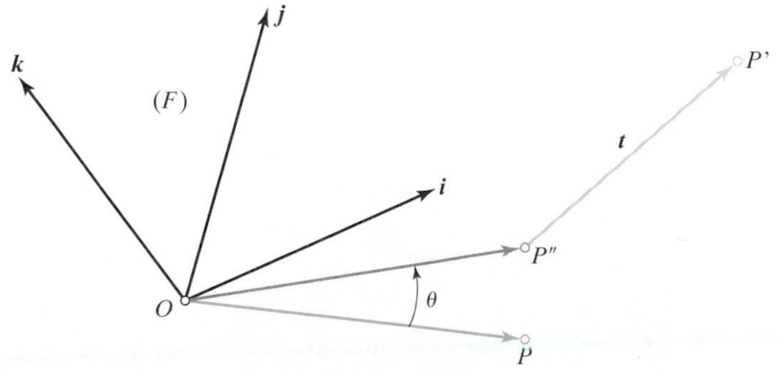

Figure 2.7 A rigid transformation maps the point P onto the point P'' through a rotation $\mathcal{R}$ before mapping P'' onto P' via a translation $\boldsymbol{t}$. In the example shown in this figure, $\mathcal{R}$ is a rotation of angle θ about the $\boldsymbol{k}$ axis of the coordinate system (F).

representing the change of coordinates between two frames is the inverse of the matrix mapping the first frame onto the second one.

What happens when $\mathcal{R}$ is replaced by an arbitrary nonsingular 3×3 matrix $\mathcal{A}$? Equation (2.8) still represents a mapping between points (or a change of coordinates between frames), but this time lengths and angles may not be preserved anymore (equivalently, the new coordinate system does not necessarily have orthogonal axes with unit length). We say that the 4×4 matrix

$$\mathcal{T} = \begin{pmatrix} \mathcal{A} & \boldsymbol{t} \\ \boldsymbol{0}^T & 1 \end{pmatrix}$$

represents an *affine transformation*. When $\mathcal{T}$ is a nonsingular but otherwise arbitrary 4×4 matrix, we say that we have a projective transformation. Affine and projective transformations also form groups. They will be given a more thorough treatment in chapters 12 and 13.

2.2 CAMERA PARAMETERS AND THE PERSPECTIVE PROJECTION

We saw in chapter 1 that the coordinates x, y, and z of a scene point P observed by a pinhole camera are related to its image coordinates x' and y' by the perspective Eq. (1.1). In reality, this equation is only valid when all distances are measured in the camera's reference frame, and image coordinates have their origin at the principal point where the axis of symmetry of the camera pierces its retina. In practice, the world and camera coordinate systems are related by a set of physical parameters, such as the focal length of the lens, the size of the pixels, the position of the principal point, and the position and orientation of the camera.

This section identifies these parameters. We distinguish the *intrinsic* parameters, which relate the camera's coordinate system to the idealized coordinate system used in chapter 1, from the *extrinsic* parameters, which relate the camera's coordinate system to a fixed world coordinate system and specify its position and orientation in space.

We ignore in the rest of this chapter the fact that, for cameras equipped with a lens, a point is only in focus when its depth and the distance between the optical center of the camera and its image plane obey the thin lens Eq. (1.6). Likewise, the nonlinear aberrations associated with real lenses are not taken into account by Eq. (1.1). We neglect these aberrations in this chapter, but revisit radial distortion in chapter 3 when we address the problem of estimating the intrinsic and extrinsic parameters of a camera (a process known as *geometric camera calibration*).

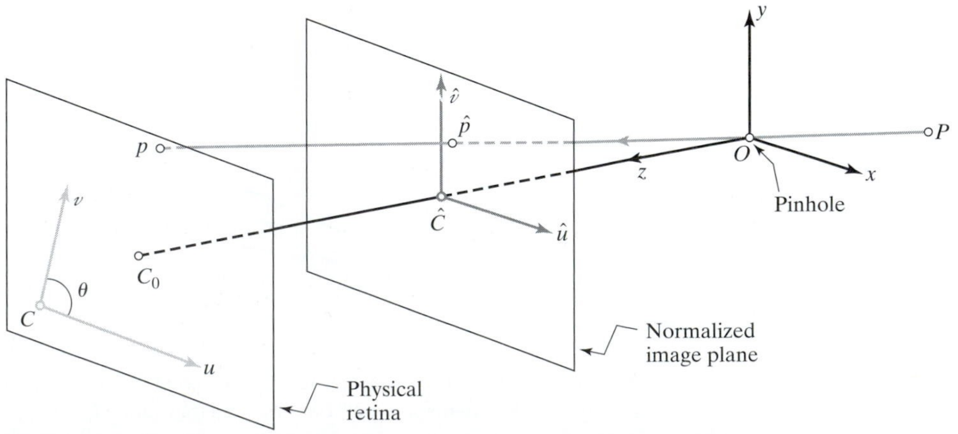

Figure 2.8 Physical and normalized image coordinate systems.

2.2.1 Intrinsic Parameters

It is possible to associate with a camera a *normalized image plane* parallel to its physical retina but located at a unit distance from the pinhole. We attach to this plane its own coordinate system with an origin located at the point $\hat{C}$ where the optical axis pierces it (Figure 2.8). The perspective projection Eq. (1.1) can be written in this normalized coordinate system as

$$\begin{cases} \hat{u} = \dfrac{x}{z} \\ \hat{v} = \dfrac{y}{z} \end{cases} \iff \hat{\boldsymbol{p}} = \frac{1}{z}(\text{Id} \quad \mathbf{0})\begin{pmatrix} \boldsymbol{P} \\ 1 \end{pmatrix}, \tag{2.9}$$

where $\hat{\boldsymbol{p}} \stackrel{\text{def}}{=} (\hat{u}, \hat{v}, 1)^T$ is the vector of homogeneous coordinates of the projection $\hat{p}$ of the point P into the normalized image plane.

The physical retina of the camera is in general different (Figure 2.8): It is located at a distance $f \neq 1$ from the pinhole,[3] and the image coordinates (u, v) of the image point p are usually expressed in pixel units (instead of, say, meters). In addition, pixels are normally rectangular instead of square, so the camera has two additional scale parameters k and l, and

$$\begin{cases} u = kf\dfrac{x}{z}, \\ v = lf\dfrac{y}{z}. \end{cases} \tag{2.10}$$

Let us talk units for a second: f is a distance, expressed in meters, for example, and a pixel has dimensions $\frac{1}{k} \times \frac{1}{l}$, where k and l are expressed in pixel $\times$ m^{-1}. The parameters k, l, and f are not independent and can be replaced by the magnifications $\alpha = kf$ and $\beta = lf$ expressed in pixel units.

In general, the origin of the camera coordinate system is at a corner C of the retina (e.g., in the case depicted in Figure 2.8, the lower left corner or sometimes the upper-left corner, when the image coordinates are the row and column indexes of a pixel) and not at its center, and the center of the CCD matrix usually does not coincide with the principal point C_0. This adds two parameters u_0 and v_0 that define the position (in pixel units) of C_0 in the retinal coordinate system, and Eq. (2.10) is replaced by

$$\begin{cases} u = \alpha\dfrac{x}{z} + u_0, \\ v = \beta\dfrac{y}{z} + v_0. \end{cases} \tag{2.11}$$

Finally, the camera coordinate system may also be skewed due to some manufacturing error, so the angle θ between the two image axes is not equal to (but of course not very different from either) 90°. In this case, it is easy to show that Eq. (2.11) transforms into

$$\begin{cases} u = \alpha\dfrac{x}{z} - \alpha\cot\theta\dfrac{y}{z} + u_0, \\ v = \dfrac{\beta}{\sin\theta}\dfrac{y}{z} + v_0. \end{cases} \tag{2.12}$$

[3]From now on, we assume that the camera is focused at infinity so the distance between the pinhole and image plane is equal to the focal length.

Combining Eqs. (2.9) and (2.12) now allows us to write the change in coordinates between the physical image frame and the normalized one as a planar affine transformation—that is,

$$p = \mathcal{K}\hat{p}, \quad \text{where} \quad p = \begin{pmatrix} u \\ v \\ 1 \end{pmatrix} \quad \text{and} \quad \mathcal{K} \stackrel{\text{def}}{=} \begin{pmatrix} \alpha & -\alpha \cot\theta & u_0 \\ 0 & \dfrac{\beta}{\sin\theta} & v_0 \\ 0 & 0 & 1 \end{pmatrix}. \tag{2.13}$$

Putting it all together, we obtain

$$p = \frac{1}{z}\mathcal{M}P, \quad \text{where} \quad \mathcal{M} \stackrel{\text{def}}{=} \begin{pmatrix} \mathcal{K} & \mathbf{0} \end{pmatrix}, \tag{2.14}$$

and $P = (x, y, z, 1)^T$ denotes this time the *homogeneous* coordinate vector of P in the camera coordinate system. In other words, homogeneous coordinates can be used to represent the perspective projection mapping by the 3×4 matrix $\mathcal{M}$.

Note that the physical size of the pixels and the skew are always fixed for a given camera and frame grabber, and in principle they can be measured during manufacturing (of course, this information may not be available—for example, in the case of stock film footage, or when the frame grabber's digitization rate is unknown). For zoom lenses, the focal length may vary with time, along with the image center when the optical axis of the lens is not exactly perpendicular to the image plane. Simply changing the focus of the camera also affects the magnification since it changes the lens-to-retina distance, but we continue to assume that the camera is focused at infinity and ignore this effect in the rest of this chapter.

2.2.2 Extrinsic Parameters

Let us now consider the case where the camera frame (C) is distinct from the world frame (W). Noting that

$$\begin{pmatrix} {}^C P \\ 1 \end{pmatrix} = \begin{pmatrix} {}^C_W\mathcal{R} & {}^C O_W \\ \mathbf{0}^T & 1 \end{pmatrix} \begin{pmatrix} {}^W P \\ 1 \end{pmatrix}$$

and substituting in Eq. (2.14) yields

$$p = \frac{1}{z}\mathcal{M}P, \quad \text{where} \quad \mathcal{M} = \mathcal{K}\begin{pmatrix} \mathcal{R} & t \end{pmatrix}, \tag{2.15}$$

$\mathcal{R} = {}^C_W\mathcal{R}$ is a rotation matrix, $t = {}^C O_W$ is a translation vector, and $P = ({}^W x, {}^W y, {}^W z, 1)^T$ denotes the *homogeneous* coordinate vector of P in the frame (W).

This is the most general form of the perspective projection equation. We can use it to determine the position of the camera's optical center O in the world coordinate system. Indeed, as shown in the exercises, its *homogenous* coordinate vector $\boldsymbol{O}$ verifies $\mathcal{M}\boldsymbol{O} = 0$. (Intuitively, this is rather obvious since the optical center is the only point whose image is not uniquely defined.) In particular, if $\mathcal{M} = \begin{pmatrix} \mathcal{A} & \boldsymbol{b} \end{pmatrix}$, where $\mathcal{A}$ is a nonsingular 3×3 matrix and $\boldsymbol{b}$ is a vector in $\mathbb{R}^3$, then the *nonhomogeneous* coordinate vector of the point O is simply $-\mathcal{A}^{-1}\boldsymbol{b}$.

It is important to understand that the depth z in Eq. (2.15) is *not* independent of $\mathcal{M}$ and $\mathcal{P}$ since, if $\boldsymbol{m}_1^T$, $\boldsymbol{m}_2^T$, and $\boldsymbol{m}_3^T$ denote the three rows of $\mathcal{M}$, it follows directly from Eq. (2.15) that $z = \boldsymbol{m}_3 \cdot \boldsymbol{P}$. In fact, it is sometimes convenient to rewrite Eq. (2.15) in the equivalent form:

$$\begin{cases} u = \dfrac{\boldsymbol{m}_1 \cdot \boldsymbol{P}}{\boldsymbol{m}_3 \cdot \boldsymbol{P}}, \\[2ex] v = \dfrac{\boldsymbol{m}_2 \cdot \boldsymbol{P}}{\boldsymbol{m}_3 \cdot \boldsymbol{P}}. \end{cases} \tag{2.16}$$

A projection matrix can be written explicitly as a function of its five intrinsic parameters (α, β, u_0, v_0, and θ) and its six extrinsic ones (the three angles defining $\mathcal{R}$ and the three coordinates of $\boldsymbol{t}$), namely,

$$\mathcal{M} = \begin{pmatrix} \alpha \boldsymbol{r}_1^T - \alpha \cot\theta \boldsymbol{r}_2^T + u_0 \boldsymbol{r}_3^T & \alpha t_x - \alpha \cot\theta t_y + u_0 t_z \\ \dfrac{\beta}{\sin\theta}\boldsymbol{r}_2^T + v_0 \boldsymbol{r}_3^T & \dfrac{\beta}{\sin\theta} t_y + v_0 t_z \\ \boldsymbol{r}_3^T & t_z \end{pmatrix}, \tag{2.17}$$

where $\boldsymbol{r}_1^T$, $\boldsymbol{r}_2^T$, and $\boldsymbol{r}_3^T$ denote the three rows of the matrix $\mathcal{R}$ and t_x, t_y, and t_z are the coordinates of the vector $\boldsymbol{t}$. If $\mathcal{R}$ is written as the product of three elementary rotations, the vectors $\boldsymbol{r}_i$ $(i = 1, 2, 3)$ can of course be written explicitly in terms of the corresponding three angles.

2.2.3 A Characterization of Perspective Projection Matrices

This section examines the conditions under which a 3×4 matrix $\mathcal{M}$ can be written in the form given by Eq. (2.17). Let us write without loss of generality $\mathcal{M} = (\mathcal{A} \quad \boldsymbol{b})$, where $\mathcal{A}$ is a 3×3 matrix and $\boldsymbol{b}$ is an element of $\mathbb{R}^3$, and let us denote by $\boldsymbol{a}_3^T$ the third row of $\mathcal{A}$. Clearly, if $\mathcal{M}$ is an instance of Eq. (2.17), then $\boldsymbol{a}_3^T$ must be a unit vector since it is equal to $\boldsymbol{r}_3^T$, the last row of a rotation matrix. Note, however, that replacing $\mathcal{M}$ by $\lambda\mathcal{M}$ in Eq. (2.16) for some arbitrary $\lambda \neq 0$ does not change the corresponding image coordinates. This leads us in the rest of this book to consider projection matrices as *homogeneous objects*, only defined up to scale, whose canonical form of Eq. (2.17) can be obtained by choosing a scale factor such that $|\boldsymbol{a}_3| = 1$. Note that the parameter z in Eq. (2.15) can only rightly be interpreted as the depth of the point P when $\mathcal{M}$ is written in this canonical form. Note also that the number of intrinsic and extrinsic parameters of a camera matches the 11 free parameters of the (homogeneous) matrix $\mathcal{M}$.

We say that a 3×4 matrix that can be written (up to scale) as Eq. (2.17) for some set of intrinsic and extrinsic parameters is a *perspective projection matrix*. It is of practical interest to put some restrictions on the intrinsic parameters of a camera since, as noted earlier, some of these parameters are fixed and may be known. In particular, we say that a 3×4 matrix is a *zero-skew perspective projection matrix* when it can be rewritten (up to scale) as Eq. (2.17) with $\theta = \pi/2$, and that it is a *perspective projection matrix with zero skew and unit aspect-ratio* when it can be rewritten (up to scale) as Eq. (2.17) with $\theta = \pi/2$ and $\alpha = \beta$. A camera with *known* nonzero skew and nonunit aspect-ratio can be transformed into a camera with zero skew and unit aspect-ratio by an appropriate change of image coordinates. Are arbitrary 3×4 matrices perspective projection matrices? The following theorem answers this question.

Theorem 1. *Let* $\mathcal{M} = (\mathcal{A} \quad \boldsymbol{b})$ *be a* 3×4 *matrix, and let* $\boldsymbol{a}_i^T$ $(i = 1, 2, 3)$ *denote the rows of the matrix* $\mathcal{A}$ *formed by the three leftmost columns of* $\mathcal{M}$.

- *A necessary and sufficient condition for* $\mathcal{M}$ *to be a perspective projection matrix is that* $\operatorname{Det}(\mathcal{A}) \neq 0$.
- *A necessary and sufficient condition for* $\mathcal{M}$ *to be a zero-skew perspective projection matrix is that* $\operatorname{Det}(\mathcal{A}) \neq 0$ *and*

$$(\boldsymbol{a}_1 \times \boldsymbol{a}_3) \cdot (\boldsymbol{a}_2 \times \boldsymbol{a}_3) = 0.$$

- *A necessary and sufficient condition for* $\mathcal{M}$ *to be a perspective projection matrix with zero skew and unit aspect-ratio is that* $\operatorname{Det}(\mathcal{A}) \neq 0$ *and*

$$\begin{cases} (\boldsymbol{a}_1 \times \boldsymbol{a}_3) \cdot (\boldsymbol{a}_2 \times \boldsymbol{a}_3) = 0, \\ (\boldsymbol{a}_1 \times \boldsymbol{a}_3) \cdot (\boldsymbol{a}_1 \times \boldsymbol{a}_3) = (\boldsymbol{a}_2 \times \boldsymbol{a}_3) \cdot (\boldsymbol{a}_2 \times \boldsymbol{a}_3). \end{cases}$$

The conditions of the theorem are clearly necessary: According to Eq. (2.15), we have $\mathcal{A} = \mathcal{KR}$, thus the determinants of $\mathcal{A}$ and $\mathcal{K}$ are the same and $\mathcal{A}$ is nonsingular. Further, a simple calculation shows that the rows of $\mathcal{KR}$ in Eq. (2.17) satisfy the conditions of the theorem under the various assumptions imposed by its statement. The theorem conditions are proved to be sufficient in Faugeras (1993) and in the exercises.

2.3 AFFINE CAMERAS AND AFFINE PROJECTION EQUATIONS

When a scene's relief is small compared with the overall distance separating it from the camera observing it, *affine projection models* can be used to approximate the imaging process. These include the *orthographic* and *weak-perspective* projection models introduced in chapter 1 as well as the parallel and paraperspective models introduced in this section. Their name is justified in chapter 12.

2.3.1 Affine Cameras

Under orthographic projection, the imaging process is simply modeled as an orthogonal projection onto the image plane. This is a reasonable approximation of perspective projection for distant objects lying at a roughly constant distance from the cameras observing them. The *parallel projection* model subsumes the orthographic one and takes into account that the objects of interest may lie off the optical axis of the camera. In this model, the viewing rays are parallel to each other, but are not necessarily perpendicular to the image plane.

The weak-perspective and paraperspective projection models generalize the orthographic and parallel projections models to allow for variations in the depth of an object relative to the camera observing it (Figure 2.9). Let O denote the optical center of this camera, and let R denote a scene reference point; the weak-perspective projection of a scene point P is constructed in two

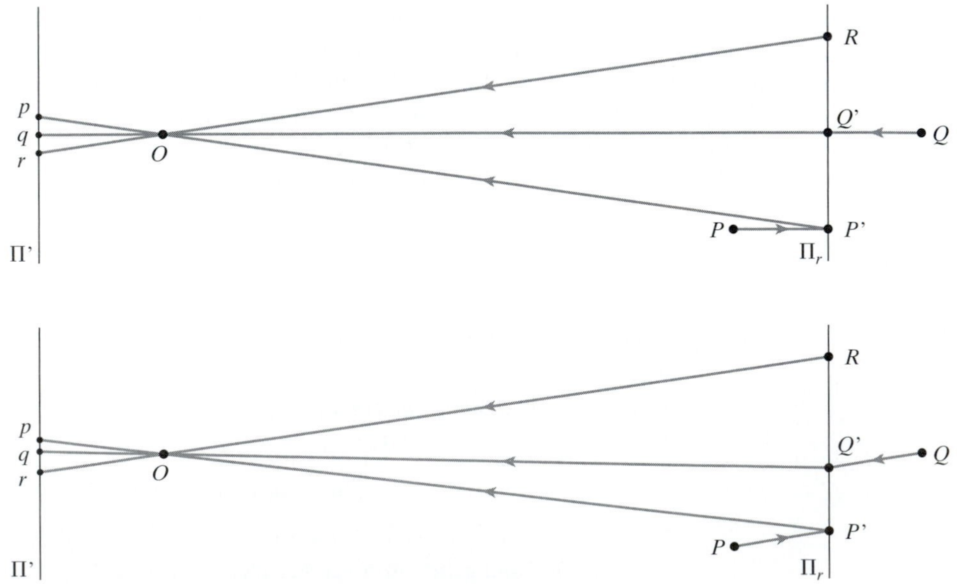

Figure 2.9 Affine projection models: (top) weak-perspective and (bottom) paraperspective projections.

steps: P is first projected orthographically onto a point P' of the plane Π_r parallel to the image plane Π' and passing through R; perspective projection is then used to map the point P' onto the image point p (top of Figure 2.9). Since Π_r is a fronto-parallel plane, the net effect of the second projection step is a scaling of the image coordinates. The paraperspective model takes into account both the distortions associated with a reference point that is off the optical axis of the camera and possible variations in depth (bottom of Figure 2.9): Using the same notation as before and denoting by Δ the line joining the optical center O to the reference point R, parallel projection in the direction of Δ is first used to map P onto a point P' of the plane Π_r; perspective projection is then used to map the point P' onto the image point p.

2.3.2 Affine Projection Equations

Let us derive the weak-perspective projection equation. If z_r denotes the depth of the reference point R, the two elementary projection stages $P \to P' \to p$ can be written in the normalized coordinate system attached to the camera as

$$\begin{pmatrix} x \\ y \\ z \end{pmatrix} \longrightarrow \begin{pmatrix} x \\ y \\ z_r \end{pmatrix} \longrightarrow \begin{pmatrix} \hat{u} \\ \hat{v} \\ 1 \end{pmatrix} = \begin{pmatrix} x/z_r \\ y/z_r \\ 1 \end{pmatrix},$$

or, in matrix form,

$$\begin{pmatrix} \hat{u} \\ \hat{v} \\ 1 \end{pmatrix} = \frac{1}{z_r} \begin{pmatrix} 1 & 0 & 0 & 0 \\ 0 & 1 & 0 & 0 \\ 0 & 0 & 0 & z_r \end{pmatrix} \begin{pmatrix} x \\ y \\ z \\ 1 \end{pmatrix}.$$

Introducing the calibration matrix $\mathcal{K}$ of the camera and its extrinsic parameters $\mathcal{R}$ and $\boldsymbol{t}$ gives the general form of the projection equation—that is,

$$\begin{pmatrix} u \\ v \\ 1 \end{pmatrix} = \frac{1}{z_r} \mathcal{K} \begin{pmatrix} 1 & 0 & 0 & 0 \\ 0 & 1 & 0 & 0 \\ 0 & 0 & 0 & z_r \end{pmatrix} \begin{pmatrix} \mathcal{R} & \boldsymbol{t} \\ \boldsymbol{0}^T & 1 \end{pmatrix} \begin{pmatrix} \boldsymbol{P} \\ 1 \end{pmatrix}, \tag{2.18}$$

where $\boldsymbol{P}$ denotes as usual the *nonhomogeneous* coordinate vector of P in the world reference frame. Finally, noting that z_r is a constant and writing

$$\mathcal{K} = \begin{pmatrix} \mathcal{K}_2 & \boldsymbol{p}_0 \\ \boldsymbol{0}^T & 1 \end{pmatrix}, \quad \text{where} \quad \mathcal{K}_2 \stackrel{\text{def}}{=} \begin{pmatrix} \alpha & -\alpha \cot\theta \\ 0 & \dfrac{\beta}{\sin\theta} \end{pmatrix} \quad \text{and} \quad \boldsymbol{p}_0 \stackrel{\text{def}}{=} \begin{pmatrix} u_0 \\ v_0 \end{pmatrix}$$

allows us to rewrite Eq. (2.18) as

$$\boldsymbol{p} = \mathcal{M} \begin{pmatrix} \boldsymbol{P} \\ 1 \end{pmatrix}, \quad \text{where} \quad \mathcal{M} = (\mathcal{A} \quad \boldsymbol{b}), \tag{2.19}$$

$\boldsymbol{p} = (u, v)^T$ is the *nonhomogeneous* coordinate vector of the point p, and $\mathcal{M}$ is a 2×4 projection matrix (compare to the general perspective Eq. [2.15]). In this expression, the 2×3 matrix $\mathcal{A}$ and the 2 vector $\boldsymbol{b}$ are, respectively, defined by

$$\mathcal{A} = \frac{1}{z_r} \mathcal{K}_2 \mathcal{R}_2 \quad \text{and} \quad \boldsymbol{b} = \frac{1}{z_r} \mathcal{K}_2 \boldsymbol{t}_2 + \boldsymbol{p}_0,$$

where $\mathcal{R}_2$ denotes the 2×3 matrix formed by the first two rows of $\mathcal{R}$ and $\boldsymbol{t}_2$ denotes the 2 vector formed by the first two coordinates of $\boldsymbol{t}$.

Note that t_z does not appear in the expression of $\mathcal{M}$, and that $\boldsymbol{t}_2$ and $\boldsymbol{p}_0$ are coupled in this expression: The projection matrix does not change when $\boldsymbol{t}_2$ is replaced by $\boldsymbol{t}_2 + \boldsymbol{a}$ and $\boldsymbol{p}_0$ is replaced by $\boldsymbol{p}_0 - \frac{1}{z_r}\mathcal{K}_2\boldsymbol{a}$. This redundancy allows us to arbitrarily choose $u_0 = v_0 = 0$. In other words, the position of the center of the image is immaterial for weak-perspective projection. Note that the values of z_r, α, and β are also coupled in the expression of $\mathcal{M}$, and that the value of z_r is a priori unknown in most applications. This allows us to write

$$\mathcal{M} = \frac{1}{z_r}\begin{pmatrix} k & s \\ 0 & 1 \end{pmatrix}\begin{pmatrix} \mathcal{R}_2 & \boldsymbol{t}_2 \end{pmatrix}, \tag{2.20}$$

where k and s denote, respectively, the aspect ratio and skew of the camera. In particular, a weak-perspective projection matrix is defined by two intrinsic parameters (k and s), five extrinsic parameters (the three angles defining $\mathcal{R}_2$ and the two coordinates of $\boldsymbol{t}_2$), and one scene-dependent *structure* parameter z_r.

It is easy to show (see Exercises) that the paraperspective projection equations can also be written in the general affine form of Eq. (2.19) with

$$\mathcal{M} = \frac{1}{z_r}\begin{pmatrix} k & s \\ 0 & 1 \end{pmatrix}\left(\begin{pmatrix} 1 & 0 & -x_r/z_r \\ 0 & 1 & -y_r/z_r \end{pmatrix}\mathcal{R} \quad \boldsymbol{t}_2\right), \tag{2.21}$$

where x_r, y_r, and z_r denote the coordinates of the reference point R in the normalized camera coordinate system. Note that Eq. (2.21) reduces (as expected) to the weak-perspective projection Eq. (2.20) when $x_r = y_r = 0$. According to Eq. (2.21), a paraperspective projection matrix is defined by two intrinsic parameters (k, s), five extrinsic parameters (the three angles defining $\mathcal{R}$ and the two coordinates of $\boldsymbol{t}_2$), and three structure parameters x_r, y_r, and z_r. In practice, the reference point R is often taken to be a point feature whose projection is observable in the image. Its coordinates x_r, y_r, and z_r cannot of course be measured in the image, but the coordinates u_r and v_r of its projection are readily available. It is easy to rewrite Eq. (2.21) as

$$\mathcal{M} = \frac{1}{z_r}\left(\begin{pmatrix} k & s & u_0 - u_r \\ 0 & 1 & v_0 - v_r \end{pmatrix}\mathcal{R} \quad \begin{pmatrix} k & s \\ 0 & 1 \end{pmatrix}\boldsymbol{t}_2\right). \tag{2.22}$$

In this formulation, the paraperspective projection matrix is defined by four intrinsic parameters (k, s, u_0, and v_0), five extrinsic parameters (the three angles defining $\mathcal{R}$ and the two coordinates of $\boldsymbol{t}_2$), and a single structure parameter z_r.

The orthographic and parallel projection equations are obtained from the weak-perspective and paraperspective ones by fixing the value of z_r to be some constant (in practice, $z_r = 1$) in Eqs. (2.20), (2.21), or (2.22). When several different orthographic (resp. parallel) cameras observe the same scene (or, equivalently, when a zooming camera films an image sequence), the actual image magnifications become relevant, and the simplified calibration matrices used in Eq. (2.20) (resp. Eqs. [2.21] or [2.22]) must be replaced by $\mathcal{K}_2$.

2.3.3 A Characterization of Affine Projection Matrices

A 2×4 matrix $\mathcal{M} = (\mathcal{A} \quad \boldsymbol{b})$, where $\mathcal{A}$ is an arbitrary rank-2 2×3 matrix and $\boldsymbol{b}$ is an arbitrary vector in $\mathbb{R}^2$, is called an *affine projection matrix*. The rank condition follows from the fact that a rank-1 matrix would project all scene points onto a single image line; also note that the matrix $\mathcal{A}$ associated with weak-perspective and paraperspective cameras has rank 2 by construction since, according to Eqs. (2.20), (2.21), and (2.22), it can be written as the product of rank-2 matrices.

Both weak-perspective and general affine projection matrices are defined by eight independent parameters. Paraperspective projection matrices, in contrast, have 10 degrees of freedom.

Weak-perspective and paraperspective projection matrices are, of course, affine ones. Conversely, a simple parameter-counting argument suggests that it should be possible to write an arbitrary affine projection matrix as a weak-perspective or a paraperspective one, but that the latter representation is not unique unless additional constraints are imposed on its form. This is confirmed by the following theorem.

Theorem 2. *An affine projection matrix can be written uniquely (up to a sign ambiguity) as a general weak-perspective projection matrix as defined by Eq. (2.20) or as a paraperspective projection matrix as defined by Eq. (2.21) or (2.22) with* $k = 1$ *and* $s = 0$.

This theorem is proven in Faugeras *et al.* (2001, Propositions 4.26 and 4.27) and the exercises. It shows that any affine projection can be written as a weak-perspective or paraperspective one, and that the geometric properties of these projection models apply to general affine projection. For example, as shown in chapter 12, weak-perspective projection preserves the parallelism of lines, and Theorem 2 implies that this property also holds for arbitrary affine projection. The fact that an arbitrary 2×4 matrix can always be written as a paraperspective projection matrix with $k = 1$ and $s = 0$ should *not* be interpreted as meaning that the aspect ratio of a paraperspective camera or its skew are irrelevant.

2.4 NOTES

Craig (1989) offers a good introduction to coordinate system representations and kinematics. Thorough presentations of geometric camera models can be found in Faugeras (1993), Hartley and Zisserman (2000), and Faugeras *et al.* (2001). The paraperspective projection model was introduced in computer vision by Ohta, Maenobu, and Sakai (1981), and its properties have been studied by Aloimonos (1990). The relationship between paraperspective and affine projection models is discussed in Basri (1996). Equations for the perspective projections of straight lines in terms of their *Plücker coordinates* are derived in Faugeras and Papadopoulo (1997) and the exercises below. The machinery introduced in this chapter is used in the next one to calibrate a camera (i.e., to compute its intrinsic and extrinsic parameters from the image positions of fiducial points). It is also a key to the methods for stereo vision and motion analysis presented in chapters 10 to 13. The main equations derived in this chapter have been collected in Table 2.1 for reference.

PROBLEMS

2.1. Write formulas for the matrices ${}^A_B\mathcal{R}$ when (B) is deduced from (A) via a rotation of angle θ about the axes $\boldsymbol{i}_A$, $\boldsymbol{j}_A$, and $\boldsymbol{k}_A$ respectively.

2.2. Show that rotation matrices are characterized by the following properties: (a) the inverse of a rotation matrix is its transpose and (b) its determinant is 1.

2.3. Show that the set of matrices associated with rigid transformations and equipped with the matrix product forms a group.

2.4. Let ${}^A\mathcal{T}$ denote the matrix associated with a rigid transformation $\mathcal{T}$ in the coordinate system (A), with

$$
{}^A\mathcal{T} = \begin{pmatrix} {}^A\mathcal{R} & {}^A\boldsymbol{t} \\ \boldsymbol{0} & 1 \end{pmatrix}.
$$

Construct the matrix ${}^B\mathcal{T}$ associated with $\mathcal{T}$ in the coordinate system (B) as a function of ${}^A\mathcal{T}$ and the rigid transformation separating (A) and (B).

2.5. Show that if the coordinate system (B) is obtained by applying to the coordinate system (A) the transformation matrix $\mathcal{T}$, then ${}^BP = \mathcal{T}^{-1}\,{}^AP$.

TABLE 2.1 Reference card: geometric camera models.

Plane equation (homogeneous)	$\mathbf{\Pi} \cdot \boldsymbol{P} = ax + by + cz - d = 0$
Quadric surface equation (homogeneous)	$\boldsymbol{P}^T \mathcal{Q} \boldsymbol{P} = 0$ with $\mathcal{Q} = \begin{pmatrix} a_{200} & \frac{1}{2}a_{110} & \frac{1}{2}a_{101} & \frac{1}{2}a_{100} \\ \frac{1}{2}a_{110} & a_{020} & \frac{1}{2}a_{011} & \frac{1}{2}a_{010} \\ \frac{1}{2}a_{101} & \frac{1}{2}a_{011} & a_{002} & \frac{1}{2}a_{001} \\ \frac{1}{2}a_{100} & \frac{1}{2}a_{010} & \frac{1}{2}a_{001} & a_{000} \end{pmatrix}$
Rotation matrix	${}^B_A\mathcal{R} = \begin{pmatrix} \boldsymbol{i}_A \cdot \boldsymbol{i}_B & \boldsymbol{j}_A \cdot \boldsymbol{i}_B & \boldsymbol{k}_A \cdot \boldsymbol{i}_B \\ \boldsymbol{i}_A \cdot \boldsymbol{j}_B & \boldsymbol{j}_A \cdot \boldsymbol{j}_B & \boldsymbol{k}_A \cdot \boldsymbol{j}_B \\ \boldsymbol{i}_A \cdot \boldsymbol{k}_B & \boldsymbol{j}_A \cdot \boldsymbol{k}_B & \boldsymbol{k}_A \cdot \boldsymbol{k}_B \end{pmatrix}$
Change of coordinates (nonhomogeneous)	${}^B P = {}^B_A\mathcal{R}\, {}^A P + {}^B O_A$
Perspective projection equation (homogeneous)	$\boldsymbol{p} = \frac{1}{z}\mathcal{M}\boldsymbol{P}$
Matrix of intrinsic parameters	$\mathcal{K} = \begin{pmatrix} \alpha & -\alpha \cot\theta & u_0 \\ 0 & \beta/\sin\theta & v_0 \\ 0 & 0 & 1 \end{pmatrix}$
Perspective projection matrix	$\mathcal{M} = \mathcal{K}(\mathcal{R} \quad \boldsymbol{t})$
Affine projection equation (nonhomogeneous)	$\boldsymbol{p} = \mathcal{M}\begin{pmatrix} \boldsymbol{P} \\ 1 \end{pmatrix} = \mathcal{A}\boldsymbol{P} + \boldsymbol{b}$
Weak-perspective projection matrix	$\mathcal{M} = (\mathcal{A} \quad \boldsymbol{b}) = \frac{1}{z_r}\begin{pmatrix} k & s \\ 0 & 1 \end{pmatrix}(\mathcal{R}_2 \quad \boldsymbol{t}_2)$
Paraperspective projection matrix I	$\mathcal{M} = \frac{1}{z_r}\begin{pmatrix} k & s \\ 0 & 1 \end{pmatrix}\left(\begin{pmatrix} 1 & 0 & -x_r/z_r \\ 0 & 1 & -y_r/z_r \end{pmatrix}\mathcal{R} \quad \boldsymbol{t}_2\right)$
Paraperspective projection matrix II	$\mathcal{M} = \frac{1}{z_r}\left(\begin{pmatrix} k & s & u_0 - u_r \\ 0 & 1 & v_0 - v_r \end{pmatrix}\mathcal{R} \quad \begin{pmatrix} k & s \\ 0 & 1 \end{pmatrix}\boldsymbol{t}_2\right)$

2.6. Show that the rotation of angle θ about the $\boldsymbol{k}$ axis of the frame (F) can be represented by

$$
{}^F P' = \mathcal{R}\, {}^F P, \quad \text{where} \quad \mathcal{R} = \begin{pmatrix} \cos\theta & -\sin\theta & 0 \\ \sin\theta & \cos\theta & 0 \\ 0 & 0 & 1 \end{pmatrix}.
$$

2.7. Show that the change of coordinates associated with a rigid transformation preserves distances and angles.

2.8. Show that when the camera coordinate system is skewed and the angle θ between the two image axes is not equal to 90°, then Eq. (2.11) transforms into Eq. (2.12).

2.9. Let $\boldsymbol{O}$ denote the *homogeneous* coordinate vector of the optical center of a camera in some reference frame, and let $\mathcal{M}$ denote the corresponding perspective projection matrix. Show that $\mathcal{M}(\boldsymbol{O}) = 0$.

2.10. Show that the conditions of Theorem 1 are necessary.

2.11. Show that the conditions of Theorem 1 are sufficient. Note that the statement of this theorem is a bit different from the corresponding theorems in Faugeras (1993) and Heyden (1995), where the condition $\text{Det}(\mathcal{A}) \neq 0$ is replaced by $\boldsymbol{a}_3 \neq 0$. Of course, $\text{Det}(\mathcal{A}) \neq 0$ implies $\boldsymbol{a}_3 \neq 0$.

2.12. If ${}^A\mathbf{\Pi}$ denotes the homogeneous coordinate vector of a plane Π in the coordinate frame (A), what is the homogeneous coordinate vector ${}^B\mathbf{\Pi}$ of Π in the frame (B)?

2.13. If ${}^A\mathcal{Q}$ denotes the symmetric matrix associated with a quadric surface in the coordinate frame (A), what is the symmetric matrix ${}^B\mathcal{Q}$ associated with this surface in the frame (B)?

2.14. Prove Theorem 2.

2.15. Line Plücker coordinates. The *exterior product* of two vectors $\boldsymbol{u}$ and $\boldsymbol{v}$ in $\mathbb{R}^4$ is defined by

$$\boldsymbol{u} \wedge \boldsymbol{v} \stackrel{\text{def}}{=} \begin{pmatrix} u_1 v_2 - u_2 v_1 \\ u_1 v_3 - u_3 v_1 \\ u_1 v_4 - u_4 v_1 \\ u_2 v_3 - u_3 v_2 \\ u_2 v_4 - u_4 v_2 \\ u_3 v_4 - u_4 v_3 \end{pmatrix}.$$

Given a fixed coordinate system and the (homogeneous) coordinates vectors $\boldsymbol{A}$ and $\boldsymbol{B}$ associated with two points A and B in $\mathbb{E}^3$, the vector $\boldsymbol{L} = \boldsymbol{A} \wedge \boldsymbol{B}$ is called the vector of Plücker coordinates of the line joining A to B.

(a) Let us write $\boldsymbol{L} = (L_1, L_2, L_3, L_4, L_5, L_6)^T$ and denote by O the origin of the coordinate system and by H its projection onto L. Let us also identify the vectors $\overrightarrow{OA}$ and $\overrightarrow{OB}$ with their non-homogeneous coordinate vectors. Show that $\overrightarrow{AB} = -(L_3, L_5, L_6)^T$ and $\overrightarrow{OA} \times \overrightarrow{OB} = \overrightarrow{OH} \times \overrightarrow{AB} = (L_4, -L_2, L_1)^T$. Conclude that the Plücker coordinates of a line obey the quadratic constraint $L_1 L_6 - L_2 L_5 + L_3 L_4 = 0$.

(b) Show that changing the position of the points A and B along the line L only changes the overall scale of the vector $\boldsymbol{L}$. Conclude that Plücker coordinates are homogeneous coordinates.

(c) Prove that the following identity holds of any vectors $\boldsymbol{x}, \boldsymbol{y}, \boldsymbol{z}$, and $\boldsymbol{t}$ in $\mathbb{R}^4$:

$$(\boldsymbol{x} \wedge \boldsymbol{y}) \cdot (\boldsymbol{z} \wedge \boldsymbol{t}) = (\boldsymbol{x} \cdot \boldsymbol{z})(\boldsymbol{y} \cdot \boldsymbol{t}) - (\boldsymbol{x} \cdot \boldsymbol{t})(\boldsymbol{y} \cdot \boldsymbol{z}).$$

(d) Use this identity to show that the mapping between a line with Plücker coordinate vector $\boldsymbol{L}$ and its image l with homogeneous coordinates $\boldsymbol{l}$ can be represented by

$$\rho \boldsymbol{l} = \tilde{\mathcal{M}} \boldsymbol{L}, \quad \text{where} \quad \tilde{\mathcal{M}} \stackrel{\text{def}}{=} \begin{pmatrix} (\boldsymbol{m}_2 \wedge \boldsymbol{m}_3)^T \\ (\boldsymbol{m}_3 \wedge \boldsymbol{m}_1)^T \\ (\boldsymbol{m}_1 \wedge \boldsymbol{m}_2)^T \end{pmatrix}, \tag{2.23}$$

and $\boldsymbol{m}_1^T, \boldsymbol{m}_2^T$, and $\boldsymbol{m}_3^T$ denote as before the rows of $\mathcal{M}$ and ρ is an appropriate scale factor.

Hint: Consider a line L joining two points A and B and denote by a and b the projections of these two points, with homogeneous coordinates $\boldsymbol{a}$ and $\boldsymbol{b}$. Use the fact that the points a and b lie on l, thus if $\boldsymbol{l}$ denote the homogeneous coordinate vector of this line, we must have $\boldsymbol{l} \cdot \boldsymbol{a} = \boldsymbol{l} \cdot \boldsymbol{b} = 0$.

(e) Given a line L with Plücker coordinate vector $\boldsymbol{L} = (L_1, L_2, L_3, L_4, L_5, L_6)^T$ and a point P with homogeneous coordinate vector $\boldsymbol{P}$, show that a necessary and sufficient condition for P to lie on L is that

$$\mathcal{L}\boldsymbol{P} = 0, \quad \text{where} \quad \mathcal{L} \stackrel{\text{def}}{=} \begin{pmatrix} 0 & L_6 & -L_5 & L_4 \\ -L_6 & 0 & L_3 & -L_2 \\ L_5 & -L_3 & 0 & L_1 \\ -L_4 & L_2 & -L_1 & 0 \end{pmatrix}.$$

(f) Show that a necessary and sufficient condition for the line L to lie in the plane Π with homogeneous coordinate vector $\mathbf{\Pi}$ is that

$$\mathcal{L}^* \mathbf{\Pi} = 0, \quad \text{where} \quad \mathcal{L}^* \stackrel{\text{def}}{=} \begin{pmatrix} 0 & L_1 & L_2 & L_3 \\ -L_1 & 0 & L_4 & L_5 \\ -L_2 & -L_4 & 0 & L_6 \\ -L_3 & -L_5 & -L_6 & 0 \end{pmatrix}.$$

3

Geometric Camera Calibration

This chapter addresses the problem of estimating the intrinsic and extrinsic parameters of a camera, a process known as *geometric camera calibration*. We assume throughout that the camera observes a set of features such as points or lines with known positions in some fixed world coordinate system (Figure 3.1): In this context, camera calibration can be modeled as an optimization process, where the discrepancy between the observed image features and their theoretical positions (as predicted by the perspective projection equations derived in chapter 2) is minimized with respect to the camera's intrinsic and extrinsic parameters.

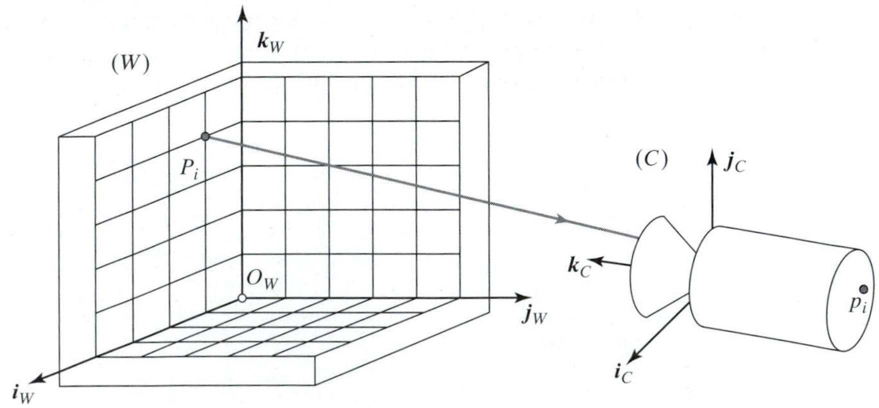

Figure 3.1 Camera calibration setup: In this example, the calibration rig is formed by three grids drawn in orthogonal planes. Other patterns could be used as well, and they may involve lines or other geometric figures.

We start with an overview of *least-squares* techniques aimed at solving this type of optimization problems before presenting several linear and nonlinear approaches to calibration. Once a camera has been calibrated, it is possible to associate with any image point a well-defined ray passing through this point and the camera's optical center as well as perform accurate three-dimensional measurements from digitized pictures. An application to mobile robot localization is briefly discussed at the end of the chapter.

3.1 LEAST-SQUARES PARAMETER ESTIMATION

As already mentioned, calibrating a camera amounts to estimating the intrinsic and extrinsic parameters that minimize the mean-squared deviation from predicted to observed image features. This section introduces a class of optimization techniques, known as least-squares methods, for solving this kind of problem. They prove useful on several other occasions in the rest of this book.

3.1.1 Linear Least-Squares Methods

Let us first consider a system of p linear equations in q unknowns:

$$\begin{cases} u_{11}x_1 + u_{12}x_2 + \cdots + u_{1q}x_q = y_1 \\ u_{21}x_1 + u_{22}x_2 + \cdots + u_{2q}x_q = y_2 \\ \ldots \\ u_{p1}x_1 + u_{p2}x_2 + \cdots + u_{pq}x_q = y_p \end{cases} \iff \mathcal{U}\boldsymbol{x} = \boldsymbol{y}. \tag{3.1}$$

In this equation,

$$\mathcal{U} = \begin{pmatrix} u_{11} & u_{12} & \cdots & u_{1q} \\ u_{21} & u_{22} & \cdots & u_{2q} \\ \ldots & \ldots & \ldots & \ldots \\ u_{p1} & u_{p2} & \cdots & u_{pq} \end{pmatrix}, \quad \boldsymbol{x} = \begin{pmatrix} x_1 \\ x_2 \\ \ldots \\ x_q \end{pmatrix} \quad \text{and} \quad \boldsymbol{y} = \begin{pmatrix} y_1 \\ y_2 \\ \ldots \\ y_p \end{pmatrix}.$$

We know from linear algebra that (in general)

- when $p < q$, the set of solutions to this equation forms a $(q - p)$-dimensional vector subspace of $\mathbb{R}^q$;
- when $p = q$, there is a unique solution;
- when $p > q$, there is no solution.

This statement is true when the *rank* (i.e., the maximum number of independent rows or columns) of $\mathcal{U}$ is maximal—that is, equal to $\min(p, q)$ (this is what we mean by *in general*). When the rank is smaller than $\min(p, q)$, the existence of solutions to Eq. (3.1) depends on the value of $\boldsymbol{y}$ and whether it belongs to the *range* of $\mathcal{U}$ (i.e., the subspace of $\mathbb{R}^p$ spanned by its columns).

Normal Equations and the Pseudoinverse The rest of this section focuses on the overconstrained case $p > q$ and assumes that $\mathcal{U}$ has maximal rank q. Since there is no exact solution in this case, we content ourselves with finding the vector $\boldsymbol{x}$ that minimizes the error measure

$$E \stackrel{\text{def}}{=} \sum_{i=1}^{p} (u_{i1}x_1 + \cdots + u_{iq}x_q - y_i)^2 = |\mathcal{U}\boldsymbol{x} - \boldsymbol{y}|^2.$$

E is proportional to the mean-squared error associated with the equations, hence the name of least-squares methods given to techniques for minimizing it.

Now, we can write $E = \boldsymbol{e} \cdot \boldsymbol{e}$, where $\boldsymbol{e} \stackrel{\text{def}}{=} \mathcal{U}\boldsymbol{x} - \boldsymbol{y}$. To find the vector $\boldsymbol{x}$ minimizing E, we write that the derivatives of this error measure with respect to the coordinates x_i $(i = 1, \dots, q)$ of $\boldsymbol{x}$ must be zero—that is,

$$\frac{\partial E}{\partial x_i} = 2\frac{\partial \boldsymbol{e}}{\partial x_i} \cdot \boldsymbol{e} = 0 \quad \text{for} \quad i = 1, \dots, q.$$

But if the columns of $\mathcal{U}$ are the vectors $\boldsymbol{c}_j = (u_{1j}, \dots, u_{mj})^T$ $(j = 1, \dots, q)$, we have

$$\frac{\partial \boldsymbol{e}}{\partial x_i} = \frac{\partial}{\partial x_i}\left[\begin{pmatrix} \boldsymbol{c}_1 & \cdots & \boldsymbol{c}_q \end{pmatrix}\begin{pmatrix} x_1 \\ \cdots \\ x_q \end{pmatrix} - \boldsymbol{y}\right] = \frac{\partial}{\partial x_i}(x_1\boldsymbol{c}_1 + \cdots + x_q\boldsymbol{c}_q - \boldsymbol{y}) = \boldsymbol{c}_i.$$

In particular, writing that $\partial E/\partial x_i = 0$ implies that $\boldsymbol{c}_i^T(\mathcal{U}\boldsymbol{x} - \boldsymbol{y}) = 0$, and stacking the constraints associated with the q coordinates of $\boldsymbol{x}$ yields the *normal equations* associated with our least-squares problem—that is,

$$\mathbf{0} = \begin{pmatrix} \boldsymbol{c}_1^T \\ \cdots \\ \boldsymbol{c}_q^T \end{pmatrix}(\mathcal{U}\boldsymbol{x} - \boldsymbol{y}) = \mathcal{U}^T(\mathcal{U}\boldsymbol{x} - \boldsymbol{y}) \Longleftrightarrow \mathcal{U}^T\mathcal{U}\boldsymbol{x} = \mathcal{U}^T\boldsymbol{y}.$$

When $\mathcal{U}$ has maximal rank q, the matrix $\mathcal{U}^T\mathcal{U}$ is easily shown to be invertible, and the solution of the normal equations is $\boldsymbol{x} = \mathcal{U}^\dagger\boldsymbol{y}$ with $\mathcal{U}^\dagger \stackrel{\text{def}}{=} [(\mathcal{U}^T\mathcal{U})^{-1}\mathcal{U}^T]$. The $q \times q$ matrix $\mathcal{U}^\dagger$ is called the *pseudoinverse* of $\mathcal{U}$. It coincides with $\mathcal{U}^{-1}$ when the matrix $\mathcal{U}$ is square and nonsingular. Linear least-squares problems can be solved without explicitly computing the pseudoinverse, using, for example, QR decomposition or singular value decomposition (more on the latter in chapter 12), which are known to be better behaved numerically.

Homogeneous Systems and Eigenvalue Problems Let us now consider a variant of our original problem, where we have again a system of p linear equations in q unknowns, but the vector $\boldsymbol{y}$ is zero—that is,

$$\begin{cases} u_{11}x_1 + u_{12}x_2 + \cdots + u_{1q}x_q = 0 \\ u_{21}x_1 + u_{22}x_2 + \cdots + u_{2q}x_q = 0 \\ \cdots \\ u_{p1}x_1 + u_{p2}x_2 + \cdots + u_{pq}x_q = 0 \end{cases} \Longleftrightarrow \mathcal{U}\boldsymbol{x} = \mathbf{0}. \tag{3.2}$$

This is a *homogeneous* equation in $\boldsymbol{x}$ (i.e., if $\boldsymbol{x}$ is a solution, so is $\lambda\boldsymbol{x}$ for any $\lambda \neq 0$). When $p = q$ and the matrix $\mathcal{U}$ is nonsingular, Eq. (3.2) admits as a unique solution $\boldsymbol{x} = \mathbf{0}$. Conversely, when $p \geq q$, nontrivial (i.e., nonzero) solutions may only exist when $\mathcal{U}$ is singular with rank strictly smaller than q. In this context, minimizing $E = |\mathcal{U}\boldsymbol{x}|^2$ only makes sense when some additional constraint is imposed on $\boldsymbol{x}$ since the value $\boldsymbol{x} = \mathbf{0}$ yields the zero global minimum of E. By homogeneity, we have $E(\lambda\boldsymbol{x}) = \lambda^2 E(\boldsymbol{x})$, and it is reasonable to choose the constraint $|\boldsymbol{x}|^2 = 1$, which avoids the trivial solution and forces the uniqueness of the result.

The error E can be rewritten as $|\mathcal{U}\boldsymbol{x}|^2 = \boldsymbol{x}^T(\mathcal{U}^T\mathcal{U})\boldsymbol{x}$. The $q \times q$ matrix $\mathcal{U}^T\mathcal{U}$ is by construction symmetric positive semidefinite (i.e., its eigenvalues are all positive or zero), and it can be diagonalized in an orthonormal basis of eigenvectors $\boldsymbol{e}_i$ $(i = 1, \dots, q)$ associated with the eigenvalues $0 \leq \lambda_1 \leq \cdots \leq \lambda_q$. Thus we can write any unit vector as $\boldsymbol{x} = \mu_1\boldsymbol{e}_1 + \cdots + \mu_q\boldsymbol{e}_q$ with $\mu_1^2 + \cdots + \mu_q^2 = 1$. In particular,

$$E(\boldsymbol{x}) - E(\boldsymbol{e}_1) = \boldsymbol{x}^T(\mathcal{U}^T\mathcal{U})\boldsymbol{x} - \boldsymbol{e}_1^T(\mathcal{U}^T\mathcal{U})\boldsymbol{e}_1 = \lambda_1^2\mu_1^2 + \cdots + \lambda_q^2\mu_q^2 - \lambda_1^2$$
$$\geq \lambda_1^2(\mu_1^2 + \cdots + \mu_q^2 - 1) = 0.$$

It follows that the unit vector $\boldsymbol{x}$ minimizing E is the eigenvector $\boldsymbol{e}_1$ associated with the minimum eigenvalue of $\mathcal{U}^T\mathcal{U}$, and the corresponding minimum value of E is λ_1^2. Various methods are available for computing the eigenvectors and eigenvalues of a symmetric matrix, including Jacobi transformations and reduction to tridiagonal form followed by QR decomposition. Singular value decomposition can also be used to compute the eigenvectors and eigenvalues without actually constructing the matrix $\mathcal{U}^T\mathcal{U}$.

Before illustrating the use of homogeneous linear least-squares techniques with an example, let us pause for a minute to consider the slightly more general problem of minimizing $|\mathcal{U}\boldsymbol{x}|^2$ under the constraint $|\mathcal{V}\boldsymbol{x}|^2 = 1$, where $\mathcal{V}$ is an $r \times q$ matrix (this reduces to homogeneous linear least squares when $\mathcal{V} = \text{Id}$). A vector $\boldsymbol{x}$ and a scalar λ such that

$$\mathcal{U}^T\mathcal{U}\boldsymbol{x} = \lambda\mathcal{V}^T\mathcal{V}\boldsymbol{x}$$

are called a *generalized eigenvector* and the corresponding *generalized eigenvalue* of the $q \times q$ symmetric matrices $\mathcal{U}^T\mathcal{U}$ and $\mathcal{V}^T\mathcal{V}$. As shown in the exercises, the solution of our constrained optimization problem is precisely the unit generalized eigenvector associated with the minimum generalized eigenvalue (which is in this case guaranteed to be positive or zero by construction). As before, effective methods for computing the generalized eigenvectors and eigenvalues of a pair of symmetric matrices are available.

Example 3.1 Fitting a line to points in a plane.

Consider n points p_i $(i = 1, \ldots, n)$ in a plane, with coordinates (x_i, y_i) in some fixed coordinate system (Figure 3.2). What is the straight line that best fits these points? To answer this question, we must first quantify how well a line δ fits a set of points or, equivalently, define some error function E measuring the discrepancy between this line and the points. The best-fitting line can then be found by minimizing E.

A reasonable choice for the error function is the mean-squared distance between the points and the line (Figure 3.2). We saw in chapter 2 that the equation of a line with unit normal $\boldsymbol{n} = (a, b)^T$ lying at a distance d from the origin is $ax + by = d$. It is in fact easy to show that the perpendicular distance between a point with coordinates $(x, y)^T$ and this line is $|ax + by - d|$. We can therefore

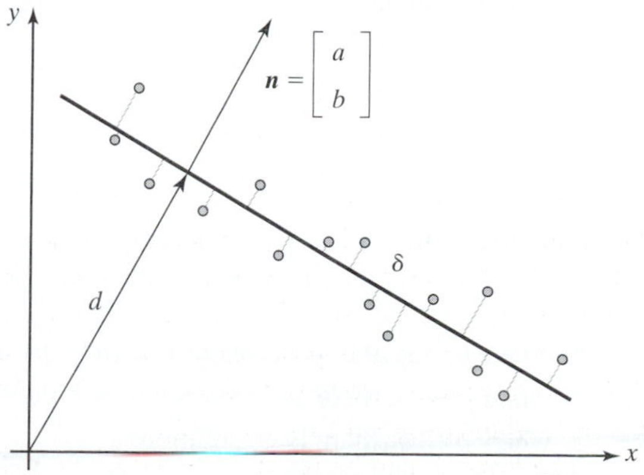

Figure 3.2 The line that best fits n points in the plane can be defined as the line δ that minimizes the mean-squared perpendicular distance to these points (i.e., in this diagram, the mean-squared length of the short parallel line segments joining δ to the points).

use

$$E(a, b, d) = \sum_{i=1}^{n} (ax_i + by_i - d)^2$$

as our error measure, and the line-fitting problem reduces to the minimization of E with respect to a, b, and d under the constraint $a^2 + b^2 = 1$. Differentiating E with respect to d shows that, at a minimum of this function, we must have $0 = \partial E/\partial d = -2\sum_{i=1}^{n}(ax_i + by_i - d)$, thus

$$d = a\bar{x} + b\bar{y}, \quad \text{where} \quad \bar{x} = \frac{1}{n}\sum_{i=1}^{n} x_i \quad \text{and} \quad \bar{y} = \frac{1}{n}\sum_{i=1}^{n} y_i, \tag{3.3}$$

and the two scalars $\bar{x}$ and $\bar{y}$ are simply the coordinates of the center of mass of the input points. Substituting this expression for d in the definition of E yields

$$E = \sum_{i=1}^{n} [a(x_i - \bar{x}) + b(y_i - \bar{y})]^2 = |\mathcal{U}\boldsymbol{n}|^2 \quad \text{where} \quad \mathcal{U} = \begin{pmatrix} x_1 - \bar{x} & y_1 - \bar{y} \\ \cdots & \cdots \\ x_n - \bar{x} & y_n - \bar{y} \end{pmatrix},$$

and our original problem finally reduces to minimizing $|\mathcal{U}\boldsymbol{n}|^2$ with respect to $\boldsymbol{n}$ under the constraint $|\boldsymbol{n}|^2 = 1$. We recognize a homogeneous linear least-squares problem, whose solution is the unit eigenvector associated with the minimum eigenvalue of the 2×2 matrix $\mathcal{U}^T\mathcal{U}$. Once a and b have been computed, the value of d is immediately obtained from Eq. (3.3). Note that $\mathcal{U}^T\mathcal{U}$ is easily shown to be equal to

$$\begin{pmatrix} \sum_{i=1}^{n} x_i^2 - n\bar{x}^2 & \sum_{i=1}^{n} x_i y_i - n\bar{x}\bar{y} \\ \sum_{i=1}^{n} x_i y_i - n\bar{x}\bar{y} & \sum_{i=1}^{n} y_i^2 - n\bar{y}^2 \end{pmatrix},$$

that is, the matrix of second moments of inertia of the points p_i. In fact, the line best fitting these points in the sense defined in this section is simply their axis of least inertia as defined in elementary mechanics.

3.1.2 Nonlinear Least-Squares Methods

Let us now consider a general system of p equations in q unknowns:

$$\begin{cases} f_1(x_1, x_2, \dots, x_q) = 0 \\ f_2(x_1, x_2, \dots, x_q) = 0 \\ \cdots \\ f_p(x_1, x_2, \dots, x_q) = 0 \end{cases} \iff \boldsymbol{f}(\boldsymbol{x}) = \boldsymbol{0}. \tag{3.4}$$

Here, f_i denotes, for $i = 1, \dots, p$, a (possibly nonlinear) differentiable function from $\mathbb{R}^p$ to $\mathbb{R}$, and we take $\boldsymbol{f} = (f_1, \dots, f_p)^T$ and $\boldsymbol{x} = (x_1, \dots, x_q)^T$. In general,

- when $p < q$, the solutions form a $(q - p)$-dimensional *subset* of $\mathbb{R}^q$;
- when $p = q$, there is a *finite set* of solutions;
- when $p > q$, there is no solution.

Let us emphasize the main differences with the linear case: In general, the dimension of the solution set is still $q - p$ in the underconstrained case, but this set does not form a vector space anymore. Its structure depends on the nature of the functions f_i. Likewise, there is usually a finite number of solutions instead of a unique one in the case $p = q$. A precise definition of the general conditions that a family of functions f_i $(i = 1, \ldots, p)$ has to satisfy for the prior statement to be true is unfortunately beyond the scope of this book.

There is no general method for finding all the solutions of Eq. (3.4) when $p = q$ or for finding the global minimum of the least-squares error

$$E(\boldsymbol{x}) \stackrel{\text{def}}{=} |\boldsymbol{f}(\boldsymbol{x})|^2 = \sum_{i=1}^{p} f_i^2(\boldsymbol{x})$$

when $p > q$. Instead, we present next a number of iterative methods that linearize the problem in hope of finding at least one suitable solution. They all rely on a first-order Taylor expansion of the functions f_i in the neighborhood of a point $\boldsymbol{x}$:

$$f_i(\boldsymbol{x} + \delta\boldsymbol{x}) = f_i(\boldsymbol{x}) + \delta x_1 \frac{\partial f_i}{\partial x_1}(\boldsymbol{x}) + \cdots + \delta x_q \frac{\partial f_i}{\partial x_q}(\boldsymbol{x}) + O(|\delta\boldsymbol{x}|^2) \approx f_i(\boldsymbol{x}) + \nabla f_i(\boldsymbol{x}) \cdot \delta\boldsymbol{x}.$$

Here, $\nabla f_i(\boldsymbol{x}) = (\partial f_i/\partial x_1, \ldots, \partial f_i/\partial x_q)^T$ is the *gradient* of f_i at the point $\boldsymbol{x}$, and we have neglected the second-order term $O(|\delta\boldsymbol{x}|^2)$. It follows immediately that

$$\boldsymbol{f}(\boldsymbol{x} + \delta\boldsymbol{x}) \approx \boldsymbol{f}(\boldsymbol{x}) + \mathcal{J}_{\boldsymbol{f}}(\boldsymbol{x})\delta\boldsymbol{x}, \tag{3.5}$$

where $\mathcal{J}_{\boldsymbol{f}}(\boldsymbol{x})$ is the *Jacobian* of $\boldsymbol{f}$—that is, the $p \times q$ matrix

$$\mathcal{J}_{\boldsymbol{f}}(\boldsymbol{x}) \stackrel{\text{def}}{=} \begin{pmatrix} \nabla f_1^T(\boldsymbol{x}) \\ \cdots \\ \nabla f_p^T(\boldsymbol{x}) \end{pmatrix} = \begin{pmatrix} \frac{\partial f_1}{\partial x_1}(\boldsymbol{x}) & \cdots & \frac{\partial f_1}{\partial x_q}(\boldsymbol{x}) \\ \cdots & \cdots & \cdots \\ \frac{\partial f_p}{\partial x_1}(\boldsymbol{x}) & \cdots & \frac{\partial f_p}{\partial x_q}(\boldsymbol{x}) \end{pmatrix}.$$

Newton's Method: Square Systems of Nonlinear Equations As mentioned earlier, Eq. (3.4) admits (in general) a finite number of solutions when $p = q$. Although there is no general method for finding all of these solutions when $\boldsymbol{f}$ is arbitrary, Eq. (3.5) can be used as the basis for a simple iterative algorithm for finding one of these solutions: Given some current estimate $\boldsymbol{x}$ of the solution, the idea is to compute a perturbation $\delta\boldsymbol{x}$ of this estimate such that $\boldsymbol{f}(\boldsymbol{x} + \delta\boldsymbol{x}) \approx \boldsymbol{0}$, or, according to Eq. (3.5),

$$\mathcal{J}_{\boldsymbol{f}}(\boldsymbol{x})\delta\boldsymbol{x} = -\boldsymbol{f}(\boldsymbol{x}).$$

When the Jacobian is nonsingular, $\delta\boldsymbol{x}$ is easily found as the solution of this $q \times q$ system of linear equations, and the process is repeated until convergence.

Newton's method converges rapidly once close to a solution: It has a *quadratic convergence rate* (i.e., the error at step $k + 1$ is proportional to the square of the error at step k). When started far from a solution, Newton's method as presented here may be unreliable. Various strategies can be used to improve its robustness, but their discussion is beyond the scope of this book.

Newton's Method: Overconstrained Systems of Nonlinear Equations When p is greater than q, we seek a local minimum of the least squares error E. Newton's method can be adapted to this case by noting that such a minimum is a zero of the error's gradient. More precisely, we introduce $\boldsymbol{F}(\boldsymbol{x}) = \frac{1}{2}\nabla E(\boldsymbol{x})$ and use Newton's method to find the desired minimum as a solution of the $q \times q$ system of nonlinear equations $\boldsymbol{F}(\boldsymbol{x}) = \boldsymbol{0}$. Differentiating E shows that

$$\boldsymbol{F}(\boldsymbol{x}) = \mathcal{J}_{\boldsymbol{f}}^T(\boldsymbol{x})\boldsymbol{f}(\boldsymbol{x}), \tag{3.6}$$

and differentiating this expression shows in turn that the Jacobian of $\boldsymbol{F}$ is

$$\mathcal{J}_{\boldsymbol{F}}(\boldsymbol{x}) = \mathcal{J}_{\boldsymbol{f}}^T(\boldsymbol{x})\mathcal{J}_{\boldsymbol{f}}(\boldsymbol{x}) + \sum_{i=1}^{p} f_i(\boldsymbol{x})\mathcal{H}_{f_i}(\boldsymbol{x}). \tag{3.7}$$

In this equation, $\mathcal{H}_{f_i}(\boldsymbol{x})$ denotes the *Hessian* of f_i—that is, the $q \times q$ matrix of second derivatives

$$\mathcal{H}_{f_i}(\boldsymbol{x}) \stackrel{\text{def}}{=} \begin{pmatrix} \frac{\partial^2 f_i}{\partial x_1^2}(\boldsymbol{x}) & \cdots & \frac{\partial^2 f_i}{\partial x_1 x_q}(\boldsymbol{x}) \\ \cdots & \cdots & \cdots \\ \frac{\partial^2 f_i}{\partial x_1 x_q}(\boldsymbol{x}) & \cdots & \frac{\partial^2 f_i}{\partial x_q^2}(\boldsymbol{x}) \end{pmatrix}.$$

The term $\delta\boldsymbol{x}$ in Newton's method satisfies $\mathcal{J}_{\boldsymbol{F}}(\boldsymbol{x})\delta\boldsymbol{x} = -\boldsymbol{F}(\boldsymbol{x})$. Equivalently, combining Eqs. (3.6) and (3.7) shows that $\delta\boldsymbol{x}$ is the solution of

$$\left[\mathcal{J}_{\boldsymbol{f}}^T(\boldsymbol{x})\mathcal{J}_{\boldsymbol{f}}(\boldsymbol{x}) + \sum_{i=1}^{p} f_i(\boldsymbol{x})\mathcal{H}_{f_i}(\boldsymbol{x})\right]\delta\boldsymbol{x} = -\mathcal{J}_{\boldsymbol{f}}^T(\boldsymbol{x})\boldsymbol{f}(\boldsymbol{x}). \tag{3.8}$$

The Gauss–Newton and Levenberg–Marquardt Algorithms Newton's method requires computing the Hessians of the functions f_i, which may be difficult and/or expensive. We discuss here two other approaches to nonlinear least-squares that do not involve the Hessians. Let us first consider the Gauss–Newton algorithm: In this approach, we use again a first-order Taylor expansion of $\boldsymbol{f}$ to minimize E, but this time we seek the value of $\delta\boldsymbol{x}$ that minimizes $E(\boldsymbol{x} + \delta\boldsymbol{x})$ for a given value of $\boldsymbol{x}$. Substituting Eq. (3.5) into Eq. (3.4) yields

$$E(\boldsymbol{x} + \delta\boldsymbol{x}) = |\boldsymbol{f}(\boldsymbol{x} + \delta\boldsymbol{x})|^2 \approx |\boldsymbol{f}(\boldsymbol{x}) + \mathcal{J}_{\boldsymbol{f}}(\boldsymbol{x})\delta\boldsymbol{x}|^2.$$

At this point, we are back in the linear least-squares setting, and the adjustment $\delta\boldsymbol{x}$ can be computed as the solution of $\mathcal{J}_{\boldsymbol{f}}^{\dagger}(\boldsymbol{x})\delta\boldsymbol{x} = -\boldsymbol{f}(\boldsymbol{x})$ or, equivalently, according to the definition of the pseudoinverse,

$$\mathcal{J}_{\boldsymbol{f}}^T(\boldsymbol{x})\mathcal{J}_{\boldsymbol{f}}(\boldsymbol{x})\delta\boldsymbol{x} = -\mathcal{J}_{\boldsymbol{f}}^T(\boldsymbol{x})\boldsymbol{f}(\boldsymbol{x}). \tag{3.9}$$

Comparing Eqs. (3.8) and (3.9), we see that the Gauss–Newton algorithm can be thought of as an approximation of Newton's method where the term involving the Hessians $\mathcal{H}_{f_i}$ has been neglected. This is justified when the values of the functions f_i at a solution (the *residuals*) are small since the matrices $\mathcal{H}_{f_i}$ are multiplied by these residuals in Eq. (3.8). In this case, the performance of the Gauss–Newton algorithm is comparable to that of Newton's method, with (nearly) quadratic convergence close to a solution. When the residuals at the solution are too large, however, it may converge slowly or not at all.

When Eq. (3.9) is replaced by

$$[\mathcal{J}_{\boldsymbol{f}}^T(\boldsymbol{x})\mathcal{J}_{\boldsymbol{f}}(\boldsymbol{x}) + \mu\text{Id}]\delta\boldsymbol{x} = -\mathcal{J}_{\boldsymbol{f}}^T(\boldsymbol{x})\boldsymbol{f}(\boldsymbol{x}), \tag{3.10}$$

where the parameter μ is allowed to vary at each iteration, we obtain the Levenberg–Marquardt algorithm, popular in computer vision circles. This is another variant of Newton's method where the term involving the Hessians is this time approximated by a multiple of the identity matrix. The Levenberg–Marqardt algorithm has convergence properties comparable to its Gauss–Newton cousin, but it is more robust: For example, unlike that algorithm, it can be used when the Jacobian $\mathcal{J}_{\boldsymbol{f}}$ does not have maximal rank and its pseudoinverse does not exist.

3.2 A LINEAR APPROACH TO CAMERA CALIBRATION

It is now time to go back to geometric camera calibration. We assume in this section that a calibration rig is observed by a camera and that the image positions (u_i, v_i) of n points P_i $(i = 1, \dots, n)$ with known homogeneous coordinate vectors $\boldsymbol{P}_i$ have been found in a picture of the rig, either automatically or by hand. We decompose the calibration process into (a) the computation of the perspective projection matrix $\mathcal{M}$ associated with the camera in this coordinate system, followed by (b) the estimation of the intrinsic and extrinsic parameters of the camera from this matrix. Degenerate point configurations for which the first step of this process may fail are identified in Section 3.2.3. As shown shortly, writing that the points p_i are the perspective images of the points P_i imposes a set of n linear constraints on the 11 independent coefficients of the corresponding projection matrix. When $n > 11$, these equations generally do not admit a common root, but the techniques introduced in Section 3.1.1 can be used to effectively construct their solution in the least-squares sense.

3.2.1 Estimation of the Projection Matrix

Let us assume that our camera has nonzero skew. According to Theorem 1 from chapter 2, the matrix $\mathcal{M}$ is not singular, but otherwise arbitrary. Clearing the denominators in the perspective projection Eq. (2.16) yields

$$\begin{cases} (\boldsymbol{m}_1 - u_i \boldsymbol{m}_3) \cdot \boldsymbol{P} = 0, \\ (\boldsymbol{m}_2 - v_i \boldsymbol{m}_3) \cdot \boldsymbol{P} = 0. \end{cases}$$

Collecting the constraints associated with our n points yields a system of $2n$ homogeneous linear equations in the twelve coefficients of the matrix $\mathcal{M}$—namely, $\mathcal{P}\boldsymbol{m} = 0$, where

$$\mathcal{P} \stackrel{\text{def}}{=} \begin{pmatrix} \boldsymbol{P}_1^T & \boldsymbol{0}^T & -u_1\boldsymbol{P}_1^T \\ \boldsymbol{0}^T & \boldsymbol{P}_1^T & -v_1\boldsymbol{P}_1^T \\ \dots & \dots & \dots \\ \boldsymbol{P}_n^T & \boldsymbol{0}^T & -u_n\boldsymbol{P}_n^T \\ \boldsymbol{0}^T & \boldsymbol{P}_n^T & -v_n\boldsymbol{P}_n^T \end{pmatrix} \quad \text{and} \quad \boldsymbol{m} \stackrel{\text{def}}{=} \begin{pmatrix} \boldsymbol{m}_1 \\ \boldsymbol{m}_2 \\ \boldsymbol{m}_3 \end{pmatrix} = 0.$$

When $n \geq 6$, homogeneous linear least-squares can be used to compute the value of the unit vector $\boldsymbol{m}$ (hence the matrix $\mathcal{M}$) that minimizes $|\mathcal{P}\boldsymbol{m}|^2$ as the solution of an eigenvalue problem.

3.2.2 Estimation of the Intrinsic and Extrinsic Parameters

Once the projection matrix $\mathcal{M}$ has been estimated, its expression in terms of the camera intrinsic and extrinsic parameters (Eq. [2.17] in chapter 2) can be used to recover these parameters as follows: We write as before $\mathcal{M} = (\mathcal{A} \quad \boldsymbol{b})$, with $\boldsymbol{a}_1^T$, $\boldsymbol{a}_2^T$, and $\boldsymbol{a}_3^T$ denoting the rows of $\mathcal{A}$. We obtain

$$\rho(\mathcal{A} \quad \boldsymbol{b}) = \mathcal{K}(\mathcal{R} \quad \boldsymbol{t}) \Longleftrightarrow \rho \begin{pmatrix} \boldsymbol{a}_1^T \\ \boldsymbol{a}_2^T \\ \boldsymbol{a}_3^T \end{pmatrix} = \begin{pmatrix} \alpha \boldsymbol{r}_1^T - \alpha \cot\theta \boldsymbol{r}_2^T + u_0 \boldsymbol{r}_3^T \\ \dfrac{\beta}{\sin\theta}\boldsymbol{r}_2^T + v_0 \boldsymbol{r}_3^T \\ \boldsymbol{r}_3^T \end{pmatrix},$$

where ρ is an unknown scale factor introduced here to account for the fact that the recovered matrix $\mathcal{M}$ has unit Frobenius form since $|\mathcal{M}| = |\boldsymbol{m}| = 1$.

In particular, using the fact that the rows of a rotation matrix have unit length and are perpendicular to each other yields immediately

$$\begin{cases} \rho = \varepsilon/|\boldsymbol{a}_3|, \\ \boldsymbol{r}_3 = \rho \boldsymbol{a}_3, \\ u_0 = \rho^2(\boldsymbol{a}_1 \cdot \boldsymbol{a}_3), \\ v_0 = \rho^2(\boldsymbol{a}_2 \cdot \boldsymbol{a}_3), \end{cases} \quad \text{where} \quad \varepsilon = \mp 1. \tag{3.11}$$

Since θ is always in the neighborhood of $\pi/2$ with a positive sine, we have

$$\begin{cases} \rho^2(\boldsymbol{a}_1 \times \boldsymbol{a}_3) = -\alpha \boldsymbol{r}_2 - \alpha \cot\theta \boldsymbol{r}_1, \\ \rho^2(\boldsymbol{a}_2 \times \boldsymbol{a}_3) = \dfrac{\beta}{\sin\theta}\boldsymbol{r}_1, \end{cases} \quad \text{and} \quad \begin{cases} \rho^2|\boldsymbol{a}_1 \times \boldsymbol{a}_3| = \dfrac{|\alpha|}{\sin\theta}, \\ \rho^2|\boldsymbol{a}_2 \times \boldsymbol{a}_3| = \dfrac{|\beta|}{\sin\theta}. \end{cases} \tag{3.12}$$

Thus,

$$\begin{cases} \cos\theta = -\dfrac{(\boldsymbol{a}_1 \times \boldsymbol{a}_3) \cdot (\boldsymbol{a}_2 \times \boldsymbol{a}_3)}{|\boldsymbol{a}_1 \times \boldsymbol{a}_3||\boldsymbol{a}_2 \times \boldsymbol{a}_3|}, \\ \alpha = \rho^2|\boldsymbol{a}_1 \times \boldsymbol{a}_3| \sin\theta, \\ \beta = \rho^2|\boldsymbol{a}_2 \times \boldsymbol{a}_3| \sin\theta, \end{cases} \tag{3.13}$$

since the sign of the magnification parameters α and β is normally known in advance and can be taken to be positive.

We can now compute $\boldsymbol{r}_1$ and $\boldsymbol{r}_2$ from the second part of Eq. (3.12) as

$$\begin{cases} \boldsymbol{r}_1 = \dfrac{\rho^2 \sin\theta}{\beta}(\boldsymbol{a}_2 \times \boldsymbol{a}_3) = \dfrac{1}{|\boldsymbol{a}_2 \times \boldsymbol{a}_3|}(\boldsymbol{a}_2 \times \boldsymbol{a}_3), \\ \boldsymbol{r}_2 = \boldsymbol{r}_3 \times \boldsymbol{r}_1. \end{cases} \tag{3.14}$$

Note that there are two possible choices for the matrix $\mathcal{R}$ depending on the value of ε. The translation parameters can now be recovered by writing $\mathcal{K}\boldsymbol{t} = \rho\boldsymbol{b}$, and hence $\boldsymbol{t} = \rho\mathcal{K}^{-1}\boldsymbol{b}$. In practical situations, the sign of t_z is often known in advance (this corresponds to knowing whether the origin of the world coordinate system is in front or behind the camera), which allows the choice of a unique solution for the calibration parameters.

3.2.3 Degenerate Point Configurations

We now examine the *degenerate configurations* of the points P_i ($i = 1, \dots, n$) that may cause the failure of the camera calibration process. We focus on the (ideal) case where the data points $\boldsymbol{p}_i$ ($i = 1, \dots, n$) can be measured with zero error, and we identify the *nullspace* of the matrix $\mathcal{P}$ (i.e., the subspace of $\mathbb{R}^{12}$ formed by the vectors $\boldsymbol{l}$ such that $\mathcal{P}\boldsymbol{l} = \boldsymbol{0}$).

Let $\boldsymbol{l}$ be such a vector. Introducing the vectors formed by successive quadruples of its coordinates (i.e., $\boldsymbol{\lambda} = (l_1, l_2, l_3, l_4)^T$, $\boldsymbol{\mu} = (l_5, l_6, l_7, l_8)^T$, and $\boldsymbol{\nu} = (l_9, l_{10}, l_{11}, l_{12})^T$) allows us to write

$$\boldsymbol{0} = \mathcal{P}\boldsymbol{l} = \begin{pmatrix} \boldsymbol{P}_1^T & \boldsymbol{0}^T & -u_1\boldsymbol{P}_1^T \\ \boldsymbol{0}^T & \boldsymbol{P}_1^T & -v_1\boldsymbol{P}_1^T \\ \dots & \dots & \dots \\ \boldsymbol{P}_n^T & \boldsymbol{0}^T & -u_n\boldsymbol{P}_n^T \\ \boldsymbol{0}^T & \boldsymbol{P}_n^T & -v_n\boldsymbol{P}_n^T \end{pmatrix} \begin{pmatrix} \boldsymbol{\lambda} \\ \boldsymbol{\mu} \\ \boldsymbol{\nu} \end{pmatrix} = \begin{pmatrix} \boldsymbol{P}_1^T\boldsymbol{\lambda} - u_1\boldsymbol{P}_1^T\boldsymbol{\nu} \\ \boldsymbol{P}_1^T\boldsymbol{\mu} - v_1\boldsymbol{P}_1^T\boldsymbol{\nu} \\ \dots \\ \boldsymbol{P}_n^T\boldsymbol{\lambda} - u_n\boldsymbol{P}_n^T\boldsymbol{\nu} \\ \boldsymbol{P}_n^T\boldsymbol{\mu} - v_n\boldsymbol{P}_n^T\boldsymbol{\nu} \end{pmatrix}. \tag{3.15}$$

Combining Eqs. (2.16) and (3.15) yields

$$\begin{cases} \boldsymbol{P}_i^T \boldsymbol{\lambda} - \dfrac{\boldsymbol{m}_1^T \boldsymbol{P}_i}{\boldsymbol{m}_3^T \boldsymbol{P}_i} \boldsymbol{P}_i^T \boldsymbol{\nu} = 0, \\ \boldsymbol{P}_i^T \boldsymbol{\mu} - \dfrac{\boldsymbol{m}_2^T \boldsymbol{P}_i}{\boldsymbol{m}_3^T \boldsymbol{P}_i} \boldsymbol{P}_i^T \boldsymbol{\nu} = 0, \end{cases} \quad \text{for} \quad i = 1, \dots, n.$$

After clearing the denominators and rearranging the terms, we finally obtain

$$\begin{cases} \boldsymbol{P}_i^T (\boldsymbol{\lambda} \boldsymbol{m}_3^T - \boldsymbol{m}_1 \boldsymbol{\nu}^T) \boldsymbol{P}_i = 0, \\ \boldsymbol{P}_i^T (\boldsymbol{\mu} \boldsymbol{m}_3^T - \boldsymbol{m}_2 \boldsymbol{\nu}^T) \boldsymbol{P}_i = 0, \end{cases} \quad \text{for} \quad i = 1, \dots, n. \tag{3.16}$$

As expected, the vector $\boldsymbol{l}$ associated with $\boldsymbol{\lambda} = \boldsymbol{m}_1$, $\boldsymbol{\mu} = \boldsymbol{m}_2$, and $\boldsymbol{\nu} = \boldsymbol{m}_3$ is a solution of these equations. Are there other solutions?

Let us first consider the case where the points P_i $(i = 1, \dots, n)$ all lie in some plane Π—that is, according to Eq. (2.2), $\boldsymbol{\Pi} \cdot \boldsymbol{P}_i = 0$ for some 4-vector $\boldsymbol{\Pi}$. Clearly, choosing $(\boldsymbol{\lambda}, \boldsymbol{\mu}, \boldsymbol{\nu})$ equal to $(\boldsymbol{\Pi}, \boldsymbol{0}, \boldsymbol{0})$, $(\boldsymbol{0}, \boldsymbol{\Pi}, \boldsymbol{0})$, or $(\boldsymbol{0}, \boldsymbol{0}, \boldsymbol{\Pi})$, or any linear combination of these vectors yields a solution of Eq. (3.16). In other words, the nullspace of $\mathcal{P}$ contains the four-dimensional vector space spanned by these vectors and $\boldsymbol{m}$. In practice, this means that the fiducial points P_i should not all lie in the same plane.

In general, for a given nonzero value of the vector $\boldsymbol{l}$, the points P_i that satisfy Eq. (3.16) must lie on the curve where the two quadric surfaces defined by the corresponding equations intersect. A closer look at Eq. (3.16) reveals that the straight line where the planes defined by $\boldsymbol{m}_3 \cdot \boldsymbol{P} = 0$ and $\boldsymbol{\nu} \cdot \boldsymbol{P} = 0$ intersect lies on both quadrics. It can be shown that the intersection curve of these two surfaces consists of this line and a *twisted cubic* curve Γ passing through the origin. A twisted cubic is entirely determined by six points lying on it, and it follows that seven points chosen at random do not fall on Γ. Since this curve passes through the origin, choosing $n \geq 6$ random points generally guarantees that the matrix $\mathcal{P}$ has rank 11 and the projection matrix can be recovered in a unique fashion.

3.3 TAKING RADIAL DISTORTION INTO ACCOUNT

We have assumed so far that our camera is equipped with a perfect lens. As shown in chapter 1, real lenses suffer from a number of aberrations. In this section, we show how to account for *radial distortion*, a type of aberration that depends on the distance separating the optical axis from the point of interest. We assume that the image center is known so that we can take $u_0 = v_0 = 0$ and model the projection process as

$$\boldsymbol{p} = \frac{1}{z} \begin{pmatrix} 1/\lambda & 0 & 0 \\ 0 & 1/\lambda & 0 \\ 0 & 0 & 1 \end{pmatrix} \mathcal{M} \boldsymbol{P}, \tag{3.17}$$

where λ is a polynomial function of the squared distance d^2 between the image center and the image point p. In most applications, it is sufficient to use a low-degree polynomial (e.g., $\lambda = 1 + \sum_{p=1}^{q} \kappa_p d^{2p}$, with $q \leq 3$) and the *distortion coefficients* κ_p $(p = 1, \dots, q)$ are normally assumed to be small. Note that d^2 is naturally expressed in terms of the *normalized* image coordinates of the point p (i.e., $d^2 = \hat{u}^2 + \hat{v}^2$). Substituting $u_0 = 0$ and $v_0 = 0$ in Eq. (2.13) allows us, after some algebraic manipulation, to rewrite d^2 as a function of u and v instead—namely,

$$d^2 = \frac{u^2}{\alpha^2} + \frac{v^2}{\beta^2} + 2\frac{uv}{\alpha\beta} \cos\theta. \tag{3.18}$$

Using Eq. (3.18) to write λ as an explicit function of u and v in Eq. (3.17) yields highly nonlinear constraints on the $q+11$ camera parameters. Although these parameters in principle can all be found using the general *nonlinear least-squares* techniques introduced in the next section, we prefer here a two-stage approach tailored to the calibration problem: Eliminating λ from Eq. (3.17) first allows us to use *linear* least squares to estimate nine of the camera parameters. The q+2 remaining ones are then computed from Eqs. (3.17) and (3.18) by a simple nonlinear process.

3.3.1 Estimation of the Projection Matrix

Geometrically, radial distortion changes the distance between the image center and the image point p, but it does not affect the direction of the vector joining these two points. This is the *radial alignment constraint* introduced by Tsai (1987*a*), and it can be expressed algebraically by writing

$$\lambda\begin{pmatrix}u\\v\end{pmatrix} = \begin{pmatrix}\dfrac{\boldsymbol{m}_1\cdot\boldsymbol{P}}{\boldsymbol{m}_3\cdot\boldsymbol{P}}\\[2ex] \dfrac{\boldsymbol{m}_2\cdot\boldsymbol{P}}{\boldsymbol{m}_3\cdot\boldsymbol{P}}\end{pmatrix} \Longrightarrow v(\boldsymbol{m}_1\cdot\boldsymbol{P}) - u(\boldsymbol{m}_2\cdot\boldsymbol{P}) = 0. \tag{3.19}$$

Given n fiducial points, we obtain n linear equations in the eight coefficients of the vectors $\boldsymbol{m}_1$ and $\boldsymbol{m}_2$—namely,

$$\mathcal{Q}\boldsymbol{n} = 0, \quad \text{where} \quad \mathcal{Q} \stackrel{\text{def}}{=} \begin{pmatrix} v_1\boldsymbol{P}_1^T & -u_1\boldsymbol{P}_1^T \\ \cdots & \cdots \\ v_n\boldsymbol{P}_n^T & -u_n\boldsymbol{P}_n^T \end{pmatrix} \quad \text{and} \quad \boldsymbol{n} = \begin{pmatrix}\boldsymbol{m}_1\\ \boldsymbol{m}_2\end{pmatrix}. \tag{3.20}$$

Note the similarity with the previous case. When $n \geq 8$, this system of equations is in general overconstrained, and a solution with unit norm can be found using linear least squares.

3.3.2 Estimation of the Intrinsic and Extrinsic Parameters

Once $\boldsymbol{m}_1$ and $\boldsymbol{m}_2$ have been estimated, we can define as before the corresponding values of $\boldsymbol{a}_1$, $\boldsymbol{a}_2$ and write

$$\rho\begin{pmatrix}\boldsymbol{a}_1^T\\ \boldsymbol{a}_2^T\end{pmatrix} = \begin{pmatrix}\alpha\boldsymbol{r}_1^T - \alpha\cot\theta\,\boldsymbol{r}_2^T + u_0\boldsymbol{r}_3^T \\ \dfrac{\beta}{\sin\theta}\boldsymbol{r}_2^T + v_0\boldsymbol{r}_3^T\end{pmatrix}.$$

Calculating the norm and dot product of the vectors $\boldsymbol{a}_1$ and $\boldsymbol{a}_2$ immediately yields the aspect ratio and the skew of the camera as

$$\frac{\beta}{\alpha} = \frac{|\boldsymbol{a}_2|}{|\boldsymbol{a}_1|} \quad \text{and} \quad \cos\theta = -\frac{\boldsymbol{a}_1\cdot\boldsymbol{a}_2}{|\boldsymbol{a}_1||\boldsymbol{a}_2|}. \tag{3.21}$$

Using the fact that $\boldsymbol{r}_2^T$ is the second row of a rotation matrix, and thus has unit norm, now yields

$$\alpha = \varepsilon\rho|\boldsymbol{a}_1|\sin\theta \quad \text{and} \quad \beta = \varepsilon\rho|\boldsymbol{a}_2|\sin\theta, \tag{3.22}$$

where, as before, $\varepsilon = \mp 1$. After some simple algebraic manipulation, we obtain

$$\begin{cases} \boldsymbol{r}_1 = \dfrac{\varepsilon}{\sin\theta}\left(\dfrac{1}{|\boldsymbol{a}_1|}\boldsymbol{a}_1 + \dfrac{\cos\theta}{|\boldsymbol{a}_2|}\boldsymbol{a}_2\right), \\ \boldsymbol{r}_2 = \dfrac{\varepsilon}{|\boldsymbol{a}_2|}\boldsymbol{a}_2. \end{cases}$$

Using these equations and $\boldsymbol{r}_3 = \boldsymbol{r}_1 \times \boldsymbol{r}_2$ allows us to recover the rotation matrix $\mathcal{R}$ up to a twofold ambiguity. Two of the translation parameters can also be recovered by writing

$$\begin{pmatrix} \alpha t_x - \alpha \cot\theta t_y \\ \dfrac{\beta}{\sin\theta} t_y \end{pmatrix} = \rho \begin{pmatrix} b_1 \\ b_2 \end{pmatrix},$$

where b_1 and b_2 are the first two coordinates of the vector $\boldsymbol{b}$, which in turn allows us to compute these parameters t_x and t_y as

$$\begin{cases} t_x = \dfrac{\varepsilon}{\sin\theta}\left(\dfrac{b_1}{|\boldsymbol{a}_1|} + \dfrac{b_2\cos\theta}{|\boldsymbol{a}_2|}\right), \\ t_y = \dfrac{\varepsilon b_2}{|\boldsymbol{a}_2|}. \end{cases}$$

Without further constraints, it is impossible to recover t_z and the absolute scale of the magnification parameters, or equivalently, the value of ρ, from the values of $\boldsymbol{m}_1$ and $\boldsymbol{m}_2$ only. To estimate these parameters, it is necessary to go back to the original projection equations: We rewrite the left side of Eq. (3.19) as

$$\begin{cases} (\boldsymbol{m}_1 - \lambda u \boldsymbol{m}_3) \cdot \boldsymbol{P} = 0, \\ (\boldsymbol{m}_2 - \lambda v \boldsymbol{m}_3) \cdot \boldsymbol{P} = 0, \end{cases} \tag{3.23}$$

Here $\boldsymbol{m}_1$ and $\boldsymbol{m}_2$ are known and, according to Eq. (2.17), $\boldsymbol{m}_3^T = (\boldsymbol{r}_3^T \quad t_z)$, where $\boldsymbol{r}_3$ is also known. Now, combining the expression for d^2 given in Eq. (3.18) with the expressions for α, β, and $\cos\theta$ given in Eqs. (3.21) and (3.22) yields

$$d^2 = \frac{1}{\rho^2}\frac{|u\boldsymbol{a}_2 - v\boldsymbol{a}_1|^2}{|\boldsymbol{a}_1 \times \boldsymbol{a}_2|^2},$$

and substituting this value in Eq. (3.23) yields a nonlinear equation in ρ, t_z, and the distortion parameters κ_p ($p = 1, \ldots, q$). Given enough data points, the nonlinear least-squares techniques that have been presented in Section 3.1.2 can be used to solve for these parameters. These methods are iterative and require initial guesses for all unknowns. Here, a reasonable estimate for ρ and t_z can be found using linear least squares by first assuming that $\lambda = 1$. Likewise, zero values are reasonable initial guesses for the distortion parameters. As before, the twofold ambiguity can be resolved when the sign of t_z is known in advance.

3.3.3 Degenerate Point Configurations

Let us determine the degenerate point configurations for which the vectors $\boldsymbol{m}_1$ and $\boldsymbol{m}_2$ cannot be uniquely determined. Given a vector $\boldsymbol{l}$ in the nullspace, we define the vectors $\boldsymbol{\lambda} = (l_1, l_2, l_3, l_4)^T$ and $\boldsymbol{\mu} = (l_5, l_6, l_7, l_8)^T$ and write

$$\mathbf{0} = \mathcal{Q}\boldsymbol{l} = \begin{pmatrix} v_1\boldsymbol{P}_1^T & -u_1\boldsymbol{P}_1^T \\ \cdots & \cdots \\ v_n\boldsymbol{P}_n^T & -u_n\boldsymbol{P}_n^T \end{pmatrix}\begin{pmatrix} \boldsymbol{\lambda} \\ \boldsymbol{\mu} \end{pmatrix} = \begin{pmatrix} v_1\boldsymbol{P}_1^T\boldsymbol{\lambda} - u_1\boldsymbol{P}_1^T\boldsymbol{\mu} \\ \cdots \\ v_n\boldsymbol{P}_n^T\boldsymbol{\lambda} - u_n\boldsymbol{P}_n^T\boldsymbol{\mu} \end{pmatrix}.$$

Taking into account the values of u_i and v_i, rearranging the terms, and clearing the denominators yields

$$\boldsymbol{P}_i^T(\boldsymbol{m}_2\boldsymbol{\lambda}^T - \boldsymbol{m}_1\boldsymbol{\mu}^T)\boldsymbol{P}_i = 0 \quad \text{for} \quad i = 1, \dots, n. \tag{3.24}$$

The vector $\boldsymbol{l}$ associated with $\boldsymbol{\lambda} = \boldsymbol{m}_1$ and $\boldsymbol{\mu} = \boldsymbol{m}_2$ is of course a solution of these equations (in the noise-free case; i.e., when all image positions are exact). When the points P_i ($i = 1, \dots, n$) all lie in some plane Π or, equivalently, $\boldsymbol{\Pi} \cdot \boldsymbol{P}_i = 0$ for some 4-vector $\boldsymbol{\Pi}$, we can choose $(\boldsymbol{\lambda}, \boldsymbol{\mu})$ equal to $(\boldsymbol{\Pi}, \boldsymbol{0})$, $(\boldsymbol{0}, \boldsymbol{\Pi})$, or any linear combination of these two vectors, and construct a solution of Eq. (3.24). The nullspace of $\mathcal{Q}$ contains the three-dimensional vector space spanned by these vectors and $\boldsymbol{l}$. Thus, as before, points that all lie in the same plane cannot be used in this calibration method.

More generally, for a given value of $\boldsymbol{\lambda}$ and $\boldsymbol{\mu}$, the points P_i form a degenerate configuration when they lie on the quadric surface defined by Eq. (3.24). Note that this surface contains the four straight lines defined by $\boldsymbol{\lambda} \cdot \boldsymbol{P} = \boldsymbol{\mu} \cdot \boldsymbol{P} = 0$, $\boldsymbol{\lambda} \cdot \boldsymbol{P} = \boldsymbol{m}_1 \cdot \boldsymbol{P} = 0$, $\boldsymbol{\mu} \cdot \boldsymbol{P} = \boldsymbol{m}_2 \cdot \boldsymbol{P} = 0$ and $\boldsymbol{m}_1 \cdot \boldsymbol{P} = \boldsymbol{m}_2 \cdot \boldsymbol{P} = 0$. Therefore, it must consist of two planes or be a cone, hyperboloid of one sheet, or hyperbolic paraboloid. In any case, for a large enough number of points in general position, our least-squares problem admits a unique solution.

3.4 ANALYTICAL PHOTOGRAMMETRY

The techniques presented so far ignore some of the constraints associated with the calibration process. For example, the camera skew was assumed to be arbitrary instead of (very close to) zero in Section 3.2. We present in this section a nonlinear approach to camera calibration that takes into account *all* the relevant constraints. This approach is borrowed from *photogrammetry*—an engineering field whose aim is to recover quantitative geometric information from one or several pictures, with applications in cartography, military intelligence, city planning, and so on. For many years, photogrammetry relied on a combination of geometric, optical, and mechanical methods to recover three-dimensional information from pictures, but the advent of computers in the 1950s has made a purely computational approach to this problem feasible. This is the domain of *analytical photogrammetry*, where the intrinsic parameters of a camera define its *interior orientation*, and the extrinsic parameters define its *exterior orientation*.

In this setting, we assume once again that we observe n fiducial points P_i ($i = 1, \dots, n$) whose positions in some world coordinate system are known, and we minimize the mean-squared distance between the measured positions (u_i, v_i) of their images and the positions $(\tilde{u}_i, \tilde{v}_i)$ predicted by the perspective projection equation with respect to a vector of camera parameters $\boldsymbol{\xi} = (\xi_1, \dots, \xi_q)^T$ ($q \geq 11$) that may include various distortion coefficients in addition to the usual intrinsic and extrinsic parameters. The least-squares error can be written as

$$E(\boldsymbol{\xi}) = \sum_{i=1}^{n} [(\tilde{u}_i(\boldsymbol{\xi}) - u_i)^2 + (\tilde{v}_i(\boldsymbol{\xi}) - v_i)^2],$$

where

$$\tilde{u}_i(\boldsymbol{\xi}) \stackrel{\text{def}}{=} \frac{\boldsymbol{m}_1(\boldsymbol{\xi}) \cdot \boldsymbol{P}_i}{\boldsymbol{m}_3(\boldsymbol{\xi}) \cdot \boldsymbol{P}_i} \quad \text{and} \quad \tilde{v}_i(\boldsymbol{\xi}) \stackrel{\text{def}}{=} \frac{\boldsymbol{m}_2(\boldsymbol{\xi}) \cdot \boldsymbol{P}_i}{\boldsymbol{m}_3(\boldsymbol{\xi}) \cdot \boldsymbol{P}_i}.$$

Contrary to the cases studied so far, the dependency of each error term on the unknown parameters $\boldsymbol{\xi}$ is not linear. Instead, it involves a combination of polynomial and trigonometric functions, and minimizing the overall error measure involves the use of the nonlinear least squares algorithms discussed in Section 3.1.2. To follow the notation introduced in that section, we rewrite our error function as

$$E(\boldsymbol{\xi}) = |\boldsymbol{f}(\boldsymbol{\xi})|^2 = \sum_{j=1}^{2n} f_j^2(\boldsymbol{\xi}), \text{ where } \begin{cases} f_{2i-1}(\boldsymbol{\xi}) = \tilde{u}_i(\boldsymbol{\xi}) - u_i \\ f_{2i}(\boldsymbol{\xi}) \quad = \tilde{v}_i(\boldsymbol{\xi}) - v_i \end{cases} \text{ for } i = 1, \dots, n.$$

The Gauss–Newton and Levenberg–Marquard techniques described in Section 3.1.2 require the gradient of the functions f_j, and Newton's method requires both their gradient and Hessian. Here we only calculate the gradient or, equivalently, the Jacobian of $\boldsymbol{f}$. Let us drop the $\boldsymbol{\xi}$ argument for conciseness and define $\tilde{x}_i = \boldsymbol{m}_1 \cdot \boldsymbol{P}_i$, $\tilde{y}_i = \boldsymbol{m}_2 \cdot \boldsymbol{P}_i$ and $\tilde{z}_i = \boldsymbol{m}_3 \cdot \boldsymbol{P}_i$ $(i = 1, \dots, n)$, so $\tilde{u}_i = \tilde{x}_i / \tilde{z}_i$ and $\tilde{v}_i = \tilde{y}_i / \tilde{z}_i$. We have

$$\begin{cases} \dfrac{\partial f_{2i-1}}{\partial \xi_j} = \dfrac{\partial \tilde{u}_i}{\partial \xi_j} = \dfrac{1}{z_i}\dfrac{\partial \tilde{x}_i}{\partial \xi_j} - \dfrac{\tilde{x}_i}{\tilde{z}_i^2}\dfrac{\partial \tilde{z}_i}{\partial \xi_j} = \dfrac{1}{\tilde{z}_i}\left(\dfrac{\partial}{\partial \xi_j}(\boldsymbol{m}_1 \cdot \boldsymbol{P}_i) - \tilde{u}_i \dfrac{\partial}{\partial \xi_j}(\boldsymbol{m}_3 \cdot \boldsymbol{P}_i)\right), \\ \dfrac{\partial f_{2i}}{\partial \xi_j} \quad = \dfrac{\partial \tilde{v}_i}{\partial \xi_j} = \dfrac{1}{\tilde{z}_i}\dfrac{\partial \tilde{y}_i}{\partial \xi_j} - \dfrac{\tilde{y}_i}{\tilde{z}_i^2}\dfrac{\partial \tilde{z}_i}{\partial \xi_j} = \dfrac{1}{\tilde{z}_i}\left(\dfrac{\partial}{\partial \xi_j}(\boldsymbol{m}_2 \cdot \boldsymbol{P}_i) - \tilde{v}_i \dfrac{\partial}{\partial \xi_j}(\boldsymbol{m}_3 \cdot \boldsymbol{P}_i)\right), \end{cases}$$

which is easily rewritten as

$$\begin{pmatrix} \dfrac{\partial f_{2i-1}}{\partial \xi_j} \\ \dfrac{\partial f_{2i}}{\partial \xi_j} \end{pmatrix} = \frac{1}{\tilde{z}_i}\begin{pmatrix} \boldsymbol{P}_i^T & \boldsymbol{0}^T & -\tilde{u}_i \boldsymbol{P}_i^T \\ \boldsymbol{0}^T & \boldsymbol{P}_i^T & -\tilde{v}_i \boldsymbol{P}_i^T \end{pmatrix} \mathcal{J}_{\boldsymbol{m}},$$

where $\boldsymbol{m}$ is as before the vector of $\mathbb{R}^{12}$ associated with $\mathcal{M}$, and $\mathcal{J}_{\boldsymbol{m}}$ denotes its Jacobian with respect to $\boldsymbol{\xi}$. We finally obtain the Jacobian of $\boldsymbol{f}$ as

$$\mathcal{J}_{\boldsymbol{f}} = \begin{pmatrix} \dfrac{1}{\tilde{z}_1}\boldsymbol{P}_1^T & \boldsymbol{0}^T & -\dfrac{\tilde{u}_1}{\tilde{z}_1}\boldsymbol{P}_1^T \\ \boldsymbol{0}^T & \dfrac{1}{\tilde{z}_1}\boldsymbol{P}_1^T & -\dfrac{\tilde{v}_1}{\tilde{z}_1}\boldsymbol{P}_1^T \\ \dots & \dots & \dots \\ \dfrac{1}{\tilde{z}_n}\boldsymbol{P}_n^T & \boldsymbol{0}^T & -\dfrac{\tilde{u}_n}{\tilde{z}_n}\boldsymbol{P}_n^T \\ \boldsymbol{0}^T & \dfrac{1}{\tilde{z}_n}\boldsymbol{P}_n^T & -\dfrac{\tilde{v}_n}{\tilde{z}_n}\boldsymbol{P}_n^T \end{pmatrix} \mathcal{J}_{\boldsymbol{m}}.$$

In this expression, $\tilde{u}_i$, $\tilde{v}_i$, $\tilde{z}_i$, and $\boldsymbol{P}_i$ depend on the point considered, but $\mathcal{J}_{\boldsymbol{m}}$ only depends on the intrinsic and extrinsic parameters of the camera. Note that this method requires an explicit parameterization of the matrix $\mathcal{R}$. Such a parameterization in terms of three elementary rotations about coordinate axes was mentioned in chapter 2. Many other parameterizations can be used as well (see Exercises and chapter 21).

3.5 AN APPLICATION: MOBILE ROBOT LOCALIZATION

The calibration methods presented in this chapter can be used in a variety of applications, from metrology to stereo vision and object localization in robotic tasks. Here we briefly describe the nonlinear approach to camera calibration proposed by Devy *et al.* (1997) and its application to mobile robot localization. Unlike the techniques discussed so far, this method uses several images (up to 20 in the experiments presented here) of a planar rectangular grid to calibrate a

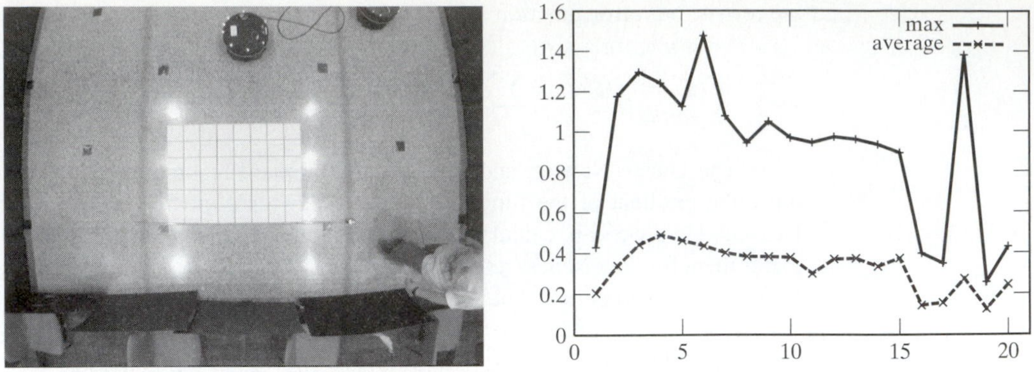

Figure 3.3 Calibration experiments. Left: One of the 20 input pictures; note the strong radial distortion. Note the mobile robot at the top of the photograph with its characteristic LED pattern. Right: Average and maximum reprojections errors (in pixels) in the 20 images. *Calibration and localization software courtesy of Michel Devy. Experiments courtesy of Fred Rothganger.*

static camera (Figure 3.3). One of these pictures is taken with the grid lying on the ground, and it is used to define the world coordinate system. After a rough manual localization, the corners of the grid are found in each picture with a precision of 1/10 pixel using a parametric model of the gray-level surface in the neighborhood of a corner.

The imaging geometry is modeled as in Section 3.3, with three radial distortion coefficients and zero skew. The calibration algorithm recovers a single set of intrinsic parameters, and one set of extrinsic parameters per input image. An initial guess for the intrinsic parameters can be obtained from information supplied by the camera and frame-grabber manufacturers. An initial guess for the extrinsic parameters can be obtained for each image using a variant of Tsai's (1987*a*) algorithm. Briefly, the projection matrix is estimated via linear least squares by choosing the z coordinate axis of the world reference frame perpendicular to the calibration grid. Accordingly, Eq. (3.20) now becomes

$$\mathcal{Q}'\boldsymbol{n}' = 0, \quad \text{where} \quad \mathcal{Q}' = \begin{pmatrix} v_1 x_1 & v_1 y_1 & v_1 & -u_1 x_1 & -u_1 y_1 & -u_1 \\ \cdots & \cdots & \cdots & \cdots & \cdots & \cdots \\ v_n x_n & v_n y_n & v_n & -u_n x_n & -u_1 y_n & -u_n \end{pmatrix}$$

and $\boldsymbol{n}' = (m_{11}, m_{12}, m_{14}, m_{21}, m_{22}, m_{24})^T$.

Note that explicitly imposing that $z_i = 0$ avoids the degeneracies encountered by the previously discussed algorithms for coplanar points. Once $\boldsymbol{n}'$ is known, since the intrinsic parameters are assumed to be (roughly) known, it is a simple matter to compute the extrinsic parameters (see Exercises). Once initial guesses for both the intrinsic and extrinsic parameters are available, nonlinear optimization (in this case, the Levenberg–Marquardt algorithm) can be used to minimize as usual the mean-squared distance between predicted and observed image features.

Figure 3.3 shows the result of experiments conducted with a 576×768 camera equipped with a 4.5 mm lens and plots the errors found when reprojecting the corners of the calibration grid model into the 20 images. Once the camera has been calibrated, it can be used to monitor the position and orientation of mobile robots in the coordinate system attached to the ground reference image. Each robot carries an array of infrared LEDs forming a distinctive pattern (Figure 3.3). During localization experiments, the camera is equipped with an infrared filter that effectively blocks out all incoming light except for that from the LEDs. Each robot is identified using a simple pattern matching algorithm, and its position and orientation are deduced from the image position of the LEDs and the camera parameters. With the camera mounted 4 m above the ground, typical localization errors within the entire field of view of the camera are below 2 cm in position and 1° in orientation, with maximum errors that may reach 5 cm and 5°.

3.6 NOTES

The linear calibration technique described in Section 3.2 is detailed in Faugeras (1993). Its variant that takes radial distortion into account is adapted from Tsai (1987*a*). Haralick and Shapiro (1992) present a concise introduction to analytical photogrammetry. The *Manual of Photogrammetry* is of course the gold standard, and newcomers (such as the authors of this book) will probably find the ingenious mechanisms and rigorous methods described in its various editions fascinating (Thompson *et al.*, 1966, Slama *et al.*, 1980). We come back to photogrammetry in the context of multiple images in chapter 10. General optimization techniques are discussed in Luenberger (1985), Bertsekas (1995), and Heath (2002), for example. An excellent survey and discussion of least-squares methods in the context of analytical photogrammetry can be found in Triggs *et al.* (2000). The output of least-squares methods admits a statistical interpretation in maximum-likelihood terms when the coordinates of the data points are modeled as random variables obeying a normal distribution. We come back to this interpretation in chapter 15, where we also revisit the problem of fitting a straight line to a set of points in the plane.

PROBLEMS

3.1. Show that the vector $\boldsymbol{x}$ that minimizes $|\mathcal{U}\boldsymbol{x}|^2$ under the constraint $|\mathcal{V}\boldsymbol{x}|^2 = 1$ is the unit generalized eigenvector associated with the minimum generalized eigenvalue of the symmetric matrices $\mathcal{U}^T\mathcal{U}$ and $\mathcal{V}^T\mathcal{V}$.

Hint: This problem is equivalent to minimizing (without constraints) the error $E = |\mathcal{U}\boldsymbol{x}|^2/|\mathcal{V}\boldsymbol{x}|^2$ with respect to $\boldsymbol{x}$.

3.2. Show that the 2×2 matrix $\mathcal{U}^T\mathcal{U}$ involved in the line-fitting example from Section 3.1.1 is the matrix of second moments of inertia of the points p_i $(i = 1, \ldots, n)$.

3.3. Extend the line-fitting method presented in Section 3.1.1 to the problem of fitting a plane to n points in $\mathbb{E}^3$.

3.4. Derive an expression for the Hessian of the functions $f_{2i-1}(\boldsymbol{\xi}) = \tilde{u}_i(\boldsymbol{\xi}) - u_i$ and $f_{2i}(\boldsymbol{\xi}) = \tilde{v}_i(\boldsymbol{\xi}) - v_i$ $(i = 1, \ldots, n)$ introduced in Section 3.4.

3.5. Euler angles. Show that the rotation obtained by first rotating about the z axis of some coordinate frame by an angle α, then rotating about the y axis of the new coordinate frame by an angle β and finally rotating about the z axis of the resulting frame by an angle γ can be represented in the original coordinate system by

$$\begin{pmatrix} \cos\alpha\cos\beta\cos\gamma - \sin\alpha\sin\gamma & -\cos\alpha\cos\beta\sin\gamma - \sin\alpha\cos\gamma & \cos\alpha\sin\beta \\ \sin\alpha\cos\beta\cos\gamma + \cos\alpha\sin\gamma & -\sin\alpha\cos\beta\sin\gamma + \cos\alpha\cos\gamma & \sin\alpha\sin\beta \\ -\sin\beta\cos\gamma & \sin\beta\sin\gamma & \cos\beta \end{pmatrix}.$$

3.6. The Rodrigues formula. Consider a rotation $\mathcal{R}$ of angle θ about the axis $\boldsymbol{u}$ (a unit vector). Show that $\mathcal{R}\boldsymbol{x} = \cos\theta\boldsymbol{x} + \sin\theta\,\boldsymbol{u} \times \boldsymbol{x} + (1 - \cos\theta)(\boldsymbol{u} \cdot \boldsymbol{x})\boldsymbol{u}$.

Hint: A rotation does not change the projection of a vector $\boldsymbol{x}$ onto the direction $\boldsymbol{u}$ of its axis and applies a planar rotation of angle θ to the projection of $\boldsymbol{x}$ into the plane orthogonal to $\boldsymbol{u}$.

3.7. Use the Rodrigues formula to show that the matrix associated with $\mathcal{R}$ is

$$\begin{pmatrix} u^2(1-c)+c & uv(1-c)-ws & uw(1-c)+vs \\ uv(1-c)+ws & v^2(1-c)+c & vw(1-c)-us \\ uw(1-c)-vs & vw(1-c)+us & w^2(1-c)+c \end{pmatrix}.$$

where $c = \cos\theta$ and $s = \sin\theta$.

3.8. Assuming that the intrinsic parameters of a camera are known, show how to compute its extrinsic parameters once the vector $\boldsymbol{n}'$ defined in Section 3.5 is known.

Hint: Use the fact that the rows of a rotation matrix have unit norm.

3.9. Assume that n fiducial lines with known Plücker coordinates are observed by a camera.

(a) Show that the line projection matrix $\tilde{\mathcal{M}}$ introduced in the exercises of chapter 2 can be recovered using linear least squares when $n \geq 9$.

(b) Show that once $\tilde{\mathcal{M}}$ is known, the projection matrix $\mathcal{M}$ can also be recovered using linear least squares.

Hint: Consider the rows $\boldsymbol{m}_i$ of $\mathcal{M}$ as the coordinate vectors of three planes Π_i and the rows $\tilde{\boldsymbol{m}}_i$ of $\tilde{\mathcal{M}}$ as the coordinate vectors of three lines, and use the incidence relationships between these planes and these lines to derive linear constraints on the vectors $\boldsymbol{m}_i$.

Programming Assignments

3.10. Use linear least-squares to fit a plane to n points $(x_i, y_i, z_i)^T$ $(i = 1, \dots, n)$ in $\mathbb{R}^3$.

3.11. Use linear least-squares to fit a conic section defined by $ax^2 + bxy + cy^2 + dx + ey + f = 0$ to n points $(x_i, y_i)^T$ $(i = 1, \dots, n)$ in $\mathbb{R}^2$.

3.12. Implement the linear calibration algorithm presented in Section 3.2.

3.13. Implement the calibration algorithm that takes into account radial distortion and that was presented in Section 3.3.

3.14. Implement the nonlinear calibration algorithm from Section 3.4.

4

Radiometry— Measuring Light

This chapter introduces a vocabulary with which we can describe the behavior of light. There are no vision algorithms, but definitions and ideas that are useful later. Some readers may find more detail here than they really want; for their benefit, there is a summary on page 67 which gives quick definitions of the main terms we use.

4.1 LIGHT IN SPACE

The measurement of light is a field known as *radiometry*. We need a series of units that describe how energy is transferred from light sources to surface patches, and what happens to the energy when it arrives at a surface. The first matter to study is the behavior of light in space.

4.1.1 Foreshortening

As a source is tilted with respect to the direction in which the illumination is traveling, it "looks smaller" to a patch of surface viewing the source. Similarly, as a patch is tilted with respect to the direction in which the illumination is traveling, it "looks smaller" to the source. This phenomenon is known as *foreshortening*.

Foreshortening is important because the effect that a distant source has on a surface depends on how that source looks from the point of view of the surface. We can see this very important fact by considering the hemisphere of directions at the relevant point of the surface (Figure 4.1). Radiation can arrive only along these directions. Furthermore, if two different sources result in exactly the same amount of radiation arriving along each incoming direction, they must have the same effect on the surface. This is because they are indistinguishable from the surface.

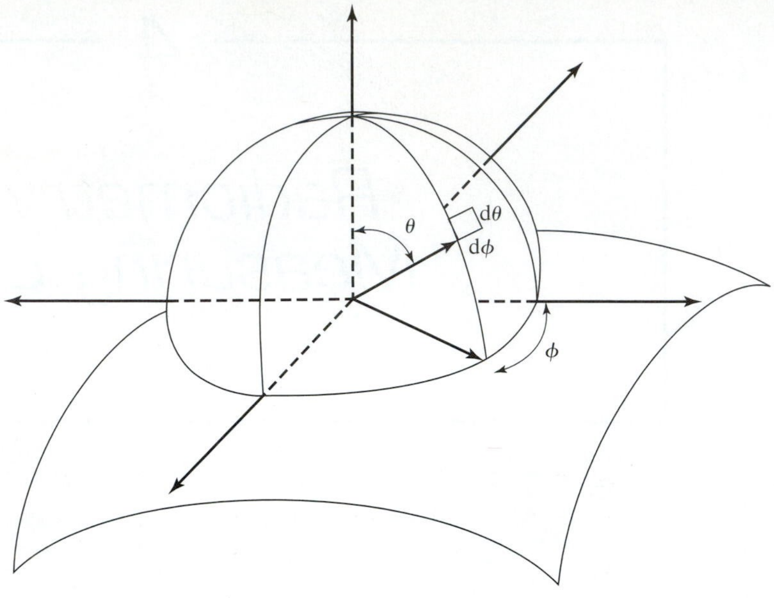

Figure 4.1 A point on a surface sees the world along a hemisphere of directions centered at the point; the surface normal is used to orient the hemisphere to obtain the θ, ϕ coordinate system that we use consistently from now on to describe angular coordinates on this hemisphere. Usually in radiation problems, we compute the brightness of the surface by summing effects due to all incoming directions, so that the fact we have given no clear way to determine the direction in which $\phi = 0$ is not a problem.

The same argument applies to surface patches seen from a source. If I substitute one patch viewing a source with another, and the two different surface patches "look" exactly the same to the source—meaning that every direction leaving the surface strikes material with the same property—then the source cannot tell the patches apart, and so must have transferred energy to each at the same rate.

4.1.2 Solid Angle

The pattern a source generates on an input hemisphere can be described by the *solid angle* that the source subtends. Solid angle is defined by analogy with angle on the plane.

The angle subtended on the plane by an infinitesimal line segment of length dl at a point $\boldsymbol{p}$ can be obtained by projecting the line segment onto the unit circle whose center is at $\boldsymbol{p}$; the length of the result is the required angle in radians (see Figure 4.2). Because the line segment is infinitesimally short, it subtends an infinitesimally small angle, which depends on the distance to the center of the circle and on the orientation of the line:

$$d\phi = \frac{dl \cos \theta_1}{r}$$

and the angle subtended by a curve can be obtained by breaking it into infinitesimal segments and summing (integration!).

Similarly, the solid angle subtended by a patch of surface at a point $\boldsymbol{x}$ is obtained by projecting the patch onto the unit sphere whose center is at $\boldsymbol{x}$; the area of the result is the required

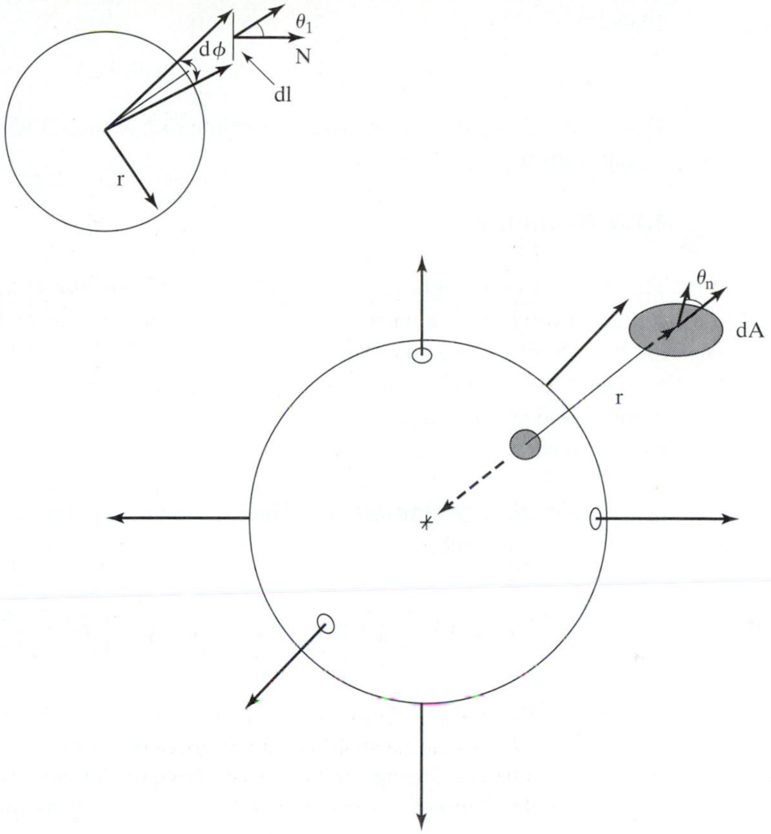

Figure 4.2 **Top:** The angle subtended by a curve segment at a particular point is obtained by projecting the curve onto the unit circle whose center is at that point and then measuring the length of the projection. For a small segment, the angle is $d\phi = (1/r)dl \cos\theta_1$. **Bottom:** A sphere, illustrating the concept of solid angle. The small circles surrounding the coordinate axes are to help you see the drawing as a 3D surface. An infinitesimal patch of surface is projected onto the unit sphere centered at the relevant point; the resulting area is the solid angle of the patch. In this case, the patch is small so that the area and hence the solid angle is $(1/r^2)\, dA \cos\theta_n$.

solid angle, whose unit is now *steradians*. Solid angle is usually denoted by the symbol ω. Notice that solid angle captures the intuition in foreshortening—patches that "look the same" on the input hemisphere subtend the same solid angle.

If the area of the patch dA is small (as suggested by the infinitesimal form), then the infinitesimal solid angle it subtends is easily computed in terms of the area of the patch and the distance to it as

$$d\omega = \frac{dA \cos\theta_n}{r^2},$$

where the terminology is given in Figure 4.2.

Solid angle can be written in terms of the usual angular coordinates on a sphere (illustrated in Figure 4.2). From Figure 4.1 and the expression for the length of circular arcs, we have that infinitesimal steps $(d\theta,\, d\phi)$ in the angles θ and ϕ cut out a region of solid angle on a sphere given by:

$$d\omega = \sin\theta d\theta d\phi.$$

Both of these expressions are worth remembering because they turn out to be useful for a variety of applications.

4.1.3 Radiance

The distribution of light in space is a function of position and direction. For example, consider shining a torch with a narrow beam in an empty room at night—we need to know where the torch is shining from and in what direction it is shining. The effect of the illumination can be represented in terms of the power an infinitesimal patch of surface would receive if it were inserted into space at a particular point and orientation. We use this approach to obtain a unit of measurement.

Definition of Radiance The appropriate unit for measuring the distribution of light in space is *radiance*, which is defined as

> the power (amount of energy per unit time) traveling at some point in a specified direction, per unit area *perpendicular to the direction of travel*, per unit solid angle.

The units of radiance are watts per square meter per steradian ($W \times m^{-2} \times sr^{-1}$). The definition of radiance may look strange, but it is consistent with the most basic phenomenon in radiometry: A small surface patch viewing a source frontally collects more energy than the same patch viewing a source radiance along a nearly tangent direction—the amount of energy a patch collects from a source depends both on how large the source looks from the patch *and* on how large the patch looks from the source. It is important to remember that the square meters in the units for radiance are *foreshortened* (i.e., perpendicular to the direction of travel) to account for this phenomenon.

Radiance is a function of position and direction (the torch with a narrow beam is a good model to keep in mind—you can both move the torch around and point the beam in different directions). Given a point P and some (possibly nonunit) vector $\boldsymbol{v}$, the radiance at P in the direction $\boldsymbol{v}$ is usually denoted $L(P, \boldsymbol{v})$. The point P may lie in free space or on a surface. In the latter case, we sometimes use $L(P, \theta, \phi)$ to denote the radiance at P in a direction with spherical coordinates θ and ϕ in some coordinate system whose z axis is along the surface normal.

Radiance Is Constant Along a Straight Line For the vast majority of important vision problems, it is safe to assume that light does not interact with the medium through which it travels—i.e., that we are in a vacuum. Radiance has the highly desirable property that, for two points P_1 and P_2 (which have a line of sight between them), the radiance leaving P_1 in the direction of P_2 is the same as the radiance arriving at P_2 from the direction of P_1.

The following proof may look vacuous at first glance; it is worth studying carefully because it is the key to a number of other computations. Figure 4.3 shows a patch of surface radiating in a particular direction. From the definition, if the radiance at the patch is $L(P_1, \theta, \phi)$, then the energy transmitted by the patch into an infinitesimal region of solid angle $d\omega$ around the direction θ, ϕ in time dt is

$$L(P_1, \theta, \phi)(\cos\theta_1 dA_1)(d\omega)\,(dt).$$

This is radiance times the foreshortened area of the patch times the solid angle into which the power is radiated times the time for which the power is radiating.

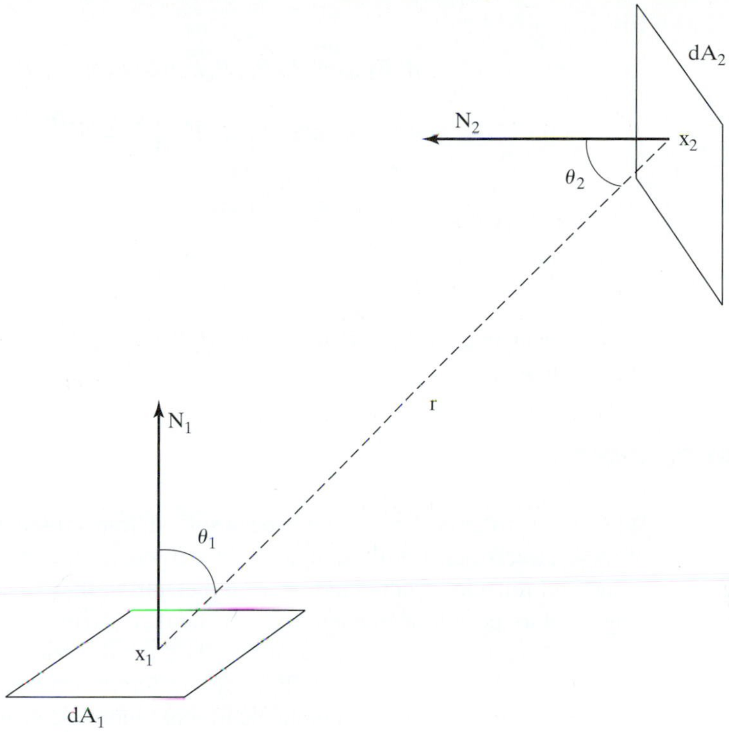

Figure 4.3 Light intensity is best measured in radiance because radiance does not go down along straight line paths in a vacuum (or, for reasonable distances, in clear air). This is shown by an energy conservation argument in the text, where one computes the energy transferred from a patch dA_1 to a patch dA_2.

Now consider two patches, one at P_1 with area dA_1 and the other at P_2 with area dA_2 (see Figure 4.3). To avoid confusion with angular coordinate systems, write the direction from P_1 to P_2 as $\overrightarrow{P_1 P_2}$. The angles θ_1 and θ_2 are as defined in Figure 4.3.

The radiance leaving P_1 in the direction of P_2 is $L(P_1, \overrightarrow{P_1 P_2})$ and the radiance arriving at P_2 from the direction of P_1 is $L(P_2, \overrightarrow{P_1 P_2})$.

This means that, in time dt, the energy leaving P_1 toward P_2 is

$$d^3 E_{1\to 2} = L(P_1, \overrightarrow{P_1 P_2})(\cos\theta_1 dA_1)(d\omega_{2(1)})\,(dt)$$

where $d\omega_{2(1)}$ is the solid angle subtended by patch 2 at patch 1 (energy emitted into this solid angle arrives at 2; all the rest disappears into the void). The notation $d^3 E_{1\to 2}$ implies that there are three infinitesimal terms involved. From the expression for solid angle above,

$$d\omega_{2(1)} = \frac{\cos\theta_2 dA_2}{r^2},$$

so the energy leaving 1 for 2 is:

$$d^3 E_{1\to 2} = L(P_1, \overrightarrow{P_1 P_2})\left(\frac{\cos\theta_1\cos\theta_2}{r^2}\right) dA_2 dA_1\, dt.$$

Because the medium is a vacuum, it does not absorb energy, so that the energy arriving at 2 from 1 is the same as the energy leaving 1 in the direction of 2. The energy arriving at 2 from

1 is

$$
\begin{aligned}
d^3E_{1\to 2} &= L(P_2, \overrightarrow{P_1P_2})(\cos\theta_2 dA_2)(d\omega_{1(2)})\,(dt) \\
&= L(P_2, \overrightarrow{P_1P_2})\left(\frac{\cos\theta_2\cos\theta_1}{r^2}\right) dA_1 dA_2\,dt \\
&= d^3E_{1\to 2} \\
&= L(P_1, \overrightarrow{P_1P_2})\left(\frac{\cos\theta_2\cos\theta_1}{r^2}\right) dA_2 dA_1\,dt,
\end{aligned}
$$

which means that $L(P_2, \overrightarrow{P_1P_2}) = L(P_1, \overrightarrow{P_1P_2})$, so that *radiance is constant along (unoccluded) straight lines.*

4.2 LIGHT AT SURFACES

When light strikes a surface, it may be absorbed, transmitted, or scattered; usually, a combination of these effects occur. For example, light arriving at skin can be scattered at various depths into tissue and reflected from blood or from melanin in there; can be absorbed; or can be scattered tangential to the skin within a film of oil and then escape at some distant point.

The picture is complicated further by the willingness of some surfaces to absorb light at one wavelength and then radiate light at a different wavelength as a result. This effect, known as *fluorescence*, is fairly common: Scorpions fluoresce visible light under x-ray illumination; human teeth fluoresce faint blue under ultraviolet light (nylon underwear tends to fluoresce, too, and false teeth generally do not—the resulting embarrassments led to the demise of uv lights in discotheques); and laundry can be made to look bright by washing powders that fluoresce under ultraviolet light. Furthermore, a surface that is warm enough emits light in the visible range.

4.2.1 Simplifying Assumptions

It is common to assume that all effects are local and can be explained with a macroscopic model with no fluorescence or emission; we call this model a **local interaction model**. This is a reasonable model for the kind of surfaces and decisions that are common in vision. In this model:

- the radiance leaving a point on a surface is due only to radiance arriving at this point (although radiance may change *direction* at a point on a surface, we assume that it does not skip from point to point);
- we assume that all light leaving a surface at a given wavelength is due to light arriving at that wavelength;
- we assume that the surfaces do not generate light internally and treat sources separately.

4.2.2 The Bidirectional Reflectance Distribution Function

We wish to describe the relationship between incoming illumination and reflected light. This is a function of both the direction in which light arrives at a surface and the direction in which it leaves.

Irradiance The appropriate unit for representing incoming power is *irradiance*, defined as:

incident power per unit area *not foreshortened*.

This definition means that a patch of surface of area dA illuminated by radiance $L_i(P, \theta_i, \phi_i)$ coming in from a differential region of solid angle $d\omega$ at angles (θ_i, ϕ_i) receives irradiance

$$(1/dA)(L_i(P, \theta_i, \phi_i))(\cos\theta_i dA)d\omega = L_i(P, \theta_i, \phi_i)\cos\theta_i d\omega.$$

This means we multiply the radiance by the foreshortening factor and by the solid angle to get irradiance. The main feature of this unit is that we could compute all the power incident on a surface at a point by summing the irradiance over the whole input hemisphere, which makes it the natural unit for *incoming* power.

The BRDF The most general model of local reflection is the *bidirectional reflectance distribution function*, usually abbreviated BRDF. The BRDF is defined as

> the ratio of the radiance in the outgoing direction to the incident irradiance

so that, if a surface illuminated by radiance $L_i(P, \theta_i, \phi_i)$ coming in from a differential region of solid angle $d\omega$ at angles (θ_i, ϕ_i) was to emit radiance $L_o(P, \theta_o, \phi_o)$, its BRDF would be

$$\rho_{bd}(\theta_o, \phi_o, \theta_i, \phi_i) = \frac{L_o(P, \theta_o, \phi_o)}{L_i(P, \theta_i, \phi_i)\cos\theta_i d\omega}.$$

The BRDF has units of inverse steradians (sr^{-1}) and could vary from 0 (no light reflected in that direction) to infinity (unit radiance in an exit direction resulting from arbitrarily small radiance in the incoming direction). The BRDF is symmetric in the incoming and outgoing direction—a fact known as the *Helmholtz reciprocity principle*.

Computing Radiance Leaving a Surface from its BRDF The radiance leaving a surface due to irradiance *in a particular direction* is easily obtained from the definition of the BRDF:

$$L_o(P, \theta_o, \phi_o) = \rho_{bd}(\theta_o, \phi_o, \theta_i, \phi_i)L_i(P, \theta_i, \phi_i)\cos\theta_i d\omega.$$

More interesting is the radiance leaving a surface due to its irradiance (whatever the direction of irradiance). We obtain this by summing over contributions from all incoming directions:

$$L_o(P, \theta_o, \phi_o) = \int_\Omega \rho_{bd}(\theta_o, \phi_o, \theta_i, \phi_i)L_i(P, \theta_i, \phi_i)\cos\theta_i d\omega.$$

Here, Ω is the incoming hemisphere.

Constraints on the BRDF The BRDF is not an arbitrary symmetric function in four variables. To see this, assume that a surface patch of area dA is subjected to a radiance of $L_i(P, \theta_i, \phi_i)$ W $\times$ m^{-2} $\times$ sr^{-1}. This means that the total energy arriving at the surface in a time interval dt is

$$\left(\int_{\Omega_i} L_i(P, \theta_i, \phi_i)\cos\theta_i d\omega_i\right) dA\, dt = \left(\int_0^{2\pi}\int_0^{\frac{\pi}{2}} L_i(P, \theta_i, \phi_i)\cos\theta_i \sin\theta_i d\theta_i d\phi_i\right) dA\, dt.$$

We have assumed that any energy leaving at the surface leaves from the same point at which it arrived and no energy is generated within the surface. This means that the total energy leaving the surface during that interval must be less than or equal to the amount arriving. So we have

$$\left(\int_{\Omega_i} L_i(P,\theta_i,\phi_i)\cos\theta_i d\omega_i\right) dA\,dt \geq \left(\int_{\Omega_o} L_o(P,\theta_o,\phi_o)\cos\theta_o d\omega_o\right) dA\,dt.$$

But the energy leaving the surface can be written as

$$\left(\int_{\Omega_o}\int_{\Omega_i} \rho_{bd}(\theta_o,\phi_o,\theta_i,\phi_i) L_i(P,\theta_i,\phi_i)\cos\theta_i d\omega_i \cos\theta_o d\omega_o\right) dA\,dt.$$

This means that, for a general BRDF, the property

$$\int_{\Omega_i} L_i(P,\theta_i,\phi_i)\cos\theta_i d\omega_i \geq \int_{\Omega_o}\int_{\Omega_i} \rho_{bd}(\theta_o,\phi_o,\theta_i,\phi_i) L_i(P,\theta_i,\phi_i)\cos\theta_i d\omega_i \cos\theta_o d\omega_o$$

must hold, *whatever the choice of* L_i. Hence, although the BRDF can be large for some pairs of incoming and outgoing angles, it can't be large for many. In fact, the average value has to be quite small: You can use this property to prove that, for a BRDF which is independent of angle, the maximum value is $1/\pi$.

4.2.3 Example: The Radiometry of Thin Lenses

To illustrate some of the concepts introduced so far, let us consider an image patch $\delta A'$ centered in P', where a thin lens concentrates the light radiating from a scene patch δA centered in P (Figure 4.4), and let us relate the object radiance L in P to the image irradiance E in P'.[1] If $\delta\omega$ denotes the solid angle subtended by δA (or $\delta A'$) from the center O of the lens, we have

$$\delta\omega = \frac{\delta A' \cos\alpha}{(z'/\cos\alpha)^2} = \frac{\delta A \cos\beta}{(z/\cos\alpha)^2}, \quad \text{thus} \quad \frac{\delta A}{\delta A'} = \frac{\cos\alpha}{\cos\beta}\left(\frac{z}{z'}\right)^2.$$

Now the area of a lens with diameter d is $\frac{\pi}{4}d^2$, and if Ω denotes the solid angle subtended by the lens from P, we have

$$\Omega = \frac{\pi}{4}\frac{d^2\cos\alpha}{(z/\cos\alpha)^2} = \frac{\pi}{4}\left(\frac{d}{z}\right)^2\cos^3\alpha.$$

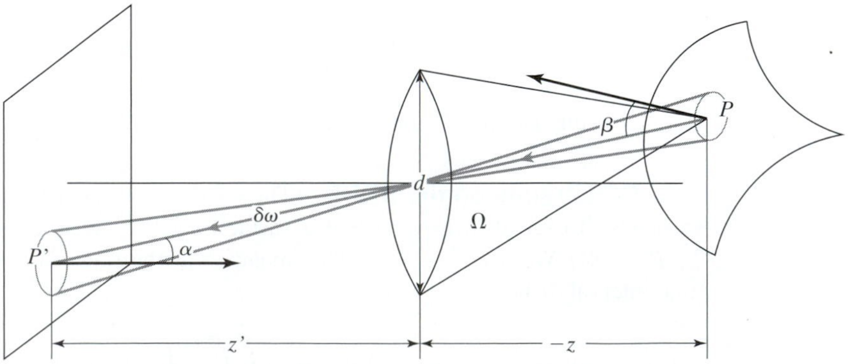

Figure 4.4 Object radiance and image irradiance.

[1] Here, we drop the radiance and irradiance parameters for conciseness, with the understanding that L denotes the radiance in P in the direction of P', and E denotes the irradiance in P'.

The power δP emitted from the patch δA and falling on the lens can be expressed in terms of the object radiance L as

$$\delta P = L\Omega\delta A \cos\beta = \frac{\pi}{4}\left(\frac{d}{z}\right)^2 L\delta A \cos^3\alpha \cos\beta.$$

The power δP is concentrated by the lens on the patch $\delta A'$ of the image plane, and it is the only power reaching this patch. Thus, the image irradiance is

$$E = \frac{\delta P}{\delta A'} = \frac{\pi}{4}\left(\frac{d}{z}\right)^2 L\frac{\delta A}{\delta A'} \cos^3\alpha \cos\beta.$$

Substituting the value of $\delta A/\delta A'$ in this equation finally yields

$$E = \left[\frac{\pi}{4}\left(\frac{d}{z'}\right)^2 \cos^4\alpha\right] L. \tag{4.1}$$

This relationship is important for several reasons: First, it shows that the image irradiance is proportional to the object radiance. In other words, what we measure (E) is proportional to what we are interested in (L). Second, the irradiance is proportional to the area of the lens and inversely proportional to the distance between its center and the image plane. The quantity $a = d/f$ is the *relative aperture* and is the inverse of the f number defined earlier. Equation (4.1) shows that E is proportional to a^2 when the lens is focused at infinity ($z' = f$ in this case). Finally, the irradiance is proportional to $\cos^4\alpha$ and falls off as the light rays deviate from the optical axis. For small values of α, this effect is hardly noticeable for both people and image analysis algorithms since the former are remarkably insensitive to smooth brightness gradients and the latter are usually more susceptible to *vignetting*—a phenomenon that dominates the $\cos^4\alpha$ fall-off in most imaging situations.

4.3 IMPORTANT SPECIAL CASES

Radiance is a fairly subtle quantity because it depends on angle. This generality is sometimes essential—for example, for describing the distribution of light in space in the torch beam example above. As another example, fix a compact disc and illuminate its underside with a torch beam. The intensity and color of light reflected from the surface depends very strongly on the angle from which the surface is viewed and on the angle from which it is illuminated. The CD example is worth trying because it illustrates how strange the behavior of reflecting surfaces can be; it also illustrates how accustomed we are to dealing with surfaces that do not behave in this way. For many surfaces—cotton cloth is one good example—the dependency of reflected light on angle is weak or nonexistent, so that a system of units that are independent of angle is useful.

4.3.1 Radiosity

If the radiance leaving a surface is independent of exit angle, there is no point describing it using a unit that explicitly depends on direction. The appropriate unit is *radiosity*, defined as

> the total power leaving a point on a surface per unit area on the surface.

Radiosity, which is usually written as $B(P)$, has units watts per square meter ($\mathrm{W} \times \mathrm{m}^{-2}$). To obtain the radiosity of a surface at a point, we can sum the radiance leaving the surface at that point over the whole exit hemisphere. Thus, if P is a point on a surface emitting radiance $L(P, \theta, \phi)$, the radiosity at that point is

$$B(P) = \int_{\Omega} L(P, \theta, \phi) \cos\theta d\omega,$$

where Ω is the exit hemisphere and the term $\cos\theta$ turns foreshortened area into area (look at the definitions again); $d\omega$ can be written in terms of θ, ϕ as before.

The Radiosity of a Surface with Constant Radiance One result to remember is the relationship between the radiosity and the radiance of a surface patch *where the radiance is independent of angle*. In this case, $L_o(x, \theta_o, \phi_o) = L_o(P)$. Now the radiosity can be obtained by summing the radiance leaving the surface over all the directions in which it leaves:

$$\begin{aligned} B(P) &= \int_{\Omega} L_o(P) \cos\theta d\omega \\ &= L_o(P) \int_0^{\frac{\pi}{2}} \int_0^{2\pi} \cos\theta \sin\theta d\phi d\theta \\ &= \pi L_o(P). \end{aligned}$$

4.3.2 Directional Hemispheric Reflectance

The BRDF is also a subtle quantity, and BRDF measurements are typically difficult, expensive, and not particularly repeatable. This is because surface dirt and aging processes can have significant effects on BRDF measurements; for example, touching a surface will transfer oil to it, typically in little ridges (from the fingertips), which can act as lenses and make significant changes in the directional behavior of the surface.

The light leaving many surfaces is largely independent of the exit angle. A natural measure of a surface's reflective properties in this case is the *directional-hemispheric reflectance*, usually termed ρ_{dh}, defined as:

> the fraction of the incident irradiance in a given direction that is reflected by the surface, whatever the direction of reflection.

The directional hemispheric reflectance of a surface is obtained by summing the radiance leaving the surface over all directions and dividing by the irradiance in the direction of illumination, which gives

$$\begin{aligned} \rho_{dh}(\theta_i, \phi_i) &= \frac{\int_{\Omega} L_o(P, \theta_o, \phi_o) \cos\theta_o d\omega_o}{L_i(P, \theta_i, \phi_i) \cos\theta_i d\omega_i} \\ &= \int_{\Omega} \left\{ \frac{L_o(P, \theta_o, \phi_o) \cos\theta_o}{L_i(P, \theta_i, \phi_i) \cos\theta_i d\omega_i} \right\} d\omega_o \\ &= \int_{\Omega} \rho_{bd}(\theta_o, \phi_o, \theta_i, \phi_i) \cos\theta_o d\omega_o. \end{aligned}$$

This property is dimensionless, and its value lies between 0 and 1.

Directional hemispheric reflectance can be computed for any surface. For some surfaces, it varies sharply with the direction of illumination. A good example is a surface with fine, symmetric triangular grooves that are black on one face and white on the other. If these grooves are sufficiently fine, it is reasonable to use a macroscopic description of the surface as flat, and with a directional hemispheric reflectance that is large along a direction pointing toward the white faces and small along that pointing toward the black.

4.3.3 Lambertian Surfaces and Albedo

For some surfaces, the directional hemispheric reflectance does not depend on illumination direction. Examples of such surfaces include cotton cloth, many carpets, matte paper and matte paints. A formal model is given by a surface whose BRDF is independent of outgoing direction (and, by the reciprocity principle, of incoming direction as well). This means the radiance leaving the surface is independent of angle. Such surfaces are known as *ideal diffuse surfaces* or *Lambertian surfaces* (after Johan Lambert, who first formalized the idea).

It is natural to use radiosity as a unit to describe the energy leaving a Lambertian surface. For Lambertian surfaces, the directional hemispheric reflectance is independent of direction. In this case the directional hemispheric reflectance is often called their *diffuse reflectance* or *albedo* and is written ρ_d. For a Lambertian surface with BRDF $\rho_{bd}(\theta_o, \phi_o, \theta_i, \phi_i) = \rho$, we have

$$\begin{aligned}\rho_d &= \int_\Omega \rho_{bd}(\theta_o, \phi_o, \theta_i, \phi_i) \cos\theta_o d\omega_o \\ &= \int_\Omega \rho \cos\theta_o d\omega_o \\ &= \rho \int_0^{\frac{\pi}{2}} \int_0^{2\pi} \cos\theta_o \sin\theta_o d\theta_o d\phi_o \\ &= \pi\rho.\end{aligned}$$

This fact is more often used in the form

$$\rho_{brdf} = \frac{\rho_d}{\pi},$$

a fact that is useful and well worth remembering.

Because our sensations of brightness correspond (roughly!) to measurements of radiance, a Lambertian surface will look equally bright from any direction, whatever the direction along which it is illuminated. This gives a rough test for when a Lambertian approximation is appropriate.

4.3.4 Specular Surfaces

A second important class of surfaces are the glossy or mirror like surfaces often known as *specular* surfaces (after the Latin word *speculum*, a mirror). An ideal specular reflector behaves like an ideal mirror. Radiation arriving along a particular direction can leave only along the *specular direction*, obtained by reflecting the direction of incoming radiation about the surface normal. Usually some fraction of incoming radiation is absorbed; on an ideal specular surface, the same fraction of incoming radiation is absorbed for every direction, the rest leaving along the specular direction. The BRDF for an ideal specular surface has a curious form (exercises) because radiation arriving in a particular direction can leave in only one direction.

Specular Lobes Relatively few surfaces can be approximated as ideal specular reflectors. A fair test of whether a flat surface can be approximated as an ideal specular reflector is whether one could safely use it as a mirror. Good mirrors are surprisingly hard to make; until recently, mirrors were made of polished metal. Typically, unless the metal is extremely highly polished and carefully maintained, radiation arriving in one direction leaves in a small lobe of directions around the specular direction. This results in a typical blurring effect. A good example is the bottom of a flat metal pie dish. If the dish is reasonably new, one can see a distorted image of one's face in the surface, but it would be difficult to use as a mirror; a more battered dish reflects a selection of distorted blobs.

Larger specular lobes mean that the specular image is more heavily distorted and is darker (because the incoming radiance must be shared over a larger range of outgoing directions). Quite commonly, it is possible to see only a specular reflection of relatively bright objects like sources. Thus, in shiny paint or plastic surfaces, one sees a bright blob—often called a *specularity*—along the specular direction from light sources, but few other specular effects. It is not often necessary to model the shape of the specular lobe. When the shape of the lobe is modeled, the most common model is the *Phong model*, which assumes that only point light sources are specularly reflected (Figure 4.5). In this model, the radiance leaving a specular surface is proportional to $\cos^n(\delta\theta) = \cos^n(\theta_o - \theta_s)$, where θ_o is the exit angle, θ_s is the specular direction and n is a parameter. Large values of n lead to a narrow lobe and small, sharp specularities; small values lead to a broad lobe and large specularities with rather fuzzy boundaries.

4.3.5 The Lambertian + Specular Model

Relatively few surfaces are either ideal diffuse or perfectly specular. The BRDF of many surfaces can be approximated as a combination of a Lambertian component and a specular component, which usually has some form of narrow lobe. Usually, the specular component is weighted by a *specular albedo*. Again, because specularities tend not to be examined in detail, the shape of this lobe is left unspecified. In this case, the surface *radiance* (because it must now depend on direction) in a given direction is typically approximated as

$$L(P, \theta_o, \phi_o) = \rho_d(P) \int_\Omega L(P, \theta_i, \phi_i) \cos\theta_i d\omega + \rho_s(P) L(P, \theta_s, \phi_s) \cos^n(\theta_s - \theta_o),$$

where θ_s, ϕ_s give the specular direction and ρ_s is the specular albedo. As we shall see, it is common not to reason about the exact magnitude of the specular radiance term.

Using this model implicitly excludes "too narrow" specular lobes because most algorithms expect to encounter occasional small, compact specularities from light sources. Surfaces with too narrow specular lobes (mirrors) produce overwhelming quantities of detail in specularities. Similarly, "too broad" lobes are excluded because the specularities would be hard to identify.

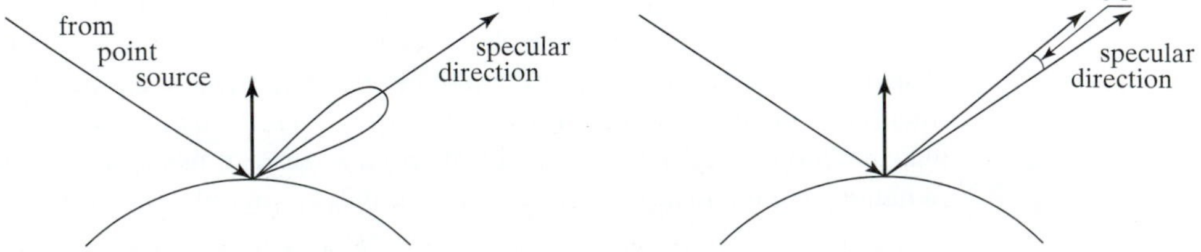

Figure 4.5 Specular surfaces commonly reflect light into a lobe of directions around the specular direction, where the intensity of the reflection depends on the direction, as shown on the left. Phong's model is used to describe the shape of this lobe in terms of the offset angle from the specular direction.

TABLE 4.1 Reference card: Radiometric Terminology

Topic	**What you must know**
Foreshortening	The fact that a large area viewed at a grazing angle will look like a small area, viewed frontally. This is important because two different receivers will receive the same amount of radiation from a source if they look exactly the same to the source; two different sources will have the same effect on a receiver if they look exactly the same to the receiver.
Radiance	**Definition:** the quantity of energy traveling at a point in a direction, per unit time, per unit area *perpendicular to the direction of travel*, per unit solid angle. **Units:** $W \times m^{-2} \times sr^{-1}$ **Use:** representing: light traveling in free space; light reflected from a surface when it depends strongly on direction
Irradiance	**Definition:** total incident power per unit surface area. **Units:** $W \times m^{-2}$ **Use:** representing light arriving at a surface
Radiosity	**Definition:** the total power leaving a point on a surface per unit area on the surface. **Units:** $W \times m^{-2}$ **Use:** representing light leaving a diffuse surface
BRDF	**Definition:** the ratio of the radiance in the outgoing direction to incident irradiance **Units:** sr^{-1} **Use:** representing reflection off general surfaces where reflection depends strongly on direction
Directional Hemispheric Reflectance	**Definition:** the fraction of the incident irradiance in a given direction that is reflected by the surface, whatever the direction of reflection. **Units:** unitless. **Use:** representing reflection off a surface where direction is unimportant
Albedo	**Definition:** directional hemispheric reflectance of a diffuse surface. **Units:** unitless. **Use:** representing a diffuse surface
Diffuse surface; Lambertian surface	**Definition:** A surface whose BRDF is constant. **Examples:** Cotton cloth; many rough surfaces; many paints and papers; surfaces whose apparent brightness doesn't change with viewing direction
Specular surface	**Definition:** A surface that behaves like a mirror **Examples:** Mirrors; polished metal
Specularity	**Definition:** Small bright patches on a surface that result from specular components of the BRDF. **Examples:** Common on plastic surfaces; on brushed metal surfaces; on some paints; and on shiny cloth. **Important feature:** they move on the surface when you move your head.

4.4 NOTES

We strongly recommend François Sillion's excellent book (Sillion 1994) for its clear account of radiometric calculations. There are a variety of more detailed publications for reference (Nayar, Ikeuchi and Kanade, 1991*b*). Our discussion of reflection is thoroughly superficial. The specular plus diffuse model appears to be originally due to Cook, Torrance, and Sparrow (Torrance and Sparrow, 1967, Cook and Torrance, 1987). A variety of modifications of this model appear in computer vision and computer graphics. Reflection models can be derived by combining a statistical description of surface roughness with electromagnetic considerations (e.g., Beckmann and Spizzichino, 1987) or by adopting scattering models (as in the work of Torrance and Sparrow, 1967, and of Cook and Torrance, 1987).

Top of the list of effects we omitted to discuss is off-specular glints, followed by specular backscatter. Off-specular glints commonly arise in brushed surfaces, where there is a large surface area oriented at a substantial angle to the macroscopic surface normal. This leads to a second specular lobe due to this region. These effects can confuse algorithms that reason about shape from specularities, if the reasoning is close enough. Specular backscatter occurs when a surface reflects light back in the source direction—usually for a similar reason that off-specular glints occur. Again, the effect is likely to confuse algorithms that reason about shape from specularities. Some classes of reflectance models that incorporate these properties are described in Tagare and de Figueiredo (1991).

It is commonly believed that rough surfaces are Lambertian. This belief has a substantial component of wishful thinking because rough surfaces often have local shadowing effects that make the radiance reflected quite strongly dependent on the illumination angle. For example, a stucco wall illuminated at a near grazing angle shows a clear pattern of light and dark regions where facets of the surface face toward the light or are shadowed. If the same wall is illuminated along the normal, this pattern largely disappears. Similar effects at a finer scale are averaged to endow rough surfaces with measurable departures from a Lambertian model (for details, see Koenderink, van Doorn, Dana and Nayar, 1999, Nayar and Oren, 1993, 1995, Oren and Nayar, 1995, and Wolff, Nayar and Oren, 1998).

Another example of an object that does not support a simple macroscopic surface model is a field of flowers. A distant viewer should be able to abstract this field as a "surface"; however, doing so leads to a surface with quite strange properties. If one views such a field along a normal direction, one sees mainly flowers; a tangential view reveals both stalks and flowers, meaning that the color changes dramatically with viewing direction (the effect is explored in Leung and Malik, 1997).

PROBLEMS

4.1. How many steradians in a hemisphere?

4.2. We have proved that radiance does not go down along a straight line *in a non-absorbing medium*, which makes it a useful unit. Show that if we were to use power per square meter of foreshortened area (which is irradiance), the unit must change with distance along a straight line. How significant is this difference?

4.3. **An absorbing medium:** Assume that the world is filled with an isotropic absorbing medium. A good, simple model of such a medium is obtained by considering a line along which radiance travels. If the radiance along the line is N at x, it is $N - (\alpha dx)N$ at $x + dx$.

- **(a)** Write an expression for the radiance transferred from one surface patch to another in the presence of this medium.
- **(b)** Now *qualitatively* describe the distribution of light in a room filled with this medium for α small and large positive numbers. The room is a cube, and the light is a single small patch in the center of the ceiling. Keep in mind that if α is large and positive, little light actually reaches the walls of the room.

4.4. Derive the relationship between the scene radiance and image irradiance for a pinhole camera with a pinhole of diameter d.

4.5. Derive the relationship between the scene radiance and image irradiance for a spherical camera with a pinhole of diameter d.

4.6. Identify common surfaces that are neither Lambertian nor specular using the underside of a CD as a working example. There are a variety of important biological examples, which are often blue in color. Give at least two different reasons that it could be advantageous to an organism to have a non-Lambertian surface.

4.7. Show that for an ideal diffuse surface the directional hemispheric reflectance is constant; now show that if a surface has constant directional hemispheric reflectance, it is ideal diffuse.

4.8. Show that the BRDF of an ideal specular surface is

$$\rho_{bd}(\theta_o, \phi_o, \theta_i, \phi_i) = \rho_s(\theta_i)\{2\delta(\sin^2\theta_o - \sin^2\theta_i)\}\{\delta(\phi_o - \phi\pi)\},$$

where $\rho_s(\theta_i)$ is the fraction of radiation that leaves.

4.9. Why are specularities brighter than diffuse reflection?

4.10. A surface has constant BRDF. What is the maximum possible value of this constant? Now assume that the surface is known to absorb 20% of the radiation incident on it (the rest is reflected); what is the value of the BRDF?

4.11. The eye responds to radiance. Explain why Lambertian surfaces are often referred to as having a brightness independent of viewing angle.

4.12. Show that the solid angle subtended by a sphere of radius ϵ at a point a distance r away from the center of the sphere is approximately $\pi(\frac{\epsilon}{r})^2$, for $r \gg \epsilon$.

5

Sources, Shadows, and Shading

Surfaces are bright or dark for two main reasons: their albedo and the amount of light they are receiving. A model of how the brightness of a surface is obtained is usually called a *shading model*. Shading models are important because with an appropriate shading model we can interpret pixel values. If the right shading model applies, it is possible to reconstruct objects and their albedos using just a few images. Furthermore, we can interpret shadows and explain their puzzling and seldom-noticed absence in most indoor scenes.

5.1 QUALITATIVE RADIOMETRY

We should like to know how "bright" surfaces are going to be under various lighting conditions, and how this "brightness" depends on local surface properties, on surface shape, and on illumination. As we saw in chapter 4, foreshortening means that different sources can have the same effect on a surface. The most powerful tool for analyzing this problem is to think about *what a source looks like from the surface*. This *qualitative radiometry* is one of these tricks that looks unsophisticated—no hard math—but is extremely powerful. In some cases, this technique gives qualitative descriptions of "brightness" without even knowing what the term means.

Recall from Section 4.1.1 and Figure 4.1 that a surface patch sees the world through a hemisphere of directions at that patch. The radiation arriving at the surface along a particular direction passes through a point on the hemisphere. If two surface patches have equivalent incoming hemispheres, they must have the same incoming radiation, *whatever the outside world looks like*. This means that any difference in "brightness" between patches with the same incoming hemisphere is a result of different surface properties. In particular, if two surface patches with the same BRDF see the same incoming hemisphere, then the radiation they output must be the same.

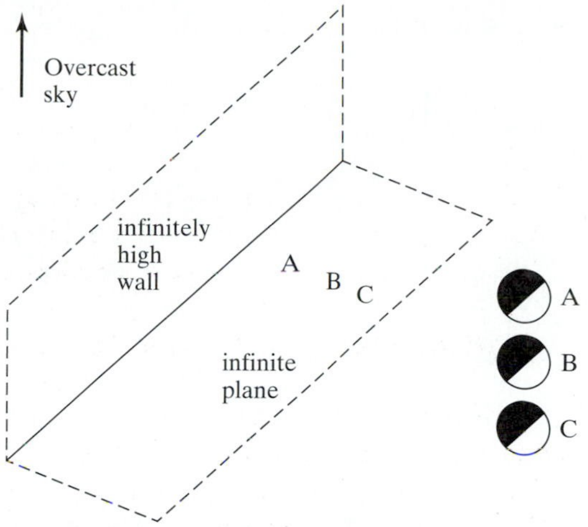

Figure 5.1 A geometry in which a qualitative radiometric solution can be obtained by thinking about what the world looks like from the point of view of a patch. We wish to know what the brightness looks like at the base of two different infinitely high walls. In this geometry, an infinitely high matte black wall cuts off the view of the overcast sky—which is a hemisphere of infinite radius and uniform "brightness". On the right, we show a representation of the directions that see or do not see the source at the corresponding points, obtained by flattening the hemisphere to a circle of directions (or, equivalently, by viewing it from above). Since each point has the same input hemisphere, the brightness must be uniform.

Lambert determined the distribution of "brightness" on a uniform plane at the base of an infinitely high black wall illuminated by an overcast sky (see Figure 5.1). In this case, every point on the plane must see the same hemisphere—half of its viewing sphere is cut off by the wall, and the other half contains the sky, which is uniform—and the plane is uniform, so every point must have the same "brightness".

A second example is somewhat trickier. We now have an infinitely thin black wall that is infinitely long in only one direction and on an infinite plane (Figure 5.2). A qualitative description would be to find what the curves of equal "brightness" look like. It is fairly easy to see that all points on any line passing through the point p in Figure 5.2 see the same input hemisphere and so must have the same "brightness". Furthermore, the distribution of "brightness" on the plane must have a symmetry about the line of the wall—we expect the brightest points to be along the extension of the line of the wall and the darkest to be at the base of the wall.

5.2 SOURCES AND THEIR EFFECTS

5.2.1 Radiometric Properties of Light Sources

We define a *light source* to be anything that emits light *that is internally generated* (i.e., not just reflected). To describe a source, we need a description of the radiance it emits in each direction. Typically, internally generated radiance is dealt with separately from reflected radiance. This is because, although a source may reflect light, the light it reflects depends on the environment, whereas the light it generates internally usually does not.

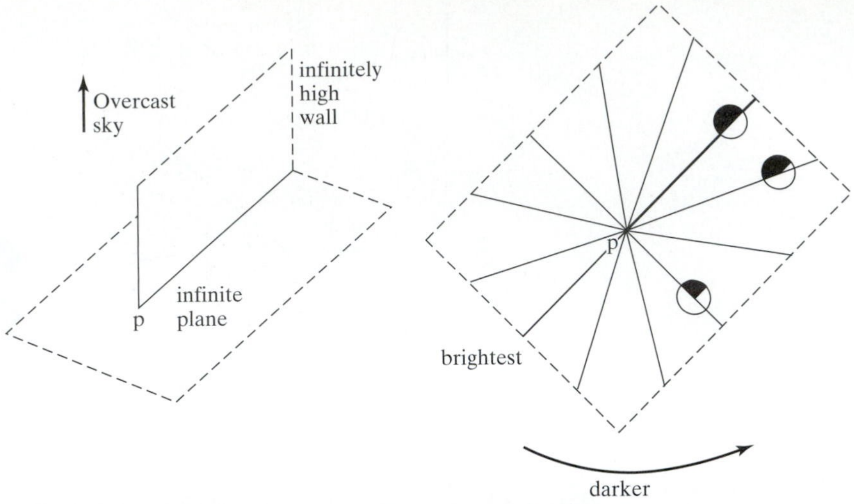

Figure 5.2 We now have a matte black, infinitely thin, half-infinite wall on an infinite white plane (shown on the **left**). This geometry also sees an overcast sky of infinite radius and uniform "brightness". In the text, we show how to determine the curves of similar "brightness" on the plane. These curves are shown on the **right**, depicted on an overhead view of the plane; the thick line represents the wall. Superimposed on these curves is a representation of the input hemisphere for some of these isophotes. Along these curves, the hemisphere is fixed (by a geometrical argument), but they change as one moves from curve to curve.

We seldom need a complete description of the radiance a source emits in each direction. It is more usual to model sources as emitting a constant radiance in each direction (possibly with a family of directions zeroed, like a spotlight). The proper quantity in this case is the *exitance*, defined as

> the internally generated energy radiated per unit time and per unit area on the radiating surface.

Exitance is similar to radiosity, and can be computed as

$$E(P) = \int_{\Omega} L_e(P, \theta_o, \phi_o) \cos\theta_o \, d\omega,$$

Together with a description of the exitance, we need a description of the geometry of the source, which has profound effects on the spatial variation of light around the source and on the shadows cast by objects near the source. Sources are usually modeled with quite simple geometries for two reasons: first, many synthetic sources can be modeled as point sources, line sources, or area sources fairly effectively; second, sources with simple geometries can still yield surprisingly complex effects.

5.2.2 Point Sources

A common approximation is to assume that the light source is an extremely small sphere, in fact, a point; such a source is known as a *point source*. It is a natural model to use because many sources are physically small compared with the environment in which they stand. We can obtain

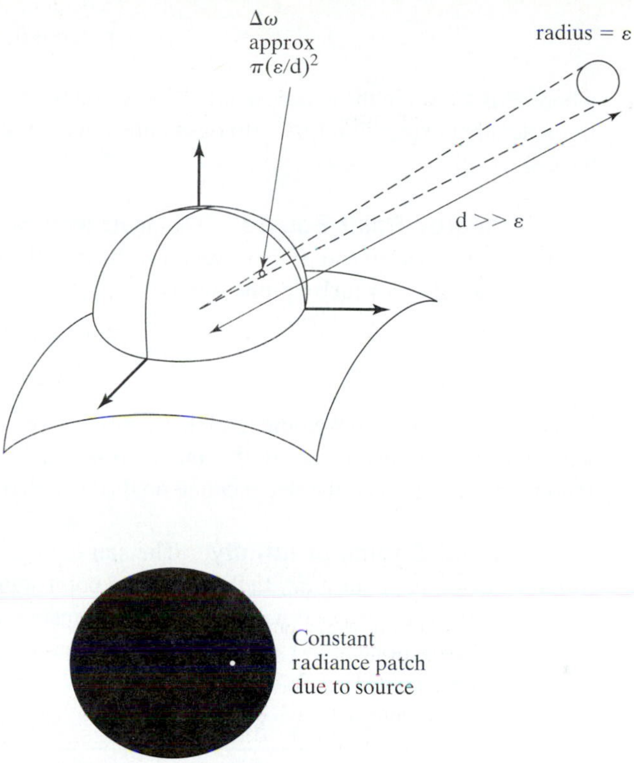

Figure 5.3 A surface patch sees a distant sphere of small radius; the sphere produces a small illuminated patch on the input hemisphere of the sphere. In the text, by reasoning about the scaling behavior of this patch as the distant sphere moves farther away or gets bigger, we obtain an expression for the behavior of the point source.

a model for the effects of a point source by modeling the source as a very small sphere that emits light at each point on the sphere, with an exitance constant over the sphere.

Assume that a surface patch is viewing a sphere of radius ϵ, at a distance r away, and that $\epsilon \ll r$ (Figure 5.3). The assumption that the sphere is far away from the patch relative to its radius almost always applies for real sources. Now the solid angle that the source subtends is Ω_s. This behaves approximately proportional to

$$\frac{\epsilon^2}{r^2}.$$

The pattern of illumination that the source creates on the hemisphere (roughly) scales, too. As the sphere moves away, the rays leaving the surface patch and striking the sphere move closer together (roughly) evenly, and the collection changes only slightly (a small set of new rays is added at the rim—the contribution from these rays must be small because they come from directions tangent to the sphere). In the limit as ϵ tends to zero, no new rays are added.

The radiosity due to the source is obtained by integrating the pattern generated by the source, times $\cos\theta_i$, over the patch of solid angle. As ϵ tends to zero, the patch shrinks and the $\cos\theta_i$ is close to constant. If ρ is the surface albedo, all this means the expression for radiosity due to the point source is

$$\rho \left(\frac{\epsilon}{r}\right)^2 E \cos\theta,$$

where E is a term in the exitance of the source integrated over the small patch. We do not need a more detailed expression for E (to determine one, we would need to actually do the integral we have shirked).

A Nearby Point Source The angle term can be written in terms of $\boldsymbol{N}(P)$ (the unit normal to the surface) and $\boldsymbol{S}(P)$ (a vector from P to the source whose length is $\epsilon^2 E$) to yield the standard model of a **nearby point source:**

$$\rho_d(P) \frac{\boldsymbol{N}(P) \cdot \boldsymbol{S}(P)}{r(P)^2}.$$

This is an extremely convenient model, because it gives an explicit relationship between radiosity and shape (the normal term). In this model, $\boldsymbol{S}$ is usually called the *source vector*. It is common (and incorrect!) to omit the dependency on distance to the source from this model.

A Point Source at Infinity The sun is far away; as a result, the terms $1/r(P)^2$ and $\boldsymbol{S}(P)$ are essentially constant. In this case, the point source is referred to as being a *point source at infinity*. If all the surface patches we are interested in are close together with respect to the distance to the source, $r(P) = r_0 + \Delta r(P)$ where $r_0 \gg \Delta r(P)$. Furthermore, $\boldsymbol{S}(P) = \boldsymbol{S}_0 + \Delta\boldsymbol{S}(P)$, where $| \boldsymbol{S}_0 | \gg | \Delta\boldsymbol{S}(P) |$. We now have

$$\frac{\boldsymbol{N} \cdot \boldsymbol{S}(P)}{r(P)^2} = \frac{\boldsymbol{N} \cdot (\boldsymbol{S}_0 + \Delta\boldsymbol{S}(P))}{(r_0 + \Delta r(P))^2} \approx \frac{\boldsymbol{N} \cdot \boldsymbol{S}_0}{r_0^2}.$$

Now both $\boldsymbol{S}_0$ and r_0 are constants, and there is no particular point in keeping them explicitly in the expression. Instead, we can write $\boldsymbol{S} = (1/r_0^2)\boldsymbol{S}_0$, and our model for the radiosity due to a point source at infinity becomes

$$B(P) = \rho_d(P)(\boldsymbol{N} \cdot \boldsymbol{S}).$$

The term $\boldsymbol{S}$ is again known as the *source vector*. Typically, this model is used by inferring a source vector that is appropriate, rather than by computing its value from the exitance of the source and the geometry.

Choosing a Point Source Model A point source at infinity is a good model for the sun, for example, because the solid angle that the sun subtends is small and essentially constant wherever it appears in the field of view (this test means that our approximation step is valid). If we use linear sensors with an unknown gain, for example, we can roll the source intensity and the unknown gain into the source vector term—this is quite often done without comment.

As you should expect from the derivation, a point source at infinity is a poor model *when the distance between objects is similar in magnitude to the distance to the source*. In this case, we cannot use the series approximation to pretend that the radiosity due to the source does not go down with distance to the source.

The heart of the problem is easy to see if we consider what the source looks like from different surface patches. It must look bigger to nearer surface patches (however small its radius); this means that the radiosity due to the source must go up. If the source is sufficiently distant—for example, the sun—we can ignore this effect because the source does not change in apparent size for any plausible motion.

However, for configurations like a light bulb in the center of a room, the solid angle subtended by the source goes up as the inverse square of the distance, meaning that the radiosity due to the source will do so too. The correct model to use in this case is the point source of Section 5.2.2. The difficulty with this model is that radiosity changes sharply over space, in a way that is inconsistent with experience. For example, if a point source is placed at the center of a cube, then the the model predicts that radiosity in the corners is roughly one third that at the center of each face —but the corners of real rooms are nowhere near as dark as that. It is quite common in practice to suppress the distance term in the nearby point source model to account for this—an activity that is radiometrically incorrect but that tends to yield a better model. The explanation of this apparent contradiction must wait until we have discussed shading models.

5.2.3 Line Sources

A *line source* has the geometry of a line—a good example is a single fluorescent light bulb. Line sources are not terribly common in natural scenes or in synthetic environments, and we discuss them only briefly. Their main interest is as an example for radiometric problems; in particular, the radiosity of patches reasonably close to a line source changes as the reciprocal of distance to the source (rather than the square of the distance). The reasoning is more interesting than the effect. We model a line source as a thin cylinder with diameter ϵ. Assume for the moment that the line source is infinitely long and that we are considering a patch that views the source frontally, as in Figure 5.4.

Figure 5.4 sketches the appearance of the source from the point of view of patch 1; now move the patch closer and consider patch 2—the width of the region on the hemisphere corresponding to the source changes, but not the length (because the source is infinitely long). In turn, because the width is approximately ϵ/r, the radiosity due to the source must go down with the reciprocal of distance. It is easy to see that with a source that is not infinitely long, this applies as long as the patch is reasonably close.

5.2.4 Area Sources

An *area source* is an area that radiates light. Area sources are important for two reasons. First, they occur quite commonly in natural scenes—an overcast sky is a good example—and in syn-

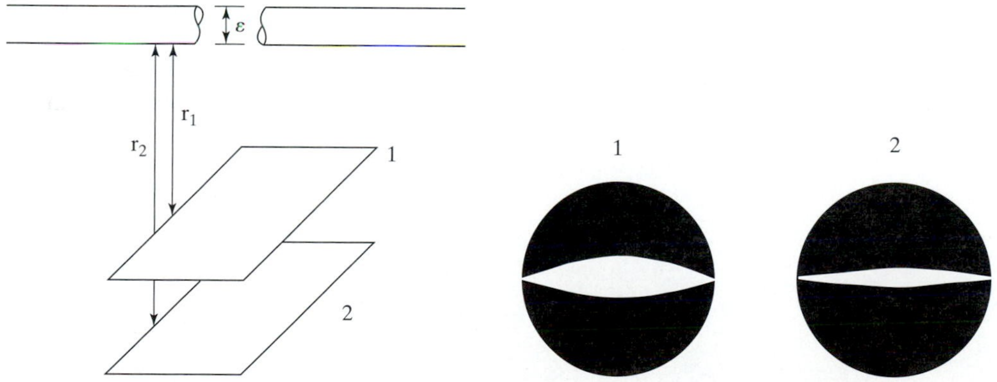

Figure 5.4 The radiosity due to a line source goes down as the reciprocal of distance for points that are reasonably close to the source. On the left, two patches viewing an infinitely long, narrow cylinder with constant exitance along its surface and diameter ϵ. On the right, the view of the source *from each patch*, drawn as the underside of the input hemisphere seen from below. Notice that the length of the source on this hemisphere does not change, but the width does (as ϵ/r). This yields the result.

thetic environments—for example, the fluorescent light boxes found in many industrial ceilings. Second, a study of area sources allows us to explain various shadowing and interreflection effects. Area sources are normally modeled as surface patches whose emitted radiance is independent of position and of direction—they can be described by their exitance.

An argument similar to that used for line sources shows that, for points not too distant from the source, the radiosity due to an area source does not change with distance to the source. This is because, if the area is large enough with respect to the distance to the source, the area subtended by the source on some input hemisphere is about the same as we move toward and away from the source. This explains the widespread use of area sources in illumination engineering—they generally yield fairly uniform illumination. For our applications, we need a more exact description of the radiosity due to an area source, so we need to write out the integral.

The Exact Radiosity due to an Area Source Assume we have a diffuse surface patch that is illuminated by an area source with exitance $E(Q)$ at the source point Q. Instead of writing angles in coordinates, we write $\overrightarrow{QP}$ for the direction from Q to P (more notation

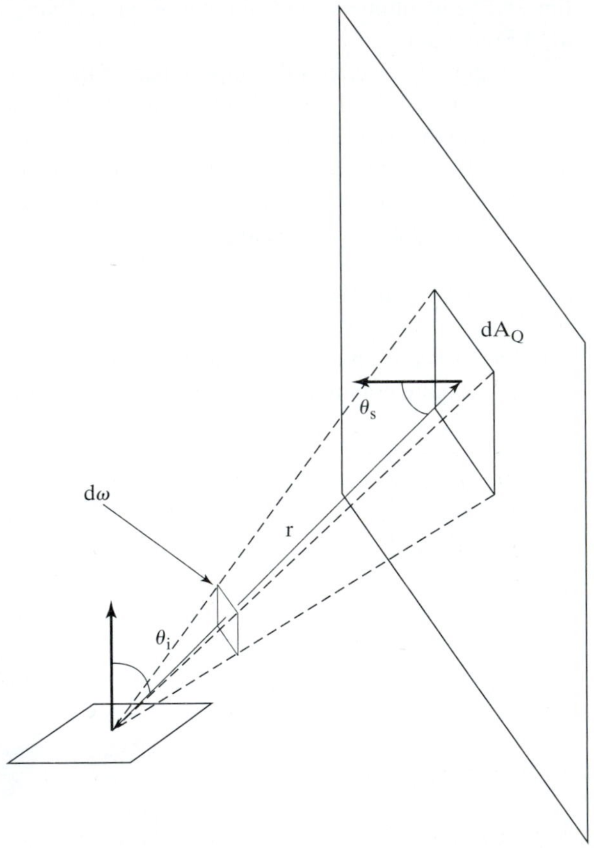

Figure 5.5 A diffuse source illuminates a diffuse surface. The source has exitance $E(Q)$, and we wish to compute the radiosity on the patch due to the source. We do this by transforming the integral of incoming radiance at the surface into an integral over the source area. This transformation is convenient because it avoids us having to use different angular domains for different surfaces. However, it still leads to an integral that is usually impossible in closed form.

is illustrated in Figure 5.5). The radiosity on the surface is obtained by summing the incoming radiance over all incoming directions. This integral can be transformed into an integral over the source as follows:

$$\begin{aligned}
B(P) &= \rho_d(P)\int_\Omega L_i(P, \overrightarrow{QP})\cos\theta_i \, d\omega \\
&= \rho_d(P)\int_\Omega L_e(Q, \overrightarrow{QP})\cos\theta_i \, d\omega \\
&= \rho_d(P)\int_\Omega \left(\frac{1}{\pi}E(Q)\right)\cos\theta_i \, d\omega \\
&= \rho_d(P)\int_{\text{Source}} \left(\frac{1}{\pi}E(Q)\right)\cos\theta_i \left(\cos\theta_s \frac{dA_Q}{r^2}\right) \\
&= \rho_d(P)\int_{\text{Source}} E(Q)\frac{\cos\theta_i\cos\theta_s}{\pi r^2} \, dA_Q.
\end{aligned}$$

The transformation works because radiance is constant along straight lines and because $E(Q) = (1/\pi)L_e(Q)$. It is useful because it means we do not have to worry about consistent angular coordinate systems. However we transform them, integrals describing the effect of area sources are generally difficult or impossible to do in closed form.

5.3 LOCAL SHADING MODELS

We have studied the physics of light because we want to know how bright things will be, and why, in the hope of extracting object information from these models. Currently, we know the radiosity at a patch *due to a source* but this is not a shading model. Radiance could arrive at surface patches in other ways (e.g., it could be reflected from other surface patches); we need to know which components to account for.

The easiest model to manipulate is a *local shading model*, which models the radiosity at a surface patch as the sum of the radiosity due only to light internally generated at sources. This means that we assume that light is not reflected from surface to surface, but instead leaves a source, arrives at some surface, and proceeds directly to the camera. This model is palpably unphysical, but is easy to analyze. The model supports a variety of algorithms and theories (see Section 5.4). Unfortunately, this model often produces wildly inaccurate predictions. Even worse, there is little reliable information about when it is safe to use this model.

An alternate model is to account for all radiation (Section 5.5). This takes into account radiance arriving from sources and that arriving from radiating surfaces. This model is physically accurate, but usually hard to manipulate.

5.3.1 Local Shading Models for Point Sources

The local shading model for a set of point sources is obtained by writing out the radiosity due to light internally generated at sources. This gives

$$B(P) = \sum_{s \in \text{sources visible from } P} B_s(P),$$

where $B_s(P)$ is the radiosity due to source s. This expression is fairly innocuous, but notice that if all the sources are point sources at infinity, the expression becomes

$$B(P) = \sum_{s \in \text{sources visible from } P} \rho_d(P) \boldsymbol{N}(P) \cdot \boldsymbol{S}_s,$$

so that if we confine our attention to a region where all points can see the same sources, we could add all the source vectors to obtain a single virtual source that had the same effects. The relationship between shape and shading is pretty direct here —the radiosity is a measurement of one component of the surface normal.

For point sources that are not at infinity, the model becomes

$$B(P) = \sum_{s \in \text{sources visible from } P} \rho_d(P) \frac{\boldsymbol{N}(P) \cdot \boldsymbol{S}(P)}{r_s(P)^2},$$

where $r_s(P)$ is the distance from the source to P; the presence of this term means that the relationship between shape and shading is somewhat more obscure.

The Appearance of Shadows In a local shading model, shadows occur when the patch can not see one or more sources. In this model, point sources produce a series of shadows with crisp boundaries; shadow regions where no source can be seen are particularly dark. Shadows cast with a single source can be crisp and black depending on the size of the source and the albedo of other nearby surfaces (which could reflect light into the shadow and soften its boundary). It was a popular 19th Century pastime to cast such shadows onto paper and then draw them, yielding the silhouettes still occasionally found in antiques shops.

The geometry of the shadow cast by a point source on a plane is analogous to the geometry of viewing in a perspective camera (Figure 5.6). Any patch on the plane is in shadow if a ray from the patch to the source passes through an object. This means that there are two kinds of shadow boundary. At *self shadow boundaries*, the surface is turning away from the light, and a ray from the patch to the source is tangent to the surface. At *cast shadow boundaries*, from the perspective of the patch, the source suddenly disappears behind an occluding object. Shadows cast onto curved surfaces can have extremely complex geometries, however.

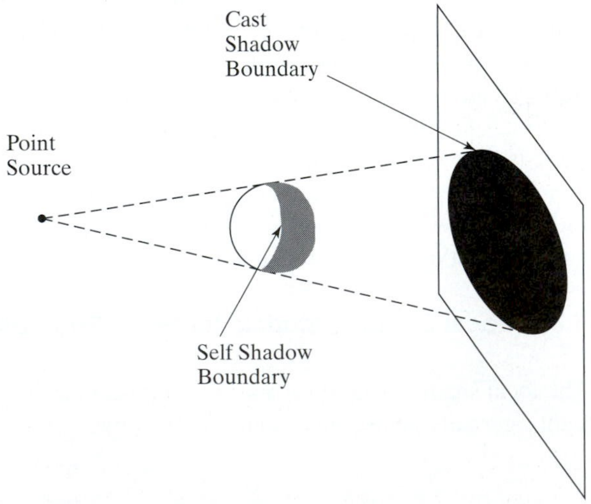

Figure 5.6 Shadows cast by point sources on planes are relatively simple. Self shadow boundaries occur when the surface turns away from the light and cast shadow boundaries occur when a distant surface occludes the view of the source.

If there are many sources, the shadows are less dark (except at points where no source is visible), and there can be many qualitatively distinct shadow regions (each source casts its own shadow—some points may not see more than one source). One example of this effect occurs in televised soccer matches. Because the stadium has multiple bright distant point-like illuminants spaced evenly around the perimeter of the stadium, there is a set of shadows radiating evenly around each player's feet. These shadows typically become brighter or darker as the player moves around usually because the illumination due to other sources and to interreflections in the region of the shadow increases or decreases.

5.3.2 Area Sources and Their Shadows

The local shading model for a set of area sources is significantly more complex because it is possible for patches to see only a portion of a given source. The model becomes

$$\begin{aligned} B(P) &= \sum_{s \in \text{ all sources}} \left\{ \int_{\text{visible component of source } s} \text{Radiosity due to source} \right\} \\ &= \sum_{s \in \text{ all sources}} \int_{\text{visible component of source } s} \left\{ E(Q) \frac{\cos\theta_Q \cos\theta_s}{\pi r^2} \, dA_Q \right\} \end{aligned}$$

using the terminology of Figure 5.5; usually, we assume that E is constant over the source.

Area sources do not produce dark shadows with crisp boundaries. This is because, from the perspective of a viewing patch, the source appears slowly from behind the occluding object (think of an eclipse of the moon—it is an exact analogy). It is common to distinguish between points in the *umbra* (a Latin word meaning "shadow")—which cannot see the source at all—and points in the *penumbra* (a compound of Latin words meaning "almost shadow")—which see part of the source. The vast majority of indoor sources are area sources of one form or another, so the effects are quite easy to see; hold an arm quite close to the wall and look at the shadow it casts. There is a dark core, which gets larger as the arm gets closer to the wall (this is the umbra), surrounded by a lighter region with a fuzzier boundary (the penumbra). Figure 5.7 illustrates the geometry.

5.3.3 Ambient Illumination

One problem with local shading models should be apparent immediately; they predict that some shadow regions are arbitrarily dark because they cannot see the source. This prediction is inaccurate in almost every case because shadows are illuminated by light from other diffuse surfaces. This effect can be significant. In rooms with light walls and area sources, it is possible to see shadows only by holding objects close to the wall or close to the source. This is because a patch on the wall sees all the other walls in the room; until an object is close to the wall, it blocks out only a small fraction of the visual hemisphere of each patch.

For some environments, the total irradiance a patch obtains from other patches is roughly constant and roughly uniformly distributed across the input hemisphere. This must be true for the interior of a sphere with a constant distribution of radiosity (by symmetry) and (by accepting a model of a cube as a sphere) is roughly true for the interior of a room with white walls. In such an environment, it is sometimes possible to model the effect of other patches by adding an *ambient illumination* term to each patch's radiosity. There are two strategies for determining this term. First, if each patch sees the same proportion of the world (e.g., the interior of a sphere), we can add the same constant term to the radiosity of each patch. The magnitude of this term is usually guessed.

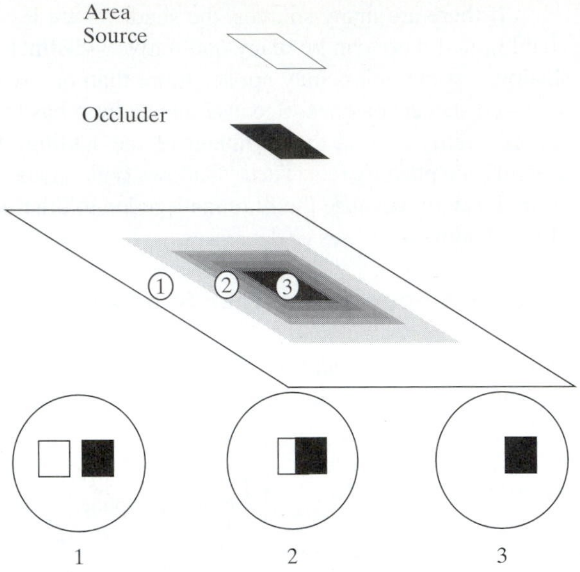

Figure 5.7 Area sources generate complex shadows with smooth boundaries, because from the point of view of a surface patch, the source disappears slowly behind the occluder. Regions where the source cannot be seen at all are known as the *umbra*; regions where some portion of the source is visible are known as the *penumbra*. A good model is to imagine lying with your back to the surface looking at the world above. At point 1, you can see all of the source; at point 2, you can see some of it; and at point 3, you can see none of it.

Second, if some patches see more or less of the world than others (this happens if regions of the world occlude a patch's view, e.g., a patch at the bottom of a groove), this can be taken into account. To do so, we need a model of the world *from the perspective of the patch under consideration*. A natural strategy is to model the world as a large, distant polygon of constant radiosity, where the view of this polygon is occluded at some patches (see Figure 5.8). The result is that the ambient term is smaller for patches that see less of the world. This model is often more accurate than adding a constant ambient term. Unfortunately, it is much more difficult to extract information from this model, possibly as difficult as for a global shading model.

5.4 APPLICATION: PHOTOMETRIC STEREO

We reconstruct a patch of surface from a series of pictures of the surface taken under different illuminants. For simplicity, we use an orthographic camera and choose a coordinate system such that the point (x, y, z) in space projects onto the point (x, y) in the image (the method we describe works for the other camera models described in chapter 1).

In this case, to measure the shape of the surface, we need to obtain the depth to the surface. This suggests representing the surface as $(x, y, f(x, y))$—a representation known as a *Monge patch* after a French military engineer who first used it (Figure 5.9). This representation is attractive because we can determine a unique point on the surface by giving the image coordinates. Notice that to obtain a measurement of a solid object, we would need to reconstruct more than one patch because we need to observe the back of the object.

Photometric stereo is a method for recovering a representation of the Monge patch from image data. The method involves reasoning about the image intensity values for several

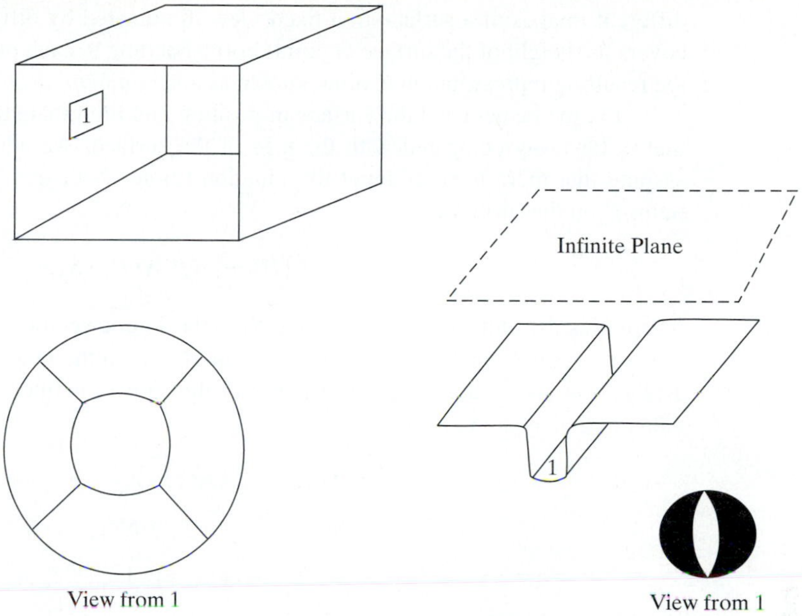

Figure 5.8 Ambient illumination is a term added to the radiosity predictions of local shading models to model the effects of radiosity from distant, reflecting surfaces. In a world like the interior of a sphere or of a cube (the case on the left), where a patch sees roughly the same thing from each point, a constant ambient illumination term is often acceptable. In more complex worlds, some surface patches see much less of the surrounding world than others. For example, the patch at the base of the groove on the right sees relatively little of the outside world, which we model as an infinite polygon of constant exitance; its input hemisphere is shown below.

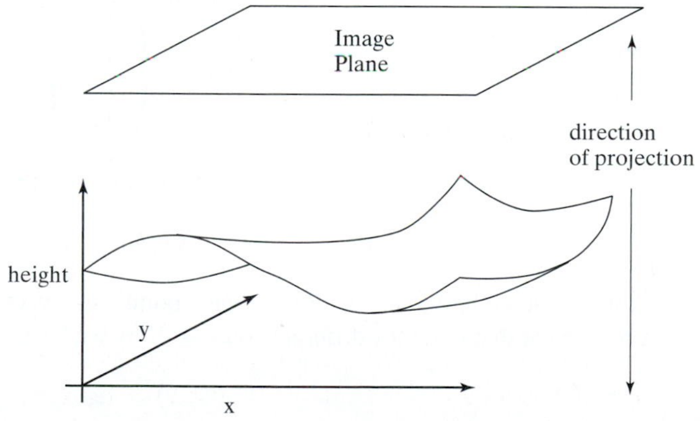

Figure 5.9 A Monge patch is a representation of a piece of surface as a height function. For the photometric stereo example, we assume that an orthographic camera—one that maps (x, y, z) in space to (x, y) in the camera—is viewing a Monge patch. This means that the shape of the surface can be represented as a function of position in the image.

different images of a surface in a fixed view illuminated by different sources. This method recovers the height of the surface at points corresponding to each pixel; in computer vision circles, the resulting representation is often known as a *height map*, *depth map*, or *dense depth map*.

Fix the camera and the surface in position and illuminate the surface using a point source that is far away compared with the size of the surface. We adopt a local shading model and assume that there is no ambient illumination (more about this later) so that the radiosity at a point P on the surface is

$$B(P) = \rho(P)\boldsymbol{N}(P) \cdot \boldsymbol{S}_1,$$

where $\boldsymbol{N}$ is the unit surface normal and $\boldsymbol{S}_1$ is the source vector. With our camera model, there is only one point P on the surface for each point (x, y) in the image, and we can write $B(x, y)$ for $B(P)$. Now we assume that the response of the camera is linear in the surface radiosity, so the value of a pixel at (x, y) is

$$\begin{aligned} I(x, y) &= kB(x, y) \\ &= k\rho(x, y)\boldsymbol{N}(x, y) \cdot \boldsymbol{S}_1 \\ &= \boldsymbol{g}(x, y) \cdot \boldsymbol{V}_1, \end{aligned}$$

where k is the constant connecting the camera response to the input radiance, $\boldsymbol{g}(x, y) = \rho(x, y)\boldsymbol{N}(x, y)$, and $\boldsymbol{V}_1 = k\boldsymbol{S}_1$.

In these equations, $\boldsymbol{g}(x, y)$ describes the surface and $\boldsymbol{V}_1$ is a property of the illumination and of the camera. We have a dot product between a vector field $\boldsymbol{g}(x, y)$ and a vector $\boldsymbol{V}_1$, which could be measured; with enough of these dot products, we could reconstruct $\boldsymbol{g}$ and so the surface.

5.4.1 Normal and Albedo from Many Views

Now if we have n sources, for each of which $\boldsymbol{V}_i$ is known, we stack each of these $\boldsymbol{V}_i$ into a known matrix $\mathcal{V}$, where

$$\mathcal{V} = \begin{pmatrix} \boldsymbol{V}_1^T \\ \boldsymbol{V}_2^T \\ \cdots \\ \boldsymbol{V}_n^T \end{pmatrix}.$$

For each image point, we stack the measurements into a vector

$$\boldsymbol{i}(x, y) = \{I_1(x, y), I_2(x, y), \dots, I_n(x, y)\}^T.$$

Notice that we have one vector per image point; each vector contains all the image brightnesses observed at that point for different sources. Now we have

$$\boldsymbol{i}(x, y) = \mathcal{V}\boldsymbol{g}(x, y),$$

and $\boldsymbol{g}$ is obtained by solving this linear system—or rather, one linear system per point in the image. Typically, $n > 3$ so that a least squares solution is appropriate. This has the advantage that the residual error in the solution provides a check on our measurements.

The difficulty with this approach is that substantial regions of the surface may be in shadow for one or the other light (see Figure 5.10). There is a simple trick that deals with shadows. If there really is no ambient illumination, then we can form a matrix from the image vector and multiply both sides by this matrix; this zeroes out any equations from points that are in shadow.

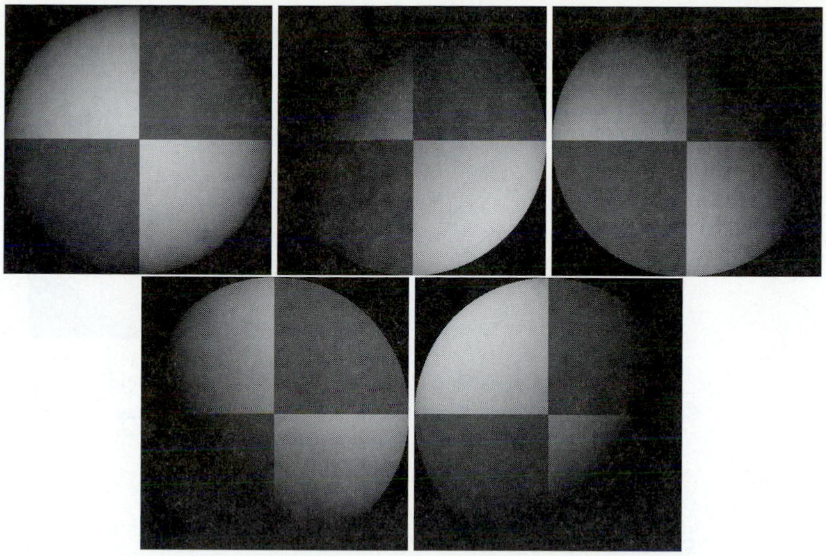

Figure 5.10 Five synthetic images of a sphere, all obtained in an orthographic view from the same viewing position. These images are shaded using a local shading model and a distant point source. This is a convex object, so the only view where there is no visible shadow occurs when the source direction is parallel to the viewing direction. The variations in brightness occuring under different sources code the shape of the surface.

We form

$$\mathcal{I}(x, y) = \begin{pmatrix} I_1(x, y) & \dots & 0 & 0 \\ 0 & I_2(x, y) & \dots & 0 \\ \dots & & & \\ 0 & 0 & \dots & I_n(x, y) \end{pmatrix}$$

and

$$\mathcal{I}\boldsymbol{i} = \mathcal{I}\mathcal{V}\boldsymbol{g}(x, y),$$

and $\mathcal{I}$ has the effect of zeroing the contributions from shadowed regions, because the relevant elements of the matrix are zero at points that are in shadow. Again, there is one linear system per point in the image; at each point, we solve this linear system to recover the $\boldsymbol{g}$ vector at that point.

Measuring Albedo We can extract the albedo from a measurement of $\boldsymbol{g}$ because $\boldsymbol{N}$ is the unit normal. This means that $|\boldsymbol{g}(x, y)| = \rho(x, y)$. This provides a check on our measurements as well. Because the albedo is in the range zero to one, any pixels where $|\boldsymbol{g}|$ is greater than one are suspect—either the pixel is not working or $\mathcal{V}$ is incorrect. Figure 5.11 shows albedo recovered using this method for the images of Figure 5.10.

Recovering Normals We can extract the surface normal from $\boldsymbol{g}$ because the normal is a unit vector

$$\boldsymbol{N}(x, y) = \frac{1}{|\boldsymbol{g}(x, y)|}\boldsymbol{g}(x, y).$$

Figure 5.12 shows normal values recovered for the images of Figure 5.10.

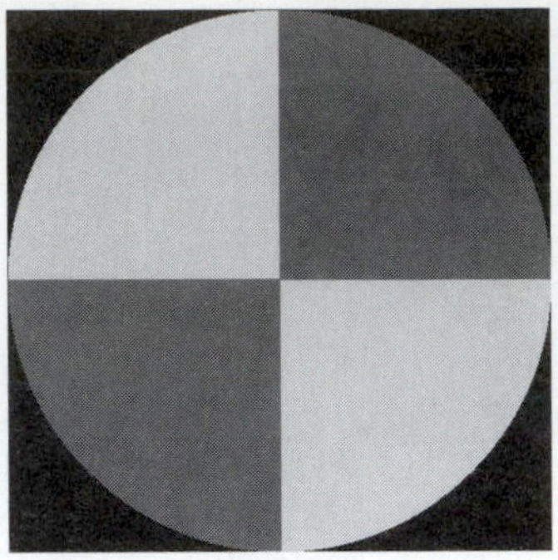

Figure 5.11 The magnitude of the vector field $\mathbf{g}(x, y)$ recovered from the input data of Figure 5.10 represented as an image—this is the reflectance of the surface.

5.4.2 Shape from Normals

The surface is $(x, y, f(x, y))$, so the normal as a function of (x, y) is

$$\mathbf{N}(x, y) = \frac{1}{\sqrt{1 + \frac{\partial f}{\partial x}^2 + \frac{\partial f}{\partial y}^2}} \left\{ -\frac{\partial f}{\partial x}, -\frac{\partial f}{\partial y}, 1 \right\}^T$$

To recover the depth map, we need to determine $f(x, y)$ from measured values of the unit normal.

Assume that the measured value of the unit normal at some point (x, y) is $(a(x, y), b(x, y), c(x, y))$. Then

$$\frac{\partial f}{\partial x} = \frac{a(x, y)}{c(x, y)} \quad \text{and} \quad \frac{\partial f}{\partial y} = \frac{b(x, y)}{c(x, y)}.$$

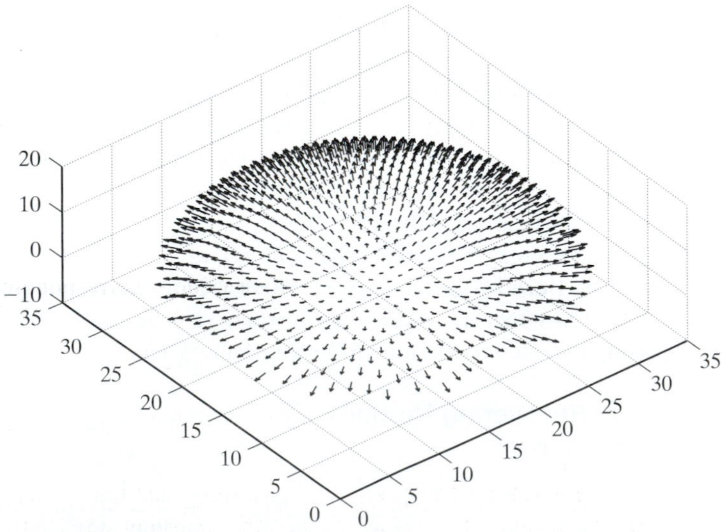

Figure 5.12 The normal field recovered from the input data of Figure 5.10.

We have another check on our data set, because

$$\frac{\partial^2 f}{\partial x \partial y} = \frac{\partial^2 f}{\partial y \partial x},$$

so we expect that

$$\frac{\partial \left(\frac{a(x,y)}{c(x,y)}\right)}{\partial y} - \frac{\partial \left(\frac{b(x,y)}{c(x,y)}\right)}{\partial x}$$

should be small at each point. In principle it should be zero, but we would have to estimate these partial derivatives numerically and so should be willing to accept small values. This test is known as a test of *integrability*, which in vision applications always boils down to checking that mixed second partials are equal.

Algorithm 5.1: Photometric Stereo

Obtain many images in a fixed view under different illuminants
Determine the matrix $\mathcal{V}$ from source and camera information
Create arrays for albedo, normal (3 components),
  p (measured value of $\frac{\partial f}{\partial x}$) and
  q (measured value of $\frac{\partial f}{\partial y}$)
For each point in the image array
  Stack image values into a vector $\boldsymbol{i}$
  Construct the diagonal matrix $\mathcal{I}$
  Solve $\mathcal{I}\mathcal{V}\boldsymbol{g} = \mathcal{I}\boldsymbol{i}$ to obtain $\boldsymbol{g}$ for this point
  Albedo at this point is $|\boldsymbol{g}|$
  Normal at this point is $\frac{\boldsymbol{g}}{|\boldsymbol{g}|}$
  p at this point is $\frac{N_1}{N_3}$
  q at this point is $\frac{N_2}{N_3}$
end
Check: is $(\frac{\partial p}{\partial y} - \frac{\partial q}{\partial x})^2$ small everywhere?

Top left corner of height map is zero
For each pixel in the left column of height map
  height value = previous height value + corresponding q value
end
For each row
  For each element of the row except for leftmost
    height value = previous height value + corresponding p value
  end
end

Shape by Integration Assuming that the partial derivatives pass this sanity test, we can reconstruct the surface up to some constant depth error. The partial derivative gives the change in surface height with a small step in either the x or the y direction. This means we can

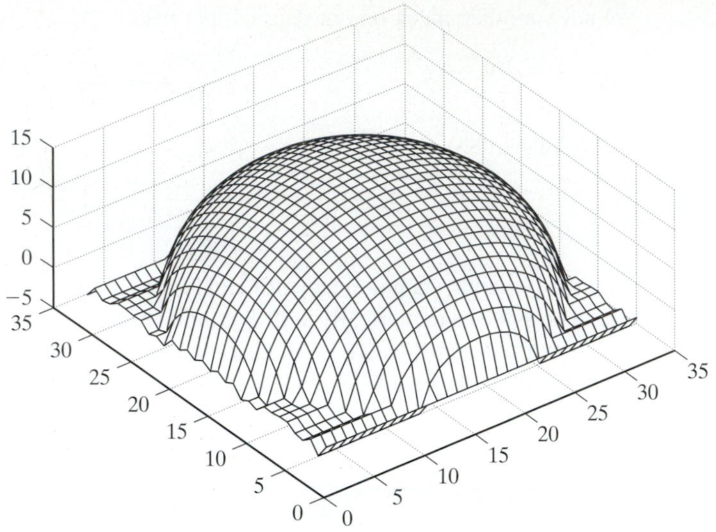

Figure 5.13 The height field obtained by integrating the normal field of Figure 5.12 using the method described in the text.

get the surface by summing these changes in height along some path. In particular, we have

$$f(x, y) = \oint_C \left(\frac{\partial f}{\partial x}, \frac{\partial f}{\partial y} \right) \cdot d\boldsymbol{l} + c,$$

where C is a curve starting at some fixed point and ending at (x, y) and c is a constant of integration, which represents the (unknown) height of the surface at the start point. The recovered surface does not depend on the choice of curve (exercises).

For example, we can reconstruct the surface at (u, v) by starting at $(0, 0)$, summing the y-derivative along the line $x = 0$ to the point $(0, v)$, and then summing the x-derivative along the line $y = v$ to the point (u, v):

$$f(u, v) = \int_0^v \frac{\partial f}{\partial y}(0, y)\, dy + \int_0^u \frac{\partial f}{\partial x}(x, v)\, dx + c.$$

This is the integration path given in Algorithm 5.1. Any other set of paths would work as well although it is probably best to use many different paths and average so as to spread around the error in the derivative estimates. Figure 5.13 shows the reconstruction obtained for the data of Figure 5.10.

Another approach to recovering shape is to choose the function $f(x, y)$ whose partial derivatives most look like the measured partial derivatives. We explore this approach for a similar problem in Section 6.5.2.

5.5 INTERREFLECTIONS: GLOBAL SHADING MODELS

Local shading models can be quite misleading. In the real world, each surface patch is illuminated not only by sources, but also by light reflected off other surface patches (a phenomenon known as *interreflection*). A model that incorporates interreflection effects is known as a *global shading model*. Interreflections lead to a variety of complex shading effects, which are still quite poorly

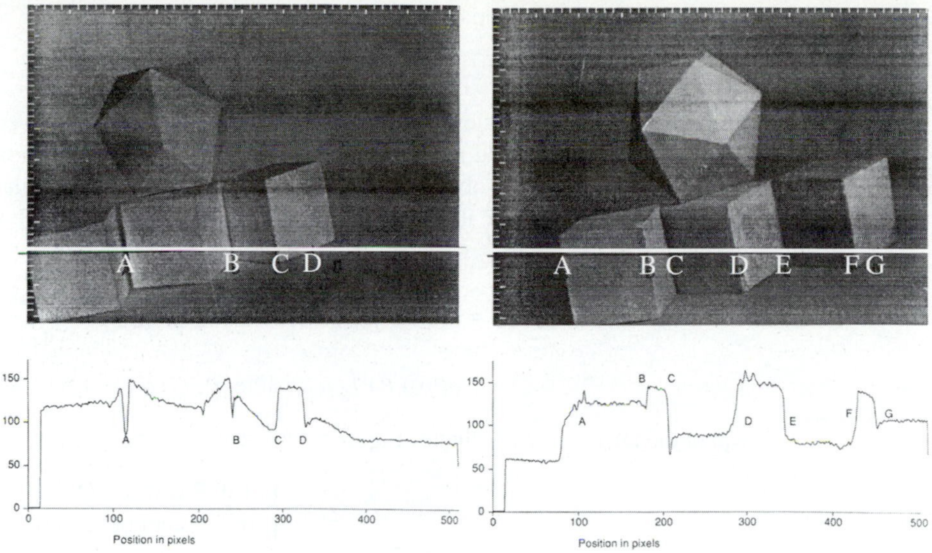

Figure 5.14 The column on the **left** shows data from a room with matte black walls and containing a collection of matte black polyhedral objects; that on the **right** shows data from a white room containing white objects. The images are qualitatively different, with darker shadows and crisper boundaries in the black room and bright reflexes in the concave corners in the white room. The graphs show sections of the image intensity along the corresponding lines in the images. *Figure from "Mutual Illumination," by D.A. Forsyth and A.P. Zisserman, Proc. CVPR, 1989, © 1989 IEEE*

understood. Unfortunately, these effects occur widely, and it is still not yet known how to simplify global shading models without losing essential qualitative properties.

For example, Figure 5.14 shows views of the interior of two rooms. One room has black walls and contains black objects. The other has white walls and contains white objects. Each is illuminated (approximately!) by a distant point source. Given that the intensity of the source is adjusted appropriately, the local shading model predicts that these pictures would be indistinguishable. In fact, the black room has much darker shadows and crisper boundaries at the creases of the polyhedra than the white room. This is because surfaces in the black room reflect less light onto other surfaces (they are darker), whereas in the white room other surfaces are significant sources of radiation. The sections of the camera response to the radiosity (these are proportional to radiosity for diffuse surfaces) shown in the figure are hugely different qualitatively. In the black room, the radiosity is constant in patches as a local shading model would predict, whereas in the white room slow image gradients are quite common—these occur in concave corners, where object faces reflect light onto one another.

This effect also explains why a room illuminated by a point light source does not show the sharp illumination gradients that a local shading model predicts (recall Section 5.2.2). The walls and floor of the room reflect illumination back, and this tends to light up the corners, which would otherwise be dark.

5.5.1 An Interreflection Model

It is well understood how to predict the radiosity on a set of diffuse surface patches. The total radiosity leaving a patch is its exitance —which is zero for all but sources—*plus* all the radiosity that is reflected from the patch:

$$B(P) = E(P) + B_{\text{refl}}(P).$$

From the point of view of our patch, there is no distinction between energy leaving another patch due to exitance and that due to reflection. This means we can take the expression for an area source and use it to obtain an expression for $B_{\text{refl}}(Q)$. In particular, from the perspective of our patch, the patch at R in the world is equivalent to an area source with exitance $B(R)$. This means

$$\begin{aligned} B_{\text{refl}}(P) &= \rho_d(P) \int_{\text{world}} \text{visible}(P, Q) B(Q) \frac{\cos\theta_P \cos\theta_Q}{\pi d_{PQ}^2} \, dA_Q \\ &= \rho_d(P) \int_{\text{world}} \text{visible}(P, Q) K(P, Q) B(Q) \, dA_Q, \end{aligned}$$

where the terminology is that of Figure 5.15 and

$$\text{visible}(P, Q) = \begin{cases} 1 \text{ if } P \text{ can see } Q, \\ 0 \text{ if } P \text{ cannot see } Q. \end{cases}$$

The term visible$(P, Q)K(P, Q)$ is usually referred to as the *interreflection kernel*. Substituting the expression for $B_{\text{refl}}(P)$ gives

$$B(P) = E(P) + \rho_d(P) \int_{\text{world}} \text{visible}(P, Q) K(P, Q) B(Q) \, dA_Q.$$

In particular, the solution appears inside the integral. Equations of this form are known as Fredholm integral equations of the second kind. This particular equation is a fairly nasty sample of the type because the interreflection kernel is generally not continuous and may have singularities. Solutions of this equation can yield quite good models of the appearance of diffuse surfaces, and the topic supports a substantial industry in the computer graphics community (good

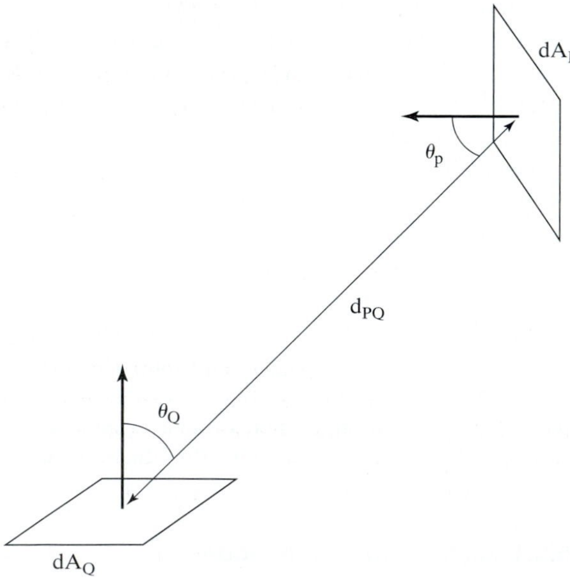

Figure 5.15 Terminology for expression derived in the text for the interreflection kernel.

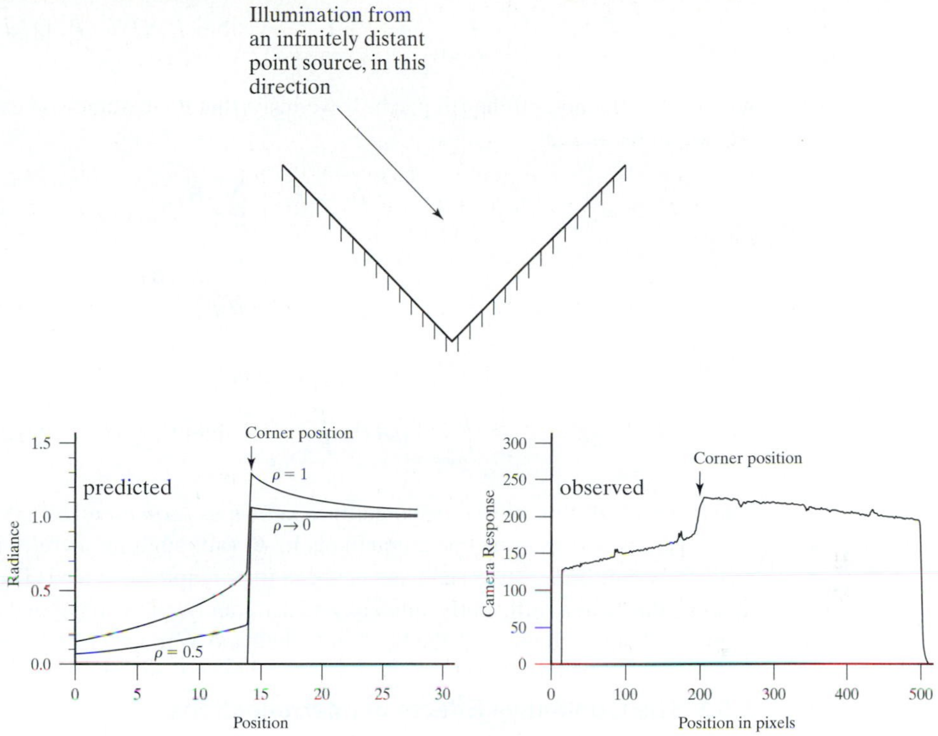

Figure 5.16 The model described in the text produces quite accurate qualitative predictions for interreflections. The **top** figure shows a concave right-angled groove illuminated by a point source at infinity where the source direction is parallel to the one face. On the **left** of the bottom row is a series of predictions of the radiosity for this configuration. These predictions have been scaled to lie on top of one another; the case $\rho \to 0$ corresponds to the local shading model. On the **right**, an observed image intensity for an image of this form for a corner made of white paper, showing the roof-like gradient in radiosity associated with the edge. A local shading model predicts a step. *Figure from "Mutual Illumination," by D.A. Forsyth and A.P. Zisserman, Proc. CVPR, 1989, © 1989 IEEE*

places to start for this topic are Cohen and Wallace (1993) or Sillion (1994)). The model produces good predictions of observed effects (Figure 5.16).

5.5.2 Solving for Radiosity

We sketch one approach to solving the global shading model to illustrate the methods. Subdivide the world into small, flat patches and approximate the radiosity as being constant over each patch. This approximation is reasonable because we could obtain an accurate representation by working with small patches. Now we construct a vector $\boldsymbol{B}$, which contains the value of the radiosity for each patch. In particular, the ith component of $\boldsymbol{B}$ is the radiosity of the ith patch.

We write the incoming radiosity at the ith patch due to radiosity on the jth patch as

$$B_{j \to i}(P) = \rho_d(P) \int_{\text{patch } j} \text{visible}(P, Q) K(P, Q) \, dA_Q B_j,$$

where P is a coordinate on the ith patch and R is a coordinate on the jth patch. Now this expression is not a constant, and so we must average it over the ith patch to get

$$\overline{B}_{j\to i} = \frac{1}{A_i}\int_{\text{patch } i} \rho_d(P) \int_{\text{patch } j} \text{visible}(P, Q) K(P, Q)\, dA_P\, dA_Q B_j,$$

where A_i is the area of the ith patch. If we insist that the exitance on each patch is constant, too, we obtain the model

$$\begin{aligned} B_i &= E_i + \sum_{\text{all } j} \overline{B}_{j\to i} \\ &= E_i + \sum_{\text{all } j} K_{ij} B_j, \end{aligned}$$

where

$$K_{ij} = \frac{1}{A_i}\int_{\text{patch } i} \rho_d(P) \int_{\text{patch } j} \text{visible}(P, Q) K(P, Q)\, dA_P\, dA_Q.$$

The elements of this matrix are sometimes known as *form factors*.

This is a system of linear equations in B_i (although an awfully big one—K_{ij} could be a million by a million matrix) and, as such, can in principle be solved. The tricks that are necessary to solve the system efficiently, quickly, and accurately are well beyond our scope; Sillion (1994) is an excellent account, as is the book of Cohen and Wallace (1993).

5.5.3 The Qualitative Effects of Interreflections

We should like to extract shape information from radiosity. This is relatively easy to do with a local model (see Section 5.4 for some details), but the model describes the world poorly, and little is known about how severely this affects the resulting shape information. Extracting shape information from a global shading model is difficult for two reasons. First, the relationship between shape and radiosity is complicated because it is governed by the interreflection kernel. Second, there are almost always surfaces that are not visible, but radiate to the objects in view. These so-called "distant surfaces" mean it is hard to account for all radiation in the scene using an interreflection model because some radiators are invisible and we may know little or nothing about them.

All this suggests that understanding qualitative, local effects of interreflection is important; armed with this understanding, we can either discount the effects of interreflection or exploit them. This topic remains largely an open research topic, but there are some things we can say.

Smoothing and Regional Properties First, interreflections have a characteristic smoothing effect. This is most obviously seen if one tries to interpret a stained glass window by looking at the pattern it casts on the floor; this pattern is almost always a set of indistinct colored blobs. The effect is seen most easily with the crude model of Figure 5.17. The geometry consists of a patch with a frontal view of an infinite plane, which is a unit distance away and carries a radiosity $\sin \omega x$. There is no reason to vary the distance of the patch from the plane because interreflection problems have scale invariant solutions—this means that the solution for a patch two units away can be obtained by reading our graph at 2ω. The patch is small enough that its contribution to the plane's radiosity can be ignored. If the patch is slanted by σ with respect to the plane, it carries radiosity that is nearly periodic, with spatial frequency $\omega \cos \sigma$. We refer to the amplitude of the component at this frequency as the gain of the patch and plot the gain in Figure 5.17. The important property of this graph is that high spatial frequencies have a difficult time jumping the gap from the plane to the patch. This means that shading effects with high

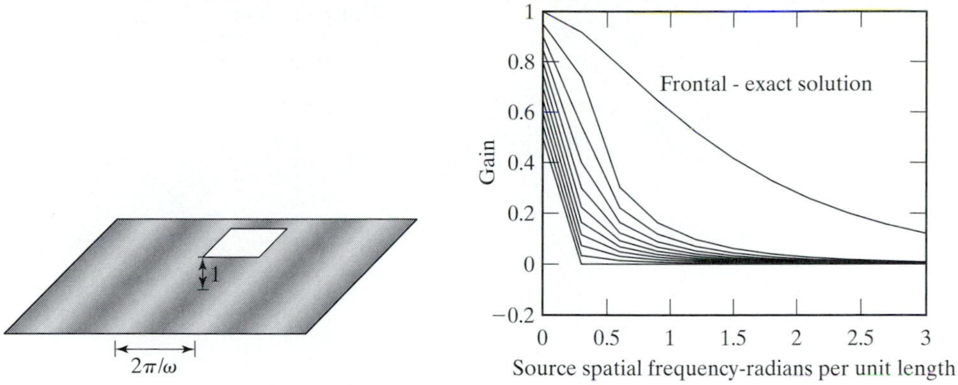

Figure 5.17 A small patch views a plane with sinusoidal radiosity of unit amplitude. This patch has a (roughly) sinusoidal radiosity due to the effects of the plane. We refer to the amplitude of this component as the *gain of the patch*. The graph shows numerical estimates of the gain for patches at 10 equal steps in slant angle, from 0 to $\pi/2$, as a function of spatial frequency *on the plane*. The gain falls extremely fast, meaning that large terms at high spatial frequencies must be regional effects, rather than the result of distant radiators. This is why it is hard to determine the pattern in a stained glass window by looking at the floor at foot of the window. *Reprinted from "Shading Primitives: Finding Folds and Shallow Grooves," by J. Haddon and D.A. Forsyth, Proc. Int. Conf. Computer Vision, 1998 © 1998 IEEE*

spatial frequency and high amplitude generally cannot come from distant surfaces (unless they are abnormally bright).

The extremely fast fall-off in amplitude with spatial frequency of terms due to distant surfaces means that, if one observes a high amplitude term at a high spatial frequency, *it is very unlikely to have resulted from the effects of distant, passive radiators* (because these effects die away quickly). There is a convention, which we see in Section 6.5.2, that classifies effects in shading as due to reflectance if they are fast ("edges") and the dynamic range is relatively low and due to illumination otherwise. We can expand this convention. There is a mid range of spatial frequencies that are largely unaffected by mutual illumination from distant surfaces because the gain is small. Spatial frequencies in this range cannot be transmitted by distant passive radiators unless these radiators have improbably high radiosity. As a result, spatial frequencies in this range can be thought of as *regional properties*, which can result only from interreflection effects within a region.

The most notable regional properties are probably *reflexes*— small bright patches that appear mainly in concave regions (illustrated in Figure 5.18 and Figure 5.19). A second important effect is *color bleeding*, where a colored surface reflects light onto another colored surface. This is a common effect that people tend not to notice unless they are consciously looking for it. It is quite often reproduced by painters.

5.6 NOTES

Shading models are handled in a quite unsystematic way in the vision literature. The point source approximation is widely abused; you should use it with care and inspect others' use of it with suspicion. We believe we are the first to draw the distinction between (a) the physical effects of sources, and (b) the shading model.

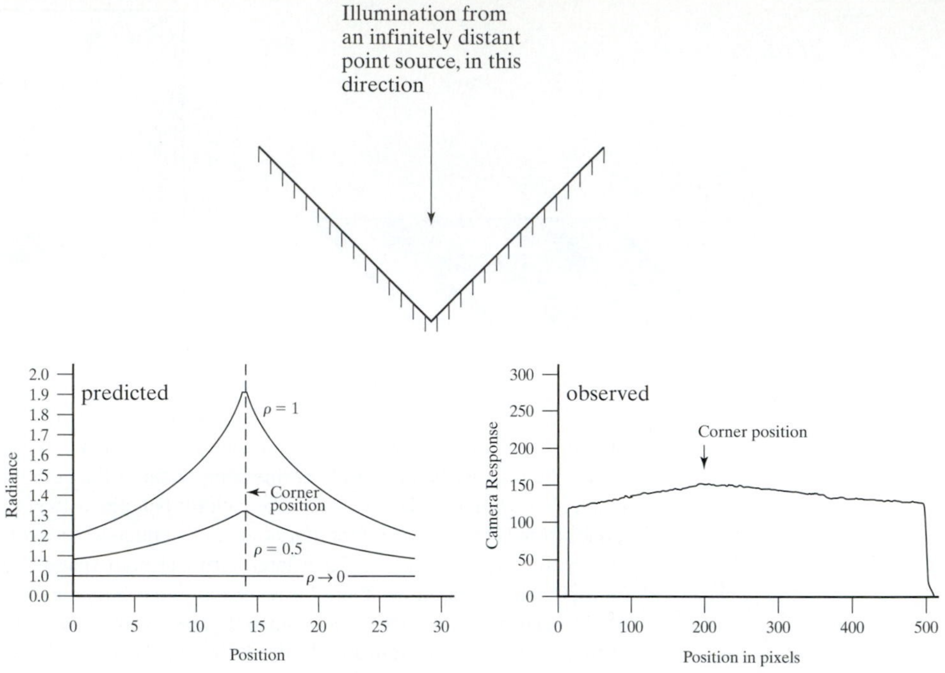

Figure 5.18 Reflexes at concave edges are a common qualitative result of interreflections. The figure on the **top** shows the situation here; a concave right-angled groove illuminated by a point light source at infinity, whose source vector is along the angle bisector. The graph on the **left** shows the intensity predictions of an interreflection model for this configuration; the case $\rho \to 0$ is a local shading model. The graphs have been lined up for easy comparison. As the surface's albedo goes up, a roof-like structure appears. The graph on the **right** shows an observation of this effect in an image of a real scene. *Figure from "Mutual Illumination," by D.A. Forsyth and A.P. Zisserman, Proc. CVPR, 1989, © 1989 IEEE*

Local Shading Models

The great virtue of local shading models is that the analysis is simple. The primary characteristic of a local shading model is that, on a surface of constant albedo, the radiosity of a surface patch is a function of the normal alone. This means that one can avoid the abstraction of reflectance and sources and instead simply code the properties of surface *and source* as a *reflectance map*. The reflectance map is a function that takes a representation of the normal, and returns the radiosity to be expected at a point with that normal.

Horn started the systematic study of shading in computer vision, with important papers on recovering shape from a local shading model using a point source (Horn, 1970, 1975), with a more recent account in Horn (1990). The methods discussed have largely fallen into disuse (at least partially because they appear unable to cope with the difficulties created by a global shading model), so we do not survey the vast literature here. A comprehensive summary is in Horn and Brooks (1989). Shape and albedo are ambiguous; with appropriate changes in albedo, surfaces of different shapes can generate the same image (Belhumeur, Kriegman and Yuille, 1999, Kriegman and Belhumeur, 1998). Because the surface normal is the key in local shading models, such models typically yield elegant links between surface shading and curvature (Koenderink and van Doorn, 1980).

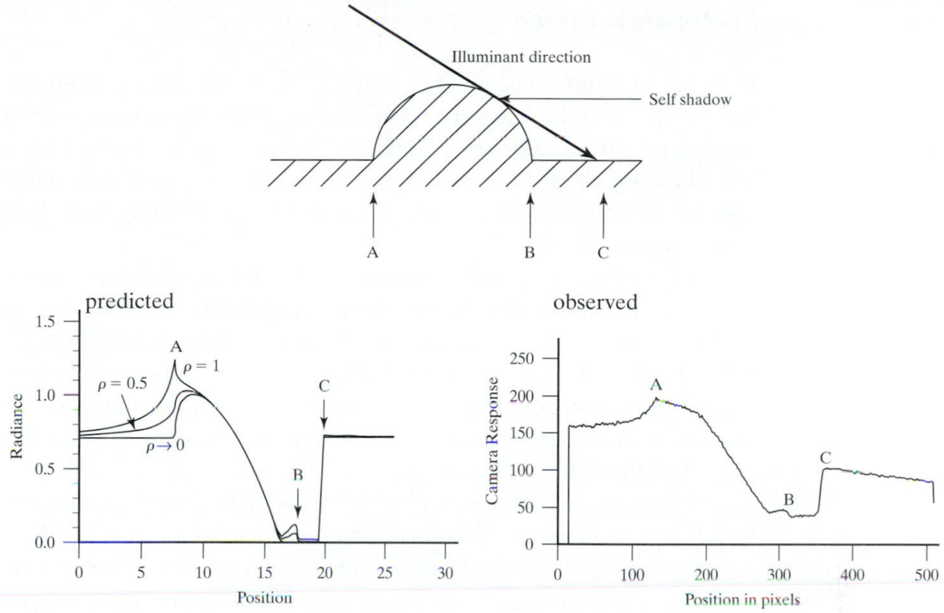

Figure 5.19 Reflexes occur quite widely; they are usually caused by a favorable view of a large reflecting surface. In the geometry shown on the **top**, the shadowed region of the cylindrical bump sees the plane background at a fairly favorable angle —if the background is large enough, near half the hemisphere of the patch at the base of the bump is a view of the plane. This means there will be a reflex with a large value attached to the edge of the bump and inside the cast shadow region (which a local model predicts as black). There is another reflex on the other side, too, as the series of solutions (again normalized for easy comparison) on the **left** show. On the **right**, an observation of this effect in a real scene. *Figure from "Mutual Illumination," by D.A. Forsyth and A.P. Zisserman, Proc. CVPR, 1989, © 1989 IEEE*

Interreflections

The effects of global shading are often ignored in the shading literature, which causes a reflex response of hostility in one of the authors. The reason to ignore interreflections is that they are extremely hard to analyze, particularly from the perspective of inferring object properties given the output of a global shading model. If interreflection effects do not change the output of a method much, then it is probably all right to ignore them. Unfortunately, this line of reasoning is seldom pursued because it is quite difficult to show that a method is stable under interreflections. The discussion of spatial frequency issues follows Haddon and Forsyth (1998*a*), after an idea of Koenderink and van Doorn (1983). Apart from this, there is not much knowledge about the overall properties of interreflected shading, which is an important gap in our knowledge. An alternative strategy is to iteratively reestimate shape using a rendering model (Nayar, Ikeuchi and Kanade, 1991*a*).

Horn is also the first author to indicate the significance of global shading effects (Horn, 1977). Koenderink and van Doorn (1983) noted that the radiosity under a global model is obtained by taking the radiosity under a local model, and applying a linear operator. One then studies that operator; in some cases, its eigenfunctions (often called *geometrical modes*) are informative. Forsyth and Zisserman (1989, 1990, 1991) then demonstrated a variety of the qualitative effects due to interreflections.

Photometric Stereo

In its original form, photometric stereo is due to Woodham. There are a number of variants of this useful idea (Horn, Woodham and Silver, 1978, Woodham, 1979, 1980, 1989, 1994). There are a variety of variations on photometric stereo. One interesting idea is to illuminate the surface with three lights of different colors (and in different positions) and use a color image. For an appropriate choice of colors, this is equivalent to obtaining three images, so the measurement process is simplified.

Generally, photometric stereo is used under circumstances where the illumination is quite easily controlled, so that it is possible to ensure that no ambient illumination is in the image. It is relatively simple to insert ambient illumination into the formulation given; we extend the matrix $\mathcal{V}$ by attaching a column of ones. In this case, $\boldsymbol{g}(x, y)$ becomes a four-dimensional vector, and the fourth component is the ambient term. However, this approach does not guarantee that the ambient term is constant over space. Instead, we would have to check that this term was constant and adjust the model if it were not.

Photometric stereo depends only on adopting a local shading model. This model need not be a Lambertian surface illuminated by a distant point source. If the radiosity of the surface is a known function of the surface normal satisfying a small number of constraints, photometric stereo is still possible. This is because the intensity of a pixel in a single view determines the normal up to a one-parameter family. This means that two views determine the normal. The simplest example of this case occurs for a surface of known albedo illuminated by a distant point source.

In fact, if the radiosity of the surface is a k-parameter function of the surface normal, photometric stereo is still possible. The intensity of the pixel in a single view determines the normal up to a $k + 1$ parameter family, and $k + 1$ views give the normal. For this approach to work, the radiosity needs to be given by a function for which our arithmetic works (e.g., if the radiosity of the surface is a constant function of the surface normal, it is not possible to infer any constraint on the normal from the radiosity). One can then recover shape and reflectance maps simultaneously (Garcia-Bermejo, Diaz Pernas and Coronado, 1996, Mukawa, 1990, Nayar, Ikeuchi and Kanade, 1990, and Tagare and de Figueiredo, 1992, 1993).

Alternative Shading Representations

Instead of trying to extract shape information from the shading signal, one might try to match it to a collection of different possible examples. This suggests studying what kinds of shaded view a surface can generate. The collection of available shadings is notably limited (Belhumeur and Kriegman, 1998). A knowledge of this collection's structure is valuable because it makes it possible to understand how to compare shaded images without being confused by illumination changes. Illumination changes are a particular problem in face finding and recognition applications (Adini, Moses and Ullman, 1997, Phillips and Vardi, 1996). Knowing the possible variations in illumination seems to help (Georghiades, Kriegman and Belhumeur, 1998, 2000, Jacobs, Belhumeur and Basri, 1998).

Another possibility is to extend the notion of qualitative analysis of interreflections to obtain a *shading primitive*—a shading pattern that is characteristic and stably linked to a shape pattern. For example, narrow grooves and deep holes in surfaces are dark, and cylinders have a characteristic extended pattern of shading. Few such primitives are known but some appear to be useful (Haddon and Forsyth, 1998*a*, 1998*b*).

PROBLEMS

5.1. What shapes can the shadow of a sphere take if it is cast on a plane and the source is a point source?

5.2. We have a square area source and a square occluder, both parallel to a plane. The source is the same size as the occluder, and they are vertically above one another with their centers aligned.
(a) What is the shape of the umbra?
(b) What is the shape of the outside boundary of the penumbra?

5.3. We have a square area source and a square occluder, both parallel to a plane. The edge length of the source is now twice that of the occluder, and they are vertically above one another with their centers aligned.
(a) What is the shape of the umbra?
(b) What is the shape of the outside boundary of the penumbra?

5.4. We have a square area source and a square occluder, both parallel to a plane. The edge length of the source is now half that of the occluder, and they are vertically above one another with their centers aligned.
(a) What is the shape of the umbra?
(b) What is the shape of the outside boundary of the penumbra?

5.5. A small sphere casts a shadow on a larger sphere. Describe the possible shadow boundaries that occur.

5.6. Explain why it is difficult to use shadow boundaries to infer shape, particularly if the shadow is cast onto a curved surface.

5.7. An infinitesimal patch views a circular area source of constant exitance frontally along the axis of symmetry of the source. Compute the radiosity of the patch due to the source exitance $E(\boldsymbol{u})$ as a function of the area of the source and the distance between the center of the source and the patch. You may have to look the integral up in tables—if you don't, you're entitled to feel pleased with yourself—but this is one of few cases that can be done in closed form. It is easier to look up if you transform it to get rid of the cosine terms.

5.8. As in Figure 5.17, a small patch views an infinite plane at unit distance. The patch is sufficiently small that it reflects a trivial quantity of light onto the plane. The plane has radiosity $B(x, y) = 1 + \sin ax$. The patch and the plane are parallel to one another. We move the patch around parallel to the plane, and consider its radiosity at various points.
(a) Show that if one translates the patch, its radiosity varies periodically with its position in x.
(b) Fix the patch's center at $(0, 0)$; determine a *closed form* expression for the radiosity of the patch at this point as a function of a. You'll need a table of integrals for this (if you don't, you're entitled to feel very pleased with yourself).

5.9. If one looks across a large bay in the daytime, it is often hard to distinguish the mountains on the opposite side; near sunset, they are clearly visible. This phenomenon has to do with scattering of light by air—a large volume of air is actually a source. Explain what is happening. We have modeled air as a vacuum and asserted that no energy is lost along a straight line in a vacuum. Use your explanation to give an estimate of the kind of scales over which that model is acceptable.

5.10. Read the book *Colour and Light in Nature*, by Lynch and Livingstone, published by Cambridge University Press, 1995.

Programming Assignments

5.11. An area source can be approximated as a grid of point sources. The weakness of this approximation is that the penumbra contains quantization errors, which can be quite offensive to the eye.
(a) Explain.
(b) Render this effect for a square source and a single occluder casting a shadow onto an infinite plane. For a fixed geometry, you should find that as the number of point sources goes up, the quantization error goes down.

(c) This approximation has the unpleasant property that it is possible to produce arbitrarily large quantization errors with any finite grid by changing the geometry. This is because there are configurations of source and occluder that produce large penumbrae. Use a square source and a single occluder, casting a shadow onto an infinite plane, to explain this effect.

5.12. Make a world of black objects and another of white objects (paper, glue and spraypaint are useful here) and observe the effects of interreflections. Can you come up with a criterion that reliably tells, *from an image*, which is which? (If you can, publish it; the problem looks easy, but isn't).

5.13. (This exercise requires some knowledge of numerical analysis.) Do the numerical integrals required to reproduce Figure 5.17. These integrals aren't particularly easy: If one uses coordinates on the infinite plane, the size of the domain is a nuisance; if one converts to coordinates on the view hemisphere of the patch, the frequency of the radiance becomes infinite at the boundary of the hemisphere. The best way to estimate these integrals is using a Monte Carlo method on the hemisphere. You should use importance sampling because the boundary contributes rather less to the integral than the top.

5.14. Set up and solve the linear equations for an interreflection solution for the interior of a cube with a small square source in the center of the ceiling.

5.15. Implement a photometric stereo system.

(a) How accurate are its measurements (i.e., how well do they compare with known shape information)? Do interreflections affect the accuracy?

(b) How repeatable are its measurements (i.e., if you obtain another set of images, perhaps under different illuminants, and recover shape from those, how does the new shape compare with the old)?

(c) Compare the minimization approach to reconstruction with the integration approach; which is more accurate or more repeatable and why? Does this difference appear in experiment?

(d) One possible way to improve the integration approach is to obtain depths by integrating over many different paths and then average these depths (you need to be a little careful about constants here). Does this improve the accuracy or repeatability of the method?

6

Color

Color is a rich and complex experience, usually caused by the vision system responding differently to different wavelengths of light (other causes include pressure on the eyeball and dreams). Although the color of objects seems to be a useful cue in identifying them, it is currently difficult to use.

6.1 THE PHYSICS OF COLOR

We extend our radiometric vocabulary to describe energy arriving in different quantities at different wavelengths and then describe typical properties of colored surfaces and colored light sources.

6.1.1 Radiometry for Colored Lights: Spectral Quantities

All of the physical units we have described can be extended with the phrase *per unit wavelength* to yield *spectral units*. These allow us to describe differences in energy, in BRDF, or in albedo with wavelength. We will ignore interactions, such as flourescence, where energy changes wavelength; thus, the definitions of Chapter 4 can be extended by adding the phrase "per unit wavelength," to obtain what are known as *spectral quantities*.

Our first spectral quantity, *spectral radiance*, is usually written as $L^{\lambda}(\boldsymbol{x}, \theta, \phi)$ and the radiance emitted in the range of wavelengths $[\lambda, \lambda + d\lambda]$ is $L^{\lambda}(\boldsymbol{x}, \theta, \phi)d\lambda$. Spectral radiance has units Watts per cubic meter per steradian ($Wm^{-3}sr^{-1}$—cubic meters because of the additional factor of the wavelength). For problems where the angular distribution of the source is unimportant, *spectral exitance* is the appropriate property; spectral exitance has units Wm^{-3}.

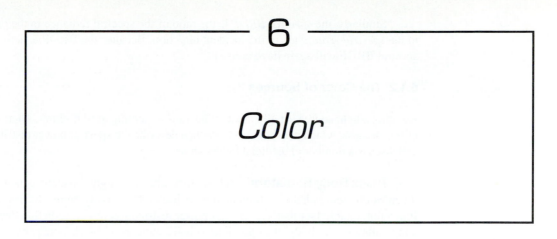

Similarly, the *spectral BRDF* is the ratio of the spectral radiance in the outgoing direction to the *spectral irradiance* in the incident direction. Because the BRDF is defined by a ratio, the spectral BRDF will again have units sr^{-1}.

6.1.2 The Color of Sources

Building a light source usually involves heating something until it glows. There is an idealization of this process, which we study first. We then describe the spectral power distribution of daylight and discuss a number of artificial light sources.

Black Body Radiators A body that reflects no light—usually called a *black body*—is the most efficient radiator of illumination. A heated black body emits electromagnetic radiation. It is a remarkable fact that the spectral power distribution of this radiation depends only on the temperature of the body. It is possible to build quite good black bodies by obtaining a hollow piece of metal and looking into the cavity through a tiny hole—very little of the light getting into the hole returns to the eye. The spectral power distribution of a hot black body can be measured by heating a cavity like this. If we write T for the temperature of the body in Kelvins, h for Planck's constant, k for Boltzmann's constant, c for the speed of light, and λ for the wavelength, we have

$$E(\lambda) \propto \frac{1}{\lambda^5} \frac{1}{(\exp(hc/k\lambda) - 1)}.$$

This means that there is one parameter family of light colors corresponding to black body radiators—the parameter being the temperature—and so we can talk about the *color temperature* of a light source. This is the temperature of the black body that looks most similar. At relatively low temperatures, black bodies are red, passing through orange to a pale yellow-white to white as the temperature increases (Figure 6.9 shows this locus).

The Sun and the Sky The most important natural light source is the sun. The sun is usually modeled as a distant, bright point. Light from the sun is scattered by the air. In particular, light can leave the sun, be scattered by the air, strike a surface and be reflected into the camera or the eye. This means the sky is an important natural light source. A crude geometrical model of the sky has it as a source consisting of a hemisphere with constant exitance. The assumption that exitance is constant is poor, however, because the sky is substantially brighter at the horizon than at the zenith. A natural model of the sky is to assume that air emits a constant amount of light per unit volume; this means that the sky is brighter on the horizon than at the zenith because a viewing ray along the horizon passes through more sky.

A patch of surface outdoors during the day is illuminated both by light that comes directly from the sun—usually called *daylight* and by light from the sun that has been scattered by the air (sometimes called *skylight* or *airlight*; the presence of clouds or snow can add other, important, phenomena). The color of daylight varies with time of day (Figure 6.1) and time of year. These effects have been widely studied.

For clear air, the intensity of radiation scattered by a unit volume depends on the fourth power of the frequency; this means that light of a long wavelength can travel much farther before being scattered than light of a short wavelength (this is known as *Rayleigh scattering*). This means that, when the sun is high in the sky, blue light is scattered out of the ray from the sun to the earth—meaning that the sun looks yellow—and can scatter from the sky into the eye—meaning that the sky looks blue. There are standard models of the spectral radiance of the sky at different times of day and latitude, too. Surprising effects occur when there are fine particles of dust in the sky (the larger particles cause much more complex scattering effects, usually modeled

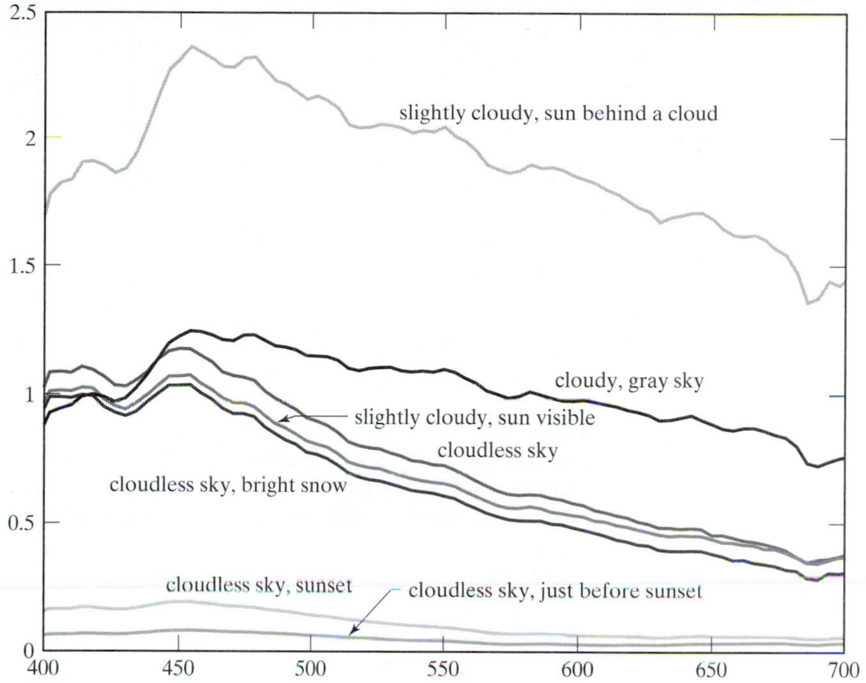

Figure 6.1 There are significant variations in the relative spectral power of daylight measured at different times of day and under different conditions. The figure shows a series of seven different daylight measurements, made by Jussi Parkkinen and Pertti Silfsten, of daylight illuminating a sample of barium sulphate (which gives a high reflectance white surface). Plot from data obtainable at `http://www.it.lut.fi/research/color/lutcs_database.html`.

rather roughly by the *Mie scattering* model, described in Lynch and Livingston, 2001, or in Minnaert, 1993). One author remembers vivid sunsets in Johannesburg caused by dust in the air from mine dumps, and there are records of blue and even green moons caused by volcanic dust in the air.

Artificial Illumination Typical artificial light sources are commonly of a small number of types:

- An *incandescent light* contains a metal filament that is heated to a high temperature. The spectrum roughly follows the black-body law, meaning that incandescent lights in most practical cases have a reddish tinge because the melting temperature of the element limits the color temperature of the light source.
- A *fluorescent light* works by generating high-speed electrons that strike gas within the bulb; this in turn releases ultraviolet radiation, which causes phosphors coating the inside of the bulb to fluoresce. Typically the coating consists of three or four phosphors, which fluoresce in quite narrow ranges of wavelengths. Most fluorescent bulbs generate light with a bluish tinge, but bulbs that mimic natural daylight are increasingly available (Figure 6.2).
- In some bulbs, an arc is struck in an atmosphere consisting of gaseous metals and inert gases. Light is produced by electrons in metal atoms dropping from an excited state to a lower energy state. Typical of such lamps is strong radiation at a small number of wavelengths, which correspond to particular state transitions. The most common cases

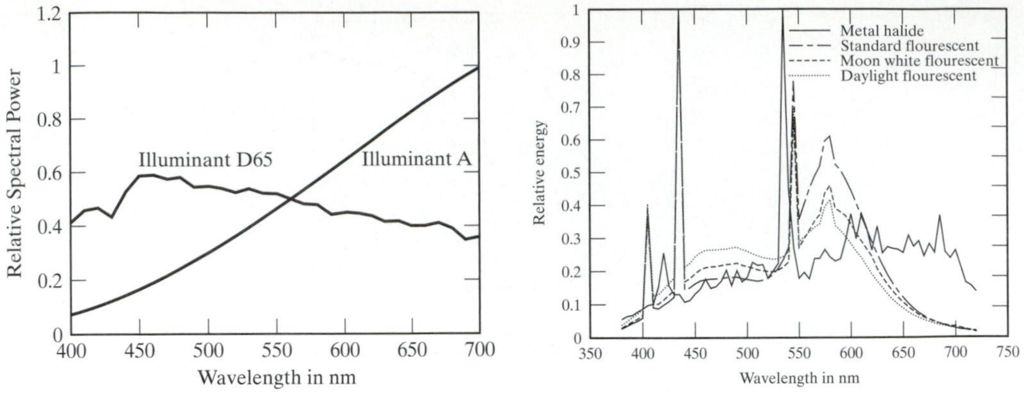

Figure 6.2 There is a variety of illuminant models; the graph on the **left** shows the relative spectral power distribution of two standard CIE models: Illuminant A—which models the light from a 100W Tungsten filament light bulb with color temperature 2800K; and Illuminant D-65—which models daylight. *The figure was plotted from data disseminated by the Color and Vision Research Laboratories database, compiled by Andrew Stockman and Lindsey Sharpe, and available at* `http://www-cvrl.ucsd.edu/index.htm`. On the **right**, we show the relative spectral power distribution of four different lamps from the Mitsubishi Electric corp. Note the bright, narrow bands that come from the flourescing phosphors in the fluorescent lamp. *The figure was plotted from data made available on the Coloring info pages* `http://colorpro.com/info/index.html`; *the data was measured by Hiroaki Sugiura.*

are *sodium arc lamps* and *mercury arc lamps*. Sodium arc lamps produce a yellow-orange light extremely efficiently and are quite commonly used for freeway lighting. Mercury arc lamps produce a blue-white light and are often used for security lighting.

Figure 6.2 shows a sample of spectra from different light bulbs.

6.1.3 The Color of Surfaces

The color of surfaces is a result of a large variety of mechanisms, including differential absorbtion at different wavelengths, refraction, diffraction, and bulk scattering (for more details, see, for example Lamb and Bourriau, 1995, Lynch and Livingston, 2001, Minnaert, 1993, or Williamson and Cummins, 1983). Usually these effects are bundled into a macroscopic BRDF model, which is typically a Lambertian plus specular approximation; the terms are now *spectral reflectance* (sometimes abbreviated to *reflectance*) or (less commonly) *spectral albedo*. Figures 6.3 and 6.4 show examples of spectral reflectances for a number of different natural objects.

The color of the light returned to the eye is affected both by the spectral radiance (color!) of the illuminant and by the spectral reflectance (color!) of the surface. If we use the Lambertian plus specular model, we have

$$E(\lambda) = \rho_{dh}(\lambda) S(\lambda) \times \text{geometric terms} + \text{specular terms},$$

where $E(\lambda)$ is the spectral radiosity of the surface, $\rho_{dh}(\lambda)$ is the spectral reflectance, and $S(\lambda)$ is the spectral irradiance. The specular terms have different colors depending on the surface—i.e., we now need a *spectral specular albedo*.

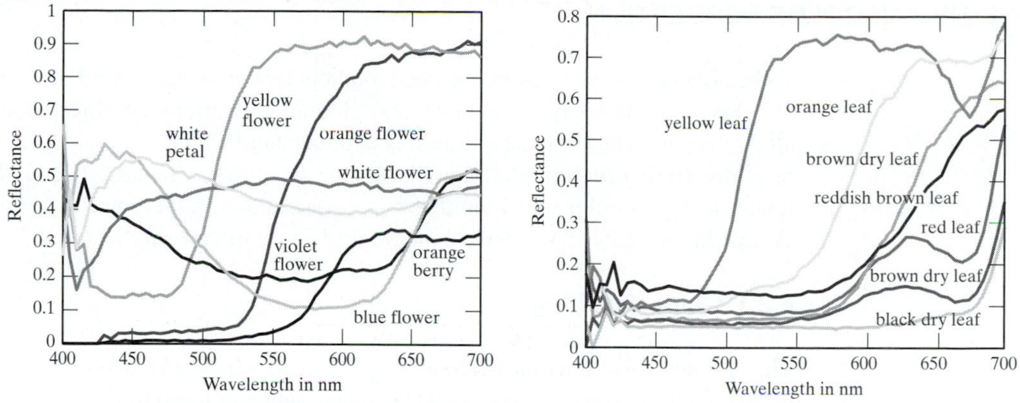

Figure 6.3 Spectral albedoes for a variety of natural surfaces measured by Esa Koivisto, Department of Physics, University of Kuopio, Finland. On the left, albedoes for a series of different natural surfaces—a color name is given for each. On the right, albedoes for different colors of leaf—again, a color name is given for each. These figures were plotted from data available at `http://www.it.lut.fi/research/color/lutcs_database.html`.

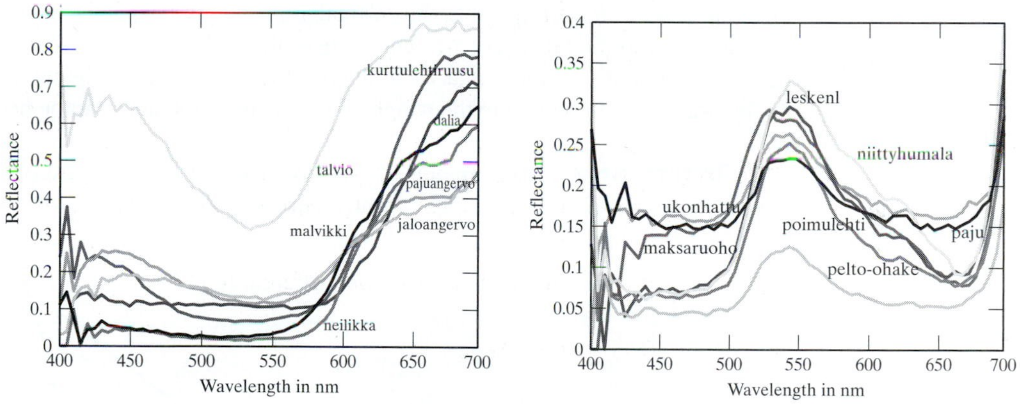

Figure 6.4 More spectral albedoes for a variety of natural surfaces measured by Esa Koivisto, Department of Physics, University of Kuopio, Finland. On the left, albedoes for a series of different red flowers. Each is given its Finnish name. On the right, albedoes for green leaves. Each is given its Finnish name. These albedoes don't vary all that much because there are relatively few mechanisms that give rise to color in plants. These figures were plotted from data available at `http://www.it.lut.fi/research/color/lutcs_database.html`.

Color and Specular Reflection Generally, metal surfaces have a specular component that is wavelength dependent—a shiny copper penny has a yellowish glint. Surfaces that do not conduct—*dielectric surfaces* —have a specular component that is independent of wavelength (e.g., the specularities on a shiny plastic object are the color of the light). Section 6.4.3 describes how these properties can be used to find specularities, and to find image regions corresponding to metal or plastic objects.

6.2 HUMAN COLOR PERCEPTION

To be able to describe colors, we need to know how people respond to them. Human perception of color is a complex function of context; illumination, memory, object identity, and emotion can all play a part. The simplest question is to understand which spectral radiances produce the same response from people under simple viewing conditions (Section 6.2.1). This yields a simple, linear theory of color matching that is accurate and extremely useful for describing colors. We sketch the mechanisms underlying the transduction of color in Section 6.2.2.

6.2.1 Color Matching

The simplest case of color perception is obtained when only two colors are in view on a black background. In a typical experiment, a subject sees a colored light—the *test light*—in one half of a split field (Figure 6.5). The subject can then adjust a mixture of lights in the other half to get it to match. The adjustments involve changing the intensity of some fixed number of *primaries* in the mixture. In this form, a large number of lights may be required to obtain a match, but many different adjustments may yield a match.

Write T for the test light, an equals sign for a match, the weights—which are non-negative—as w_i, and the primaries P_i. A match can then written in an algebraic form as

$$T = w_1 P_1 + w_2 P_2 + \cdots,$$

meaning that test light T matches the particular mixture of primaries given by $(w_1, w_2, \ldots)$. The situation is simplified if *subtractive matching* is allowed: In subtractive matching, the viewer can add some amount of some primaries to the *test light* instead of to the match. This can be written in algebraic form by allowing the weights in the expression above to be negative.

Trichromacy It is a matter of experimental fact that for most observers only three primaries are required to match a test light. There are some caveats. First, subtractive matching

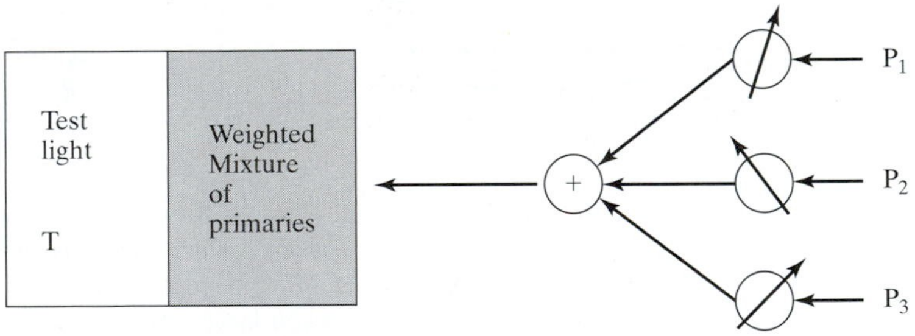

Figure 6.5 Human perception of color can be studied by asking observers to mix colored lights to match a test light shown in a split field. The drawing shows the outline of such an experiment. The observer sees a test light T and can adjust the amount of each of three primaries in a mixture displayed next to the test light. The observer is asked to adjust the amounts so that the mixture looks the same as the test light. The mixture of primaries can be written as $w_1 P_1 + w_2 P_2 + w_3 P_3$; if the mixture matches the test light, then we write $T = w_1 P_1 + w_2 P_2 + w_3 P_3$. It is a remarkable fact that for most people three primaries are sufficient to achieve a match for many colors and for all colors if we allow subtractive matching (i.e., some amount of some of the primaries is mixed with the test light to achieve a match). Some people require fewer primaries. Furthermore, most people choose the same mixture weights to match a given test light.

must be allowed; second, the primaries must be independent, meaning that no mixture of two of the primaries may match a third. This phenomenon is known as the principle of *trichromacy*. It is often explained by assuming that there are three distinct types of color transducer in the eye. Recently, evidence has emerged from genetic studies to support this view (Nathans, Piantanida, Eddy, Shows and Hogness, 1986*a*, Nathans, Thomas and Hogness, 1986*b*). Given the same primaries and test light, most observers select the *same* mixture of primaries to match that test light. This phenomenon is usually explained by assuming that the three distinct types of color transducers are common to most people. The direct evidence from genetic studies seems to support this view, too.

Grassman's Laws Under the circumstances described, matching is (to an accurate approximation) linear. This yields *Grassman's laws*.

First, if we mix two test lights, then mixing the matches will match the result—that is, if

$$T_a = w_{a1}P_1 + w_{a2}P_2 + w_{a3}P_3$$

and

$$T_b = w_{b1}P_1 + w_{b2}P_2 + w_{b3}P_3,$$

then

$$T_a + T_b = (w_{a1} + w_{b1})P_1 + (w_{a2} + w_{b2})P_2 + (w_{a3} + w_{b3})P_3.$$

Second, if two test lights can be matched with the same set of weights, then they will match each other—that is, if

$$T_a = w_1P_1 + w_2P_2 + w_3P_3$$

and

$$T_b = w_1P_1 + w_2P_2 + w_3P_3,$$

then

$$T_a = T_b.$$

Finally, matching is linear: if

$$T_a = w_1P_1 + w_2P_2 + w_3P_3,$$

then

$$kT_a = (kw_1)P_1 + (kw_2)P_2 + (kw_3)P_3$$

for non-negative k.

Exceptions Given the same test light and set of primaries, most people use the same set of weights to match the test light. This, trichromacy and Grassman's laws are about as true as any law covering biological systems can be. The exceptions include the following:

- people with aberrant color systems as a result of genetic ill fortune (who may be able to match everything with fewer primaries);

- people with aberrant color systems as a result of neural ill-fortune (who may display all sorts of effects, including a complete absence of the sensation of color);
- some elderly people (whose choice of weights differ from the norm because of the development of macular pigment in the eye);
- very bright lights (whose hue and saturation look different from less bright versions of the same light);
- and very dark conditions (where the mechanism of color transduction is somewhat different than in brighter conditions).

6.2.2 Color Receptors

Trichromacy suggests that there are profound constraints on the way color is transduced in the eye. One hypothesis that satisfactorily explains this phenomenon is to assume that there are three distinct types of receptor in the eye that mediate color perception. Each of these receptors turns incident light into neural signals. It is possible to reason about the sensitivity of these receptors from color matching experiments. If two test lights that have different spectra look the same, they must have the same effect on these receptors.

The Principle of Univariance The *principle of univariance* states that the activity of these receptors is of one kind (i.e., they respond strongly or weakly, but do not signal the wavelength of the light falling on them). Experimental evidence can be obtained by carefully dissecting light-sensitive cells and measuring their responses to light at different wavelengths or by reasoning backward from color matches. Univariance is a powerful idea because it gives us a good and simple model of human reaction to colored light: Two lights will match if they produce the same receptor responses, *whatever their spectral radiances*.

Because the system of matching is linear, the receptors must be linear. Let us write p_k for the response of the kth type of receptor, $\sigma_k(\lambda)$ for its sensitivity, $E(\lambda)$ for the light arriving at the receptor, and Λ for the range of visible wavelengths. We can obtain the overall response of a receptor by adding up the response to each separate wavelength in the incoming spectrum so that

$$p_k = \int_\Lambda \sigma_k(\lambda) E(\lambda) d\lambda.$$

Rods and Cones Anatomical investigation of the retina shows two types of cell that are sensitive to light differentiated by their shape. The light-sensitive region of a *cone* has a roughly conical shape, whereas that in a *rod* is roughly cylindrical. Cones largely dominate color vision and completely dominate the fovea. Cones are somewhat less sensitive to light than rods are, meaning that in low light, color vision is poor and it is impossible to read (one doesn't have sufficient spatial precision because the fovea isn't working).

Studies of the genetics of color vision support the idea that there are three types of cone differentiated by their sensitivity (in the large; there is some evidence that there are slight differences from person to person within each type). The sensitivities of the three different kinds of receptor to different wavelengths can be obtained by comparing color matching data for normal observers with color matching data for observers lacking one type of cone. Sensitivities obtained in this fashion are shown in Figure 6.6. The three types of cone are properly called *S cones*, *M cones* and *L cones* (for their peak sensitivity being to short, medium, and long wavelength light, respectively). They are occasionally called blue, green and red cones; however, this is bad practice, because the sensation of red is definitely not caused by the stimulation of red cones and so on.

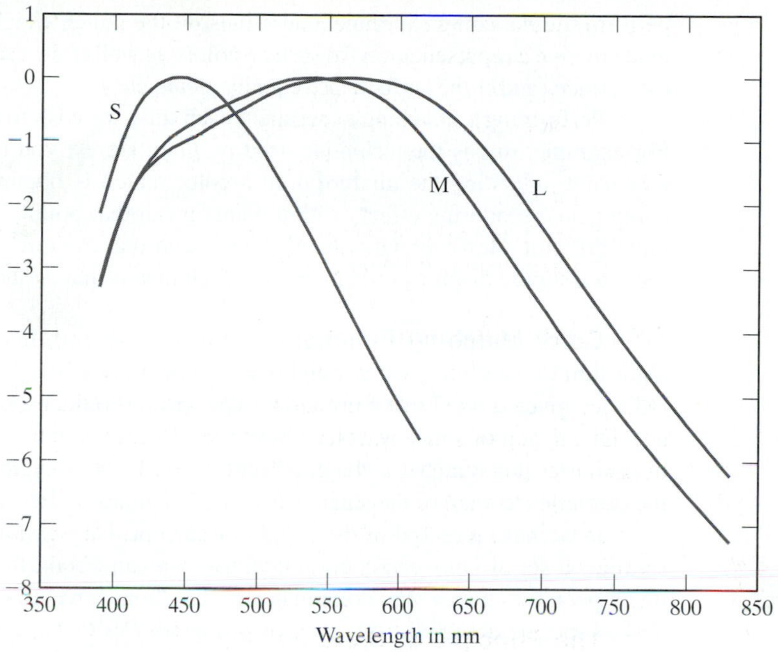

Figure 6.6 There are three types of color receptor in the human eye, usually called *cones*. These receptors respond to all photons in the same way, but in different amounts. The figure shows the log of the relative spectral sensitivities of the three kinds of color receptor in the human eye. The first two receptors—sometimes called the *red* and *green* cones respectively, but more properly named the *long-* and *medium-wavelength* receptors—have peak sensitivities at quite similar wavelengths. The third receptor (*blue* cone or, more properly, *short-wavelength* receptor) has a different peak sensitivity. The response of a receptor to incoming light can be obtained by summing the product of the sensitivity and the spectral radiance of the light over all wavelengths. *Figures plotted from data disseminated by the Color and Vision Research Laboratories database, compiled by Andrew Stockman and Lindsey Sharpe, and available at* `http://www-cvrl.ucsd.edu/index.htm`.

6.3 REPRESENTING COLOR

Describing colors accurately is a matter of great commercial importance. Many products are closely associated with specific colors—for example, the golden arches, the color of various popular computers and the color of photographic film boxes—and manufacturers are willing to go to a great deal of trouble to ensure that different batches have the same color. This requires a standard system for talking about color. Simple names are insufficient because relatively few people know many color names, and most people are willing to associate a large variety of colors with a given name.

6.3.1 Linear Color Spaces

There is a natural mechanism for representing color: Agree on a standard set of primaries and then describe any colored light by the three values of weights that people would use to match the light using those primaries. In principle, this is easy to use. To describe a color, we set up and perform the matching experiment and transmit the match weights. Of course, this approach extends to give a representation for surface colors as well if we use a standard light for illuminating the surface (and if the surfaces are equally clean, etc.).

Performing a matching experiment each time we wish to describe a color can be practical. For example, this is the technique used by paint stores; you take in a flake of paint, and they mix paint, adjusting the mixture until a color match is obtained. Paint stores do this because complicated scattering effects within paints mean that predicting the color of a mixture can be quite difficult. However, Grassman's laws mean that mixtures of colored lights—at least those seen in a simple display—mix *linearly*, which means that a much simpler procedure is available.

Color Matching Functions When colors mix linearly, we can construct a simple algorithm to determine which weights would be used to match a source of some known spectral radiance given a fixed set of primaries. The spectral radiance of the source can be thought of as a weighted sum of single wavelength sources. Because color matching is linear, the combination of primaries that matches a weighted sum of single wavelength sources is obtained by matching the primaries to each of the single wavelength sources and then adding up these match weights.

If we have a record of the weight of each primary required to match a single wavelength source—a set of *color matching functions*—we can obtain the weights used to match an arbitrary spectral radiance. The color matching functions—which we shall write as $f_1(\lambda)$, $f_2(\lambda)$, and $f_3(\lambda)$—can be obtained from a set of primaries P_1, P_2 and P_3 by experiment. Essentially, we tune the weight of each primary to match a unit radiance source at every wavelength. We then obtain a set of weights, one for each wavelength, for matching a unit radiance source $U(\lambda)$. We can write this process as

$$U(\lambda) = f_1(\lambda)P_1 + f_2(\lambda)P_2 + f_3(\lambda)P_3$$

(i.e., at each wavelength λ, $f_1(\lambda)$, $f_2(\lambda)$, and $f_3(\lambda)$ give the weights required to match a unit radiance source at that wavelength).

The source—which we write $S(\lambda)$—is a sum of a vast number of single wavelength sources, each with a different intensity. We now match the primaries to each of the single wavelength sources and then add up these match weights, obtaining

$$\begin{aligned} S(\lambda) &= w_1 P_1 + w_2 P_2 + w_3 P_3 \\ &= \left\{\int_\Lambda f_1(\lambda)S(\lambda)d\lambda\right\} P_1 + \left\{\int_\Lambda f_2(\lambda)S(\lambda)d\lambda\right\} P_2 + \left\{\int_\Lambda f_3(\lambda)S(\lambda)d\lambda\right\} P_3. \end{aligned}$$

General Issues for Linear Color Spaces Linear color naming systems can be obtained by specifying primaries—which imply color matching functions—or by specifying color matching functions—which imply primaries. It is an inconvenient fact of life that, if the primaries are real lights, at least one of the color matching functions is negative for some wavelengths. This is not a violation of natural law; it just implies that subtractive matching is required to match some lights, whatever set of primaries is used. It is a nuisance, though.

One way to avoid this problem is to specify color matching functions that are everywhere positive (which guarantees that the primaries are imaginary because for some wavelengths their spectral radiance is negative).

Although this looks like a problem—how would one create a real color with imaginary primaries?—it isn't, because color naming systems are hardly ever used that way. Usually, we would simply compare weights to tell whether colors are similar, and for that purpose it is enough to know the color matching functions. A variety of different systems have been standardised by the CIE (the *commission international d'éclairage*, which exists to make standards on such things).

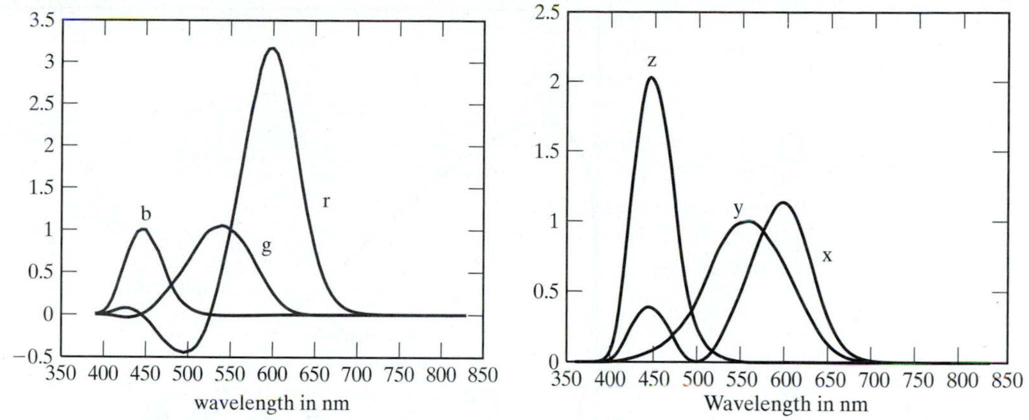

Figure 6.7 On the left, color matching functions for the primaries for the RGB system. The negative values mean that subtractive matching is required to match lights at that wavelength with the RGB primaries. On the right, color matching functions for the CIE X, Y, and Z primaries; the color matching functions are everywhere positive, but the primaries are not real. *Figures plotted from data disseminated by the Color and Vision Research Laboratories database, compiled by Andrew Stockman and Lindsey Sharpe, and available at* `http://www-cvrl.ucsd.edu/index.htm`.

The CIE XYZ Color Space The *CIE XYZ color space* is one quite popular standard. The color matching functions were chosen to be everywhere positive, so that the coordinates of any real light are always positive. It is not possible to obtain CIE X, Y, or Z primaries because for some wavelengths the value of their spectral radiance is negative. However, given color matching functions alone, one can specify the XYZ coordinates of a color and hence describe it.

Linear color spaces allow a number of useful graphical constructions that are more difficult to draw in three dimensions than in two, so it is common to intersect the XYZ space with the

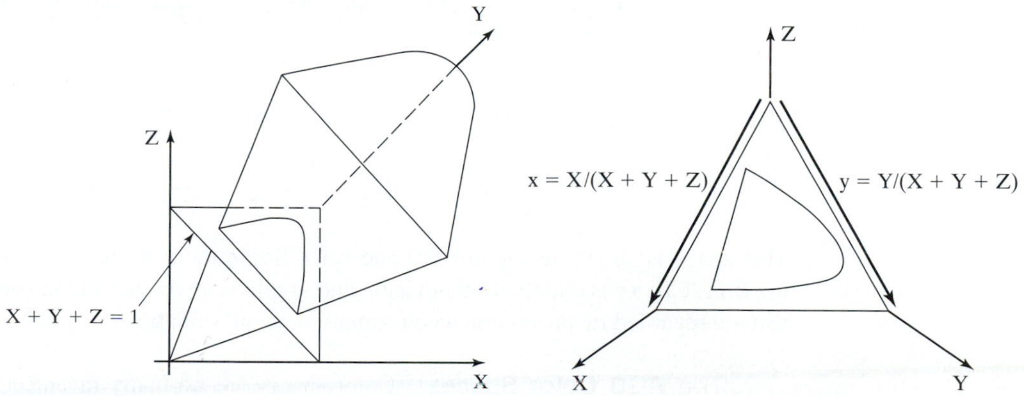

Figure 6.8 The volume of all visible colors in CIE XYZ coordinate space is a cone whose vertex is at the origin. Usually it is easier to suppress the brightness of a color, which we can do because to a good approximation perception of color is linear, and we do this by intersecting the cone with the plane $X + Y + Z = 1$ to get the CIE xy space shown in Figures 6.9 and 6.10

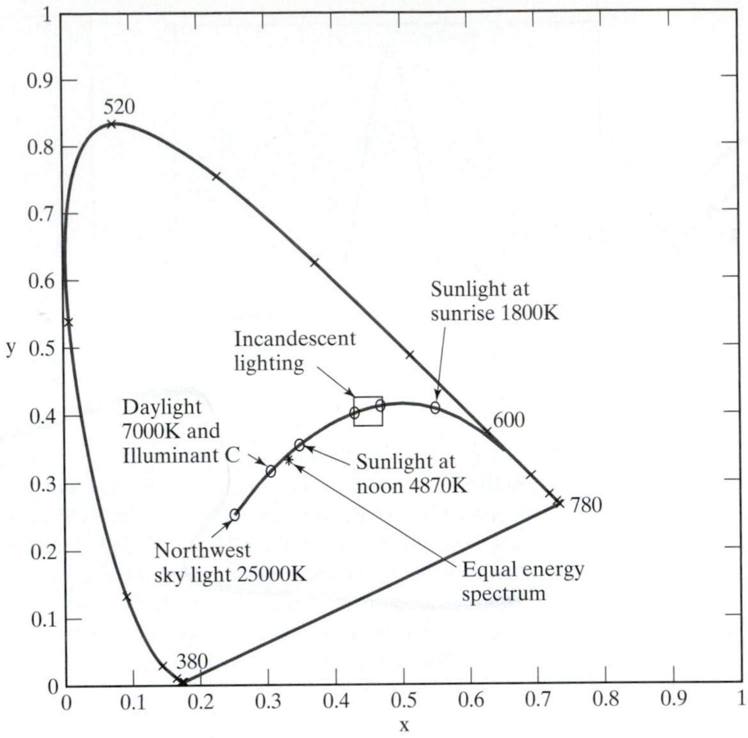

Figure 6.9 The figure shows a constant brightness section of the standard 1931 standard CIE xy color space. This space has two coordinate axes. The curved boundary of the figure is often known as the *spectral locus*; it represents the colors experienced when lights of a single wavelength are viewed. The figure shows a locus of colors due to black-body radiators at different temperatures and a locus of different sky colors. Near the center of the diagram is the neutral point, the color whose weights are equal for all three primaries. CIE selected the primaries so that this light appears achromatic. Generally, colors that lie farther away from the neutral point are more saturated—the difference between deep red and pale pink—and hue—the difference between green and red—as one moves around the neutral point.

plane $X + Y + Z = 1$ (as shown in Figure 6.8) and draw the resulting figure using coordinates

$$(x, y) = \left(\frac{X}{X + Y + Z}, \frac{Y}{X + Y + Z} \right).$$

This space is shown in Figures 6.9 and 6.10. Some more useful constructions appear in Figure 6.11. CIE xy is widely used in vision and graphics textbooks and in some applications, but is usually regarded by professional colorimetrists as out of date.

The RGB Color Spaces Color spaces are normally invented for practical reasons, and so a wide variety exist. The *RGB color space* is a linear color space that formally uses single wavelength primaries (645.16 nm for R, 526.32nm for G, and 444.44nm for B; see Figure 6.7). Informally, RGB uses whatever phosphors a monitor has as primaries. Available colors are usually represented as a unit cube—usually called the *RGB cube*—whose edges represent the R, G, and B weights. The cube is drawn in Figure 6.12.

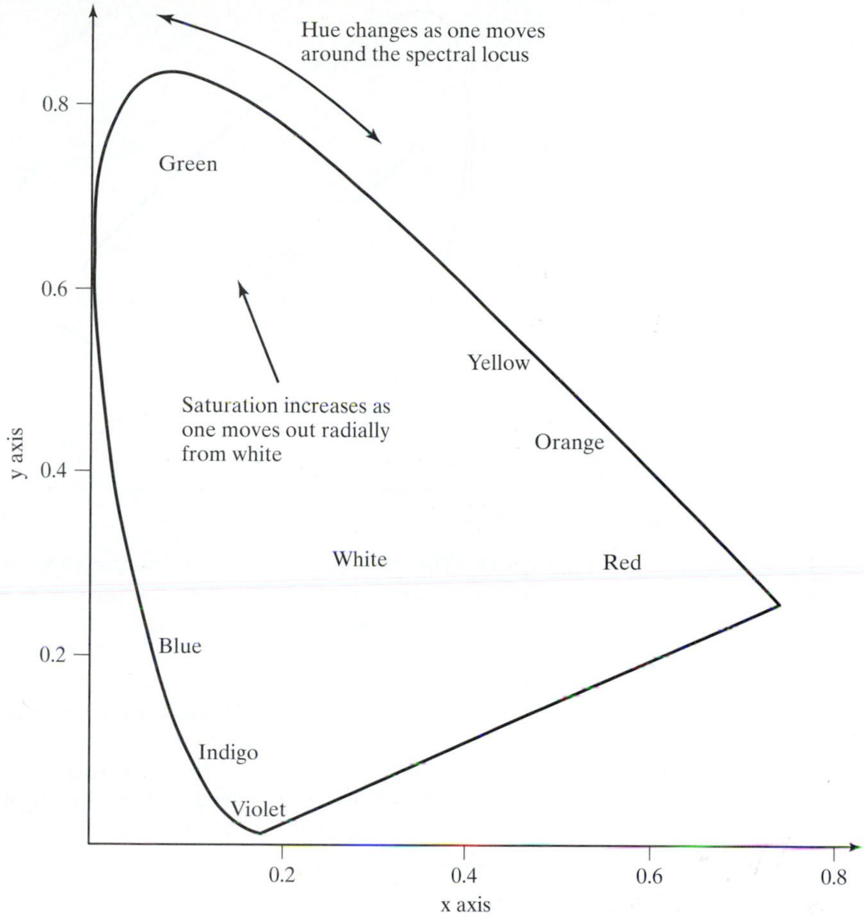

Figure 6.10 The figure shows a constant brightness section of the standard 1931 standard CIE xy color space, with color names marked on the diagram. Generally, colors that lie farther away from the neutral point are more saturated—the difference between deep red and pale pink—and hue—the difference between green and red—as one moves around the neutral point.

CMY and Black Intuition from one's finger-painting days suggests that the primary colors should be red, yellow and blue, and that red and green mix to make yellow. The reason this intuition doesn't apply to monitors is that it is about pigments—which mix subtractively—rather than about lights. Pigments remove color from incident light, which is reflected from paper. Thus, red ink is really a dye that absorbs green and blue light—incident red light passes through this dye and is reflected from the paper.

Color spaces for this kind of subtractive matching can be quite complicated. In the simplest case, mixing is linear (or reasonably close to linear) and the *CMY space* applies. In this space, there are three primaries: *cyan* (a blue-green color); *magenta* (a purplish color), and *yellow*. These primaries should be thought of as subtracting a light primary from white light; cyan is $W - R$ (white − red); magenta is $W - G$ (white − green), and yellow is $W - B$ (white − blue). Now the appearance of mixtures may be evaluated by reference to the RGB color space. For example, cyan and magenta mixed give

$$(W - R) + (W - G) = R + G + B - R - G = B,$$

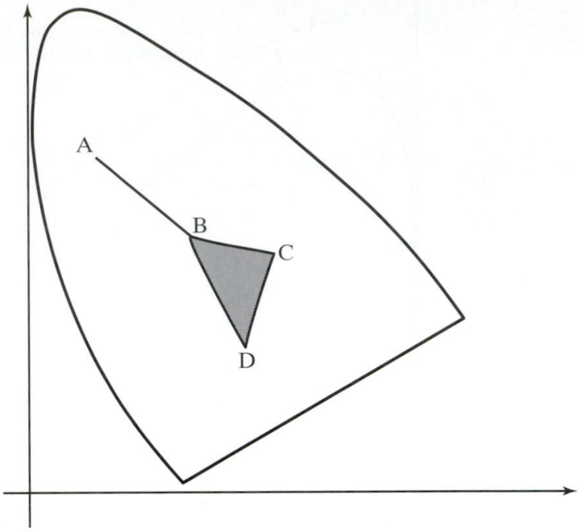

Figure 6.11 The linear model of the color system allows a variety of useful constructions. If we have two lights whose CIE coordinates are ***A*** and ***B***, all the colors that can be obtained from non-negative mixtures of these lights are represented by the line segment joining ***A*** and ***B***. In turn, given ***B***, ***C***, and ***D***, the colors that can by obtained by mixing them lie in the triangle formed by the three points. This is important in the design of monitors—each monitor has only three phosphors, and the more saturated the color of each phosphor, the bigger the set of colors that can be displayed. This also explains why the same colors can look quite different on different monitors. The curvature of the spectral locus gives the reason that no set of three real primaries can display all colors without subtractive matching.

that is, blue. Notice that $W + W = W$ because we assume that ink cannot cause paper to reflect more light than it does when uninked. Practical printing devices use at least four inks (cyan, magenta, yellow, and black) because mixing color inks leads to a poor black, it is difficult to ensure good enough registration between the three color inks to avoid colored haloes around text, and color inks tend to be more expensive than black inks. Getting really good results from a color printing process is still difficult: Different inks have significantly different spectral properties, different papers have different spectral properties too, and inks can mix nonlinearly.

6.3.2 Non-linear Color Spaces

The coordinates of a color in a linear space may not necessarily encode properties that are common in language or are important in applications. Useful color terms include: *hue*—the property of a color that varies in passing from red to green; *saturation*—the property of a color that varies in passing from red to pink; and *brightness* (sometimes called *lightness* or *value*)—the property that varies in passing from black to white. For example, if we are interested in checking whether a color lies in a particular range of reds, we might wish to encode the hue of the color directly.

Another difficulty with linear color spaces is that the individual coordinates do not capture human intuitions about the topology of colors; it is a common intuition that hues form a circle, in the sense that hue changes from red through orange to yellow and then green and from there to cyan, blue, purple, and then red again. Another way to think of this is to think of local hue relations: Red is next to purple and orange; orange is next to red and yellow; yellow is next

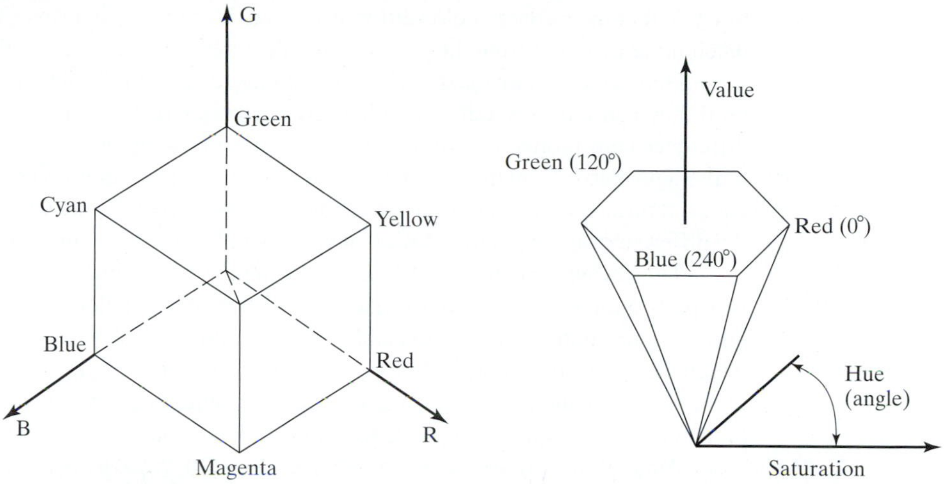

Figure 6.12 On the left, we see the RGB cube; this is the space of all colors that can be obtained by combining three primaries (R, G, and B—usually defined by the color response of a monitor) with weights between zero and one. It is common to view this cube along its neutral axis—the axis from the origin to the point (1, 1, 1)—to see a hexagon, shown in the middle. This hexagon codes hue (the property that changes as a color is changed from green to red) as an angle, which is intuitively satisfying. On the right, we see a cone obtained from this cross-section, where the distance along a generator of the cone gives the value (or brightness) of the color, angle around the cone gives the hue, and distance out gives the saturation of the color.

to orange and green; green is next to yellow and cyan; cyan is next to green and blue; blue is next to cyan and purple; and purple is next to blue and red. Each of these local relations works, and globally they can be modeled by laying hues out in a circle. This means that no individual coordinate of a linear color space can model hue because that coordinate has a maximum value that is far away from the minimum value.

Hue, Saturation, and Value A standard method for dealing with this problem is to construct a color space that reflects these relations by applying a nonlinear transformation to the RGB space. There are many such spaces. One, called *HSV space* (for hue, saturation, and value), is obtained by looking down the center axis of the RGB cube. Because RGB is a linear space, brightness—called *value* in HSV—varies with scale out from the origin. We can flatten the RGB cube to get a 2D space of constant value and for neatness deform it to be a hexagon. This gets the structure shown in Figure 6.12, where hue is given by an angle that changes as one goes round the neutral point and saturation changes as one moves away from the neutral point.

There are a variety of other possible changes of coordinate from between linear color spaces, or from linear to nonlinear color spaces (the recent book of Fairchild, 1998 is a good reference). There is no obvious advantage to using one set of coordinates over another (particularly if the difference between coordinate systems is just a one–one transformation) unless one is concerned with coding, bit rates, and the like, or with perceptual uniformity.

Uniform Color Spaces Usually one cannot reproduce colors exactly. This means it is important to know whether a color difference would be noticeable to a human viewer; it is generally useful to compare the significance of small color differences. It is usually dangerous to try and compare large color differences; consider trying to answer the question, "Is the blue patch more different from the yellow patch than the red patch is from the green patch?".

One can determine *just noticeable differences* by modifying a color shown to an observer until they can only just tell it has changed in a comparison with the original color. When these differences are plotted on a color space, they form the boundary of a region of colors that are indistinguishable from the original colors. Usually ellipses are fitted to the just noticeable differences. It turns out that in CIE xy space these ellipses depend quite strongly on where in the space the difference occurs, as the Macadam ellipses in Figure 6.13 illustrate.

This means that the size of a difference in (x, y) coordinates, given by $((\Delta x)^2 + (\Delta y)^2)^{(1/2)}$, is a poor indicator of the significance of a difference in color (if it was a good indicator, the ellipses representing indistinguishable colors would be circles). A *uniform color space* is one in which the distance in coordinate space is a fair guide to the significance of the difference between two colors—in such a space, if the distance in coordinate space were below some threshold, a human observer would not be able to tell the colors apart.

A more uniform space can be obtained from CIE XYZ by using a projective transformation to skew the ellipses; this yields the *CIE u'v' space*, illustrated in Figure 6.14. The coordinates are:

$$(u', v') = \left(\frac{4X}{X + 15Y + 3Z}, \frac{9Y}{X + 15Y + 3Z} \right).$$

Generally, the distance between coordinates in u', v' space is a fair indicator of the significance of the difference between two colors. Of course, this omits differences in brightness.

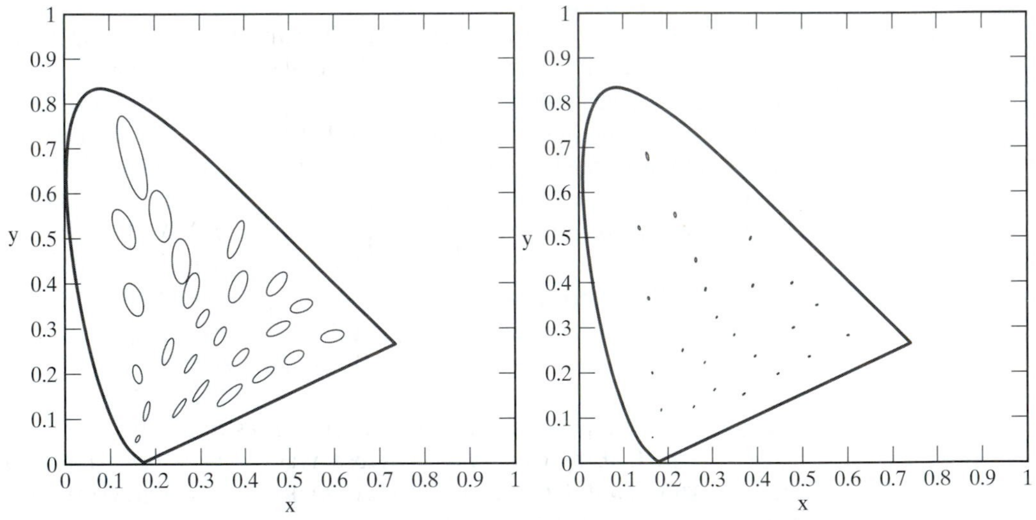

Figure 6.13 This figure shows variations in color matches on a CIE x, y space. At the center of the ellipse is the color of a test light; the size of the ellipse represents the scatter of lights that the human observers tested would match to the test color; the boundary shows where the just noticeable difference is. The ellipses in the figure on the **left** have been magnified 10x for clarity; on the **right** they are plotted to scale. The ellipses are known as MacAdam ellipses after their inventor. Notice that the ellipses at the top are larger than those at the bottom of the figure, and that they rotate as they move up. This means that the magnitude of the difference in x, y coordinates is a poor guide to the difference in color. Ellipses are plotted using data from MacAdam (1942).

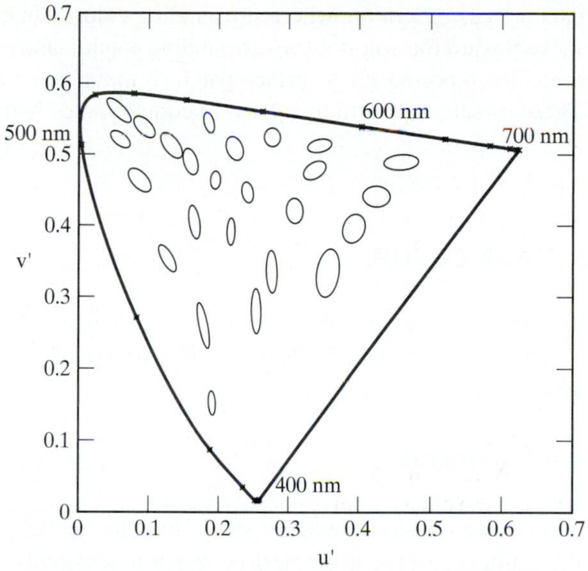

Figure 6.14 This figure shows the CIE 1976 u', v' space, which is obtained by a projective transformation of CIE x, y space. The intention is to make the MacAdam ellipses uniformly circles—this would yield a uniform color space. A variety of nonlinear transforms can be used to make the space more uniform (see Fairchild (1998) for details)

CIE LAB is now almost universally the most popular uniform color space. Coordinates of a color in LAB are obtained as a nonlinear mapping of the XYZ coordinates:

$$L^* = 116\left(\frac{Y}{Y_n}\right)^{\frac{1}{3}} - 16$$

$$a^* = 500\left[\left(\frac{X}{X_n}\right)^{\frac{1}{3}} - \left(\frac{Y}{Y_n}\right)^{\frac{1}{3}}\right]$$

$$b^* = 200\left[\left(\frac{Y}{Y_n}\right)^{\frac{1}{3}} - \left(\frac{Z}{Z_n}\right)^{\frac{1}{3}}\right]$$

Here X_n, Y_n, and Z_n are the X, Y, and Z coordinates of a reference white patch. The reason to care about the LAB space is that it is substantially uniform. In some problems, it is important to understand how different two colors will look *to a human observer*, and differences in LAB coordinates give a good guide.

6.3.3 Spatial and Temporal Effects

Predicting the appearance of complex displays of color (i.e., a stimulus that is more interesting than a pair of lights) is difficult. If the visual system has been exposed to a particular illuminant for some time, this causes the color system to adapt—a process known as *chromatic adaptation*. Adaptation causes the color diagram to skew, in the sense that two observers, adapted to different illuminants, can report that spectral radiosities with quite different chromaticities have the same color. Adaptation can be caused by surface patches in view. Other mechanisms that are significant are *assimilation*—where surrounding colors cause the color reported for a surface patch to move toward the color of the surrounding patch—and *contrast*—where surrounding colors cause the color reported for a surface patch to move away from the color of the surrounding patch. These effects appear to be related to coding issues within the optic nerve and to color constancy (Section 6.5).

6.4 A MODEL FOR IMAGE COLOR

To interpret the color values reported by a camera, we need some understanding of what cameras do and of what physical effects we wish to model. Our model supports several quite simple and powerful inference algorithms.

6.4.1 Cameras

Most color cameras contain a single imaging device. At each sensory element, there is one of three filters to give it the desired spectral sensitivity function (roughly, red, green, and blue). These filters are arranged in a mosaic; a variety of different patterns is used. The output of the CCD is then processed to reconstruct full red, green, and blue images. Of course, some signal information is lost in this process; what is lost depends on the details of the camera and of the mosaic. Typically, these losses do not affect the perceived quality of the image to a viewer, but they can significantly affect various spatial computations—for example, the ability of an edge detector to localize an edge (because the spatial resolution in intensity may not be what it seems).

It is desirable to have the system of CCD element and filter have a spectral sensitivity that is within a linear transform of the CIE XYZ color matching functions. This property would mean that the camera would match colors in the same way that people did. The requirement seems to be quite hard to meet in practice because experimental evidence suggests that most cameras don't really have this desirable property.

CCDs are intrinsically linear devices. However, most users are used to film, which tends to compress the incoming dynamic range (brightness differences at the top end of the range are reduced, as are those at the bottom end of the range). The output of a linear device tends to look too harsh (the darks are too dark and the lights are too light), so that manufacturers apply various forms of compression to the output. The most common is called *gamma correction*. This is a form of compression originally intended to account for nonlinearities within monitors. Typically, the intensity of a monitor goes as V_{in}^{γ}, where V_{in} is the input voltage at the electron gun ($\gamma = 2.2$ for CRT monitors). In most computer display devices, the voltage supplied to the electron gun is a linear function of the value in the framebuffer. This means that, if we desire an intensity I, the value to use in the framebuffer is proportional to $I^{1/\gamma}$. Typically, cameras are gamma corrected, meaning that one can take the value reported by a camera and put it directly in the framebuffer to get the right intensity. This means that the output of the camera is *not* a linear function of the input.

There was a time when CCD cameras came with a separate box of control electronics, which had a switch that turned off the nonlinearities; those happy days are now past. Typically, the input–output relationship of a camera needs to be calibrated because there are a variety of possible nonlinearities (e.g., Barnard and Funt, 2002, Holst, 1998, or Vora, Farell, Tietz and Brainard, 1997). Extremists occasionally tinker with the camera electronics, a solution not for the faint of heart. In what follows, we assume that the input–output relationship of the camera is known and has been accounted for in a way that makes the camera seem linear.

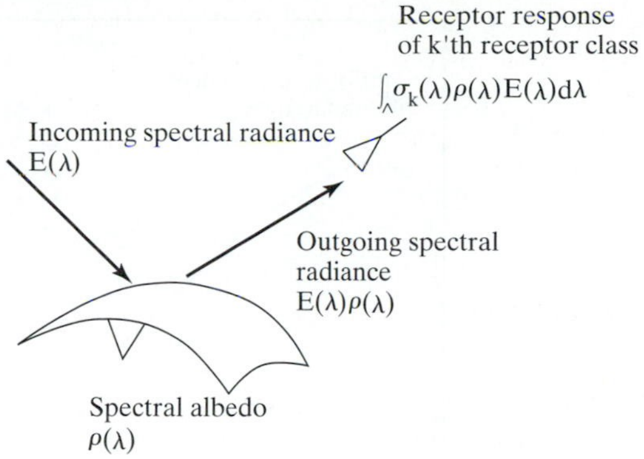

Figure 6.15 If a patch of perfectly diffuse surface with diffuse spectral reflectance $\rho(\lambda)$ is illuminated by a light whose spectrum is $E(\lambda)$, the spectrum of the reflected light is $\rho(\lambda)E(\lambda)$ (multiplied by some constant to do with surface orientation, which we have already decided to ignore). Thus, if a linear photoreceptor of the kth type sees this surface patch, its response is $p_k = \int_\Lambda \sigma_k(\lambda)\rho(\lambda)E(\lambda)d\lambda$, where Λ is the range of all relevant wavelengths and $\sigma_k(\lambda)$ is the sensitivity of the kth photoreceptor.

6.4.2 A Model for Image Color

The color of light arriving at a camera is determined by two factors: first, the spectral reflectance of the surface that the light is leaving; and second, the spectral radiance of the light falling on that surface. If a patch of perfectly diffuse surface with diffuse spectral reflectance $\rho(\lambda)$ is illuminated by a light whose spectrum is $E(\lambda)$, the spectrum of the reflected light is $\rho(\lambda)E(\lambda)$ (multiplied by some constant to do with surface orientation, which we have already decided to ignore). Thus, if a linear photoreceptor of the kth type sees this surface patch, its response is:

$$p_k = \int_\Lambda \sigma_k(\lambda)\rho(\lambda)E(\lambda)d\lambda,$$

where Λ is the range of all relevant wavelengths and $\sigma_k(\lambda)$ is the sensitivity of the kth photoreceptor.

The color of the light falling on surfaces can vary widely—from blue fluorescent light indoors, to warm orange tungsten lights, to orange or even red light at sunset—so that the color of the light arriving at the camera can be quite a poor representation of the color of the surfaces being viewed (Figures 6.16, 6.17, 6.18 and 6.19).

By suppressing details in the physical models of chapter 5, we can model the value at a camera pixel as

$$\boldsymbol{C}(\boldsymbol{x}) = g_d(\boldsymbol{x})\boldsymbol{d}(\boldsymbol{x}) + g_s(\boldsymbol{x})\boldsymbol{s}(\boldsymbol{x}) + \boldsymbol{i}(\boldsymbol{x}).$$

In this model,

- $\boldsymbol{d}(\boldsymbol{x})$ is the *image* color of an equivalent *flat* frontal surface viewed under the same light;
- $g_d(\boldsymbol{x})$ is a term that varies over space and accounts for the change in brightness due to the orientation of the surface;
- $\boldsymbol{s}(\boldsymbol{x})$ is the *image* color of the specular reflection from an equivalent *flat* frontal surface;
- $g_s(\boldsymbol{x})$ is a term that varies over space and accounts for the change in the amount of energy specularly reflected;

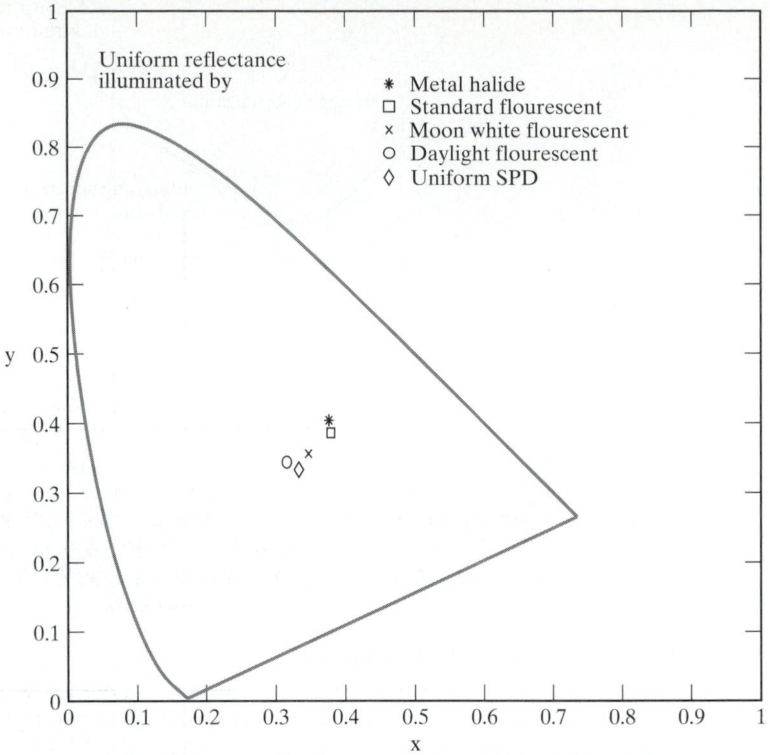

Figure 6.16 Light sources can have quite widely varying colors. This figure shows the color of the four light sources of Figure 6.2, compared with the color of a uniform spectral power distribution, plotted in CIE x, y coordinates.

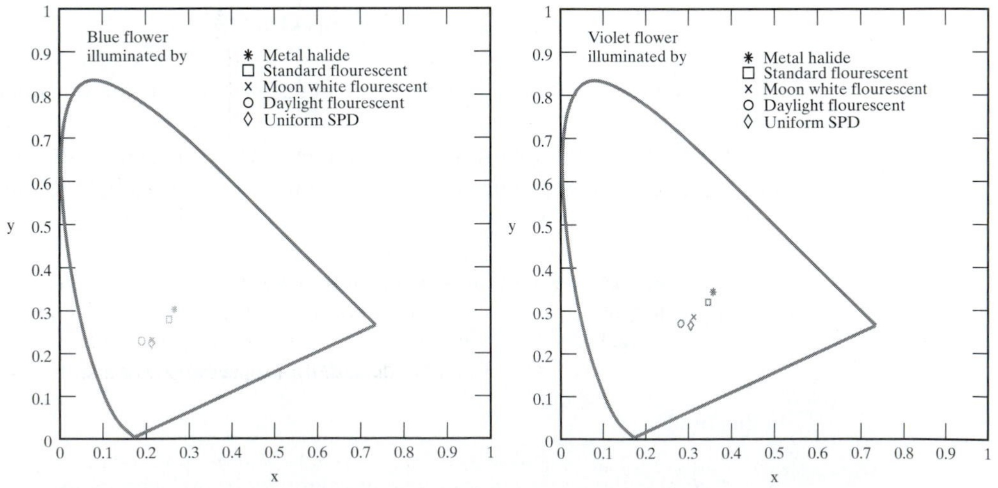

Figure 6.17 Surfaces have significantly different colors when viewed under different lights. These figures show the colors taken on by the blue and violet flowers of Figure 6.3 when viewed under the four different sources of Figure 6.2 and under a uniform spectral power distribution.

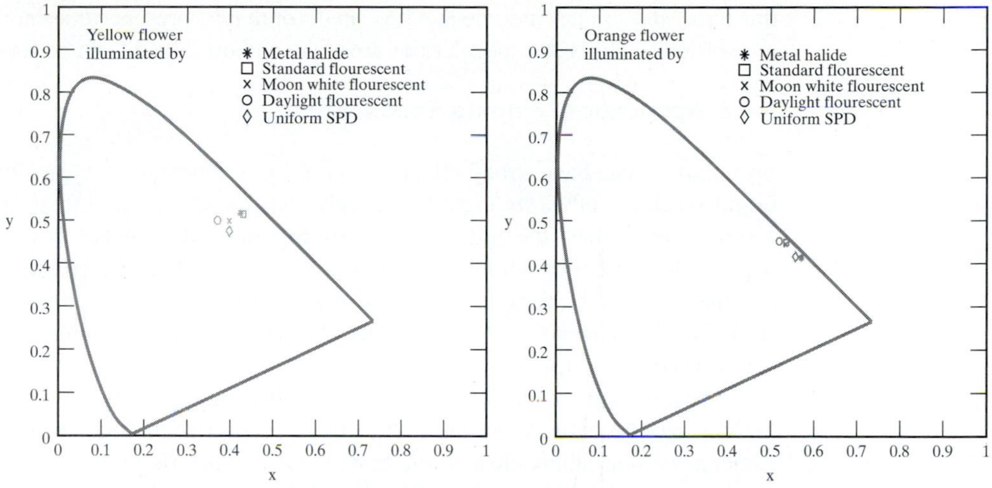

Figure 6.18 Surfaces have significantly different colors when viewed under different lights. These figures show the colors taken on by the yellow and orange flowers of Figure 6.3 when viewed under the four different sources of Figure 6.2 and under a uniform spectral power distribution.

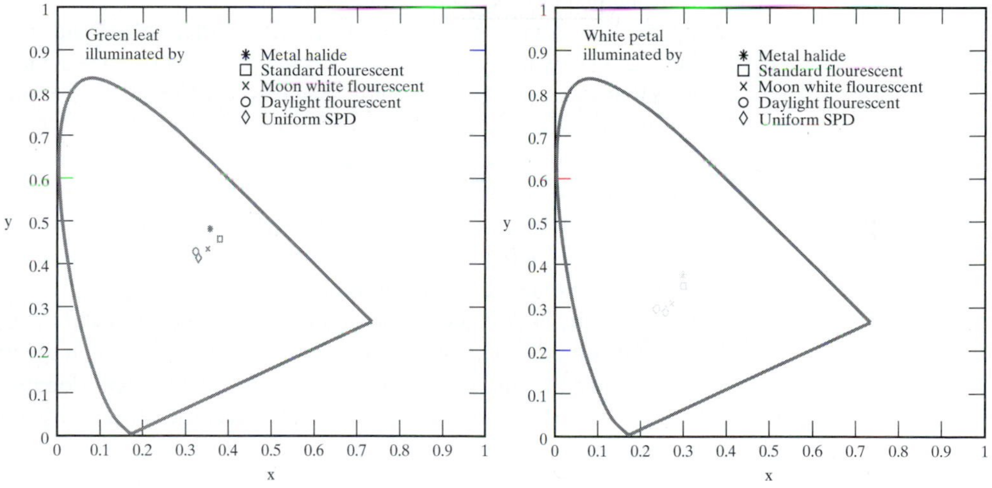

Figure 6.19 Surfaces have significantly different colors when viewed under different lights. These figures show the colors taken on by the white petal of Figure 6.3 and one of the leaves of Figure 6.4 when viewed under the four different sources of Figure 6.2 and under a uniform spectral power distribution.

- and $\boldsymbol{i}(\boldsymbol{x})$ is a term that accounts for colored interreflections, spatial changes in illumination, and the like.

We are primarily interested in information that can be extracted from color at a local level, and so we are ignoring the detailed structure of the terms $g_d(\boldsymbol{x})$ and $\boldsymbol{i}(\boldsymbol{x})$. Nothing is known about how to extract information from $\boldsymbol{i}(\boldsymbol{x})$; all evidence suggests that this is difficult. The term can sometimes be quite small with respect to other terms and usually changes quite slowly over space. We ignore this term, and so must assume that it is small (or that its presence does not disrupt our algorithms too severely). However, specularities are small and bright, and can be found.

6.4.3 Application: Finding Specularities

Specularities can have strong effects on an object's appearance. Typically, they appear as small, bright patches, called *highlights*. Highlights have a substantial effect on human perception of a surface properties; the addition of small, highlight-like patches to a figure makes the object depicted look glossy or shiny. Specularities are often sufficiently bright to saturate the camera so that the color can be hard to measure. However, because the appearance of a specularity is quite strongly constrained, there are a number of effective schemes for marking them, and the results can be used as a shape cue.

The dynamic range of practically available albedoes is relatively small. Surfaces with very high or very low albedo are difficult to make. Uniform illumination is common too, and most cameras are reasonably close to linear within their operating range. This means that very bright patches cannot be due to diffuse reflection; they must be either sources (of one form or another—perhaps a stained glass window with the light behind it) or specularities. Furthermore, specularities tend to be small. Thus, looking for small, bright patches can be an effective way to find specularities (Brelstaff and Blake, 1988).

In color images, specularities on dielectric and conductive materials often look quite different. This link to conductivity occurs because electric fields cannot penetrate conductors (the electrons inside just move around to cancel the field), so that light striking a metal surface can be either absorbed or specularly reflected. Dull metal surfaces look dull because of surface roughness effects, and shiny metal surfaces have shiny patches that have a characteristic color because the conductor absorbs energy in different amounts at different wavelengths. However, light striking a dielectric surface can penetrate it.

Many dielectric surfaces can be modeled as a clear matrix with randomly embedded pigments; this is a particularly good model for plastics and some paints. In this model, there are two components of reflection that correspond to our specular and diffuse notions: *body reflection*, which comes from light penetrating the matrix, striking various pigments, and then leaving; and *surface reflection*, which comes from light specularly reflected from the surface. Assuming the pigment is randomly distributed (small, not on the surface, etc.) and the matrix is reasonable, the body reflection component behaves like a diffuse component with a spectral albedo that depends on the pigment and the surface component is independent of wavelength.

Assume we are looking at a single object dielectric object with a single color. We expect that the interreflection term can be ignored and our model of camera pixel brightnesses becomes

$$\boldsymbol{p}(\boldsymbol{x}) = g_d(\boldsymbol{x})\boldsymbol{d} + g_s(\boldsymbol{x})\boldsymbol{s},$$

where $\boldsymbol{s}$ is the color of the source and $\boldsymbol{d}$ is the color of the diffuse reflected light, $g_d(\boldsymbol{x})$ is a geometric term that depends on the orientation of the surface, and $g_s(\boldsymbol{x})$ is a term that gives the extent of the specular reflection. If the object is curved, then $g_s(\boldsymbol{x})$ is small over much of the surface and large only around specularities; $g_d(\boldsymbol{x})$ varies more slowly with the orientation of the surface. We now map the colors produced by this surface in receptor response space and look at the structures that appear there (Figure 6.20).

The term $g_d(\boldsymbol{x})\boldsymbol{d}$ produces a line that should extend to pass through the origin because it represents the same vector of receptor responses multiplied by a constant that varies over space. If there is a specularity, then we expect to see a second line due to $g_s(\boldsymbol{x})\boldsymbol{s}$. This does not, in general, pass through the origin (because of the diffuse term). This is a line, rather than a planar region, because $g_s(\boldsymbol{x})$ is large over only a small range of surface normals. We expect that, because the surface is curved, this corresponds to a small region of surface. The term $g_d(\boldsymbol{x})$

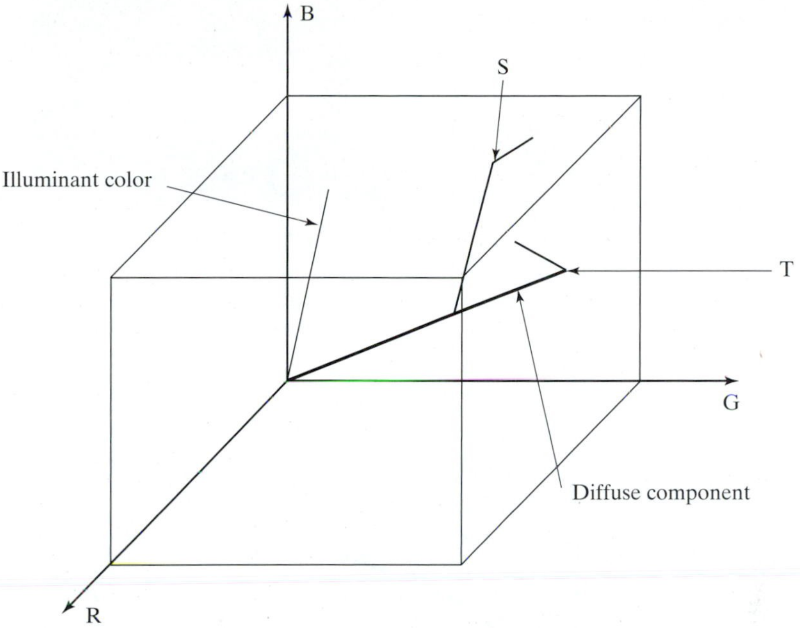

Figure 6.20 Assume we have a picture of a single uniformly colored surface. Our model of reflected light should lead to a gamut that looks like the drawing. We are assuming that reflected light consists of the diffuse term plus a specular term, and the specular term is the color of the light source. Most points on the surface do not have a significant specular term and instead are brighter or darker versions of the same diffuse surface color. At some points, the specular term is large, and this leads to a "dog-leg" in the gamut caused by adding the diffuse term to the source term. If the diffuse reflection is bright, one or another color channel might saturate (point T); similarly, if the specular reflection is bright, one or another color channel might saturate (point S).

should be approximately constant in this region. We expect a line, rather than an isolated pixel value, because we expect surfaces to have (possibly narrow) specular lobes, meaning that the specular coefficient has a range of values. This second line may collide with a face of the color cube and get clipped.

The resulting dog-leg pattern leads pretty much immediately to a specularity marking algorithm—find the pattern and then find the specular line. All the pixels on this line are specular pixels, and the specular and diffuse components can be estimated easily. For the approach to work effectively, we need to be confident that only one object is represented in the collection of pixels. This is helped by using local image windows as illustrated by Figure 6.21. The observations underlying the method hold even if the surface is not monochrome—a coffee mug with a picture on it, for example—but finding the resulting structures in the color space now becomes something of a nuisance and, to our knowledge, has not been demonstrated.

6.5 SURFACE COLOR FROM IMAGE COLOR

It would be attractive to have a *color constancy* algorithm that could take an image, discount the effect of the light, and report the actual color of the surfaces being viewed. Color constancy is

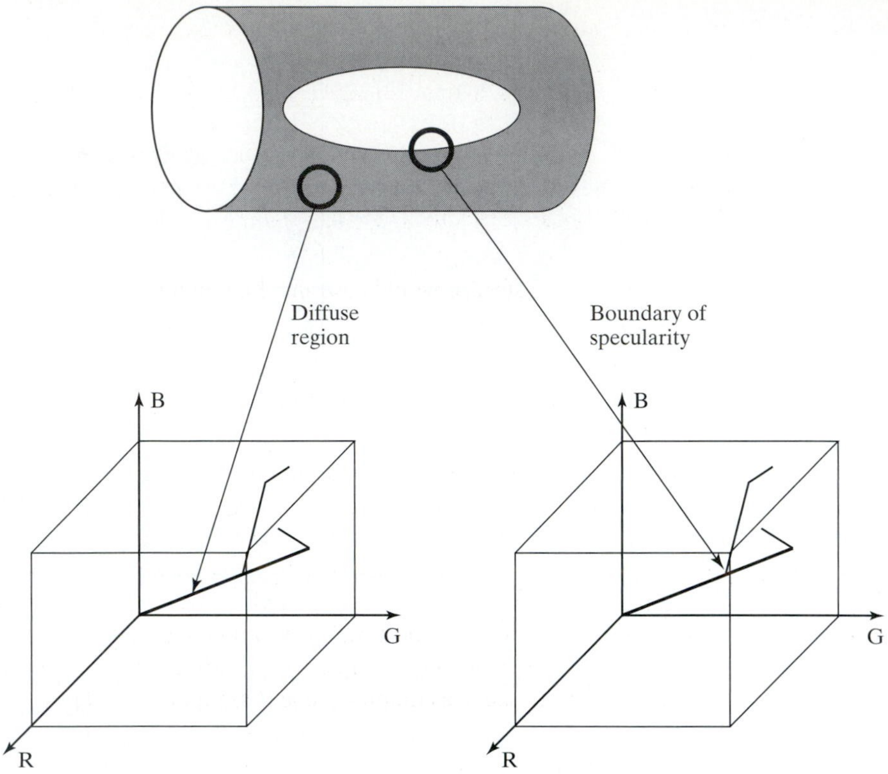

Figure 6.21 The linear clusters produced by specularities on plastic objects can be found by reasoning about windows of image pixels. In a world of plastic objects on a black background, a background window produces a region of pixels that are point-like in color space—all pixels have the same color. A window that lies along the body produces a line-like cluster of points in color space, because the intensity varies, but the color does not. At the boundary of a specularity, windows produce plane-like clusters because points are a weighted combination of two different colors (the specular and the body color). Finally, at the interior of a specular region, the windows can produce volume-like clusters, because the camera saturates, and the extent of the window can include both the boundary-style window and saturated points. Whether a region is line-like, plane-like, or volume-like can be determined easily by looking at the eigenvalues of the covariance of the pixels.

an interesting subproblem that has the flavour of a quite general vision problem: We are determining some parameters of the world from ambiguous image measurements, we need to use a model to disentangle these measurements, and we should like to report some representation of the uncertainty resulting from that measurement (perhaps confidence intervals, the covariance of the posterior, or a series of representative solutions).

6.5.1 Surface Color Perception in People

There is some form of color constancy algorithm in the human vision system. People are often unaware of this, and inexperienced photographers are sometimes surprised that a scene photographed indoors under fluorescent lights has a blue cast, whereas the same scene photographed outdoors may have a warm orange cast.

It is common to distinguish between color constancy, which is usually thought of in terms of intensity-independent descriptions of color like hue and saturation, and *lightness constancy*, which is the skill that allows humans to report whether a surface is white, gray, or black (the *lightness* of the surface) despite changes in the intensity of illumination (the *brightness*). Color constancy is neither perfectly accurate nor unavoidable. Humans can report the following:

- the color a surface would have in white light (often called *surface color*);
- color of the light arriving at the eye, a skill that allows artists to paint surfaces illuminated by colored lighting; and
- sometimes the color of the light falling on the surface.

All of these reports could be by-products of a color constancy process.

The colorimetric theories of Section 6.3 can predict the color an observer will perceive when shown an isolated spot of light of a given power spectral distribution. The human color constancy algorithm appears to obtain cues from the structure of complex scenes, meaning that predictions from colorimetric theories can be wildly inaccurate if the spot of light is part of a larger, complex scene. Demonstrations by Land and McCann (1971) (which are illustrated in Figure 6.22), give convincing examples of this effect. It is surprisingly difficult to predict what colors a human will see in a complex scene (Fairchild, 1998, Helson, 1938*a*, 1938*b*, 1934, 1940).

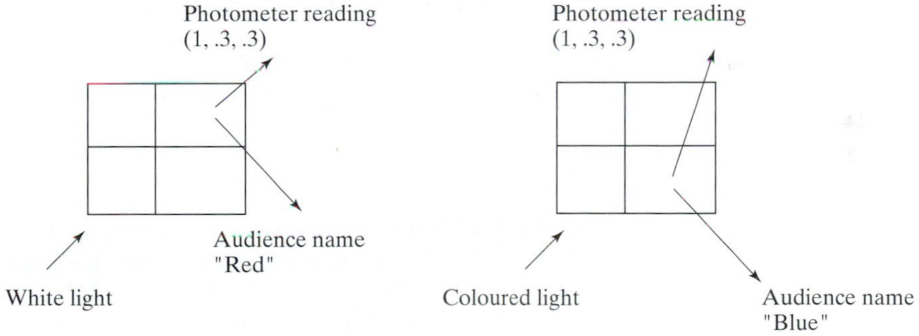

Figure 6.22 Land showed an audience a quilt of rectangles of flat colored papers—since known as a Mondrian for a purported resemblance to the work of that artist—illuminated using three slide projectors, casting red, green and blue light respectively. He used a photometer to measure the energy leaving a particular spot in three different channels, corresponding to the three classes of receptor in the eye. He recorded the measurement, and asked the audience to name the patch. Assume the answer was "red" (on the **left**). Land then adjusted the slide projectors so that some other patch reflected light that gave the same photometer measurements, and asked the audience to name that patch. The reply would describe the patch's color in white light—if the patch looked blue in white light, the answer would be "blue" (on the **right**). In later versions of this demonstration, Land put wedge-shaped neutral density filters into the slide projectors so that the color of the light illuminating the quilt of papers would vary slowly across the quilt. Again, although the photometer readings vary significantly from one end of a patch to another, the audience sees the patch as having a constant color.

This is one of the many difficulties that make it hard to produce really good color reproduction systems.

Human competence at color constancy is surprisingly poorly understood. The main experiments on humans (Arend and Reeves, 1986, McCann, McKee and Taylor, 1976) do not explore all circumstances. For example, it is not known how robust color constancy is or the extent to which high-level cues contribute to color judgements. Color constancy clearly fails sometimes—otherwise there would be no film industry—but the circumstances under which it fails are not well understood. There is a large body of data on surface lightness perception for achromatic stimuli. Since the brightness of a surface varies with its orientation as well as with the intensity of the illuminant, one would expect that human lightness constancy would be poor. In fact, it is extremely good over a wide range of illuminant variation (Jacobsen and Gilchrist, 1988).

6.5.2 Inferring Lightness

There is a lot of evidence that human lightness constancy involves two processes: One compares the brightness of various image patches and uses this comparison to determine which patches are lighter and which darker; the second establishes some form of absolute standard to which these comparisons can be referred (e.g. Gilchrist, Kossyfidis, Bonato, Agostini, Cataliotti, Li, Spehar, Annan and Economou, 1999). We describe lightness algorithms first because they tend to be simpler than color constancy algorithms.

A Simple Model of Image Brightness The radiance arriving at a pixel depends on the illumination of the surface being viewed, its BRDF, its configuration with respect to the source, and the camera responses. The situation is considerably simplified by assuming that the scene is plane and frontal, that surfaces are Lambertian, and that the camera responds linearly to radiance.

This yields a model of the camera response C at a point $\boldsymbol{X}$ as the product of an illumination term, an albedo term, and a constant that comes from the camera gain:

$$C(\boldsymbol{x}) = k_c I(\boldsymbol{x})\rho(\boldsymbol{x}).$$

If we take logarithms, we get

$$\log C(\boldsymbol{x}) = \log k_c + \log I(\boldsymbol{x}) + \log \rho(\boldsymbol{x}).$$

We now make a second set of assumptions:

- First, we assume that albedoes change only quickly over space—this means that a typical set of albedoes will look like a collage of papers of different grays. This assumption is quite easily justified: There are relatively few continuous changes of albedo in the world (the best example occurs in ripening fruit), and changes of albedo often occur when one object occludes another (so we would expect the change to be fast). This means that spatial derivatives of the term $\log \rho(\boldsymbol{x})$ are either zero (where the albedo is constant) or large (at a change of albedo).
- Second, illumination changes only slowly over space. This assumption is somewhat realistic. For example, the illumination due to a point source will change relatively slowly unless the source is very close—so the sun is a source that is particularly good for this example. As another example, illumination inside rooms tends to change very slowly because the white walls of the room act as area sources. This assumption fails dramatically at shadow boundaries, however. We have to see these as a special case and assume that either there are no shadow boundaries or that we know where they are.

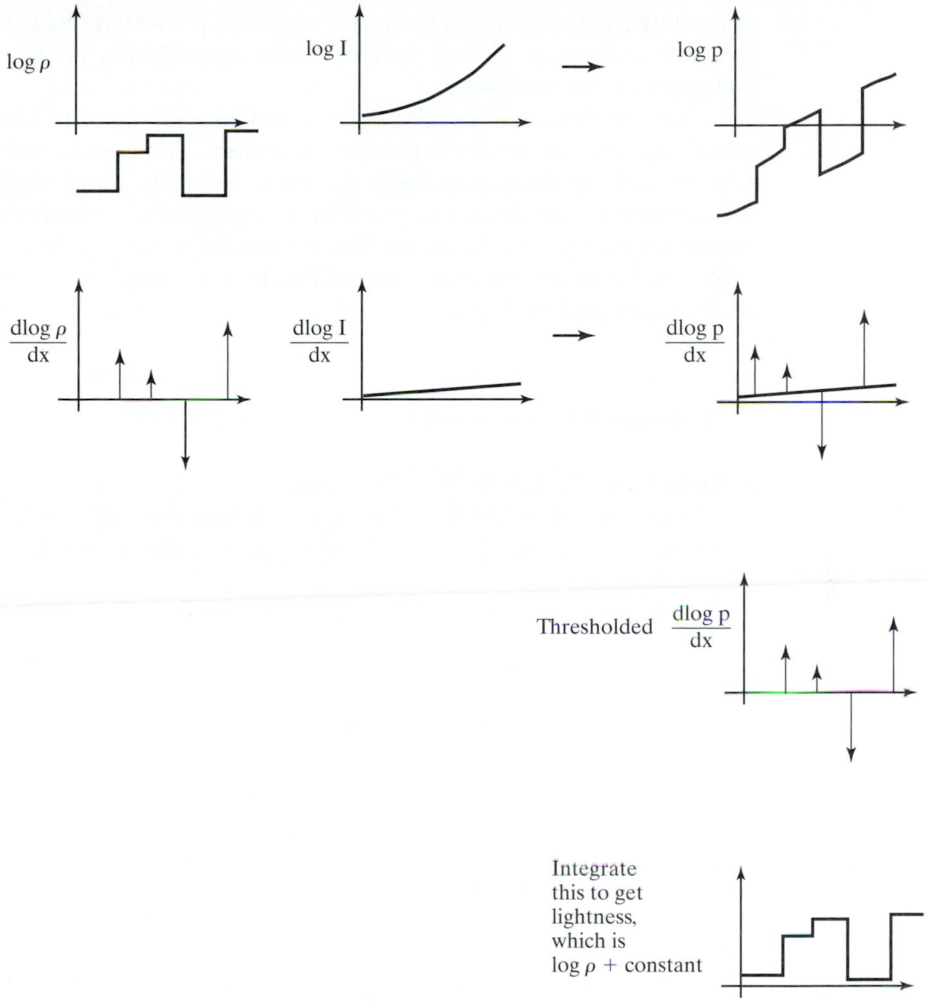

Figure 6.23 The lightness algorithm is easiest to illustrate for a 1D image. In the **top row**, the graph on the **left** shows $\log \rho(x)$, that on the **center** $\log I(x)$, and that on the **right** their sum, which is $\log C$. The log of image intensity has large derivatives at changes in surface reflectance and small derivatives when the only change is due to illumination gradients. Lightness is recovered by differentiating the log intensity, thresholding to dispose of small derivatives, and integrating at the cost of a missing constant of integration.

Recovering Lightness from the Model It is relatively easy to build algorithms that use our model. The earliest algorithm, the Retinex algorithm of Land and McCann (1971), has fallen into disuse. A natural approach is to differentiate the log transform, throw away small gradients, and integrate the results (Horn, 1974). There is a constant of integration missing, so lightness ratios are available, but absolute lightness measurements are not. Figure 6.23 illustrates the process for a one-dimensional example, where differentiation and integration are easy.

This approach can be extended to two dimensions as well. Differentiating and thresholding is easy: At each point, we estimate the magnitude of the gradient; if the magnitude is less than some threshold, we set the gradient vector to zero or else we leave it alone. The difficulty is in integrating these gradients to get the log albedo map. The thresholded gradients may not be the gradients of an image because the mixed second partials may not be equal (integrability again; compare with Section 5.4.2).

The problem can be rephrased as a minimization problem: choose the log albedo map whose gradient is most like the thresholded gradient. This is a relatively simple problem because computing the gradient of an image is a linear operation. The x-component of the thresholded gradient is scanned into a vector $\boldsymbol{p}$ and the y-component is scanned into a vector $\boldsymbol{q}$. We write the vector representing log-albedo as $\boldsymbol{l}$. Now the process of forming the x derivative is linear, and so there is some matrix $\mathcal{M}_x$, such that $\mathcal{M}_x \boldsymbol{l}$ is the x derivative; for the y derivative, we write the corresponding matrix $\mathcal{M}_y$.

Algorithm 6.1: Determining the Lightness of Image Patches

Form the gradient of the log of the image
At each pixel, if the gradient magnitude is below a threshold, replace that gradient with zero
Reconstruct the log-albedo by solving the minimization problem described in the text
Obtain a constant of integration
Add the constant to the log-albedo, and exponentiate

The problem becomes to find the vector $\boldsymbol{l}$ that minimizes

$$| \mathcal{M}_x \boldsymbol{l} - \boldsymbol{p} |^2 + | \mathcal{M}_y \boldsymbol{l} - \boldsymbol{q} |^2 .$$

This is a quadratic minimization problem, and the answer can be found by a linear process. Some special tricks are required because adding a constant vector to $\boldsymbol{l}$ cannot change the derivatives, so the problem does not have a unique solution. We explore the minimization problem in the exercises.

The constant of integration needs to be obtained from some other assumption. There are two obvious possibilities:

- we can assume that the *brightest patch is white*;
- we can assume that the *average lightness is constant*.

We explore the consequences of these models in the exercises.

Finite-Dimensional Linear Models Now the model of image color in Section 6.4.2 is

$$\boldsymbol{C}(\boldsymbol{x}) = g_d(\boldsymbol{x})\boldsymbol{d}(\boldsymbol{x}) + g_s(\boldsymbol{x})\boldsymbol{s}(\boldsymbol{x}) + \boldsymbol{i}(\boldsymbol{x}).$$

We decided to ignore the interreflection term $\boldsymbol{i}(\boldsymbol{x})$. In principle, we could use the methods of Section 6.4.3 to generate new images without specularities. This brings us to the term $g_d(\boldsymbol{x})\boldsymbol{d}(\boldsymbol{x})$. Assume that $g_d(\boldsymbol{x})$ is a constant, so we are viewing a flat, frontal surface.

The resulting term, $\boldsymbol{d}(\boldsymbol{x})$, models the world as a collage of flat, frontal diffuse colored surfaces. We assume that there is a single illuminant that has a constant color over the whole image. This term is a conglomeration of illuminant, receptor, and reflectance information. It is impossible to disentangle completely in a realistic world. However, current algorithms can make quite usable estimates of surface color from image colors given a well-populated world of colored surfaces and a reasonable illuminant.

The term $\boldsymbol{d}(\boldsymbol{x})$ results from interactions among the spectral irradiance of the source, the spectral albedo of the surfaces, and the camera sensitivity. We need a model to account for these interactions. Recall from Section 6.4.2 that if a patch of perfectly diffuse surface with diffuse spectral reflectance $\rho(\lambda)$ is illuminated by a light whose spectrum is $E(\lambda)$, the spectrum of the reflected light is $\rho(\lambda)E(\lambda)$ (multiplied by some constant to do with surface orientation, which we have already decided to ignore).

Thus, if a linear photoreceptor of the kth type sees this surface patch, its response is:

$$p_k = \int_\Lambda \sigma_k(\lambda)\rho(\lambda)E(\lambda)d\lambda,$$

where Λ is the range of all relevant wavelengths and $\sigma_k(\lambda)$ is the sensitivity of the kth photoreceptor (Figure 6.15).

This response is linear in the surface reflectance and linear in the illumination, which suggests using linear models for the families of possible surface reflectances and illuminants. A *finite-dimensional linear model* models surface spectral albedoes and illuminant spectral irradiance as a weighted sum of a finite number of basis functions. We need not use the same bases for reflectances and for illuminants.

If a finite-dimensional linear model of surface reflectance is a reasonable description of the world, any surface reflectance can be written as

$$\rho(\lambda) = \sum_{j=1}^{n} r_j \phi_j(\lambda),$$

where the $\phi_j(\lambda)$ are the basis functions for the model of reflectance, and the r_j vary from surface to surface.

Similarly, if a finite-dimensional linear model of the illuminant is a reasonable model, any illuminant can be written as

$$E(\lambda) = \sum_{i=1}^{m} e_i \psi_i(\lambda),$$

where the $\psi_i(\lambda)$ are the basis functions for the model of illumination.

When both models apply, the response of a receptor of the kth type is

$$\begin{aligned} p_k &= \int \sigma_k(\lambda) \left(\sum_{j=1}^{n} r_j \phi_j(\lambda) \right) \left(\sum_{i=1}^{m} e_i \psi_i(\lambda) \right) d\lambda \\ &= \sum_{i=1,j=1}^{m,n} e_i r_j \left(\int \sigma_k(\lambda)\phi_j(\lambda)\psi_i(\lambda) \right) d\lambda \\ &= \sum_{i=1,j=1}^{m,n} e_i r_j g_{ijk}, \end{aligned}$$

where we expect that the

$$g_{ijk} = \int \sigma_k(\lambda)\phi_j(\lambda)\psi_i(\lambda)d\lambda$$

are known, as they are components of the world model (they can be learned from observations; see the exercises).

6.5.3 Surface Color from Finite-Dimensional Linear Models

Each of the indexed terms can be interpreted as components of a vector, and we use the notation $\boldsymbol{p}$ for the vector with kth component p_k. We could represent surface color either directly by the vector of coefficients $\boldsymbol{r}$ or more indirectly by computing $\boldsymbol{r}$ and then determining what the surfaces would look like under white light. The latter representation is more useful in practice; among other things, the results are easy to interpret.

Normalizing Average Reflectance Assume that the spatial average of reflectance in all scenes is constant and known (e.g., we might assume that all scenes have a spatial average of reflectance that is dull gray). In the finite-dimensional basis for reflectance, we can write this average as

$$\sum_{j=1}^{n} \overline{r_j}\phi_j(\lambda).$$

Now if the average reflectance is constant, the average of the receptor responses must be constant too (if the imaging process is linear—see the discussion), and the average of the response of the kth receptor can be written as:

$$\overline{p_k} = \sum_{i=1,j=1}^{m,n} e_i g_{ijk} \overline{r_j}.$$

If $\overline{\boldsymbol{p}}$ is the vector with kth component $\overline{p}_k$ (using the notation above) and $\mathcal{A}$ is the matrix with k, ith component

$$\sum_{j=1}^{n} \overline{r}_j g_{ijk},$$

then we can write the previous expression as:

$$\overline{\boldsymbol{p}} = \mathcal{A}\boldsymbol{e}.$$

For reasonable choices of receptors, the matrix $\mathcal{A}$ will have full rank, meaning that we can determine $\boldsymbol{e}$, which gives the illumination, *if* the finite-dimensional linear model for illumination has the same dimension as the number of receptors. Of course, once the illumination is known, we can report the surface reflectance at each pixel or correct the image to look as if it were taken under white light.

Algorithm 6.2: Color Constancy from Known Average Reflectance

Compute the average color $\overline{\boldsymbol{p}}$ for the image
Compute $\boldsymbol{e}$ from $\overline{\boldsymbol{p}} = \mathcal{A}\boldsymbol{e}$
To obtain a version of the image under white light, $\boldsymbol{e}^w$
  For each pixel, compute $\boldsymbol{r}$ from $p_k = \sum_{i=1,j=1m,n} e_i g_{ijk} r_j$
  Replace the pixel value with $p_k^w = \sum_{i=1,j=1m,n} e_i^w g_{ijk} r_j$

The underlying assumption that average reflectance is a known constant is dangerous, however, because it is usually not even close to being true. For example, if we assume that the average reflectance is a medium gray (a popular choice; see e.g., Buchsbaum, 1980, Gershon, Jepson and Tsotsos, 1986), an image of a leafy forest glade is reported as a collection of objects of various grays illuminated by green light. One way to avoid this problem is to change the average for different kinds of scenes (Gershon, Jepson and Tsotsos, 1986), but how do we decide what average to use? Another approach is to compute an average that is not a pure spatial average. For example, one might average the colors that were represented by 10 or more pixels, but without weighting them by the number of pixels present.

Normalizing the Gamut Not every possible pixel value can be obtained by taking images of real surfaces under white light. It is usually impossible to obtain values where one channel responds strongly and others do not (e.g., 255 in the red channel and 0 in the green and blue channels). This means that the gamut of an image—the collection of all pixel values—contains information about the light source. For example, if one observes a pixel that has value (255, 0, 0), then the light source is likely to be red in color.

If an image gamut contains two pixel values, say $\boldsymbol{p}_1$ and $\boldsymbol{p}_2$, then it must be possible to take an image *under the same illuminant* that contains the value $t\boldsymbol{p}_1 + (1-t)\boldsymbol{p}_2$ for $0 \leq t \leq 1$ (because we could mix the colorants on the surfaces). This means that the convex hull of the image gamut contains illuminant information. These constraints can be exploited to constrain the color of the illuminant.

Write G for the convex hull of the gamut of the given image, W for the convex hull of the gamut of an image of many different surfaces under white light, and $\mathcal{M}_{\boldsymbol{e}}$ for the map that takes an image seen under illuminant $\boldsymbol{e}$ to an image seen under white light. Then the only illuminants we need to consider are those such that $\mathcal{M}_{\boldsymbol{e}}(G) \in W$. This is most helpful if the family $\mathcal{M}_{\boldsymbol{e}}$ has a reasonable structure; one natural example is to assume that elements of $\mathcal{M}_{\boldsymbol{e}}$ are diagonal matrices.

In the case of finite-dimensional linear models, $\mathcal{M}_{\boldsymbol{e}}$ depends linearly on $\boldsymbol{e}$ so that the family of illuminants that satisfy the constraint is also convex. This family can be constructed by intersecting a set of convex hulls, each corresponding to the family of maps that takes a hull vertex of G to some point inside W (or we could write a long series of linear constraints on $\boldsymbol{e}$).

Algorithm 6.3: Color Constancy by Gamut Mapping

Obtain the gamut W of many images of many different colored surfaces under white light (this is the convex hull of all image pixel values)
Obtain the gamut G of the image (this is the convex hull of all image pixel values)
Obtain every element of the family of illuminant maps $\mathcal{M}_{\boldsymbol{e}}$ such that $\mathcal{M}_{\boldsymbol{e}}G \in W$ this represents all possible illuminants
Choose some element of this family, and apply it to every pixel in the image

Once we have formed this family, it remains to find an appropriate illuminant. There are a variety of possible strategies. If something is known about the likelihood of encountering particular illuminants, then one might choose the most likely. Assuming that most pictures contain many different colored surfaces leads to the choice of illuminant that makes the restored gamut the largest. One might use other constraints on illuminants (for example, all the illuminants must have non-negative energy at all wavelengths) to constrain the set even further (Finlayson and Hordley, 2000).

6.6 NOTES

The use of color in computer vision is surprisingly primitive. One difficulty is some legitimate uncertainty about what it is good for. John Mollon's remark that the primate color system could be seen as an innovation of some kinds of fruiting tree (Mollon, 1995) is one explanation, but it is not much help.

There are a number of important general resources on the use of color. We recommend Hardin and Maffi (1997), Lamb and Bourriau (1995), Lynch and Livingston (2001), Minnaert (1993), Trussell, Allebach, Fairchild, Funt and Wong (1997), Williamson and Cummins (1983). Wyszecki and Stiles (1982) contains an enormous amount of helpful information.

Trichromacy and Color Spaces

Until quite recently, there was no conclusive explanation of why trichromacy applied, although it was generally believed to be due to the presence of three different types of color receptor in the eye. Work on the genetics of photoreceptors by Nathans *et al.* can be interpreted as confirming this hunch (see Nathans, Piantanida, Eddy, Shows and Hogness, 1986*a*, and Nathans, Thomas and Hogness, 1986*b*), although a full explanation is still far from clear because this work can also be interpreted as suggesting many individuals have more than three types of photoreceptor (Mollon, 1995).

There is an astonishing number of color spaces and color appearance models available. The important issue is not in what coordinate system one measures color, but how one counts the difference, so color metrics may still bear some thought.

Color metrics are an old topic; usually, one fits a metric tensor to MacAdam ellipses. The difficulty with this approach is that a metric tensor carries the strong implication that you can measure differences over large ranges by integration, whereas it is very hard to see large-range color comparisons as meaningful. Another concern is that the weight observers place on a difference in a Maxwellian view and the semantic significance of a difference in image colors are two very different things.

Specularity Finding

The specularity finding method we describe is due to Shafer (1985*b*), with improvements due to Klinker, Shafer and Kanade (1987*a,b*), (1990), and to Maxwell and Shafer (2000). Specularities can also be detected because they are small and bright (Brelstaff and Blake, 1988) or because they differ in color and motion from the background (Lee and Bajcsy, 1992*a*, Lee and Bajcsy, 1992*b*).

Lightness

Land reported a variety of color vision experiments (Land, 1959*a*, 1959*b*, 1959*c*, 1983). There has not been much recent study of lightness constancy algorithms. The basic idea is due to Land and McCann (1971). His work was formalized for the computer vision community by Horn (1974). A variation on Horn's algorithm was constructed by Blake (1985). This is the lightness algorithm we describe. It appeared originally in a slightly different form, where it was called the *Retinex* algorithm (Land and McCann, 1971). Retinex was originally intended as a color constancy algorithm. It is surprisingly difficult to analyze (Brainard and Wandell, 1986).

Lightness techniques are not as widely used as they should be, particularly given that there is some evidence that they produce useful information on real images (Brelstaff and Blake, 1987). Classifying illumination versus albedo simply by looking at the magnitude of the gradient is crude, and ignores important cues. One of these cues is that large changes must be illumination, however fast they occur. Another is that color changes at shadow boundaries are different from color changes at albedo boundaries. The question of which changes are albedo and which are illuminant looks like an inference problem to us, and a do-able one at that (it's fairly easy to write down a likelihood model; the priors are what one worries about). Some attempts to study this problem using a local model have been made (e.g., Freeman, Pasztor and Carmichael, 2000).

Color Constancy

Finite-dimensional linear models for spectral reflectances can be supported by an appeal to surface physics as spectral absorbtion lines are thickened by solid state effects. The main experimental justifications for finite-dimensional linear models of surface reflectance are measurements, by Cohen (1964), of the surface reflectance of a selection of standard reference surfaces known as *Munsell chips*, and measurements of a selection of natural objects by Krinov (1947). Cohen (1964) performed a principal axis decomposition of his data to obtain a set of basis functions, and Maloney (1984) fitted weighted sums of these functions to Krinov's date to get good fits with patterned deviations. The first three principal axes explained in each case a high percentage of the sample variance (near 99 %), and hence a linear combination of these functions fitted all the sampled functions rather well. More recently, Maloney (1986) fitted Cohen's (1964) basis vectors to a large set of data, including Krinov's (1947) data, and further data on the surface reflectances of Munsell chips, and concluded that the dimension of an accurate model of surface reflectance was on the order of five or six.

Finite-dimensional linear models are an important tool in color constancy. There is a large collection of algorithms that follow rather naturally from the approach. Some algorithms exploit the properties of the linear spaces involved (Maloney, 1984, Maloney and Wandell, 1986, Wandell, 1987). On surfaces like plastics, the specular component of the reflected light is the same color as the illuminant. If we can identify specular regions from such objects in the image, the color of the illuminant is known. This idea has been popular for a long time—Judd (1940) writing in 1960 about early German work in surface color perception refers to it as "a more usual view". The idea has been popular recently, too (D'Zmura and Lennie, 1986, Flock, 1984, Klinker, Shafer and Kanade, 1987*a,b*, Lee, 1986). Assuming that the average color is constant is another popular approach (Buchsbaum, 1980, Gershon, 1987).

Gamut mapping methods are due to Forsyth (1990). The method has been extensively enhanced (e.g. Barnard, 2000, Finlayson and Hordley, 1999, 2000). The structure of the family of maps associated with a change in illumination, $\mathcal{M}_e$, has been studied quite extensively. The first work is due to Von Kries (who didn't think about it quite the way we do). He assumed that color constancy was, in essence, the result of independent lightness calculations in each channel, meaning that one can rectify an image by scaling each channel independently. This practice is known as Von Kries' law. In our notation, the law boils down to assuming that $\mathcal{M}_e$ consists of diagonal matrices. Von Kries' law has proved to be a remarkably good law (Finlayson, Drew and Funt, 1994*a*). Current best practice involves applying a linear transformation to the channels and then scaling the result using diagonal maps (Finlayson et al., 1994*a*, 1994*b*).

There is surprisingly little work on color constancy that unifies a study of the spatial variation in illumination with solutions for surface color, which is why we were reduced to ignoring a number of terms in our color model. Ideally, one would work in shadows and surface orientation, too. Again, the whole thing looks like an inference problem to us, but a subtle one. The main papers on this extremely important topic are Barnard, Finlayson and Funt (1997), Funt and Drew (1988). There is substantial room for research here, too.

Interreflections between colored surfaces lead to a phenomenon called *color bleeding*, where each surface reflects colored light onto the other. The phenomenon can be surprisingly large in practice. People seem to be quite good at ignoring it entirely, to the extent that most people don't realize that the phenomenon occurs at all. Discounting color bleeding probably uses spatial cues. Some skill is required to spot really compelling examples. The best known to the authors is occasionally seen in southern California, where there are many large hedges of white oleander by the roadside. White oleander has dark leaves and white flowers. Occasionally, in bright sunlight, one sees a hedge with yellow oleander flowers; a moment's thought attributes the color to the yellow service truck parked by the road reflecting yellow light onto the white flowers. One's ability to discount color bleeding effects seems to have been disrupted by the dark leaves of the plant breaking up the spatial pattern. Color bleeding contains cues to surface color that are quite difficult to disentangle (see Drew and Funt, 1990, Funt and Drew, 1993, and Funt, Drew and Ho, 1991 for studies).

It is possible to formulate and attack color constancy as an inference problem (Forsyth, 1999, Freeman and Brainard, 1997). The advantage of this approach is that the algorithm could report a range of possible surface colors given data.

Color in Recognition

As future chapters show, it is quite tricky to build systems that use object color to help in recognition. Color constancy is conceived of as a process that should improve matching because we can deal in object properties, rather than properties of the particular view (Finlayson, Chatterjee and Funt, 1996, Funt, Barnard and Martin, 1998, Funt and Finlayson, 1995). Uniform color spaces offer some help here if we are willing to swallow a fairly loose evolutionary argument: It is worth understanding the color differences that humans recognize because they are adapted to useful measurements.

ASSIGNMENTS

PROBLEMS

6.1. Sit down with a friend and a packet of colored papers, and compare the color names that you use. You need a large packet of papers—one can very often get collections of colored swatches for paint, or for the Pantone color system very cheaply. The best names to try are basic color names—the terms *red*, *pink*, *orange*, *yellow*, *green*, *blue*, *purple*, *brown*, *white*, *gray* and *black*, which (with a small number of other terms) have remarkable canonical properties that apply widely across different languages (the papers in Hardin and Maffi, 1997 give a good summary of current thought on this issue). You will find it surprisingly easy to disagree on which colors should be called blue and which green, for example.

6.2. Derive the equations for transforming from RGB to CIE XYZ and back. This is a linear transformation. It is sufficient to write out the expressions for the elements of the linear transformation—you don't have to look up the actual numerical values of the color matching functions.

6.3. Linear color spaces are obtained by choosing primaries and then constructing color matching functions for those primaries. Show that there is a linear transformation that takes the coordinates of a color in one linear color space to those in another; the easiest way to do this is to write out the transformation in terms of the color matching functions.

6.4. Exercise 3 means that, in setting up a linear color space, it is possible to choose primaries arbitrarily, but there are constraints on the choice of color matching functions. Why? What are these constraints?

6.5. Two surfaces that have the same color under one light and different colors under another are often referred to as *metamers*. An *optimal color* is a spectral reflectance or radiance that has value 0 at some wavelengths and 1 at others. Although optimal colors don't occur in practice, they are a useful device (due to Ostwald) for explaining various effects.

(a) Use optimal colors to explain how metamerism occurs.

(b) Given a particular spectral albedo, show that there are an infinite number of metameric spectral albedoes.

(c) Use optimal colors to construct an example of surfaces that look different under one light (say, red and green) and the same under another.

(d) Use optimal colors to construct an example of surfaces that swop apparent color when the light is changed (i.e., surface one looks red and surface two looks green under light one, and surface one looks green and surface two looks red under light two).

6.6. You have to map the gamut for a printer to that of a monitor. There are colors in each gamut that do not appear in the other. Given a monitor color that can't be reproduced exactly, you could choose the printer color that is closest. Why is this a bad idea for reproducing images? Would it work for reproducing "business graphics" (bar charts, pie charts, and the like, which all consist of many differernt large blocks of a single color)?

6.7. *Volume color* is a phenomenon associated with translucent materials that are colored—the most attractive example is a glass of wine. The coloring comes from different absorption coefficients at different wavelengths. Explain (a) why a small glass of sufficiently deeply colored red wine (a good Cahors or Gigondas) looks black (b) why a big glass of lightly colored red wine also looks black. Experimental work is optional.

6.8. (This exercise requires some knowledge of numerical analysis.) In Section 6.5.2, we set up the problem of recovering the log albedo for a set of surfaces as one of minimizing

$$| \mathcal{M}_x \boldsymbol{l} - \boldsymbol{p} |^2 + | \mathcal{M}_y \boldsymbol{l} - \boldsymbol{q} |^2,$$

where $\mathcal{M}_x$ forms the x derivative of $\boldsymbol{l}$ and $\mathcal{M}_y$ forms the y derivative (i.e., $\mathcal{M}_x \boldsymbol{l}$ is the x-derivative).

(a) We asserted that $\mathcal{M}_x$ and $\mathcal{M}_y$ existed. Use the expression for forward differences (or central differences, or any other difference approximation to the derivative) to form these matrices. Almost every element is zero.

(b) The minimization problem can be written in the form

$$\text{choose } \boldsymbol{l} \text{ to minimize } (\mathcal{A}\boldsymbol{l} + \boldsymbol{b})^T(\mathcal{A}\boldsymbol{l} + b).$$

Determine the values of $\mathcal{A}$ and $\boldsymbol{b}$, and show how to solve this general problem. You will need to keep in mind that $\mathcal{A}$ does not have full rank, so you can't go inverting it.

6.9. In Section 6.5.2, we mentioned two assumptions that would yield a constant of integration.

(a) Show how to use these assumptions to recover an albedo map.

(b) For each assumption, describe a situation where it fails, and describe the nature of the failure. Your examples should work for cases where there are many different albedoes in view.

6.10. Read the book *Colour: Art and Science*, by Lamb and Bourriau, Cambridge University Press, 1995.

Programming Assignments

6.11. Spectra for illuminants and for surfaces are available on the web (try `http://www.it.lut.fi/research/color/lutcs_database.html`). Fit a finite-dimensional linear model to a set of illuminants and surface reflectances using principal components analysis, render the resulting models, and compare your rendering with an exact rendering. Where do you get the most significant errors? Why?

6.12. Print a colored image on a color inkjet printer using different papers and compare the result. It is particularly informative to (a) ensure that the driver knows what paper the printer will be printing on, and compare the variations in colors (which are ideally imperceptible), and (b) deceive the driver about what paper it is printing on (i.e., print on plain paper and tell the driver it is printing on photographic paper). Can you explain the variations you see? Why is photographic paper glossy?

6.13. Fitting a finite-dimensional linear model to illuminants and reflectances separately is somewhat ill-advised because there is no guarantee that the *interactions* will be represented well (they're not accounted for in the fitting error). It turns out that one can obtain g_{ijk} by a fitting process that sidesteps the use of basis functions. Implement this procedure (which is described in detail in Marimont and Wandell (1992)), and compare the results with those obtained from the previous assignment.

6.14. Build a color constancy algorithm that uses the assumption that the spatial average of reflectance is constant. Use finite-dimensional linear models. You can get values of g_{ijk} from your solution to Exercise 3.

6.15. We ignore color interreflections in our surface color model. Do an experiment to get some idea of the size of color shifts possible from color interreflections (which are astonishingly big). Humans seldom interpret color interreflections as surface color. Speculate as to why this might be the case, using the discussion of the lightness algorithm as a guide.

6.16. Build a specularity finder along the lines described in Section 6.4.3.

PART II
Early Vision: Just One Image

7

Linear Filters

Pictures of zebras and of dalmatians have black and white pixels, and in about the same number, too. The differences between the two have to do with the characteristic appearance of small groups of pixels, rather than individual pixel values. In this chapter, we introduce methods for obtaining descriptions of the appearance of a small group of pixels.

Our main strategy is to use weighted sums of pixel values using different patterns of weights to find different image patterns. Despite its simplicity, this process is extremely useful. It allows us to smooth noise in images, and to find edges and other image patterns.

7.1 LINEAR FILTERS AND CONVOLUTION

Many important effects can be modeled with a simple model. Construct a new array, the same size as the image. Fill each location of this new array with a weighted sum of the pixel values from the locations surrounding the corresponding location in the image *using the same set of weights each time*. Different sets of weights could be used to represent different processes. One example is computing a local average taken over a fixed region. We could average all pixels within a $2k+1 \times 2k+1$ block of the pixel of interest. For an input image $\mathcal{F}$, this gives an output

$$\mathcal{R}_{ij} = \frac{1}{(2k+1)^2} \sum_{u=i-k}^{u=i+k} \sum_{v=j-k}^{v=j+k} \mathcal{F}_{uv}.$$

The weights in this example are simple (each pixel is weighted by the same constant), but we could use a more interesting set of weights. For example, we could use a set of weights that was large at the center and fell off sharply as the distance from the center increased to model the kind of smoothing that occurs in a defocused lens system.

Whatever the weights chosen, the output of this procedure is *shift-invariant*—meaning that the value of the output depends on the pattern in an image neighborhood, rather than the position of the neighborhood—and *linear*—meaning that the output for the sum of two images is the same as the sum of the outputs obtained for the images separately. The procedure is known as **linear filtering**.

7.1.1 Convolution

We introduce some notation at this point. The pattern of weights used for a linear filter is usually referred to as the *kernel* of the filter. The process of applying the filter is usually referred to as *convolution*. There is a catch: For reasons that will appear later (Section 7.2.1), it is convenient to write the process in a non-obvious way. In particular, given a filter kernel $\mathcal{H}$, the convolution of the kernel with image $\mathcal{F}$ is an image $\mathcal{R}$. The i, jth component of $\mathcal{R}$ is given by

$$R_{ij} = \sum_{u,v} H_{i-u,j-v} F_{u,v}.$$

This process defines convolution—we say that $\mathcal{H}$ has been convolved with $\mathcal{F}$ to yield $\mathcal{R}$. You should look closely at this expression—the "direction" of the dummy variable u (resp. v) has been reversed compared with correlation. This is important, because if you forget that it is there you compute the wrong answer. The reason for the reversal emerges from the derivation of Section 7.2.1. We carefully avoid inserting the range of the sum; in effect, we assume that the sum is over a large enough range of u and v that all nonzero values are taken into account. Furthermore, we assume that any values that haven't been specified are zero; this means that we can model the kernel as a small block of nonzero values in a sea of zeros. We use this convention, which is common, regularly in what follows.

Example 7.1 Smoothing by Averaging.

Images typically have the property that the value of a pixel is usually similar to that of its neighbor. Assume that the image is affected by noise of a form where we can reasonably expect that this property is preserved. For example, there might be occasional dead pixels, or small random numbers with zero mean might have been added to the pixel values. It is natural to attempt to reduce the effects of this noise by replacing each pixel with a weighted average of its neighbors, a process often referred to as *smoothing* or *blurring*.

Replacing each pixel with an unweighted average computed over some fixed region centered at the pixel is the same as convolution with a kernel that is a block of ones multiplied by a constant. You can (and should) establish this point by close attention to the range of the sum. This process is a poor model of blurring—its output does not look like that of a defocused camera (Figure 7.1). The reason is clear. Assume that we have an image in which every point but the center point was zero, and the center point was one. If we blur this image by forming an unweighted average at each point, the result looks like a small bright box, but this is not what defocused cameras do. We want a blurring process that takes a small bright dot to a circularly symmetric region of blur, brighter at the center than at the edges and fading slowly to darkness. As Figure 7.1 suggests, a set of weights of this form produces a much more convincing defocus model.

Example 7.2 Smoothing with a Gaussian.

A good formal model for this fuzzy blob is the *symmetric Gaussian kernel*

$$G_\sigma(x, y) = \frac{1}{2\pi\sigma^2} \exp\left(-\frac{(x^2 + y^2)}{2\sigma^2}\right)$$

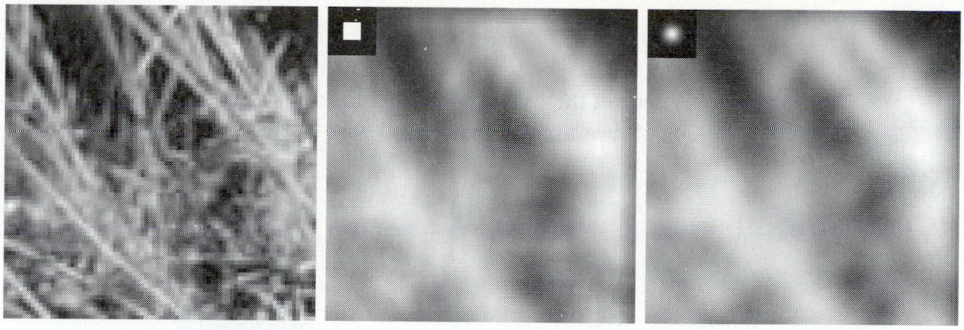

Figure 7.1 Although a uniform local average may seem to give a good blurring model, it generates effects not usually seen in defocusing a lens. The images above compare the effects of a uniform local average with weighted average. The image on the **left** shows a view of grass. In the **center**, the result of blurring this image using a uniform local model and on the **right**, the result of blurring this image using a set of Gaussian weights. The degree of blurring in each case is about the same, but the uniform average produces a set of narrow vertical and horizontal bars—an effect often known as *ringing*. The bottom row shows the weights used to blur the image, themselves rendered as an image; bright points represent large values and dark points represent small values (in this example, the smallest values are zero).

illustrated in Figure 7.2. σ is referred to as the *standard deviation* of the Gaussian (or its "sigma!"); the units are interpixel spaces, usually referred to as *pixels*. The constant term makes the integral over the whole plane equal to one and is often ignored in smoothing applications. The name comes from the fact that this kernel has the form of the probability density for a 2D normal (or Gaussian) random variable with a particular covariance.

This smoothing kernel forms a weighted average that weights pixels at its center much more strongly than at its boundaries. One can justify this approach qualitatively: Smoothing suppresses noise by enforcing the requirement that pixels should look like their neighbors. By downweighting distant neighbors in the average, we can ensure that the requirement that a pixel look like its neighbors is less strongly imposed for distant neighbors. A qualitative analysis gives the following:

- If the standard deviation of the Gaussian is very small—say smaller than one pixel—the smoothing will have little effect because the weights for all pixels off the center will be very small;
- For a larger standard deviation, the neighboring pixels will have larger weights in the weighted average, which means in turn that the average will be strongly biased toward a consensus of the neighbors—this will be a good estimate of a pixel's value, and the noise will largely disappear at the cost of some blurring;
- Finally, a kernel that has a large standard deviation will cause much of the image detail to disappear along with the noise.

Figure 7.3 illustrates these phenomena. You should notice that Gaussian smoothing can be effective at suppressing noise.

In applications, a discrete smoothing kernel is obtained by constructing a $2k+1 \times 2k+1$ array whose i, jth value is

$$H_{ij} = \frac{1}{2\pi\sigma^2} \exp\left(-\frac{((i-k-1)^2 + (j-k-1)^2)}{2\sigma^2}\right).$$

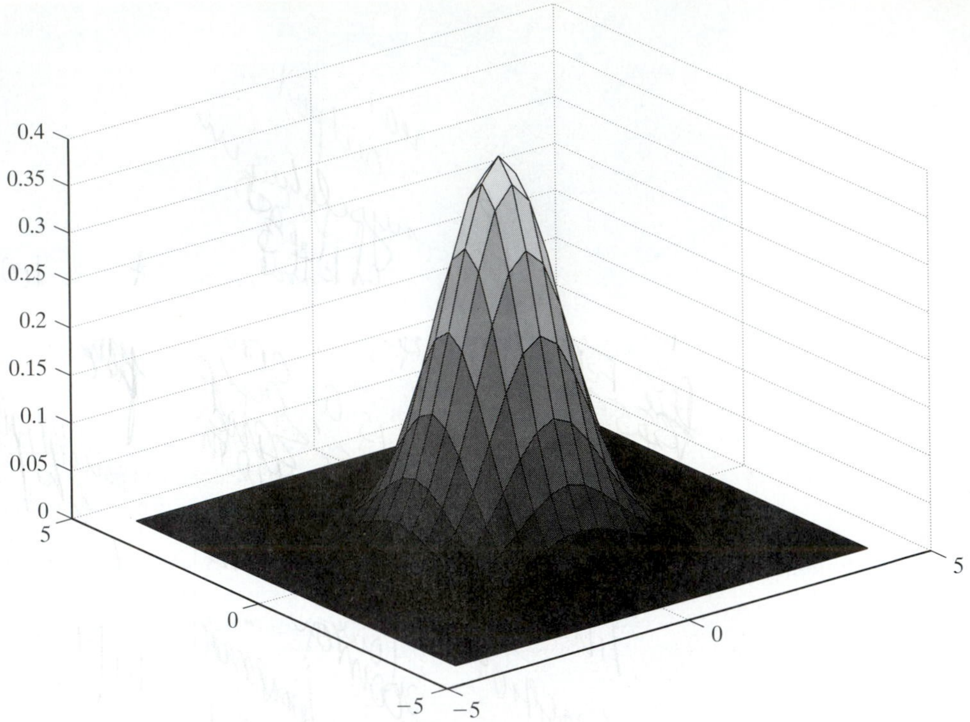

Figure 7.2 The symmetric Gaussian kernel in 2D. This view shows a kernel scaled so that its sum is equal to one; this scaling is quite often omitted. The kernel shown has $\sigma = 1$. Convolution with this kernel forms a weighted average that stresses the point at the center of the convolution window and incorporates little contribution from those at the boundary. Notice how the Gaussian is qualitatively similar to our description of the point spread function of image blur; it is circularly symmetric, has strongest response in the center, and dies away near the boundaries.

Notice that some care must be exercised with σ; if σ is too small, then only one element of the array will have a nonzero value. If σ is large, then k must be large, too, otherwise we are ignoring contributions from pixels that should contribute with substantial weight.

Example 7.3 Derivatives and Finite Differences.

Image derivatives can be approximated using another example of a convolution process. Because

$$\frac{\partial f}{\partial x} = \lim_{\epsilon \to 0} \frac{f(x+\epsilon, y) - f(x, y)}{\epsilon},$$

we might estimate a partial derivative as a symmetric *finite difference*:

$$\frac{\partial h}{\partial x} \approx h_{i+1,j} - h_{i-1,j}.$$

This is the same as a convolution, where the convolution kernel is

$$\mathcal{H} = \left\{ \begin{array}{ccc} 0 & 0 & 0 \\ 1 & 0 & -1 \\ 0 & 0 & 0 \end{array} \right\}.$$

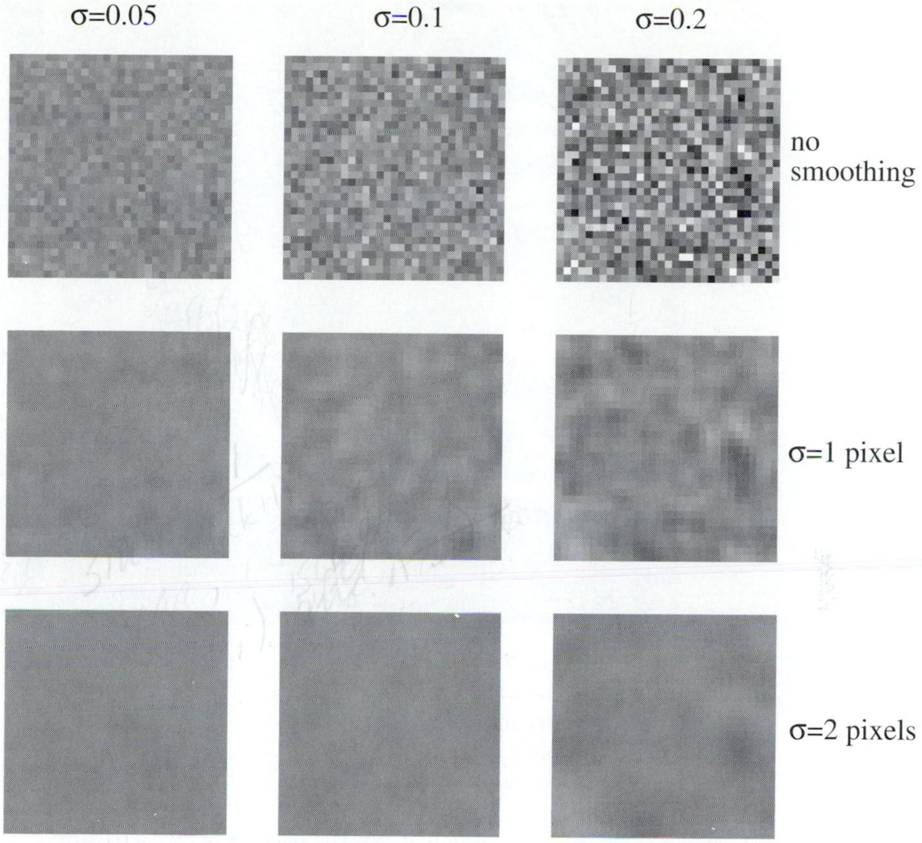

Figure 7.3 The **top row** shows images of a constant mid-gray level corrupted by additive Gaussian noise. In this noise model, each pixel has a zero-mean normal random variable added to it. The range of pixel values is from zero to one, so that the standard deviation of the noise in the first column is about 1/20 of full range. The **center row** shows the effect of smoothing the corresponding image in the top row with a Gaussian filter of σ one pixel. Notice the annoying overloading of notation here; there is Gaussian noise and Gaussian filters, and both have σ's. One uses context to keep these two straight, although this is not always as helpful as it could be because Gaussian filters are particularly good at suppressing Gaussian noise. This is because the noise values at each pixel are independent, meaning that the expected value of their average is going to be the noise mean. The **bottom row** shows the effect of smoothing the corresponding image in the top row with a Gaussian filter of σ two pixels.

Notice that this kernel could be interpreted as a template: It gives a large positive response to an image configuration that is positive on one side and negative on the other, and a large negative response to the mirror image.

As Figure 7.4 suggests, finite differences give a most unsatisfactory estimate of the derivative. This is because finite differences respond strongly (i.e., have an output with large magnitude) at fast changes, and fast changes are characteristic of noise. Roughly, this is because image pixels tend to look like one another. For example, if we had bought a discount camera with some pixels that were stuck at either black or white, the output of the finite difference process would be large at those pixels

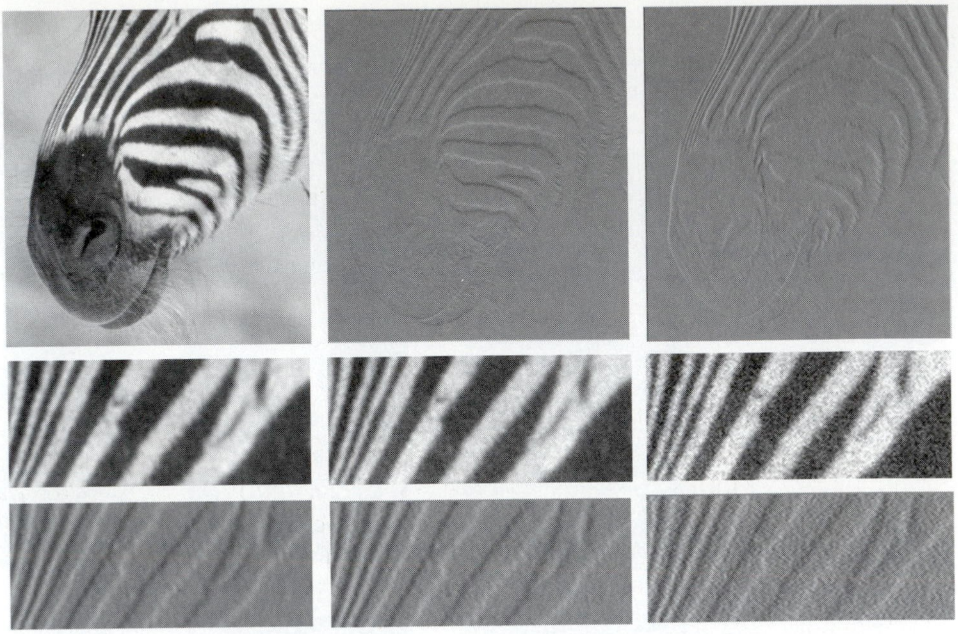

Figure 7.4 The **top row** shows estimates of derivatives obtained by finite differences. The image at the **left** shows a detail from a picture of a zebra. The **center** image shows the partial derivative in the y-direction— which responds strongly to horizontal stripes and weakly to vertical stripes—and the **right** image shows the partial derivative in the x-direction— which responds strongly to vertical stripes and weakly to horizontal stripes. However, finite differences respond strongly to noise. The image at **center left** shows a detail from a picture of a zebra; the next image in the row is obtained by adding a random number with zero mean and normal distribution ($\sigma = 0.03$—the darkest value in the image is 0, and the lightest 1) to each pixel; and the third image is obtained by adding a random number with zero mean and normal distribution ($\sigma = 0.09$) to each pixel. The **bottom row** shows the partial derivative in the x-direction of the image at the head of the row. Notice how strongly the differentiation process emphasizes image noise—the derivative figures look increasingly grainy. In the derivative figures, a mid-gray level is a zero value, a dark gray level is a negative value, and a light gray level is a positive value.

because they are, in general, substantially different from their neighbors. All this suggests that some form of smoothing is appropriate before differentiation; the details appear in Sections 8.1 and 8.2.

7.2 SHIFT INVARIANT LINEAR SYSTEMS

Convolution represents the effect of a large class of system. In particular, most imaging systems have, to a good approximation, three significant properties:

- **Superposition:** We expect that

$$R(f+g) = R(f) + R(g);$$

that is, the response to the sum of stimuli is the sum of the individual responses.

- **Scaling:** The response to a zero input is zero. Taken with superposition, we have that the response to a scaled stimulus is a scaled version of the response to the original stimulus—that is,

$$R(kf) = kR(f).$$

A device that exhibits superposition and scaling is *linear*.

- **Shift invariance:** In a *shift invariant* system, the response to a translated stimulus is just a translation of the response to the stimulus. This means that, for example, if a view of a small light aimed at the center of the camera is a small bright blob, then if the light is moved to the periphery, we should see the same small bright blob, only translated.

A device that is linear and shift invariant is known as a *shift invariant linear system*, or often just as a *system*.

The response of a shift invariant linear system to a stimulus is obtained by convolution. We demonstrate this first for systems that take discrete inputs—say vectors or arrays—and produce discrete outputs. We then use this to describe the behavior of systems that operate on continuous functions of the line or the plane, and from this analysis we obtain some useful facts about convolution.

7.2.1 Discrete Convolution

In the 1D case, we have a shift invariant linear system that takes a vector and responds with a vector. This case is the easiest to handle because there are fewer indices to look after. The 2D case, a system that takes an array and responds with an array, follows easily. In each case, we assume that the input and output are infinite dimensional. This allows us to ignore some minor issues that arise at the boundaries of the input. We deal with these in Section 7.2.3.

Discrete Convolution in One Dimension We have an input vector, $\boldsymbol{f}$. For convenience, we assume that the vector is infinite, and its elements are indexed by the integers (i.e., there is an element with index -1, say). The ith component of this vector is f_i. Now $\boldsymbol{f}$ is a weighted sum of basis elements. A convenient basis is a set of elements that have a one in a single component and zeros elsewhere. We write

$$\boldsymbol{e}_0 = \ldots 0, 0, 0, 1, 0, 0, 0, \ldots$$

This is a data vector that has a 1 in the zeroth place, and zeros elsewhere. Define a shift operation, which takes a vector to a shifted version of that vector. In particular, the vector $\texttt{Shift}(\boldsymbol{f}, i)$ has, as its jth component, the $j - i$th component of $\boldsymbol{f}$. For example, $\texttt{Shift}(\boldsymbol{e}_0, 1)$ has a zero in the first component. Now we can write

$$\boldsymbol{f} = \sum_i f_i \, \texttt{Shift}(\boldsymbol{e}_0, i).$$

We write the response of our system to a vector $\boldsymbol{f}$ as

$$R(\boldsymbol{f}).$$

Now because the system is shift invariant, we have

$$R(\texttt{Shift}(\boldsymbol{f}, k)) = \texttt{Shift}(R(\boldsymbol{f}), k).$$

Furthermore, because it is linear, we have

$$R(k\boldsymbol{f}) = kR(\boldsymbol{f}).$$

This means that

$$\begin{aligned} R(\boldsymbol{f}) &= R\left(\sum_i f_i \, \mathtt{Shift}(\boldsymbol{e}_0, i)\right) \\ &= \sum_i R(f_i \, \mathtt{Shift}(\boldsymbol{e}_0, i)) \\ &= \sum_i f_i R(\mathtt{Shift}(\boldsymbol{e}_0, i)) \\ &= \sum_i f_i \, \mathtt{Shift}(R(\boldsymbol{e}_0), i)). \end{aligned}$$

This means that, to obtain the system's response to any data vector, we need to know only its response to $\boldsymbol{e}_0$. This is usually called the system's *impulse response*. Assume that the impulse response can be written as $\boldsymbol{g}$. We have

$$R(\boldsymbol{f}) = \sum_i f_i \, \mathtt{Shift}(\boldsymbol{g}, i) = \boldsymbol{g} * \boldsymbol{f}.$$

This defines an operation—the 1D, discrete version of convolution—which we write with a $*$.

This is all very well, but it doesn't give us a particularly easy expression for the output. If we consider the jth element of $R(\boldsymbol{f})$, which we write as R_i, we must have

$$R_j = \sum_i g_{j-i} f_i,$$

which conforms to (and explains the origin of) the form used in Section 7.1.1.

Discrete Convolution in Two Dimensions We now use an array of values and write the i, jth element of the array $\mathcal{D}$ as D_{ij}. The appropriate analogy to an impulse response is the response to a stimulus that looks like

$$\mathcal{E}_{00} = \begin{matrix} \ldots & \ldots & \ldots & \ldots & \ldots \\ \ldots & 0 & 0 & 0 & \ldots \\ \ldots & 0 & 1 & 0 & \ldots \\ \ldots & 0 & 0 & 0 & \ldots \\ \ldots & \ldots & \ldots & \ldots & \ldots \end{matrix}$$

If $\mathcal{G}$ is the response of the system to this stimulus, the same considerations as for 1D convolution yield a response to a stimulus $\mathcal{F}$—that is,

$$R_{ij} = \sum_{u,v} G_{i-u,j-v} F_{uv},$$

which we write as

$$\mathcal{R} = \mathcal{G} * * \mathcal{H}.$$

7.2.2 Continuous Convolution

There are shift invariant linear systems that produce a continuous response to a continuous input; for example, a camera lens takes a set of radiances and produces another set, and many lenses are approximately shift invariant. A brief study of these systems allows us to study the information lost by approximating a continuous function—the incoming radiance values across an image plane—by a discrete function—the value at each pixel.

The natural description is in terms of the system's response to a rather unnatural function, the δ-function, which is not a function in formal terms. We do the derivation first in one dimension to make the notation easier.

Convolution in One Dimension We obtain an expression for the response of a continuous shift invariant linear system from our expression for a discrete system. We can take a discrete input and replace each value with a box straddling the value; this gives a continuous input function. We then make the boxes narrower and consider what happens in the limit.

Our system takes a function of one dimension and returns a function of one dimension. Again, we write the response of the system to some input $f(x)$ as $R(f)$; when we need to emphasize that f is a function, we write $R(f(x))$. The response is also a *function*; occasionally, when we need to emphasize this fact, we write $R(f)(u)$. We can express the linearity property in this notation by writing

$$R(kf) = kR(f)$$

(for k some constant) and the shift invariance property by introducing a `Shift` operator, which takes functions to functions:

$$\texttt{Shift}(f, c) = f(u - c).$$

With this `Shift` operator, we can write the shift invariance property as

$$R(\texttt{Shift}(f, c)) = \texttt{Shift}(R(f), c).$$

We define the *box* function as:

$$box_\epsilon(x) = \begin{cases} 0 & abs(x) > \frac{\epsilon}{2} \\ 1 & abs(x) < \frac{\epsilon}{2} \end{cases} .$$

The value of $box_\epsilon(\epsilon/2)$ does not matter for our purposes. The input function is $f(x)$. We construct an even grid of points x_i, where $x_{i+1} - x_i = \epsilon$. We now construct a vector $\boldsymbol{f}$ whose ith component (written f_i) is $f(x_i)$. This vector can be used to represent the function.

We obtain an approximate representation of f by $\sum_i f_i \, \texttt{Shift}(box_\epsilon, x_i)$. We apply this input to a shift invariant linear system; the response is a weighted sum of shifted responses to box functions. This means that

$$\begin{aligned} R(\sum_i f_i \, \texttt{Shift}(box_\epsilon, x_i)) &= \sum_i R(f_i \, \texttt{Shift}(box_\epsilon, x_i)) \\ &= \sum_i f_i R(\texttt{Shift}(box_\epsilon, x_i)) \\ &= \sum_i f_i \, \texttt{Shift}\left(R\left(\frac{box_\epsilon}{\epsilon}\epsilon\right), x_i\right) \\ &= \sum_i f_i \, \texttt{Shift}\left(R\left(\frac{box_\epsilon}{\epsilon}\right), x_i\right)\epsilon. \end{aligned}$$

So far, everything has followed our derivation for discrete functions. We now have something that looks like an approximate integral if $\epsilon \to 0$.

We introduce a new device, called a δ-function, to deal with the term box_ϵ/ϵ. Define

$$d_\epsilon(x) = \frac{box_\epsilon(x)}{\epsilon}.$$

The δ-function is:

$$\delta(x) = \lim_{\epsilon \to 0} d_\epsilon(x).$$

We don't attempt to evaluate this limit, so we need not discuss the value of $\delta(0)$. One interesting feature of this function is that, for practical shift invariant linear systems, the response of the system to a δ-function exists and has *compact support* (i.e., is zero except on a finite number of intervals of finite length). For example, a good model of a δ-function in 2D is an extremely small, extremely bright light. If we make the light smaller and brighter while ensuring the total energy is constant, we expect to see a small but finite spot due to the defocus of the lens. The δ-function is the natural analogue for $\boldsymbol{e}_0$ in the continuous case.

This means that the expression for the response of the system,

$$\sum_i f_i \, \texttt{Shift}\left(R\left(\frac{box_\epsilon}{\epsilon}\right), x_i \right) \epsilon,$$

turns into an integral as ϵ limits to zero. We obtain

$$\begin{aligned} R(f) &= \int \left\{ R(\delta)(u - x') \right\} f(x')\, dx' \\ &= \int g(u - x') f(x')\, dx', \end{aligned}$$

where we have written $R(\delta)$—which is usually called the **impulse response** of the system—as g and have omitted the limits of the integral. These integrals could be from $-\infty$ to ∞, but more stringent limits could apply if g and h have compact support. This operation is called **convolution** (again), and we write the foregoing expression as

$$R(f) = (g * f).$$

Convolution is *symmetric*, meaning

$$(g * h)(x) = (h * g)(x).$$

Convolution is *associative*, meaning that

$$(f * (g * h)) = ((f * g) * h).$$

This latter property means that we can find a single shift invariant linear system that behaves like the composition of two different systems. This comes in useful when we discuss sampling.

Convolution in Two Dimensions The derivation of convolution in two dimensions requires more notation. A box function is now given by $box_{\epsilon^2}(x, y) = box_\epsilon(x) box_\epsilon(y)$; we now have

$$d_\epsilon(x, y) = \frac{box_{\epsilon^2}(x, y)}{\epsilon^2}.$$

The δ-function is the limit of $d_\epsilon(x, y)$ function as $\epsilon \to 0$. Finally, there are more terms in the sum. All this activity results in the expression

$$R(h)(x, y) = \int\int g(x - x', y - y')h(x', y')\,dx\,dy$$
$$= (g * *h)(x, y),$$

where we have used two $*$s to indicate a two-dimensional convolution. Convolution *in 2D* is *symmetric*, meaning that

$$(g * *h) = (h * *g)$$

and *associative*, meaning that

$$((f * *g) * *h) = (f * *(g * *h)).$$

A natural model for the impulse response of a two-dimensional system is to think of the pattern seen in a camera viewing a very small, distant light source (which subtends a very small viewing angle). In practical lenses, this view results in some form of fuzzy blob, justifying the name **point spread function**, which is often used for the impulse response of a 2D system. The point spread function of a linear system is often known as its *kernel*.

7.2.3 Edge Effects in Discrete Convolutions

In practical systems, we cannot have infinite arrays of data. This means that when we compute the convolution, we need to contend with the edges of the image; at the edges, there are pixel locations where computing the value of the convolved image requires image values that don't exist. There are a variety of strategies we can adopt:

- **Ignore these locations**—this means that we report only values for which every required image location exists. This has the advantage of probity, but the disadvantage that the output is smaller than the input. Repeated convolutions can cause the image to shrink quite drastically.
- **Pad the image with constant values**—this means that, as we look at output values closer to the edge of the image, the extent to which the output of the convolution depends on the image goes down. This is a convenient trick because we can ensure that the image doesn't shrink, but it has the disadvantage that it can create the appearance of substantial gradients near the boundary.
- **Pad the image in some other way**—for example, we might think of the image as a doubly periodic function so that if we have an $n \times m$ image, then column $m + 1$—required for the purposes of convolution—would be the same as column $m - 1$. This can create the appearance of substantial second derivative values near the boundary.

7.3 SPATIAL FREQUENCY AND FOURIER TRANSFORMS

We have used the trick of thinking of a signal $g(x, y)$ as a weighted sum of a large (or infinite) number of small (or infinitely small) box functions. This model emphasizes that a signal is an element of a vector space—the box functions form a convenient basis, and the weights are coefficients on this basis. We need a new technique to deal with two related problems so far left open:

- Although it is clear that a discrete image version cannot represent the full information in a signal, we have not yet indicated what is lost;

- It is clear that we cannot shrink an image simply by taking every kth pixel—this could turn a checkerboard image all white or all black—and we should like to know how to shrink an image safely.

All of these problems are related to the presence of fast changes in an image. For example, shrinking an image is most likely to miss fast effects because they could slip between samples; similarly, the derivative is large at fast changes.

These effects can be studied by *a change of basis*. We change the basis to be a set of sinusoids and represent the signal as an infinite weighted sum of an infinite number of sinusoids. This means that fast changes in the signal are obvious, because they correspond to large amounts of high-frequency sinusoids in the new basis.

7.3.1 Fourier Transforms

The change of basis is effected by a *Fourier transform*. We define the Fourier transform of a signal $g(x, y)$ to be

$$\mathcal{F}(g(x, y))(u, v) = \iint_{-\infty}^{\infty} g(x, y)e^{-i2\pi(ux+vy)}\, dx\, dy.$$

Assume that appropriate technical conditions are true to make this integral exist. It is sufficient for all moments of g to be finite; a variety of other possible conditions are available (Bracewell, 1995). The process takes a complex valued function of x, y and returns a complex valued function of u, v (images are complex valued functions with zero imaginary component).

For the moment, fix u and v, and let us consider the meaning of the value of the transform at that point. The exponential can be rewritten

$$e^{-i2\pi(ux+vy)} = \cos(2\pi(ux + vy)) + i\sin(2\pi(ux + vy)).$$

These terms are sinusoids on the x, y plane, whose orientation and frequency are given by u, v. For example, consider the real term, which is constant when $ux + vy$ is constant (i.e., along a straight line in the x, y plane whose orientation is given by $\tan\theta = v/u$). The gradient of this term is perpendicular to lines where $ux + vy$ is constant, and the frequency of the sinusoid is $\sqrt{u^2 + v^2}$. These sinusoids are often referred to as *spatial frequency components*; a variety are illustrated in Figure 7.5.

The integral should be seen as a dot product. If we fix u and v, the value of the integral is the dot product between a sinusoid in x and y and the original function. This is a useful analogy because dot products measure the amount of one vector in the direction of another.

In the same way, the value of the transform at a particular u and v can be seen as measuring the amount of the sinusoid with given frequency and orientation in the signal. The transform takes a function of x and y to the function of u and v whose value at any particular (u, v) is the amount of that particular sinusoid in the original function. This view justifies the model of a Fourier transform as a change of basis.

Linearity The Fourier transform is linear:

$$\mathcal{F}(g(x, y) + h(x, y)) = \mathcal{F}(g(x, y)) + \mathcal{F}(h(x, y))$$

and

$$\mathcal{F}(kg(x, y)) = k\mathcal{F}(g(x, y)).$$

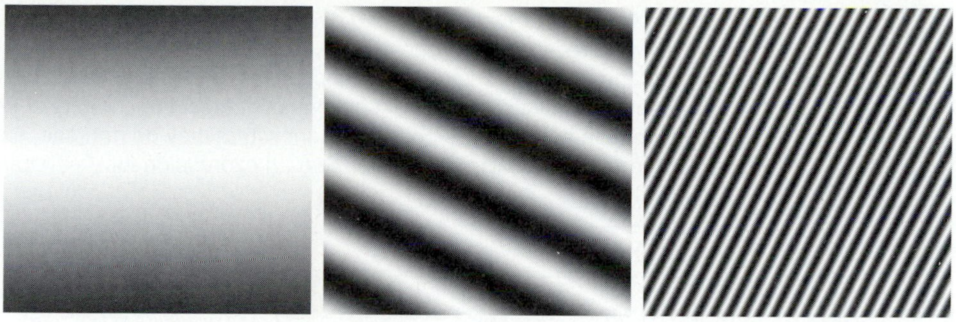

Figure 7.5 The real component of Fourier basis elements shown as intensity images. The brightest point has value one, and the darkest point has value zero. The domain is $[-1, 1] \times [-1, 1]$, with the origin at the center of the image. On the left, $(u, v) = (0, 0.4)$; in the center, $(u, v) = (1, 2)$ and on the right $(u, v) = (10, -5)$. These are sinusoids of various frequencies and orientations described in the text.

The Inverse Fourier Transform It is useful to recover a signal from its Fourier transform. This is another change of basis with the form

$$g(x, y) = \int\int_{-\infty}^{\infty} \mathcal{F}(g(x, y))(u, v)e^{i2\pi(ux+vy)}\, du\, dv.$$

Fourier Transform Pairs Fourier transforms are known in closed form for a variety of useful cases; a large set of examples appears in Bracewell (1995). We list a few in Table 7.1 for reference. The last line of Table 7.1 contains the *convolution theorem*; convolution in the signal domain is the same as multiplication in the Fourier domain. We use this important fact several times in what follows (Section 9.2.2).

Phase and Magnitude The Fourier transform consists of a real and a complex component:

$$\begin{aligned}
\mathcal{F}(g(x, y))(u, v) &= \int\int_{-\infty}^{\infty} g(x, y)\cos(2\pi(ux + vy))\, dx\, dy \\
&\quad + i\int\int_{-\infty}^{\infty} g(x, y)\sin(2\pi(ux + vy))\, dx\, dy \\
&= \Re(\mathcal{F}(g)) + i * \Im(\mathcal{F}(g)) \\
&= \mathcal{F}_R(g) + i * \mathcal{F}_I(g).
\end{aligned}$$

It is usually inconvenient to draw complex functions of the plane. One solution is to plot $\mathcal{F}_R(g)$ and $\mathcal{F}_I(g)$ separately; another is to consider the *magnitude* and *phase* of the complex functions, and to plot these instead. These are then called the *magnitude spectrum* and *phase spectrum*, respectively.

The value of the Fourier transform of a function at a particular u, v point depends on the whole function. This is obvious from the definition because the domain of the integral is the whole domain of the function. It leads to some subtle properties, however. First, a local change in the function (e.g., zeroing out a block of points) is going to lead to a change *at every point* in the Fourier transform. This means that the Fourier transform is quite difficult to use as a

TABLE 7.1 A variety of functions of two dimensions and their Fourier transforms. This table can be used in two directions (with appropriate substitutions for u, v and x, y) because the Fourier transform of the Fourier transform of a function is the function. Observant readers may suspect that the results on infinite sums of δ functions contradict the linearity of Fourier transforms. By careful inspection of limits, it is possible to show that they do not (see, e.g., Bracewell, 1995). Observant readers may also have noted that an expression for $\mathcal{F}(\frac{\partial f}{\partial y})$ can be obtained by combining two lines of this table.

Function	Fourier transform
$g(x, y)$	$\int\int_{-\infty}^{\infty} g(x, y)e^{-i2\pi(ux+vy)}\, dx\, dy$
$\int\int_{-\infty}^{\infty} \mathcal{F}(g(x, y))(u, v)e^{i2\pi(ux+vy)}\, du\, dv$	$\mathcal{F}(g(x, y))(u, v)$
$\delta(x, y)$	1
$\frac{\partial f}{\partial x}(x, y)$	$u\mathcal{F}(f)(u, v)$
$0.5\delta(x + a, y) + 0.5\delta(x - a, y)$	$\cos 2\pi au$
$e^{-\pi(x^2+y^2)}$	$e^{-\pi(u^2+v^2)}$
$box_1(x, y)$	$\frac{\sin u}{u}\frac{\sin v}{v}$
$f(ax, by)$	$\frac{\mathcal{F}(f)(u/a, v/b)}{ab}$
$\sum_{i=-\infty}^{\infty}\sum_{j=-\infty}^{\infty}\delta(x - i, y - j)$	$\sum_{i=-\infty}^{\infty}\sum_{j=-\infty}^{\infty}\delta(u - i, v - j)$
$(f * *g)(x, y)$	$\mathcal{F}(f)\mathcal{F}(g)(u, v)$
$f(x - a, y - b)$	$e^{-i2\pi(au+bv)}\mathcal{F}(f)$
$f(x\cos\theta - y\sin\theta, x\sin\theta + y\cos\theta)$	$\mathcal{F}(f)(u\cos\theta - v\sin\theta, u\sin\theta + v\cos\theta)$

representation (e.g., it might be very difficult to tell if a pattern was present in an image just by looking at the Fourier transform). Second, the magnitude spectra of images tends to be similar. This appears to be a fact of nature, rather than something that can be proven axiomatically. As a result, the magnitude spectrum of an image is surprisingly uninformative (see Figure 7.6 for an example).

7.4 SAMPLING AND ALIASING

The crucial reason to discuss Fourier transforms is to get some insight into the difference between discrete and continuous images. In particular, it is clear that some information has been lost when we work on a discrete pixel grid, but what? A good, simple example to think about comes from an image of a checkerboard, and is given in Figure 7.7. The problem has to do with the number of samples relative to the function; we can formalize this rather precisely given a sufficiently powerful model.

Figure 7.6 The second image in each row shows the log of the magnitude spectrum for the first image in the row; the third image shows the phase spectrum scaled so that $-\pi$ is dark and π is light. The final images are obtained by swapping the magnitude spectra. Although this swap leads to substantial image noise, it doesn't substantially affect the interpretation of the image, suggesting that the phase spectrum is more important for perception than the magnitude spectrum.

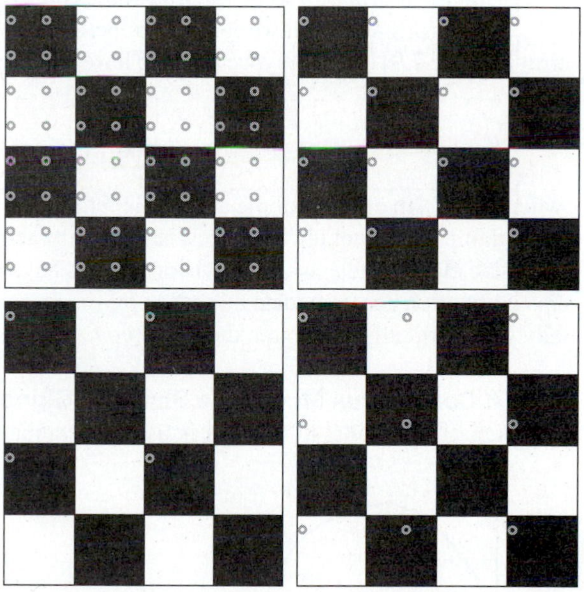

Figure 7.7 The two checkerboards on the **top** illustrate a sampling procedure that appears to be successful (whether it is or not depends on some details that we will deal with later). The grey circles represent the samples; if there are sufficient samples, then the samples represent the detail in the underlying function. The sampling procedures shown on the **bottom** is unequivocally unsuccessful; the samples suggest that there are fewer checks than there are. This illustrates two important phenomena: First, successful sampling schemes sample data often enough; second, unsuccessful sampling schemes cause high-frequency information to appear as lower frequency information.

7.4.1 Sampling

Passing from a continuous function—like the irradiance at the back of a camera system—to a collection of values on a discrete grid—like the pixel values reported by a camera—is referred to as *sampling*. We construct a model that allows us to obtain a precise notion of what is lost in sampling.

Sampling in One Dimension Sampling in one dimension takes a function and returns a discrete set of values. The most important case involves sampling on a uniform discrete grid, and we assume that the samples are defined at integer points. This means we have a process that takes some function and returns a vector of values:

$$\texttt{sample}_{1D}(f(x)) = \boldsymbol{f}.$$

We model this sampling process by assuming that the elements of this vector are the values of the function $f(x)$ at the sample points and allowing negative indices to the vector (Figure 7.8). This means that the ith component of $\boldsymbol{f}$ is $f(x_i)$.

Sampling in Two Dimensions Sampling in 2D is very similar to sampling in 1D. Although sampling can occur on nonregular grids (the best example being the human retina), we proceed on the assumption that samples are drawn at points with integer coordinates. This yields a uniform rectangular grid, which is a good model of most cameras. Our sampled images are then rectangular arrays of finite size (all values outside the grid being zero).

In the formal model, we sample a function of two dimensions, instead of one, yielding an array (Figure 7.9). This array we allow to have negative indices in both dimensions, and can then write

$$\texttt{sample}_{2D}(F(x, y)) = \mathcal{F},$$

where the i, jth element of the array $\mathcal{F}$ is $F(x_i, y_j) = F(i, j)$.

Samples are not always evenly spaced in practical systems. This is quite often due to the pervasive effect of television; television screens have an aspect ratio of 4:3 (width:height). Cameras quite often accommodate this effect by spacing sample points slightly farther apart horizontally than vertically (in jargon, they have *non-square pixels*).

A Continuous Model of a Sampled Signal We need a continuous model of a sampled signal. Generally, this model is used to evaluate integrals—in particular, taking a Fourier

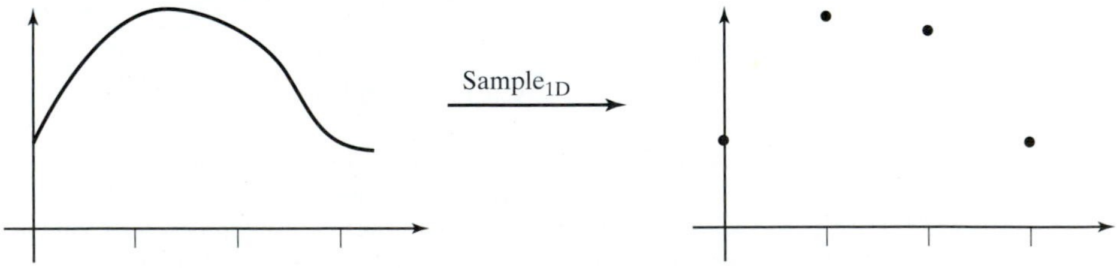

Figure 7.8 Sampling in 1D takes a function and returns a vector whose elements are values of that function at the sample points, as the top figures show. For our purposes, it is enough that the sample points be integer values of the argument. We allow the vector to be infinite dimensional and have negative as well as positive indices.

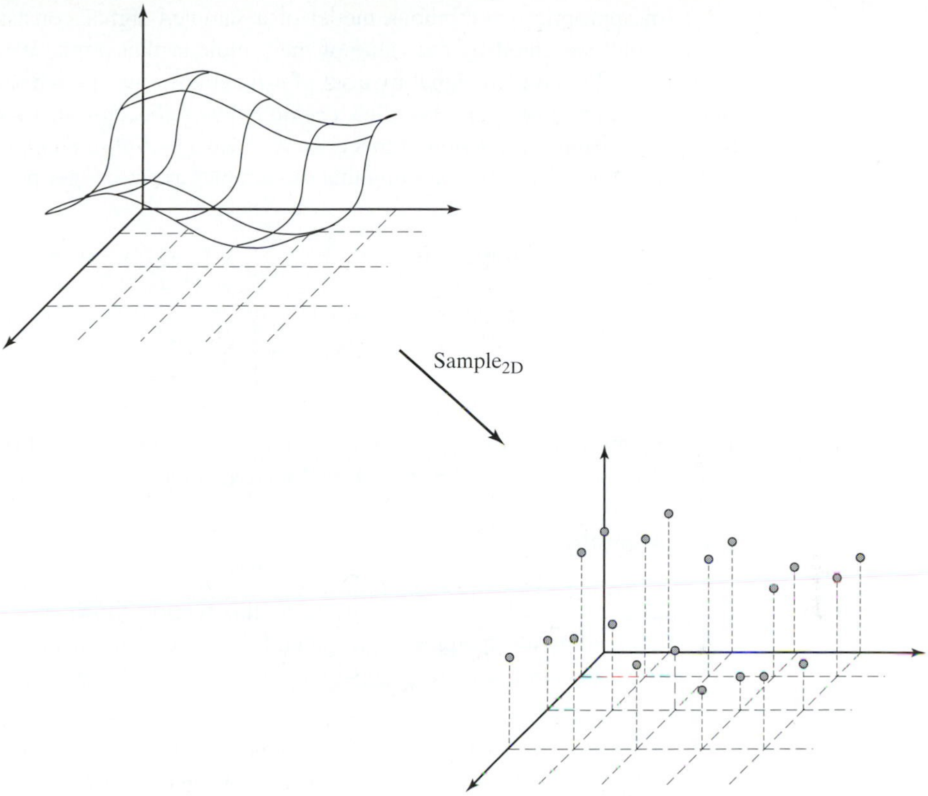

Figure 7.9 Sampling in 2D takes a function and returns an array; again, we allow the array to be infinite dimensional and to have negative as well as positive indices.

transform involves integrating the product of our model with a complex exponential. It is clear how this integral should behave—the value of the integral should be obtained by adding up values at each integer point. This means we cannot model a sampled signal as a function that is zero everywhere except at integer points (where it takes the value of the signal) because this model has a zero integral.

An appropriate continuous model of a sampled signal relies on an important property of the δ function:

$$\begin{aligned}\int_{-\infty}^{\infty} a\delta(x) f(x)\, dx &= a \lim_{\epsilon\to 0} \int_{-\infty}^{\infty} d(x;\epsilon) f(x)\, dx \\ &= a \lim_{\epsilon\to 0} \int_{-\infty}^{\infty} \frac{bar(x;\epsilon)}{\epsilon}(f(x))\, dx \\ &= a \lim_{\epsilon\to 0} \sum_{i=-\infty}^{\infty} \frac{bar(x;\epsilon)}{\epsilon}(f(i\epsilon) bar(x - i\epsilon;\epsilon))\epsilon \\ &= af(0).\end{aligned}$$

Here we have used the idea of an integral as the limit of a sum of small strips.

An appropriate continuous model of a sampled signal consists of a δ-function at each sample point weighted by the value of the sample at that point. We can obtain this model by multiplying the sampled signal by a set of δ-functions, one at each sample point. In one dimension, a function of this form is called a *comb function* (because that's what the graph looks like). In two dimensions, a function of this form is called a *bed-of-nails function* (for the same reason).

Working in 2D and assuming that the samples are at integer points, this procedure gets

$$\begin{aligned}\texttt{sample}_{2D}(f) &= \sum_{i=-\infty}^{\infty}\sum_{j=-\infty}^{\infty} f(i,j)\delta(x-i, y-j)\\ &= f(x,y)\left\{\sum_{i=-\infty}^{\infty}\sum_{j=-\infty}^{\infty}\delta(x-i,y-j)\right\}.\end{aligned}$$

This function is zero except at integer points (because the δ-function is zero except at integer points), and its integral is the sum of the function values at the integer points.

7.4.2 Aliasing

Sampling involves a loss of information. As this section shows, a signal sampled too slowly is misrepresented by the samples; high spatial frequency components of the original signal appear as low spatial frequency components in the sampled signal—an effect known as *aliasing*.

The Fourier Transform of a Sampled Signal A sampled signal is given by a product of the original signal with a bed-of-nails function. By the convolution theorem, the Fourier transform of this product is the convolution of the Fourier transforms of the two functions. This means that the Fourier transform of a sampled signal is obtained by convolving the Fourier transform of the signal with another bed-of-nails function.

Now convolving a function with a shifted δ-function merely shifts the function (see exercises). This means that the Fourier transform of the sampled signal is the sum of a collection of shifted versions of the Fourier transforms of the signal, that is,

$$\begin{aligned}\mathcal{F}(\texttt{sample}_{2D}(f(x,y))) &= \mathcal{F}\left(f(x,y)\left\{\sum_{i=-\infty}^{\infty}\sum_{j=-\infty}^{\infty}\delta(x-i,y-j)\right\}\right)\\ &= \mathcal{F}(f(x,y)) **\mathcal{F}\left(\left\{\sum_{i=-\infty}^{\infty}\sum_{j=-\infty}^{\infty}\delta(x-i,y-j)\right\}\right)\\ &= \sum_{i=-\infty}^{\infty} F(u-i, v-j),\end{aligned}$$

where we have written the Fourier transform of $f(x, y)$ as $F(u, v)$.

If the support of these shifted versions of the Fourier transform of the signal does not intersect, we can easily reconstruct the signal from the sampled version. We take the sampled signal, Fourier transform it, and cut out one copy of the Fourier transform of the signal and Fourier transform this back (Figure 7.10).

However, if the support regions *do* overlap, we are not able to reconstruct the signal because we can't determine the Fourier transform of the signal in the regions of overlap, where different copies of the Fourier transform will add. This results in a characteristic effect, usually called **aliasing**, where high spatial frequencies appear to be low spatial frequencies (see Figure 7.12

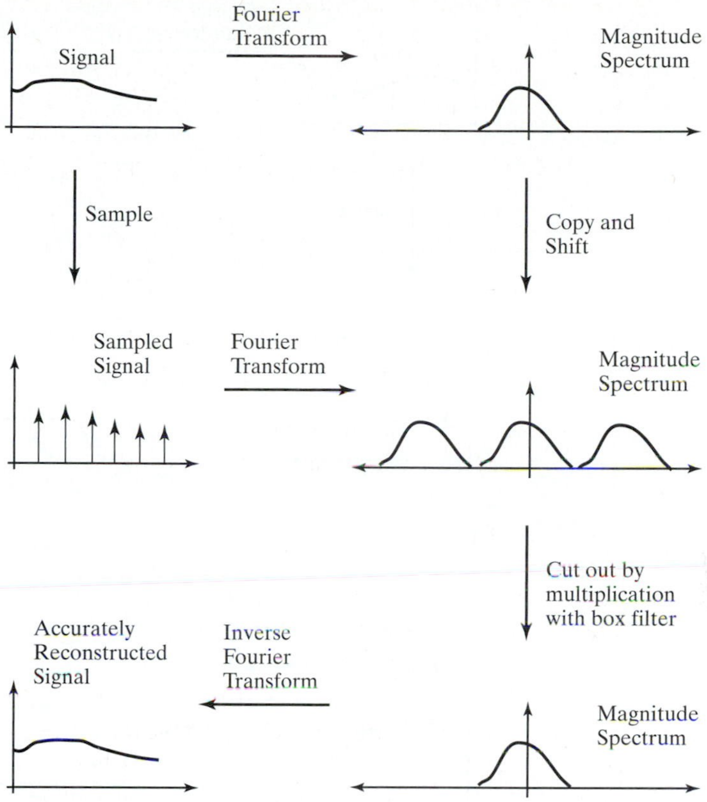

Figure 7.10 The Fourier transform of the sampled signal consists of a sum of copies of the Fourier transform of the original signal, shifted with respect to each other by the sampling frequency. Two possibilities occur. If the shifted copies do not intersect with each other (as in this case), the original signal can be reconstructed from the sampled signal (we just cut out one copy of the Fourier transform and inverse transform it). If they do intersect (as in Figure 7.11), the intersection region is added, and so we cannot obtain a separate copy of the Fourier transform, and the signal has aliased.

and exercises). Our argument also yields *Nyquist's theorem*—the sampling frequency must be at least twice the highest frequency present for a signal to be reconstructed from a sampled version.

7.4.3 Smoothing and Resampling

Nyquist's theorem means it is dangerous to shrink an image by simply taking every kth pixel (as Figure 7.12 confirms). Instead, we need to filter the image so that spatial frequencies above the new sampling frequency are removed. We could do this exactly by multiplying the image Fourier transform by a scaled 2D bar function, which would act as a low-pass filter. Equivalently, we would convolve the image with a kernel of the form $(\sin x \sin y)/(xy)$. This is a difficult and expensive (a polite way of saying *impossible*) convolution because this function has infinite support.

The most interesting case occurs when we want to halve the width and height of the image. We assume that the sampled image has no aliasing (because if it did, there would be nothing we

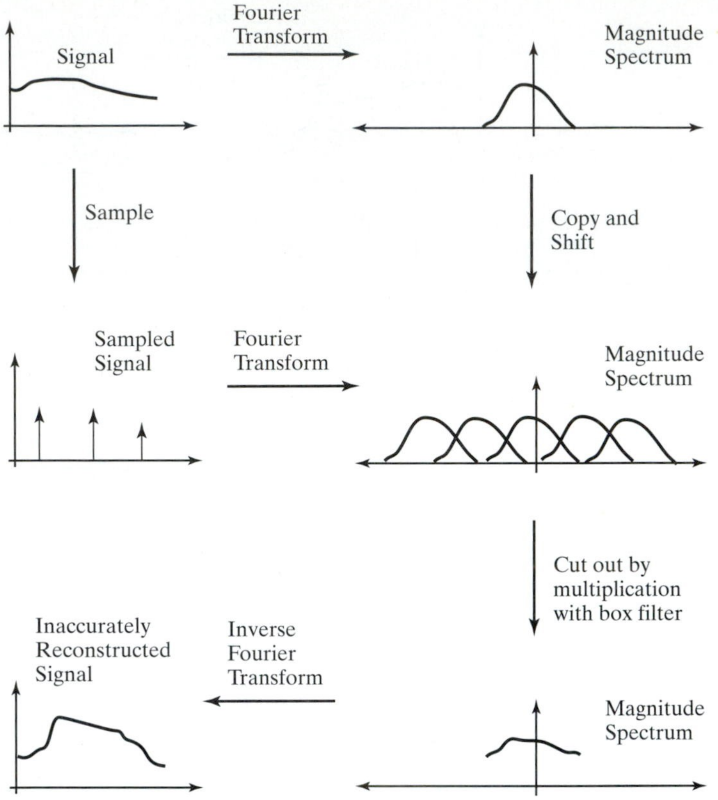

Figure 7.11 The Fourier transform of the sampled signal consists of a sum of copies of the Fourier transform of the original signal, shifted with respect to each other by the sampling frequency. Two possibilities occur. If the shifted copies do not intersect with each other (as in Figure 7.10), the original signal can be reconstructed from the sampled signal (we just cut out one copy of the Fourier transform and inverse transform it). If they do intersect (as in this figure), the intersection region is added, and so we cannot obtain a separate copy of the Fourier transform, and the signal has aliased. This also explains the tendency of high spatial frequencies to alias to lower spatial frequencies.

could do about it anyway; once an image has been sampled, any aliasing that is going to occur has happened, and there's not much we can do about it without an image model). This means that the Fourier transform of the sampled image is going to consist of a set of copies of some Fourier transform, with centers shifted to integer points in u, v space.

If we resample this signal, the copies now have centers on the half-integer points in u, v space. This means that, to avoid aliasing, we need to apply a filter that strongly reduces the content of the original Fourier transform outside the range $|u| < 1/2$, $|v| < 1/2$. Of course, if we reduce the content of the signal *inside* this range, we may lose information, too. Now the Fourier transform of a Gaussian is a Gaussian, and Gaussians die away fairly quickly. Thus, if we were to convolve the image with a Gaussian—or multiply its Fourier transform by a Gaussian, which is the same thing—we could achieve what we want.

The choice of Gaussian depends on the application; if σ is large, there is less aliasing (because the value of the kernel outside our range is very small), but information is lost because the kernel is not flat within our range; similarly, if σ is small, less information is lost within the

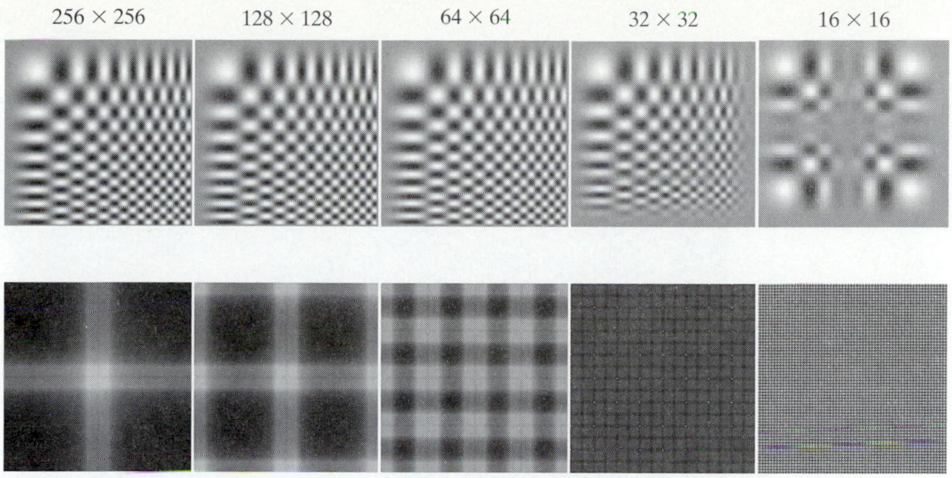

Figure 7.12 The **top row** shows sampled versions of an image of a grid obtained by multiplying two sinusoids with linearly increasing frequency—one in x and one in y. The other images in the series are obtained by resampling by factors of two without smoothing (i.e., the next is a 128x128, then a 64x64, etc., all scaled to the same size). Note the substantial aliasing; high spatial frequencies alias down to low spatial frequencies, and the smallest image is an extremely poor representation of the large image. The **bottom row** shows the magnitude of the Fourier transform of each image displayed as a log to compress the intensity scale. The constant component is at the center. Notice that the Fourier transform of a resampled image is obtained by scaling the Fourier transform of the original image and then tiling the plane. Interference between copies of the original Fourier transform means that we cannot recover its value at some points; this is the mechanism underlying aliasing.

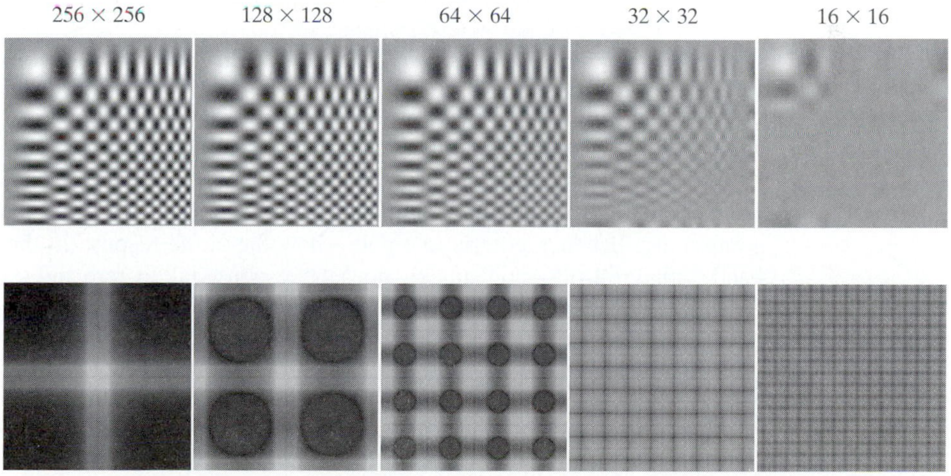

Figure 7.13 **Top:** Resampled versions of the image of Figure 7.12, again by factors of two, but this time each image is smoothed with a Gaussian of σ one pixel before resampling. This filter is a low-pass filter, and so suppresses high spatial frequency components, reducing aliasing. **Bottom:** The effect of the low-pass filter is easily seen in these log-magnitude images; the low-pass filter suppresses the high spatial frequency components so that components interfere less to reduce aliasing.

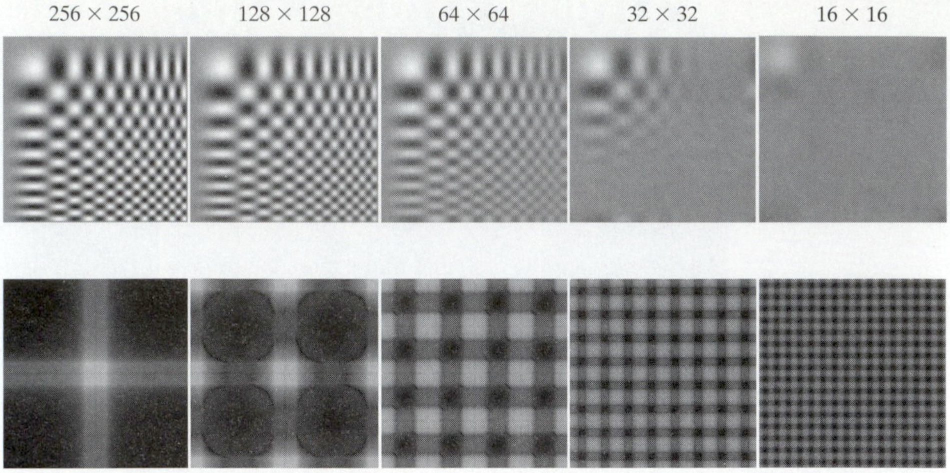

Figure 7.14 **Top:** Resampled versions of the image of Figure 7.12, again by factors of two, but this time each image is smoothed with a Gaussian of σ two pixels before resampling. This filter suppresses high spatial frequency components more aggressively than that of Figure 7.13. **Bottom:** The effect of the low-pass filter is easily seen in these log-magnitude images; the low-pass filter suppresses the high spatial frequency components so that components interfere less, to reduce aliasing.

range, but aliasing can be more substantial. Figures 7.13 and 7.14 illustrate the effects of different choices of σ.

We have been using a Gaussian as a low-pass filter because its response at high spatial frequencies is low and at low spatial frequencies is high. In fact, the Gaussian is not a particularly good low-pass filter. What one wants is a filter whose response is pretty close to constant for some range of low spatial frequencies— the pass band—and whose response is also pretty close to constant—and zero—for higher spatial frequencies—the stop band. It is possible to design low-pass filters that are significantly better than Gaussians. The design process involves detailed compromise between criteria of ripple—how flat is the respnse in the pass band and the stop band?—and roll-off—how quickly does the response fall to zero and stay there? The basic steps for resampling an image are given in Algorithm 7.1.

Algorithm 7.1: Subsampling an Image by a Factor of Two

Apply a low-pass filter to the original image
  (a Gaussian with a σ of between one
  and two pixels is usually an acceptable choice).
Create a new image whose dimensions on edge are half
  those of the old image
Set the value of the i, jth pixel of the new image to the value
  of the $2i$, $2j$th pixel of the filtered image

7.5 FILTERS AS TEMPLATES

It turns out that filters offer a natural mechanism for finding simple patterns because filters respond most strongly to pattern elements that look like the filter. For example, smoothed derivative filters are intended to give a strong response at a point where the derivative is large. At these points, the kernel of the filter looks like the effect it is intended to detect. The x-derivative filters look like a vertical light blob next to a vertical dark blob (an arrangement where there is a large x-derivative), and so on.

It is generally the case that filters intended to give a strong response to a pattern look like that pattern (Figure 7.15). This is a simple geometric result.

7.5.1 Convolution as a Dot Product

Recall from Section 7.1.1 that, for $\mathcal{G}$ the kernel of some linear filter, the response of this filter to an image $\mathcal{H}$ is given by:

$$R_{ij} = \sum_{u,v} G_{i-u,j-v} H_{uv}.$$

Now consider the response of a filter at the point where i and j are zero. This is

$$R = \sum_{u,v} G_{-u,-v} H_{u,v}.$$

This response is obtained by associating image elements with filter kernel elements, multiplying the associated elements, and summing. We could scan the image into a vector, and the filter kernel into another vector, in such a way that associated elements are in the same component. By inserting zeros as needed, we can ensure that these two vectors have the same dimension. Once this is done, the process of multiplying associated elements and summing is precisely the same as taking a dot product.

This is a powerful analogy because this dot product, like any other, achieves its largest value when the vector representing the image is parallel to the vector representing the filter

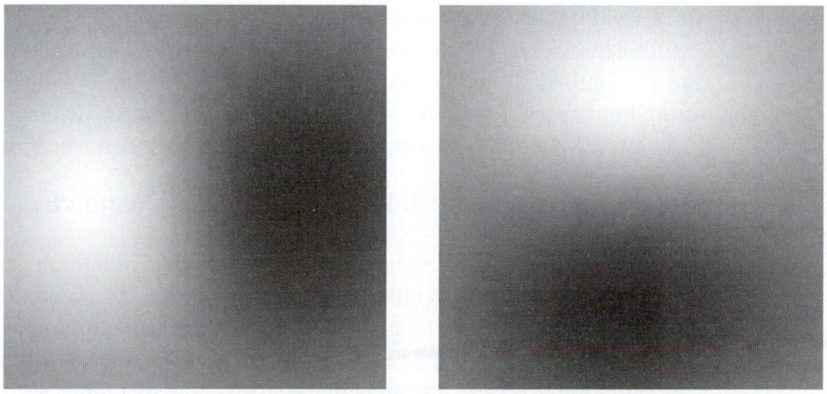

Figure 7.15 Filter kernels look like the effects they are intended to detect. On the **left**, a smoothed derivative of Gaussian filter that looks for large changes in the x-direction (such as a dark blob next to a light blob); on the **right**, a smoothed derivative of Gaussian filter that looks for large changes in the y-direction.

kernel. This means that a filter responds most strongly when it encounters an image pattern that looks like the filter. The response of a filter gets stronger as a region gets brighter, too.

Now consider the response of the image to a filter at some other point. Nothing significant about our model has changed. Again, we can scan the image into one vector and the filter kernel into another vector, such that associated elements lie in the same components. Again, the result of applying this filter is a dot product. There are two useful ways to think about this dot product.

7.5.2 Changing Basis

We can think of convolution as a dot product between the image and *a different vector at each point* (because we have moved the filter kernel to lie over some other point in the image). The new vector is obtained by rearranging the old one so that the elements lie in the right components to make the sum work out. This means that, by convolving an image with a filter, we are representing the image on a new *basis* of the vector space of images—the basis given by the different shifted versions of the filter. The original basis elements were vectors with a zero in all slots except one. The new basis elements are shifted versions of a single pattern.

For many of the kernels discussed, we expect that this process will *lose* information—for the same reason that smoothing suppresses noise—so that the coefficients on this basis are redundant. While the coefficients may be redundant, they expose image structure in a useful way. This basis transformation is valuable in texture analysis. Typically, we choose a basis that consists of small, useful pattern components. Large values of the basis coefficients suggest that a pattern component is present, and texture can be represented by representing the relationships between these pattern components, usually with some form of probability model.

7.6 TECHNIQUE: NORMALIZED CORRELATION AND FINDING PATTERNS

We can think of convolution as comparing a filter with a patch of image centered at the point whose response we are looking at. In this view, the image neighborhood corresponding to the filter kernel is scanned into a vector that is compared with the filter kernel. By itself, this dot product is a poor way to find features because the value may be large simply because the image region is bright. By analogy with vectors, we are interested in the cosine of the angle between the filter vector and the image neighborhood vector; this suggests computing the root sum of squares of the relevant image region (the image elements that would lie under the filter kernel) and dividing the response by that value.

This yields a value that is large and positive when the image region looks like the filter kernel, and small and negative when the image region looks like a contrast-reversed version of the filter kernel. This value could be squared if contrast reversal doesn't matter. This is a cheap and effective method for finding patterns, often called *normalized correlation*.

7.6.1 Controlling the Television by Finding Hands by Normalized Correlation

It would be nice to have systems that could respond to human gestures. For example, you might wave at the light to get the room illuminated, point at the air conditioning to get the room temperature changed, or make an appropriate gesture at an annoying politician on television and get a change in channel. In typical consumer applications, there are quite strict limits to the amount of computation available, meaning that it is essential that the gesture recognition system be simple. However, such systems are usually quite limited in what they need to do, too.

Controlling the Television Typically, a user interface is in some state—perhaps a menu is displayed—and an event occurs—perhaps a button is pressed on a remote control. This event causes the interface to change state—a new menu item is highlighted, say—and the whole process continues. In some states, some events cause the system to perform some action—the channel might change. All this means that a state machine is a natural model for a user interface.

One way for vision to fit into this model is to provide events. This is good because there are generally few different kinds of event, and we know what kinds of event the system should care about in any particular state. As a result, the vision system needs only to determine whether either nothing or one of a small number of known kinds of event has occurred. It is quite often possible to build systems that meet these constraints.

A relatively small set of events is required to simulate a remote control; one needs events that look like button presses (e.g., to turn the television on or off), and events that look like pointer motion (e.g., to increase the volume; it is possible to do this with buttons, too). With these events, the television can be turned on, and an on-screen menu system navigated.

Finding Hands Freeman, Anderson and et al. (1998) produced an interface where an open hand turns the television on. This can be robust because all the system needs to do is determine whether there is a hand in view. Furthermore, the user will cooperate by holding their hand up and open. Because the user is expected to be a fairly constant distance from the camera—so the size of the hand is roughly known, and there is no need to search over scales—and in front of the television, the image region that needs to be searched to determine whether there is a hand is quite small.

The hand is held up in a fairly standard configuration and orientation to turn the television set on, and it usually appears at about the same distance from the television (so we know what it looks like). This means that a normalized correlation score is sufficient to find the hand. Any points in the correlation image where the score is high enough correspond to hands. This approach can be used to control volume and so on, as well as turn the television on and off. To do so, we need some notion of where the hand is going—to one side turns the volume up, to the other turns it down—and this can be obtained by comparing the position in the previous frame with that in the current frame. The system displays an iconic representation of its interpretation of hand position so the user has some feedback as to what the system is doing (Figure 7.16). Notice that an attractive feature of this approach is that it could be self-calibrating. In this approach, when you install your television set, you sit in front of it and show it your hand a few times to allow it to get an estimate of the scale at which the hand appears.

7.7 TECHNIQUE: SCALE AND IMAGE PYRAMIDS

Images look quite different at different scales. For example, the zebra's muzzle in Figure 7.17 can be described in terms of individual hairs—which might be coded in terms of the response of oriented filters that operate at a scale of a small number of pixels—or in terms of the stripes on the zebra. In the case of the zebra, we would not want to apply large filters to find the stripes. This is because these filters are inclined to spurious precision—we don't wish to represent the disposition of each hair on the stripe—inconvenient to build, and slow to apply. A more practical approach than applying large filters is to apply smaller filters to smoothed and resampled versions of the image.

7.7.1 The Gaussian Pyramid

An *image pyramid* is a collection of representations of an image. The name comes from a visual analogy. Typically, each layer of the pyramid is half the width and half the height of the previous layer; if we were to stack the layers on top of each other, a pyramid would result. In a **Gaussian**

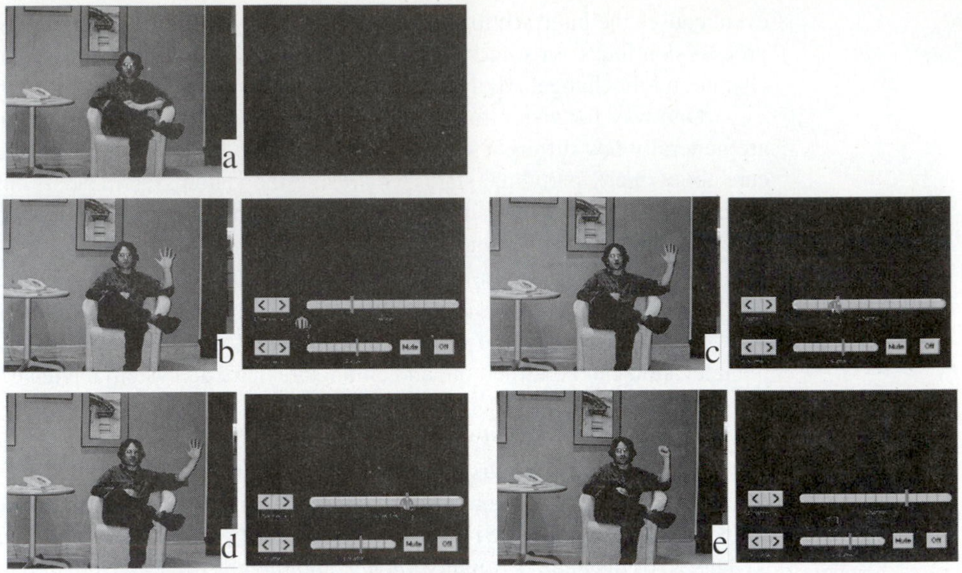

Figure 7.16 Examples of Freeman *et al.*'s system controlling a television set. Each state is illustrated with what the television sees on the **left** and what the user sees on the **right**. In **(a)**, the television is asleep, but a process is watching the user. An open hand causes the television to come on and show its user interface panel **(b)**. Focus on the panel tracks the movement of the user's open hand in **(c)**, and the user can change channel by using this tracking to move an icon on the screen in **(d)**. Finally, the user displays a closed hand in **(e)** to turn off the set. *Reprinted from "Computer Vision for Interactive Computer Graphics," by W.Freeman et al., IEEE Computer Graphics and Applications, 1998 © 1998 IEEE*

Algorithm 7.2: Forming a Gaussian Pyramid

Set the finest scale layer to the image
For each layer, going from next to finest to coarsest
  Obtain this layer by smoothing the next finest
  layer with a Gaussian, and then subsampling it
end

pyramid, each layer is smoothed by a symmetric Gaussian kernel and resampled to get the next layer (Figure 7.17). These pyramids are most convenient if the image dimensions are a power of two or a multiple of a power of two. The smallest image is the most heavily smoothed; the layers are often referred to as *coarse scale* versions of the image.

With a little notation, we can write simple expressions for the layers of a Gaussian pyramid. The operator $S^{\downarrow}$ downsamples an image; in particular, the j, kth element of $S^{\downarrow}(\mathcal{I})$ is the $2j, 2k$th element of $\mathcal{I}$. The nth level of a pyramid $P(\mathcal{I})$ is denoted $P(\mathcal{I})_n$. With this notation, we have

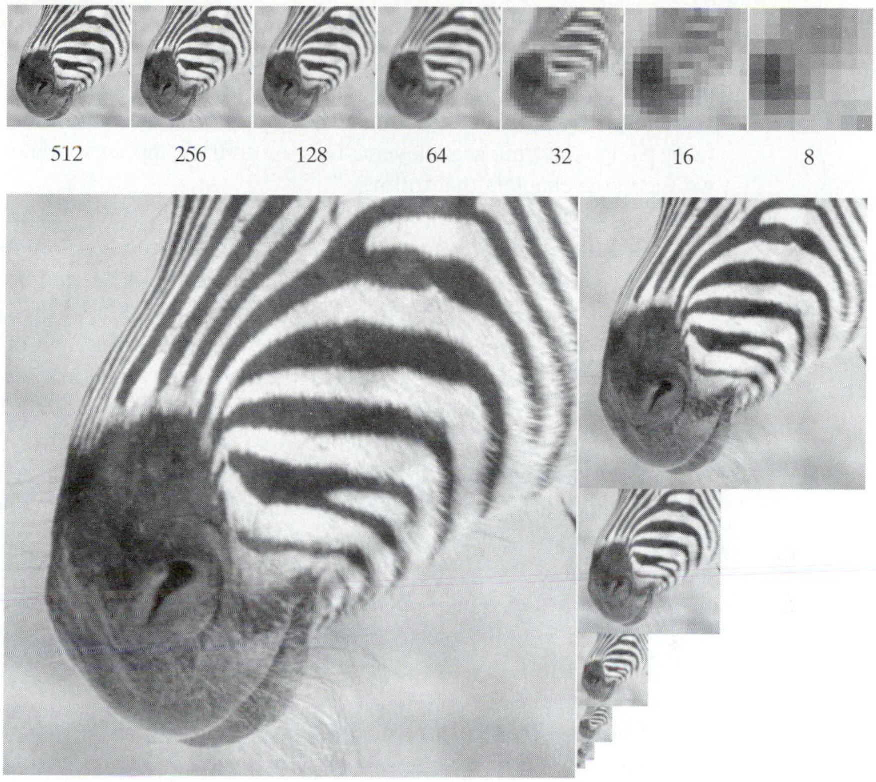

Figure 7.17 A Gaussian pyramid of images running from 512x512 to 8x8. On the top row, we have shown each image at the same size (so that some have bigger pixels than others), and the lower part of the figure shows the images to scale. Notice that if we convolve each image with a fixed size filter, it responds to quite different phenomena. An 8x8 pixel block at the finest scale might contain a few hairs; at a coarser scale, it might contain an entire stripe; and at the coarsest scale, it contains the animal's muzzle.

$$\begin{aligned} P_{\text{Gaussian}}(\mathcal{I})_{n+1} &= S^{\downarrow}(G_{\sigma} ** P_{\text{Gaussian}}(\mathcal{I})_n) \\ &= S^{\downarrow} G_{\sigma}(P_{\text{Gaussian}}(\mathcal{I})_n) \end{aligned}$$

(where we have written G_{σ} for the linear operator that takes an image to the convolution of that image with a Gaussian). The finest scale layer is the original image:

$$P_{\text{Gaussian}}(\mathcal{I})_1 = \mathcal{I}.$$

7.7.2 Applications of Scaled Representations

Gaussian pyramids are useful because they make it possible to extract representations of different types of structure in an image. There are three standard applications.

Search over Scale Numerous objects can be represented as small image patterns. A standard example is a frontal view of a face. Typically, at low resolution, frontal views of faces have a quite distinctive pattern: The eyes form dark pools, under a dark bar (the eyebrows), separated by a lighter bar (specular reflections from the nose), and above a dark bar (the mouth). There are various methods for finding faces that exploit these properties (see chapter 22). These methods all assume that the face lies in a small range of scales. All other faces are found by searching a pyramid. To find bigger faces, we look at coarser scale layers, and to find smaller faces we look at finer scale layers. This useful trick applies to many different kinds of feature, as we see in the chapters that follow.

Spatial Search One application is spatial search, a common theme in computer vision. Typically, we have a point in one image and are trying to find a point in a second image that corresponds to it. This problem occurs in stereopsis—where the point has moved because the two images are obtained from different viewing positions—and in motion analysis—where the image point has moved either because the camera moved or because it is on a moving object.

Searching for a match in the original pairs of images is inefficient because we may have to wade through a great deal of detail. A better approach, which is now pretty much universal, is to look for a match in a heavily smoothed and resampled image and then refine that match by looking at increasingly detailed versions of the image. For example, we might reduce 1024×1024 images down to 4×4 versions, match those, and then look at 8×8 versions (because we know a rough match, it is easy to refine it); we then look at 16×16 versions, and so on, all the way up to 1024×1024. This gives an extremely efficient search because a step of a single pixel in the 4×4 version is equivalent to a step of 256 pixels in the 1024×1024 version. This strategy is known as *coarse-to-fine matching*.

Feature Tracking Most features found at coarse levels of smoothing are associated with large, high-contrast image events because for a feature to be marked at a coarse scale a large pool of pixels need to agree that it is there. Typically, finding coarse scale phenomena misestimates both the size and location of a feature. For example, a single pixel error in a coarse-scale image represents a multiple pixel error in a fine-scale image.

At fine scales, there are many features, some of which are associated with smaller, low-contrast events. One strategy for improving a set of features obtained at a fine scale is to track features across scales to a coarser scale and accept only the fine scale features that have identifiable parents at a coarser scale. This strategy, known as *feature tracking* in principle, can suppress features resulting from textured regions (often referred to as noise) and features resulting from real noise.

7.8 NOTES

We don't claim to be exhaustive in our treatment of linear systems, but it wouldn't be possible to read the literature on filters in vision without a grasp of the ideas in this chapter. We have given a fairly straightforward account here; more details on these topics can be found in the excellent books by Bracewell (1995), (2000).

Real Imaging Systems versus Shift Invariant Linear Systems

Imaging systems are only approximately linear. Film is not linear—it does not respond to weak stimuli, and it saturates for bright stimuli—but one can usually get away with a linear model within a reasonable range. CCD cameras are linear within a working range. They give a small, but non zero response to a zero input as a result of thermal noise (which is why astronomers cool their cameras) and they saturate for very bright stimuli. CCD cameras often contain electronics that transforms their output to make them behave more like film because consumers are used to film. Shift invariance is approximate as well because lenses tend to distort responses near the image boundary. Some lenses—fish-eye lenses are a good example—are not shift invariant.

Scale

There is a large body of work on scale space and scaled representations. The origins appear to lie with Witkin (1983) and the idea was developed by Koenderink and van Doorn (1986). Since then, a huge literature has sprung up (one might start with ter Haar Romeny, Florack, Koenderink and Viergever, 1997 or Nielsen, Johansen, Olsen and Weickert, 1999). We have given only the briefest picture here because the analysis tends to be quite tricky. The usefulness of the techniques is currently hotly debated, too.

Anisotropic Scaling

One important difficulty with scale space models is that the symmetric Gaussian smoothing process tends to blur out edges rather too aggressively for comfort. For example, if we have two trees near one another on a skyline, the large-scale blobs corresponding to each tree may start merging before all the small-scale blobs have finished. This suggests that we should smooth differently at edge points than at other points. For example, we might make an estimate of the magnitude and orientation of the gradient: For large gradients, we would then use an oriented smoothing operator that smoothed aggressively perpendicular to the gradient and little along the gradient; for small gradients, we might use a symmetric smoothing operator. This idea used to be known as *edge-preserving smoothing*.

In the modern, more formal version, due to Perona and Malik (1990*a,b*), we notice the scale space representation family is a solution to the *diffusion equation*

$$\begin{aligned}\frac{\partial \Phi}{\partial \sigma} &= \frac{\partial^2 \Phi}{\partial x^2} + \frac{\partial^2 \Phi}{\partial y^2} \\ &= \nabla^2 \Phi,\end{aligned}$$

with the initial condition

$$\Phi(x, y, 0) = \mathcal{I}(x, y)$$

If this equation is modified to have the form

$$\begin{aligned}\frac{\partial \Phi}{\partial \sigma} &= \nabla \cdot (c(x, y, \sigma)\nabla \Phi) \\ &= c(x, y, \sigma)\nabla^2 \Phi + (\nabla c(x, y, \sigma)) \cdot (\nabla \Phi)\end{aligned}$$

with the same initial condition, then if $c(x, y, \sigma) = 1$, we have the diffusion equation we started with, and if $c(x, y, \sigma) = 0$ there is no smoothing. We assume that c does not depend on σ. If we knew where the edges were in the image, we could construct a mask that consisted of regions where $c(x, y) = 1$, isolated by patches along the edges where $c(x, y) = 0$; in this case, a solution would smooth *inside* each separate region, but not over the edge. Although we do not know where the edges are — the exercise would be empty if we did—we can obtain reasonable choices of $c(x, y)$ from the magnitude of the image gradient. If the gradient is large, then c should be small and vice versa. There is a substantial literature dealing with this approach; a good place to start is ter Haar Romeny (1994).

ASSIGNMENTS

PROBLEMS

7.1. Show that forming unweighted local averages, which yields an operation of the form

$$\mathcal{R}_{ij} = \frac{1}{(2k+1)^2} \sum_{u=i-k}^{u=i+k} \sum_{v=j-k}^{v=j+k} \mathcal{F}_{uv}$$

is a convolution. What is the kernel of this convolution?

7.2. Write $\mathcal{E}_0$ for an image that consists of all zeros with a single one at the center. Show that convolving this image with the kernel

$$H_{ij} = \frac{1}{2\pi\sigma^2} \exp\left(-\frac{((i-k-1)^2 + (j-k-1)^2)}{2\sigma^2} \right)$$

(which is a discretised Gaussian) yields a circularly symmetric fuzzy blob.

7.3. Show that convolving an image with a discrete, separable 2D filter kernel is equivalent to convolving with two 1D filter kernels. Estimate the number of operations saved for an NxN image and a $2k + 1 \times 2k + 1$ kernel.

7.4. Show that convolving a function with a δ function simply reproduces the original function. Now show that convolving a function with a shifted δ function shifts the function.

7.5. We said that convolving the image with a kernel of the form $(\sin x \sin y)/(xy)$ is impossible because this function has infinite support. Why would it be impossible to Fourier transform the image, multiply the Fourier transform by a box function, and then inverse-Fourier transform the result? Hint: Think support.

7.6. Aliasing takes high spatial frequencies to low spatial frequencies. Explain why the following effects occur:

(a) In old cowboy films that show wagons moving, the wheel often seems to be stationary or moving in the wrong direction (i.e., the wagon moves from left to right and the wheel seems to be turning counterclockwise).

(b) White shirts with thin dark pinstripes often generate a shimmering array of colors on television.

(c) In ray-traced pictures, soft shadows generated by area sources look blocky.

Programming Assignments

7.7. One way to obtain a Gaussian kernel is to convolve a constant kernel with itself many times. Compare this strategy with evaluating a Gaussian kernel.

(a) How many repeated convolutions do you need to get a reasonable approximation? (You need to establish what a reasonable approximation is; you might plot the quality of the approximation against the number of repeated convolutions).

(b) Are there any benefits that can be obtained like this? (Hint: Not every computer comes with an FPU.)

7.8. Write a program that produces a Gaussian pyramid from an image.

7.9. A sampled Gaussian kernel must alias because the kernel contains components at arbitrarily high spatial frequencies. Assume that the kernel is sampled on an infinite grid. As the standard deviation gets smaller, the aliased energy must increase. Plot the energy that aliases against the standard deviation of the Gaussian kernel in pixels. Now assume that the Gaussian kernel is given on a 7x7 grid. If the aliased energy must be of the same order of magnitude as the error due to truncating the Gaussian, what is the smallest standard deviation that can be expressed on this grid?

8

Edge Detection

Sharp changes in image brightness are interesting for many reasons. First, object boundaries often generate sharp changes in brightness—a light object may lie on a dark background or a dark object may lie on a light background. Second, reflectance changes often generate sharp changes in brightness, which can have quite distinctive patterns—zebras have stripes and leopards have spots. Cast shadows can also generate sharp changes in brightness. Finally, sharp changes in surface orientation are often associated with sharp changes in image brightness.

Points in the image where brightness changes particularly sharply are often called *edges* or **edge points**. We should like edge points to be associated with the boundaries of objects and other kinds of meaningful changes. It is hard to define precisely the changes we would like to mark—is the region of a pastoral scene where the leaves give way to the sky the boundary of an object? Typically, it is hard to tell a semantically meaningful edge from a nuisance edge, and to do so requires a great deal of high-level information. Nonetheless, experience building vision systems suggests that interesting things often happen at an edge in an image and it is worth knowing where the edges are.

8.1 NOISE

A primary problem in edge detection is image noise. This is because edge detectors are constructed to respond strongly to sharp changes; but one way to get sharp changes in an image is to add noise to the pixels (because the noise values at each pixel are typically uncorrelated, meaning they can be very different). As Section 7.3 indicated, noise makes finite difference estimates of image derivatives unusable. We use this observation as an impetus to study image noise in general.

The term *noise* usually means image measurements from which we do not know how to extract information or from which we do not care to extract information; all the rest is *signal*. It is wrong to believe that noise does not contain information—for example, we should be able to extract some estimate of the camera temperature by taking pictures in a dark room with the lens cap on. Furthermore, since we cannot say anything meaningful about noise without a noise model, it is wrong to say that noise is not modeled. Noise is everything we don't wish to use, and that's all there is to it.

8.1.1 Additive Stationary Gaussian Noise

In the *additive stationary Gaussian noise* model, each pixel has added to it a value chosen independently from the same Gaussian probability distribution. Almost always the mean of this distribution is zero. The standard deviation is a parameter of the model. The model is intended to describe thermal noise in cameras. It is illustrated in Figure 8.1.

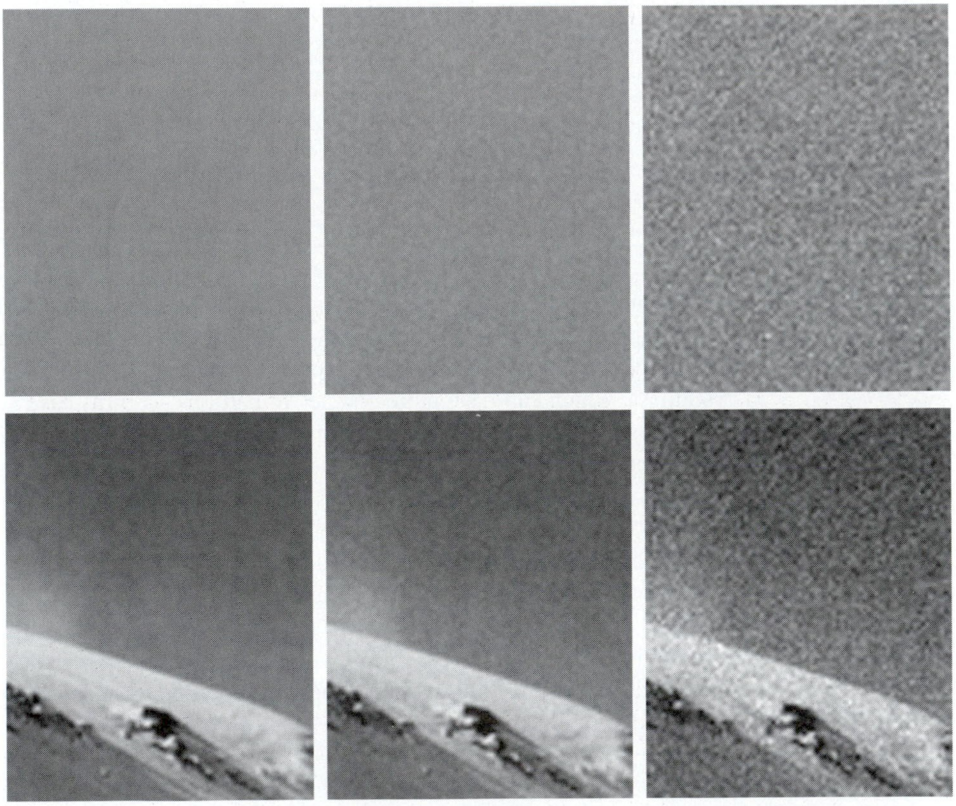

Figure 8.1 The top row shows three realizations of a stationary additive Gaussian noise process. We have added half the range of brightnesses to these images to show both negative and positive values of noise. From left to right, the noise has standard deviation 1/256, 4/256, and 16/256 of the full range of brightness, respectively. This corresponds roughly to bits zero, two, and five of a camera that has an output range of eight bits per pixel. The lower row shows this noise added to an image. In each case, values below zero or above the full range have been adjusted to zero or the maximum value accordingly.

Linear Filter Response to Additive Gaussian Noise Assume we have a discrete linear filter whose kernel is $\mathcal{G}$ and we apply it to a noise image $\mathcal{N}$ consisting of stationary additive Gaussian noise with mean μ and standard deviation σ. The response of the filter at some point i, j will be

$$R(\mathcal{N})_{i,j} = \sum_{u,v} G_{i-u,j-v} N_{u,v}.$$

Because the noise is stationary, the expectations that we compute do not depend on the point, and we assume that i and j are zero and dispense with the subscript. Assume the kernel has finite support so that only some subset of the noise variables contributes to the expectation; write this subset as $n_{0,0}, \ldots, n_{r,s}$. The expected value of this response must be

$$\begin{aligned} \mathrm{E}[R(\mathcal{N})] &= \int_{-\infty}^{\infty} \{R(\mathcal{N})\} p(N_{0,0}, \ldots, N_{r,s}) dN_{0,0} \ldots dN_{r,s} \\ &= \sum_{u,v} G_{-u,-v} \left\{ \int_{-\infty}^{\infty} N_{u,v} p(N_{u,v}) dN_{u,v} \right\}, \end{aligned}$$

where we have done some aggressive moving around of variables and integrated out all the variables that do not appear in each expression in the sum. Since all the $N_{u,v}$ are independent and identically distributed Gaussian random variables with mean μ, we have

$$\mathrm{E}[R(\mathcal{N})] = \mu \sum_{u,v} G_{i-u,j-v}.$$

The variance of the noise response is obtained as easily. We want to determine

$$\mathrm{E}[\{R(\mathcal{N})_{i,j} - \mathrm{E}[R(\mathcal{N})_{i,j}]\}^2],$$

and this is the same as

$$\int \{R(\mathcal{N})_{i,j} - \mathrm{E}[R(\mathcal{N})_{i,j}]\}^2 p(N_{0,0}, \ldots, N_{r,s}) dN_{0,0} \ldots dN_{r,s},$$

which expands to

$$\int \left\{ \sum_{u,v} G_{-u,-v}(N_{u,v} - \mu) \right\}^2 p(N_{0,0}, \ldots, N_{r,s}) dN_{0,0} \ldots dN_{r,s},$$

This expression expands into a sum of two kinds of integral. Terms of the form

$$\int G^2_{-u,-v}(N_{u,v} - \mu)^2 p(N_{0,0}, \ldots, N_{r,s}) dN_{0,0} \ldots dN_{r,s}$$

(for some u, v) can be integrated easily because each $N_{u,v}$ is independent; the integral is $\sigma^2 G^2_{-u,-v}$, where σ is the standard deviation of the noise. Terms of the form

$$\int G_{-u,-v} G_{-a,-b}(N_{u,v} - \mu)(N_{a,b} - \mu) p(N_{0,0}, \ldots, N_{r,s}) dN_{0,0} \ldots dN_{r,s}$$

(for some u, v and a, b) integrate to zero again because each noise term is independent. We now have

$$\mathrm{E}\left[\{R(\mathcal{N})_{i,j} - \mathrm{E}[R(\mathcal{N})_{i,j}]\}^2\right] = \sigma^2 \sum G^2_{u,v}$$

Difficulties with the Additive Stationary Gaussian Noise Model Taken literally, the additive stationary Gaussian noise model is a poor model of image noise. First, the model allows positive (and, more alarmingly, *negative!*) pixel values of arbitrary magnitude. With appropriate choices of standard deviation for typical current cameras operating indoors or in daylight, this doesn't present much of a problem because these pixel values are extremely unlikely to occur in practice. In rendering noise images, the problematic pixels that do occur are fixed at zero or full output, respectively.

Second, noise values are completely independent, so this model does not capture the possibility of groups of pixels that have correlated responses perhaps because of the design of the camera electronics or because of hot spots in the camera integrated circuit. This problem is harder to deal with, because noise models that do model this effect tend to be difficult to deal with analytically. Finally, this model does not describe *dead pixels* (pixels that consistently report no incoming light or are consistently saturated) terribly well. If the standard deviation is quite large and we threshold pixel values, then dead pixels will occur, but the standard deviation may be too large to model the rest of the image well. A crucial advantage of additive Gaussian noise is that it is easy to estimate the response of filters to this noise model. In turn, this gives us some idea of how effective the filter is at responding to signal and ignoring noise.

8.1.2 Why Finite Differences Respond to Noise

Our discussion of the response of linear filters to additive stationary Gaussian noise offers some insight into the noise behavior of finite differences. Assume we have an image of stationary Gaussian noise of zero mean, and consider the variance of the response to a finite difference filter that estimates derivatives of increasing order. We use the kernel

$$\begin{matrix} 0 & 0 \\ 1 & -1 \\ 0 & 0 \end{matrix}$$

to estimate the first derivative. Now a second derivative is simply a first derivative applied to a first derivative, so the kernel is

$$\begin{matrix} 0 & 0 & 0 \\ 1 & -2 & 1 \\ 0 & 0 & 0 \end{matrix}\ .$$

With a little thought, you can convince yourself that, under this scheme, the kernel coefficients of a kth derivative come from the $k+1$th row of Pascal's triangle with appropriate flips of sign. For each of these derivative filters, the mean response to Gaussian noise is zero, but the variance of this response goes up sharply; for the kth derivative, it is the sum of squares of the $k+1$th row of Pascal's triangle times the standard deviation. Figure 8.2 illustrates this result.

There is an alternative explanation. From Table 7.1, differentiating a function is the same as multiplying its Fourier transform by a frequency variable; this means that the high spatial frequency components are heavily emphasized at the expense of the low-frequency components. This is intuitively plausible—differentiating a function must set the constant component to zero, and the amplitude of the derivative of a sinusoid goes up with its frequency. Furthermore, this property is the reason we are interested in derivatives; we are discussing the derivative precisely because fast changes (which generate high spatial frequencies) have large derivatives.

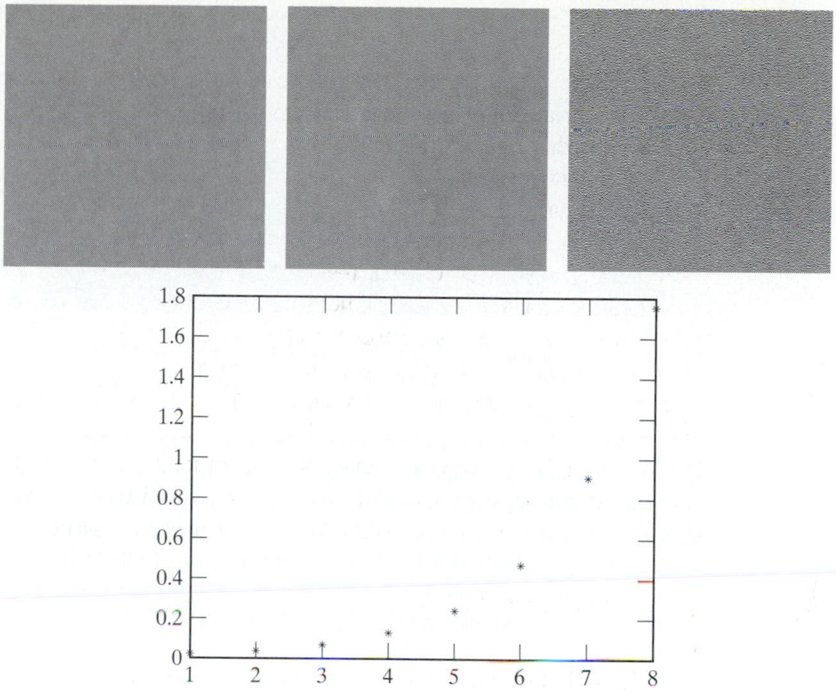

Figure 8.2 Finite differences can accentuate additive Gaussian noise substantially following the argument in Section 8.1.2. On the **top left** an image of zero mean Gaussian noise with standard deviation 4/256 of the full range. The **top center** figure shows a finite difference estimate of the third derivative in the x direction, and the **top right** shows the sixth derivative in the x direction. In each case, the image has been centered by adding half the full range to show both positive and negative deviations. The images are shown using the same gray-level scale; in the case of the sixth derivative, some values exceed the range of this scale. The graph on the **bottom** shows the standard deviations of these noise images for the first eight derivatives (estimated using the argument based around Pascal's triangle).

8.2 ESTIMATING DERIVATIVES

As Figure 7.4 indicates, simple finite difference filters tend to give strong responses to noise so that applying two finite difference filters (one in each direction) is a poor way to estimate a gradient. The way to deal with this problem is to smooth the image and then differentiate it (we could also smooth the derivative). In practice, the image is almost always smoothed with a Gaussian filter—in fact, the finite difference operator is smoothed. We discuss this practice first, and then for those who want more information, we discuss why smoothing helps and why a Gaussian is a good choice of smoothing filter.

8.2.1 Derivative of Gaussian Filters

Smoothing an image and then differentiating it is the same as convolving it with the derivative of a smoothing kernel. This fact is most easily seen by thinking about continuous convolution.

First, differentiation is linear and shift invariant. This means that there is some kernel—we dodge the question of what it looks like—that differentiates. That is, given a function $I(x, y)$,

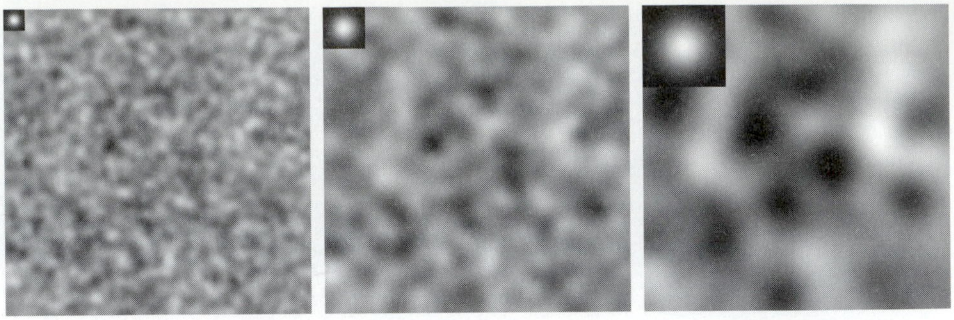

Figure 8.3 Smoothing stationary additive Gaussian noise results in signals where pixel values tend to be increasingly similar to the value of neighboring pixels. This occurs at about the scale of the filter kernel because the filter kernel causes the correlations. The figures show noise smoothed with increasingly large Gaussian smoothing kernels. Gray pixels have zero value, darker values are negative, and brighter values are positive. The kernels are shown in the top right-hand corners of the figures to indicate the spatial scale of the kernel (we have scaled the brightness of the kernels, which are Gaussians, so that the center pixel is white and the boundary pixels are black). Smoothed noise tends to look like natural texture as the figures indicate.

$$\frac{\partial I}{\partial x} = K_{(\partial/\partial x)} * * I.$$

Now we want the derivative of a smoothed function. We write the convolution kernel for the smoothing as S. Recalling that convolution is associative, we have

$$(K_{(\partial/\partial x)} * *(S * * I)) = (K_{(\partial/\partial x)} * * S) * * I = \left(\frac{\partial S}{\partial x}\right) * * I$$

This fact appears in its most commonly used form when the smoothing function is a Gaussian; we can then write

$$\frac{\partial\left(G_\sigma * * I\right)}{\partial x} = \left(\frac{\partial G_\sigma}{\partial x}\right) * * I,$$

that is, we need only convolve with the derivative of the Gaussian, rather than convolve and then differentiate. Smoothing results in much smaller noise responses from the derivative estimates (Figure 8.4).

8.2.2 Why Smoothing Helps

In general, any change of significance to us has effects over a pool of pixels. For example, the contour of an object can result in a long chain of points where the image derivative is large. For many kinds of noise model, large image derivatives due to noise are an essentially local event. This means that smoothing a differentiated image tends to pool support for the changes we are interested in and to suppress the effects of noise. An alternative interpretation of the point is that the changes we are interested in will not be suppressed by some smoothing, which tends to suppress the effects of noise.

There is an alternative explanation as to why smoothing may help. Assume we smooth a noisy image and then differentiate it. First, the variance of the noise tends to be reduced by a

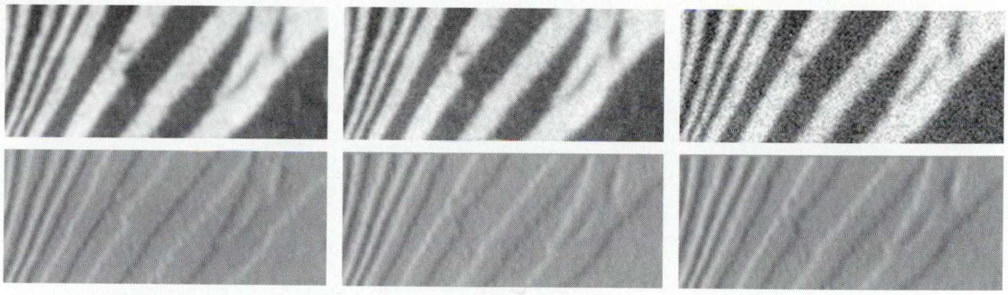

Figure 8.4 Derivative of Gaussian filters are less extroverted in their response to noise than finite difference filters. The image at **top left** shows a detail from a picture of a zebra; **top center** shows the same image corrupted by zero mean stationary additive Gaussian noise, with $\sigma = 0.03$ (pixel values range from 0 to 1). **Top right** shows the same image corrupted by zero mean stationary additive Gaussian noise, with $\sigma = 0.09$. The second row shows the partial derivative in the x-direction of each image, in each case estimated by a derivative of Gaussian filter with σ one pixel. Notice how the smoothing helps reduce the impact of the noise.

smoothing kernel. This is because we tend to use smoothing kernels which are positive and for which

$$\sum_{uv} G_{uv} = 1,$$

which means that

$$\sum_{uv} G_{uv}^2 \leq 1.$$

Second, pixels have a greater tendency to look like neighboring pixels—if we take stationary additive Gaussian noise and smooth it, the pixel values of the resulting signal *are no longer independent*. In some sense, this is what smoothing was about— recall that we introduced smoothing as a method to predict a pixel's value from the values of its neighbors. However, if pixels tend to look like their neighbors, then derivatives must be smaller (because they measure the tendency of pixels to look different from their neighbors).

Another approach is to reason in terms of spatial frequencies. It is possible to show that stationary additive Gaussian noise has uniform energy at each frequency. If we differentiate the noise, we emphasize the high frequencies. If we do not attempt to ameliorate this situation, the gradient magnitude map is likely to have occasional large values due to noise. Filtering with a Gaussian filter suppresses these high spatial frequencies as it does for resampling (Section 7.4.3).

Smoothed noise has applications. As Figure 8.3 indicates, smoothed noise tends to look like some kinds of natural texture, and smoothed noise is widely used as a source of textures in computer graphics applications (Ebert, Musgrave, Peachey, Worley and Perlin, 1998, Perlin, 1985).

8.2.3 Choosing a Smoothing Filter

The smoothing filter can be chosen by taking a model of an edge and using some set of criteria to choose a filter that gives the best response to that model. It is difficult to pose this problem as a two-dimensional problem because edges in 2D can be curved. Conventionally, the smoothing filter is chosen by formulating a one-dimensional problem and then using a rotationally symmetric version of the filter in 2D.

The one-dimensional filter must be obtained from a model of an edge. The usual model is a step function of unknown height in the presence of stationary additive Gaussian noise and is given by

$$edge(x) = AU(x) + n(x),$$

where

$$U(x) = \begin{cases} 0 & \text{if } x < 0 \\ 1 & \text{if } x > 0 \end{cases}.$$

(the value of $U(0)$ is irrelevant to our purpose). A is usually referred to as the *contrast* of the edge. In the 1D problem, finding the gradient magnitude is the same as finding the square of the derivative response. For this reason, we usually seek a derivative estimation filter rather than a smoothing filter (which can then be reconstructed from the derivative estimation filter).

Canny (1986) established the practice of choosing a derivative estimation filter by using the continuous model to optimize a combination of three criteria:

- **Signal to noise ratio**—the filter should respond more strongly to the edge *at* $x = 0$ than to noise.
- **Localization**—the filter response should reach a maximum close to $x = 0$.
- **Low false positives**—there should be only one maximum of the response in a reasonable neighborhood of $x = 0$.

Once a continuous filter has been found, it is discretized. The criteria can be combined in a variety of ways yielding a variety of somewhat different filters. It is a remarkable fact that the optimal smoothing filters derived by most combinations of these criteria tend to look a great deal like Gaussians—this is intuitively reasonable because the smoothing filter must place strong weight on center pixels and less weight on distant pixels rather like a Gaussian. In practice, optimal smoothing filters are usually replaced by a Gaussian, with no particularly important degradation in performance.

The choice of σ used in estimating the derivative is often called the *scale* of the smoothing. Scale has a substantial effect on the response of a derivative filter. Assume we have a narrow bar on a constant background, rather like the zebra's whisker. Smoothing on a scale smaller than the width of the bar means that the filter responds on each side of the bar, and we are able to resolve the rising and falling edges of the bar. If the filter width is much greater, the bar is smoothed into the background and the bar generates little or no response (Figure 8.5).

8.2.4 Why Smooth with a Gaussian?

Although a Gaussian is not the only possible blurring kernel, it is convenient because it has a number of important properties. First, if we convolve a Gaussian with a Gaussian, the result is another Gaussian:

$$G_{\sigma_1} ** G_{\sigma_2} = G_{\sqrt{\sigma_1^2+\sigma_2^2}}.$$

This means that it is possible to obtain heavily smoothed images by resmoothing smoothed images. This is a significant property because discrete convolution can be an expensive operation (particularly if the kernel of the filter is large), and it is common to want versions of an image smoothed by different amounts.

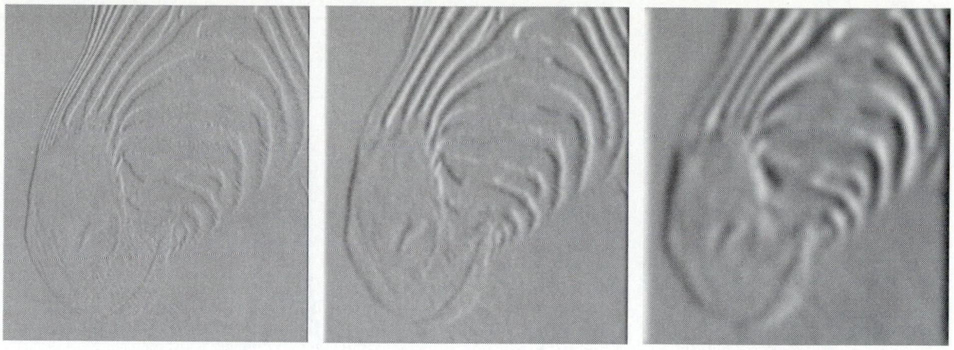

Figure 8.5 The scale (i.e., σ) of the Gaussian used in a derivative of Gaussian filter has significant effects on the results. The three images show estimates of the derivative in the x direction of an image of the head of a zebra obtained using a derivative of Gaussian filter with σ one pixel, three pixels, and seven pixels (moving to the right). Note how images at a finer scale show some hair, the animal's whiskers disappear at a medium scale, and the fine stripes at the top of the muzzle disappear at the coarser scale.

Efficiency Consider convolving an image with a Gaussian kernel with σ one pixel. Although the Gaussian kernel is nonzero over an infinite domain, for most of that domain it is extremely small because of the exponential form. For σ one pixel, points outside a 5×5 integer grid centered at the origin have values less than $e^{-4} = 0.0184$, and points outside a 7×7 integer grid centered at the origin have values less than $e^{-9} = 0.0001234$. This means that we can ignore their contributions and represent the discrete Gaussian as a small array (5×5 or 7×7 according to taste and the number of bits you allocate to represent the kernel).

However, if σ is 10 pixels, we may need a 50×50 array or worse. A back of the envelope count of operations should convince you that convolving a reasonably sized image with a 50×50 array is an unattractive prospect. The alternative—convolving repeatedly with a much smaller kernel— is much more efficient *because we don't need to keep every pixel in the interim.* This is because a smoothed image is, to some extent, redundant (most pixels contain a significant component of their neighbors' values). As a result, some pixels can be discarded. We then have a quite efficient strategy: smooth, subsample, smooth, subsample, and so on. The result is an image that has the same information as a heavily smoothed image, but is much smaller and easier to obtain. We explore the details of this approach in Section 7.7.1.

The Central Limit Theorem Gaussians have another significant property that we do not prove but illustrate in Figure 8.6. For an important family of functions, convolving any member of that family of functions with itself repeatedly eventually yields a Gaussian. With the associativity of convolution, this implies that if we choose a different smoothing kernel and apply it repeatedly to the image, the result eventually looks like we smoothed the image with a Gaussian.

Gaussians are Separable Finally, an isotropic Gaussian can be factored as

$$\begin{aligned} G_\sigma(x, y) &= \frac{1}{2\pi\sigma^2} \exp\left(-\frac{(x^2+y^2)}{2\sigma^2}\right) \\ &= \left(\frac{1}{\sqrt{2\pi}\sigma} \exp\left(-\frac{(x^2)}{2\sigma^2}\right)\right) \times \left(\frac{1}{\sqrt{2\pi}\sigma} \exp\left(-\frac{(y^2)}{2\sigma^2}\right)\right), \end{aligned}$$

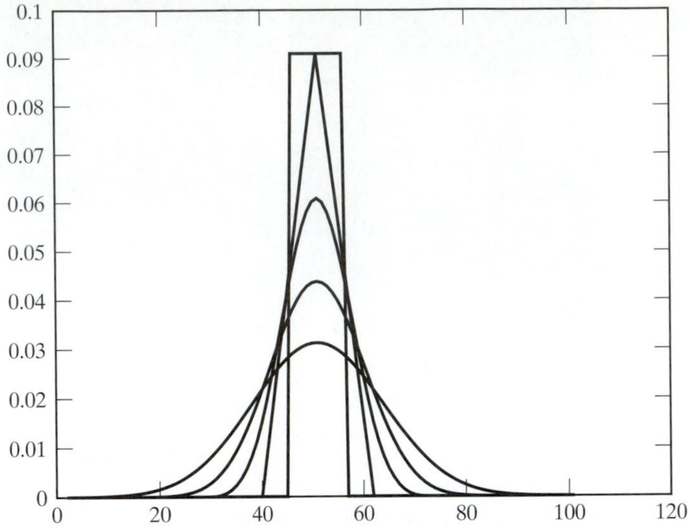

Figure 8.6 The central limit theorem states that repeated convolution of a positive kernel with itself eventually limits toward a kernel that is a scaling of a Gaussian. The graph illustrates this effect for 1D convolution; the triangle is obtained by convolving a box function with itself; each succeeding stage is obtained by convolving the previous stage with itself.

and this is a product of two 1D Gaussians. Generally, a function $f(x, y)$ that factors as $f(x, y) = g(x)h(y)$ is referred to as a *tensor product*. It is common to refer to filter kernels that are tensor products as *separable kernels*. Separability is a useful property indeed. In particular, convolving with a filter kernel that is separable is the same as convolving with two 1D kernels—one in the x direction and another in the y direction (exercises).

Many other kernels are separable. Separable filter kernels result in discrete representations that factor as well. In particular, if $\mathcal{H}$ is a discretized separable filter kernel, there are some vectors $\boldsymbol{f}$ and $\boldsymbol{g}$ such that

$$H_{ij} = f_i g_j.$$

It is possible to identify this property using techniques from numerical linear algebra because the rank of the matrix $\mathcal{H}$ must be one. Commercial convolution packages often test the kernel to see whether it is separable before applying it to the image. The cost of this test is easily paid off by the savings if the kernel does turn out to be separable. Many kernels can be approximated in a useful way as a sum of separable kernels. If the number of kernels is sufficiently small, then the approximation can represent a practical saving in convolution. This is a particularly attractive strategy if one wishes to convolve an image with many different filters; in this case, one tries to obtain a representation of each of these filters as a weighted sum of separable kernels, which are tensor products of a small number of basis elements. It is then possible to convolve the images with the basis elements and then form different weighted sums of the result to obtain convolutions of the image with different filters.

Aliasing in Subsampled Gaussians The discussion of aliasing gives us some insight into available smoothing parameters. Any Gaussian kernel that we use is a sampled approximation to a Gaussian sampled on a single pixel grid. This means that, for the original kernel to be reconstructed from the sampled approximation, it should contain no components of spatial frequency greater than 0.5pixel^{-1}. This isn't possible with a Gaussian because its Fourier transform is also Gaussian, and hence isn't bandlimited. The best we can do is insist that the quantity of energy in the signal that is aliased is below some threshold—in turn, this implies a minimum value of σ that is available for a smoothing filter on a discrete grid (for values lower than this minimum, the smoothing filter is badly aliased; see the exercises).

8.3 DETECTING EDGES

The two main strategies for detecting edges both model edges as fast changes in brightness. In the first, we observe that the fastest change occurs when a 2D analogue of the second derivative vanishes (Section 8.3.1). Although this approach is historically important, it is no longer popular. The alternative is to explicitly search for points where the magnitude of the gradient is extremal (Section 8.3.2).

8.3.1 Using the Laplacian to Detect Edges

In one dimension, the second derivative of a signal is zero when the derivative magnitude is extremal. This means that, if we wish to find large changes, a good place to look is where the second derivative is zero. This approach extends to two dimensions. We now need a sensible analogue to the second derivative. This needs to be rotationally invariant. It is not hard to show that the *Laplacian* has this property. The Laplacian of a function in 2D is defined as

$$(\nabla^2 f)(x, y) = \frac{\partial^2 f}{\partial x^2} + \frac{\partial^2 f}{\partial y^2}.$$

It is natural to smooth the image before applying a Laplacian. Notice that the Laplacian is a linear operator (if you're not sure about this, you should check), meaning that we could represent taking the Laplacian as convolving the image with some kernel (which we write as K_{∇^2}. Because convolution is associative, we have that

$$(K_{\nabla^2} ** (G_\sigma ** I)) = (K_{\nabla^2} ** G_\sigma) ** I = (\nabla^2 G_\sigma) ** I.$$

The reason this is important is that, just as for first derivatives, smoothing an image and then applying the Laplacian is the same as convolving the image with the Laplacian of the kernel used for smoothing. Figure 8.7 shows the resulting kernel.

This leads to a simple and historically important edge detection strategy illustrated in Figure 8.8. We convolve an image with a *Laplacian of Gaussian* at some scale, and mark the points where the result has value zero—the *zero crossings*. These points should be checked to ensure that the gradient magnitude is large. The method is due to Marr and Hildreth (1980).

The response of a Laplacian of Gaussian filter is positive on one side of an edge and negative on another. This means that adding some percentage of this response back to the original image yields a picture in which edges have been sharpened and detail is more easy to see. This observation dates back to a photographic developing technique called *unsharp masking*, where a blurred positive is used to increase visibility of detail in bright areas by subtracting a local average of the brightness in that area. This is roughly the same as filtering the image with a difference of Gaussians, multiplying the result by a small constant, and adding this back to the original image. Now the difference between two Gaussian kernels looks similar to a Laplacian of Gaussian kernel, and it is quite common to replace one with the other. This means that unsharp masking adds an edge term back to the image.

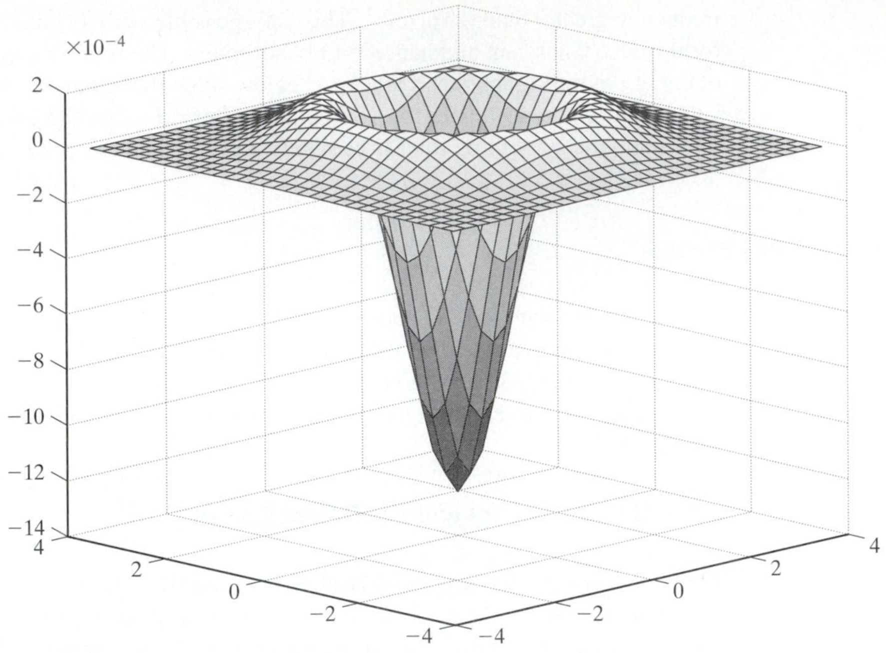

Figure 8.7 The Laplacian of Gaussian filter kernel, shown here for σ one pixel, can be thought of as subtracting the center pixel from a weighted average of the surround (hence the analogy with unsharp masking described in the text). It is quite common to replace this kernel with the difference of two Gaussians—one with a small value of σ and the other with a large value of σ.

Laplacian of Gaussian edge detectors have fallen into some disfavor. Because the Laplacian of Gaussian filter is not oriented, its response is composed of an average across an edge and one along the edge. This means that the behavior at corners—where the direction along the edge changes—is poor. They mark the boundaries of sharp corners quite inaccurately. Furthermore, at trihedral or greater vertices, they have difficulty recording the topology of the corner correctly, as Figure 8.9 illustrates. Second, the components along the edge tend to contribute to the response of the filter to noise but not necessarily to an edge; this means that zero crossings may not lie exactly on an edge.

8.3.2 Gradient-Based Edge Detectors

In a gradient-based edge detector, we compute some estimate of the gradient magnitude—almost always using a Gaussian as a smoothing filter—and use this estimate to determine the position of edge points. Typically, the gradient magnitude can be large along a thick trail in an image (Figure 8.10). Object outlines are curves, however, and we should like to obtain a curve of the most distinctive points on this trail.

A natural approach is to look for points where the gradient magnitude is a maximum along the direction perpendicular to the edge. For this approach, the direction perpendicular to the edge can be estimated using the direction of the gradient (Figure 8.11). These considerations yield Algorithm 8.1. Most current edgefinders follow these lines, but there remain substantial debates about the proper execution of the details.

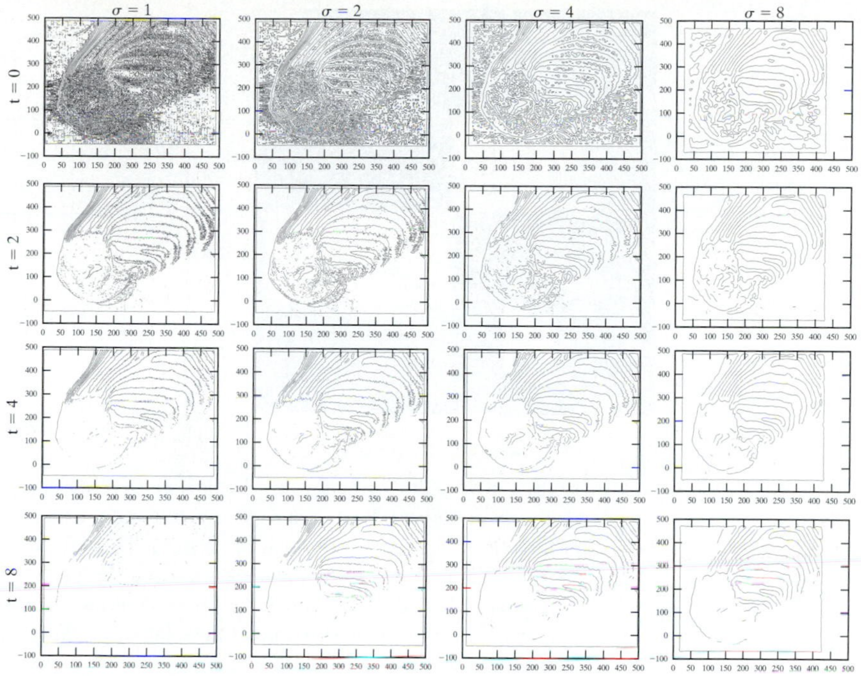

Figure 8.8 Zero crossings of the Laplacian of Gaussian for various scales and at various gradient magnitude thresholds. Each column shows a fixed scale, with t, the threshold on gradient magnitude increasing as one moves down (by a factor of two from image to image). Each row shows a fixed t, with scale increasing from σ one pixel to σ eight pixels by factors of two. Notice that the fine-scale, low-threshold edges contain a quantity of detailed information that may or may not be useful (depending on one's interest in the hairs on the zebra's nose). As the scale increases, the detail is suppressed; as the threshold increases, small regions of edge drop out. No scale or threshold gives the outline of the zebra's head; all respond to its stripes, although as the scale increases, the narrow stripes on the top of the muzzle are no longer resolved.

Algorithm 8.1: Gradient-Based Edge Detection

Form an estimate of the image gradient
Obtain the gradient magnitude from this estimate
Identify image points where the value
  of the gradient magnitude is maximal
  in the direction perpendicular to the edge
  and also large; these points are edge points

Nonmaximum Suppression Given estimates of gradient magnitude, we would like to obtain edge points. Again, there is clearly no objective definition, and we proceed by reasonable intuition. The gradient magnitude can be thought of as a chain of low hills. Marking local extrema would mark isolated points—the hilltops in the analogy. A better criterion is to slice

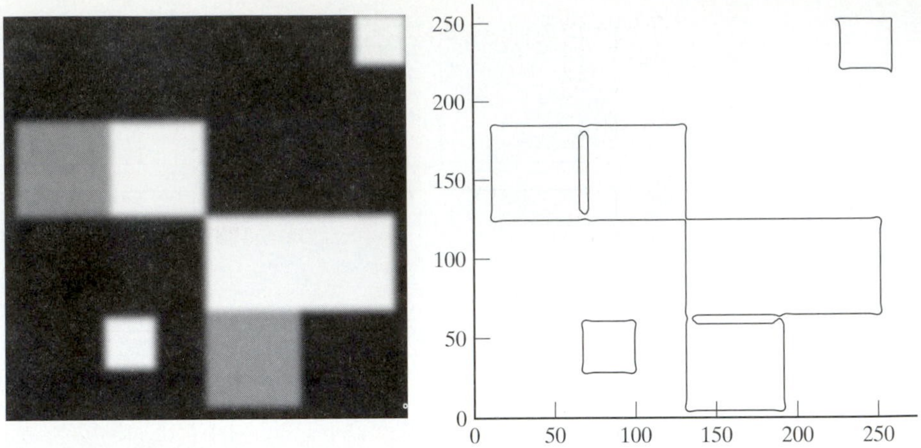

Figure 8.9 Zero crossings of Laplacian of Gaussian output can behave strangely at corners. First, at a right angled corner, the zero crossing bulges out at the corner (but passes through the vertex). This effect is not due to digitization or quantization, but can be shown to occur in the continuous case as well. At corners where three or more edges meet, contours behave strangely, with the details depending on the structure of the contour marking algorithm—this algorithm (the one shipped with Matlab) produces curious loops. This effect can be mitigated with careful design of the contour marking process, which needs to incorporate a fairly detailed vertex model.

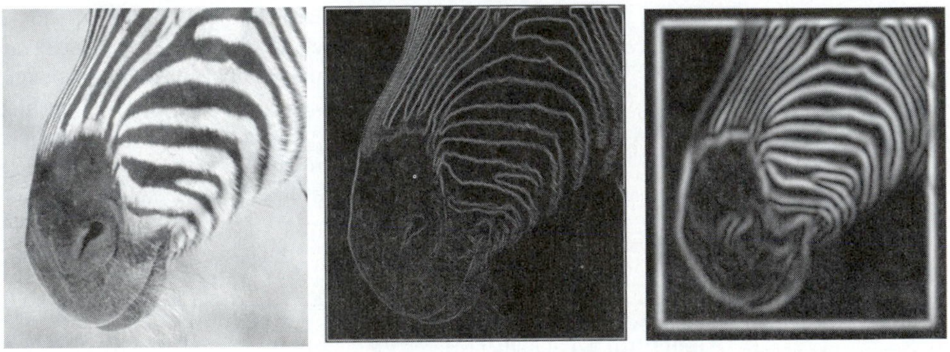

Figure 8.10 The gradient magnitude can be estimated by smoothing an image and then differentiating it. This is equivalent to convolving with the derivative of a smoothing kernel. The extent of the smoothing affects the gradient magnitude; in this figure, we show the gradient magnitude for the figure of a zebra at different scales. At the **center**, gradient magnitude estimated using the derivatives of a Gaussian with $\sigma = 1$ pixel; and on the **right** gradient magnitude estimated using the derivatives of a Gaussian with $\sigma = 2$ pixel. Notice that large values of the gradient magnitude form thick trails.

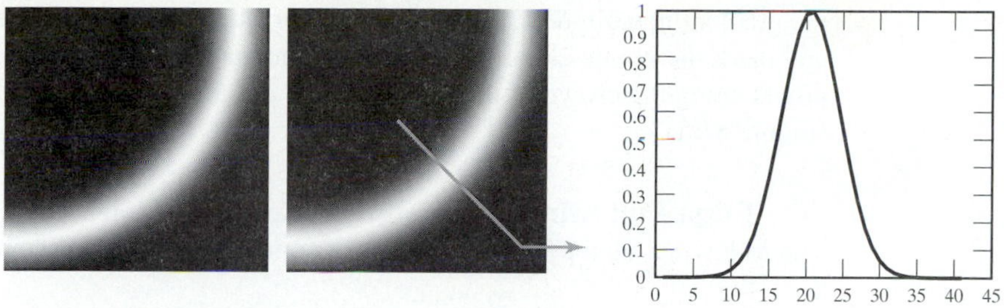

Figure 8.11 The gradient magnitude tends to be large along thick trails in an image. Typically, we would like to condense these trails into curves of representative edge points. A natural way to do this is to cut the trail perpendicular to its direction and look for a peak. We use the gradient direction as an estimate of the direction in which to cut. The figure on the **left** shows a trail of large gradient magnitude; the figure at the **center** shows an appropriate cutting direction; the figure on the **right** shows the peak in this direction.

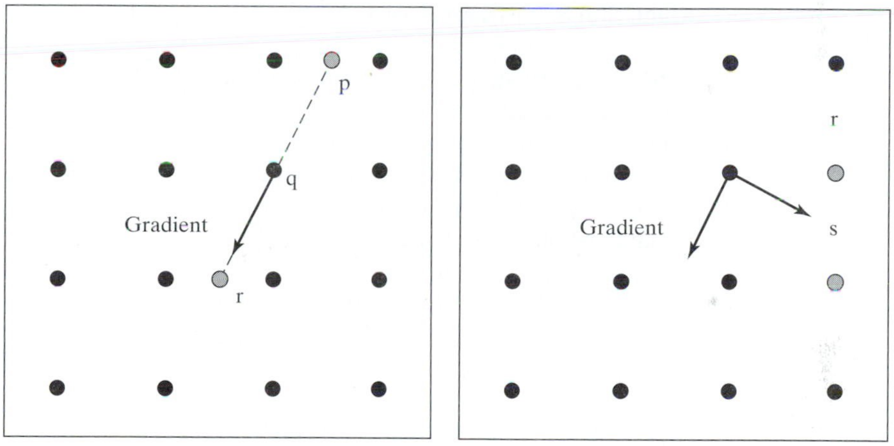

Figure 8.12 Nonmaximum suppression obtains points where the gradient magnitude is at a maximum *along the direction of the gradient*. The figure on the left shows how we reconstruct the gradient magnitude. The dots are the pixel grid. We are at pixel $\boldsymbol{q}$, attempting to determine whether the gradient is at a maximum; the gradient direction through $\boldsymbol{q}$ does not pass through any convenient pixels in the forward or backward direction, so we must interpolate to obtain the values of the gradient magnitude at $\boldsymbol{p}$ and $\boldsymbol{r}$; if the value at $\boldsymbol{q}$ is larger than both, $\boldsymbol{q}$ is an edge point. Typically, the magnitude values are reconstructed with a linear interpolate, which in this case would use the pixels to the left and right of $\boldsymbol{p}$ and $\boldsymbol{r}$, respectively, to interpolate values at those points. On the right, we sketch how to find candidates for the next edge point given that $\boldsymbol{q}$ is an edge point; an appropriate search direction is perpendicular to the gradient, so that points $\boldsymbol{s}$ and $\boldsymbol{t}$ should be considered for the next edge point. Notice that, in principle, we don't need to restrict ourselves to pixel points on the image grid because we know where the predicted position lies between $\boldsymbol{s}$ and $\boldsymbol{t}$. Hence, we could again interpolate to obtain gradient values for points off the grid.

the gradient magnitude along the gradient direction, which should be perpendicular to the edge, and mark the points along the slice where the magnitude is maximal. This would get a chain of points along the crown of the hills in our chain; the process is called *nonmaximum suppression* (Figure 8.12).

Edge Following Typically, we expect edge points to occur along curve like chains. The following are the significant steps in nonmaximum suppression:

- determining whether a given point is an edge point;
- and, if it is, finding the next edge point.

Once these steps are understood, it is easy to enumerate all edge chains. We find the first edge point, mark it, expand all chains through that point exhaustively, marking all points along those chains, and continue to do this for all unmarked edge points.

Algorithm 8.2: Nonmaximum Suppression

```
While there are points with high gradient
that have not been visited
   Find a start point that is a local maximum in the
      direction perpendicular to the gradient
      erasing points that have been checked
   While possible, expand a chain through
      the current point by:
         1) predicting a set of next points, using
            the direction perpendicular to the gradient
         2) finding which (if any) is a local maximum
            in the gradient direction
         3) testing if the gradient magnitude at the
            maximum is sufficiently large
         4) leaving a record that the point and
            neighbors have been visited
         record the next point, which becomes the current point
   end
end
```

The two main steps are simple. For the moment, assume that edges are to be marked at pixel locations (rather than, say, at some finer subdivision of the pixel grid). We can determine whether the gradient magnitude is maximal at any pixel by comparing it with values at points some way backward and forward *along the gradient direction* (Figure 8.11). This is a function of distance along the gradient; typically we step forward to the next row (or column) of pixels and backward to the previous to determine whether the magnitude at our pixel is larger (Figure 8.12). The gradient direction does not usually pass through the next pixel, so we must interpolate to determine the value of the gradient magnitude at the points we are interested in; a linear interpolate is usual.

If the pixel turns out to be an edge point, the next edge point in the curve can be guessed by taking a step perpendicular to the gradient. In general, this step does not end on a pixel; a natural

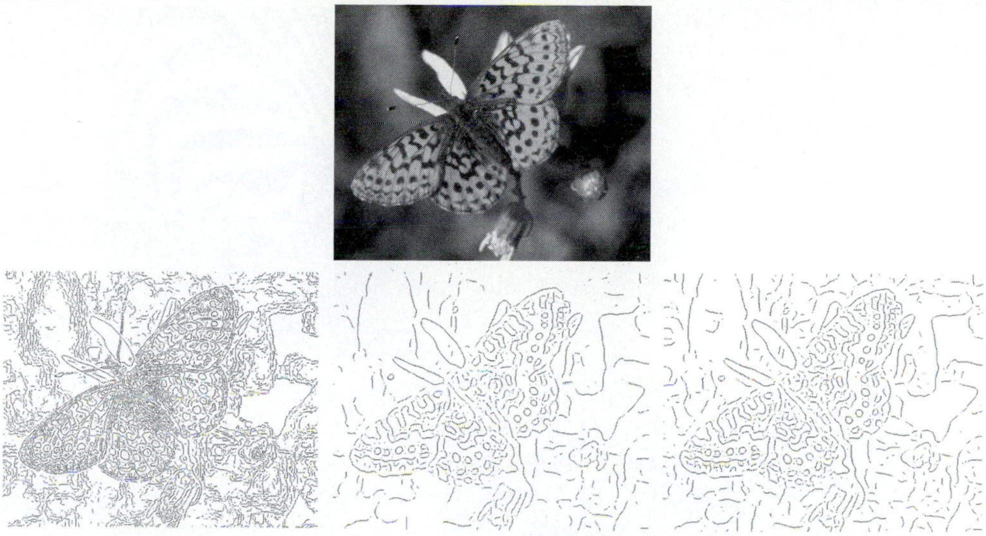

Figure 8.13 Edge points marked on the pixel grid for the image shown on the **top**. The edge points on the **left** are obtained using a Gaussian smoothing filter at σ one pixel, and gradient magnitude has been tested against a high threshold to determine whether a point is an edge point. The edge points on the **center** are obtained using a Gaussian smoothing filter at σ four pixels, and gradient magnitude has been tested against a high threshold to determine whether a point is an edge point. The edge points on the **right** are obtained using a Gaussian smoothing filter at σ four pixels, and gradient magnitude has been tested against a low threshold to determine whether a point is an edge point. At a fine scale, fine detail at high contrast generates edge points, which disappear at the coarser scale. When the threshold is high, curves of edge points are often broken because the gradient magnitude dips below the threshold; for the low threshold, a variety of new edge points of dubious significance are introduced.

strategy is to look at the neighboring pixels that lie close to that direction (see Figure 8.12). This approach leads to a set of curves that can be represented by rendering them in black on a white background, as in Figures 8.13, 8.14 and 8.15.

Hysteresis There are too many of these curves to come close to being a reasonable representation of object boundaries. This is, in part, because we have marked maxima of the gradient magnitude without regard to how large these maxima are. It is more usual to apply a threshold test to ensure that the maxima are greater than some lower bound. This in turn leads to broken edge curves (look closely at Figures 8.13 to 8.15). The usual trick for dealing with this is to use *hysteresis*; we have two thresholds and refer to the *larger* when starting an edge chain and the *smaller* while following it. The trick often results in an improvement in edge outputs (see Exercises).

8.3.3 Technique: Orientation Representations and Corners

Edge detectors notoriously fail at corners because the assumption that estimates of the partial derivatives in the x and y direction suffice to estimate an oriented gradient becomes unsupportable. At sharp corners or unfortunately oriented corners, these partial derivative estimates are poor because their support will cross the corner. There are a variety of specialized corner detec-

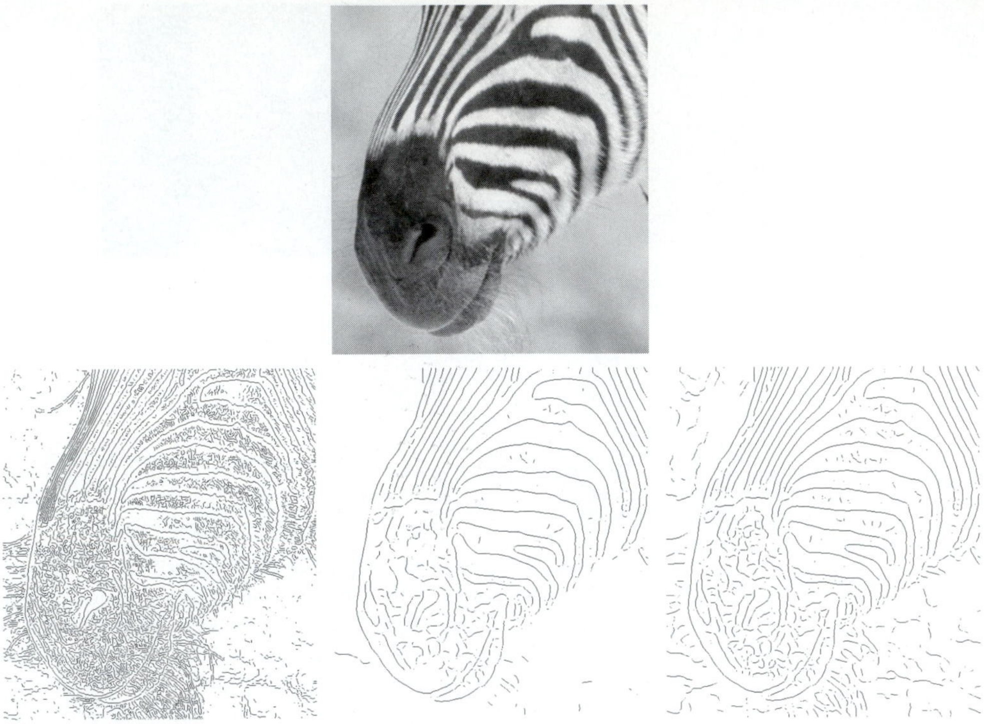

Figure 8.14 Edge points marked on the pixel grid for the image shown on the **top**. The edge points on the **left** are obtained using a Gaussian smoothing filter at σ one pixel, and gradient magnitude has been tested against a high threshold to determine whether a point is an edge point. The edge points on the **center** are obtained using a Gaussian smoothing filter at σ four pixels, and gradient magnitude has been tested against a high threshold to determine whether a point is an edge point. The edge points on the **right** are obtained using a Gaussian smoothing filter at σ four pixels, and gradient magnitude has been tested against a low threshold to determine whether a point is an edge point. At a fine scale, fine detail at high contrast generates edge points, which disappear at the coarser scale. When the threshold is high, curves of edge points are often broken because the gradient magnitude dips below the threshold; for the low threshold, a variety of new edge points of dubious significance are introduced.

tors, which look for image neighborhoods where the gradient swings sharply (Figure 8.16). More generally, the statistics of the gradient in an image neighborhood yields quite a useful description of the neighborhood. There is a rough taxonomy of four qualitative types of image window:

- *constant windows*, where the gray level is approximately constant;
- *edge windows*, where there is a sharp change in image brightness that runs along a single direction within the window;
- *flow windows*, where there are several fine parallel stripes—say hair or fur—within the window;
- and *2D windows*, where there is some form of 2D texture—say spots or a corner—within the window.

Figure 8.15 Edge points marked on the pixel grid for the image shown on the **top**. The edge points on the **left** are obtained using a Gaussian smoothing filter at σ one pixel, and gradient magnitude has been tested against a high threshold to determine whether a point is an edge point. The edge points on the **center** are obtained using a Gaussian smoothing filter at σ four pixels, and gradient magnitude has been tested against a high threshold to determine whether a point is an edge point. The edge points on the **right** are obtained using a Gaussian smoothing filter at σ four pixels, and gradient magnitude has been tested against a low threshold to determine whether a point is an edge point. At a fine scale, fine detail at high contrast generates edge points, which disappear at the coarser scale. When the threshold is high, curves of edge points are often broken because the gradient magnitude dips below the threshold; for the low threshold, a variety of new edge points of dubious significance are introduced.

These cases correspond to different kinds of behavior on the part of the image gradient. In constant windows, the gradient vector is short; in edge windows, there is a small number of long gradient vectors all pointing in a single direction; in flow windows, there are many gradient vectors pointing in two directions; and in 2D windows, the gradient vector swings.

These distinctions can be quite easily drawn by looking at variations in orientation within a window. In particular, the matrix

Figure 8.16 An image of a joshua tree on the **left** and its orientations shown as vectors superimposed on the image on the **right**. The orientation is superimposed on the image as small vectors. Notice that around corners and in textured regions, the orientation vector swings sharply.

$$\mathcal{H} = \sum_{window} \left\{ (\nabla I)(\nabla I)^T \right\}$$

$$\approx \sum_{window} \left\{ \begin{array}{cc} \left(\frac{\partial G_\sigma}{\partial x} ** \mathcal{I}\right)\left(\frac{\partial G_\sigma}{\partial x} ** \mathcal{I}\right) & \left(\frac{\partial G_\sigma}{\partial x} ** \mathcal{I}\right)\left(\frac{\partial G_\sigma}{\partial y} ** \mathcal{I}\right) \\ \left(\frac{\partial G_\sigma}{\partial x} ** \mathcal{I}\right)\left(\frac{\partial G_\sigma}{\partial y} ** \mathcal{I}\right) & \left(\frac{\partial G_\sigma}{\partial y} ** \mathcal{I}\right)\left(\frac{\partial G_\sigma}{\partial y} ** \mathcal{I}\right) \end{array} \right\}$$

gives a good idea of the behavior of the orientation in a window. In a constant window, both eigenvalues of this matrix are small because all terms are small. In an edge window, we expect to see one large eigenvalue associated with gradients at the edge and one small eigenvalue because few gradients run in other directions. In a flow window, we expect the same properties of the eigenvalues, except that the large eigenvalue is likely to be larger because many edges contribute. Finally, in a 2D window, both eigenvalues are large.

The behavior of this matrix is most easily understood by plotting the ellipses

$$(x, y)^T \mathcal{H}^{-1} (x, y) = \epsilon$$

for some small constant ϵ. These ellipses are superimposed on the image windows. Their major and minor axes are along the eigenvectors of $\mathcal{H}$, and the extent of the ellipses along their major or minor axes corresponds to the size of the eigenvalues; this means that a large circle corresponds to an edge window and a narrow extended ellipse indicates an edge window (as in Figure 8.17 and Figure 8.18). Thus, corners could be marked by marking points where the area of this ellipse is extremal and large. The localization accuracy of this approach is limited by the size of the window and the behavior of the gradient. More accurate localization can be obtained at the price of providing a more detailed model of the corner sought (see, for example, Harris and Stephens, 1988 or Schmid, Mohr and Bauckhage, 2000).

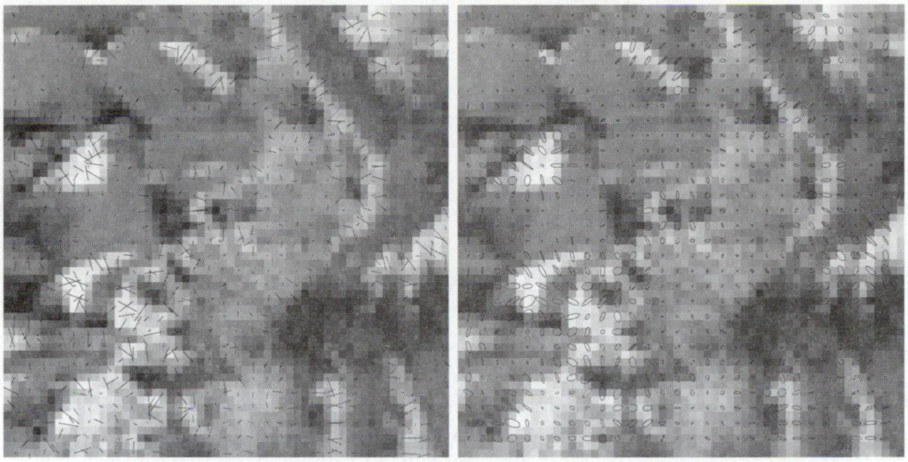

Figure 8.17 The orientation field for a detail of the joshua tree picture. On the **left**, the orientations shown as vectors and superimposed on the image. Orientations have been censored to remove those where the gradient magnitude is too small. The **right** figure shows the ellipses described in the text for a 3x3 window.

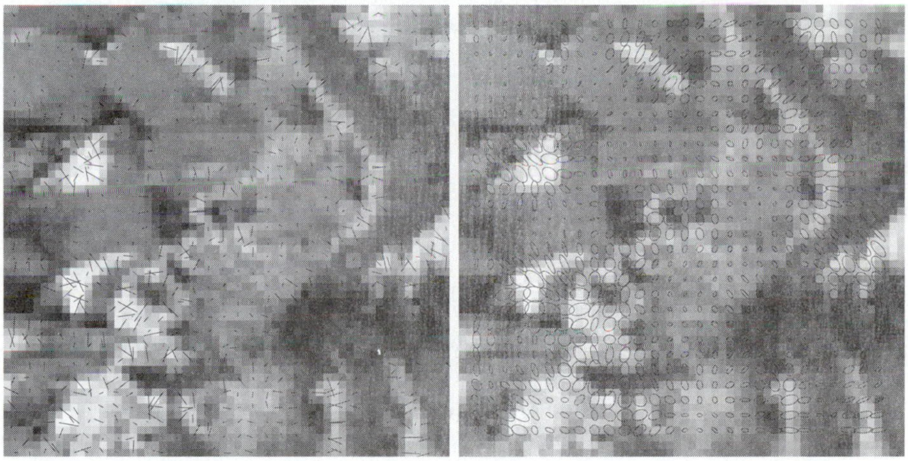

Figure 8.18 The orientation field for a detail of the joshua tree picture. On the **left**, the orientations shown as vectors and superimposed on the image. Orientations have been censored to remove those where the gradient magnitude is too small. The **right** figure shows the ellipses described in the text, for a 5x5 window.

8.4 NOTES

There is a huge edge detection literature. The earliest paper of which we are aware is Julez (1959) (yes, 1959!). Those wishing to be acquainted with the early literature in detail should start with a 1975 survey by Davis (1975); Herskovits and Binford (1970); Horn (1971); and Hueckel (1971), who models edges and then detects the model.

Edge detection is a subject alive with controversy, much of it probably empty. We have hardly scratched the surface. There are many optimality criteria for edge detectors, and rather more "optimal" edge detectors. The key paper in this literature is by Canny (1986); significant variants are due to Deriche (1987) and to Spacek (1986). Faugeras' textbook contains a detailed and accessible exposition of the main issues (1993). At the end of the day, most variants boil down to smoothing the image with something that looks a lot like a Gaussian before measuring the gradient.

Object boundaries are not the same as sharp changes in image values. First, objects may not have a strong contrast with their backgrounds through sheer bad luck. Second, objects are often covered with texture or markings that generate edges of their own—so many that it is often hard to wade through them to find the relevant pieces of object boundary. Finally, shadows and the like may generate edges that have no relation to object boundaries. There are some strategies for dealing with these difficulties.

First, some applications allow management of illumination; if it is possible to choose the illumination, a careful choice can make a tremendous difference in the contrast and eliminate shadows. Second, by setting smoothing parameters large and contrast thresholds high, it is often possible to ensure that edges due to texture are smoothed over and not marked. This is a dubious business, because it can be hard to choose reliable values of the smoothing and the thresholds, and because it is perverse to regard texture purely as a nuisance, rather than a source of information.

There are other ways to handle the uncomfortable distinction between edges and object boundaries. First, one might work to make better edge detectors. This approach is the root of a huge literature, dealing with matters like localization, corner topology, and the like. We incline to the view that returns are diminishing rather sharply in this endeavor; we can provide only some pointers to this (vast) literature. The reader could start with Bergholm (1987), Deriche (1990), Elder and Zucker (1998), Fleck (1992*a*), Kube and Perona (1996), Olson (1998), Perona and Malik (1990*a,b*), or Torre and Poggio (1986).

Second, one might deny the usefulness of edge detection entirely. This approach is rooted in the observation that some stages of edge detection, particularly nonmaximum suppression, discard information that is awfully difficult to retrieve. This is because a hard decision—testing against a threshold—has been made. Instead, the argument proceeds, one should keep this information around in a "soft" (a propaganda term for probabilistic) way. Attactive as these arguments sound, we are inclined to discount this view because there are currently no practical mechanisms for handling the volumes of soft information so obtained.

Finally, one might regard this as an issue to be dealt with by overall questions of system architecture—the fatalist view that almost every visual process is going to have obnoxious features, and the correct approach to this problem is to understand the integration of visual information well enough to construct vision systems that are tolerant to this. Although it sweeps a great deal of dust under the carpet (precisely *how* does one construct such architectures?) we find this approach most attractive and discuss it again and again.

All edge detectors behave badly at corners; only the details vary. In the case of zero crossings of the Laplacian of Gaussian, the problem is well understood (Berzins, 1984). This bad behavior has resulted in two lively strands in the literature (What goes wrong? What to do about it?). There are a variety of quite sophisticated corner detectors, mainly because corners make quite good point features for correspondence algorithms supporting such activities as stereopsis, reconstruction, or structure from motion. This has led to quite detailed practical knowledge of the localisation properties of corner detectors (e.g., Schmid, Mohr and Bauckhage, 2000).

Another lively strand in the literature is to determine how well edge detectors do. One may study localization accuracy (e.g., Kakarala and Hero, 1992, Lyvers and Mitchell, 1988) or stability (e.g., Cho, Meer and Cabrera, 1997, 1998); one may compare with human preferences (e.g., Bowyer, Kranenburg and Dougherty, 1999, Dougherty and Bowyer, 1998, Heath, Sarkar, Sanocki and Bowyer, 1997) or compare performance in the context of a fixed task, such as structure from motion (e.g., Shin, Goldgof and Bowyer, 1998) or recognition (e.g., Shin, Goldgof and Bowyer, 1999). All edge detectors share some difficulties (e.g., at corners Fleck, 1992*b*).

The edges that our edge detectors respond to are sometimes called *step edges* because they consist of a sharp, "discontinuous" change in value that is sometimes modeled as a step. A variety of other forms of edge have been studied. The most commonly cited example is the *roof edge*, which consists of a rising segment meeting a falling segment, rather like some of the reflexes that can result from the effects of interreflections (Figure 5.16). Another example that also results from interreflections is a composite of a step and a roof. It is possible to find these phenomena by using essentially the same steps as outlined before (find an "optimal" filter, and do nonmaximum suppression on its outputs) (Canny, 1986, Perona and Malik, 1990*a,b*). In practice, this is seldom done. There appear to be two reasons. First, there is no comfortable basis in theory (or practice) for the models that are adopted. What particular composite edges are worth looking for? The easy answer—those for which optimal filters are reasonably easy to derive—is most unsatisfactory. Second, the semantics of roof edges and more complex composite edges is even vaguer than that of step edges. There is little notion of what one would *do* with roof edge once it had been found.

Edges are poorly defined and usually hard to detect, but one can solve problems with the output of an edge detector. Roof edges are similarly poorly defined and similarly hard to detect; we have never seen problems solved with the output of a roof edge detector. The real difficulty here is that there seems to be no reliable mechanism for predicting, in advance, what is worth detecting. We scratch the surface of this very difficult problem in what follows.

ASSIGNMENTS

PROBLEMS

8.1. Each pixel value in 500×500 pixel image $\mathcal{I}$ is an independent, normally distributed random variable with zero mean and standard deviation one. Estimate the number of pixels that, where the absolute value of the x derivative, estimated by forward differences (i.e., $|I_{i+1,j} - I_{i,j}|$), is greater than 3.

8.2. Each pixel value in 500×500 pixel image $\mathcal{I}$ is an independent, normally distributed random variable with zero mean and standard deviation one. $\mathcal{I}$ is convolved with the $2k + 1 \times 2k + 1$ kernel $\mathcal{G}$. What is the covariance of pixel values in the result? There are two ways to do this; on a case-by-case basis (e.g., at points that are greater than $2k + 1$ apart in either the x or y direction, the values are clearly independent) or in one fell swoop. Don't worry about the pixel values at the boundary.

8.3. We have a camera that can produce output values that are integers in the range from 0 to 255. Its spatial resolution is 1024 by 768 pixels, and it produces 30 frames a second. We point it at a scene that, in the absence of noise, would produce the constant value 128. The output of the camera is subject to noise that we model as zero mean stationary additive Gaussian noise with a standard deviation of 1. How long must we wait before the noise model predicts that we should see a pixel with a negative value? (Hint: You may find it helpful to use logarithms to compute the answer as a straightforward evaluation of $\exp(-128^2/2)$ will yield 0; the trick is to get the large positive and large negative logarithms to cancel.)

8.4. We said a sensible 2D analogue to the 1D second derivative must be rotationally invariant in Section 8.3.1. Why is this true?

Programming Assignments

8.5. Why is it necessary to check that the gradient magnitude is large at zero crossings of the Laplacian of an image? Demonstrate a series of edges for which this test is significant.

8.6. The Laplacian of a Gaussian looks similar to the difference between two Gaussians at different scales. Compare these two kernels for various values of the two scales. Which choices give a good approximation? How significant is the approximation error in edge finding using a zero-crossing approach?

8.7. Obtain an implementation of Canny's edge detector (you could try the vision home page; MATLAB has an implementation in the image processing toolbox, too) and make a series of images indicating the effects of scale and contrast thresholds on the edges that are detected. How easy is it to set up the edge detector to mark only object boundaries? Can you think of applications where this would be easy?

8.8. It is quite easy to defeat hysteresis in edge detectors that implement it—essentially, one sets the lower and higher thresholds to have the same value. Use this trick to compare the behavior of an edge detector with and without hysteresis. There are a variety of issues to look at:

(a) What are you trying to do with the edge detector output? It is sometimes helpful to have linked chains of edge points. Does hysteresis help significantly here?

(b) Noise suppression: We often wish to force edge detectors to ignore some edge points and mark others. One diagnostic that an edge is useful is high contrast (it is by no means reliable). How reliably can you use hysteresis to suppress low-contrast edges without breaking high-contrast edges?

9

Texture

Texture is a phenomenon that is widespread, easy to recognise and hard to define. Typically, whether an effect is referred to as texture or not depends on the scale at which it is viewed. A leaf that occupies most of an image is an object, but the foliage of a tree is a texture. Texture arises from a number of different sources. Firstly, views of large numbers of small objects are often best thought of as textures. Examples include grass, foliage, brush, pebbles and hair. Secondly, many surfaces are marked with orderly patterns that look like large numbers of small objects. Examples include: the spots of animals like leopards or cheetahs; the stripes of animals like tigers or zebras; the patterns on bark, wood and skin.

There are three standard problems to do with texture:

- **Texture segmentation** is the problem of breaking an image into components within which the texture is constant. Texture segmentation involves both representing a texture, and determining the basis on which segment boundaries are to be determined. In this chapter, we deal only with the question of how textures should be *represented* (Section 9.1); chapters 14 and 16 show how to segment textured images using this representation.
- **Texture synthesis** seeks to construct large regions of texture from small example images. We do this by using the example images to build probability models of the texture, and then drawing on the probability model to obtain textured images. There are a variety of methods for building a probability model; three successful current methods are described in Section 9.3.
- **Shape from texture** involves recovering surface orientation or surface shape from image texture. We do this by assuming that texture "looks the same" at different points on a surface; this means that the deformation of the texture from point to point is a cue to

Figure 9.1 A set of texture examples, used in experiments with human subjects to tell how easily various types of textures can be discriminated. Note that these textures are made of quite stylized subelements, repeated in a meaningful way. *Reprinted from A Computational Model of Texture Segmentation, J. Malik and P. Perona, Proc. Computer Vision and Pattern Recognition, 1989, © 1989, IEEE*

the shape of the surface. In Section 9.4, we describe the main lines of reasoning in this (rather technical) area.

9.1 REPRESENTING TEXTURE

Image textures generally consist of organised patterns of quite regular subelements (sometimes called *textons*). For example, one texture in Figure 9.1 consists of triangles. Similarly, another texture in that figure consists of arrows. One natural way to try and represent texture is to find the textons, and then describe the way in which they are laid out.

The difficulty with this approach is that there is no known canonical set of textons, meaning that it isn't clear what one should look for. Instead of looking for patterns at the level of arrow-

Figure 9.2 Typical textured images. For materials such as brush, grass, foliage and water, our perception of what the material is is quite intimately related to the texture (for the figure on the **left**, what would the surface feel like if you ran your fingers over it? is it wet?, etc.). Notice how much information you are getting about the type of plants, their shape, etc. from the textures in the figure on the **right**. These textures are also made of quite stylized subelements, arranged in a rough pattern.

heads and triangles, we could look for even simpler pattern elements—dots and bars, say—and then reason about their spatial layout. The advantage of this approach is that it is easy to look for simple pattern elements by filtering an image.

9.1.1 Extracting Image Structure with Filter Banks

In Section 7.5, we saw that convolving an image with a linear filter yields a representation of the image on a different basis. The advantage of transforming an image to the new basis given by convolving it with a filter, is that the process makes the local structure of the image clear. This is because there is a strong response when the image pattern in a neighborhood looks similar to the filter kernel, and a weak response when it doesn't.

This suggests representing image textures in terms of the response of a collection of filters. The collection of different filters would consist of a series of patterns—spots and bars are usual—at a collection of scales (to identify bigger or smaller spots or bars, say). The value at a point in a derived image represents the local "spottiness" ("barriness", etc.) at a particular scale at the corresponding point in the image. While this representation is now heavily redundant, it exposes structure ("spottiness", "barriness", etc.), in a way that has proven helpful.

Generally, spot filters are useful because they respond strongly to small regions that differ from their neighbors (for example, on either side of an edge, or at a spot). The other attraction is that they detect non-oriented structure. Bar filters, on the other hand, are oriented, and tend to respond to oriented structure.

Spots and Bars by Weighted Sums of Gaussians But what filters should we use? There is no canonical answer. A variety of answers have been tried. By analogy with the human visual cortex, it is usual to use at least one spot filter and a collection of oriented bar filters at different orientations, scales and *phases*. The phase of the bar refers to the phase of a cross-section perpendicular to the bar, thought of as a sinusoid (i.e., if the cross section passes through zero at the origin, then the phase is $0°$).

One way to obtain these filters is to form a weighted difference of Gaussian filters at different scales; this technique was used for the filters of Figure 9.3. The filters for this example consist of

- **A spot**, given by a weighted sum of three concentric, symmetric Gaussians, with weights $1, -2$ and 1, and corresponding sigmas 0.62, 1 and 1.6.
- **Another spot**, given by a weighted sum of two concentric, symmetric Gaussians, with weights 1 and -1, and corresponding sigmas 0.71 and 1.14.
- **A series of oriented bars**, consisting of a weighted sum of three oriented Gaussians, which are offset with respect to one another. There are six versions of these bars; each is a rotated version of a horizontal bar. The Gaussians in the horizontal bar have weights $-1, 2$ and -1. They have different sigma's in the x and in the y directions; the σ_x values are all 2, and the σ_y values are all 1. The centers are offset along the y axis, lying at $(0, 1)$, $(0, 0)$ and $(0, -1)$.

You should understand that the details of the choice of filter are almost certainly immaterial. There is a body of experience that suggests that there should be a series of spots and bars at various scales and orientations—which is what this collection provides—but very little reason to believe that optimizing the choice of filters produces any major advantage.

Figures 9.4 and 9.5 illustrate the absolute value of the responses of this bank of filters to an input image of a butterfly. Notice that, while the bar filters are not completely reliable bar

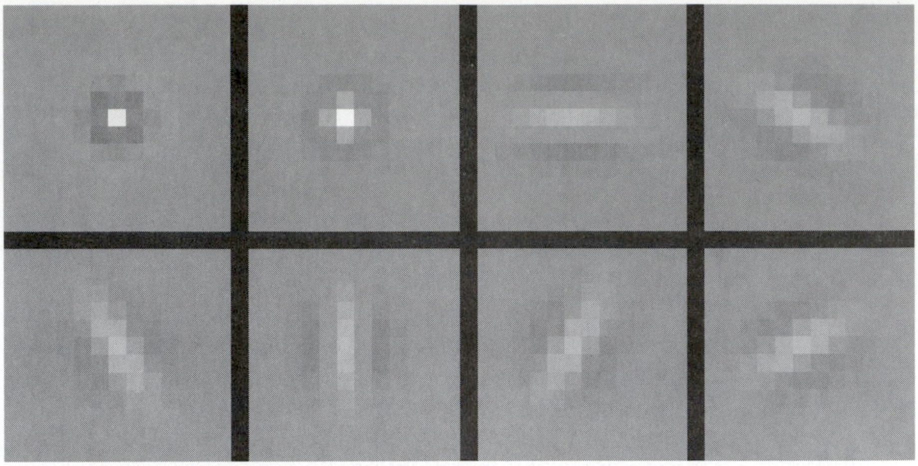

Figure 9.3 A set of eight filters used for expanding images into a series of responses. These filters are shown at a fixed scale, with zero represented by a mid-grey level, lighter values being positive and darker values being negative. They represent two distinct spots, and six bars; the set of filters is that used by Malik and Perona (1990).

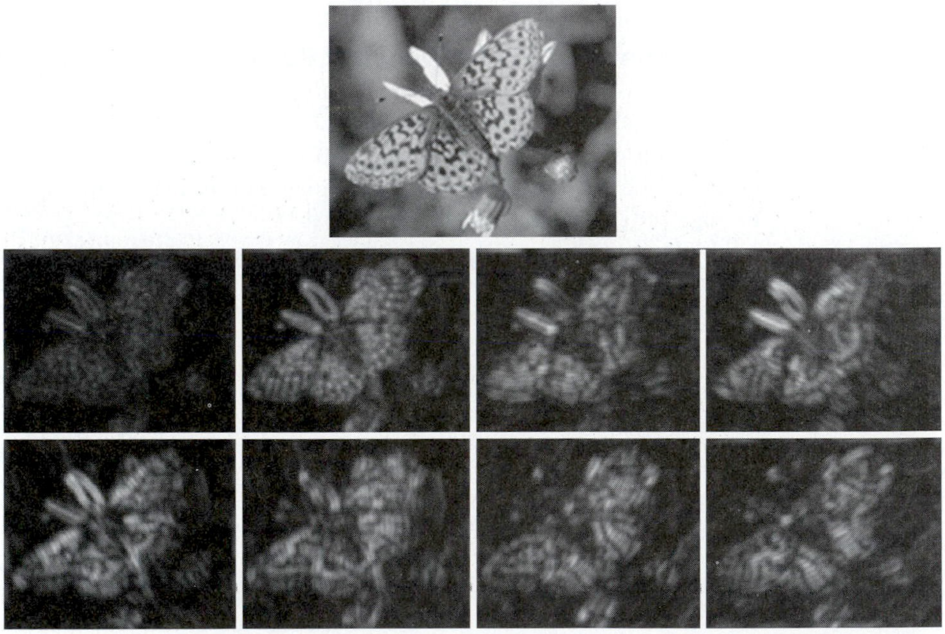

Figure 9.4 At the top, an image of a butterfly at a fine scale, and below, the result of applying each of the filters of Figure 9.3 to that image. The results are shown as absolute values of the output, lighter pixels representing stronger responses, and the images are laid out corresponding to the filter position in the top row.

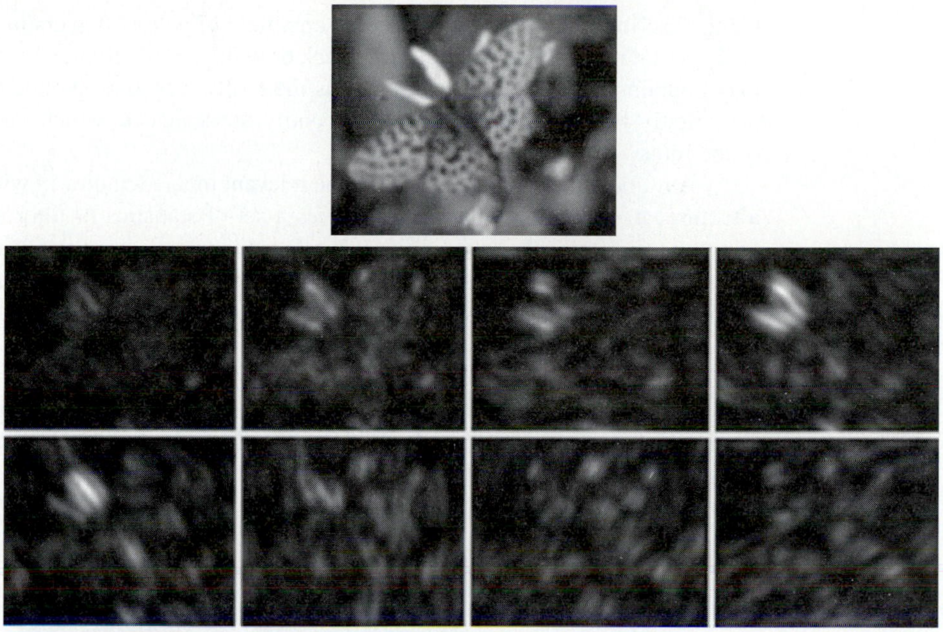

Figure 9.5 The input image of a butterfly and responses of the filters of Figure 9.3 at a coarser scale than that of Figure 9.4. Notice that the oriented bars respond to the bars on the wings, the antennae, and the edges of the wings; the fact that one bar has responded does not mean that another will not, but the size of the response is a cue to the orientation of the bar in the image.

detectors (because a bar filter at a particular orientation responds to bars of a variety of sizes and orientations), the filter outputs give a reasonable representation of the image data. Generally, bar filters respond strongly to oriented bars and weakly to other patterns, and the spot filter responds to isolated spots.

How Many Filters and at What Orientation? It is not known just how many filters are "best" for useful texture algorithms. Perona (1995) lists the number of scales and orientation used in a variety of systems; numbers run from four to eleven scales and from two to eighteen orientations. The number of orientations varies from application to application and does not seem to matter much, as long as there are at least about six orientations. Typically, the "spot" filters are Gaussians and the "bar" filters are obtained by differentiating oriented Gaussians.

Similarly, there does not seem to be much benefit in using more complicated sets of filters than the basic spot and bar combination. There is a tension here: using more filters leads to a more detailed (and more redundant representation of the image); but we must also convolve the image with all these filters, which can be expensive. One way to simplify the process is to control the amount of redundant information we deal with, by building a pyramid.

9.1.2 Representing Texture Using the Statistics of Filter Outputs

A set of filtered images, in itself, is not a representation of texture, because we need some representation of the overall distribution of texture elements. For example, a field of yellow flowers may consist of many small yellow spots, with some vertical green bars; a zebra may consist of black stripes on a white background. We are implicitly assuming a scale over which the texture is being described. A small image window on a field of yellow flowers may contain only a flower; on a zebra it may consist of a constant black or white region. Similarly, a window that is too large may contain some background as well as the relevant texture. Notice that there are two scales here: firstly, the scale of the filters and secondly, the scale over which we consider the distribution of the filters.

Assume that we know the size of the relevant image window in which we wish to represent a texture. A typical representation involves a set of statistics of filter outputs for that window. Outputs are commonly squared (among other things, this has the advantage of counting black next to white stripes in the same way as white next to black stripes). For example, in Figure 9.6, we illustrate a putative representation in terms of horizontal and vertical textures. This representation is obtained by taking the outputs of horizontal (resp. vertical) bar filters and squaring them. We then smooth the result at a coarse scale. This smoothing is equivalent to estimating the mean of the squared filter outputs in some window. Finally, the smoothed outputs are passed to a classifier that describes the texture. In the example of Figure 9.6, the texture is placed in one of four classes, depending on whether the vertical or the horizontal output or both or neither are large.

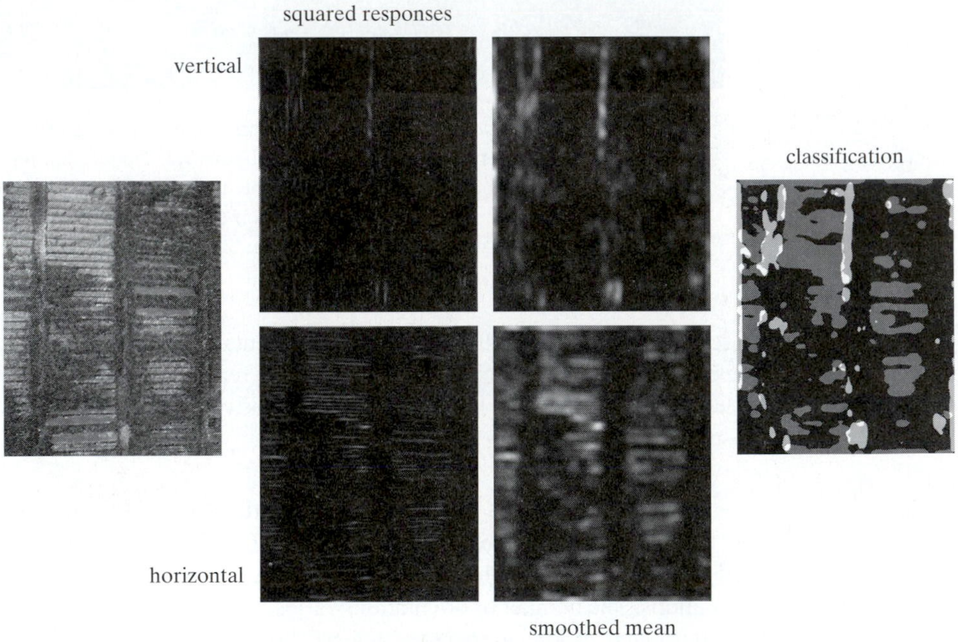

Figure 9.6 A putative texture representation in terms of filter outputs. We have sharply reduced the number of filters (there are two derivative filters, one vertical and one horizontal). The image on the **left** is the input; notice it has components that could reasonably be described as horizontal, vertical and fuzzy. The images in the **center left** column show the squared values of the filter outputs (which have been squared so that black-to-white transitions count the same as white-to-black transitions). The values are shown on the same linear scale, with lighter points indicating stronger responses. These have been smoothed to yield the images on the **center right** (which can be interpreted as the mean of the squared response over a small window). The mean response to vertical stripes is strong in the vertical map, and that to horizontal stripes in the horizontal map. Finally, we have thresholded these two images and combined them to get the image on the **right** (black values are neither horizontal nor vertical; dark grey values are horizontal; light grey values are vertical; and white values are "both").

The Choice of Statistic The question of *what* statistics should be collected depends to some extent on what we intend to represent. However, work on texture synthesis has indicated some constraints on appropriate choices of model, which is why we spend so much ink on the topic in Section 9.3. Assume, for the moment, that the scale of the window over which statistics should be collected has been set. One strategy is to compute the mean of the squared filter outputs for a range of filters (Malik and Perona, 1989). A window is then described by a vector of numbers, each of which is the mean of the squared response of some filter over the window. This approach can tell, for example, spotty windows—where the mean response of the spot filters will be high—from stripey windows—where the mean response of the bar filters will be high. This is the approach of Figure 9.6, but with a richer set of filters.

An alternative approach is to compute the mean and standard deviation of the filter outputs over the window, and use these for the feature vector (Ma and Manjunath, 1996). Texture descriptions of this form can be used to recover image windows based on examples (Figure 9.7). This is useful, because in a satellite image, whether a region depicts housing or vegetation can be determined from the texture. This means that, if we can match textures, we can find all regions of,

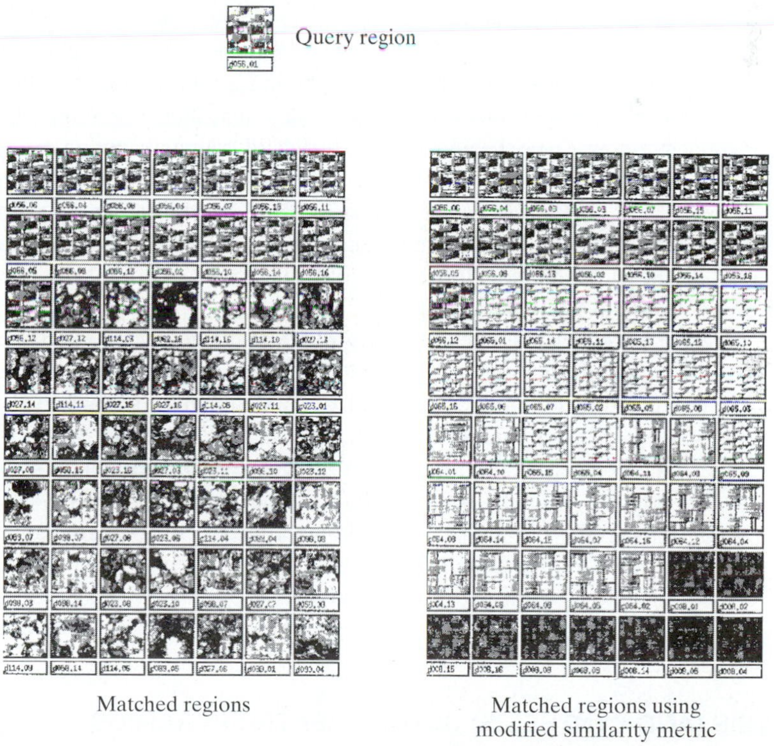

Figure 9.7 Textures can be represented as the mean and standard deviation of filter outputs taken over a window. If we use a collection of different filters, this yields a vector of numbers to represent the window; spotty textures will have large mean spot filter outputs, stripey textures will have large mean bar filter outputs, and so on. This means that an image window can be compared to others by computing a distance based on the feature vector. A pure Euclidean distance yields acceptable results (**left**), but modifying the distance function can yield very good results (**right**). *Reprinted from "Texture Features and Learning Similarity," by W.Y. Ma and B.S. Manjunath, Proc. IEEE Conf. Computer Vision and Pattern Recognition, 1996, © 1996, IEEE*

say, vegetation in a satellite image. Two textures with quite different feature vectors may appear to be similar; this can be dealt with by modifying the way that we measure differences between feature vectors.

Neither mean nor mean and standard deviation is an ideal representation, because the relationships between filter responses can be significant. For example, imagine a texture that consists of small spots arranged in stripes—an aerial view of a field of cabbages, say. In a texture like this, the small spot filter will respond strongly and the large bar filter will respond strongly, but *the responses are correlated*—the large bar filter responds where the small spot filter responds. In the case of a texture consisting of many large bars scattered around with small spots in the background, the small spot filter will respond strongly and the large bar filter will respond strongly, but the responses may not be correlated. We could try and record the covariance of the filter outputs—which would handle the example of the field of cabbages—but there will generally be too many terms to form an accurate estimate. Instead, one typically identifies some covariance terms that may be useful in a particular application and uses those.

The Choice of Scale Another question that is typically dealt with as a practical matter is the choice of the scale to use in representing a texture. Generally, one chooses a small window at the point of interest and then increases the size of the window until an increase does not cause a significant change. For example, imagine selecting a pixel on an image of a zebra. In a very small window around that pixel, the image has a constant value, say black. As the window gets slightly larger, there is a sharp change and there are some black and some white pixels in the window. Once the window is somewhat larger and contains several stripes, enlarging the window further will produce no significant change—it will just be a bigger, stripey window.

One statistic that can be used to determine when to stop expanding the window is the *polarity*. We first determine the *dominant orientation* of the window—the average direction of the gradient. Now for each gradient vector, we form the dot product between the gradient vector and the dominant orientation. We then form a smoothed average of the positive dot products and a smoothed average of the magnitude of the negative dot products, and take the difference of the two. This measures the extent to which gradients in a region point along the dominant orientation (positive dot products) vs. against the dominant orientation (negative dot products).

We can measure this statistic for any particular window scale. We do so for a range of window sizes, and then start at the finest scale and look at increasingly large windows until the polarity has not changed when the scale changed. Notice that there is some possibility that this criterion is not unique. For example, imagine a very high resolution photograph of a zebra. If we start with a sufficiently small window (which may span a single hair), this criterion will select a scale at which a window contains several hairs. If we start with a somewhat larger window, containing many hairs, the criterion should select a scale that encompasses several stripes.

9.2 ANALYSIS (AND SYNTHESIS) USING ORIENTED PYRAMIDS

Representing texture using the statistics of a series of filter outputs requires convolving the image with many filters at many scales. There are quite good methods for doing this systematically, which we describe in this section. Many readers may be content to leave this issue unexamined, and such readers should skip to Section 9.3.

The process of convolving an image with a range of filters is referred to as *analysis*; convolving an image with a collection of oriented filters is sometimes, rather loosely, described as *analyzing orientation* or *representing orientation*. The Gaussian pyramid (Section 7.7.1) is an example of image analysis by a bank of filters—in this case, smoothing filters. The Gaussian pyramid handles scale systematically by subsampling the image once it has been smoothed. This means that generating the next coarsest scale is easier, because we don't process redundant information.

In fact, the Gaussian pyramid is a highly redundant representation because each layer is a low pass filtered version of the previous layer—this means that we are representing the lowest spatial frequencies many times. A layer of the Gaussian pyramid is a prediction of the appearance of the next finer scale layer—this prediction isn't exact, but it means that it is unnecessary to store all of the next finer scale layer. We need keep only a record of the errors in the prediction. This is the motivating idea behind the *Laplacian pyramid*.

The Laplacian pyramid will yield a representation of various different scales that has fairly low redundancy, but it doesn't immediately deal with orientation. By thinking about pyramids in the Fourier domain, we obtain a method for encoding orientation as well (Section 9.2.2). In Section 9.2.3, we will sketch a method that obtains a representation of orientation as well.

9.2.1 The Laplacian Pyramid

The Laplacian pyramid makes use of the fact that a coarse layer of the Gaussian pyramid predicts the appearance of the next finer layer. If we have an upsampling operator that can produce a version of a coarse layer of the same size as the next finer layer, then we need only store the difference between this prediction and the next finer layer itself.

Clearly, we cannot create image information, but we can expand a coarse scale image by replicating pixels. This involves an upsampling operator $S^{\uparrow}$ which takes an image at level $n+1$ to an image at level n. In particular, $S^{\uparrow}(\mathcal{I})$ takes an image, and produces an image twice the size in each dimension. The four elements of the output image at $(2j-1, 2k-1)$; $(2j, 2k-1)$; $(2j-1, 2k)$; and $(2j, 2k)$ all have the same value as the j, kth element of $\mathcal{I}$.

Analysis—Building a Laplacian Pyramid from an Image The coarsest scale layer of a Laplacian pyramid is the same as the coarsest scale layer of a Gaussian pyramid. Each of the finer scale layers of a Laplacian pyramid is a difference between a layer of the Gaussian pyramid and a prediction obtained by upsampling the next coarsest layer of the Gaussian pyramid. This means that:

$$P_{\text{Laplacian}}(\mathcal{I})_m = P_{\text{Gaussian}}(\mathcal{I})_m$$

(where m is the coarsest level) and

$$\begin{aligned} P_{\text{Laplacian}}(\mathcal{I})_k &= P_{\text{Gaussian}}(\mathcal{I})_k - S^{\uparrow}(P_{\text{Gaussian}}(\mathcal{I})_{k+1}) \\ &= (Id - S^{\uparrow} S^{\downarrow} G_{\sigma}) P_{\text{Gaussian}}(\mathcal{I})_k \end{aligned}$$

All this yields Algorithm 9.1. While the name "Laplacian" is somewhat misleading—there are no differential operators here—it is not outrageous, because each layer is approximately the result of a difference of Gaussian filter.

Each layer of the Laplacian pyramid can be thought of as the response of a band-pass filter. This is because we are taking the image at a particular resolution, and subtracting the components that can be predicted by a coarser resolution version—which corresponds to the low spatial frequency components of the image. This means in turn that we expect that an image of a set of stripes at a particular spatial frequency would lead to strong responses at one level of the pyramid and weak responses at other levels (Figure 9.8).

Because different levels of the pyramid represent different spatial frequencies, the Laplacian pyramid can be used as a reasonably effective image compression scheme.

Algorithm 9.1: Building a Laplacian Pyramid from an Image

Form a Gaussian pyramid
Set the coarsest layer of the Laplacian pyramid to be
  the coarsest layer of the Gaussian pyramid
For each layer, going from next to coarsest to finest
  Obtain this layer of the Laplacian pyramid by
    upsampling the next coarser layer, and subtracting
    it from this layer of the Gaussian pyramid
end

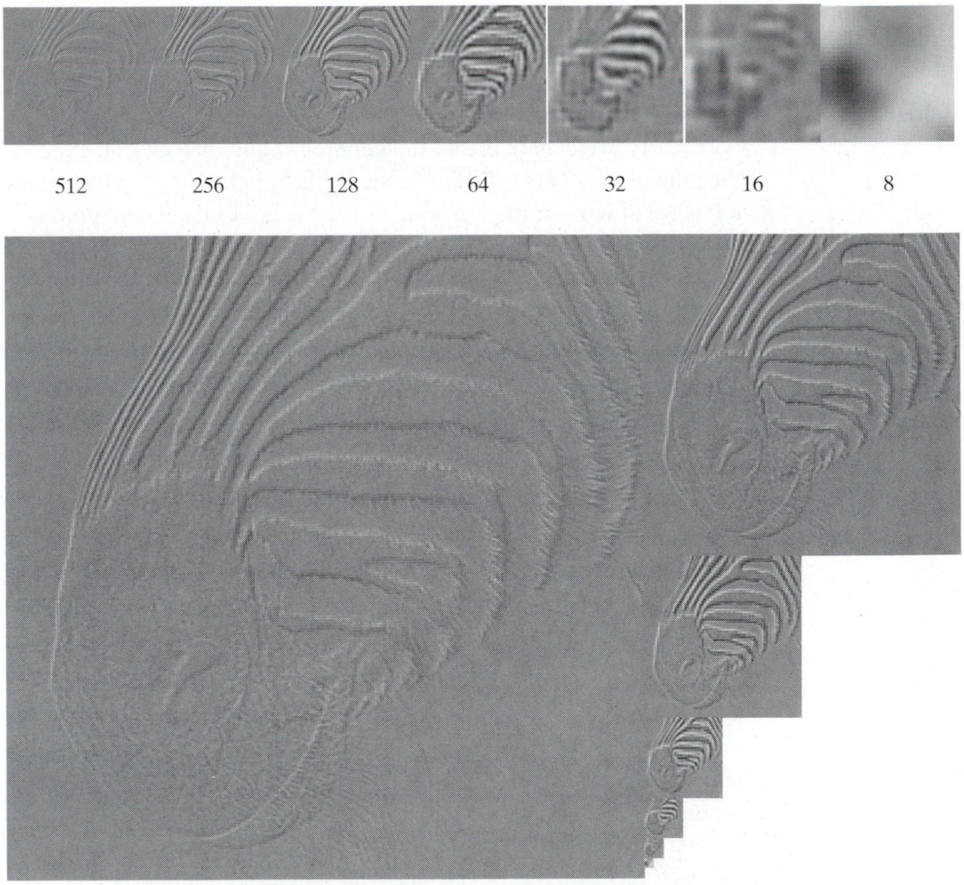

Figure 9.8 A Laplacian pyramid of images, running from 512x512 to 8×8. A zero response is coded with a mid-grey; positive values are lighter and negative values are darker. Notice that the stripes give stronger responses at particular scales, because each layer corresponds (roughly) to the output of a band-pass filter.

Synthesis—Recovering an Image from its Laplacian Pyramid Laplacian pyramids have one important feature. It is easy to recover an image from its Laplacian pyramid. We do this by recovering the Gaussian pyramid from the Laplacian pyramid, and then taking the finest scale of the Gaussian pyramid (which is the image). It is easy to get to the Gaussian pyramid from the Laplacian. Firstly, the coarsest scale of the Gaussian pyramid is the same as the coarsest scale of the Laplacian pyramid. The next-to-coarsest scale of the Gaussian pyramid is obtained by taking the coarsest scale, upsampling it, and adding the next-to-coarsest scale of the Laplacian pyramid (and so on up the scales). This process is known as *synthesis* and is described in Algorithm 9.2.

Algorithm 9.2: Synthesis: Obtaining an Image from a Laplacian Pyramid

```
Set the working image to be the coarsest layer
For each layer, going from next to coarsest to finest
  Upsample the working image and add the current layer
    to the result
  Set the working image to be the result of this operation
end
The working image now contains the original image
```

9.2.2 Filters in the Spatial Frequency Domain

The convolution theorem (that convolution in the spatial domain is the same as multiplication in the Fourier domain) yields some intuition about what filters do and what information pyramids contain. We shall illustrate this theorem by showing a natural analogy between smoothing and low-pass filtering; that some kinds of band-pass filter naturally respond to oriented structure; and that a form of local spatial frequency analysis can be obtained using a particular family of filters.

Smoothing and Low-Pass Filters The convolution theorem yields that convolving an image with an isotropic Gaussian with standard deviation σ is the same as multiplying the Fourier transform of the image by an isotropic Gaussian of standard deviation $1/\sigma$. Now a Gaussian falls off quite quickly, particularly if its standard deviation is large. This means that the Fourier transform of the result will have relatively little energy at high spatial frequencies, where a high spatial frequency is a few multiples of $1/\sigma$. We can interpret this as a *low-pass filter*—one that has a high gain for low spatial frequencies and a low gain for high spatial frequencies. This is quite a satisfactory interpretation: if we smooth with a Gaussian with a very small standard deviation, all but the highest spatial frequencies are preserved; and if we smooth with a Gaussian with a very large standard deviation, the result will be pretty much the average value of the image. This means that a Gaussian pyramid is, in essence, a set of low-pass filtered versions of the image.

Band-Pass Filters and Orientation Selective Operators A *band-pass filter* is one that has high gain for some range of spatial frequencies and a low gain for higher and for lower spatial frequencies. One type of band-pass filter is insensitive to orientation. A natural example of such a filter is to smooth an image with the difference of two isotropic Gaussians; one with a small standard deviation and one with a large standard deviation. In the Fourier domain, the kernel of this filter looks like an annulus of large values (the left half of Figure 9.9); this means that it selects a range of spatial frequencies, but is not selective to orientations (because points at

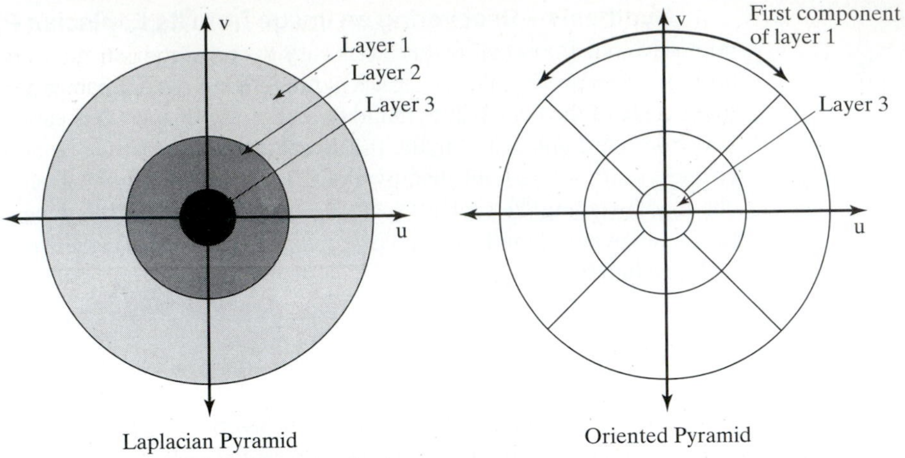

Figure 9.9 Each layer of the Laplacian pyramid consists the elements of a smoothed and resampled image that are not represented by the next smoother layer. Assuming that a Gaussian is a sufficiently good smoothing filter, each layer can be thought of as representing the image components within a range of spatial frequencies—this means that the Fourier transform of each layer is an annulus of values from the Fourier transform space (u, v) space (recall that the magnitude of (u, v) gives the spatial frequency). The sum of these annuluses is the Fourier transform of the image, so that each layer cuts an annulus out of the image's Fourier transform. An oriented pyramid cuts each annulus into a set of wedges. If (u, v) space is represented in polar coordinates, each wedge corresponds to an interval of radius values and an interval of angle values (recall that $\arctan(u/v)$ gives the orientation of the Fourier basis element).

the same distance from the origin in Fourier space refer to basis elements of the frequency, but at different orientations). While an ideal bandpass filter would have a unit value within the annulus and a zero value outside, such a filter would have infinite spatial support—making it difficult to work with—and the difference of Gaussians appears to be a satisfactory practical choice. Of course, this difference of Gaussians is the filter used to obtain the Laplacian pyramid, so the Laplacian pyramid consists of a set of band-pass filtered versions of the image.

An alternative type of band-pass filter has a Fourier transform that is large within a wedge of the annulus, and small outside (the right half of Figure 9.9)—this filter is *orientation selective*, meaning that it responds most strongly to signals that have a particular range of spatial frequencies *and* orientations.

Local Spatial Frequency Analysis and Gabor Filters One difficulty with the Fourier transform is that Fourier coefficients depend on the entire image; the value of the Fourier transform for some particular (u, v) is computed using every image pixel. This is an inconvenient way to think of images, because we have lost all spatial information. For example, the stripes of Figure 9.12 get wider as one moves across the image. If we think in terms of spatial frequency only locally defined, then we can think of this phenomenon in terms of the spatial frequency content of the image changing as we move across it. In some window around a point, the narrow stripes look like high spatial frequency terms and the wide stripes look like low spatial frequency terms.

Gabor filters achieve this. The kernels look like Fourier basis elements that are multiplied by Gaussians, meaning that a Gabor filter responds strongly at points in an image where there are components that *locally* have a particular spatial frequency and orientation. Gabor filters come in pairs, often referred to as *quadrature pairs*; one of the pair recovers symmetric components in a particular direction, and the other recovers antisymmetric components. The mathematical form of the symmetric kernel is

$$G_{\text{symmetric}}(x, y) = \cos\left(k_x x + k_y y\right) \exp - \left\{ \frac{x^2 + y^2}{2\sigma^2} \right\}$$

and the antisymmetric kernel has the form

$$G_{\text{antisymmetric}}(x, y) = \sin\left(k_0 x + k_1 y\right) \exp - \left\{ \frac{x^2 + y^2}{2\sigma^2} \right\}$$

The filters are illustrated in Figures 9.10 and 9.11; (k_x, k_y) give the spatial frequency to which the filter responds most strongly, and σ is referred to as the *scale* of the filter. In principle, by applying a very large number of Gabor filters at different scales, orientations and spatial frequencies, one can analyze an image into a detailed local description. There is an analogy between Gabor filtering with $\sigma = \infty$ and a Fourier transform; this explains why there are two types of filter, and indicates why we can think of a Gabor filter as performing a local spatial frequency analysis.

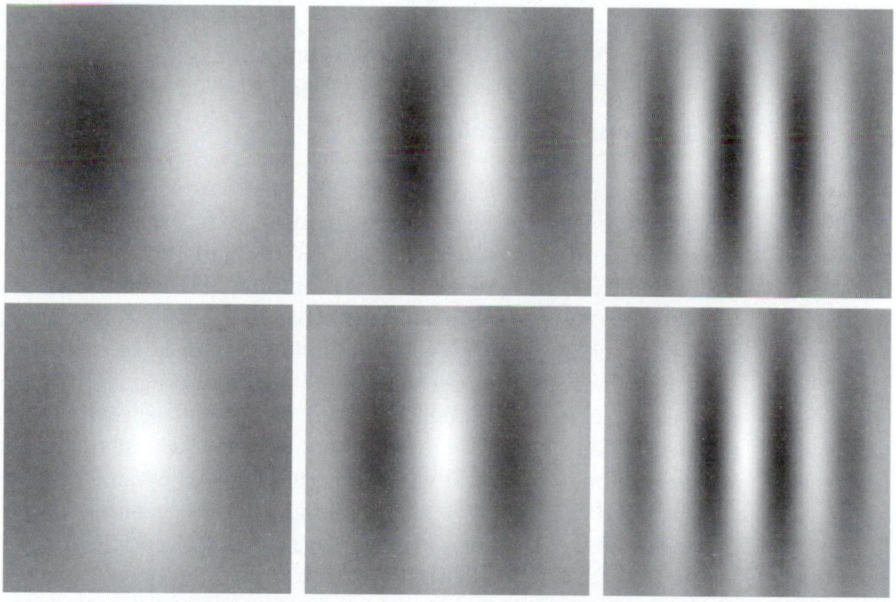

Figure 9.10 Gabor filter kernels are the product of a symmetric Gaussian with an oriented sinusoid; the form of the kernels is given in the text. The images show Gabor filter kernels as images, with mid-grey values representing zero, darker values representing negative numbers and lighter values representing positive numbers. The top row shows the antisymmetric component, and the bottom row shows the symmetric component. The symmetric and antisymmetric components have a phase difference of $\pi/2$ radians, because a cross-section perpendicular to the bar (horizontally, in this case) gives sinusoids that have this phase difference. The scale of these filters is constant, and they are shown for three different spatial frequencies. Figure 9.11 shows Gabor filters at a finer scale.

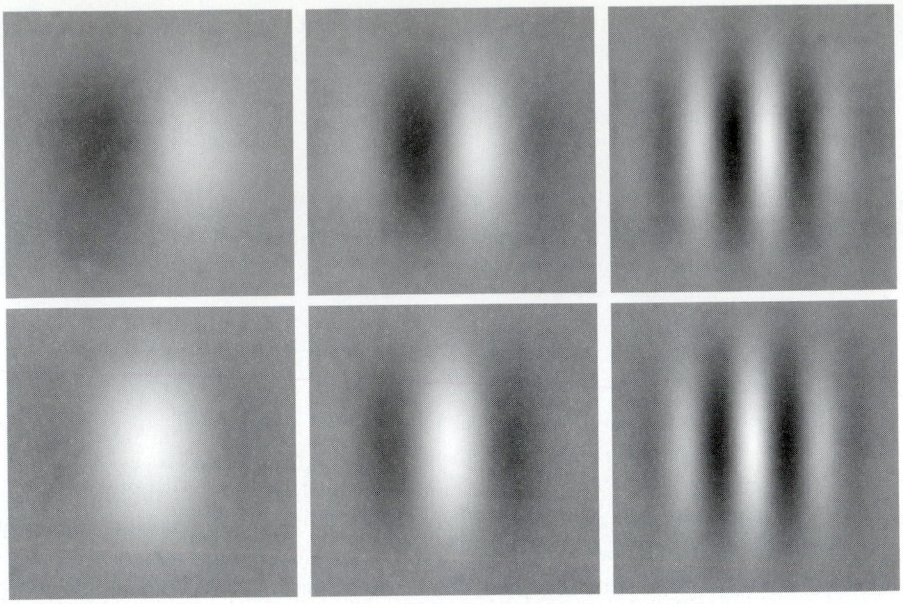

Figure 9.11 The images shows Gabor filter kernels as images, with mid-grey values representing zero, darker values representing negative numbers and lighter values representing positive numbers. The top row shows the antisymmetric component, and the bottom row shows the symmetric component. The scale of these filters is constant, and they are shown for three different spatial frequencies. These filters are shown at a finer scale than those of Figure 9.10.

9.2.3 Oriented Pyramids

A Laplacian pyramid does not contain enough information to reason about image texture, because there is no explicit representation of the orientation of the stripes. A natural strategy for dealing with this is to take each layer and decompose it further, to obtain a set of components each of which represents a energy at a distinct orientation. Each component can be thought of as the response of an oriented filter at a particular scale and orientation. The result is a detailed analysis of the image, known as an *oriented pyramid* (Figure 9.13).

A comprehensive discussion of the design of oriented pyramids would take us out of our way. The first design constraint is that the filter should select a small range of spatial frequencies and orientations, as in Figure 9.9. There is a second design constraint for our analysis filters: synthesis should be easy. If we think of the oriented pyramid as a decomposition of the Laplacian pyramid (Figure 9.14), then synthesis involves reconstructing each layer of the Laplacian pyramid, and then synthesizing the image from the Laplacian pyramid. The ideal strategy is to have a set of filters that have oriented responses *and* where synthesis is easy. It is possible to produce a set of filters such that reconstructing a layer from its components involves filtering the image a second time with the same filter (as Figure 9.15 suggests). An efficient implementation of these pyramids is available at `http://www.cis.upenn.edu/~eero/steerpyr.html`. The design process is described in detail in Karasaridis and Simoncelli (1996) and Simoncelli and Freeman (1995).

9.3 APPLICATION: SYNTHESIZING TEXTURES FOR RENDERING

Renderings of object models look more realistic if they are textured (it's worth thinking about why this should be true, even though the point is widely accepted as obvious). There are a va-

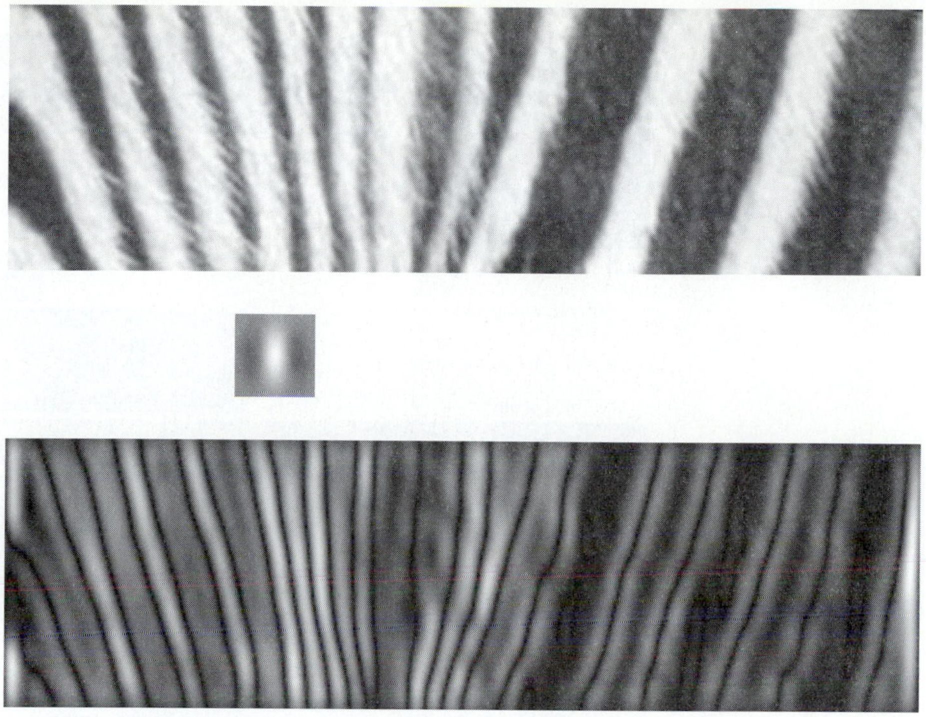

Figure 9.12 The image on the **top** shows a detail from an image of a zebra, chosen because it has a stripes at somewhat different scales and orientations. This has been convolved with the kernel in the center, which is a Gabor filter kernel. The image at the **bottom** shows the absolute value of the result; notice that the response is large when the spatial frequency of the bars roughly matches that windowed by the Gaussian in the Gabor filter kernel (i.e., the stripes in the kernel are about as wide as, and at about the same orientation as, the three stripes in the kernel). When the stripes are larger or smaller, the response falls off; thus, the filter is performing a kind of local spatial frequency analysis. This filter is one of a quadrature pair (it is the symmetric component). The response of the anti-symmetric component is similarly frequency selective. The two responses can be seen as the two components of the (complex valued) local Fourier transform, so that magnitude and phase information can be extracted from them.

riety of techniques for texture mapping; the basic idea is that when an object is rendered, the reflectance value used to shade a pixel is obtained by reference to a **texture map**. Some system of coordinates is adopted on the surface of the object to associate the elements of the texture map with points on the surface. Different choices of coordinate system yield renderings that look quite different, and it is not always easy to ensure that the texture lies on a surface in a natural way (for example, consider painting stripes on a zebra—where should the stripes go to yield a natural pattern?). Despite this issue, texture mapping seems to be an important trick for making rendered scenes look more realistic.

Texture mapping demands textures, and texture mapping a large object may require a substantial texture map. This is particularly true if the object is close to the view, meaning that the texture on the surface is seen at a high resolution, so that problems with the resolution of the texture map will become obvious. Tiling texture images can work poorly, because it can be difficult to obtain images that tile well—the borders have to line up, and even if they did, the

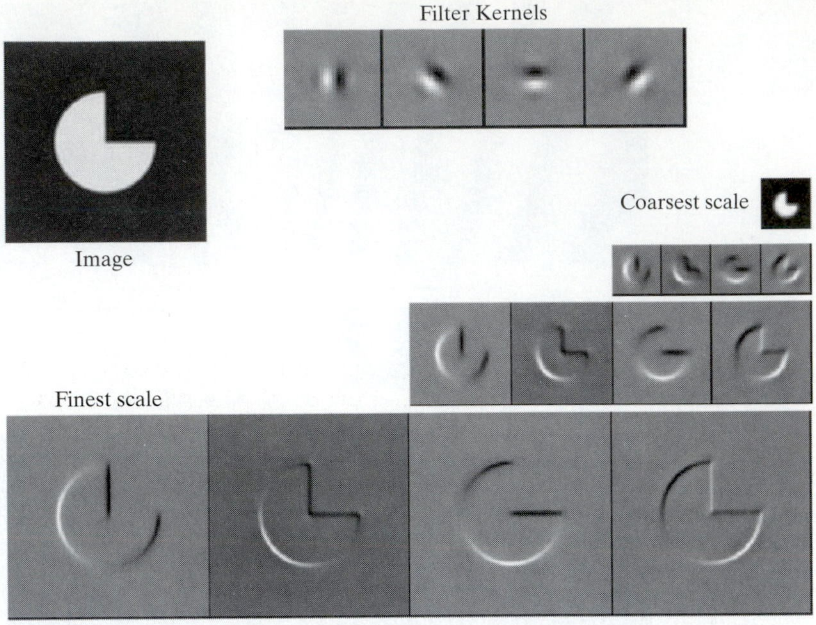

Figure 9.13 An oriented pyramid, formed from the image at the **top left**, with four orientations per layer. This is obtained by firstly decomposing an image into subbands which represent bands of spatial frequency (as with the Laplacian pyramid), and then applying oriented filters (**top right**) to these subbands to decompose them into a set of distinct images, each of which represents the amount of energy at a particular scale and orientation in the image. Notice how the orientation layers have strong responses to the edges in particular directions, and weak responses at other directions. Code for constructing oriented pyramids, written and distributed by Eero Simoncelli, can be found at `http://www.cis.upenn.edu/~eero/steerpyr.html`. *Reprinted from "Shiftable MultiScale Transforms," by Simoncelli et al., IEEE Transactions on Information Theory, 1992, © 1992, IEEE*

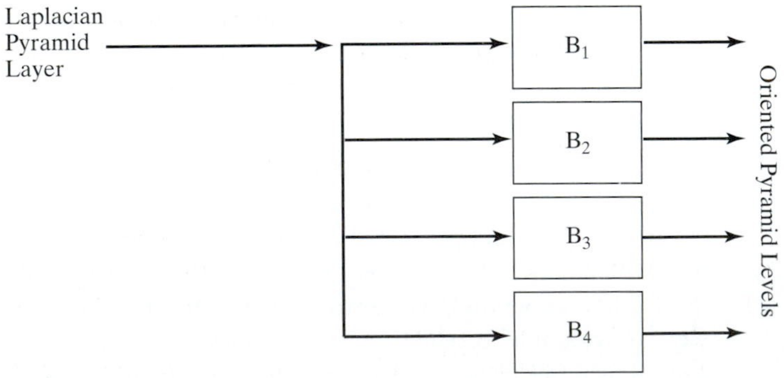

Figure 9.14 The oriented pyramid is obtained by taking layers of the Laplacian pyramid, and then applying oriented filters (represented in this schematic drawing by boxes). Each layer of the Laplacian pyramid represents a range of spatial frequencies; the oriented filters decompose this range of spatial frequencies into a set of orientations.

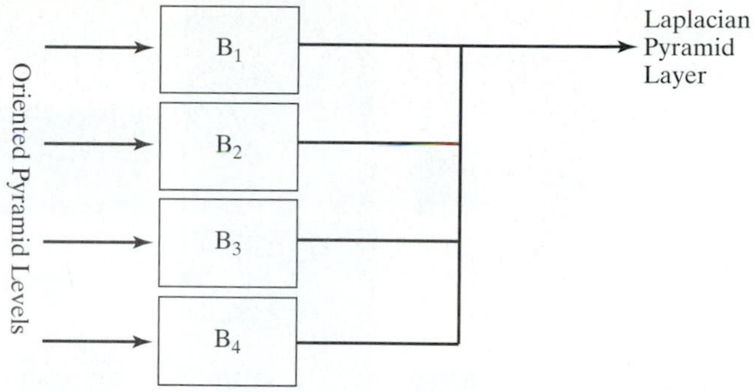

Figure 9.15 In the oriented pyramid, synthesis is possible by refiltering the layers and then adding them, as this schematic indicates. This property is obtained by appropriate choice of filters.

resulting periodic structure can be annoying. It is possible to buy image textures from a variety of sources, but an ideal would be to have a program that can generate large texture images from a small example. Quite sophisticated programs of this form can be built, and they illustrate the usefulness of representing textures by filter outputs.

9.3.1 Homogeneity

The general strategy for texture synthesis is to think of a texture as a sample from some probability distribution and then to try and obtain other samples from that same distribution. To make this approach practical, we need to obtain a probability model from the sample texture. The first thing to do is assume that the texture is *homogenous*. This means that local windows of the texture "look the same", from wherever in the texture they were drawn. More formally, the probability distribution on values of a pixel is determined by the properties of some neighborhood of that pixel, rather than by, say, the position of the pixel.

An assumption of homogeneity means that we can construct a model for the texture outside the boundaries of our example region, based on the properties of our example region. The assumption often applies to natural textures over a reasonable range of scales. For example, the stripes on a zebra's back are homogenous, but remember that those on its back are vertical and those on its legs, horizontal. We can use the example texture to obtain the probability model for the synthesized texture in various ways; we describe only one here.

9.3.2 Synthesis by Sampling Local Models

As Efros and Leung (1999) point out, the example image can serve as a a probability model. Assume for the moment that we have every pixel in the synthesized image, except one. To obtain a probability model for the value of that pixel, we could match a neighborhood of the pixel to the example image. Every matching neighborhood in the example image has a possible value for the pixel of interest. This collection of values is a conditional histogram for the pixel of interest. By drawing a sample uniformly and at random from this collection, we obtain the value that is consistent with the example image.

Finding Matching Image Neighbourhoods The essence of the matter is to take some form of neighbourhood around the pixel of interest, and to compare it to neighbourhoods in the example image. The size and shape of this neighbourhood is significant, because it codes

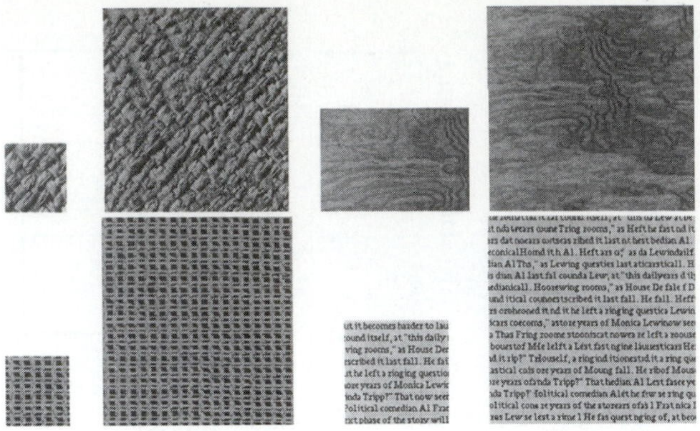

Figure 9.16 Efros' texture synthesis algorithm (Algorithm 9.3) matches neighbourhoods of the image being synthesized to the example image, and then chooses at random amongst the possible values reported by matching neighbourhoods. This means that the algorithm can reproduce complex spatial structures, as these examples indicate. The small block on the **left** is the example texture; the algorithm synthesizes the block on the **right**. Note that the synthesized text looks like text; it appears to be constructed of words of varying lengths that are spaced like text; and each word looks as though it is composed of letters (though this illusion fails as one looks closely). *Figure from Texture Synthesis by Non-parametric Sampling, A. Efros and T.K. Leung, Proc. Int. Conf. Computer Vision, 1999 © 1999, IEEE*

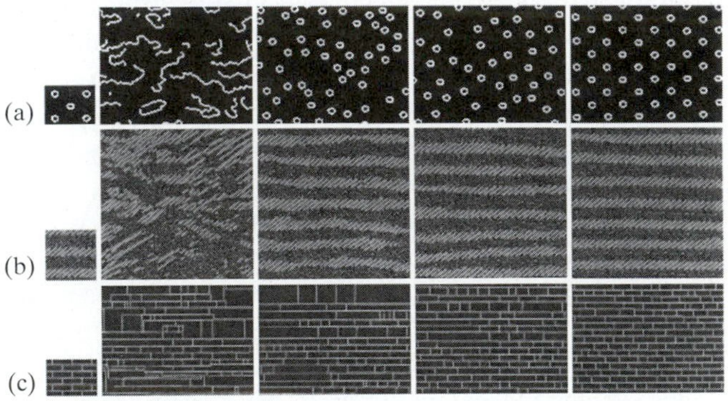

Figure 9.17 The size of the image neighbourhood to be matched makes a significant difference in Algorithm 9.3. In the figure, the textures at the right are synthesized from the small blocks on the **left**, using neighbourhoods that are increasingly large as one moves to the **right**. If very small neighbourhoods are matched, then the algorithm cannot capture large scale effects easily. For example, in the case of the spotty texture, if the neighbourhood is too small to capture the spot structure (and so sees only pieces of curve), the algorithm synthesizes a texture consisting of curve segments. As the neighbourhood gets larger, the algorithm can capture the spot structure, but not the even spacing. With very large neighbourhoods, the spacing is captured as well. *Figure from Texture Synthesis by Non-parametric Sampling, A. Efros and T.K. Leung, Proc. Int. Conf. Computer Vision, 1999 © 1999, IEEE*

the range over which pixels can affect one another's values directly (see Figure 9.17). Efros uses a square neighborhood, centered at the pixel of interest.

The similarity between two image neighbourhoods can be measured by forming the sum of squared differences of corresponding pixel values. This value is small when the neighbourhoods are similar, and large when they are different (it is essentially the length of the difference vector). Of course, the value of the pixel to be synthesized is not counted in the sum of squared differences.

Synthesizing Textures using Neighbourhoods Now we know how to obtain the value of a single missing pixel: choose uniformly and at random amongst the values of pixels in the example image whose neighborhoods match the neighbourhood of our pixel (i.e., where the sum of squared differences between the two neighbour hoods is smaller than some threshold).

Generally, we need to synthesize more than just one pixel. Usually, the values of some pixels in the neighborhood of the pixel to be synthesized are not known—these pixels need to be synthesized too. One way to obtain a collection of examples for the pixel of interest is to count only the known values in computing the sum of squared differences, and to adjust the threshold pro rata. The synthesis process can be started by choosing a block of pixels at random from the example image, yielding Algorithm 9.3.

Algorithm 9.3: Non-parametric Texture Synthesis

```
Choose a small square of pixels at random from the example image
Insert this square of values into the image to be synthesized
Until each location in the image to be synthesized has a value
  For each unsynthesized location on
    the boundary of the block of synthesized values
    Match the neighborhood of this location to the
      example image, ignoring unsynthesized
      locations in computing the matching score
    Choose a value for this location uniformly and at random
      from the set of values of the corresponding locations in the
      matching neighborhoods
  end
end
```

9.4 SHAPE FROM TEXTURE

A patch of texture of viewed frontally looks very different from a same patch viewed at a glancing angle, because foreshortening causes the texture elements (and the gaps between them!) to shrink more in some directions than in others. This suggests that we can recover some shape information from texture, at the cost of supplying a texture model. This is a task at which humans excel (Figure 9.18). Remarkably, quite general texture models appear to supply enough information to infer shape. This is most easily seen for planes (Section 9.4.1); while the details remain opaque in the case of curved surfaces, the general issues remain the same.

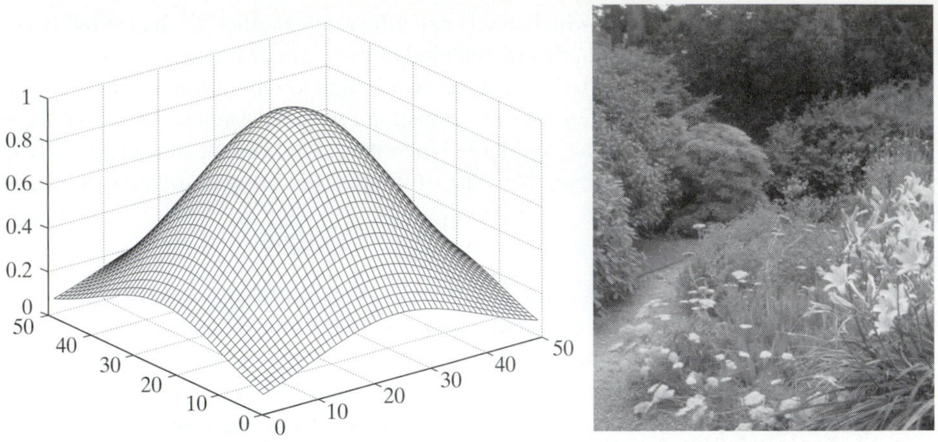

Figure 9.18 Humans obtain information about the shape of surfaces in space from the appearance of the texture on the surface. The figure on the **left** shows one common use for this effect—away from the contour regions, our only source of information about the surface depicted is the distortion of the texture on the surface. On the **right**, the texture of the bushes gives a sense they form rounded surfaces.

9.4.1 Shape from Texture for Planes

If we know we are viewing a plane, shape from texture boils down to determine the configuration of the plane relative to the camera. Assume that we hypothesize a configuration; we can then project the image texture back onto that plane. If we have some model of the "uniformity" of the texture, then we can test that property for the backprojected texture. We now obtain the plane with the "best" backprojected texture on it. This general strategy works for a variety of texture models. We will confine our discussion to the case of an orthographic camera. If the camera is not orthographic, the arguments we use will go through, but require substantially more work and more notation. We discuss other cases in the commentary.

Representing a Plane Now assume that we are viewing a single textured plane in an orthographic camera. Because the camera is orthographic, there is no way to measure the depth to the plane. However, we can think about the orientation of the plane. Let us work in terms of the camera coordinate system. We need to know firstly, the angle between the normal of the textured plane and the viewing direction—sometimes called the *slant*—and secondly, the angle the projected normal makes in the camera coordinate system—sometimes called the *tilt* (Figure 9.19). In an image of a plane, there is a *tilt direction*—the direction in the plane parallel to the projected normal.

Isotropy Assumptions An *isotropic* texture is one where the probability of encountering a texture element does not depend on the orientation of that element. This means that a probability model for an isotropic texture need not depend on the orientation of the coordinate system on the textured plane.

If we assume that the texture is isotropic, both slant and tilt can be read from the image. We could synthesize an orthographic view of a textured plane by first rotating the coordinate system by the tilt and then secondly contracting along one coordinate direction by the cosine of the slant—call this process a **viewing transformation**. The easiest way to see this is to assume

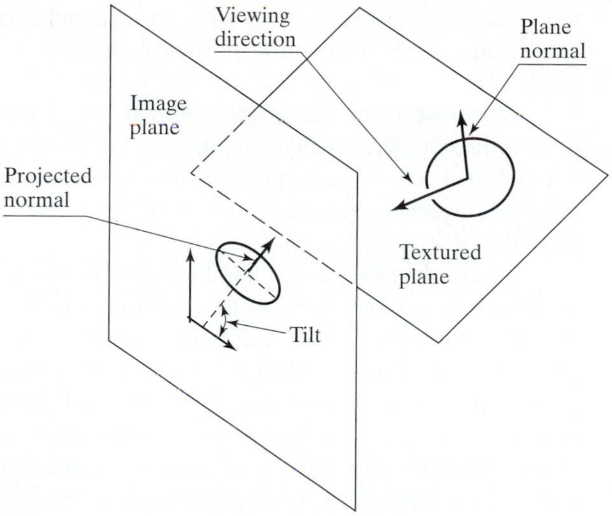

Figure 9.19 The orientation of a plane with respect to the camera plane can be given by the slant—which is the angle between the normal of the textured plane and the viewing direction—and the tilt—which is the angle the projected normal makes with the camera coordinate system. The figure illustrates the tilt, and shows a circle projecting to an ellipse.

that the texture consists of a set of circles, scattered about the plane. In an orthographic view, these circles will project to ellipses, whose minor axes will give the tilt, and whose aspect ratios will give the slant (see the exercises and Figure 9.19).

An orthographic view of an isotropic texture is *not* isotropic (unless the plane is parallel to the image plane). This is because the contraction in the slant direction interferes with the isotropy of the texture. Elements that point along the contracted direction get shorter. Furthermore, elements that have a component along the contracted direction have that component shrunk. Now corresponding to a viewing transformation is an **inverse viewing transformation** (which turns an image plane texture into the object plane texture, given a slant and tilt). This yields a strategy for determining the orientation of the plane: find an inverse viewing transformation that turns the image texture into an isotropic texture, and recover the slant and tilt from that inverse viewing transformation.

There are variety of ways to find this viewing transformation. One natural strategy is to use the energy output of a set of oriented filters. This is the squared response, summed over the image. For an isotropic texture, we would expect the energy output to be the same for each orientation at any given scale, because the probability of encountering a pattern does not depend on its orientation. Thus, a measure of isotropy is the standard deviation of the energy output as a function of orientation. We could sum this measure over scales, perhaps weighting the measure by the total energy in the scale. The smaller the measure, the more isotropic the texture. We now find the inverse viewing transformation that makes the image looks most isotropic by this measure, using standard methods from optimization.

Notice that this approach immediately extends to perspective projection, spherical projection, and other types of viewing transformation. We simply have to search over a larger family of transformations for the transformation that makes the image texture look most isotropic. One does need to be careful, however. For example, scaling an isotropic texture will lead to another isotropic texture, meaning that it isn't possible to recover a scaling parameter, and it's a bad idea to try. The main difficulty with using an assumption of isotropy to recover the orientation of a plane is that there are very few isotropic textures in the world.

Homogeneity Assumptions It isn't possible to recover the orientation of a plane in an orthographic view by assuming that the texture is homogeneous (the definition is in Section 9.3.1). This is because the viewing transformation takes one homogeneous texture into another homogeneous texture. However, if we assume that the view is perspective, it becomes possible.

To see this, first notice that homogeneity means that, if a large even grid is imposed on the plane, the number of events that occur in each box should be (approximately) the same. For example, if a texture consists of a homogenous pattern of spots, the expected number of spots per box is the same for any box. However, if we were to see a perspective view of a textured plane with a grid superimposed, then some grid elements would project to large quadrilaterals and others to very small quadrilaterals (unless the view is frontal). In turn, this means that the projected texture cannot be homogenous *in the image plane*—because some elements in a grid of boxes on the image plane will have many projected quadrilaterals and hence many texture events in them, and others will have few. For the example of the spotted plane, this just means that the spots that project close to the plane's horizon appear small. The appropriate strategy is now to choose a transformation that will make the image plane texture "most homogenous"; notice that we can determine the orientation of the plane with respect to the camera plane, but not its depth, because a frontal view of a homogenous texture is homogenous—everything scales by the same amount.

9.5 NOTES

We have aggressively compressed the texture literature in this chapter. Over the years, there have been a wide variety of techniques for representing image textures, typically looking at the statistics of how patterns lie with respect to one another. The disagreements are in how a pattern should be described, and what statistics to look at. While it is a bit early to say that the approach that represents patterns using linear filters is correct, it is currently dominant, mainly because it is very easy to solve problems with this strategy. Readers who are seriously interested in texture will probably most resent our omission of the Markov Random Field model, a choice based on the amount of mathematics required to develop the model and the absence of satisfactory inference algorithms for MRF's. We refer the interested reader to Chellappa and Jain (1993), Cross and Jain (1983), Manjunath and Chellappa (1991), or Speis and Healey (1996).

Another important omission is the discussion of wavelet methods for representing texture. While these methods follow the rather rough lines given above—represent a texture by thinking about the output of a lot of filters—there is a comprehensive theory behind those filters. We refer the interested reader to Ma and Manjunath (1995), (1996) or Manjunath and Ma (1996*b,c*).

Filters, Pyramids and Efficiency

If we are to represent texture with the output of a large range of filters at many scales and orientations, then we need to be efficient at filtering. This is a topic that has attracted much attention; the usual approach is to try and construct a tensor product basis that represents the available families of filters well. With an appropriate construction, we need to convolve the image with a small number of separable kernels, and can estimate the responses of many different filters by combining the results in different ways (hence the requirement that the basis be a tensor product). Significant papers include Freeman and Adelson (1991), Greenspan, Belongie, Perona, Goodman, Rakshit and Anderson (1994), Hel-Or and Teo (1996), Perona (1992), (1995), Simoncelli and Farid (1995), and Simoncelli and Freeman (1995).

Texture Synthesis

Texture synthesis exhausted us long before we could exhaust it. The most significant omission, apart from MRF's, is the work of Zhu, Wu and Mumford (1998), which uses sophisticated entropy criteria to firstly choose filters by which to represent a texture and secondly construct probability models for that texture.

Shape from Texture

There are surprisingly few methods for recovering a surface model from a projection of a texture field that is assumed to lie on that surface. **Global methods** attempt to recover an entire surface model, using assumptions about the distribution of texture elements. Appropriate assumptions are **isotropy** (Witkin, 1981) (the disadvantage of this method is that there are relatively few natural isotropic textures) or **homogeneity** (Aloimonos, 1986, Blake and Marinos, 1990). Methods based around homogeneity assume that texels are the result of a homogenous Poisson point process on a plane; the gradient of the density of the texel centers then yields the plane's parameters. However, deformation of individual texture elements is not accounted for.

Local methods recover some differential geometric parameters at a point on a surface (typically, normal and curvatures). This class of methods, which is due to Garding (1992), has been successfully demonstrated for a variety of surfaces by Malik and Rosenholtz (1997) and Rosenholtz and Malik (1997); a reformulation in terms of wavelets is due to Clerc and Mallat (1999). The methods have a crucial flaw; it is necessary either to know that texture element coordinate frames form a frame field that is locally parallel around the point in question, or to know the differential rotation of the frame field (see Garding, 1995 for this point, which is emphasized by the choice of textures displayed in Rosenholtz and Malik, 1997; the assumption is known as **texture stationarity**). For example, if one were to use these methods to recover the curvature of a doughnut dipped in chocolate sprinkles, it would be necessary to ensure that the sprinkles were all parallel on the surface (or that the field of angles from sprinkle to sprinkle was known). As a result, the method can be demonstrated to work only on quite a small class of textured surfaces. A second, important, difficulty lies in the data recovered; these methods all make local estimates of normal *and curvature*. But curvature is a derivative of the normal; as a result, while one local estimate may be helpful, there is no reason to believe that a collection of local estimates will be consistent. This is a problem of **integrability**. Surface interpolation methods have largely fallen out of fashion in computer vision, due to the uncertainty regarding the semantic status of surface patches in regions where data is absent. Shape from texture is a problem where an interpolate has an unquestionably useful role—it expresses the fact that, because one has a prior belief that surfaces are relatively slowly changing, incomplete local measurements of the surface normal can constrain one another and lead to good global estimates of the normal at some points.

PROBLEMS

9.1. Show that a circle appears as an ellipse in an orthographic view, and that the minor axis of this ellipse is the tilt direction. What is the aspect ratio of this ellipse?

9.2. We will study measuring the orientation of a plane in an orthographic view, given the texture consists of points laid down by a homogenous Poisson point process. Recall that one way to generate points according to such a process is to sample the x and y coordinate of the point uniformly and at random. We assume that the points from our process lie within a unit square.

(a) Show that the probability that a point will land in a particular set is proportional to the area of that set.

(b) Assume we partition the area into disjoint sets. Show that the number of points in each set has a multinomial probability distribution.

We will now use these observations to recover the orientation of the plane. We partition the *image texture* into a collection of disjoint sets.

(c) Is the area of each set, *backprojected onto the textured plane*, a function of the orientation of the plane?

(d) Is it possible to determine the plane's orientation using this information? Use the result of (c).

Programming Assignments

9.3. **Texture synthesis:** Implement the non-parametric texture synthesis algorithm of Section 9.3.2. Use your implementation to study:

(a) the effect of window size on the synthesized texture;

(b) the effect of window shape on the synthesized texture;

(c) the effect of the matching criterion on the synthesized texture (i.e., using weighted sum of squares instead of sum of squares, etc.).

9.4. **Texture representation:** Implement a texture classifier that can distinguish between at least six types of texture; use the scale selection mechanism of Section 9.1.2, and compute statistics of filter outputs. We recommend that you use at least the mean and covariance of the outputs of about six oriented bar filters and a spot filter. You may need to read up on classification in chapter 22; use a simple classifier (nearest neighbor using Mahalanobis distance should do the trick).

PART III
Early Vision: Multiple Images

10

The Geometry of Multiple Views

Despite the wealth of information contained in a photograph, the depth of a scene point along the corresponding projection ray is not directly accessible in a single image. With at least two pictures, depth can be measured through triangulation. This is of course one of the reasons that most animals have at least two eyes and/or move their head when looking for friend or foe, as well as the motivation for equipping an autonomous robot with a stereo or motion analysis system. Before building such a system, we must understand how several views of the same scene constrain its three-dimensional structure as well as the corresponding camera configurations. This is the goal of this chapter. In particular, we elucidate the geometric and algebraic constraints that hold among two, three, or more views of the same scene. In the familiar setting of binocular stereo vision, we show that the first image of any point must lie in the plane formed by its second image and the optical centers of the two cameras. This *epipolar constraint* can be represented algebraically by a 3×3 matrix, called the *essential matrix* when the intrinsic parameters of the cameras are known and the *fundamental matrix* otherwise. Three pictures of the same line introduce a different constraint—namely, that the intersection of the planes formed by their preimages be degenerate. Algebraically, this geometric relationship can be represented by a $3 \times 3 \times 3$ *trifocal tensor*. More images introduce additional constraints, for example four projections of the same point satisfy certain quadrilinear relations whose coefficients are captured by the *quadrifocal tensor*. Remarkably, the equations satisfied by multiple pictures of the same scene feature can be set up without any knowledge of the cameras or the scene they observe, and a number of methods for estimating their parameters directly from image data are presented in this chapter.

Computer vision is not the only scientific field concerned with the geometry of multiple views: The goal of *photogrammetry*, already mentioned in chapter 3, is precisely to recover quantitative geometric information from multiple pictures. Applications of the epipolar and trifocal constraints to the classical photogrammetry problem of *transfer* (i.e., the prediction of the position of a point in an image given its position in a number of reference pictures) are briefly discussed in this chapter, along with some examples. Many more applications in the domains of stereo and motion analysis are presented in latter chapters.

10.1 TWO VIEWS

10.1.1 Epipolar Geometry

Consider the images p and p' of a point P observed by two cameras with optical centers O and O'. These five points all belong to the *epipolar plane* defined by the two intersecting rays OP and $O'P$ (Figure 10.1). In particular, the point p' lies on the line l' where this plane and the retina Π' of the second camera intersect. The line l' is the *epipolar line* associated with the point p, and it passes through the point e' where the *baseline* joining the optical centers O and O' intersects Π'. Likewise, the point p lies on the epipolar line l associated with the point p', and this line passes through the intersection e of the baseline with the plane Π.

The points e and e' are called the *epipoles* of the two cameras. The epipole e' is the projection of the optical center O of the first camera in the image observed by the second camera and vice versa. As noted before, if p and p' are images of the same point, then p' must lie on the epipolar line associated with p. This *epipolar constraint* plays a fundamental role in stereo vision and motion analysis.

Let us assume, for example, that we know the intrinsic and extrinsic parameters of the two cameras of a *stereo rig*. As shown in chapter 11, the most difficult part of stereo data analysis is establishing correspondences between the two images (i.e., deciding which points in the second picture match the points in the first one). The epipolar constraint greatly limits the search for these correspondences: Indeed, since we assume that the rig is calibrated, the coordinates of the point p completely determine the ray joining O and p, and thus the associated epipolar plane $OO'p$ and epipolar line l'. The search for matches can be restricted to this line instead of the whole image (Figure 10.2). In two-frame motion analysis, each camera may be internally calibrated, but the rigid transformation separating the two camera coordinate systems is unknown. In this case, the epipolar geometry obviously constrains the set of possible motions. The next sections explore several variants of this situation.

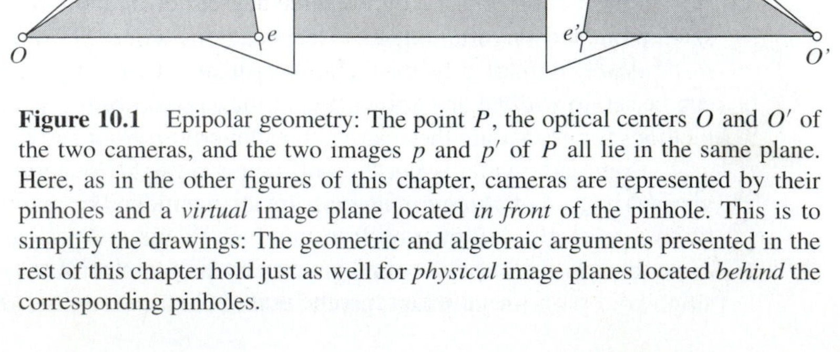

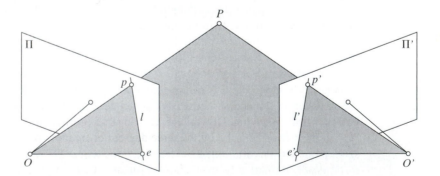

Figure 10.1 Epipolar geometry: The point P, the optical centers O and O' of the two cameras, and the two images p and p' of P all lie in the same plane. Here, as in the other figures of this chapter, cameras are represented by their pinholes and a *virtual* image plane located *in front* of the pinhole. This is to simplify the drawings: The geometric and algebraic arguments presented in the rest of this chapter hold just as well for *physical* image planes located *behind* the corresponding pinholes.

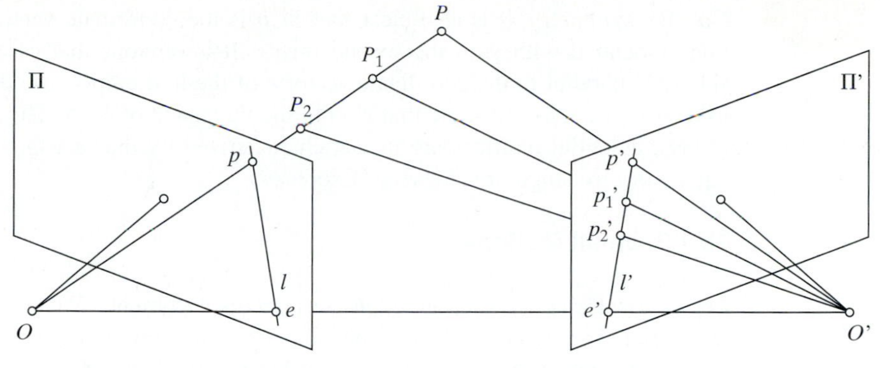

Figure 10.2 Epipolar constraint: Given a calibrated stereo rig, the set of possible matches for the point p is constrained to lie on the associated epipolar line l'.

10.1.2 The Calibrated Case

Here we assume that the intrinsic parameters of each camera are known, so $\boldsymbol{p} = \hat{\boldsymbol{p}}$. Clearly, the epipolar constraint implies that the three vectors $\overrightarrow{Op}$, $\overrightarrow{O'p'}$, and $\overrightarrow{OO'}$ are coplanar. Equivalently, one of them must lie in the plane spanned by the other two, or

$$\overrightarrow{Op} \cdot [\overrightarrow{OO'} \times \overrightarrow{O'p'}] = 0.$$

We can rewrite this coordinate-independent equation in the coordinate frame associated to the first camera as

$$\boldsymbol{p} \cdot [\boldsymbol{t} \times (\mathcal{R}\boldsymbol{p}')], \tag{10.1}$$

where $\boldsymbol{p} = (u, v, 1)^T$ and $\boldsymbol{p}' = (u', v', 1)^T$ denote the homogeneous image coordinate vectors of p and p', $\boldsymbol{t}$ is the coordinate vector of the translation $\overrightarrow{OO'}$ separating the two coordinate systems, and $\mathcal{R}$ is the rotation matrix such that a free vector with coordinates $\boldsymbol{w}'$ in the second coordinate system has coordinates $\mathcal{R}\boldsymbol{w}'$ in the first one. In this case, the two projection matrices are given in the coordinate system attached to the first camera by $(\text{Id} \quad \boldsymbol{0})$ and $(\mathcal{R}^T \quad -\mathcal{R}^T\boldsymbol{t})$.

Equation (10.1) can finally be rewritten as

$$\boldsymbol{p}^T \mathcal{E} \boldsymbol{p}' = 0, \tag{10.2}$$

where $\mathcal{E} = [\boldsymbol{t}_\times]\mathcal{R}$, and $[\boldsymbol{a}_\times]$ denotes the skew-symmetric matrix such that $[\boldsymbol{a}_\times]\boldsymbol{x} = \boldsymbol{a} \times \boldsymbol{x}$ is the cross-product of the vectors $\boldsymbol{a}$ and $\boldsymbol{x}$. The matrix $\mathcal{E}$ is called the *essential matrix*, and it was first introduced by Longuet–Higgins (1981). Its nine coefficients are only defined up to scale, and they can be parameterized by the three degrees of freedom of the rotation matrix $\mathcal{R}$ and the two degrees of freedom defining the direction of the translation vector $\boldsymbol{t}$.

Note that $\mathcal{E}\boldsymbol{p}'$ can be interpreted as the coordinate vector representing the epipolar line associated with the point p' in the first image: Indeed, an image line l can be defined by its equation $au + bv + c = 0$, where (u, v) denote the coordinates of a point on the line, (a, b) is the unit normal to the line, and $-c$ is the (signed) distance between the origin and l. Alternatively, we can define the line equation in terms of the homogeneous coordinate vector $\boldsymbol{p} = (u, v, 1)^T$ of a point on the line and the vector $\boldsymbol{l} = (a, b, c)^T$ by $\boldsymbol{l} \cdot \boldsymbol{p} = 0$, in which case the constraint $a^2 + b^2 = 1$ is relaxed since the equation holds independently of any scale change applied to $\boldsymbol{l}$. In this context, Eq. (10.2) expresses the fact that the point p lies on the epipolar line associated with the vector

$\mathcal{E}p'$. By symmetry, it is also clear that $\mathcal{E}^T p$ is the coordinate vector representing the epipolar line associated with p in the second image. It is obvious that essential matrices are singular since $\boldsymbol{t}$ is parallel to the coordinate vector $\boldsymbol{e}$ of the first epipole, so that $\mathcal{E}^T \boldsymbol{e} = -\mathcal{R}^T[\boldsymbol{t}_\times]\boldsymbol{e} = 0$. Likewise, it is easy to show that $\boldsymbol{e}'$ is in the nullspace of $\mathcal{E}$. As shown by Huang and Faugeras (1989), essential matrices are in fact characterized by the fact that they are singular with two equal nonzero singular values (see Exercises).

10.1.3 Small Motions

Let us now turn our attention to *infinitesimal* displacements. We consider a moving camera with translational velocity $\boldsymbol{v}$ and rotational velocity $\boldsymbol{\omega}$ and rewrite Eq. (10.2) for two frames separated by a small time interval δt. Let us denote by $\dot{\boldsymbol{p}} = (\dot{u}, \dot{v}, 0)^T$ the velocity of the point p or *motion field*. Using the *exponential representation* of rotations (see Exercises), it is possible to show that (to first order)

$$\begin{cases} \boldsymbol{t} = \delta t\, \boldsymbol{v}, \\ \mathcal{R} = \mathrm{Id} + \delta t\, [\boldsymbol{\omega}_\times], \\ \boldsymbol{p}' = \boldsymbol{p} + \delta t\, \dot{\boldsymbol{p}}. \end{cases} \tag{10.3}$$

Substituting in Eq. (10.2) and neglecting all terms of order two or greater in δt yields:

$$\boldsymbol{p}^T([\boldsymbol{v}_\times][\boldsymbol{\omega}_\times])\boldsymbol{p} - (\boldsymbol{p} \times \dot{\boldsymbol{p}}) \cdot \boldsymbol{v} = 0. \tag{10.4}$$

Equation (10.4) is simply the instantaneous form of the Longuet–Higgins relation (10.2), which captures the epipolar geometry in the discrete case. Note that in the case of pure translation, we have $\boldsymbol{\omega} = 0$, thus $(\boldsymbol{p} \times \dot{\boldsymbol{p}}) \cdot \boldsymbol{v} = 0$. In other words, the three vectors $\boldsymbol{p} = \overrightarrow{op}$, $\dot{\boldsymbol{p}}$, and $\boldsymbol{v}$ must be coplanar. If e denotes the infinitesimal epipole or *focus of expansion* (i.e., the point where the line passing through the optical center and parallel to the velocity vector $\boldsymbol{v}$ pierces the image plane), we obtain the well-known result that the motion field points toward the focus of expansion under pure translational motion (Figure 10.3).

10.1.4 The Uncalibrated Case

The Longuet–Higgins relation holds for *internally calibrated* cameras. When the intrinsic parameters are unknown (*uncalibrated* cameras), we can write $\boldsymbol{p} = \mathcal{K}\hat{\boldsymbol{p}}$ and $\boldsymbol{p}' = \mathcal{K}'\hat{\boldsymbol{p}}'$, where $\mathcal{K}$

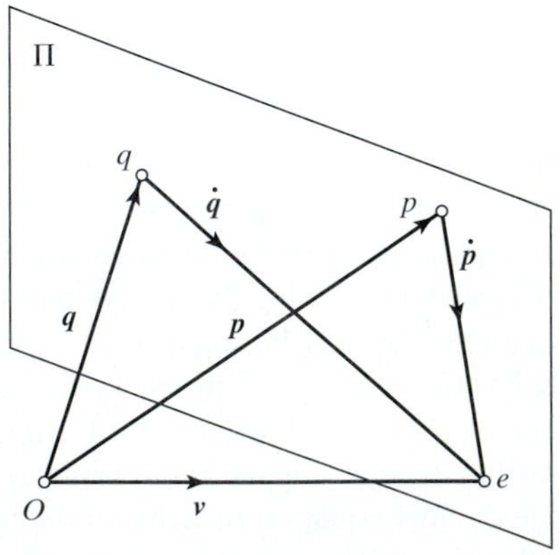

Figure 10.3 Focus of expansion: Under pure translation, the motion field at every point in the image points toward the focus of expansion.

and $\mathcal{K}'$ are 3×3 calibration matrices and $\hat{\boldsymbol{p}}$ and $\hat{\boldsymbol{p}}'$ are normalized image coordinate vectors. The Longuet–Higgins relation holds for these vectors, and we obtain

$$\boldsymbol{p}^T \mathcal{F} \boldsymbol{p}' = 0, \tag{10.5}$$

where the matrix $\mathcal{F} = \mathcal{K}^{-T} \mathcal{E} \mathcal{K}'^{-1}$, called the *fundamental matrix*, is not, in general, an essential matrix. It has again rank two, and the eigenvector of $\mathcal{F}$ (resp. $\mathcal{F}^T$) corresponding to its zero eigenvalue is as before the position $\boldsymbol{e}'$ (resp. $\boldsymbol{e}$) of the epipole. Note that $\mathcal{F}\boldsymbol{p}'$ (resp. $\mathcal{F}^T \boldsymbol{p}$) represents the epipolar line corresponding to the point $\boldsymbol{p}'$ (resp. $\boldsymbol{p}$) in the first (resp. second) image.

The rank two constraint means that the fundamental matrix only admits seven independent parameters. Several choices of parameterization are possible, but the most natural one is in terms of the coordinate vectors $\boldsymbol{e} = (\alpha, \beta)^T$ and $\boldsymbol{e}' = (\alpha', \beta')^T$ of the two epipoles, and of the *epipolar transformation* that maps one set of epipolar lines onto the other one. We examine in chapter 13 the properties of this transformation in the context of structure from motion. For the time being, let us just note (without proof) that it can be parameterized (up to scale) by four numbers—a, b, c, d—and that the fundamental matrix can be written as

$$\mathcal{F} = \begin{pmatrix} b & a & -a\beta - b\alpha \\ -d & -c & c\beta + d\alpha \\ d\beta' - b\alpha' & c\beta' - a\alpha' & -c\beta\beta' - d\beta'\alpha + a\beta\alpha' + b\alpha\alpha' \end{pmatrix}. \tag{10.6}$$

10.1.5 Weak Calibration

As mentioned earlier, the essential matrix is defined up to scale by five independent parameters. It is therefore possible (at least in principle) to calculate it by writing Eq. (10.2) for five point correspondences. Likewise, the fundamental matrix is defined by seven independent coefficients (the parameters a, b, c, d in Eq. (10.6) are only defined up to scale) and can in principle be estimated from seven point correspondences. Methods for estimating the essential and fundamental matrices from a minimal number of parameters indeed exist (see Notes), but they are far too involved to be described here. This section addresses the simpler problem of estimating the epipolar geometry from a redundant set of point correspondences between two images taken by cameras with unknown intrinsic parameters—a process known as *weak calibration*.

Note that Eq. (10.5) is linear in the nine coefficients of the fundamental matrix $\mathcal{F}$:

$$(u, v, 1) \begin{pmatrix} F_{11} & F_{12} & F_{13} \\ F_{21} & F_{22} & F_{23} \\ F_{31} & F_{32} & F_{33} \end{pmatrix} \begin{pmatrix} u' \\ v' \\ 1 \end{pmatrix} = 0. \tag{10.7}$$

Since this equation is homogeneous in the coefficients of $\mathcal{F}$, we can set $F_{33} = 1$ and use eight point correspondences $p_i \leftrightarrow p_i'$ $(i = 1, \ldots, 8)$ to rewrite the corresponding instances of Eq. (10.7) as an 8×8 system of nonhomogeneous linear equations:

$$\begin{pmatrix} u_1u_1' & u_1v_1' & u_1 & v_1u_1' & v_1v_1' & v_1 & u_1' & v_1' \\ u_2u_2' & u_2v_2' & u_2 & v_2u_2' & v_2v_2' & v_2 & u_2' & v_2' \\ u_3u_3' & u_3v_3' & u_3 & v_3u_3' & v_3v_3' & v_3 & u_3' & v_3' \\ u_4u_4' & u_4v_4' & u_4 & v_4u_4' & v_4v_4' & v_4 & u_4' & v_4' \\ u_5u_5' & u_5v_5' & u_5 & v_5u_5' & v_5v_5' & v_5 & u_5' & v_5' \\ u_6u_6' & u_6v_6' & u_6 & v_6u_6' & v_6v_6' & v_6 & u_6' & v_6' \\ u_7u_7' & u_7v_7' & u_7 & v_7u_7' & v_7v_7' & v_7 & u_7' & v_7' \\ u_8u_8' & u_8v_8' & u_8 & v_8u_8' & v_8v_8' & v_8 & u_8' & v_8' \end{pmatrix} \begin{pmatrix} F_{11} \\ F_{12} \\ F_{13} \\ F_{21} \\ F_{22} \\ F_{23} \\ F_{31} \\ F_{32} \end{pmatrix} = - \begin{pmatrix} 1 \\ 1 \\ 1 \\ 1 \\ 1 \\ 1 \\ 1 \\ 1 \end{pmatrix}.$$

Using this system to estimate the fundamental matrix gives the *eight-point algorithm* originally proposed by Longuet–Higgins (1981) in the case of calibrated cameras. It fails when the associated 8×8 matrix is singular. As shown in Faugeras (1993) and the exercises, this only happens when the eight points and two optical centers lie on a quadric surface. Fortunately, this is quite unlikely since a quadric surface is completely determined by nine points, which means that there is generally no quadric that passes through 10 arbitrary points.

When $n > 8$ correspondences are available, $\mathcal{F}$ can be estimated using linear least squares by minimizing

$$\sum_{i=1}^{n} (\boldsymbol{p}_i^T \mathcal{F} \boldsymbol{p}_i')^2 \tag{10.8}$$

with respect to the coefficients of $\mathcal{F}$ under the constraint that the vector formed by these coefficients has unit norm.

Note that both the eight-point algorithm and its least-squares version ignore the rank two property of fundamental matrices.[1] To enforce this constraint, Luong *et al.* (1993, 1996) proposed to use the matrix $\mathcal{F}$ output by the eight-point algorithm as the basis for a two-step estimation process: First, use linear least squares to find the epipoles $\boldsymbol{e}$ and $\boldsymbol{e}'$ that minimize $|\mathcal{F}^T \boldsymbol{e}|^2$ and $|\mathcal{F}\boldsymbol{e}'|^2$; second, substitute the coordinates of these points in Eq. (10.6): This yields a linear parameterization of the fundamental matrix by the coefficients of the epipolar transformation, which can now be estimated by minimizing Eq. (10.8) via linear least squares.

The least-squares version of the eight-point algorithm minimizes the mean-squared *algebraic distance* associated with the epipolar constraint (i.e., the mean-squared value of $e(\boldsymbol{p}, \boldsymbol{p}') = \boldsymbol{p}^T \mathcal{F} \boldsymbol{p}'$ calculated over all point correspondences). This error function admits a geometric interpretation: In particular, we have

$$e(\boldsymbol{p}, \boldsymbol{p}') = \lambda \mathrm{d}(\boldsymbol{p}, \mathcal{F}\boldsymbol{p}') = \lambda' \mathrm{d}(\boldsymbol{p}', \mathcal{F}^T \boldsymbol{p}),$$

where $d(\boldsymbol{p}, \boldsymbol{l})$ denotes the (signed) Euclidean distance between the point $\boldsymbol{p}$ and the line $\boldsymbol{l}$, and $\mathcal{F}\boldsymbol{p}$ and $\mathcal{F}^T\boldsymbol{p}'$ are the epipolar lines associated with $\boldsymbol{p}$ and $\boldsymbol{p}'$. The scale factors λ and λ' are simply the norms of the vectors formed by the first two components of $\mathcal{F}\boldsymbol{p}'$ and $\mathcal{F}^T\boldsymbol{p}$, and their dependence on the pair of data points observed may bias the estimation process.

It is of course possible to eliminate the scale factors and directly minimize the mean-squared *geometric* distance between the image points and the corresponding epipolar lines—that is,

$$\sum_{i=1}^{n} \left[\mathrm{d}^2(\boldsymbol{p}_i, \mathcal{F}\boldsymbol{p}_i') + \mathrm{d}^2(\boldsymbol{p}_i', \mathcal{F}^T \boldsymbol{p}_i) \right].$$

This is a nonlinear problem regardless of the parameterization chosen for the fundamental matrix, but the minimization can be initialized with the result of the eight-point algorithm. This method was first proposed by Luong *et al.* (1993), and it has been shown to provide results vastly superior to those obtained using the eight-point method. As an alternative, Hartley (1995) proposed to *normalize* the linear eight-point algorithm. This approach is based on the observation that the poor performance of the original technique is due, for the most part, to poor numerical conditioning. This suggests translating and scaling the data so they are centered at the origin and the average distance to the origin is $\sqrt{2}$ pixel. In practice, this normalization dramatically improves the conditioning of the linear least-squares estimation process. Concretely, the algorithm is divided into four steps: First, transform the image coordinates using appropriate translation and scaling operators $\mathcal{T} : \boldsymbol{p}_i \to \tilde{\boldsymbol{p}}_i$ and $\mathcal{T}' : \boldsymbol{p}_i' \to \tilde{\boldsymbol{p}}_i'$. Second, use linear least squares to compute

[1]The original algorithm proposed by Longuet–Higgins ignores that essential matrices have rank two and two equal singular values as well.

the matrix $\tilde{\mathcal{F}}$ minimizing

$$\sum_{i=1}^{n} (\tilde{\boldsymbol{p}}_i^T \tilde{\mathcal{F}} \tilde{\boldsymbol{p}}_i')^2.$$

Third, enforce the rank two constraint; this can be done using the two-step method of Luong *et al.* described earlier, but Hartley uses instead a technique suggested by Tsai and Huang (1984) in the calibrated case, which constructs the *singular value decomposition* $\tilde{\mathcal{F}} = \mathcal{U}\mathcal{S}\mathcal{V}^T$ of $\tilde{\mathcal{F}}$. Singular value decomposition is formally defined in chapter 12. Let us just note here that $\mathcal{S} = \text{diag}(r, s, t)$ is a diagonal 3×3 matrix with entries $r \geq s \geq t$, $\mathcal{U}, \mathcal{V}$ are orthogonal 3×3 matrices, and, as shown in chapter 12, the rank two matrix $\bar{\mathcal{F}}$ minimizing the Frobenius norm of $\tilde{\mathcal{F}} - \bar{\mathcal{F}}$ is simply $\bar{\mathcal{F}} = \mathcal{U}\text{diag}(r, s, 0)\mathcal{V}^T$. The last step of the algorithm sets $\mathcal{F} = \mathcal{T}^T \bar{\mathcal{F}} \mathcal{T}'$ as the final estimate of the fundamental matrix.

Figure 10.4 shows weak-calibration experiments using as input data a set of 37 point correspondences between two images of a toy house. The data points are shown in the figure as small discs, and the recovered epipolar lines are shown as short line segments. Figure 10.4(a) shows the output of the least-squares version of the plain eight-point algorithm, and Figure 10.4(b) shows the results obtained using Hartley's variant of this method. As expected, the results are much better in the second case and, in fact, extremely close to those obtained using the geometric distance criterion of Luong *et al.* (1993, 1996).

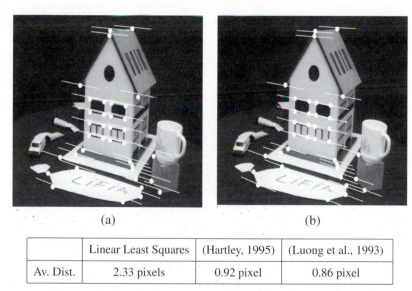

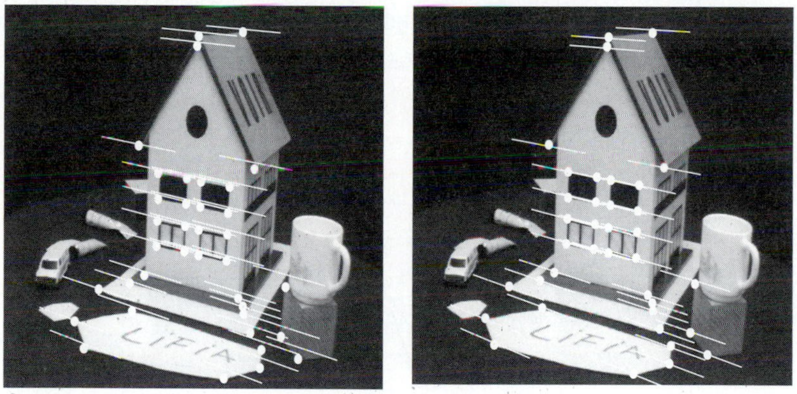

(a) (b)

	Linear Least Squares	(Hartley, 1995)	(Luong et al., 1993)
Av. Dist.	2.33 pixels	0.92 pixel	0.86 pixel

Figure 10.4 Weak-calibration experiment using 37 point correspondences between two images of a toy house. The figure shows the epipolar lines found by (a) the least-squares version of the eight-point algorithm, and (b) the normalized variant of this method proposed by Hartley (1995). Note, for example, the much larger error in (a) for the feature point close to the bottom of the mug. Quantitative comparisons are given in the table, where the average distances between the data points and corresponding epipolar lines are shown for both techniques as well as the nonlinear algorithm of Luong *et al.* (1993). *Data courtesy of Boubakeur Boufama and Roger Mohr.*

10.2 THREE VIEWS

Let us now go back to the calibrated case where $\boldsymbol{p} = \hat{\boldsymbol{p}}$ as we study the geometric constraints associated with three views of the same scene. Consider three perspective cameras observing the same point P, whose images are denoted by p_1, p_2, and p_3 (Figure 10.5). The optical centers O_1, O_2, and O_3 of the cameras define a *trifocal plane* that intersects their retinas along three *trifocal lines* t_1, t_2, and t_3. Each one of these lines passes through the associated epipoles (e.g., the line t_2 associated with the second camera passes through the projections e_{12} and e_{32} of the optical centers of the two other cameras).

Each pair of cameras defines an epipolar constraint—that is,

$$\begin{cases} \boldsymbol{p}_1^T \mathcal{E}_{12} \boldsymbol{p}_2 = 0, \\ \boldsymbol{p}_2^T \mathcal{E}_{23} \boldsymbol{p}_3 = 0, \\ \boldsymbol{p}_3^T \mathcal{E}_{31} \boldsymbol{p}_1 = 0, \end{cases} \tag{10.9}$$

where $\mathcal{E}_{ij}$ denotes the essential matrix associated with the image pairs $i \leftrightarrow j$. These three constraints are not independent since we must have $\boldsymbol{e}_{31}^T \mathcal{E}_{12} \boldsymbol{e}_{32} = \boldsymbol{e}_{12}^T \mathcal{E}_{23} \boldsymbol{e}_{13} = \boldsymbol{e}_{23}^T \mathcal{E}_{31} \boldsymbol{e}_{21} = 0$ (to see why, consider the epipoles e_{31} and e_{32}; they are the first and second images of the optical center O_3 of the third camera and are therefore in epipolar correspondence).

Any two of the equations in Eq. (10.9) are independent. In particular, when the essential matrices are known, it is possible to predict the position of the point p_1 from the positions of the two corresponding points p_2 and p_3: Indeed, the first and third constraints in Eq. (10.9) form

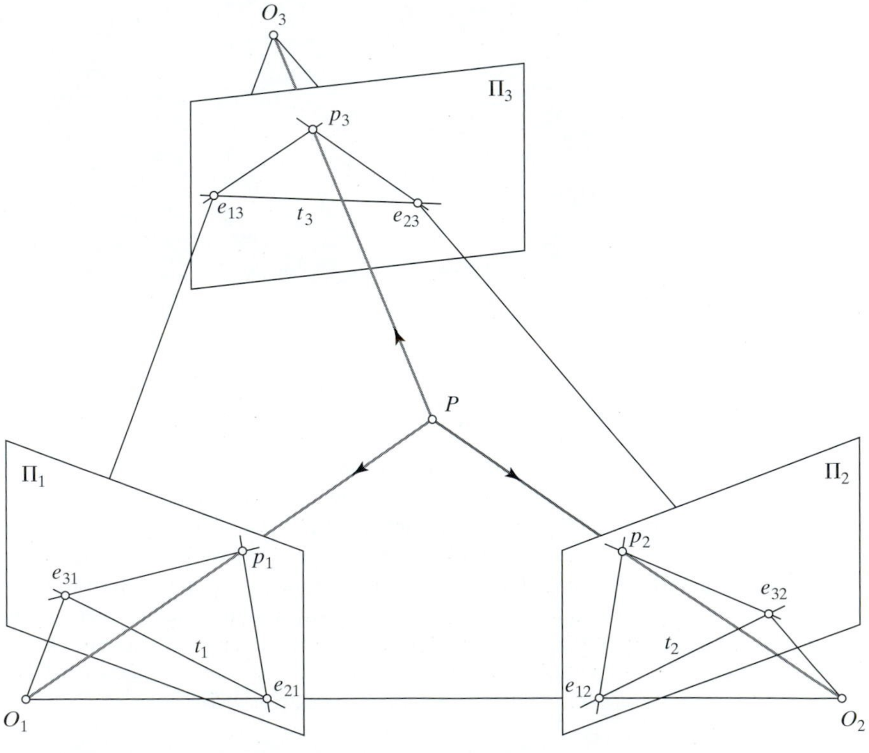

Figure 10.5 Trinocular epipolar geometry. Note that the point P does *not* lie, in general, in the trifocal plane defined by the points O_1, O_2, and O_3.

a system of two linear equations in the two unknown coordinates of p_1. Geometrically, p_1 is found as the intersection of the epipolar lines associated with p_2 and p_3 (Figure 10.5). Thus, the trinocular epipolar geometry offers a solution to the problem of transfer mentioned in the introduction.

10.2.1 Trifocal Geometry

A second set of constraints can be obtained by considering three images of a line instead of a point: The set of points that project onto an image line l is the plane L that contains the line and pinhole. We can characterize this plane as follows: If $\mathcal{M}$ denotes a 3×4 projection matrix, a point P in L projects onto the point p on l when $z\boldsymbol{p} = \mathcal{M}\boldsymbol{P}$, or

$$\boldsymbol{l}^T \mathcal{M}\boldsymbol{P} = 0, \tag{10.10}$$

where $\boldsymbol{P} = (x, y, z, 1)^T$ is the 4-vector of homogeneous coordinates of P and $\boldsymbol{l} = (a, b, c)^T$ is the 3-vector of homogeneous coordinates of l. Equation (10.10) is, of course, the equation of the plane L that contains both the optical center O of the camera and the line l, and $\boldsymbol{L} = \mathcal{M}^T\boldsymbol{l}$ is the coordinate vector of this plane.

Two images l_1 and l_2 of the same line do not constrain the relative position and orientation of the associated cameras since the corresponding planes L_1 and L_2 always intersect (unless they are parallel, in which case they can be thought of as intersecting *at infinity*; more on this in chapter 13). Let us now consider three images l_i, l_2, and l_3 of the same line l and denote by L_1, L_2, and L_3 the associated planes (Figure 10.6). The intersection of these planes forms a line instead of being reduced to a point in the generic case. Algebraically, this means that the system

$$\begin{pmatrix} \boldsymbol{L}_1^T \\ \boldsymbol{L}_2^T \\ \boldsymbol{L}_3^T \end{pmatrix} \boldsymbol{P} = \boldsymbol{0}$$

of three equations in three unknowns x, y, and z must be degenerate, or, equivalently, the rank of the 3×4 matrix

$$\mathcal{L} \stackrel{\text{def}}{=} \begin{pmatrix} \boldsymbol{l}_1^T \mathcal{M}_1 \\ \boldsymbol{l}_2^T \mathcal{M}_2 \\ \boldsymbol{l}_3^T \mathcal{M}_3 \end{pmatrix}$$

must be 2, which in turn implies that the determinants of all its 3×3 minors must be zero. These determinants are clearly trilinear combinations of the coordinate vectors $\boldsymbol{l}_1$, $\boldsymbol{l}_2$, and $\boldsymbol{l}_3$. As shown next, only two of the four determinants are independent.

10.2.2 The Calibrated Case

To obtain an explicit formula for the trilinear constraints, we pick the coordinate system attached to the first camera as the world reference frame, which allows us to write the projection matrices as $\mathcal{M}_1 = (\text{Id} \quad \boldsymbol{0})$, $\mathcal{M}_2 = (\mathcal{R}_2 \quad \boldsymbol{t}_2)$, and $\mathcal{M}_3 = (\mathcal{R}_3 \quad \boldsymbol{t}_3)$, and to rewrite $\mathcal{L}$ as

$$\mathcal{L} = \begin{pmatrix} \boldsymbol{l}_1^T & 0 \\ \boldsymbol{l}_2^T \mathcal{R}_2 & \boldsymbol{l}_2^T \boldsymbol{t}_2 \\ \boldsymbol{l}_3^T \mathcal{R}_3 & \boldsymbol{l}_3^T \boldsymbol{t}_3 \end{pmatrix}. \tag{10.11}$$

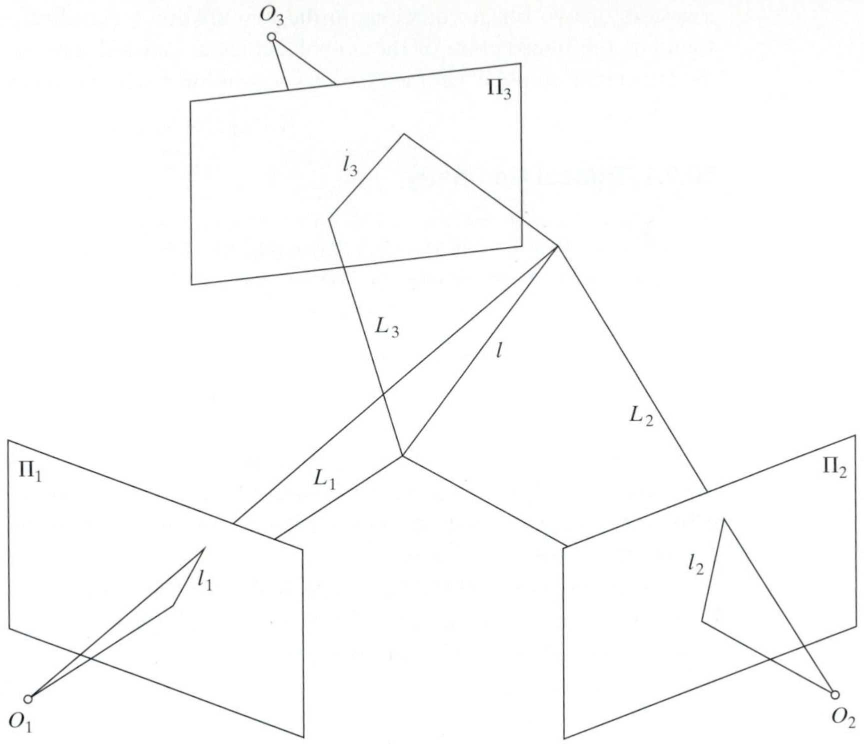

Figure 10.6 Three images of a line define it as the (degenerate) intersection of three planes.

As shown in the exercises, three of the minor determinants can be written together as

$$l_1 \times \begin{pmatrix} l_2^T \mathcal{G}_1^1 l_3 \\ l_2^T \mathcal{G}_1^2 l_3 \\ l_2^T \mathcal{G}_1^3 l_3 \end{pmatrix} = \mathbf{0}, \tag{10.12}$$

where

$$\mathcal{G}_1^i = t_2 R_3^{iT} - R_2^i t_3^T \quad \text{for} \quad i = 1, 2, 3, \tag{10.13}$$

and R_2^i and R_3^i ($i = 1, 2, 3$) denote the columns of $\mathcal{R}_2$ and $\mathcal{R}_3$. The fourth determinant is equal to $|l_1 \ \ \mathcal{R}_2 l_2 \ \ \mathcal{R}_3 l_3|$, and it is zero when the normals to the planes L_1, L_2, and L_3 are coplanar. The corresponding equation can be written as a linear combination of the three determinants in Eq. (10.12) (see Exercises). Only two of those are linearly independent of course.

The three 3×3 matrices $\mathcal{G}_1^i$ define the $3 \times 3 \times 3$ *trifocal tensor* with 27 coefficients (or 26 up to scale). (A *tensor* is the multidimensional array of coefficients associated with a multilinear form, in the same way that matrices are associated with bilinear forms.) Since O_1 is the origin of the coordinate system in which all projection equations are expressed, the vectors t_2 and t_3 can be interpreted as the homogeneous image coordinates of the epipoles e_{12} and e_{13}. In particular, it follows from Eq. (10.13) that $l_2^T \mathcal{G}_1^i l_3 = 0$ for any pair of matching epipolar lines l_2 and l_3.

Equation (10.12) can be rewritten as

$$\boldsymbol{l}_1 \propto \begin{pmatrix} \boldsymbol{l}_2^T \mathcal{G}_1^1 \boldsymbol{l}_3 \\ \boldsymbol{l}_2^T \mathcal{G}_1^2 \boldsymbol{l}_3 \\ \boldsymbol{l}_2^T \mathcal{G}_1^3 \boldsymbol{l}_3 \end{pmatrix}, \tag{10.14}$$

where we use $\boldsymbol{a} \propto \boldsymbol{b}$ to indicate that two vectors $\boldsymbol{a}$ and $\boldsymbol{b}$ only differ by a (nonzero) scale factor. It follows that the trifocal tensor also constrains the positions of three corresponding points: Indeed, suppose that P is a point on l. Its first image lies on l_1, so $\boldsymbol{p}_1^T \boldsymbol{l}_1 = 0$. In particular,

$$\boldsymbol{p}_1^T \begin{pmatrix} \boldsymbol{l}_2^T \mathcal{G}_1^1 \boldsymbol{l}_3 \\ \boldsymbol{l}_2^T \mathcal{G}_1^2 \boldsymbol{l}_3 \\ \boldsymbol{l}_2^T \mathcal{G}_1^3 \boldsymbol{l}_3 \end{pmatrix} = 0. \tag{10.15}$$

Given three point correspondences $p_1 \leftrightarrow p_2 \leftrightarrow p_3$ (Figure 10.7), we obtain four independent constraints by rewriting Eq. (10.15) for independent pairs of lines passing through p_2 and p_3 (e.g., $\boldsymbol{l}_i' = [1, 0, -u_i]^T$ and $\boldsymbol{l}_i'' = [0, 1, -v_i]^T$ for $i = 2, 3$). These constraints are trilinear in the coordinates of the points p_1, p_2, and p_3. When the tensor is known, it can thus be used to predict the position of, say, p_1 from the positions of p_2 and p_3 in the other images, giving a second solution to the transfer problem.

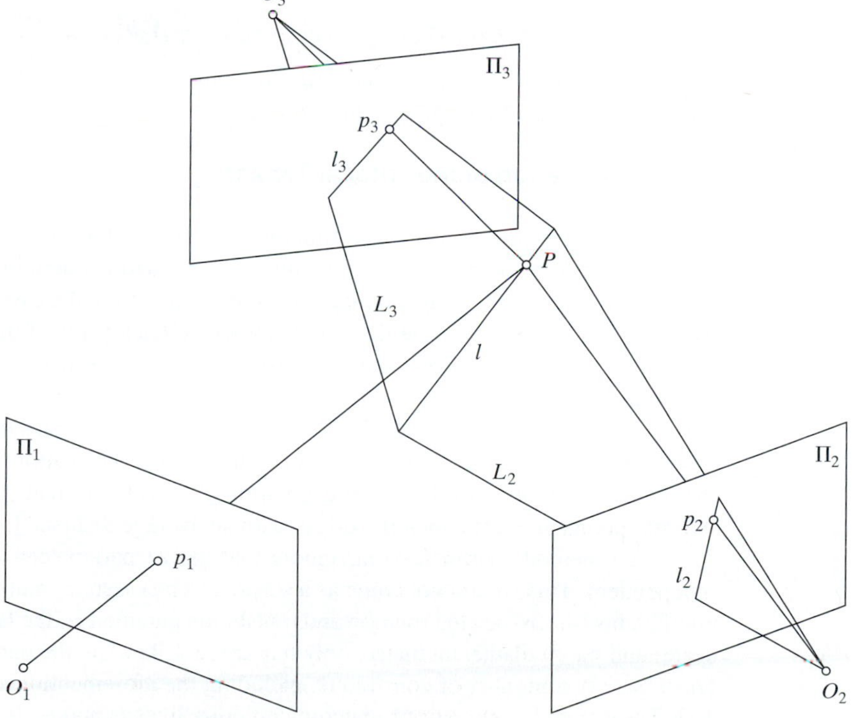

Figure 10.7 Given three images p_1, p_2, and p_3 of the same point P, and two arbitrary image lines l_2 and l_3 passing through p_2 and p_3, the ray passing through O_1 and p_1 must intersect the line where the planes L_2 and L_3 projecting onto l_2 and l_3 meet in space.

10.2.3 The Uncalibrated Case

We can still derive trilinear constraints in the image line coordinates when the intrinsic parameters of the three cameras are unknown. Since in this case $\boldsymbol{p} = \mathcal{K}\hat{\boldsymbol{p}}$ and the image line associated with the vector $\boldsymbol{l}$ is defined by $\boldsymbol{l}^T\boldsymbol{p} = 0$, we immediately obtain $\boldsymbol{l} = \mathcal{K}^{-T}\hat{\boldsymbol{l}}$ or, equivalently, $\hat{\boldsymbol{l}} = \mathcal{K}^T\boldsymbol{l}$.

In particular, Eq. (10.11) holds when $\boldsymbol{p}_i = \hat{\boldsymbol{p}}_i$ and $\boldsymbol{l}_i = \hat{\boldsymbol{l}}_i$. In the general case, we have

$$\mathcal{L} = \begin{pmatrix} \boldsymbol{l}_1^T\mathcal{K}_1 & 0 \\ \boldsymbol{l}_2^T\mathcal{K}_2\mathcal{R}_2 & \boldsymbol{l}_2^T\mathcal{K}_2\boldsymbol{t}_2 \\ \boldsymbol{l}_3^T\mathcal{K}_3\mathcal{R}_3 & \boldsymbol{l}_3^T\mathcal{K}_3\boldsymbol{t}_3 \end{pmatrix},$$

and

$$\text{Rank}(\mathcal{L}) = 2 \Longleftrightarrow \text{Rank}\left[\mathcal{L}\begin{pmatrix} \mathcal{K}_1^{-1} & 0 \\ 0 & 1 \end{pmatrix}\right] = \text{Rank}\begin{pmatrix} \boldsymbol{l}_1^T & 0 \\ \boldsymbol{l}_2^T\mathcal{A}_2 & \boldsymbol{l}_2^T\boldsymbol{b}_2 \\ \boldsymbol{l}_3^T\mathcal{A}_3 & \boldsymbol{l}_3^T\boldsymbol{b}_3 \end{pmatrix} = 2,$$

where $\mathcal{A}_i \stackrel{\text{def}}{=} \mathcal{K}_i\mathcal{R}_i\mathcal{K}_1^{-1}$ and $\boldsymbol{b}_i \stackrel{\text{def}}{=} \mathcal{K}_i\boldsymbol{t}_i$ for $i = 2, 3$. Note that the projection matrices associated with our three cameras are now $\mathcal{M}_1 = (\mathcal{K}_1 \quad \boldsymbol{0})$, $\mathcal{M}_2 = (\mathcal{A}_2\mathcal{K}_1 \quad \boldsymbol{b}_2)$, and $\mathcal{M}_3 = (\mathcal{A}_3\mathcal{K}_1 \quad \boldsymbol{b}_3)$. In particular, $\boldsymbol{b}_2$ and $\boldsymbol{b}_3$ can still be interpreted as the homogeneous image coordinates of the epipoles e_{12} and e_{13}, and the trilinear constraints of Eqs. (10.14) and (10.15) still hold when, this time,

$$\mathcal{G}_1^i = \boldsymbol{b}_2\boldsymbol{A}_3^{iT} - \boldsymbol{A}_2^i\boldsymbol{b}_3^T,$$

where $\boldsymbol{A}_2^i$ and $\boldsymbol{A}_3^i$ $(i = 1, 2, 3)$ denote the columns of $\mathcal{A}_2$ and $\mathcal{A}_3$. As before, we have $\boldsymbol{l}_2^T\mathcal{G}_1^i\boldsymbol{l}_3 = 0$ for any pair of matching epipolar lines l_2 and l_3.

10.2.4 Estimation of the Trifocal Tensor

We now address the problem of estimating the trifocal tensor from point and line correspondences established across triples of pictures. The equations defining the tensor are linear in its coefficients and depend only on image measurements. As in the case of weak calibration, we can use linear methods to estimate these 26 parameters. Each triple of matching points provides four independent linear equations, and every triple of matching lines provides two additional linear constraints. Thus, the tensor coefficients can be computed from p points and l lines granted that $2p + l \geq 13$. For example, 7 triples of points or 13 triples of lines do the trick, as do 3 triples of points and 7 triples of lines, and so on. As in the case of weak calibration, it is possible to improve the numerical stability of the tensor estimation process by normalizing the image coordinates so the data points are centered at the origin with an average distance from the origin of $\sqrt{2}$ pixel.

The methods outlined so far ignore that the 26 parameters of the trifocal tensor are *not* independent. This should not come as a surprise: The essential matrix only has five independent coefficients (the associated rotation and translation parameters, the latter being only defined up to scale) and the fundamental matrix only has seven. Likewise, the parameters defining the trifocal tensor satisfy a number of constraints, including the aforementioned equations $\boldsymbol{l}_2^T\mathcal{G}_1^i\boldsymbol{l}_3 = 0$ $(i = 1, 2, 3)$ satisfied by any pair of matching epipolar lines l_2 and l_3. It is also easy to show that the matrices $\mathcal{G}_1^i$ are singular—a property we come back to in chapter 13. Faugeras and Mourrain (1995) showed that the coefficients of the trifocal tensor of an uncalibrated trinocular stereo rig satisfy eight independent constraints, reducing the total number of independent parameters to 18. The method described in Hartley (1995) enforces these constraints *a posteriori* by recovering the epipoles e_{12} and e_{13} (or equivalently the vectors $\boldsymbol{t}_2$ and $\boldsymbol{t}_3$ in Eq. [10.13]) from the linearly estimated trifocal tensor, then recovering in a linear fashion a set of tensor coefficients that satisfy the constraints.

10.3 MORE VIEWS

What about four views? In this section, we follow Faugeras and Mourrain (1995) and first note that clearing the denominators in the perspective projection Eq. (2.16) derived in chapter 2 yields

$$\begin{pmatrix} u\mathcal{M}^3 - \mathcal{M}^1 \\ v\mathcal{M}^3 - \mathcal{M}^2 \end{pmatrix} \boldsymbol{P} = 0, \tag{10.16}$$

where $\mathcal{M}^1$, $\mathcal{M}^2$, and $\mathcal{M}^3$ denote the three rows of the matrix $\mathcal{M}$. (Note that we depart here from our habit of denoting the rows of a projection matrix by $\boldsymbol{m}_1^T$, $\boldsymbol{m}_2^T$, and $\boldsymbol{m}_3^T$. This is to avoid possible confusions between the different rows of different matrices in the rest of this section. It should be clear that $\mathcal{M}^i$ and $\boldsymbol{m}_i^T$ denote the same row vector.)

Suppose now that we have four views, with associated projection matrices $\mathcal{M}_j$ ($j = 1, 2, 3, 4$). Writing Eq. (10.16) for each one of these yields

$$\mathcal{Q}\boldsymbol{P} = 0, \quad \text{where} \quad \mathcal{Q} \stackrel{\text{def}}{=} \begin{pmatrix} u_1\mathcal{M}_1^3 - \mathcal{M}_1^1 \\ v_1\mathcal{M}_1^3 - \mathcal{M}_1^2 \\ u_2\mathcal{M}_2^3 - \mathcal{M}_2^1 \\ v_2\mathcal{M}_2^3 - \mathcal{M}_2^2 \\ u_3\mathcal{M}_3^3 - \mathcal{M}_3^1 \\ v_3\mathcal{M}_3^3 - \mathcal{M}_3^2 \\ u_4\mathcal{M}_4^3 - \mathcal{M}_4^1 \\ v_4\mathcal{M}_4^3 - \mathcal{M}_4^2 \end{pmatrix}. \tag{10.17}$$

This system of eight homogeneous equations in four unknowns admits a nontrivial solution. It follows that the rank of the corresponding 8×4 matrix $\mathcal{Q}$ is at most 3, or, equivalently, all its 4×4 minors must have zero determinants. Geometrically, each pair of equations in Eq. (10.17) represents the ray R_i ($i = 1, 2, 3, 4$) associated with the image point p_i, and $\mathcal{Q}$ must have rank 3 for these rays to intersect at a point P (Figure 10.8).

The matrix $\mathcal{Q}$ has three kinds of 4×4 minors:

1. Those that involve two rows from one projection matrix and two rows from another one. The equations associated with the six minors of this type include, for example,[2]

$$\text{Det} \begin{pmatrix} u_1\mathcal{M}_1^3 - \mathcal{M}_1^1 \\ v_1\mathcal{M}_1^3 - \mathcal{M}_1^2 \\ u_2\mathcal{M}_2^3 - \mathcal{M}_2^1 \\ v_2\mathcal{M}_2^3 - \mathcal{M}_2^2 \end{pmatrix} = 0. \tag{10.18}$$

These determinants yield bilinear constraints on the position of the associated image points. It is easy to show (see Exercises) that the corresponding equations reduce to the epipolar constraints of Eq. (10.2) when we take $\mathcal{M}_1 = (\text{Id} \quad \boldsymbol{0})$ and $\mathcal{M}_2 = (\mathcal{R}^T \quad -\mathcal{R}^T\boldsymbol{t})$.

[2]General formulas can be obtained by using, for example (u^1, u^2), instead of (u, v) and playing around with indexes and *tensorial notation*. We abstain from this worthy exercise here.

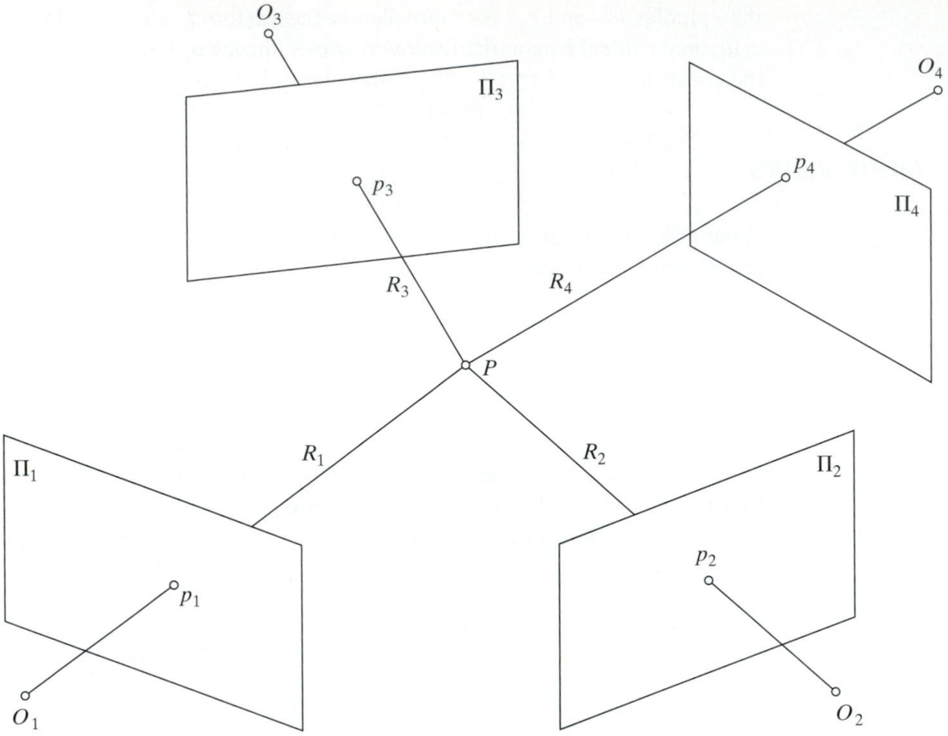

Figure 10.8 Four images p_1, p_2, p_3, and p_4 of the same point P define this point as the intersection of the corresponding rays R_i $(i = 1, 2, 3, 4)$.

2. The second type of minors involves two rows from one projection matrix and one row from each of two other matrices. There are 48 of those, and the associated equations include, for example,

$$\text{Det}\begin{pmatrix} u_1\mathcal{M}_1^3 - \mathcal{M}_1^1 \\ v_1\mathcal{M}_1^3 - \mathcal{M}_1^2 \\ u_2\mathcal{M}_2^3 - \mathcal{M}_2^1 \\ v_3\mathcal{M}_3^3 - \mathcal{M}_3^2 \end{pmatrix} = 0. \tag{10.19}$$

These minors yield trilinear constraints on the corresponding image positions. It is easy to show (see Exercises) that the corresponding equations reduce to the trifocal constraints of Eq. (10.15) when we take $\mathcal{M}_1 = (\text{Id} \quad \mathbf{0})$. In particular, they can be expressed in terms of the matrices $\mathcal{G}_1^i$ $(i = 1, 2, 3)$. Note that this completes the geometric interpretation of the trifocal constraints that express here that the rays associated with three images of the same point must intersect in space.

3. The last type of determinant involves one row of each matrix. The equations associated with the 16 minors of this form include, for example,

$$\text{Det}\begin{pmatrix} v_1\mathcal{M}_1^3 - \mathcal{M}_1^2 \\ u_2\mathcal{M}_2^3 - \mathcal{M}_2^1 \\ v_3\mathcal{M}_3^3 - \mathcal{M}_3^2 \\ v_4\mathcal{M}_4^3 - \mathcal{M}_4^2 \end{pmatrix} = 0. \tag{10.20}$$

These equations yield *quadrilinear constraints* on the position of the points $\boldsymbol{p}_i$ ($i = 1, 2, 3, 4$). Geometrically, each row of the matrix $\mathcal{Q}$ is associated with an image line or equivalently with a plane passing through the optical center of the corresponding camera. Thus each quadrilinearity expresses the fact that the four associated planes intersect in a point (instead of not intersecting at all in the generic case).

Let us focus on the quadrilinear equations. Developing determinants such as Eq. (10.20) with respect to the image coordinates reveals immediately that the coefficients of the quadrilinear constraints can be written as

$$\varepsilon_{ijkl} \operatorname{Det} \begin{pmatrix} \mathcal{M}_1^i \\ \mathcal{M}_2^j \\ \mathcal{M}_3^k \\ \mathcal{M}_4^l \end{pmatrix}, \tag{10.21}$$

where $\varepsilon_{ijkl} = \mp 1$ and i, j, k, and l are indexed between 1 and 3 (see Exercises). These coefficients determine the *quadrifocal tensor* (Triggs, 1995).

Like its trifocal cousin, this tensor can be interpreted geometrically using both points and lines. In particular, consider four pictures p_i ($i = 1, 2, 3, 4$) of a point P and four arbitrary image lines l_i passing through these points. The four planes L_i ($i = 1, 2, 3, 4$) formed by the preimages of the lines must intersect in P, which implies in turn that the 4×4 matrix

$$\mathcal{L} \stackrel{\text{def}}{=} \begin{pmatrix} \boldsymbol{l}_1^T \mathcal{M}_1 \\ \boldsymbol{l}_2^T \mathcal{M}_2 \\ \boldsymbol{l}_3^T \mathcal{M}_3 \\ \boldsymbol{l}_4^T \mathcal{M}_4 \end{pmatrix}$$

must have rank 3 and, in particular, that its determinant must be zero. This obviously provides a quadrilinear constraint on the coefficients of the four lines l_i ($i = 1, 2, 3, 4$). In addition, since each row $\boldsymbol{L}_i^T = \boldsymbol{l}_i^T \mathcal{M}_i$ of $\mathcal{L}$ is a linear combination of the rows of the associated matrix $\mathcal{M}_i$, the coefficients of the quadrilinearities obtained by developing $\operatorname{Det}(\mathcal{L})$ with respect to the coordinates of the lines l_i are simply the coefficients of the quadrifocal tensor as defined by Eq. (10.21).

Finally, note that since $\operatorname{Det}(\mathcal{L})$ is linear in the coordinates of $\boldsymbol{l}_1$, the vanishing of this determinant can be written as $\boldsymbol{l}_1 \cdot \boldsymbol{q}(\boldsymbol{l}_2, \boldsymbol{l}_3, \boldsymbol{l}_4) = 0$, where $\boldsymbol{q}$ is a (trilinear) function of the coordinates of the lines l_i ($i = 2, 3, 4$). Since this relationship holds for any line l_1 passing through p_1, it follows that $\boldsymbol{p}_1 \propto \boldsymbol{q}(\boldsymbol{l}_2, \boldsymbol{l}_3, \boldsymbol{l}_4)$. Geometrically, this means that the ray passing through O_1 and p_1 must also pass through the intersection of the planes formed by the preimages of l_2, l_3, and l_4 (Figure 10.9). Algebraically, this means that, given the quadrifocal tensor and arbitrary lines passing through three images of a point, we can predict the position of this point in a fourth image. This provides yet another method for transfer.

Note that the quadrifocal constraints are valid in both the calibrated and uncalibrated cases since we have made no assumption on the form of the matrices $\mathcal{M}_i$. The quadrifocal tensor is defined by 81 coefficients (or 80 up to scale), but it can be shown that these coefficients satisfy 51 independent constraints, reducing the total number of independent parameters to 29. It can also be shown that, although each quadruple of images of the same point yields 16 independent constraints like Eq. (10.20) on the 80 tensor coefficients, there exists a linear dependency among the 32 equations associated with each pair of points. Thus, six point correspondences are necessary to estimate the quadrifocal tensor in a linear fashion. Algorithms for performing this task and enforcing the 51 constraints associated with actual quadrifocal tensors can be found in Hartley (1998). Finally, Faugeras and Mourrain (1995) have shown that the quadrilinear tensor is

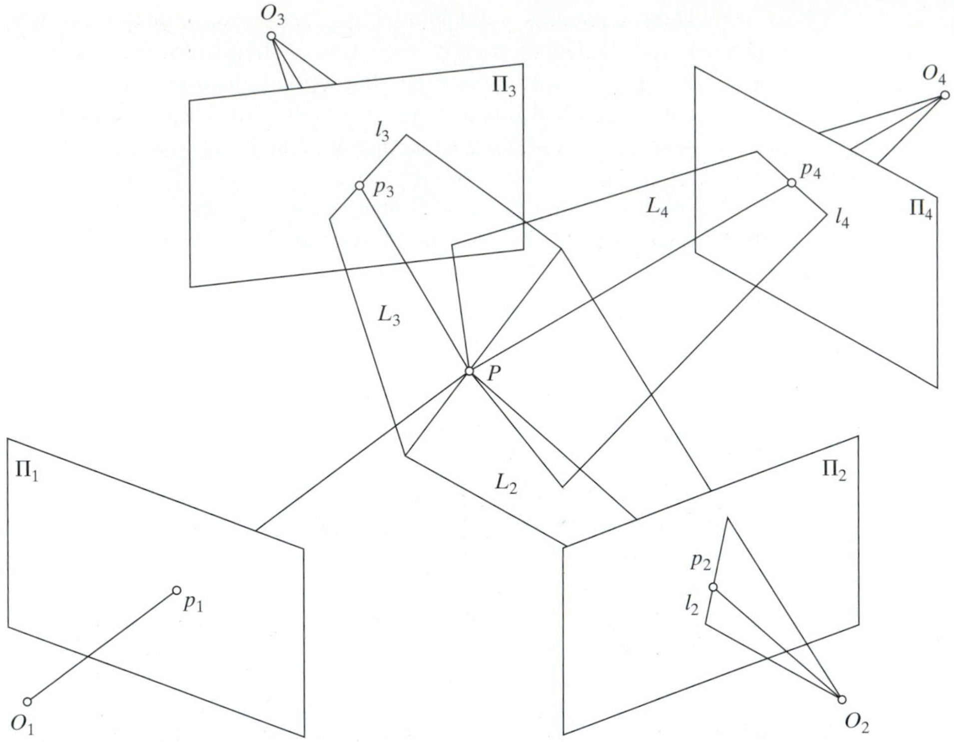

Figure 10.9 Given four images p_1, p_2, p_3, and p_4 of some point P and three arbitrary image lines l_2, l_3, and l_4 passing through the points p_2, p_3, and p_4, the ray passing through O_1 and p_1 must also pass through the point where the three planes L_2, L_3, and L_4 formed by the preimages of these lines intersect.

algebraically dependent on the associated essential/fundamental matrices and trifocal tensor, and thus does not add independent new constraints. Likewise, it can be shown that additional views do not add independent constraints either.

10.4 NOTES

The essential matrix as an algebraic form of the epipolar constraint was discovered by Longuet–Higgins (1981), and its properties have been elucidated by Huang and Faugeras (1989). The fundamental matrix was introduced by Luong and Faugeras (1992, 1996). Robust methods for estimating the fundamental matrix from point correspondences include Zhang *et al.* (1995). We come back to the properties of the fundamental matrix and of the epipolar transformation in chapter 13, when we adress the problem of recovering the structure of a scene and the motion of a camera from a sequence of perspective images. The instantaneous version of the epipolar constraint given by Eq. (10.4) and derived in Section 10.1.3 is only valid for calibrated cameras. See Viéville and Faugeras (1995) for the case of cameras with varying intrinsic parameters. The trilinear constraints associated with three views of a line were introduced independently by Spektakis and Aloimonos (1990) and Weng, Huang and Ahuja (1992) in the context of motion analysis for internally calibrated cameras. They were extended by Shashua (1995) and Hartley (1997) to the uncalibrated case. The quadrifocal tensor was introduced by Triggs (1995). Its properties are investigated in Faugeras and Mourrain (1995), Faugeras and Papadopoulo (1997), Hartley (1998), and Heyden (1998).

The introduction mentioned that photogrammetry is concerned with the extraction of quantitative information from multiple pictures. In this context, binocular and trinocular geometric constraints are regarded as the source of *condition equations* that determine the intrinsic and extrinsic parameters (called *interior* and *exterior orientation* parameters in photogrammetry) of a stereo pair or triple. In particular, the Longuet–Higgins relation appears, in a slightly disguised form, as the *coplanarity condition equation*, and trinocular constraints yield *scale-restraint condition equations* that take calibration and image measurement errors into account (Thompson *et al.*, 1966, chapter X). In this case, the rays associated with three images of the same point are not guaranteed to intersect anymore (Figure 10.10).

The setup is as follows: If the rays R_1 and R_i ($i = 2, 3$) associated with the image points p_1 and p_i do not intersect, the minimum distance between them is reached at the points P_1 and P_i, such that the line joining these points is perpendicular to both R_1 and R_i. Algebraically, this can be written as

$$\overrightarrow{O_1 P_1} = z_1^i \overrightarrow{O_1 p_1} = \overrightarrow{O_1 O_i} + z_i \overrightarrow{O_i p_i} + \lambda_i (\overrightarrow{O_1 p_1} \times \overrightarrow{O_i p_i}) \quad \text{for} \quad i = 2, 3. \tag{10.22}$$

Assuming that the cameras are internally calibrated so the projection matrices associated with the second and third cameras are $(\mathcal{R}_2^T \quad -\mathcal{R}_2^T \boldsymbol{t}_2)$ and $(\mathcal{R}_3^T \quad -\mathcal{R}_3^T \boldsymbol{t}_3)$, Eq. (10.22) can be rewritten in the coordinate system attached to the first camera as

$$z_1^i \boldsymbol{p}_1 = \boldsymbol{t}_i + z_i \mathcal{R}_i \boldsymbol{p}_i + \lambda_i (\boldsymbol{p}_1 \times \mathcal{R}_i \boldsymbol{p}_i) \quad \text{for} \quad i = 2, 3. \tag{10.23}$$

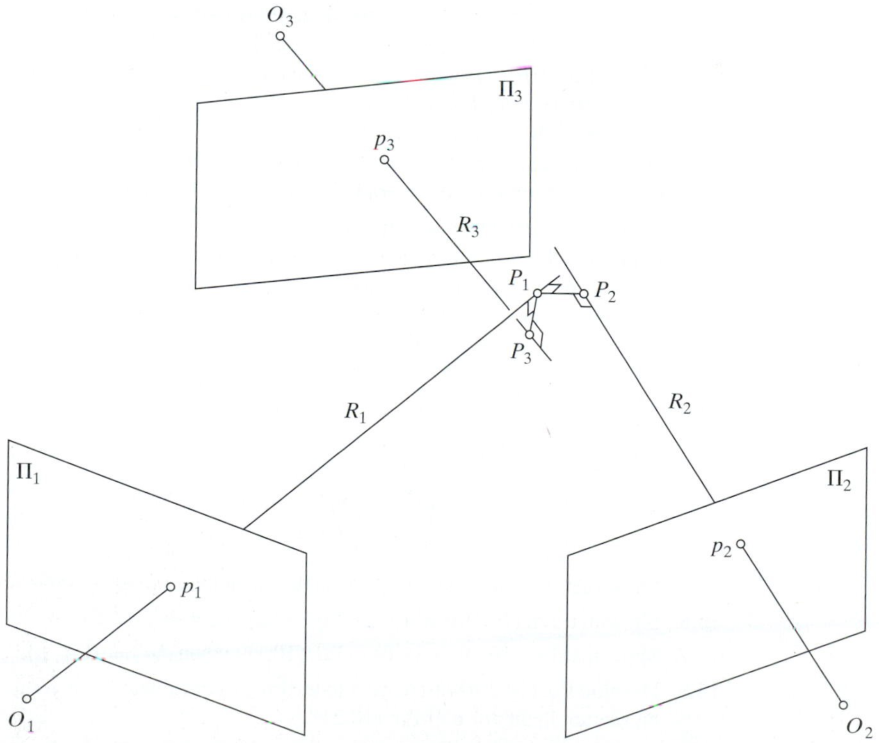

Figure 10.10 Trinocular constraints in the presence of calibration or measurement errors: The rays R_1, R_2, and R_3 may not intersect.

Note that a similar equation could be written as well for completely uncalibrated cameras by including terms depending on the (unknown) intrinsic parameters. In either case, Eq. (10.23) can be used to calculate the unknowns z_i, λ_i, and z_1^i in terms of $\boldsymbol{p}_1$, $\boldsymbol{p}_i$ and the projection matrices associated with the cameras (see Exercises). The scale-restraint condition is then written as $z_1^2 = z_1^3$. Although it is more complex than the trifocal constraint (in particular, it is not trilinear in the coordinates of the points p_1, p_2, and p_3), this condition does not involve the coordinates of the observed point, and it can be used (in principle) to estimate the trifocal geometry directly from image data. A potential advantage is that the error function $z_1^2 - z_1^3$ has a clear geometric meaning: It is the difference between the estimates of the depth of P obtained using the pairs of cameras $1 \leftrightarrow 2$ and $1 \leftrightarrow 3$. It would be interesting to further investigate the relationship between the trifocal tensor and the scale-constraint condition, as well as its practical application to the estimation of the trifocal geometry.

PROBLEMS

10.1. Show that one of the singular values of an essential matrix is 0 and the other two are equal. (Huang and Faugeras, 1989 have shown that the converse is also true—that is, any 3×3 matrix with one singular value equal to 0 and the other two equal to each other is an essential matrix.)

Hint: The singular values of $\mathcal{E}$ are the eigenvalues of $\mathcal{E}\mathcal{E}^T$.

10.2. Exponential representation of rotation matrices. The matrix associated with the rotation whose axis is the unit vector $\boldsymbol{a}$ and whose angle is θ can be shown to be equal to $e^{\theta[\boldsymbol{a}_\times]} \stackrel{\text{def}}{=} \sum_{i=0}^{+\infty} \frac{1}{i!}(\theta[\boldsymbol{a}_\times])^i$. Use this representation to derive Eq. (10.3).

10.3. The infinitesimal epipolar constraint of Eq. (10.4) was derived by assuming that the observed scene was static and the camera was moving. Show that when the camera is fixed and the scene is moving with translational velocity $\boldsymbol{v}$ and rotational velocity $\boldsymbol{\omega}$, the epipolar constraint can be rewritten as $\boldsymbol{p}^T([\boldsymbol{v}_\times][\boldsymbol{\omega}_\times])\boldsymbol{p} + (\boldsymbol{p} \times \dot{\boldsymbol{p}}) \cdot \boldsymbol{v} = 0$. Note that this equation is now the sum of the two terms appearing in Eq. (10.4) instead of their difference.

Hint: If $\mathcal{R}$ and $\boldsymbol{t}$ denote the rotation matrix and translation vectors appearing in the definition of the essential matrix for a moving camera, show that the object displacement that yields the same motion field for a static camera is given by the rotation matrix $\mathcal{R}^T$ and the translation vector $-\mathcal{R}^T\boldsymbol{t}$.

10.4. Show that when the 8×8 matrix associated with the eight-point algorithm is singular, the eight points and the two optical centers lie on a quadric surface (Faugeras, 1993).

Hint: Use the fact that when a matrix is singular, there exists some nontrivial linear combination of its columns that is equal to zero. Also take advantage of the fact that the matrices representing the two projections in the coordinate system of the first camera are in this case $(\text{Id} \quad \boldsymbol{0})$ and $(\mathcal{R}^T \quad -\mathcal{R}^T\boldsymbol{t})$.

10.5. Show that three of the determinants of the 3×3 minors of

$$\mathcal{L} = \begin{pmatrix} \boldsymbol{l}_1^T & 0 \\ \boldsymbol{l}_2^T\mathcal{R}_2 & \boldsymbol{l}_2^T\boldsymbol{t}_2 \\ \boldsymbol{l}_3^T\mathcal{R}_3 & \boldsymbol{l}_3^T\boldsymbol{t}_3 \end{pmatrix} \quad \text{can be written as} \quad \boldsymbol{l}_1 \times \begin{pmatrix} \boldsymbol{l}_2^T\mathcal{G}_1^1\boldsymbol{l}_3 \\ \boldsymbol{l}_2^T\mathcal{G}_1^2\boldsymbol{l}_3 \\ \boldsymbol{l}_2^T\mathcal{G}_1^3\boldsymbol{l}_3 \end{pmatrix} = \boldsymbol{0}.$$

Show that the fourth determinant can be written as a linear combination of these.

10.6. Show that Eq. (10.18) reduces to Eq. (10.2) when $\mathcal{M}_1 = (\text{Id} \quad \boldsymbol{0})$ and $\mathcal{M}_2 = (\mathcal{R}^T \quad -\mathcal{R}^T\boldsymbol{t})$.

10.7. Show that Eq. (10.19) reduces to Eq. (10.15) when $\mathcal{M}_1 = (\text{Id} \quad \boldsymbol{0})$.

10.8. Develop Eq. (10.20) with respect to the image coordinates, and verify that the coefficients can indeed be written in the form of Eq. (10.21).

10.9. Use Eq. (10.23) to calculate the unknowns z_i, λ_i, and z_1^i in terms of $\boldsymbol{p}_1$, $\boldsymbol{p}_i$, $\mathcal{R}_i$, and $\boldsymbol{t}_i$ $(i = 2, 3)$. Show that the value of λ_i is directly related to the epipolar constraint, and characterize the degree of the dependency of $z_1^2 - z_1^3$ on the data points.

Programming Assignments

10.10. Implement the eight-point algorithm for weak calibration from binocular point correspondences.

10.11. Implement the linear least-squares version of that algorithm with and without Hartley's preconditioning step.

10.12. Implement an algorithm for estimating the trifocal tensor from point correspondences.

10.13. Implement an algorithm for estimating the trifocal tensor from line correspondences.

11

Stereopsis

Fusing the pictures recorded by our two eyes and exploiting the difference (or *disparity*) between them allows us to gain a strong sense of depth. This chapter is concerned with the design and implementation of algorithms that mimic our ability to perform this task, known as *stereopsis*. Reliable computer programs for stereoscopic perception are of course invaluable in visual robot navigation (Figure 11.1), cartography, aerial reconnaissance, and close-range photogrammetry.

Figure 11.1 Left: The Stanford cart sports a single camera moving in discrete increments along a straight line and providing multiple snapshots of outdoor scenes. Right: The INRIA mobile robot uses three cameras to map its environment. As shown by these examples, although two eyes are sufficient for stereo fusion, mobile robots are sometimes equipped with three (or more) cameras. The bulk of this chapter is concerned with binocular perception but stereo algorithms using multiple cameras are discussed in Section 11.4. *Photos courtesy of Hans Moravec and Olivier Faugeras.*

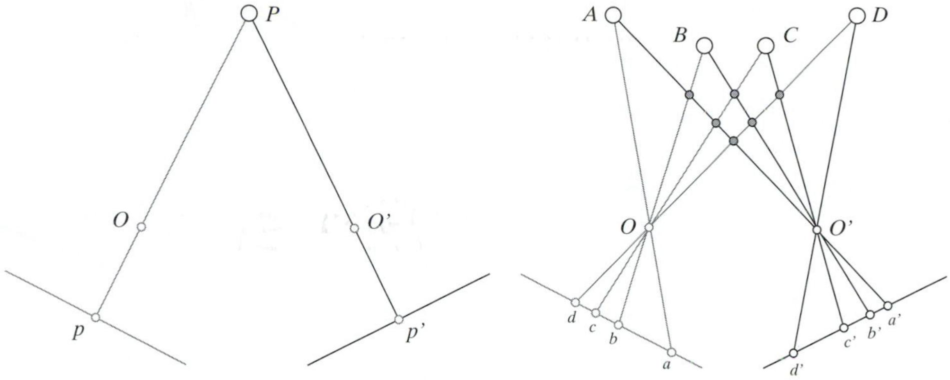

Figure 11.2 The binocular fusion problem: In the simple case of the diagram shown on the left, there is no ambiguity, and stereo reconstruction is a simple matter. In the more usual case shown on the right, any of the four points in the left picture may, a priori, match any of the four points in the right one. Only four of these correspondences are correct; the other ones yield the incorrect reconstructions shown as small gray discs.

They are also of great interest in tasks such as image segmentation for object recognition or the construction of three-dimensional scene models for computer graphics applications.

Stereo vision involves two processes: The *fusion* of features observed by two (or more) eyes and the *reconstruction* of their three-dimensional preimage. The latter is relatively simple: The preimage of matching points can (in principle) be found at the intersection of the rays passing through these points and the associated pupil centers (or pinholes; see Figure 11.2, left). Thus, when a single image feature is observed at any given time, stereo vision is easy. However, each picture consists of hundreds of thousands of pixels, with tens of thousands of image features such as edge elements, and some method must be devised to establish the correct correspondences and avoid erroneous depth measurements (Figure 11.2, right). The epipolar constraint plays a fundamental role in this process since it restricts the search for image correspondences to matching epipolar lines.

We assume in the rest of this chapter that all cameras have been carefully calibrated so their intrinsic and extrinsic parameters are precisely known relative to some fixed world coordinate system (this implies of course that the essential matrices and/or trifocal tensors associated with pairs or triples of cameras are known as well). The case of uncalibrated cameras is examined in the context of structure from motion in chapters 12 and 13.

11.1 RECONSTRUCTION

Given a calibrated stereo rig and two matching image points p and p', it is in principle straightforward to reconstruct the corresponding scene point by intersecting the two rays $R = Op$ and $R' = O'p'$. However, the rays R and R' will never, in practice, actually intersect due to calibration and feature localization errors (Figure 11.3). In this context, various reasonable approaches to the reconstruction problem can be adopted. For example, we can construct the line segment perpendicular to R and R' that intersects both rays: Its mid-point P is the closest point to the two rays and can be taken as the preimage of p and p'. It should be noted that a similar construction was used at the end of chapter 10 to characterize algebraically the geometry of multiple views in the presence of calibration or measurement errors. Equations (10.22) and (10.23) derived in that

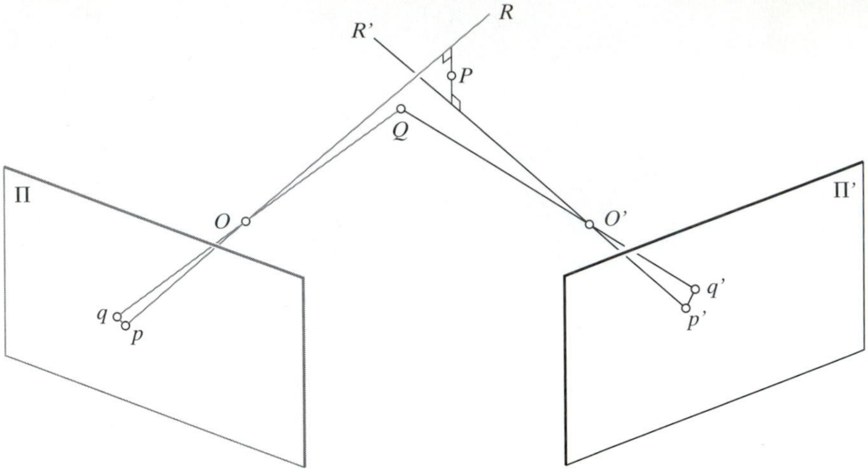

Figure 11.3 Triangulation in the presence of measurement errors. See text for details.

chapter are readily adapted to the calculation of the coordinates of P in the frame attached to the first camera.

Alternatively, we can reconstruct a scene point using a purely algebraic approach: Given the projection matrices $\mathcal{M}$ and $\mathcal{M}'$ and the matching points p and p', we can rewrite the constraints $z\boldsymbol{p} = \mathcal{M}\boldsymbol{P}$ and $z'\boldsymbol{p}' = \mathcal{M}\boldsymbol{P}$ as

$$\begin{cases} \boldsymbol{p} \times \mathcal{M}\boldsymbol{P} = 0 \\ \boldsymbol{p}' \times \mathcal{M}'\boldsymbol{P} = 0 \end{cases} \iff \begin{pmatrix} [\boldsymbol{p}_\times]\mathcal{M} \\ [\boldsymbol{p}'_\times]\mathcal{M}' \end{pmatrix} \boldsymbol{P} = 0.$$

This is an overconstrained system of four independent linear equations in the coordinates of P that is easily solved using the linear least-squares techniques introduced in chapter 3. Unlike the previous approach, this reconstruction method does not have an obvious geometric interpretation, but generalizes readily to the case of three or more cameras, each new picture simply adding two additional constraints.

Finally, we can reconstruct the scene point associated with p and p' as the point Q with images q and q' that minimizes $d^2(p, q) + d^2(p', q')$ (Figure 11.3). Unlike the two other methods presented in this section, this approach does not allow the closed-form computation of the reconstructed point, which must be estimated via nonlinear least-squares techniques such as those introduced in chapter 3. The reconstruction obtained by either of the other two methods can be used as a reasonable guess to initialize the optimization process. This nonlinear approach also readily generalizes to the case of multiple images.

11.1.1 Image Rectification

The calculations associated with stereo algorithms are often considerably simplified when the images of interest have been *rectified* (i.e., replaced by two equivalent pictures with a common image plane parallel to the baseline joining the two optical centers; see Figure 11.4). The rectification process can be implemented by projecting the original pictures onto the new image plane. With an appropriate choice of coordinate system, the rectified epipolar lines are scanlines of the new images, and they are also parallel to the baseline. There are two degrees of freedom involved in the choice of the rectified image plane: (a) the distance between this plane and the baseline, which is essentially irrelevant since modifying it only changes the scale of the recti-

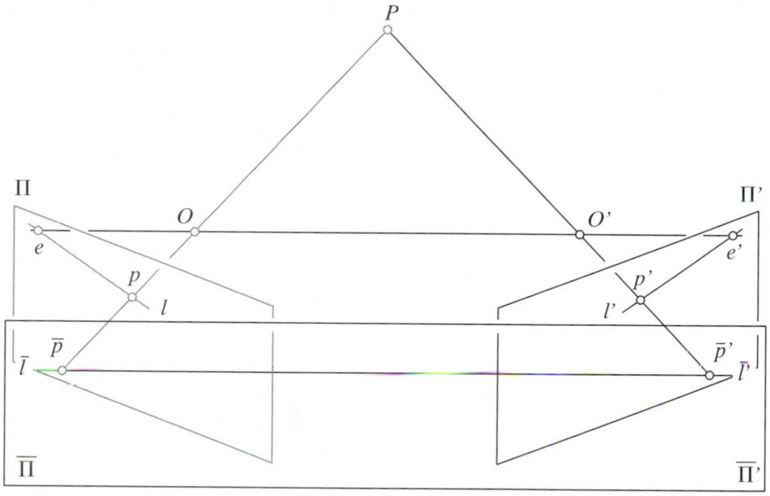

Figure 11.4 A rectified stereo pair: The two image planes Π and Π' are reprojected onto a common plane $\bar{\Pi} = \bar{\Pi}'$ parallel to the baseline. The epipolar lines l and l' associated with the points p and p' in the two pictures map onto a common scanline $\bar{l} = \bar{l}'$ also parallel to the baseline and passing through the reprojected points $\bar{p}$ and $\bar{p}'$. With modern computer graphics hardware and software, the rectified images are easily constructed by considering each input image as a polyhedral mesh and using texture mapping to render the projection of this mesh into the plane $\bar{\Pi} = \bar{\Pi}'$.

fied pictures—an effect easily balanced by an inverse scaling of the image coordinate axes; and (b) the direction of the rectified plane normal in the plane perpendicular to the baseline; natural choices include picking a plane parallel to the line where the two original retinas intersect and minimizing the distortion associated with the reprojection process.

In the case of rectified images, the notion of disparity introduced earlier takes a precise meaning: Given two points p and p' located on the same scanline of the left and right images, with coordinates (u, v) and (u', v), the disparity is defined as the difference $d = u' - u$. Let us assume from now on normalized image coordinates. As shown in the exercises, if B denotes the distance between the optical centers, also called baseline in this context, the depth of P in the (normalized) coordinate system attached to the first camera is $z = -B/d$. In particular, the coordinate vector of the point P in the frame attached to the first camera is $\boldsymbol{P} = -(B/d)\boldsymbol{p}$, where $\boldsymbol{p} = (u, v, 1)^T$ is the vector of normalized image coordinates of p. This provides yet another reconstruction method for rectified stereo pairs.

11.2 HUMAN STEREOPSIS

Before moving on to algorithms for establishing binocular correspondences, let us pause for a moment to discuss the mechanisms underlying human stereopsis. First, it should be noted that, unlike the cameras rigidly attached to a passive stereo rig, the two eyes of a person can rotate in their sockets. At each instant, they *fixate* on a particular point in space (i.e., they rotate so that the corresponding images form in the centers of their foveas).

Figure 11.5 illustrates a simplified, two-dimensional situation. If l and r denote the (counterclockwise) angles between the vertical planes of symmetry of two eyes and two rays passing through the same scene point, we define the corresponding disparity as $d = r - l$ (Figure 11.5).

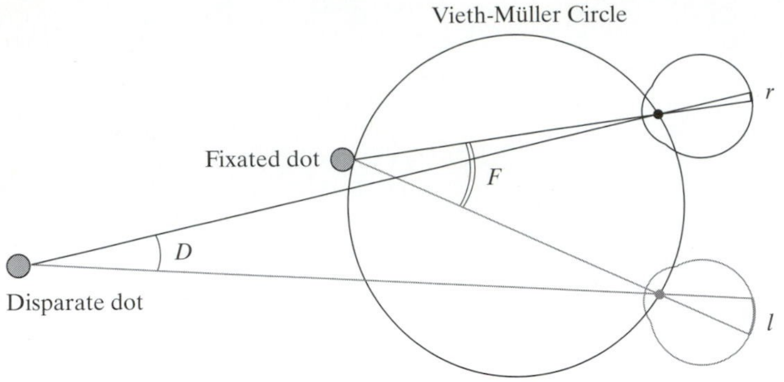

Figure 11.5 In this diagram, the close-by dot is fixated by the eyes, and it projects onto the center of their foveas with no disparity. The two images of the far dot deviate from this central position by different amounts, indicating a different depth.

It is an elementary exercise in trigonometry to show that $d = D - F$, where D denotes the angle between these rays, and F is the angle between the two rays passing through the fixated point. Points with zero disparity lie on the *Vieth–Müller circle* that passes through the fixated point and the anterior nodal points of the eyes. Points lying inside this circle have a positive disparity, points lying outside it have, as in Figure 11.5, a negative disparity, and the locus of all points having a given disparity d forms, as d varies, the family of all circles passing through the two eyes' nodal points. This property is clearly sufficient to rank order in depth dots that are near the fixation point. However, it is also clear that the *vergence angles* between the vertical *median plane* of symmetry of the head and the two fixation rays must be known to reconstruct the absolute position of scene points.

The three-dimensional case is naturally more complicated, the locus of zero-disparity points becoming a surface, the *horopter*, but the general conclusion is the same, and absolute positioning requires the vergence angles. As demonstrated by Wundt and Helmholtz (1909) nearly 100 years ago, there is strong evidence that these angles cannot be measured accurately by our nervous system. However, *relative* depth, or rank ordering of points along the line of sight, can be judged quite accurately. For example, it is possible to decide which one of two targets near the horopter is closer to an observer for disparities of a few seconds of arc (*stereoacuity threshold*), which matches the minimum separation that can be measured with one eye (*monocular hyperacuity threshold*).

Concerning the construction of correspondences between the left and right images, Julesz (1960) asks the following question: Is the basic mechanism for binocular fusion a monocular process (where local brightness patterns [micropatterns] or higher organizations of points into objects [macropatterns] are identified *before* being fused), a binocular one (where the two images are combined into a single field where all further processing takes place), or a combination of both? Some anecdotal evidence hints at a binocular mechanism. To quote Julesz: "In aerial reconnaissance it is known that objects camouflaged by a complex background are very difficult to detect but jump out if viewed sterescopically." To gather more conclusive data and settle the matter, Julesz introduces a new device, the *random dot stereogram*, a pair of synthetic images obtained by randomly spraying black dots on white objects, typically a small square plate floating over a larger one (Figure 11.6). To quote him again: "When viewed monocularly, the images appear completely random. But when viewed stereoscopically, the image pair gives the impression of a square markedly in front of (or behind) the surround." The conclusion is clear:

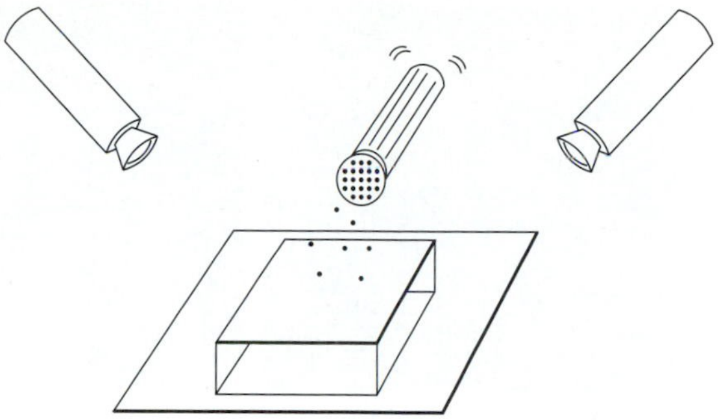

Figure 11.6 Random dot stereograms: Shaking (virtual) pepper over two plates.

Human binocular fusion cannot be explained by peripheral processes directly associated with the physical retinas. Instead, it must involve the central nervous system and an imaginary *cyclopean retina* that combines the left and right image stimuli as a single unit.

The *dipole* model of stereopsis proposed by Julesz is *cooperative*, with neighboring matches influencing each other to avoid ambiguities and promote a global analysis of the scene. The approach proposed by Marr and Poggio (1976) is another instance of a cooperative process that performs quite well on random dot stereograms. Their algorithm relies on three constraints: (a) *compatibility* (black dots can only match black dots, or, more generally, two image features can only match if they have possibly arisen from the same physical marking), (b) *uniqueness* (a black dot in one image matches at most one black dot in the other picture), and (c) *continuity* (the disparity of matches varies smoothly almost everywhere in the image). Given a number of black dots on a pair of corresponding epipolar lines, Marr and Poggio build a graph that reflects possible correspondences (Figure 11.7).

The nodes of the graph are pairs of black dots within some disparity range, reflecting the compatibility constraint; vertical and horizontal arcs represent inhibitory connections associated

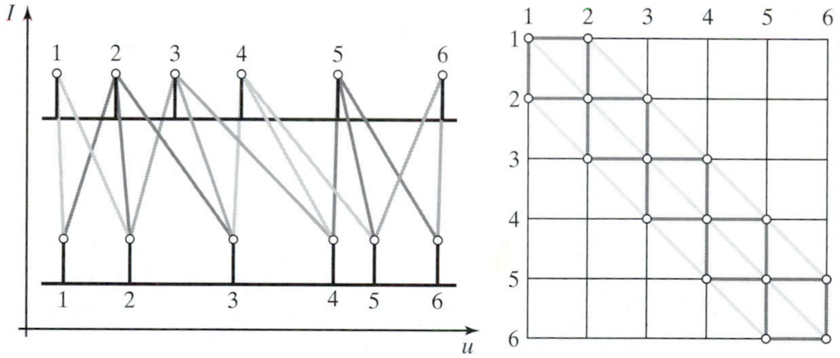

Figure 11.7 A cooperative approach to stereopsis: The Marr–Poggio (1976) algorithm. The left part of the figure shows two intensity profiles along the same scanline of two images. The spikes correspond to black dots. The line segments joining the two profiles indicate possible matches between dots given some maximum disparity range. These matches are also shown in the right part of the figure, where they form the nodes of a graph. The vertical and horizontal arcs of this graph join nodes associated with the same dot in the left or right image. The diagonal arcs join nodes with similar disparities.

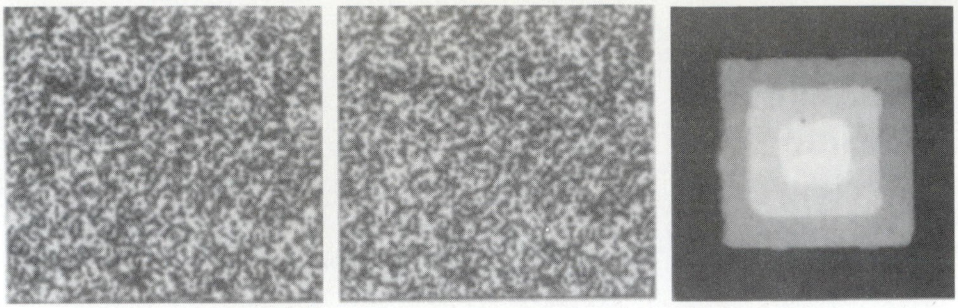

Figure 11.8 From left to right: A random dot stereogram depicting four planes at varying depth (a "wedding cake") and the disparity map obtained after 14 iterations of the Marr–Poggio cooperative algorithm. *Reprinted from VISION: A COMPUTATIONAL INVESTIGATION INTO THE HUMAN REPRESENTATION AND PROCESSING OF VISUAL INFORMATION by David Marr, © 1982 by David Marr. Reprinted by permission of Henry Holt and Company, LLC.*

with the uniqueness constraint (any match between two dots should discourage any other match for both the left dot—horizontal inhibition—and the right one—vertical inhibition—in the pair); and diagonal arcs represent excitory connections associated with the continuity constraint (any match should favor nearby matches with similar disparities).

In this approach, a quality measure is associated with each node. It is initialized to 1 for every pair of potential matches within some disparity range. The matching process is iterative and parallel, each node being assigned at each iteration a weighted combination of its neighbors' values. Excitory connections are assigned weights equal to 1, and inhibitory ones weights equal to $-w$ (where w is a suitable weighting factor). A node is assigned a value of 1 when the corresponding weighted sum exceeds some threshold and a value of 0 otherwise. This approach works quite reliably on random dot stereograms (Figure 11.8), but not on natural images. As we return to computer vision in the next section, we present a number of techniques that perform better on most real pictures, but the original Marr–Poggio algorithm and its implementation retain the interest of offering an early example of a theory of stereopsis that allows the fusion of random dot stereograms.

11.3 BINOCULAR FUSION

11.3.1 Correlation

Correlation methods find pixel-wise image correspondences by comparing intensity profiles in the neighborhood of potential matches, and they are among the first techniques ever proposed to solve the binocular fusion problem (Kelly *et al.*, 1977, Gennery, 1980). Concretely, let us consider a rectified stereo pair and a point (u, v) in the first image. We associate with the window of size $p = (2m + 1) \times (2n + 1)$ centered in (u, v) the vector $\mathbf{w}(u, v) \in \mathbb{R}^p$ obtained by scanning the window values one row at a time (the order is in fact irrelevant as long as it is fixed). Now, given a potential match $(u + d, v)$ in the second image, we can construct a second vector $\mathbf{w}'(u + d, v)$ and define the corresponding *normalized correlation function* as

$$C(d) = \frac{1}{|\mathbf{w} - \bar{\mathbf{w}}|} \frac{1}{|\mathbf{w}' - \bar{\mathbf{w}}'|} \left[(\mathbf{w} - \bar{\mathbf{w}}) \cdot (\mathbf{w}' - \bar{\mathbf{w}}') \right],$$

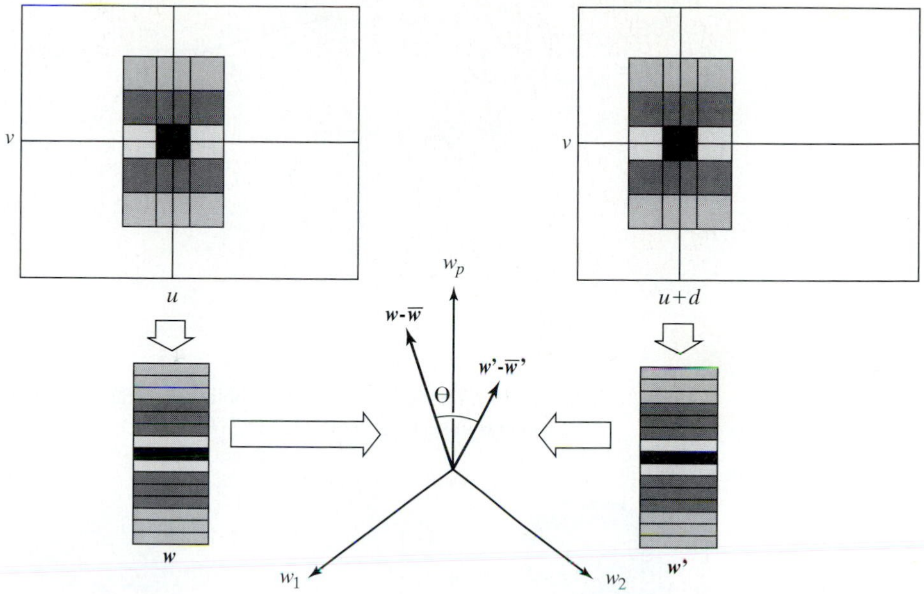

Figure 11.9 Correlation of two 3×5 windows along corresponding epipolar lines. The second window position is separated from the first one by an offset d. The two windows are encoded by vectors $\boldsymbol{w}$ and $\boldsymbol{w}'$ in $\mathbb{R}^{15}$, and the correlation function measures the cosine of the angle θ between the vectors $\boldsymbol{w}-\bar{\boldsymbol{w}}$ and $\boldsymbol{w}'-\bar{\boldsymbol{w}}'$ obtained by subtracting from the components of $\boldsymbol{w}$ and $\boldsymbol{w}'$ the average intensity in the corresponding windows.

where the u, v and d indexes have been omitted for the sake of conciseness and $\bar{\boldsymbol{a}}$ denotes the vector whose coordinates are all equal to the mean of the coordinates of $\boldsymbol{a}$ (Figure 11.9).

The normalized correlation function C clearly ranges from -1 to $+1$. It reaches its maximum value when the image brightnesses of the two windows are related by an affine transformation $I' = \lambda I + \mu$ for some constants λ and μ with $\lambda > 0$ (see Exercises). In other words, maxima of this function correspond to image patches separated by a constant offset and a positive scale factor, and stereo matches can be found by seeking the maximum of the C function over some predetermined range of disparities.[1]

At this point, let us make a few remarks about matching methods based on correlation. First, it is easily shown (see Exercises) that maximizing the normalized correlation function is equivalent to minimizing

$$\left| \frac{1}{|\boldsymbol{w} - \bar{\boldsymbol{w}}|}(\boldsymbol{w} - \bar{\boldsymbol{w}}) - \frac{1}{|\boldsymbol{w}' - \bar{\boldsymbol{w}}'|}(\boldsymbol{w}' - \bar{\boldsymbol{w}}') \right|^2,$$

or equivalently the sum of the squared differences between the pixel values of the two windows after they have been submitted to the corresponding normalization process. Second, although the calculation of the normalized correlation function at every pixel of an image for some range of disparities is computationally expensive, it can be implemented efficiently using recursive techniques (see Exercises). Finally, a major problem with correlation-based techniques for estab-

[1]The invariance of C to affine transformations of the brightness function affords correlation-based matching techniques some degree of robustness in situations where the observed surface is not quite Lambertian or the two cameras have different gains or lenses with different f numbers.

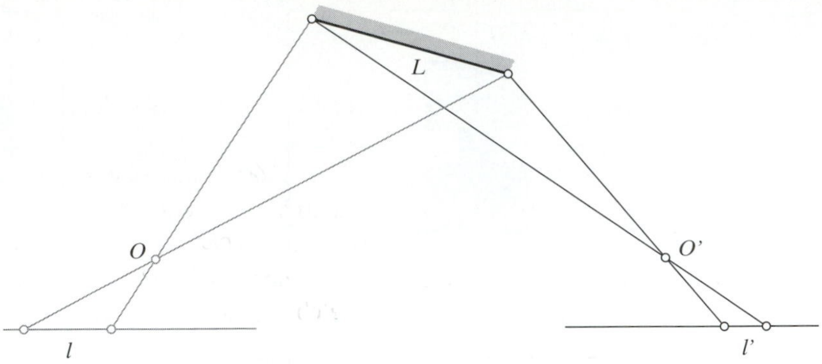

Figure 11.10 The foreshortening of (oblique) surfaces depends on the position of the cameras observing them: $l/L \neq l'/L$.

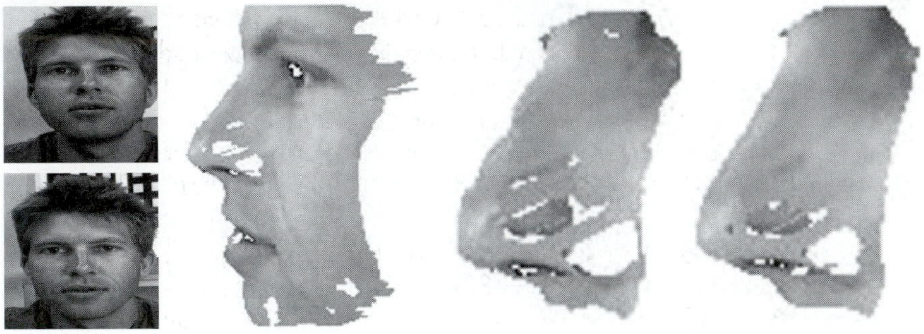

Figure 11.11 Correlation-based stereo matching: (a) a pair of stereo pictures; (b) a texture-mapped view of the reconstructed surface; (c) comparison of the regular (left) and refined (right) correlation methods in the nose region. The latter clearly gives better results. *Reprinted from "Computing Differential Properties of 3D Shapes from Stereopsis Without 3D Models," by F. Devernay and O.D. Faugeras, Proc. IEEE Conference on Computer Vision and Pattern Recognition, (1994).* © *1994 IEEE.*

lishing stereo correspondences is that they implicitly assume that the observed surface is (locally) parallel to the two image planes (Figure 11.10).

This suggests a two-pass algorithm where initial estimates of the disparity are used to warp the correlation windows to compensate for inequal amounts of foreshortening in the two pictures. Figure 11.11 shows an example, where a warped window is defined in the right image for each rectangle in the left image using the disparity in the center of the rectangle and its derivatives (Devernay & Faugeras, 1994). An optimization process is used to find the values of the disparity and its derivatives that maximize the correlation between the left rectangle and the right window, using interpolation to retrieve appropriate values in the right image.

11.3.2 Multi-Scale Edge Matching

Slanted surfaces pose problems to correlation-based matchers. Other arguments against correlation can be found in Julesz (1960) and Marr (1982), suggesting that correspondences should be found at a variety of scales, with matches between (hopefully) physically significant image

Algorithm 11.1: The Marr–Poggio (1979) Multiscale Binocular Fusion Algorithm.

1. Convolve the two (rectified) images with $\nabla^2 G_\sigma$ filters of increasing standard deviations $\sigma_1 < \sigma_2 < \sigma_3 < \sigma_4$.
2. Find zero crossings of the Laplacian along horizontal scanlines of the filtered images.
3. For each filter scale σ, match zero crossings with the same parity and roughly equal orientations in a $[-w_\sigma, +w_\sigma]$ disparity range, with $w_\sigma = 2\sqrt{2}\sigma$.
4. Use the disparities found at larger scales to offset the images in the neighborhood of matches and cause unmatched regions at smaller scales to come into correspondence.

features such as edges preferred to matches between raw pixel intensities. These principles are implemented in Algorithm 11.11.1, which is due to Marr and Poggio (1979).

Matches are sought at each scale in the $[-w_\sigma, w_\sigma]$ disparity range, where $w_\sigma = 2\sqrt{2}\sigma$ is the width of the central negative portion of the $\nabla^2 G_\sigma$ filter. This choice is motivated by psychophysical and statistical considerations. In particular, assuming that the convolved images are white Gaussian processes, Grimson (1981*a*) showed that the probability of a false match occurring in the $[-w_\sigma, +w_\sigma]$ disparity range of a given zero crossing is only 0.2 when the orientations of the matched features are within 30° of each other. A simple mechanism can be used to disambiguate the multiple potential matches that may still occur within the matching range (see Grimson (1981*a*) for details). Of course, limiting the search for matches to the $[-w_\sigma, +w_\sigma]$ range prevents the algorithm from matching *correct* pairs of zero crossings whose disparity falls outside this interval. Since w_σ is proportional to the scale σ at which matches are sought, eye movements (or equivalently image offsets) controlled by the disparities found at large scales must be used to bring large-disparity pairs of zero crossings within matchable range at a fine scale. This process occurs in Step 4 of the algorithm and is illustrated by Figure 11.12. Once matches have been found, the corresponding disparities can be stored in a buffer called the $2\frac{1}{2}$-*dimensional sketch* by Marr and Nishihara (1978). This algorithm has been implemented by Grimson (1981*a*), and extensively tested on random dot stereograms and natural images. An example appears in Figure 11.12 (bottom).

11.3.3 Dynamic Programming

It is reasonable to assume that the order of matching image features along a pair of epipolar lines is the inverse of the order of the corresponding surface attributes along the curve where the epipolar plane intersects the observed object's boundary (Figure 11.13, left). This is the so-called *ordering constraint* introduced in the early 1980s (Baker and Binford, 1981; Ohta and Kanade, 1985). Interestingly enough, it may not be satisfied by real scenes, in particular when small solids occlude parts of larger ones (Figure 11.13, right) or, more rarely at least in robot vision, when transparent objects are involved.

Despite these reservations, the ordering constraint remains a reasonable one, and it can be used to devise efficient algorithms relying on *dynamic programming* (Forney, 1973; Aho *et al.*, 1974) to establish stereo correspondences (Figure 11.14). Specifically, let us assume that a number of feature points (say edgels) have been found on corresponding epipolar lines. Our objective here is to match the intervals separating those points along the two intensity profiles (Figure 11.14, left). According to the ordering constraint, the order of the feature points must be the same, although the occasional interval in either image may be reduced to a single point corresponding to missing correspondences associated with occlusion and/or noise.

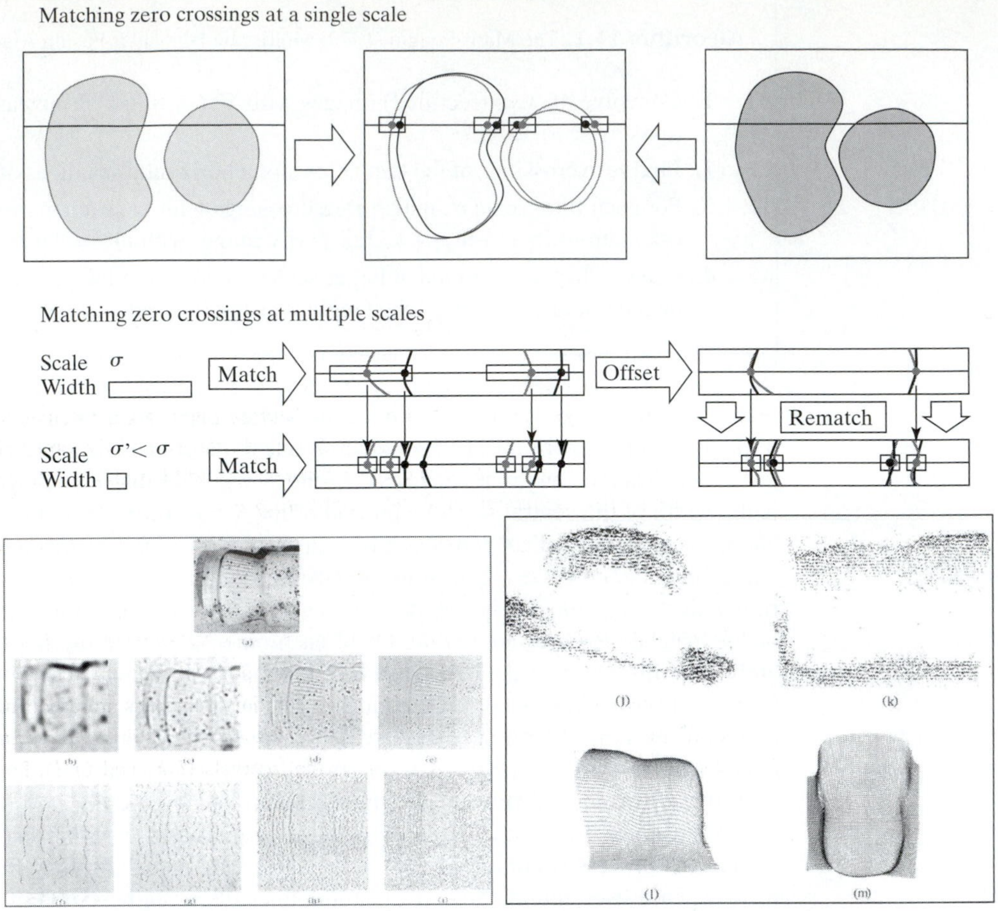

Figure 11.12 Top: Singlescale matching. Middle: Multiscale matching. Bottom: Results. Bottom left: The input data (including one of the input pictures, the output of four $\nabla^2 G_\sigma$ filters, and the corresponding zero crossings). Bottom right: Two views of the depth map constructed by the matching process and two views of the surface obtained by interpolating the reconstructed points. *Reprinted from VISION: A COMPUTATIONAL INVESTIGATION INTO THE HUMAN REPRESENTATION AND PROCESSING OF VISUAL INFORMATION by David Marr, © 1982 by David Marr. Reprinted by permission of Henry Holt and Company, LLC.*

This setting allows us to restate the matching problem as the optimization of a path's cost over a graph whose nodes correspond to pairs of left and right image features; arcs represent matches between left and right intensity profile intervals bounded by the features of the corresponding nodes (Figure 11.14, right). The cost of an arc measures the discrepancy between the corresponding intervals (e.g., the squared difference of the mean intensity values). This optimization problem can be solved using dynamic programming as shown in Algorithm 11.11.2.

As given, Algorithm 11.11.2 has a computational complexity of $O(mn)$, where m and n, respectively, denote the number of edge points on the matched left and right scanlines.[2] Variants

[2]Our version of the algorithm assumes that all edges are matched. To account for noise and edge detection errors, it is reasonable to allow the matching algorithm to skip a bounded number of edges, but this does not change its asymptotic complexity (Ohta and Kanade, 1985).

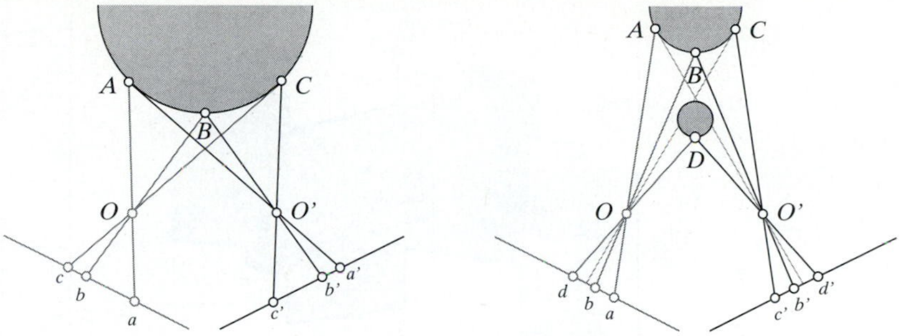

Figure 11.13 Ordering constraints. In the (usual) case shown in the left part of the diagram, the order of feature points along the two (oriented) epipolar lines is the same. In the case shown in the right part of the figure, a small object lies in front of a larger one. Some of the surface points are not visible in one of the images (e.g., A is not visible in the right image), and the order of the image points is not the same in the two pictures: b is on the right of d in the left image, but b' is on the left of d' in the right image.

Algorithm 11.2: A dynamic-programming algorithm for establishing stereo correspondences between two corresponding scanlines with m and n edge points, respectively (the endpoints of the scanlines are included for convenience). Two auxiliary functions are used: Inferior-Neighbors(k, l) returns the list of neighbors (i, j) of the node (k, l) such that $i \leq k$ and $j \leq l$, and Arc-Cost(i, j, k, l) evaluates and returns the cost of matching the intervals (i, k) and (j, l). For correctness, $C(1, 1)$ should be initialized with a value of zero.

```
    % Loop over all nodes (k, l) in ascending order.
for k = 1 to m do
  for l = 1 to n do
    % Initialize optimal cost C(k, l) and backward pointer B(k, l).
    C(k, l) ← +∞; B(k, l) ← nil;
    % Loop over all inferior neighbors (i, j) of (k, l).
    for (i, j) ∈ Inferior-Neighbors(k, l) do
      % Compute new path cost and update backward pointer if necessary.
      d ← C(i, j) + Arc-Cost(i, j, k, l);
      if d < C(k, l) then C(k, l) ← d; B(k, l) ← (i, j) endif;
      endfor;
    endfor;
  endfor;
% Construct optimal path by following backward pointers from (m, n).
P ← {(m, n)}; (i, j) ← (m, n);
while B(i, j) ≠ nil do (i, j) ← B(i, j); P ← {(i, j)} ∪ P endwhile.
```

of this approach have been implemented by Baker and Binford (1981), who combine a coarse-to-fine intra-scanline search procedure with a cooperative process for enforcing interscanline consistency, and Ohta and Kanade (1985), who use dynamic programming for both intra- and inter-scanline optimization, the latter procedure being conducted in a three-dimensional search space. Figure 11.15 shows a sample result taken from (Ohta and Kanade, 1985).

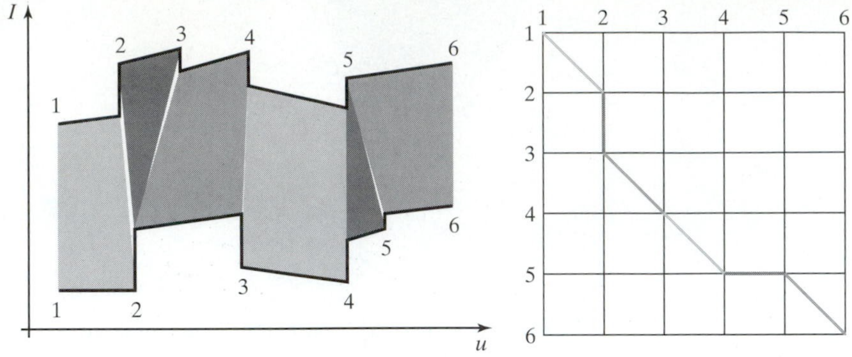

Figure 11.14 Dynamic programming and stereopsis: The left part of the figure shows two intensity profiles along matching epipolar lines. The polygons joining the two profiles indicate matches between successive intervals (some of the matched intervals may have zero length). The right part of the diagram represents the same information in graphical form: An arc (thick line segment) joins two nodes (i, i') and (j, j') when the intervals (i, j) and (i', j') of the intensity profiles match each other.

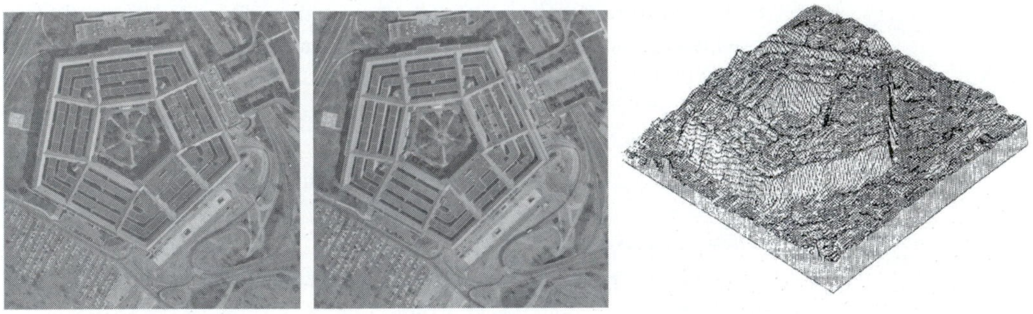

Figure 11.15 Two images of the Pentagon and an isometric plot of the disparity map computed by the dynamic-programming algorithm of Ohta and Kanade (1985). *Reprinted from "Stereo by Intra- and Inter-Scanline Search," by Y. Ohta and T. Kanade, IEEE Transactions on Pattern Analysis and Machine Intelligence, 7(2):139–154, (1985).*

11.4 USING MORE CAMERAS

11.4.1 Three Cameras

Adding a third camera eliminates (in large part) the ambiguity inherent in two-view point matching. In essence, the third image can be used to check hypothetical matches between the first two pictures (Figure 11.16): The three-dimensional point associated with such a match is first reconstructed then reprojected into the third image. If no compatible point lies nearby, then the match must be wrong. In fact, the reconstruction/reprojection process can be avoided by noting, as in chapter 10, that given three weakly (and a fortiori strongly) calibrated cameras and two images of a point one can always predict its position in a third image by intersecting the corresponding epipolar lines.

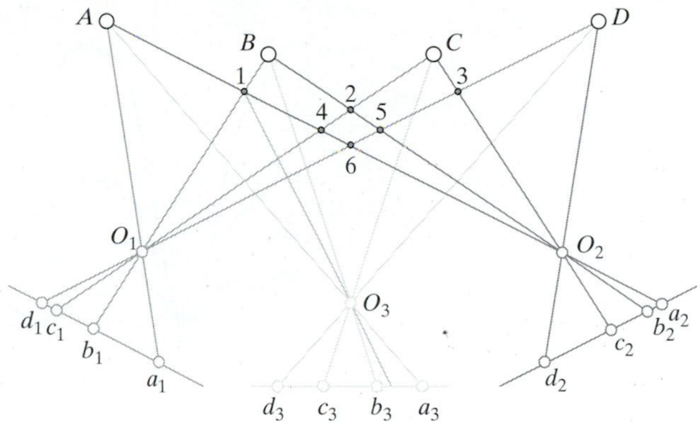

Figure 11.16 The small gray discs indicate the incorrect reconstructions associated with the left and right images of four points. The addition of a central camera removes the matching ambiguity: None of the corresponding rays intersects any of the six discs. Alternatively, matches between points in the first two images can be checked by reprojecting the corresponding three-dimensional point in the third image. For example, the match between b_1 and a_2 is obviously wrong since there is no feature point in the third image near the reprojection of the hypothetical reconstruction numbered 1 in the diagram.

11.4.2 Multiple Cameras

In most trinocular stereo algorithms, potential correspondences are hypothesized using two of the images, then confirmed or rejected using the third one. In contrast, Okutami and Kanade (1993) have proposed a multicamera method where matches are found using all pictures at the same time. The basic idea is simple, but elegant: Assuming that all the images have been rectified, the search for the correct disparities is replaced by a search for the correct depth, or rather its inverse. Of course, the inverse depth is proportional to the disparity for each camera, but the disparity varies from camera to camera, and the inverse depth can be used as a common search index. Picking the first image as a reference, Okutami and Kanade add the sums of squared differences associated with all other cameras into a global evaluation function E (as shown earlier, this is of course equivalent to adding the correlation functions associated with the images).

Figure 11.17 plots the value of E as a function of inverse depth for various sets of cameras. It should be noted that the corresponding images contain a repetitive pattern and that using only two or three cameras does not yield a single, well-defined minimum. However, adding more cameras provides a clear minimum corresponding to the correct match.

Figure 11.18 shows a sequence of 10 rectified images and a plot of the surface reconstructed by the algorithm.

11.5 NOTES

The fact that disparity gives rise to stereopsis in human beings was first demonstrated by Wheatstone's (1838) invention of the stereoscope. That disparity is sufficient for stereopsis without eye movements was demonstrated shortly afterward by Dove (1841) with illumination provided by an electric spark too brief for eye vergence to take place (Helmholtz, 1909, p. 455). Human stereopsis is further discussed in the classical book of Helmholtz (1909), an amazing read for anyone interested in the history of the field, as well as the books by Julesz (1960, 1971), Frisby (1980)

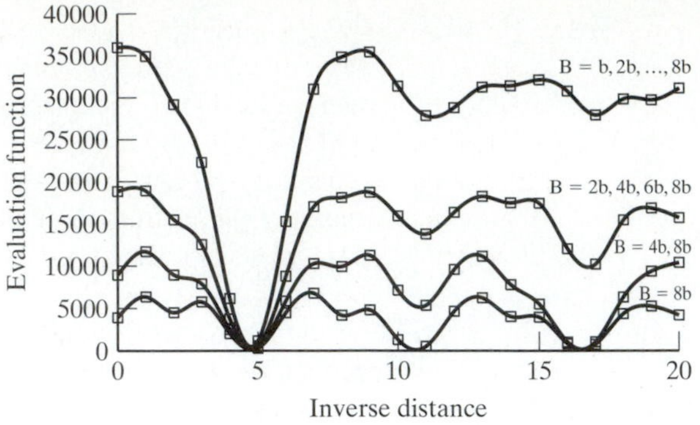

Figure 11.17 Combining multiple views: The sum of squared differences is plotted here as a function of the inverse depth for various numbers of input pictures. The data are taken from a scanline near the top of the images shown in Figure 11.18, whose intensity is nearly periodic. The diagram clearly shows that the minimum of the function becomes less and less ambiguous as more images are added. *Reprinted from "A Multiple-Baseline Stereo System," by M. Okutami and T. Kanade, IEEE Transactions on Pattern Analysis and Machine Intelligence, 15(4):353–363, (1993). © 1993 IEEE.*

and Marr (1982). Theories of human binocular perception not presented in this chapter for lack of space include Koenderink and Van Doorn (1976*a*), Pollard *et al.* (1970), McKee *et al.* (1990), and Anderson and Nakayama (1994).

Excellent treatments of machine stereopsis can be found in the books of Grimson (1981*b*), Marr (1982), Horn (1986), and Faugeras (1993). Marr focuses on the computational aspects of human stereo vision, whereas Horn's account emphasizes the role of photogrammetry in artificial stereo systems. Grimson and Faugeras emphasize the geometric and algorithmic aspects of stereopsis. The constraints associated with stereo matching are discussed in (Binford, 1984). Early techniques for line matching in binocular stereo include Medioni and Nevatia (1984) and

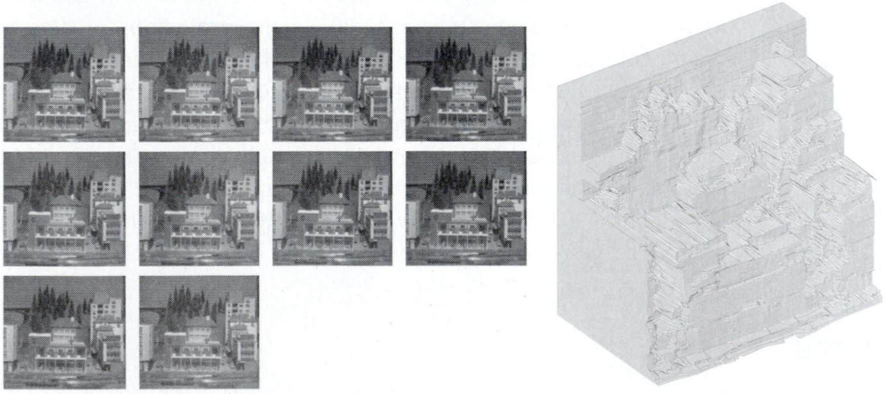

Figure 11.18 A series of 10 images and the corresponding reconstruction. The gridboard near the top of the images is the source for the nearly periodic brightness signal giving rise to ambiguities in Figure 11.17. *Reprinted from "A Multiple-Baseline Stereo System," by M. Okutami and T. Kanade, IEEE Transactions on Pattern Analysis and Machine Intelligence, 15(4):353–363, (1993). © 1993 IEEE.*

Ayache and Faugeras (1987). Algorithms for trinocular fusion include Milenkovic and Kanade (1985), Yachida *et al.* (1986), Ayache and Lustman (1987), and Robert and Faugeras (1991). As shown in Robert and Faugeras (1991) and the exercises, the trifocal tensor introduced in chapter 10 can be used to also predict the tangent and curvature along some image curve in one image given the corresponding quantities measured in in the other images. This fact can be used to effectively match and reconstruct curves from three images.

As noted earlier, image edges are often used as the basis for establishing binocular correspondences, at least in part because they can (in principle) be identified with physical properties of the imaging process, corresponding for example to albedo, color, or occlusion boundaries. A point rarely taken into account by stereo-matching algorithms is that binocular fusion *always* fails along the contours of solids bounded by smooth surfaces (Figure 11.19). Indeed, in this case, the corresponding image edges are viewpoint dependent, and matching them yields erroneous reconstructions. As shown in Arbogast and Mohr (1991), Vaillant and Faugeras (1992), Cipolla and Blake (1992), and Boyer and Berger (1996), three cameras are sufficient in this case to reconstruct a local second-degree surface model.

It is not quite clear at this point whether feature-based matching is preferable to gray-level matching. The former is accurate near surface markings, but only yields a sparse set of measurements, whereas the latter may give poor results in uniform regions but provides dense correspondences in textured areas. In this context, the topic of dense surface interpolation from sparse samples is important, although it has hardly been mentioned in this chapter. The interested reader is referred to Grimson (1981*b*) and Terzopoulos (1984) for more details.

A different approach to stereo vision that we have also failed to discuss for lack of space involves higher level interpretation processes—for example, prediction/verification methods operating on graphical image descriptions (Ayache and Faverjon, 1997) or hierarchical techniques matching curves, surfaces, and volumes found in two images (Lim and Binford, 1988).

All of the algorithms presented in this chapter (implicitly) assume that the images being fused are quite similar. This is equivalent to considering a short baseline. An effective algorithm for dealing with wide baselines can be found in Pritchett and Zisserman (1998). Another, model-based approach is discussed in chapter 26.

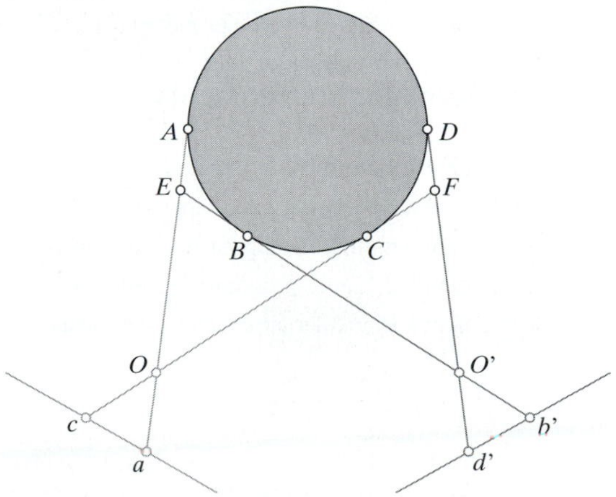

Figure 11.19 Stereo matching fails at smooth object boundaries: For narrow baselines, the pairs (c, d') and (a, b') are easily matched by most edge-based algorithms, yielding the fictitious points F and E as the corresponding three-dimensional reconstructions.

Finally, we have limited our attention to stereo rigs with fixed intrinsic and extrinsic parameters. *Active vision* is concerned with the construction of vision systems capable of dynamically modifying these parameters (e.g., changing camera zoom and vergence angles, and taking advantage of these capabilities in perceptual and robotic tasks; see Aloimonos *et al.*, 1987, Bajcsy, 1988, Ahuja and Abott 1993, Brunnström *et al.*, 1996).

PROBLEMS

11.1. Show that, in the case of a rectified pair of images, the depth of a point P in the normalized coordinate system attached to the first camera is $z = -B/d$, where B is the baseline and d is the disparity.

11.2. Use the definition of disparity to characterize the accuracy of stereo reconstruction as a function of baseline and depth.

11.3. Give reconstruction formulas for verging eyes in the plane.

11.4. Give an algorithm for generating an ambiguous random dot stereogram that can depict two different planes hovering over a third one.

11.5. Show that the correlation function reaches its maximum value of 1 when the image brightnesses of the two windows are related by the affine transform $I' = \lambda I + \mu$ for some constants λ and μ with $\lambda > 0$.

11.6. Prove the equivalence of correlation and sum of squared differences for images with zero mean and unit Frobenius norm.

11.7. Recursive computation of the correlation function.

(a) Show that $(\boldsymbol{w} - \bar{\boldsymbol{w}}) \cdot (\boldsymbol{w}' - \bar{\boldsymbol{w}}') = \boldsymbol{w} \cdot \boldsymbol{w}' - (2m+1)(2n+1)\bar{I}\bar{I}'$.

(b) Show that the average intensity $\bar{I}$ can be computed recursively, and estimate the cost of the incremental computation.

(c) Generalize the prior calculations to all elements involved in the construction of the correlation function, and estimate the overall cost of correlation over a pair of images.

11.8. Show how a first-order expansion of the disparity function for rectified images can be used to warp the window of the right image corresponding to a rectangular region of the left one. Show how to compute correlation in this case using interpolation to estimate right-image values at the locations corresponding to the centers of the left window's pixels.

11.9. Show how to use the trifocal tensor to predict the tangent line along an image curve from tangent line measurements in two other pictures.

Programming Assignments

11.10. Implement the rectification process.

11.11. Implement a correlation-based approach to stereopsis.

11.12. Implement a multiscale approach to stereopsis.

11.13. Implement a dynamic-programming approach to stereopsis.

11.14. Implement a trinocular approach to stereopsis.

12

Affine Structure from Motion

This chapter revisits the problem of estimating the three-dimensional shape of a scene from multiple pictures. In the context of stereopsis, the cameras used to acquire the input images are normally calibrated so their intrinsic parameters are known and their extrinsic ones have been determined relative to some fixed world coordinate system. This greatly simplifies the reconstruction process and explains the emphasis put on the binocular (or trinocular) fusion problem in conventional stereo vision systems. We consider in this chapter a more difficult setting where the cameras' positions and possibly their intrinsic parameters are a priori unknown and may change over time. This is typical of *image-based rendering* applications, where a video clip recorded by a hand-held camcorder, possibly zooming during the shoot, is used to capture the shape of an object and render it under new viewing conditions (chapter 26). This is also relevant for active vision systems whose calibration parameters vary dynamically and planetary robot probes for which these parameters may change due to the large accelerations at take-off and landing. Recovering the cameras' positions is of course just as important as estimating the scene shape in the context of mobile robot navigation.

We ignore the correspondence problem in the rest of this chapter, assuming that the projections of n points have been matched across m pictures.[1] We focus instead on the purely geometric *structure-from-motion* problem of using image matches to estimate both the three-dimensional positions of the corresponding scene points in some fixed coordinate system (i.e., the scene *structure*) and the projection matrices associated with the cameras observing them (or, equivalently, the *motion* of the points relative to the cameras). This chapter is concerned with scenes whose relief is small compared with their overall depth relative to the cameras observing them, so perspective projection can be approximated by the simpler *affine* models of the imaging process

[1]Methods for establishing such correspondences across both continuous image sequences and scattered views of a scene are discussed in chapters 17 and 23.

introduced in chapters 1 and 2. The full perspective structure-from-motion problem is discussed in the next chapter. Concretely, given n *fixed* points P_j ($j = 1, \dots, n$) observed by m affine cameras and the corresponding mn (nonhomogeneous) coordinate vectors $\boldsymbol{p}_{ij}$ of their images, we rewrite the affine projection Eq. (2.19) as

$$\boldsymbol{p}_{ij} = \mathcal{M}_i \begin{pmatrix} \boldsymbol{P}_j \\ 1 \end{pmatrix} = \mathcal{A}_i \boldsymbol{P}_j + \boldsymbol{b}_i \quad \text{for} \quad i = 1, \dots, m \quad \text{and} \quad j = 1, \dots, n, \tag{12.1}$$

and define *affine structure from motion* as the problem of estimating the m 2×4 matrices $\mathcal{M}_i = (\mathcal{A}_i \quad \boldsymbol{b}_i)$ and the n positions $\boldsymbol{P}_j$ of the points P_j in some fixed coordinate system from the mn image correspondences $\boldsymbol{p}_{ij}$.

When the projection matrices $\mathcal{M}_i$ are allowed to take an arbitrary form (i.e., when the intrinsic and extrinsic parameters of the cameras are unknown, see chapter 2), Eq. (12.1) provides $2mn$ constraints on the $8m + 3n$ unknown coefficients defining the matrices $\mathcal{M}_i$ and the point positions $\boldsymbol{P}_j$. Since $2mn$ is greater than $8m + 3n$ for large enough values of m and n, it is thus clear that a sufficient number of views of a sufficient number of points allows the recovery of the corresponding structure and motion parameters via, say, the least-squares techniques presented in chapter 3. However, it is important to understand that, if $\mathcal{M}_i$ and $\boldsymbol{P}_j$ are solutions of Eq. (12.1), so are $\mathcal{M}'_i$ and $\boldsymbol{P}'_j$, where

$$\mathcal{M}'_i = \mathcal{M}_i \mathcal{Q} \quad \text{and} \quad \begin{pmatrix} \boldsymbol{P}'_j \\ 1 \end{pmatrix} = \mathcal{Q}^{-1} \begin{pmatrix} \boldsymbol{P}_j \\ 1 \end{pmatrix} \tag{12.2}$$

and $\mathcal{Q}$ is an arbitrary *affine transformation* matrix—that is, it can be written (see chapter 2 and next section) as

$$\mathcal{Q} = \begin{pmatrix} \mathcal{C} & \boldsymbol{d} \\ \boldsymbol{0}^T & 1 \end{pmatrix} \quad \text{with} \quad \mathcal{Q}^{-1} = \begin{pmatrix} \mathcal{C}^{-1} & -\mathcal{C}^{-1}\boldsymbol{d} \\ \boldsymbol{0}^T & 1 \end{pmatrix}, \tag{12.3}$$

where $\mathcal{C}$ is a nonsingular 3×3 matrix and $\boldsymbol{d}$ is a vector in $\mathbb{R}^3$. In other words, any solution of the affine structure-from-motion problem can *only be defined up to an affine transformation ambiguity*. Taking into account the 12 parameters defining a general affine transformation, we should thus expect a finite number of solutions as soon as $2mn \geq 8m + 3n - 12$. For $m = 2$, this suggests that four point correspondences should be sufficient to determine (up to an affine transformation) the two projection matrices and the three-dimensional position of any other point. This is confirmed formally in Section 12.2.

When the intrinsic parameters of the cameras are known so the corresponding calibration matrices can be taken equal to the identity, the parameters of the projection matrices $\mathcal{M}_i = (\mathcal{A}_i \quad \boldsymbol{b}_i)$ must obey additional constraints. For example, according to Eq. (2.20), the matrix $\mathcal{A}_i$ associated with a (calibrated) weak-perspective camera is formed by the first two rows of a rotation matrix, scaled by the inverse of the depth of the corresponding reference point. As shown in Section 12.4, constraints such as these can be used to eliminate the affine ambiguity when enough images are available. This suggests decomposing the solution of the affine structure-from-motion problem into two steps: (a) first use at least two views of the scene to construct a unique (up to an arbitrary affine transformation) three-dimensional representation of the scene, called its *affine shape*; then (b) use additional views and the constraints associated with known camera calibration parameters and specific affine models to uniquely determine the rigid Euclidean structure of the scene. The first stage of this approach yields the essential part of the solution: The affine shape is a full-fledged three-dimensional representation of the scene, which, as shown in chapter 26, can be used in its own right to synthesize new views of the scene. The second step simply amounts to finding a *Euclidean upgrade* of the scene (i.e., to computing a single affine transformation that account for its rigidity and map its affine shape onto a Euclidean one).

Using three or more images overconstrains the structure-from-motion problem and leads to more robust least-squares solutions. Accordingly, a significant portion of this chapter is devoted to the problem of recovering the affine shape of a scene from several (possibly many) pictures. We conclude with techniques for segmenting a set of data points into objects undergoing different motions.

12.1 ELEMENTS OF AFFINE GEOMETRY

Let us start by introducing some elementary notions of affine geometry. The corresponding geometric and algebraic tools allow us to state and prove the fundamental properties of affine projection models. They also serve as building blocks for the structure-from-motion algorithms introduced in the rest of this chapter.

As noted in Snapper and Troyer (1989), affine geometry is, roughly speaking, what is left after all ability to measure lengths, areas, and angles has been removed from Euclidean geometry. The concept of parallelism remains, however, as well as the ability to measure the ratio of distances between collinear points. Giving a rigorous axiomatic introduction to affine geometry would be out of place here. Instead, we remain quite informal and recall the basic facts about real affine spaces that are necessary to understand the rest of this chapter. The reader familiar with notions such as barycentric combinations, affine coordinate systems, and affine transformations may safely proceed to the next section.

12.1.1 Affine Spaces and Barycentric Combinations

A *real affine space* is a set X of *points*, together with a real vector space $\vec{X}$, and an *action* ϕ of the additive group of $\vec{X}$ on X. The vector space $\vec{X}$ is said to *underlie* the affine space X. Informally, the action of a group on a set maps the elements of this group onto bijections of the set. Here, the action ϕ associates with every vector $\boldsymbol{u} \in \vec{X}$ a bijection $\phi_{\boldsymbol{u}} : X \to X$ such that, for any $\boldsymbol{u}, \boldsymbol{v}$ in $\vec{X}$ and any point P in X, $\phi_{\boldsymbol{u}+\boldsymbol{v}}(P) = \phi_{\boldsymbol{u}} \circ \phi_{\boldsymbol{v}}(P)$, $\phi_{\boldsymbol{0}}(P) = P$, and for any pair of points P, Q in X, there exists a unique vector $\boldsymbol{u}$ in $\vec{X}$ such that $\phi_{\boldsymbol{u}}(P) = Q$. These definitions may sound a bit abstract, so let us give some concrete examples. A familiar affine space is, of course, $\mathbb{E}^3$, where X is the set of physical points and $\vec{X}$ is the set of translations of X onto itself. Another affine space can be constructed by choosing both X and $\vec{X}$ to be equal to $\mathbb{R}^n$, with the action ϕ defined by $\phi_{\boldsymbol{u}}(P) = P + \boldsymbol{u}$, where P and $\boldsymbol{u}$ are both elements of $\mathbb{R}^n$ and "+" denotes the addition in that vector space.

Example 12.1 **$\mathbb{R}^2$ as an affine plane.**

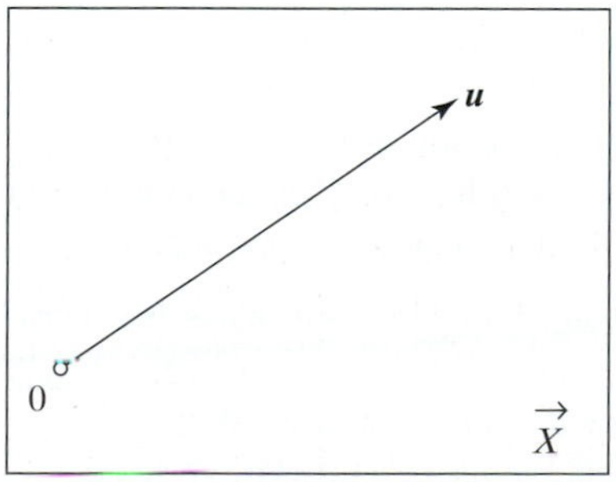

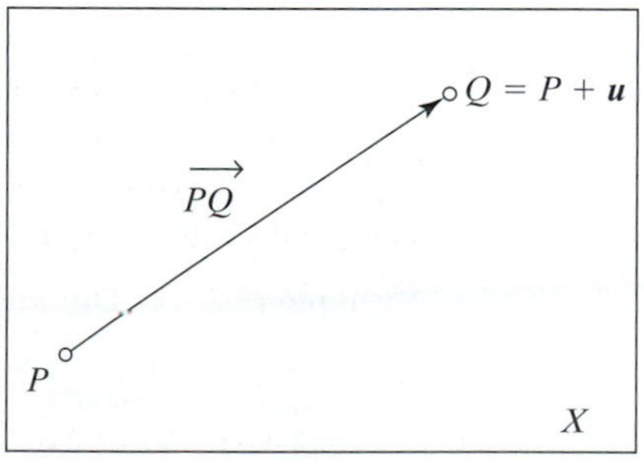

The vector space $\mathbb{R}^2$ can be considered as an affine space by choosing $X = \vec{X} = \mathbb{R}^2$. Given $P = (x, y)^T$ and $\boldsymbol{u} = (a, b)^T$, we define $\phi_{\boldsymbol{u}}(P) \stackrel{\text{def}}{=} P + \boldsymbol{u} = (x + a, y + b)^T$. Given $P = (x, y)^T$ and

$Q = (x', y')^T$, the unique vector $\boldsymbol{u}$ such that $P + \boldsymbol{u} = Q$ is of course $\boldsymbol{u} = Q - P = \overrightarrow{PQ} \stackrel{\text{def}}{=} (x' - x, y' - y)^T$.

Following this example, we denote from now on the point $\phi_{\boldsymbol{u}}(P)$ by $P + \boldsymbol{u}$ and the vector $\boldsymbol{u}$ such that $\phi_{\boldsymbol{u}}(P) = Q$ by $\overrightarrow{PQ}$ or, equivalently, by $Q - P$. This is justified by the fact that choosing a point O as the *origin* of X allows us to identify every other point P with the vector $\boldsymbol{u} = \overrightarrow{OP}$ such that $\phi_{\boldsymbol{u}}(O) = P$. Indeed,

$$Q = P + \overrightarrow{PQ} \Longleftrightarrow \overrightarrow{OQ} = \overrightarrow{OP} + \overrightarrow{PQ} \quad \text{and} \quad Q - P = \overrightarrow{PQ} \Longleftrightarrow \overrightarrow{OQ} - \overrightarrow{OP} = \overrightarrow{PQ}.$$

The introduction of an origin is often useful for beginners who want to keep their affine notation straight. It should be absolutely clear, however, that the point $P + \boldsymbol{u}$ and the vector $\overrightarrow{PQ} = Q - P$ are *totally independent* of the choice of any origin whatsoever. Likewise, the symbols "+" and "−" in these expressions are used for notational convenience and *do not* convey their usual meaning of addition and subtraction in the additive group of a vector space.

Although it is possible to "add" a vector to a point and to "subtract" two points, it is not possible to "add" two points or to "multiply" a point by a scalar (see Exercises). However, a restricted kind of "linear combination" of points can be defined: Consider $m+1$ points $A_0, A_1 \dots, A_m$ and $m + 1$ weights $\alpha_0, \alpha_1, \dots, \alpha_m$ such that $\alpha_0 + \alpha_1 + \dots + \alpha_m = 1$; the corresponding *barycentric combination* of the points A_0 to A_m is the point

$$\sum_{i=0}^{m} \alpha_i A_i \stackrel{\text{def}}{=} A_j + \sum_{i=0, i \neq j}^{m} \alpha_i (A_i - A_j), \tag{12.4}$$

where j is an integer between 0 and m. The right-hand side of this equation defines a point by adding a vector (a linear combination of the vectors $A_i - A_j$) to a point (A_j). It is easily shown that this definition is *independent* of the value of j (see Exercises), which justifies the symmetrical role played by the points A_i ($i = 0, \dots, m$) in the notation $\sum_{i=0}^{m} \alpha_i A_i$. This notation is further justified by introducing an origin O and noting that $\sum_{i=0}^{m} \alpha_i \overrightarrow{OA_i} = \overrightarrow{OA_j} + \sum_{i=0, i \neq j}^{m} \alpha_i (\overrightarrow{OA_i} - \overrightarrow{OA_j})$ when $\alpha_0 + \alpha_1 + \dots + \alpha_m = 1$. However, the definition of barycentric combinations by Eq. (12.4) is preferable since it is obviously independent of any choice of origin.

A familiar example of barycentric combination is the *center of mass* of $m + 1$ points, corresponding to the case where all weights are equal to $1/(m + 1)$. Any other set of weight values adding to 1 yields a valid barycentric combination.

12.1.2 Affine Subspaces and Affine Coordinates

An *affine subspace* of X is defined by a point O and a vector subspace U of $\vec{X}$ as the set of points $O + U \stackrel{\text{def}}{=} \{O + \boldsymbol{u}, \boldsymbol{u} \in U\}$. Its *dimension* is the dimension of the associated vector subspace. Two affine subspaces $O' + U'$ and $O'' + U''$, such that U' is a subspace of U'', or U'' is a subspace of U' are said to be parallel. Affine subspaces of dimension 1 and 2 are, respectively, called *lines* and *planes*. When $\vec{X}$ is of finite dimension n, its affine subspaces of dimension $n - 1$ are called *hyperplanes*. Affine lines, planes, and hyperplanes take their usual meaning in the affine spaces associated with physical three-dimensional space and $\mathbb{R}^n$.

Example 12.2 The intersection of two affine subspaces is either empty or an affine subspace.

Consider two subspaces $Y' = O' + U'$ and $Y'' = O'' + U''$ of some affine space X, and denote by Z their intersection. Let P_0 denote some point in Z. We have by definition $P_0 = O' + \boldsymbol{u}'_0 = O'' + \boldsymbol{u}''_0$ for some vectors $\boldsymbol{u}'_0$ in U' and $\boldsymbol{u}''_0$ in U''. Likewise, given any other point P in Z, we can write $P = O' + \boldsymbol{u}' = O'' + \boldsymbol{u}''$ for some vectors $\boldsymbol{u}'$ in U' and $\boldsymbol{u}''$ in U''. In particular, we must have

$$P = P_0 + \boldsymbol{u}' - \boldsymbol{u}'_0 = P_0 + \boldsymbol{u}'' - \boldsymbol{u}''_0,$$

which implies that (a) $\boldsymbol{u}' - \boldsymbol{u}'_0 = \boldsymbol{u}'' - \boldsymbol{u}''_0$ is an element of $U' \cap U''$, and (b) P is an element of $P_0 + U' \cap U''$. Conversely, any point P in $P_0 + U' \cap U''$ can be written as $P = P_0 + \boldsymbol{u}$ for some vector $\boldsymbol{u}$ in $U' \cap U''$; thus,

$$P = O' + (P_0 - O') + \boldsymbol{u} = O' + \boldsymbol{u}' + \boldsymbol{u} = O'' + (P_0 - O'') + \boldsymbol{u} = O'' + \boldsymbol{u}'' + \boldsymbol{u},$$

which implies that P is an element of Z. We finally conclude that $Z = P_0 + U' \cap U''$. Note that the intersection of two affine spaces may be empty. For example two parallel lines do not intersect. Neither do two skew lines in space of course, although they are not parallel to each other.

Affine subspaces can also be defined purely in terms of points: Let $S(A_0, A_1 \dots, A_m)$ denote the set of all barycentric combinations of $m+1$ points $A_0, A_1, \dots, A_m$. It is easy to verify that $S(A_0, A_1, \dots, A_m)$ is indeed an affine subspace (see Exercises), and that its dimension is at most m (e.g., two distinct points define a line, three points define [in general] a plane, etc.). We say that $m+1$ points are independent if they do not lie in a subspace of dimension at most $m-1$, so $m+1$ independent points define (or *span*) an m-dimensional subspace.

Example 12.3 Two complementary definitions of an affine plane.

Consider three noncollinear points A_0, A_1, and A_2 in $\mathbb{R}^3$ viewed as an affine space. These points define the plane $\Pi = A_0 + U$ of $\mathbb{R}^3$ associated with the point A_0 and the vector plane U spanned by the two vectors $\boldsymbol{u}_1 = \overrightarrow{A_0A_1}$ and $\boldsymbol{u}_2 = \overrightarrow{A_0A_2}$.

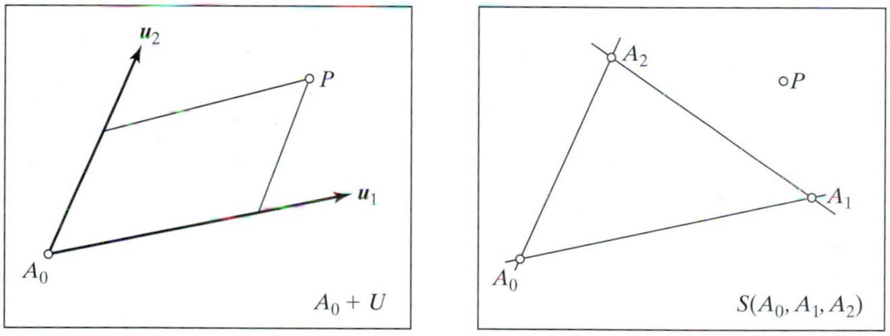

Equivalently, the plane Π can be viewed as the affine subspace $S(A_0, A_1, A_2)$ of $\mathbb{R}^3$, and any point P in Π can be represented as a barycentric combination of the points A_0, A_1, and A_2.

An *affine coordinate system* for $O + U$ consists of a point A_0 (called its *origin*) in $O + U$ and a coordinate system $(\boldsymbol{u}_1, \dots, \boldsymbol{u}_m)$ for U. The *affine coordinates* of a point P in $O + U$ are defined as the coordinates of the vector $\overrightarrow{A_0P}$ in the coordinate system $(\boldsymbol{u}_1, \dots, \boldsymbol{u}_m)$. It is crucial to understand that the (Euclidean) point coordinates used in chapter 2 and in conventional Euclidean geometry are just affine coordinates. The vectors $\boldsymbol{u}_i$ $(i = 1, \dots, m)$ used to define the corresponding coordinate systems simply have the additional property of having unit length and being orthogonal to each other. This property is not required for general affine coordinate systems, and indeed the notions of lengths and angles may not be defined in general affine spaces.

Example 12.4 Affine coordinate changes.

Given some coordinate system $(F) = (O, \boldsymbol{u}, \boldsymbol{v}, \boldsymbol{w})$ for the affine space $\mathbb{E}^3$ and a point P of $\mathbb{E}^3$ such that $\overrightarrow{OP} = x\boldsymbol{u} + y\boldsymbol{v} + z\boldsymbol{w}$, we can define, using the same notation as in chapter 2, the (affine) coordinate vector of P as ${}^F P = (x, y, z)^T$.

Given two affine coordinate systems $(A) = (O_A, \boldsymbol{u}_A, \boldsymbol{v}_A, \boldsymbol{w}_A)$ and $(B) = (O_B, \boldsymbol{u}_B, \boldsymbol{v}_B, \boldsymbol{w}_B)$ for the affine space $\mathbb{E}^3$, let us define the 3×3 matrix

$${}^B_A\mathcal{C} = \left({}^B\boldsymbol{u}_A \quad {}^B\boldsymbol{v}_A \quad {}^B\boldsymbol{w}_A\right),$$

where ${}^{B}\boldsymbol{a}$ denotes the coordinate vector of the vector $\boldsymbol{a}$ in the (vector) coordinate system $(\boldsymbol{u}_A, \boldsymbol{v}_A, \boldsymbol{w}_A)$. It is easy to show that ${}^{B}P = {}^{B}_{A}\mathcal{C}\,{}^{A}P + {}^{B}O_A$, or, in (affine) homogeneous coordinates,

$$\begin{pmatrix} {}^{B}P \\ 1 \end{pmatrix} = {}^{B}_{A}\mathcal{T} \begin{pmatrix} {}^{A}P \\ 1 \end{pmatrix}, \quad \text{where} \quad {}^{B}_{A}\mathcal{T} = \begin{pmatrix} {}^{B}_{A}\mathcal{C} & {}^{B}O_A \\ \mathbf{0}^T & 1 \end{pmatrix}.$$

Note the obvious similarity with the formula for a change of Euclidean coordinate system in chapter 2. Here, however, the basis vectors of the two coordinate frames do not form orthonormal bases, so ${}^{B}_{A}\mathcal{C}$ is an ordinary nonsingular 3×3 matrix instead of a rotation matrix, and ${}^{B}_{A}\mathcal{T}$ is an affine transformation matrix.

An alternative way of defining a coordinate system for an n-dimensional affine space X is to pick $n+1$ independent points $A_0, A_1 \dots, A_n$ in X. The *barycentric coordinates* α_i $(i = 0, 1, \dots, n)$ of a point P in Y are uniquely defined by $P = \alpha_0 A_0 + \alpha_1 A_1 + \cdots + \alpha_n A_n$. They are related to affine coordinates in a simple way: Choosing $j = 0$ in Eq. (12.4) yields

$$P = \alpha_0 A_0 + \alpha_1 A_1 + \cdots + \alpha_n A_n = A_0 + \alpha_1 (A_1 - A_0) + \cdots + \alpha_n (A_n - A_0),$$

showing that the affine coordinates of P in the basis formed by the points A_i $(i = 0, 1, \dots, m)$ are $\alpha_1, \dots, \alpha_m$.

When an n-dimensional affine space X has been equipped with an affine basis, a necessary and sufficient condition for $m+1$ points A_i to define a p-dimensional affine subspace of X (with $m \geq p$ and $n \geq p$) is for the $(n+1) \times (m+1)$ matrix

$$\mathcal{D} = \begin{pmatrix} x_{01} & x_{11} & \dots & x_{m1} \\ \dots & \dots & \dots & \dots \\ x_{0n} & x_{1n} & \dots & x_{mn} \\ 1 & 1 & \dots & 1 \end{pmatrix}$$

formed by their coordinate vectors $(x_{i1}, \dots, x_{in})^T$ $(i = 0, 1, \dots, m)$ to have rank $p+1$. Indeed, a rank lower than $p+1$ means that any column of this matrix is a barycentric combination of at most p of its columns, and a rank higher than $p+1$ implies that at least $p+2$ of the points are independent.

Example 12.5 The equation of a line in the plane.

Consider three points A_0, A_1, and A_2 in an affine plane, with coordinate vectors $(x_0, y_0)^T$, $(x_1, y_1)^T$, and $(x_2, y_2)^T$ in some basis of this plane. According to the previous paragraph, a necessary and sufficient condition for these points to lie in an affine subspace of dimension 1 (i.e., be collinear) is that the rank of the matrix

$$\mathcal{D} = \begin{pmatrix} x_0 & x_1 & x_2 \\ y_0 & y_1 & y_2 \\ 1 & 1 & 1 \end{pmatrix}$$

be equal to 2, or, equivalently, that its determinant be equal to zero. Note that

$$\text{Det}(\mathcal{D}) = x_1 y_2 - x_2 y_1 + x_2 y_0 - x_0 y_2 + x_0 y_1 - x_1 y_0 = \begin{pmatrix} x_1 - x_0 \\ y_1 - y_0 \end{pmatrix} \times \begin{pmatrix} x_2 - x_0 \\ y_2 - y_0 \end{pmatrix},$$

where "$\times$" denotes here the operator that associates with two vectors in $\mathbb{R}^2$ the determinants of their coordinates. Thus $\text{Det}(\mathcal{D}) = 0$ is indeed equivalent to $\overrightarrow{A_0 A_1}$ and $\overrightarrow{A_0 A_2}$ being parallel or to the three points being collinear. When the points A_0 and A_1 are fixed, $\text{Det}(\mathcal{D}) = 0$ can be seen as an equation defining the line passing through A_0 and A_1 in terms of the coordinates of A_2, and it has of course the form $ax_2 + by_2 + c = 0$. This method can be generalized to affine subspaces defined by arbitrary numbers of points: The corresponding equations are simply obtained by writing that the appropriate minors of the matrix $\mathcal{D}$ have zero determinants.

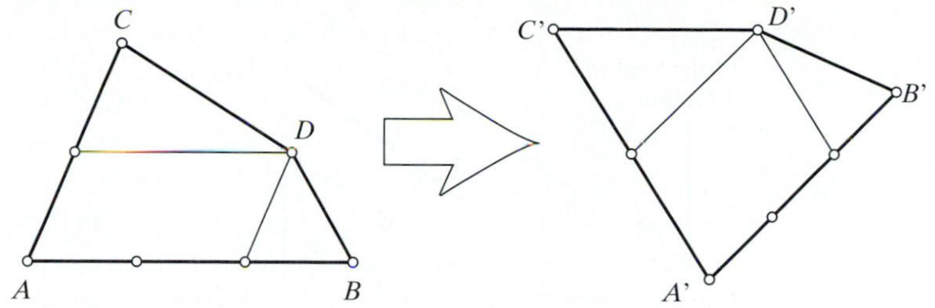

Figure 12.1 An affine transformation of the plane. The points A, B, C, and D are transformed into the points A', B', C', and D'. The affine coordinates of D in the basis of the plane formed by A, B, and C are the same as those of D' in the basis formed by A', B', and C'—namely 2/3 and 1/2.

12.1.3 Affine Transformations and Affine Projection Models

An *affine transformation* between two affine spaces X and Y is a bijection from X onto Y that maps m-dimensional subspaces of X onto m-dimensional subspaces of Y, maps parallel subspaces onto parallel subspaces, and preserves barycentric combinations (or, equivalently, affine coordinates; Figure 12.1). It can be shown that affine transformations can also be characterized by the (seemingly weaker) property of mapping lines onto lines and preserving the *ratio of the signed lengths of parallel line segments*.

An affine transformation between two affine spaces X and Y of dimension m is completely defined by the images $B_0, \ldots, B_m$ of $m+1$ independent points $A_0, \ldots, A_m$. Indeed, the image of any other point with affine coordinates α_i $(i = 0, \ldots, m)$ in the basis of X formed by the points A_i have the same coordinates in the basis of Y formed by the points B_i. Conversely, it can be shown that given any independent points $B_0, \ldots, B_m$ in Y, there is a unique affine transformation mapping the points A_i onto the points B_i. It is thus clear that affine transformations do not preserve angles or distances—a fact confirmed by Figure 12.1. In fact, it can also be shown that affine transformations of $\mathbb{R}^3$ can always be written as the combination of a translation, rotation, nonuniform scaling, and shear.

The relationship between vector and affine spaces induces a relationship between linear and affine transformations. In particular, it is easy to show (see Exercises) that an affine transformation $\psi : X \to Y$ between two affine subspaces X and Y associated with the vector spaces $\vec{X}$ and $\vec{Y}$ can be written as

$$\psi(P) = \psi(O) + \vec{\psi}(P - O),$$

where O is some arbitrarily chosen origin, and $\vec{\psi} : \vec{X} \to \vec{Y}$ is a linear mapping from $\vec{X}$ onto $\vec{Y}$ that is independent of the choice of O. When X and Y are of (finite) dimension m and an affine coordinate system with origin O is chosen, this yields the familiar expression

$$\psi(\boldsymbol{P}) = \boldsymbol{d} + \mathcal{C}\boldsymbol{P} = \mathcal{C}\boldsymbol{P} + \boldsymbol{d},$$

where $\boldsymbol{P}$ denotes the coordinate vector of P in the chosen basis, $\boldsymbol{d}$ denotes the coordinate vector of $\psi(O)$, and $\mathcal{C}$ is the $m \times m$ matrix representing $\vec{\psi}$ in the same coordinate system. Thus, affine transformations as defined in chapter 2 are indeed affine transformations as defined in this chapter.

A fundamental property of parallel projections is that they induce affine transformations from planes onto their images. Let us first show that they preserve the ratio of signed distances between collinear points: The triangles OAa, OBb, and OCc in Figure 12.2(left) are similar,

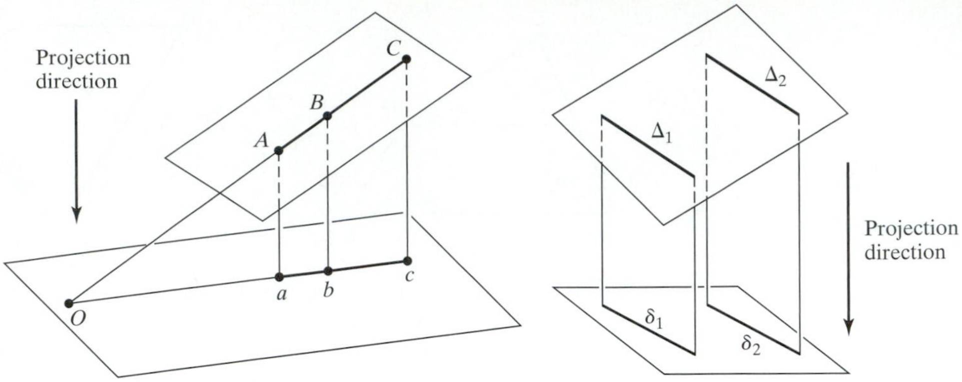

Figure 12.2 Parallel projection preserves: (left) the ratio of signed distances between collinear points and (right) the parallelism of lines.

and it follows that $\overline{AB}/\overline{BC} = \overline{ab}/\overline{bc}$ for any orientation of the lines OC and Oc. To show that parallel projections preserve the parallelism of lines, we use the fact that the intersection of a plane with two parallel planes consists of two parallel lines (see Exercises). Now consider the situation depicted in Figure 12.2(right), where two parallel lines Δ_1 and Δ_2 are projected onto a plane. The planes defined respectively by these two lines and the parallel projection direction are parallel to each other and therefore intersect the image plane along two parallel lines δ_1 and δ_2.

Weak- and paraperspective projections from one plane onto another are also affine transformations. This follows immediately from the fact that they can always be written as the composition of a parallel projection and an affine transformation of the image plane that compounds the effects of the inverse-depth scaling and intrinsic camera parameters. As shown by Theorem 2 in chapter 2, a general affine projection can always be written as a weak-perspective one, thus affine projections from one plane onto another are indeed affine transformations.

It follows immediately that affine projections preserve parallel lines and barycentric combinations. In particular, the center of mass of a set of scene points projects onto the center of mass of their images (which gives a simple method for selecting the reference point of a paraperspective camera; see chapter 2), and the ratio of signed distances between collinear points is an affine-projection invariant (which is useful in the object recognition context; see chapter 23, for example).

12.1.4 Affine Shape

We say that two (possibly infinite) point sets S and S' in some affine space X are *affinely equivalent* when there exists an affine transformation $\psi : X \to X$, such that S' is the image of S under ψ. It is easy to show that affine equivalence is an equivalence relation, and we define the *affine shape* of a point set S in X as the equivalence class of all affinely equivalent point sets. Affine structure from motion can thus be seen as the problem of recovering the affine shape of the observed scene (and/or the equivalence classes formed by the corresponding projection matrices) from features matched in an image sequence. We now have the right tools for solving this problem.

12.2 AFFINE STRUCTURE AND MOTION FROM TWO IMAGES

Let us start with the case where two affine images of the same scene are available (the case of multiple pictures is addressed in the following section). The two structure-from-motion tech-

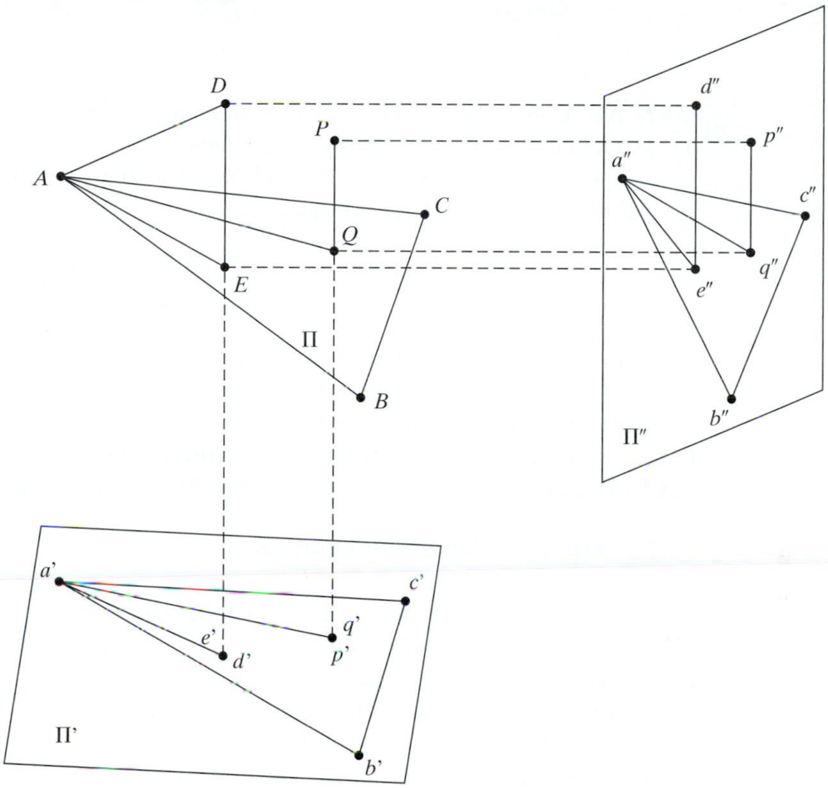

Figure 12.3 Geometric construction of the affine coordinates of a point P in the basis formed by the four points A, B, C, and D. This diagram illustrates the parallel projection case, but the reasoning used in this section is valid in the general affine setting.

niques discussed in this section are complementary: The first one uses geometric reasoning to uncover the affine shape of the scene (from which the projection matrices can be found if needed), whereas the second one uses simple algebraic manipulations to estimate the projection matrices (from which the positions of the scene points are easily calculated).

12.2.1 Geometric Scene Reconstruction

We already mentioned that two affine views of four points A, B, C, D should be sufficient to compute the affine coordinates of any other point P in the basis (A, B, C, D). This is indeed the case, and we now present the constructive proof from (Koenderink and Van Doorn, 1990). Remember that the affine projection of a plane onto another plane is an affine transformation. In particular, when the point P belongs to the plane Π that contains the triangle ABC, its affine coordinates in the basis of Π formed by these three points can be directly measured in either of the two images. Now let E (resp. Q) denote the intersection of the line passing through the points D and d' (resp. P and p') with the plane Π (Figure 12.3). The projections e'' and q'' of the points E and Q onto the plane Π'' have the same affine coordinates in the basis (a'', b'', c'') as the points d' and p' in the basis (a', b', c').

In addition, since the two segments ED and QP are parallel to the first projection direction, the two line segments $e''d''$ and $q''p''$ are also parallel, and we can measure the ratio

$$\lambda = \frac{\overline{q''p''}}{\overline{e''d''}} = \frac{\overline{QP}}{\overline{ED}},$$

where $\overline{AB}$ denotes the signed distance between the two points A and B for some arbitrary (but fixed) orientation of the line joining these points.

If we now denote by $(\alpha_{d'}, \beta_{d'})$ and $(\alpha_{p'}, \beta_{p'})$ the coordinates of the points $d' = e'$ and $p' = q'$ in the basis (a', b', c'), we can write

$$\begin{aligned}\overrightarrow{AP} &= \overrightarrow{AQ} + \overrightarrow{QP} = \alpha_{p'}\overrightarrow{AB} + \beta_{p'}\overrightarrow{AC} + \lambda\overrightarrow{ED}\\ &= (\alpha_{p'} - \lambda\alpha_{d'})\overrightarrow{AB} + (\beta_{p'} - \lambda\beta_{d'})\overrightarrow{AC} + \lambda\overrightarrow{AD}.\end{aligned}$$

In other words, the affine coordinates of P in the (A, B, C, D) basis are $(\alpha_{p'} - \lambda\alpha_{d'}, \beta_{p'} - \lambda\beta_{d'}, \lambda)$. This is the *affine structure-from-motion theorem*: Given two affine views of four noncoplanar points, the affine shape of the scene is uniquely determined (Koenderink and Van Doorn, 1990). Figure 12.4 shows three projections of the synthetic face used in Koenderink and Van Doorn's experiments, along with an affine profile view computed from two of the images.

12.2.2 Algebraic Motion Estimation

Let us now explore a completely different approach, where geometric insight is somewhat neglected in favor of simple algebraic manipulations that exploit the affine ambiguity of structure from motion to simplify the form of the projection matrices. The outcome is an extremely simple technique for recovering these matrices and the corresponding affine shape.

Let us start by introducing the affine equivalent of the epipolar constraint. We consider two affine images and rewrite the corresponding projection equations

$$\begin{cases} \boldsymbol{p} = \mathcal{A}\boldsymbol{P} + \boldsymbol{b} \\ \boldsymbol{p}' = \mathcal{A}'\boldsymbol{P} + \boldsymbol{b}' \end{cases} \quad \text{as} \quad \begin{pmatrix} \mathcal{A} & \boldsymbol{p} - \boldsymbol{b} \\ \mathcal{A}' & \boldsymbol{p}' - \boldsymbol{b}' \end{pmatrix} \begin{pmatrix} \boldsymbol{P} \\ -1 \end{pmatrix} = \boldsymbol{0},$$

and a necessary and sufficient condition for these equations to admit a nontrivial solution is that

$$\text{Det}\begin{pmatrix} \mathcal{A} & \boldsymbol{p} - \boldsymbol{b} \\ \mathcal{A}' & \boldsymbol{p}' - \boldsymbol{b}' \end{pmatrix} = 0,$$

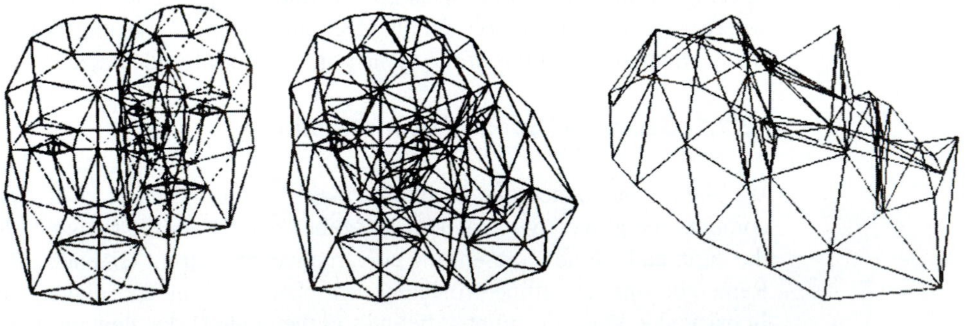

Figure 12.4 Affine reconstruction from two views. Left and middle: three views of a face; Images 0 and 1 are overlaid on the left, and Images 1 and 2 are overlaid in the middle part of the figure. Right: A profile view of the affine face computed from Images 0 and 1 (the third picture is used in Section 12.4 to turn this affine reconstruction into a Euclidean one). *Reprinted with permission from "Affine Structure from Motion," by J.J. Koenderink and A.J. Van Doorn, Journal of the Optical Society of America A, 8:377–385, (1990). © 1990 Optical Society of America.*

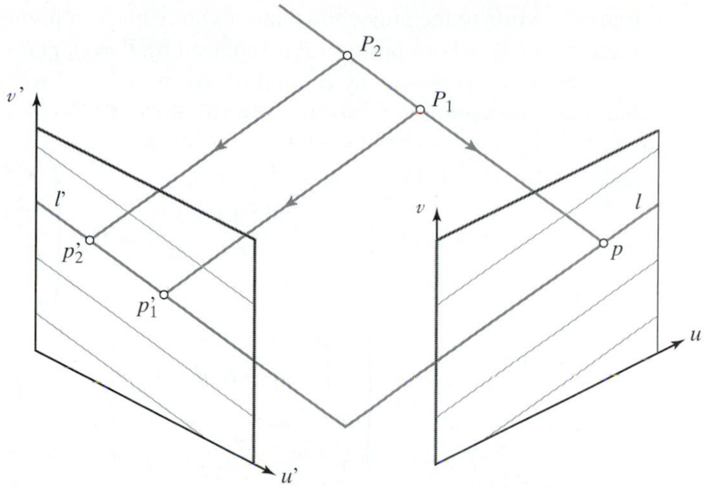

Figure 12.5 Affine epipolar geometry: Given two parallel-projection images, a point p in the first image and the two projection directions define an epipolar plane that intersects the second image along the epipolar line l'. As in the perspective case, any match p' for p is constrained to belong to this line.

or

$$\alpha u + \beta v + \alpha' u' + \beta' v' + \delta = 0, \tag{12.5}$$

where α, β, α', β', and δ are constants depending on $\mathcal{A}$, $\boldsymbol{b}$, $\mathcal{A}'$, and $\boldsymbol{b}'$. This is the *affine epipolar constraint.* Indeed, given a point p in the first image, the position of the matching point p' is constrained by Eq. (12.5) to lie on the line l' defined by $\alpha' u' + \beta' v' + \gamma' = 0$, where $\gamma' = \alpha u + \beta v + \delta$ and vice versa (Figure 12.5).

Note that the epipolar lines associated with each image are parallel to each other: For example, moving p changes γ' or, equivalently, the distance from the origin to the epipolar line l', but does not modify the direction of l'.

The affine epipolar constraint can be rewritten in the familiar form

$$(u, v, 1)\mathcal{F}\begin{pmatrix} u' \\ v' \\ 1 \end{pmatrix} = 0, \quad \text{where} \quad \mathcal{F} \stackrel{\text{def}}{=} \begin{pmatrix} 0 & 0 & \alpha \\ 0 & 0 & \beta \\ \alpha' & \beta' & \delta \end{pmatrix}$$

is the *affine fundamental matrix*. This suggests that the affine epipolar geometry can be seen as the limit of the perspective one. Indeed, it can be shown that an affine picture is the limit of a sequence of images taken by a perspective camera that zooms in on the scene as it backs away from it (see Exercises for details).

Let us now show that the projection matrices can be estimated from the epipolar constraint. The inherent affine ambiguity of affine structure from motion actually allows us to simplify the calculations: According to Eqs. (12.2) and (12.3), if $\mathcal{M} = (\mathcal{A} \quad \boldsymbol{b})$ and $\mathcal{M}' = (\mathcal{A}' \quad \boldsymbol{b}')$ are solutions of our problem, so are $\tilde{\mathcal{M}} = \mathcal{M}\mathcal{Q}$ and $\tilde{\mathcal{M}}' = \mathcal{M}'\mathcal{Q}$, where

$$\mathcal{Q} = \begin{pmatrix} \mathcal{C} & \boldsymbol{d} \\ \boldsymbol{0}^T & 1 \end{pmatrix}$$

is an arbitrary affine transformation. The new projection matrices can be written as $\tilde{\mathcal{M}} = (\mathcal{A}\mathcal{C} \quad \mathcal{A}\boldsymbol{d} + \boldsymbol{b})$ and $\tilde{\mathcal{M}}' = (\mathcal{A}'\mathcal{C} \quad \mathcal{A}'\boldsymbol{d} + \boldsymbol{b}')$. Note that, according to Eq. (12.3), applying this transformation to the projection matrices amounts to applying the inverse transformation to every scene point P, whose position $\boldsymbol{P}$ is replaced by $\tilde{\boldsymbol{P}} = \mathcal{C}^{-1}(\boldsymbol{P} - \boldsymbol{d})$.

Now let us denote by $\boldsymbol{a}_1^T$ and $\boldsymbol{a}_2^T$ (resp. $\boldsymbol{a}_1'^T$ and $\boldsymbol{a}_2'^T$) the two rows of $\mathcal{A}$ (resp. $\mathcal{A}'$) and introduce the vectors $\boldsymbol{b} = (b_1, b_2)^T$ and $\boldsymbol{b}' = (b_1', b_2')^T$. We can rewrite the epipolar constraint as

$$0 = \text{Det}\begin{pmatrix} \mathcal{A}\mathcal{C} & \boldsymbol{p} - \mathcal{A}\boldsymbol{d} - \boldsymbol{b} \\ \mathcal{A}'\mathcal{C} & \boldsymbol{p}' - \mathcal{A}'\boldsymbol{d} - \boldsymbol{b}' \end{pmatrix} = \text{Det}\left(\begin{array}{c|c} \boldsymbol{a}_1^T\mathcal{C} & u - \boldsymbol{a}_1^T\boldsymbol{d} - b_1 \\ \boldsymbol{a}_2^T\mathcal{C} & v - \boldsymbol{a}_2^T\boldsymbol{d} - b_2 \\ \hline \boldsymbol{a}_1'^T\mathcal{C} & u' - \boldsymbol{a}_1'^T\boldsymbol{d} - b_1' \\ \boldsymbol{a}_2'^T\mathcal{C} & v' - \boldsymbol{a}_2'^T\boldsymbol{d} - b_2' \end{array}\right)$$

$$= \text{Det}\left(\begin{array}{c|c} \boldsymbol{a}_1^T\mathcal{C} & u - \boldsymbol{a}_1^T\boldsymbol{d} - b_1 \\ \boldsymbol{a}_2^T\mathcal{C} & v - \boldsymbol{a}_2^T\boldsymbol{d} - b_2 \\ \boldsymbol{a}_1'^T\mathcal{C} & u' - \boldsymbol{a}_1'^T\boldsymbol{d} - b_1' \\ \hline \boldsymbol{a}_2'^T\mathcal{C} & v' - \boldsymbol{a}_2'^T\boldsymbol{d} - b_2' \end{array}\right) = \text{Det}\begin{pmatrix} \mathcal{S}\mathcal{C} & \boldsymbol{q} - \mathcal{S}\boldsymbol{d} - \boldsymbol{r} \\ \boldsymbol{c}^T & v' - d \end{pmatrix},$$

where

$$\mathcal{S} = \begin{pmatrix} \boldsymbol{a}_1^T \\ \boldsymbol{a}_2^T \\ \boldsymbol{a}_1'^T \end{pmatrix}, \quad \boldsymbol{q} = \begin{pmatrix} u \\ v \\ u' \end{pmatrix}, \quad \boldsymbol{r} = \begin{pmatrix} b_1 \\ b_2 \\ b_1' \end{pmatrix}, \quad \boldsymbol{c} = \mathcal{C}^T\boldsymbol{a}_2' \quad \text{and} \quad d = \boldsymbol{a}_2'^T\boldsymbol{d} + b_2'.$$

When $\mathcal{S}$ is nonsingular, we can choose $\mathcal{C} = \mathcal{S}^{-1}$ and $\boldsymbol{d} = -\mathcal{S}^{-1}\boldsymbol{r}$. If $\boldsymbol{c} = (a, b, c)^T$, this reduces the two projection matrices to the canonical forms

$$\tilde{\mathcal{M}} = \begin{pmatrix} 1 & 0 & 0 & 0 \\ 0 & 1 & 0 & 0 \end{pmatrix} \quad \text{and} \quad \tilde{\mathcal{M}}' = \begin{pmatrix} 0 & 0 & 1 & 0 \\ a & b & c & d \end{pmatrix}, \tag{12.6}$$

and allows us to rewrite the epipolar constraint as

$$\text{Det}\begin{pmatrix} 1 & 0 & 0 & u \\ 0 & 1 & 0 & v \\ 0 & 0 & 1 & u' \\ a & b & c & v' - d \end{pmatrix} = -au - bv - cu' + v' - d = 0,$$

where the coefficients a, b, c, and d are related to the parameters α, β, α', β', and δ by $a : \alpha = b : \beta = c : \alpha' = -1 : \beta' = d : \delta$.

Given enough point correspondences, the coefficients a, b, c, and d can be estimated via linear least squares, similar to the perspective case studied in chapter 10. Once these parameters have been found, the two projection matrices are known, and the position of any point can be estimated from its image coordinates by using once again linear least squares to solve the corresponding system of four equations,

$$\begin{pmatrix} 1 & 0 & 0 & u \\ 0 & 1 & 0 & v \\ 0 & 0 & 1 & u' \\ a & b & c & v' - d \end{pmatrix}\begin{pmatrix} \tilde{\boldsymbol{P}} \\ -1 \end{pmatrix} = 0, \tag{12.7}$$

for the three unknown coordinates of $\tilde{\boldsymbol{P}}$.

Note that the first three equations in Eq. (12.7) are in principle sufficient to solve for $\tilde{\boldsymbol{P}}$ as $(u, v, u')^T$ without estimating the coefficients a, b, c, and d and without requiring a minimum number of matches. This is not as surprising as one may originally think: In the case of two calibrated orthographic cameras with perpendicular projection directions and parallel v axes, taking $x = u$, $y = v$, and $z = u'$ does yield the correct Euclidean reconstruction (have another look at Figure 12.5, assuming orthographic projection and imagining that the epipolar lines are parallel to the u and u' axes). In practice, of course, using all four equations may yield more accurate results. The proposed method reduces the first row of $\mathcal{A}'$ to $(0, 0, 1)$ via the affine transformation $\mathcal{Q}$. When $\mathcal{S}$ is (close to) singular, it is possible to apply instead the same reduction to the second row of $\mathcal{A}'$. When both $\mathcal{S}$ and the matrix constructed in that fashion are singular, the two image planes are parallel and the scene structure cannot be recovered.

12.3 AFFINE STRUCTURE AND MOTION FROM MULTIPLE IMAGES

The methods presented in the previous section are aimed at recovering the affine scene structure and/or the corresponding projection matrices from a minimum number of images. We now address the problem of estimating the same information from a potentially large number of pictures. We first show that any *fixed* set of affine images of a scene exhibits an affine structure; then we use this property to derive the factorization method of Tomasi and Kanade (1992) for estimating the affine structure and motion of a scene from an image sequence.

12.3.1 The Affine Structure of Affine Image Sequences

We suppose in this section and the next that we observe a static scene with a fixed set of m affine cameras and denote by $p_1, \dots, p_m$ the m projections of the scene point P. Stacking the corresponding m instances of Eq. (12.1) yields

$$\boldsymbol{q} = \boldsymbol{r} + \mathcal{A}\boldsymbol{P},$$

where

$$\boldsymbol{q} \stackrel{\text{def}}{=} \begin{pmatrix} \boldsymbol{p}_1 \\ \dots \\ \boldsymbol{p}_m \end{pmatrix}, \quad \boldsymbol{r} \stackrel{\text{def}}{=} \begin{pmatrix} \boldsymbol{b}_1 \\ \dots \\ \boldsymbol{b}_m \end{pmatrix} \quad \text{and} \quad \mathcal{A} \stackrel{\text{def}}{=} \begin{pmatrix} \mathcal{A}_1 \\ \dots \\ \mathcal{A}_m \end{pmatrix}.$$

If I denotes the set of all images taken by the m cameras, we have

$$I = \{\boldsymbol{r} + \mathcal{A}\boldsymbol{P} | \boldsymbol{P} \in \mathbb{R}^3\} = \boldsymbol{r} + V_{\mathcal{A}},$$

where $V_{\mathcal{A}}$ denotes the *range* of the $2m \times 3$ matrix $\mathcal{A}$ (i.e., the three-dimensional vector subspace of $\mathbb{R}^{2m}$ spanned by its column vectors. In other words, I is a three-dimensional subspace of the affine space $\mathbb{R}^{2m}$). In particular, if we consider as before n points $P_1, \dots, P_n$ observed by m cameras, we can define the $(2m + 1) \times n$ data matrix

$$\mathcal{D} = \begin{pmatrix} \boldsymbol{q}_1 & \dots & \boldsymbol{q}_n \\ 1 & \dots & 1 \end{pmatrix},$$

and it follows from Section 12.1 that this matrix has (at most) rank 4.

12.3.2 A Factorization Approach to Affine Structure from Motion

Tomasi and Kanade (1992) exploited the affine structure of affine images in a robust factorization method for estimating the structure of a scene and the corresponding camera motion through singular value decomposition (see insert).

Technique: Singular Value Decomposition

Let $\mathcal{A}$ be an $m \times n$ matrix, with $m \geq n$, then $\mathcal{A}$ can always be written as

$$\mathcal{A} = \mathcal{U}\mathcal{W}\mathcal{V}^T,$$

where

- $\mathcal{U}$ is an $m \times n$ column-orthogonal matrix (i.e., $\mathcal{U}^T\mathcal{U} = \mathrm{Id}_n$),
- $\mathcal{W}$ is a diagonal matrix whose diagonal entries w_i ($i = 1, \dots, n$) are the singular values of $\mathcal{A}$ with $w_1 \geq w_2 \geq \cdots \geq w_n \geq 0$,
- and $\mathcal{V}$ is an $n \times n$ orthogonal matrix, i.e., $\mathcal{V}^T\mathcal{V} = \mathcal{V}\mathcal{V}^T = \mathrm{Id}_n$.

This is the *singular value decomposition* (*SVD*) of the matrix $\mathcal{A}$, and it can be computed using the algorithm described in Wilkinson and Reich (1971).

As shown by the following theorem, the singular value decomposition of a matrix is related to the eigenvalues and eigenvectors of its square.

Theorem 3. *The singular values of the matrix $\mathcal{A}$ are the eigenvalues of the matrix $\mathcal{A}^T\mathcal{A}$ and the columns of the matrix $\mathcal{V}$ are the corresponding eigenvectors.*

This theorem can be used to solve overconstrained homogeneous linear equations of the form $\mathcal{A}\boldsymbol{x} = \mathbf{0}$ as defined in chapter 3 without explicitly computing the corresponding matrix $\mathcal{A}^T\mathcal{A}$. The solution is simply the column vector of the matrix $\mathcal{V}$ in the singular value decomposition of $\mathcal{A}$ that is associated with the smallest singular value.

The SVD of a matrix can also be used to characterize matrices that are rank-deficient. Suppose that $\mathcal{A}$ has rank $p < n$. Then the matrices $\mathcal{U}$, $\mathcal{W}$, and $\mathcal{V}$ can be written as

$$\mathcal{U} = \begin{array}{|c|c|}\hline \mathcal{U}_p & \mathcal{U}_{n-p} \\ \hline\end{array} \quad \mathcal{W} = \begin{array}{|c|c|}\hline \mathcal{W}_p & 0 \\ \hline 0 & 0 \\ \hline\end{array} \quad \text{and} \quad \mathcal{V}^T = \begin{array}{|c|}\hline \mathcal{V}_p^T \\ \hline \mathcal{V}_{n-p}^T \\ \hline\end{array},$$

and

- the columns of $\mathcal{U}_p$ form an orthonormal basis of the space spanned by the columns of $\mathcal{A}$ (i.e., its *range*),
- and the columns of $\mathcal{V}_{n-p}$ for a basis of the space spanned by the solutions of $A\boldsymbol{x} = 0$ (i.e., the *null space* of this matrix).

The $m \times p$ and $n \times p$ matrices $\mathcal{U}_p$ and $\mathcal{V}_p$ are both column-orthogonal, and we have of course $\mathcal{A} = \mathcal{U}_p\mathcal{W}_p\mathcal{V}_p^T$.

The following theorem shows that singular value decomposition also provides a valuable *approximation* procedure. In both cases, $\mathcal{U}_p$ and $\mathcal{V}_p$ denote as before the matrices formed by the p leftmost columns of the matrices $\mathcal{U}$ and $\mathcal{V}$, and $\mathcal{W}_p$ is the $p \times p$ diagonal matrix formed by the p largest singular values. This time, however, $\mathcal{A}$ may have maximal rank n, and the remaining singular values may be nonzero.

Theorem 4. *When $\mathcal{A}$ has a rank greater than p, $\mathcal{U}_p\mathcal{W}_p\mathcal{V}_p^T$ is the best possible rank-p approximation of $\mathcal{A}$ in the sense of the Frobenius norm.*

This theorem plays a fundamental role in the factorization approach to structure from motion presented in this chapter.

Assuming that the origin of the object coordinate system is one of the observed points or their center of mass, say P_0, we can translate the origin of the image coordinate system to the corresponding image point, say p_0. The transformation $\boldsymbol{p} \rightarrow \boldsymbol{p} - \boldsymbol{p}_0$ freezes the origin of the set of images I, which becomes the three-dimensional *vector space* $V_{\mathcal{A}}$. In other words, we can write, for any point P, and for $i = 1, \ldots, m$, that $\boldsymbol{p}_i = \mathcal{A}_i \boldsymbol{P}$. Equivalently, $\boldsymbol{q} = \mathcal{A}\boldsymbol{P}$, and

$$I = \{\mathcal{A}\boldsymbol{P} | \boldsymbol{P} \in \mathbb{R}^3\} = V_{\mathcal{A}}.$$

Given m images of n points $P_1, \ldots, P_n$, we can now define the $2m \times n$ data matrix

$$\mathcal{D} \stackrel{\text{def}}{=} (\boldsymbol{q}_1 \quad \ldots \quad \boldsymbol{q}_n) = \mathcal{A}\mathcal{P}, \quad \text{with} \quad \mathcal{P} \stackrel{\text{def}}{=} (\boldsymbol{P}_1 \quad \ldots \quad \boldsymbol{P}_n).$$

As the product of a $2m \times 3$ matrix and a $3 \times n$ matrix, $\mathcal{D}$ has, in general, rank 3. If $\mathcal{U}\mathcal{W}\mathcal{V}^T$ is its singular value decomposition, this means that only three of the singular values are nonzero, thus $\mathcal{D} = \mathcal{U}_3\mathcal{W}_3\mathcal{V}_3^T$, where $\mathcal{U}_3$ and $\mathcal{V}_3$ denote the $2m \times 3$ and $3 \times n$ matrices formed by the three leftmost columns of the matrices $\mathcal{U}$ and $\mathcal{V}$, and $\mathcal{W}_3$ is the 3×3 diagonal matrix formed by the corresponding nonzero singular values.

We claim that we can take $\mathcal{A}_0 = \mathcal{U}_3$ and $\mathcal{P}_0 = \mathcal{W}_3\mathcal{V}_3^T$ as representative of the true (affine) camera motion and scene shape. Indeed, the columns of $\mathcal{A}$ form by definition a basis for the range $V_{\mathcal{A}}$ of $\mathcal{D}$, whereas the columns of $\mathcal{A}_0$ form by construction another basis for this vector space. This implies that there exists a 3×3 matrix $\mathcal{Q}$ such that $\mathcal{A} = \mathcal{A}_0\mathcal{Q}$ and, thus, $\mathcal{P} = \mathcal{Q}^{-1}\mathcal{P}_0$. Conversely, $\mathcal{D} = (\mathcal{A}_0\mathcal{Q})(\mathcal{Q}^{-1}\mathcal{P}_0)$ for any invertible 3×3 matrix $\mathcal{Q}$. Adding to this linear ambiguity the degrees of freedom corresponding to the position of the origin of the world coordinate system confirms once again the affine ambiguity of the structure-from-motion problem, and the fact that singular value decomposition provides representative estimates of the affine motion and scene structure.

Our reasoning so far is only valid in an idealized, noiseless situation. In practice, due to image noise, errors in localization of feature points, and to the mere fact that actual cameras are not affine, the equation $\mathcal{D} = \mathcal{A}\mathcal{P}$ does not hold exactly, and the matrix $\mathcal{D}$ has (in general) full rank. Let us show that singular value decomposition still yields a reasonable estimate of the affine structure and motion in this case: the best we can hope for is to minimize

$$E \stackrel{\text{def}}{=} \sum_{i,j} |\boldsymbol{p}_{ij} - \mathcal{A}_i\boldsymbol{P}_j|^2 = \sum_j |\boldsymbol{q}_j - \mathcal{A}\boldsymbol{P}_j|^2 = |\mathcal{D} - \mathcal{A}\mathcal{P}|^2$$

Algorithm 12.1: The Tomasi–Kanade factorization algorithm for affine shape from motion. Note that the original algorithm, proposed in Tomasi and Kanade (1992) uses $\mathcal{A}_0 = \mathcal{U}_3\sqrt{\mathcal{W}_3}$ and $\mathcal{P}_0 = \sqrt{\mathcal{W}_3}\mathcal{V}_3^T$. Both solutions are mathematically and numerically equivalent.

1. Compute the singular value decomposition $\mathcal{D} = \mathcal{U}\mathcal{W}\mathcal{V}^T$.
2. Construct the matrices $\mathcal{U}_3$, $\mathcal{V}_3$, and $\mathcal{W}_3$ formed by the three leftmost columns of the matrices $\mathcal{U}$ and $\mathcal{V}$, and the corresponding 3×3 submatrix of $\mathcal{W}$.
3. Define

$$\mathcal{A}_0 = \mathcal{U}_3 \quad \text{and} \quad \mathcal{P}_0 = \mathcal{W}_3\mathcal{V}_3^T;$$

the $2m \times 3$ matrix $\mathcal{A}_0$ is an estimate of the camera motion, and the $3 \times n$ matrix $\mathcal{P}_0$ is an estimate of the scene structure.

with respect to the matrices $\mathcal{A}_i$ $(i = 1, \ldots, m)$ and vectors $\boldsymbol{P}_j$ $(j = 1, \ldots, m)$ or, equivalently, with respect to the matrices $\mathcal{A}$ and $\mathcal{P}$.

According to Theorem 4, the matrix $\mathcal{A}_0\mathcal{P}_0$ is the closest rank-3 approximation to $\mathcal{D}$. Since the rank of $\mathcal{AP}$ is 3 for any rank-3 $2m \times 3$ matrix $\mathcal{A}$ and rank-3 $3 \times n$ matrix $\mathcal{P}$, the minimum value of E is thus reached for $\mathcal{A} = \mathcal{A}_0$ and $\mathcal{P} = \mathcal{P}_0$, which confirms that $\mathcal{A}_0$ and $\mathcal{P}_0$ are the optimal estimates of the true camera motion and scene structure. This does not contradict the inherent ambiguity of affine structure from motion: All affinely equivalent solutions yield the same value for E. In particular, singular value decomposition can be used to estimate the affine structure and motion from the data matrix $\mathcal{D}$ as shown in Algorithm 12.12.1.

12.4 FROM AFFINE TO EUCLIDEAN IMAGES

Let us assume that a rigid scene is observed by two calibrated orthographic cameras so the image points are represented by their normalized coordinate vectors. In this case, the transformation between the coordinate systems attached to the cameras goes from affine to Euclidean (i.e., it can be written as the composition of a rotation and a translation). Under orthographic projection, a translation in depth has no effect, and a translation in the image plane (frontoparallel translation) is easily eliminated by aligning the two projections of some scene point A. Any rotation about the viewing direction is also easily identified and discarded. At this stage, the two views differ by a rotation about some axis in a frontoparallel plane passing through the projection of A. Koenderink and Van Doorn (1990) showed that there exists a one-parameter family of such rotations, determining the shape up to a depth scaling and a shear, and that the addition of a third view finally restricts the solution to one or two pairs related through a reflection in the frontoparallel plane (Figure 12.4). The details of this construction are a bit too involved to be included here. Instead, we introduce in the rest of this section a simple method for going from affine to Euclidean structure when the cameras' affine projection matrices have been estimated.

12.4.1 Euclidean Constraints and Calibrated Affine Cameras

Let us first have another look at the orthographic, weak-perspective, and paraperspective models of the imaging process (we do not detail the parallel projection case since it is rarely used in practice), assuming that the cameras have been calibrated. Obviously, the affine projection Eq. (12.1) still holds in this case, but this time there are some constraints on the components of the projection matrix $\mathcal{M} = (\mathcal{A} \quad \boldsymbol{b})$.

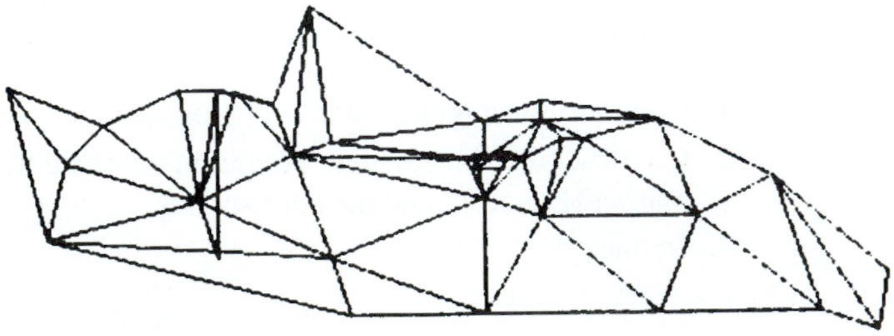

Figure 12.6 Euclidean reconstruction from the three views of a face shown in Figure 12.4. *Reprinted with permission from "Affine Structure from Motion," by J.J. Koenderink and A.J. Van Doorn, Journal of the Optical Society of America A, 8:377–385, (1990). © 1990 Optical Society of America.*

Recall from Eq. (2.20) in chapter 2 that a weak-perspective projection matrix can be written as

$$\mathcal{M} = \frac{1}{z_r}\begin{pmatrix} k & s \\ 0 & 1 \end{pmatrix}(\mathcal{R}_2 \quad \boldsymbol{t}_2),$$

where $\mathcal{R}_2$ is the 2×3 matrix formed by the first two rows of a rotation matrix and $\boldsymbol{t}_2$ is a vector in $\mathbb{R}^2$. When the camera is calibrated, we can use normalized image coordinates and take $k = 1$ and $s = 0$. The projection matrix becomes

$$\hat{\mathcal{M}} = (\hat{\mathcal{A}} \quad \hat{\boldsymbol{b}}) = \frac{1}{z_r}(\mathcal{R}_2 \quad \boldsymbol{t}_2). \tag{12.8}$$

An orthographic camera is a weak-perspective camera with $z_r = 1$, and it follows from Eq. (12.8) that the matrix $\hat{\mathcal{A}}$ is part of a rotation matrix, with unit row vectors $\hat{\boldsymbol{a}}_1^T$ and $\hat{\boldsymbol{a}}_2^T$ orthogonal to each other. In other words, an orthographic camera is an affine camera with the additional constraints

$$\hat{\boldsymbol{a}}_1 \cdot \hat{\boldsymbol{a}}_2 = 0 \quad \text{and} \quad |\hat{\boldsymbol{a}}_1|^2 = |\hat{\boldsymbol{a}}_2|^2 = 1. \tag{12.9}$$

The general weak-perspective case is similar, but the rows of the matrix $\hat{\mathcal{A}}$ are not unit vectors anymore. It follows that a weak-perspective camera is an affine camera with the two constraints

$$\hat{\boldsymbol{a}}_1 \cdot \hat{\boldsymbol{a}}_2 = 0 \quad \text{and} \quad |\hat{\boldsymbol{a}}_1|^2 = |\hat{\boldsymbol{a}}_2|^2. \tag{12.10}$$

Finally, it is easy to use the parameterization of paraperspective cameras given by Eq. (2.22) in chapter 2 to show (see Exercises) that a paraperspective camera is an affine camera that satisfies the constraints

$$\hat{\boldsymbol{a}}_1 \cdot \hat{\boldsymbol{a}}_2 = \frac{u_r v_r}{2(1+u_r^2)}|\hat{\boldsymbol{a}}_1|^2 + \frac{u_r v_r}{2(1+v_r^2)}|\hat{\boldsymbol{a}}_2|^2 \quad \text{and} \quad \frac{|\hat{\boldsymbol{a}}_1|^2}{(1+u_r^2)} = \frac{|\hat{\boldsymbol{a}}_2|^2}{(1+v_r^2)}, \tag{12.11}$$

where (u_r, v_r) denote the coordinates of the perspective projection of the reference point R associated with the paraperspective projection model.

12.4.2 Computing Euclidean Upgrades from Multiple Views

Let us focus on orthographic projection and assume that we have recovered the affine shape of a scene and the projection matrix $\mathcal{M}$ associated with each view. We already know that all solutions of the structure-from-motion problem are the same up to an affine ambiguity. In particular, if the position of a scene point in a *Euclidean* coordinate system is $\hat{\boldsymbol{P}}$ and the corresponding projection matrix is $\hat{\mathcal{M}} = (\hat{\mathcal{A}} \quad \hat{\boldsymbol{b}})$, there must exist some affine transformation

$$\mathcal{Q} = \begin{pmatrix} \mathcal{C} & \boldsymbol{d} \\ \boldsymbol{0}^T & 1 \end{pmatrix}$$

such that $\hat{\mathcal{M}} = \mathcal{M}\mathcal{Q}$ and $\hat{\boldsymbol{P}} = \mathcal{C}^{-1}(\tilde{\boldsymbol{P}} - \boldsymbol{d})$. Such a transformation is called a *Euclidean upgrade* because it maps the affine shape of a scene onto its Euclidean one.

Let us now show how compute such an upgrade when $m \geq 3$ orthographic images are available. Let $\mathcal{M}_i = (\mathcal{A}_i \quad \boldsymbol{b}_i)$ denote the corresponding projection matrices, estimated using the factorization method of Section 12.3.2, for example. If $\hat{\mathcal{M}}_i = \mathcal{M}_i\mathcal{Q}$, we can rewrite the orthographic constraints of Eq. (12.9) as

$$\begin{cases} \hat{\boldsymbol{a}}_{i1} \cdot \hat{\boldsymbol{a}}_{i2} = 0, \\ |\hat{\boldsymbol{a}}_{i1}|^2 = 1, \\ |\hat{\boldsymbol{a}}_{i2}|^2 = 1, \end{cases} \iff \begin{cases} \boldsymbol{a}_{i1}^T \mathcal{C}\mathcal{C}^T \boldsymbol{a}_{i2} = 0, \\ \boldsymbol{a}_{i1}^T \mathcal{C}\mathcal{C}^T \boldsymbol{a}_{i1} = 1, \\ \boldsymbol{a}_{i2}^T \mathcal{C}\mathcal{C}^T \boldsymbol{a}_{i2} = 1, \end{cases} \quad \text{for} \quad i = 1, \ldots, m, \tag{12.12}$$

where $\boldsymbol{a}_{i1}^T$ and $\boldsymbol{a}_{i2}^T$ denote the rows of the matrix $\mathcal{A}_i$. This overconstrained system of $3m$ quadratic equations in the coefficients of $\mathcal{C}$ can be solved via nonlinear least squares. An alternative is to consider Eq. (12.12) as a set of *linear* constraints on the matrix $\mathcal{D} \stackrel{\text{def}}{=} \mathcal{C}\mathcal{C}^T$. The coefficients of $\mathcal{D}$ can be found in this case via linear least squares, and $\mathcal{C}$ can then be computed as $\sqrt{\mathcal{D}}$ using Cholesky decomposition. It should be noted that this requires that the recovered matrix $\mathcal{D}$ be positive definite, which is not guaranteed in the presence of noise. Note also that the solution of Eq. (12.12) is only defined up to an arbitrary rotation. To determine $\mathcal{Q}$ uniquely and simplify the calculations, it is possible to map $\mathcal{M}_1$ (and possibly $\mathcal{M}_2$) to its canonical form and essentially follow the procedure given in the previous section.

Figure 12.7 shows an example, including four pictures in a video sequence of a house, a view of the recovered scene structure, and a real picture taken from a similar viewpoint for comparison.

The computation of a Euclidean upgrade for weak- and paraperspective projections follows a similar path, except for the fact that the two constraints of Eq. (12.10) or Eq. (12.11) written for m images replace the $3m$ constraints of Eq. (12.12). Note that in these cases it is not possible to determine the absolute scale of the scene since the Euclidean constraints of Eqs. (12.10) and

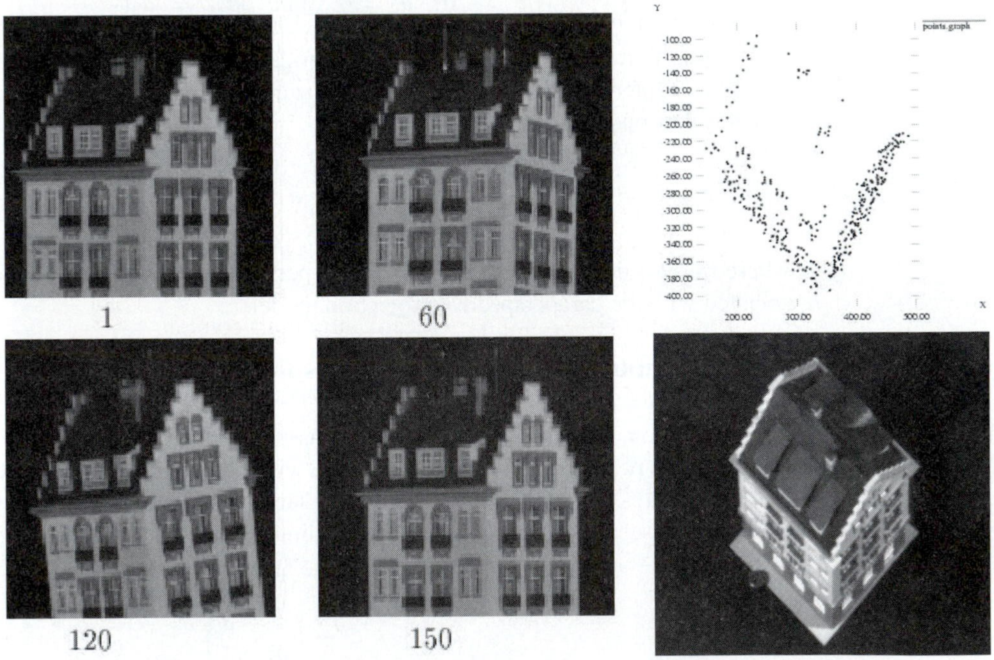

Figure 12.7 Euclidean structure from motion—experimental results. Left: Sample images of a house in a 150-frame sequence. Right: A view of the reconstructed structure (top) and a real picture of the house (bottom) taken from a similar viewpoint. *Reprinted from "Factoring Image Sequences into Shape and Motion," by C. Tomasi and T. Kanade, Proc. IEEE Workshop on Visual Motion, (1991). © 1991 IEEE.*

(12.11) are homogeneous. In other words, the structure of the scene can only be recovered up to an arbitrary *similarity* (e.g., a rigid transformation followed by an isotropic scaling). Accordingly, we now take *Euclidean shape* to mean the equivalence class formed by point sets related by similarities (some authors use instead the term *metric shape* to emphasize the scale ambiguity).

12.5 AFFINE MOTION SEGMENTATION

We have assumed so far that the n points observed all undergo the same motion. What happens if these points belong instead to k objects undergoing different motions? This section presents two methods for segmenting the data points into such independently moving objects.

12.5.1 The Reduced Row-Echelon Form of the Data Matrix

Exactly as in Section 12.3.1, we can define the data matrix

$$\mathcal{D} = \begin{pmatrix} \boldsymbol{p}_{11} & \dots & \boldsymbol{p}_{1n} \\ \dots & \dots & \dots \\ \boldsymbol{p}_{m1} & \dots & \boldsymbol{p}_{mn} \\ 1 & \dots & 1 \end{pmatrix}.$$

This time, however, $\mathcal{D}$ does not have rank 4 anymore. Instead, the columns of the data matrix corresponding to each object define a four-dimensional subspace D_i $(i = 1, \dots, k)$ of its range, and the overall rank of $\mathcal{D}$ is (at most) $4k$. As remarked by Gear (1998), constructing the *reduced row-echelon form (RREF)* of $\mathcal{D}$ identifies the subspaces D_i and the column vectors that lie in them, providing a segmentation of the input points into rigid objects (or, more precisely, into objects that may undergo affine deformations).

The RREF of a matrix $\mathcal{U}$ is a matrix $\mathcal{V}$ whose rows are linear combinations of the rows of $\mathcal{U}$ and that satisfies the following conditions:

1. all rows consisting entirely of zeros are at its bottom;
2. the first nonzero entry in each row is a 1, called the *leading* 1;
3. the leading 1 in each row is to the right of all leading 1s in rows above it; and
4. each leading 1 is the only nonzero entry in its column.

A *base column* is a column that contains a leading 1. By construction, the only nonzero entries of any nonbase column are in rows in which exactly one base column has a 1. In addition, any nonbase column $\boldsymbol{v}$ lies in the subspace spanned by the base columns $\boldsymbol{v}_{j_1}, \dots, \boldsymbol{v}_{j_k}$ associated with its nonzero entries $\alpha_1, \dots, \alpha_k$, and $\boldsymbol{v}$ can be written as $\alpha_1 \boldsymbol{v}_{j_1} + \dots + \alpha_k \boldsymbol{v}_{j_k}$. The number of base columns gives the rank r of the matrix.

Let us illustrate these properties with a sample 7×6 matrix $\mathcal{U}$ and its RREF $\mathcal{V}$ (the entries of $\mathcal{U}$ have been chosen to give $\mathcal{V}$ a simple form):

$$\mathcal{U} = \begin{pmatrix} 1 & 0 & 1 & -5 & 2 & -9 \\ 2 & 4 & 10 & 0 & 1 & 1 \\ -1 & 1 & 1 & 3 & 0 & 1 \\ 0 & 1 & 2 & -1 & 3 & -10 \\ 3 & -2 & -1 & 0 & 1 & 3 \\ 0 & 5 & 10 & 2 & -2 & 8 \\ -2 & 3 & 4 & 1 & 0 & -3 \end{pmatrix} \longrightarrow \mathcal{V} = \begin{pmatrix} 1 & 0 & 1 & 0 & 0 & 2 \\ 0 & 1 & 2 & 0 & 0 & 0 \\ 0 & 0 & 0 & 1 & 0 & 1 \\ 0 & 0 & 0 & 0 & 1 & -3 \\ 0 & 0 & 0 & 0 & 0 & 0 \\ 0 & 0 & 0 & 0 & 0 & 0 \\ 0 & 0 & 0 & 0 & 0 & 0 \end{pmatrix}.$$

Let us denote by $\boldsymbol{u}_i$ and $\boldsymbol{v}_i$ $(i = 1, \ldots, 6)$ the columns of the matrices $\mathcal{U}$ and $\mathcal{V}$. There are four base columns, $\boldsymbol{v}_1$, $\boldsymbol{v}_2$, $\boldsymbol{v}_4$, and $\boldsymbol{v}_5$, so the rank of $\mathcal{U}$ is 4. The nonzero entries of $\boldsymbol{v}_3$ are in the same rows as the leading 1s of base columns $\boldsymbol{v}_1$ and $\boldsymbol{v}_2$, indicating that $\boldsymbol{v}_3$ lies in the subspace of $\mathbb{R}^7$ spanned by $\boldsymbol{v}_1$ and $\boldsymbol{v}_2$. The values of these entries are 1 and 2, implying that $\boldsymbol{v}_3 = \boldsymbol{v}_1 + 2\boldsymbol{v}_2$. Likewise, the nonzero entries of $\boldsymbol{v}_6$ are in the same rows as the leading 1s of $\boldsymbol{v}_1$, $\boldsymbol{v}_4$, and $\boldsymbol{v}_5$, indicating that $\boldsymbol{v}_6$ lies in the subspace of $\mathbb{R}^7$ spanned by $\boldsymbol{v}_1$, $\boldsymbol{v}_4$, and $\boldsymbol{v}_5$, and the values of these entries are 2, 1, and -3, showing that $\boldsymbol{v}_6 = 2\boldsymbol{v}_1 + \boldsymbol{v}_4 - 3\boldsymbol{v}_5$. In fact, these properties also hold for the original matrix $\mathcal{U}$ (i.e., $\boldsymbol{u}_3 = \boldsymbol{u}_1 + 2\boldsymbol{u}_2$ and $\boldsymbol{u}_6 = 2\boldsymbol{u}_1 + \boldsymbol{u}_4 - 3\boldsymbol{u}_5$, as immediately confirmed by inspection of $\mathcal{U}$). This is due to the fact that the rows of $\mathcal{V}$ are linear combinations of the rows of $\mathcal{U}$.

The same properties hold for arbitrary matrices and their RREFs, and this shows that the RREF of the data matrix $\mathcal{D}$ can, in theory, be used for affine motion segmentation. Indeed, it identifies a basis for the range of $\mathcal{D}$ (the columns of this matrix corresponding to base columns in its RREF) as well as all the column vectors that lie in the subspaces spanned by subsets of this basis. When the four-dimensional subspaces D_i associated with each object only intersect at the origin (which is expected to be true for large enough values of m), the corresponding groups of points form connected components of the graph whose nodes are the columns of the RREF and whose arcs link pairs of columns with nonzero entries in at least one common row.

Unfortunately, the situation is more complicated in practice due to noise and numerical errors. A plain implementation of the RREF using, say, Gauss–Jordan elimination with pivoting, normally results in a full-rank matrix, with none of the nonbase columns lying in a four-dimensional subspace of the range (see Exercises). Gear (1998) gives several "robustified" methods for computing the RREF of a matrix, including Gauss–Jordan elimination with a test for discarding small pivot values and QR reduction followed by Gauss–Jordan elimination applied to the corresponding triangular matrix $\mathcal{R}$, and presents successful segmentation experiments involving both synthetic and real image sequences.

12.5.2 The Shape Interaction Matrix

The approach presented in the previous section relies only on the affine structure of affine images. Costeira and Kanade (1998) have proposed a different method, based on a factorization of the data matrix. We present this technique in the case of two groups of points undergoing different motions. The generalization to an arbitrary number of independently moving objects is straightforward.

In the setting of motion segmentation, it is not possible to define a rank-3 data matrix for each object since the centroid of the corresponding points is unknown. Instead, let us assume noiseless data and define the data matrices $\mathcal{D}^{(i)}$ $(i = 1, 2)$ by

$$\mathcal{D}^{(i)} \stackrel{\text{def}}{=} \begin{pmatrix} \boldsymbol{p}_{11}^{(i)} & \cdots & \boldsymbol{p}_{1n_i}^{(i)} \\ \cdots & \cdots & \cdots \\ \boldsymbol{p}_{m1}^{(i)} & \cdots & \boldsymbol{p}_{mn_i}^{(i)} \end{pmatrix},$$

where n_i is the number of points associated with object number i and $n_1 + n_2 = n$. Each data matrix has rank 4 since it can be rewritten as $\mathcal{D}^{(i)} = \mathcal{M}^{(i)}\mathcal{P}^{(i)}$ where, this time,

$$\mathcal{M}^{(i)} \stackrel{\text{def}}{=} \begin{pmatrix} \mathcal{M}_1^{(i)} \\ \cdots \\ \mathcal{M}_m^{(i)} \end{pmatrix} \quad \text{and} \quad \mathcal{P}^{(i)} \stackrel{\text{def}}{=} \begin{pmatrix} \boldsymbol{P}_1^{(i)} & \cdots & \boldsymbol{P}_{n_i}^{(i)} \\ 1 & \cdots & 1 \end{pmatrix}.$$

Let us define the $2m \times n$ composite data matrix $\mathcal{D} \stackrel{\text{def}}{=} (\mathcal{D}^{(1)} \quad \mathcal{D}^{(2)})$ as well as the composite $2m \times 8$ (motion) and $8 \times n$ (structure) matrices

$$\mathcal{M} \stackrel{\text{def}}{=} \begin{pmatrix} \mathcal{M}^{(1)} & \mathcal{M}^{(2)} \end{pmatrix} \quad \text{and} \quad \mathcal{P} \stackrel{\text{def}}{=} \begin{pmatrix} \mathcal{P}^{(1)} & \mathbf{0} \\ \mathbf{0} & \mathcal{P}^{(2)} \end{pmatrix}.$$

With this notation, we have $\mathcal{D} = \mathcal{M}\mathcal{P}$, which shows that $\mathcal{D}$ has (at most) rank 8. Now the rows of the matrix $\mathcal{P}$ form a basis for the 8-dimensional subspace of $\mathbb{R}^{2m}$ spanned by the rows of the matrix $\mathcal{D}$. As shown in Strang (1980) for example, the operator that maps any vector onto its orthogonal projection into the space spanned by the columns of a matrix $\mathcal{A}$ can be represented by the matrix $\mathcal{Z} \stackrel{\text{def}}{=} \mathcal{A}(\mathcal{A}^T\mathcal{A})^{-1}\mathcal{A}^T$. In particular, the matrix $\mathcal{Z}$ associated with the rows of $\mathcal{D}$ (or equivalently the columns of $\mathcal{D}^T$) is by construction block diagonal since $\mathcal{P}^T$ is also block diagonal.

Of course, $\mathcal{P}$ is unknown in our case, but any other matrix whose rows form a basis for the row space of $\mathcal{D}$ can be used as well. For example, if the rank-8 SVD of $\mathcal{D}$ is $\mathcal{U}_8\mathcal{W}_8\mathcal{V}_8^T$, we can use the rows of $\mathcal{V}_8^T$ as a basis, and we obtain $\mathcal{Z} = \mathcal{V}_8(\mathcal{V}_8^T\mathcal{V}_8)^{-1}\mathcal{V}_8^T = \mathcal{V}_8\mathcal{V}_8^T$ since $\mathcal{V}_8$ is orthogonal. The matrix $\mathcal{Z}$ constructed in this fashion is called the *shape interaction matrix* by Costeira and Kanade (1998), and it is once again block diagonal.

The above construction assumes that the data points are ordered consistently with the object they belong to. In general, of course, this is not the case. It can be shown that the values of the entries of the matrix $\mathcal{Z}$ are independent of the order of the points. Changing this order just swaps the columns of $\mathcal{D}$ and swap the rows and columns of $\mathcal{Z}$ accordingly. Thus, recovering the correct point ordering (and the corresponding segmentation into objects) amounts to finding the row and column swaps of the matrix $\mathcal{Z}$ that reduces it to block-diagonal form.

Costeira and Kanade have proposed several methods for finding the correct swaps in the presence of noise. One possibility is to minimize the sum of the squares of the off-diagonal block entries over all rows and column permutations (see Costeira and Kanade, 1998 for details). Figure

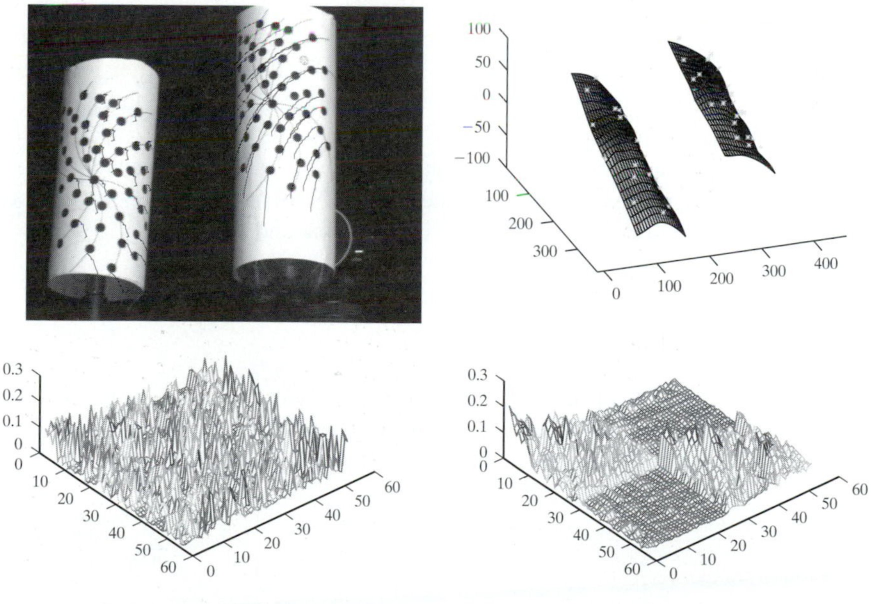

Figure 12.8 Motion segmentation—experimental results. Top-left: One frame from a sequence of pictures of two cylinders, including feature tracks. Top-right: The recovered shapes after motion segmentation. Bottom-left: The shape interaction matrix. Bottom-right: The matrix after sorting. *Reprinted from "A Multi-Body Factorization Method for Motion Analysis," by J. Costeira and T. Kanade, Proc. International Conference on Computer Vision, 1995.* © *1995 IEEE.*

12.8 shows experimental results, including the images of two objects and the corresponding feature tracks, a plot of the corresponding shape interaction matrix before and after sorting, and the corresponding segmentation results.

12.6 NOTES

The structure-from-motion problem was first studied in the calibrated orthographic setting by Ullman (1979). Its first solution in the affine setting is due to Koenderink and Van Doorn (1990). The factorization algorithm discussed in Section 12.3.2 is due to Tomasi and Kanade (1992). As shown in this chapter, the decomposition of structure from motion into an affine and a Euclidean stage affords simple and robust methods for shape reconstruction from image sequences. In essence, this *linearizes* the structure and/or motion estimation process, delaying the introduction of the nonlinear Euclidean constraints until the affine scene shape has been reconstructed. The affine stage is also valuable by itself since it is the basis for the motion-based segmentation methods introduced by Gear (1998) and Costeira and Kanade (1998) and discussed in Section 12.5; see Boult and Brown (1991) for another other approache to the same problem. As shown in chapter 26, other applications include interactive image synthesis in the augmented reality domain. Variations of the affine structure of affine images or, equivalently, of the rank 4 property of the data matrix associated with an affine motion sequence include the facts that an affine image is the linear combination of three model images (Ullman and Basri, 1991), and that the image trajectories of a scene point are linear combinations of the trajectories of three reference points (Weinshall and Tomasi, 1995). The nonlinear least-squares method for computing the Euclidean upgrade matrix $\mathcal{Q}$ is due to Tomasi and Kanade (1992). The Cholesky approach to the same problem is due to Poelman and Kanade (1997); see Weinshall and Tomasi (1995) for another variant. Various extensions of the approach presented in this chapter have been proposed recently, including the incremental recovery of structure and motion (Weinshall and Tomasi, 1995; Morita and Kanade, 1997), the extension of the affine/Euclidean decomposition to a projective/affine/Euclidean stratification (Faugeras, 1995), along with corresponding projective shape estimation algorithms (Faugeras, 1992; Hartley *et al.*, 1992; see also next chapter), and the generalization of the factorization approach of Tomasi and Kanade (1992) to the perspective case (Sturm and Triggs, 1996) and various other computer vision problems that have a natural bilinear structure (Koenderink and Van Doorn, 1997).

PROBLEMS

12.1. Explain why any definition of the "addition" of two points or of the "multiplication" of a point by a scalar is necessarily coordinate dependent.

12.2. Show that the definition of a barycentric combination as

$$\sum_{i=0}^{m} \alpha_i A_i \stackrel{\text{def}}{=} A_j + \sum_{i=0, i\neq j}^{m} \alpha_i (A_i - A_j),$$

is independent of the choice of j.

12.3. Prove that

$${}^{B}P = {}^{B}_{A}\mathcal{C}\,{}^{A}P + {}^{B}O_A \iff \begin{pmatrix} {}^{B}P \\ 1 \end{pmatrix} = \begin{pmatrix} {}^{B}_{A}\mathcal{C} & {}^{B}O_A \\ \mathbf{0}^T & 1 \end{pmatrix} \begin{pmatrix} {}^{A}P \\ 1 \end{pmatrix}.$$

12.4. Show that the set of barycentric combinations of $m+1$ points $A_0, \dots, A_m$ in X is indeed an affine subspace of X, and show that its dimension is at most m.

12.5. Derive the equation of a line defined by two points in $\mathbb{R}^3$. (Hint: You actually need *two* equations.)

12.6. Show that the intersection of a plane with two parallel planes consists of two parallel lines.

12.7. Show that an affine transformation $\psi : X \to Y$ between two affine subspaces X and Y associated with the vector spaces $\vec{X}$ and $\vec{Y}$ can be written as $\psi(P) = \psi(O) + \vec{\psi}(P - O)$, where O is some arbitrarily chosen origin, and $\vec{\psi} : \vec{X} \to \vec{Y}$ is a linear mapping from $\vec{X}$ onto $\vec{Y}$ that is independent of the choice of O.

12.8. Show that affine cameras (and the corresponding epipolar geometry) can be viewed as the limit of a sequence of perspective images with increasing focal length receding away from the scene.

12.9. Generalize the notion of multilinearities introduced in chapter 10 to the affine case.

12.10. Prove Theorem 3.

12.11. Show that a calibrated paraperspective camera is an affine camera that satisfies the constraints

$$\hat{\boldsymbol{a}}_1 \cdot \hat{\boldsymbol{a}}_2 = \frac{u_r v_r}{2(1+u_r^2)}|\hat{\boldsymbol{a}}_1|^2 + \frac{u_r v_r}{2(1+v_r^2)}|\hat{\boldsymbol{a}}_2|^2 \quad \text{and} \quad \frac{|\hat{\boldsymbol{a}}_1|^2}{(1+u_r^2)} = \frac{|\hat{\boldsymbol{a}}_2|^2}{(1+v_r^2)},$$

where (u_r, v_r) denote the coordinates of the perspective projection of the point R.

12.12. What do you expect the RREF of an $m \times n$ matrix with random entries to be when $m \geq n$? What do you expect it to be when $m < n$? Why?

Programming Assignments

12.13. Implement the Koenderink–Van Doorn approach to affine shape from motion.

12.14. Implement the estimation of affine epipolar geometry from image correspondences and the estimation of scene structure from the corresponding projection matrices.

12.15. Implement the Tomasi–Kanade approach to affine shape from motion.

12.16. Add random numbers uniformly distributed in the [0, 0.0001] range to the entries of the matrix $\mathcal{U}$ used to illustrate the RREF and compute its RREF (using, e.g., the `rref` routine in MATLAB); then compute again the RREF using a "robustified" version of the reduction algorithm (using, e.g., `rref` with a nonzero tolerance). Comment on the results.

13

Projective Structure from Motion

This chapter addresses once again the recovery of scene structure and/or camera motion from correspondences established by matching the images of n points in m pictures. This time, however, we assume a perspective projection model. Given n *fixed* points P_j $(j = 1, \ldots, n)$ observed by m cameras and the corresponding mn *homogeneous* coordinate vectors $\boldsymbol{p}_{ij} = (u_{ij}, v_{ij}, 1)^T$ of their images, let us write the corresponding perspective projection equations as

$$\begin{cases} u_{ij} = \dfrac{\boldsymbol{m}_{i1} \cdot \boldsymbol{P}_j}{\boldsymbol{m}_{i3} \cdot \boldsymbol{P}_j} \\ v_{ij} = \dfrac{\boldsymbol{m}_{i2} \cdot \boldsymbol{P}_j}{\boldsymbol{m}_{i3} \cdot \boldsymbol{P}_j} \end{cases} \quad \text{for} \quad i = 1, \ldots, m \quad \text{and} \quad j = 1, \ldots, n, \tag{13.1}$$

where $\boldsymbol{m}_{i1}^T$, $\boldsymbol{m}_{i2}^T$, and $\boldsymbol{m}_{i3}^T$ denote the rows of the 3×4 projection matrix $\mathcal{M}_i$ associated with camera number i in some fixed coordinate system, and $\boldsymbol{P}_j$ denotes the *homogeneous* coordinate vector of the point P_j in that coordinate system. We define *projective structure from motion* as the problem of estimating the m matrices $\mathcal{M}_i$ and the n vectors $\boldsymbol{P}_j$ from the mn image correspondences $\boldsymbol{p}_{ij}$.

When $\mathcal{M}_i$ and $\boldsymbol{P}_j$ are solutions of Eq. (13.1), so are of course $\lambda_i \mathcal{M}_i$ and $\mu_j \boldsymbol{P}_j$ for any nonzero values of λ_i and μ_j. In particular, as already noted in chapter 2, the matrices $\mathcal{M}_i$ satisfying Eq. (13.1) are only defined up to scale, with 11 independent parameters, and so are the vectors $\boldsymbol{P}_j$, with 3 independent parameters (when necessary, these can be reduced to the canonical form $(x_j, y_j, z_j, 1)^T$ as long as their fourth coordinate is not zero, which is the generic case). Like its affine cousin, projective structure from motion suffers from a deeper ambiguity that justifies its name: When the camera calibration parameters are unknown, the projection matrices $\mathcal{M}_i$ are, according to Theorem 1 (chapter 2), arbitrary rank-3 3×4 matrices. Hence, if $\mathcal{M}_i$ and

$\boldsymbol{P}_j$ are solutions of Eq. (13.1), so are $\mathcal{M}_i' = \mathcal{M}_i \mathcal{Q}$ and $\boldsymbol{P}_j' = \mathcal{Q}^{-1}\boldsymbol{P}_j$, where $\mathcal{Q}$ is a *projective transformation matrix* (i.e., an arbitrary nonsingular 4×4 matrix). The matrix $\mathcal{Q}$ is only defined up to scale, with 15 free parameters, since multiplying it by a nonzero scalar simply amounts to applying inverse scalings to $\mathcal{M}_i$ and $\boldsymbol{P}_j$. Since Eq. (13.1) provides $2mn$ constraints on the $11m$ parameters of the matrices $\mathcal{M}_i$ and the $3n$ parameters of the vectors $\boldsymbol{P}_j$, taking into account the *projective ambiguity* of structure from motion suggests that this problem admits a finite number of solutions as soon as $2mn \geq 11m + 3n - 15$. For $m = 2$, seven point correspondences should thus be sufficient to determine (up to a projective transformation) the two projection matrices and the position of any other point. This is confirmed formally in Sections 13.2 and 13.3.

In the rest of this chapter, *projective geometry* plays the role that affine geometry played in chapter 12, and it affords a similar overall methodology. Once again, ignoring (at first) the Euclidean constraints associated with calibrated cameras allows us to linearize the recovery of the *projective* scene structure and camera motion from point correspondences. We then exploit the geometric constraints associated with (partially or fully) calibrated perspective cameras to upgrade the projective reconstruction to a Euclidean one.

13.1 ELEMENTS OF PROJECTIVE GEOMETRY

The means of measurement available in projective geometry are even more primitive than those available in affine geometry. The affine notion of ratios of lengths along parallel lines and, in fact, the notion of parallelism are gone. The concepts of points, lines, and planes remain, however, as well as a new, weaker scalar measure of the arrangement of collinear points—the *cross-ratio*. As in the affine case, a rigorous axiomatic introduction to projective geometry would be out of place in this book, and we remain rather informal in the rest of this section.

13.1.1 Projective Spaces

Let us consider a real vector space $\vec{X}$ of dimension $n + 1$. If $\boldsymbol{v}$ is a nonzero element of $\vec{X}$, the set $\mathbb{R}\boldsymbol{v}$ of all vectors proportional to $\boldsymbol{v}$ is called a *ray*, and it is uniquely characterized by any one of its nonzero elements. The *real projective space* $X = P(\vec{X})$ of dimension n associated with $\vec{X}$ is the set of rays in $\vec{X}$ or, equivalently, the quotient of the set $\vec{X}\backslash 0$ of nonzero vectors in $\vec{X}$ under the equivalence relation "$\boldsymbol{v} \sim \boldsymbol{v}'$ if and only if $\boldsymbol{v} = k\boldsymbol{v}'$ for some $k \in \mathbb{R}$". Elements of X are called *points*, and we say that a family of points are linearly dependent (resp. independent) when representative vectors for the corresponding rays are linearly dependent (resp. independent). The map $p : \vec{X}\backslash 0 \to P(\vec{X})$ associates with any nonzero element $\boldsymbol{v}$ of $\vec{X}$ the corresponding point $p(\boldsymbol{v})$ of X.

Example 13.1 A Model of $P(\mathbb{R}^3)$.

Consider an affine plane Π of $\mathbb{R}^3$. The rays of $\mathbb{R}^3$ that are not parallel to Π are in one-to-one correspondence with the points of this plane. For example, the rays R_A, R_B, and R_C associated with the vectors $\boldsymbol{v}_A$, $\boldsymbol{v}_B$, and $\boldsymbol{v}_C$ below can be mapped onto the points A, B and C where they intersect Π. The vectors $\boldsymbol{v}_A$, $\boldsymbol{v}_B$, and $\boldsymbol{v}_C$ are linearly independent, and so are (by definition) the points A, B, and C.

As a ray gets close to being parallel to Π, the point where it intersects this plane recedes to infinity, and in fact it can be shown that a model of the projective plane $P(\mathbb{R}^3)$ (i.e., a projective space $\hat{\Pi}$ of dimension 2 isomorphic to $P(\mathbb{R}^3)$) can be constructed by adding to Π a one-dimensional set of *points at infinity* associated with the rays parallel to this plane. Here, for example, the ray R_D parallel to Π maps onto the point at infinity D of $\hat{\Pi}$.

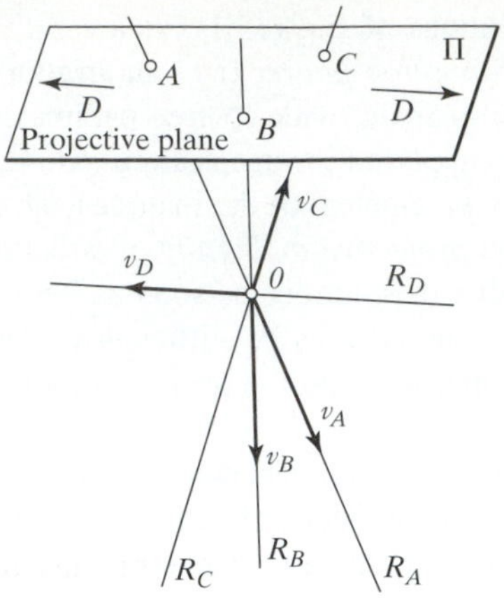

Since any affine plane can be mapped onto $\mathbb{R}^2$ by choosing some affine coordinate system, Example 13.1 suggests that affine planes, and for that matter $\mathbb{E}^3$ or any other affine space, can somehow be embedded in projective spaces, an appropriate choice of points at infinity completing the embedding. Such a completion process is presented in Section 13.1.3.

13.1.2 Projective Subspaces and Projective Coordinates

Consider an $(m+1)$-dimensional vector subspace $\vec{Y}$ of $\vec{X}$. The set Let $Y = P(\vec{Y})$ of rays in $\vec{Y}$ is called a *projective subspace* of X, and its *dimension* is m. Given a basis $(\boldsymbol{e}_0, \boldsymbol{e}_1, \dots, \boldsymbol{e}_m)$ for $\vec{Y}$, we can associate with each point P in Y a one-parameter family of elements of $\mathbb{R}^{m+1}$—namely, the coordinate vectors $(x_0, x_1, \dots, x_m)^T$ of the vectors $\boldsymbol{v} \in \vec{Y}$ such that $P = p(\boldsymbol{v})$. These tuples are proportional to one another, and a representative tuple is called a set of *homogeneous projective coordinates* of the point P.

Homogeneous coordinates can also be characterized intrinsically in terms of families of points in Y: Consider $m+1$ $(m \le n)$ linearly independent points $A_0, A_1, \dots, A_m$ and $m+1$ vectors $\boldsymbol{v}_i$ $(i = 0, 1, \dots, m)$ representative of the corresponding rays. If an additional point A^* linearly depends on the points A_i and $\boldsymbol{v}^*$ is a representative vector of the corresponding ray, we can write

$$\boldsymbol{v}^* = \mu_0 \boldsymbol{v}_0 + \mu_1 \boldsymbol{v}_1 + \cdots + \mu_m \boldsymbol{v}_m.$$

The coefficients μ_i are not uniquely determined since *each* vector $\boldsymbol{v}_i$ is only defined up to a nonzero scale factor. However, when none of the coefficients μ_i vanishes (i.e., when $\boldsymbol{v}^*$ does not lie in the vector subspace spanned by any m vectors $\boldsymbol{v}_i$ or, equivalently, when the corresponding points are linearly independent), we can uniquely define the $m+1$ nonzero vectors $\boldsymbol{e}_i = \mu_i \boldsymbol{v}_i$ such that

$$\boldsymbol{v}^* = \boldsymbol{e}_0 + \boldsymbol{e}_1 + \cdots + \boldsymbol{e}_m.$$

In particular, any vector $\boldsymbol{v}$ linearly dependent on the vectors $\boldsymbol{v}_i$ can now be written *uniquely* as

$$\boldsymbol{v} = x_0 \boldsymbol{e}_0 + x_1 \boldsymbol{e}_1 + \cdots + x_m \boldsymbol{e}_m.$$

This defines a one-to-one correspondence between the rays $\mathbb{R}(x_0, x_1, \dots, x_m)^T$ of $\mathbb{R}^{m+1}$ and a projective subspace S_m of X. S_m is, in fact, the projective space Y associated with the vector subspace $\vec{Y}$ of $\vec{X}$ spanned by the vectors $\boldsymbol{v}_i$ (or, equivalently, by the vectors $\boldsymbol{e}_i$). If $P = p(\boldsymbol{v})$ is the point of S_m associated with the ray $\mathbb{R}\boldsymbol{v}$, the numbers $x_0, x_1, \dots, x_m$ are called the homogeneous (projective) coordinates of P in the *projective coordinate system* determined by the $m+1$ *fundamental points* A_i and the *unit point* A^*. Note that, since the vector $\boldsymbol{v}$ associated with a ray is only defined up to scale, so are the homogeneous coordinates of a point.

It is a simple matter to verify that the coordinate vectors of the fundamental and unit points in the corresponding projective frame have a particularly simple form—namely,

$$\boldsymbol{A}_0 = \begin{pmatrix} 1 \\ 0 \\ \vdots \\ 0 \end{pmatrix}, \quad \boldsymbol{A}_1 = \begin{pmatrix} 0 \\ 1 \\ \vdots \\ 0 \end{pmatrix}, \quad \ldots, \quad \boldsymbol{A}_m = \begin{pmatrix} 0 \\ 0 \\ \vdots \\ 1 \end{pmatrix} \quad \text{and } \boldsymbol{A}^* = \begin{pmatrix} 1 \\ 1 \\ \vdots \\ 1 \end{pmatrix}.$$

It should be clear that the two notions of homogeneous coordinates that have been introduced in this section coincide. The only difference is in the choice of the coordinate vectors $\boldsymbol{e}_0, \boldsymbol{e}_1, \ldots, \boldsymbol{e}_m$, that are given a priori in the former case and constructed from the points forming a given projective frame in the latter one.

Example 13.2 Projective Coordinate Changes.

Given some coordinate system $(A) = (A_0, A_1, A_2, A_3, A^*)$ for the three-dimensional projective space X, we can define the (homogeneous projective) coordinate vector of any point P as ${}^AP = ({}^Ax_0, {}^Ax_1, {}^Ax_2, {}^Ax_3)^T$. Let us now consider a second projective frame $(B) = (B_0, B_1, B_2, B_3, B^*)$ for X. It can easily be shown (see Exercises) that the corresponding change of coordinates can be written as

$$\rho\, {}^BP = {}^B_A\mathcal{T}\, {}^AP, \tag{13.2}$$

where ${}^B_A\mathcal{T}$ is a 4×4 projective transformation matrix defined up to scale, and ρ is a scalar chosen so the scales of the two sides of the equations are the same. Let us now show how to compute this matrix. Writing Eq. (13.2) for the points defining the frame (A) yields

$$\rho_0\, {}^BA_0 = {}^B_A\mathcal{T} \begin{pmatrix} 1 \\ 0 \\ 0 \\ 0 \end{pmatrix}, \quad \rho_1\, {}^BA_1 = {}^B_A\mathcal{T} \begin{pmatrix} 0 \\ 1 \\ 0 \\ 0 \end{pmatrix}, \quad \rho_2\, {}^BA_2 = {}^B_A\mathcal{T} \begin{pmatrix} 0 \\ 0 \\ 1 \\ 0 \end{pmatrix}, \quad \rho_3\, {}^BA_3 = {}^B_A\mathcal{T} \begin{pmatrix} 0 \\ 0 \\ 0 \\ 1 \end{pmatrix}$$

and

$$\rho^*\, {}^BA^* = {}^B_A\mathcal{T} \begin{pmatrix} 1 \\ 1 \\ 1 \\ 1 \end{pmatrix}.$$

Since the matrix ${}^B_A\mathcal{T}$ is only defined up to a scale factor, we can choose $\rho^* = 1$, and it follows that

$${}^B_A\mathcal{T} = \begin{pmatrix} \rho_0\, {}^BA_0 & \rho_1\, {}^BA_1 & \rho_2\, {}^BA_2 & \rho_3\, {}^BA_3 \end{pmatrix},$$

where the scalars ρ_i are the solutions of the linear system

$$\begin{pmatrix} {}^BA_0 & {}^BA_1 & {}^BA_2 & {}^BA_3 \end{pmatrix} \begin{pmatrix} \rho_0 \\ \rho_1 \\ \rho_2 \\ \rho_3 \end{pmatrix} = {}^BA^*.$$

Note the obvious similarity with the formulas for changes of Euclidean or affine coordinate systems in chapters 2 and 12. Similar formulas for coordinate changes can be written for arbitrary projective spaces of finite dimension.

A projective subspace S_1 of dimension 1 of X is called a *line*. Linear subspaces of dimension 2 and $n-1$ are, respectively, called *planes* and *hyperplanes*. A hyperplane S_{n-1} consists of the set of points P linearly dependent on n linearly independent points $P_0, P_1, \ldots, P_{n-1}$.

Example 13.3 Projective lines and planes.

A projective line is uniquely determined by two distinct points A and B lying on it, but defining a projective frame requires three distinct points A_0, A_1, and A^* on that line. Likewise, a plane is uniquely determined by three points A, B, and C lying in it, but defining a projective frame requires four points in that plane: Three fundamental points A_0, A_1, and A_2 forming a nondegenerate triangle and a unit point A^* not lying on one of the edges of this triangle.

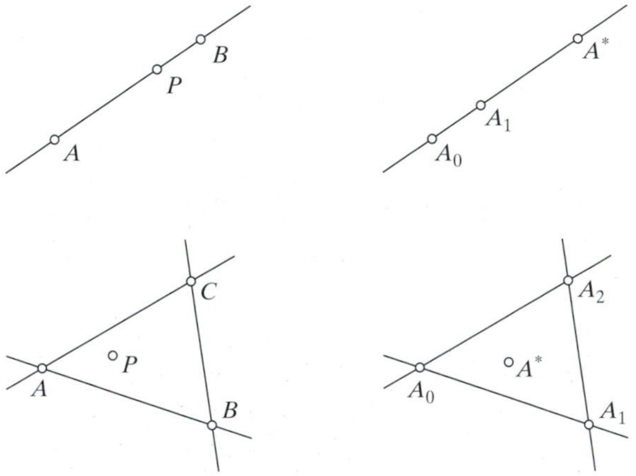

However, if $\boldsymbol{A}$ and $\boldsymbol{B}$ denote the coordinate vectors of two distinct points in some projective frame for the line passing through these points, the coordinate vector $\boldsymbol{P}$ of any point on that line can be written uniquely as $\boldsymbol{P} = \lambda\boldsymbol{A} + \mu\boldsymbol{B}$. This follows immediately from the fact that the rays R_A and R_B associated with distinct points A and B are linearly independent, but the ray R_P associated with a point P on the same line lies in the vector plane defined by R_A and R_B. Likewise, if $\boldsymbol{A}$, $\boldsymbol{B}$, and $\boldsymbol{C}$ denote the coordinate vectors of three noncollinear points in some projective frame for the plane they lie in, the coordinate vector $\boldsymbol{P}$ of any point in that plane can be written uniquely as $\boldsymbol{P} = \lambda\boldsymbol{A}+\mu\boldsymbol{B}+\nu\boldsymbol{C}$.

13.1.3 Affine and Projective Spaces

Example 13.1 introduced (informally) the idea of embedding an affine plane into a projective one with the addition of a one-dimensional set of points at infinity. More generally, it is possible to construct the *projective closure* $\tilde{X}$ of an affine space X of dimension n by adding to it a set of points at infinity associated with the directions of its lines. These points form a hyperplane of $\tilde{X}$ called the *hyperplane at infinity* and denoted by ∞_X.

Let us pick some point A in X and introduce $\tilde{X} \stackrel{\text{def}}{=} P(\vec{X} \times \mathbb{R})$, where $\vec{X}$ is the vector space underlying X. We can embed X into $\tilde{X}$ via the injective map $J_A : X \to \tilde{X}$ defined by $J_A(P) = p(\overrightarrow{AP}, 1)$ (Figure 13.1).[1] The complement of $J_A(X)$ in $\tilde{X}$ is the hyperplane at infinity $\infty_X \stackrel{\text{def}}{=} P(\vec{X} \times \{0\})$ mentioned earlier.

[1]Here we identify X and the underlying vector space $\vec{X}$ by identifying each point P in X with the vector $\overrightarrow{AP}$. This *vectorialization* process is of course dependent on the choice of the origin A, but it can easily be shown that $\tilde{X}$ is indeed independent of that choice. A more rigorous approach to the projective completion process is to introduce the *universal vector space* associated with an affine space, but it would be out of place here. See Berger (1987, chapter 5) for details. Note also the abuse of notation in writing $p(\boldsymbol{v}, \lambda)$ for $p((\boldsymbol{v}, \lambda))$.

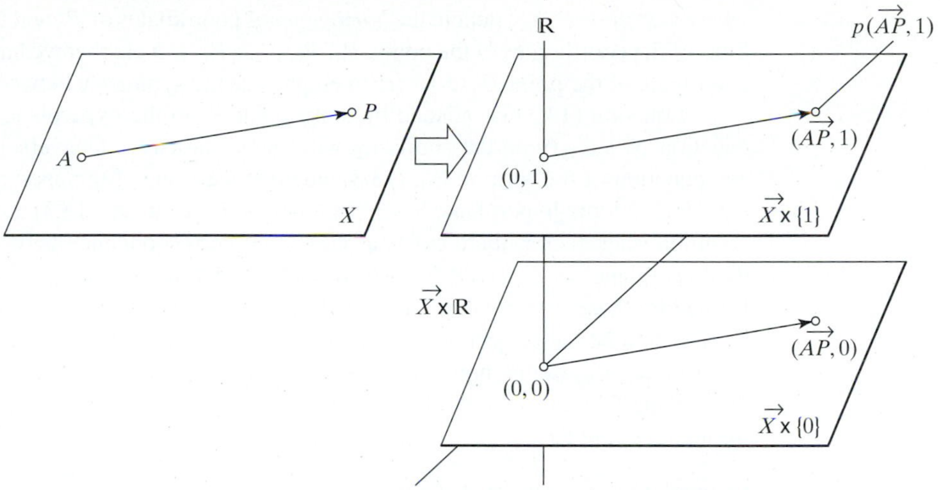

Figure 13.1 The projective completion of an affine space.

Now consider a fixed affine frame $(A_0, A_1, \ldots, A_n)$ of X and embed X into $\tilde{X}$ using J_{A_0}. The vectors $\overrightarrow{A_0A_i}$ $(i = 1, \ldots, n)$ form a basis of $\vec{X}$, thus the $n + 1$ vectors $\boldsymbol{e}_i \stackrel{\text{def}}{=} (\overrightarrow{A_0A_i}, 0)$ $(i = 1, \ldots, n)$ and $\boldsymbol{e}_{n+1} \stackrel{\text{def}}{=} (\boldsymbol{0}, 1)$ form a basis of $\vec{X} \times \mathbb{R}$. In particular, if $(x_1, \ldots, x_n)$ denote the affine coordinates of P in the basis $(A_0, A_1, \ldots, A_n)$ of X, we have

$$\begin{aligned} J_{A_0}(P) &= p(\overrightarrow{A_0P}, 1) = p(x_1\overrightarrow{A_0A_1} + \cdots + x_n\overrightarrow{A_0A_n}, 1) \\ &= p(x_1\boldsymbol{e}_1 + \cdots + x_n\boldsymbol{e}_n + \boldsymbol{e}_{n+1}), \end{aligned}$$

and the homogeneous projective coordinates of $J_{A_0}(P)$ associated with the basis of $\vec{X} \times \mathbb{R}$ formed by the vectors $(\boldsymbol{e}_1, \ldots, \boldsymbol{e}_{n+1})$ are thus $(x_1, \ldots, x_n, 1)$. The coordinates of points in ∞_X, on the other hand, have the form $(x_1, \ldots, x_n, 0)$. In particular, the projective completion process justifies, at long last, the representation of image and scene points by homogeneous coordinates introduced in chapter 2 and used throughout this book.

The introduction of points at infinity frees projective geometry from the numerous exceptions encountered in the affine case. For example, parallel lines in some affine plane Π do not intersect unless they coincide. In contrast, any two distinct lines in a projective plane intersect in exactly one point (this is because the associated vector spaces intersect along a ray), with pairs of parallel lines in Π intersecting at the point at infinity in $\tilde{\Pi}$ that is associated with their common direction (see Exercises).

13.1.4 Hyperplanes and Duality

As mentioned before, two distinct lines of a projective plane have exactly one common point. Likewise, two distinct points belong to exactly one line. These two statements can actually be taken as *incidence axioms*, leading to a purely axiomatic construction of the projective plane. Points and lines play a symmetric or, more precisely, *dual* role in these statements.

To introduce *duality* a bit more generally, let us equip the n-dimensional projective space X with a fixed projective frame and consider $n + 1$ points $P_0, P_1, \ldots, P_n$ lying in some hyperplane S_{n-1} of X. Since these points are by construction linearly dependent, the $(n + 1) \times (n + 1)$ matrix formed by collecting their coordinate vectors is singular. Expanding the determinant of this matrix with respect to its last column yields

$$u_0x_0 + u_1x_1 + \cdots + u_nx_n = 0, \tag{13.3}$$

where $(x_0, x_1, \ldots, x_n)$ denote the *homogeneous* coordinates of P_n and $(u_0, u_1, \ldots, u_n)$ are functions of the coordinates of the points $P_0, P_1, \ldots, P_{n-1}$. Note that we have refrained to set the last coordinate of the point P_n to 1 here to emphasize the symmetry between the scalars u_i and x_i.

Equation (13.3) is satisfied by every point P_n in the hyperplane S_{n-1}, and it is called the equation of S_{n-1} (note the similarity with the affine case). Conversely, it is easily shown that any equation of the form in Eq. (13.3) where at least one of the coefficients u_i is nonzero is the equation of some hyperplane. Since the coefficients u_i in Eq. (13.3) are only defined up to some common scale factor, there exists a one-to-one correspondence between the rays of $\mathbb{R}^{n+1}$ and the hyperplanes of X, and it follows that we can define a second projective space $X^* = P(\vec{X}^*)$ formed by these hyperplanes and called the *dual* of $\vec{X}$ (this is justified by the fact that X^* can be shown to be the projective space associated with the dual vector space $\vec{X}^*$ of $\vec{X}$). The scalars $(u_0, u_1, \ldots, u_n)$ define homogeneous projective coordinates for the point corresponding to the hyperplane S_{n-1} in X^*, and Eq. (13.3) can also be seen as defining the set of hyperplanes passing through the point P_n.

Example 13.4 The Dual of a Line.

Points and lines are dual notions in $\mathbb{P}^2 \stackrel{\text{def}}{=} \tilde{\mathbb{E}}^2$, points and planes are dual in $\mathbb{P}^3 \stackrel{\text{def}}{=} \tilde{\mathbb{E}}^3$, but points and lines are *not* dual in $\mathbb{P}^3$. In general, what is the dual of a line in X? A line is a one-dimensional linear subspace of X whose elements are linearly dependent on two points on the line. Likewise, a line of X^* is a one-dimensional subspace of the dual, called a *pencil of hyperplanes*, whose elements are linearly dependent on two hyperplanes in the pencil. In the plane, the dual of a line is a pencil of lines intersecting at a common point.

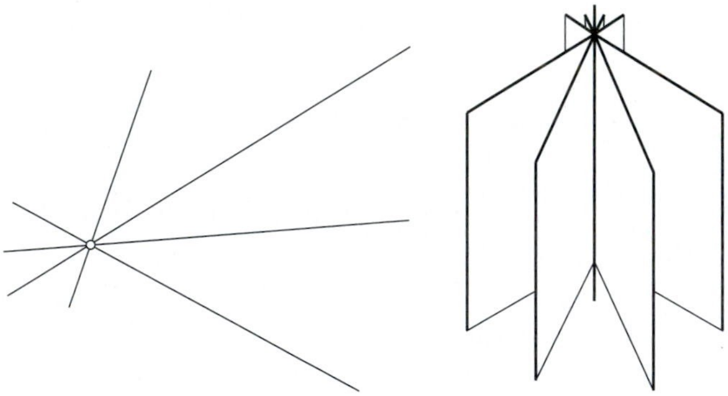

In three dimensions, the dual of a line is a pencil of planes that intersect along a common line.

Let us close this section by noting (without proof) that any geometric theorem that holds for points in X induces a corresponding theorem for hyperplanes (i.e., points in X^*) and vice versa, the two theorems being said to be dual of each other.

13.1.5 Cross-Ratios and Projective Coordinates

This section focuses on the three-dimensional projective space $\tilde{\mathbb{E}}^3$. The *nonhomogeneous* projective coordinates of a point can be defined geometrically in terms of *cross-ratios*. In the affine case, given four collinear points A, B, C, D such that A, B, and C are distinct, we define the cross-ratio of these points as

$$\{A, B; C, D\} \stackrel{\text{def}}{=} \frac{\overline{CA}}{\overline{CB}} \times \frac{\overline{DB}}{\overline{DA}},$$

where $\overline{PQ}$ denotes the signed distance between two points P and Q for some choice of orientation of the line Δ joining them. The orientation of this line is fixed but arbitrary since reversing it obviously does not change the cross-ratio. Note that $\{A, B; C, D\}$ is, a priori, only defined when $D \neq A$ since its calculation involves a division by zero when $D = A$. We extend the definition of the cross-ratio to the whole affine line by using the symbol ∞ to denote the ratio formed by dividing any nonzero real number by zero and to the whole projective line $\tilde{\Delta}$ by defining $\{A, B; C, \infty_\Delta\} = \overline{CA}/\overline{CB}$. Alternatively, given three points A, B, and C on a *projective* line Δ, it can be shown that there exists a unique projective transformation $h : \Delta \to \tilde{\mathbb{R}}$ mapping Δ onto the projective completion $\tilde{\mathbb{R}} = \mathbb{R} \cup \infty$ of the real line such that $h(A) = \infty$, $h(B) = 0$ and $h(C) = 1$. The cross-ratio can also be defined by $\{A, B; C, D\} \stackrel{\text{def}}{=} h(D)$.

Given a projective frame (A_0, A_1, A^*) for a line Δ and a point P lying on Δ with homogeneous coordinates (x_0, x_1) in that frame, we can define a nonhomogeneous coordinate for P as $k_0 = x_0/x_1$. The scalar k_0 is sometimes called *projective parameter* of P, and it is easy to show that $k_0 = \{A_0, A_1; A_2, P\}$.

As noted earlier, a set of lines passing through the same point O is called a *pencil of lines*. The cross-ratio of four *coplanar* lines Δ_1, Δ_2, Δ_3 and Δ_4 in some pencil is defined as the cross-ratio of the intersections of these lines with any other line Δ in the same plane that does not pass through O, and it is easily shown to be independent of the choice of Δ (Figure 13.2a).

Consider now four planes Π_1, Π_2, Π_3, and Π_4 in the same pencil, and denote by Δ their common line. The cross-ratio of these planes is defined as the cross-ratio of the pencil of lines formed by their intersection with any other plane Π not containing Δ (Figure 13.2b). Once again, the cross-ratio is easily shown to be independent of the choice of Π.

In the plane, the nonhomogeneous projective coordinates (k_0, k_1) of the point P in the basis (A_0, A_1, A_2, A^*) are defined by $k_0 = x_0/x_2$ and $k_1 = x_1/x_2$, and it can be shown that

$$\begin{cases} k_0 = \{A_1A_0,\ A_1A_2;\ A_1A^*,\ A_1P\}, \\ k_1 = \{A_0A_1,\ A_0A_2;\ A_0A^*,\ A_0P\}, \end{cases}$$

where MN denotes the line joining the points M and N, and $\{\Delta_1, \Delta_2; \Delta_3, \Delta_4\}$ denotes the cross-ratio of the pencil of lines Δ_1, Δ_2, Δ_3, Δ_4.

Similarly, the nonhomogeneous projective coordinates (k_0, k_1, k_2) of the point P in the basis $(A_0, A_1, A_2, A_3, A^*)$ are defined by $k_0 = x_0/x_3$, $k_1 = x_1/x_3$, and $k_2 = x_2/x_3$, and it can be

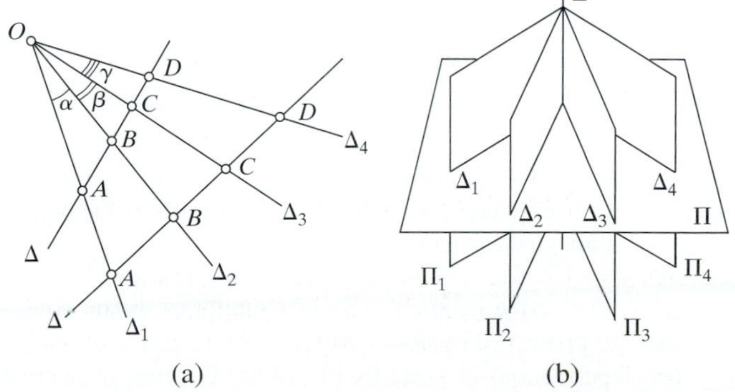

Figure 13.2 Definition of the cross-ratio of: (a) four lines, and (b) four planes. As shown in the exercises, the cross-ratio $\{A, B; C, D\}$ only depends on the three angles α, β, and γ. In particular, we have $\{A, B; C, D\} = \{A', B'; C', D'\}$.

shown that

$$\begin{cases} k_0 = \{A_1A_2A_0,\ A_1A_2A_3;\ A_1A_2A^*,\ A_1A_2P\}, \\ k_1 = \{A_2A_0A_1,\ A_2A_0A_3;\ A_2A_0A^*,\ A_2A_0P\}, \\ k_2 = \{A_0A_1A_2,\ A_0A_1A_3;\ A_0A_1A^*,\ A_0A_1P\}, \end{cases}$$

where LMN denotes the plane spanned by the three points L, M, and N, and $\{\Pi_1, \Pi_2; \Pi_3, \Pi_4\}$ denotes the cross-ratio of the pencil of planes $\Pi_1, \Pi_2, \Pi_3, \Pi_4$.

13.1.6 Projective Transformations

Consider a bijective linear map $\vec{\psi} : \vec{X} \to \vec{X}'$ between two vector spaces $\vec{X}$ and $\vec{X}'$. By linearity, $\vec{\psi}$ maps rays of $\vec{X}$ onto rays of $\vec{X}'$. Since it is bijective, it also maps nonzero vectors onto nonzero vectors, and we can define the induced map $\psi : P(\vec{X}) \to P(\vec{X}')$ by $\psi(p(\boldsymbol{v})) \stackrel{\text{def}}{=} p(\vec{\psi}(\boldsymbol{v}))$ for any $\boldsymbol{v} \neq 0$ in $\vec{X}$. The map ψ is bijective and is called a *projective transformation* (or *homography*). It is easy to show that projective transformations form a group under the law of composition of maps. When $\vec{X}' = \vec{X}$, this group is called the *projective group* of $X = P(\vec{X})$.

Example 13.5 Projective correspondence between coplanar points and their pictures.

Consider two planes and a point O lying outside these planes in $\mathbb{E}^3$. As shown in the exercises, the perspective projection mapping any point A in the (projective closure of the) first plane onto the intersection of the line AO with the (projective closure of the) second plane is a projective transformation.

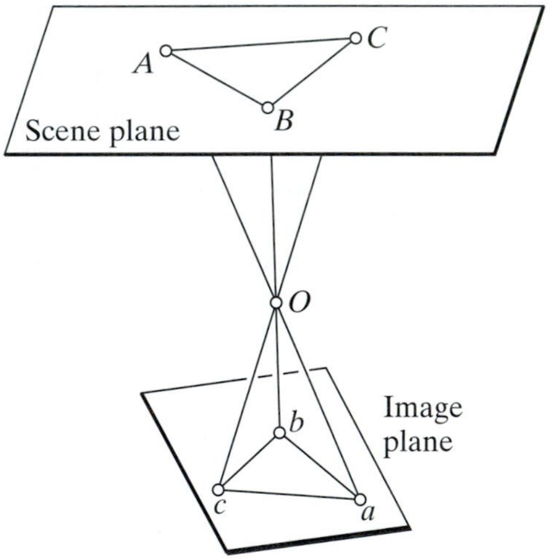

This property should not come as a surprise since, following Example 13.1, the two (projective) planes can be thought of as models of the projective spaces associated with the set of rays through the point O.

Projective geometry can be thought of as the study of the properties of projective spaces that are preserved by homographies. An example of such an *invariant* is the linear independence (or dependence) of a family of points. Given a projective transformation $\psi : X \to X'$, let us consider $m+1$ linearly independent vectors $\boldsymbol{v}_0, \boldsymbol{v}_1, \ldots, \boldsymbol{v}_m$ in $\vec{X}$ and the corresponding points $A_0, A_1, \ldots, A_m$ in X. Since $\vec{\psi}$ is bijective, the vectors $\vec{\psi}(\boldsymbol{v}_i)$ are linearly independent and so are the points $A'_i = \psi(A_i)$. It follows immediately that if $(A) = (A_0, A_1, \ldots, A_{n+1})$ is a projective frame for the n-dimensional projective space X, so is $(A') = (A'_0, A'_1, \dots, A'_{n+1})$ for X'. Conversely, given two n-dimensional projective spaces X and X', equipped respectively with the bases $(A_0, A_1, \dots, A_{n+1})$ and $(A'_0, A'_1, \dots, A'_{n+1})$, it can be shown that there exists a unique homography $\psi : X \to X'$ such that $\psi(A_i) = A'_i$ for $i = 0, 1, \dots, n+1$.

Projective coordinates form a second invariant. Indeed, due to the linearity of the underlying map $\vec{\psi}$, if the point P has coordinates $(x_0, x_1, \dots, x_n)$ in the projective frame $(A_0, A_1, \dots, A_{n+1})$ of X, the point $\psi(P)$ has the same coordinates in the coordinate frame of X' formed by the points $A'_i = \psi(A_i)$. In fact, projective transformations can be characterized as mappings that transform lines into lines and preserve cross-ratios (thus projective coordinates). Coming back to Example 13.5, it follows that an image of a set of coplanar points completely determines the projective coordinates of these points relative to the frame formed by four of them. This proves useful in designing invariant-based recognition systems in later chapters.

Like a rigid or an affine transformation, a homography ψ between two projective spaces of dimension n can conveniently be represented by an $(n+1) \times (n+1)$ matrix once coordinate systems (F) and (F') for these spaces have been chosen: This is again due to the linearity of the underlying operator $\vec{\psi}$. Thus, if $P' = \psi(P)$, we can write ${}^{F'}P' = \mathcal{Q}\,{}^{F}P$, where $\mathcal{Q}$ is a nonsingular $(n+1)\times(n+1)$ matrix defined up to scale since homogeneous projective coordinates are only defined up to scale.

Example 13.6 Parameterizing the Fundamental Matrix.

Let us revisit the problem of determining the epipolar geometry of uncalibrated cameras. This problem was introduced in chapter 10, where we gave without proof an explicit parameterization of the fundamental matrix. We now construct this parameterization. Let us define the *epipolar transformation* as the mapping from one set of epipolar lines onto the other one. As shown by the diagram, this transformation is a homography.

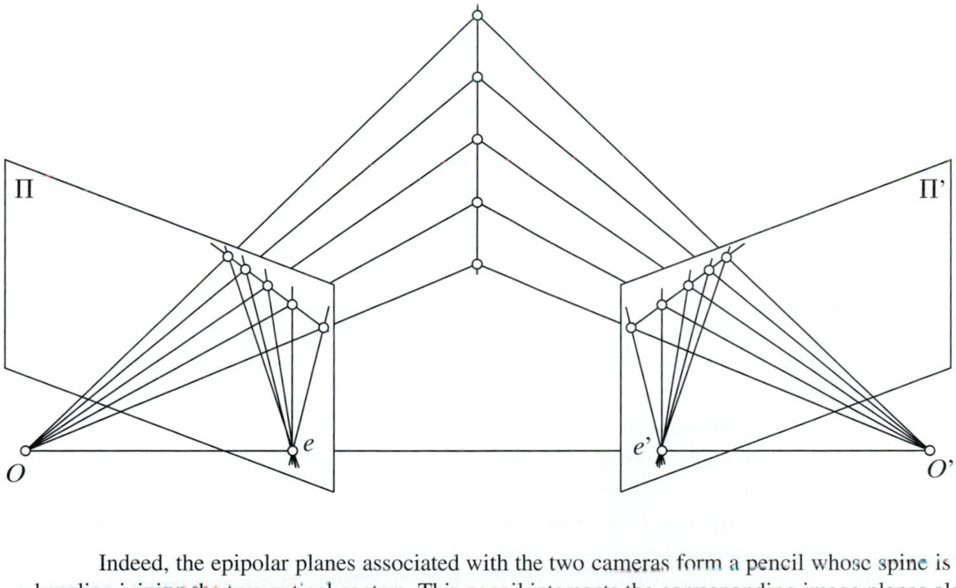

Indeed, the epipolar planes associated with the two cameras form a pencil whose spine is the baseline joining the two optical centers. This pencil intersects the corresponding image planes along the two families of epipolar lines, and the cross-ratio of any quadruple of lines in either family is of course the same as the cross-ratio of the corresponding planes. In turn, this means that the epipolar transformation preserves the cross-ratio and is therefore a projective transformation.

Let us denote by $(\alpha, \beta)^T$ and $(\alpha', \beta')^T$ the (affine) coordinates of the two epipoles e and e' in the corresponding image coordinate systems, and let us use $(u, v)^T$ and $(u', v')^T$ to denote the coordinates of points on matching epipolar lines l and l'. Using the fact that the linear map associated with the epipolar transformation maps the ray $\mathbb{R}(u-\alpha, v-\beta)^T$ onto the ray $\mathbb{R}(u'-\alpha', v'-\beta')^T$, it is easy to show (see Exercises) that the slopes τ and τ' of the lines l and l' satisfy

$$\tau' = \frac{a\tau + b}{c\tau + d}, \quad \text{with} \quad \tau \stackrel{\text{def}}{=} \frac{v-\beta}{u-\alpha} \quad \text{and} \quad \tau' \stackrel{\text{def}}{=} \frac{v'-\beta'}{u'-\alpha'}. \tag{13.4}$$

Clearing the denominators in Eq. (13.4) yields a bilinear expression in u, v and u', v', easily rewritten as $\boldsymbol{p}^T \mathcal{F} \boldsymbol{p}' = 0$, where $\mathcal{F}$ is written in the form given without proof in chapter 10—that is,

$$\mathcal{F} = \begin{pmatrix} b & a & -a\beta - b\alpha \\ -d & -c & c\beta + d\alpha \\ d\beta' - b\alpha' & c\beta' - a\alpha' & -c\beta\beta' - d\beta'\alpha + a\beta\alpha' + b\alpha\alpha' \end{pmatrix}.$$

13.1.7 Projective Shape

Following the approach used in the affine case, we say that two point sets S and S' in some projective space X are *projectively equivalent* when there exists a projective transformation $\psi : X \to X$ such that S' is the image of S under ψ. As in the affine case, it is easy to show that projective equivalence is an equivalence relation, and the *projective shape* of a point set S in X is defined as the equivalence class of all projectively equivalent point sets. Likewise, projective structure from motion can now be redefined as the problem of recovering the projective shape of the observed scene (and/or the equivalence classes formed by the corresponding projection matrices) from features matched in an image sequence.

13.2 PROJECTIVE STRUCTURE AND MOTION FROM BINOCULAR CORRESPONDENCES

The rest of this chapter is concerned with the recovery of the three-dimensional projective structure of a scene assuming that n points have been tracked in m images of this scene. This section focuses on the case of two images. Structure and motion estimation from three or more views are addressed in the next two sections. We assume that the epipoles are known, which, as shown in chapter 10, requires establishing at least seven point correspondences.

13.2.1 Geometric Scene Reconstruction

Let us start with a geometric method for estimating the projective shape of a scene when the epipoles are known. The inherent ambiguity of projective structure from motion simplifies our task by allowing us to choose appropriate points as a projective frame.

Let us assume that we observe four noncoplanar points A, B, C, D with a weakly-calibrated stereo rig (Figure 13.3). Let O' (resp. O'') denote the position of the optical center of the first (resp. second) camera. For any point P let p' (resp. p'') denote the position of the projection of P into the first (resp. second) image and let P' (resp. P'') denote the intersection of the ray $O'P$ (resp. $O''P$) with the plane ABC. The epipoles are e' and e'', and the baseline intersects the plane ABC in E. (Clearly, $E' = E'' = E$, $A' = A'' = A$, etc.)

We choose A, B, C, O', O'' as a basis for projective three-space, and our goal is to reconstruct the position of D. Choosing a', b', c', e' as a basis for the first image plane, we can measure the coordinates of d' in this basis and reconstruct the point D' in the basis A, B, C, E of the plane ABC. Similarly, we can reconstruct the point D'' from the projective coordinates of d'' in the basis a'', b'', c'', e'' of the second image plane. The point D is finally reconstructed as the intersection of the two lines $O'D'$ and $O''D''$.

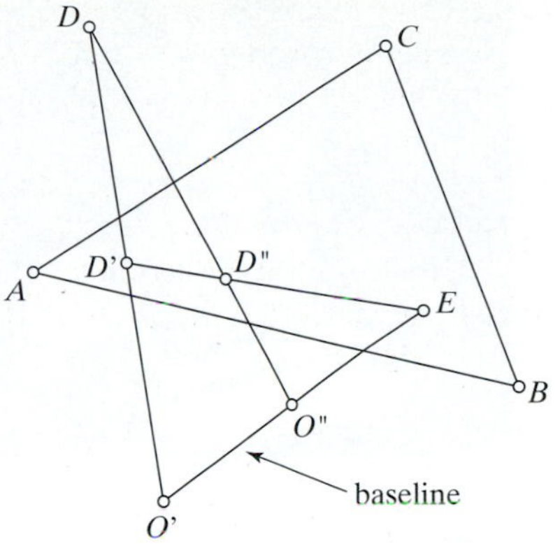

Figure 13.3 Geometric construction of the projective coordinates of the point D in the basis formed by the five points A, B, C, O', and O''.

We can now express this geometric construction in algebraic terms. It turns out to be simpler to reorder the points of our projective frame and calculate the nonhomogeneous projective coordinates of D in the basis formed by the tetrahedron A, O'', O', B and the unit point C. These coordinates are defined by the following three cross-ratios:

$$\begin{cases} k_0 = \{O''O'A,\ O''O'B;\ O''O'C,\ O''O'D\}, \\ k_1 = \{O'AO'',\ O'AB;\ O'AC,\ O'AD\}, \\ k_2 = \{AO''O',\ AO''B;\ AO''C,\ AO''D\}. \end{cases}$$

By intersecting the corresponding pencils of planes with the two image planes, we immediately obtain the values of k_0, k_1, k_2 as cross-ratios directly measurable in the two images:

$$\begin{cases} k_0 = \{e'a',\ e'b';\ e'c',\ e'd'\} = \{e''a'',\ e''b'';\ e''c'',\ e''d''\}, \\ k_1 = \{a'e',\ a'b';\ a'c',\ a'd'\}, \\ k_2 = \{a''e'',\ a''b'';\ a''c'',\ a''d''\}. \end{cases}$$

Figure 13.4 illustrates this method with data consisting of 46 point correspondences established between two images taken by weakly calibrated cameras. Figure 13.4(a) shows the input images and point matches. Figure 13.4(b) shows a view of the corresponding projective scene reconstruction, the raw projectives coordinates being used for rendering purposes. Since this form of display is not particularly enlightening, we also show in Figure 13.4(c) the reconstruction obtained by applying to the scene points the projective transformation mapping the three reference points (shown as small circles) and the camera centers onto their calibrated Euclidean positions. The true point positions are displayed as well for comparison.

13.2.2 Algebraic Motion Estimation

This section presents a purely algebraic approach to the problem of estimating the projective shape of a scene from binocular point correspondences, assuming once again that the stereo rig has been weakly calibrated. The perspective projection Eq. (2.15) introduced in chapter 2 extends naturally to the projective completion of $\mathbb{E}^3$ and maintains the same form in arbitrary projective frames for that space. Indeed, if we rewrite Eq. (2.15) as $z\boldsymbol{p} = \mathcal{M}\boldsymbol{P}$ in some Euclidean coordinate system (F), we obtain a similar equation $z\boldsymbol{p} = \mathcal{M}'\boldsymbol{P}'$ in a projective frame (F'), where $\boldsymbol{P}' = {}^{F'}P = {}^{F'}_{F}\mathcal{T}\boldsymbol{P}$, and $\mathcal{M}' = \mathcal{M}\ {}^{F'}_{F}\mathcal{T}^{-1}$.

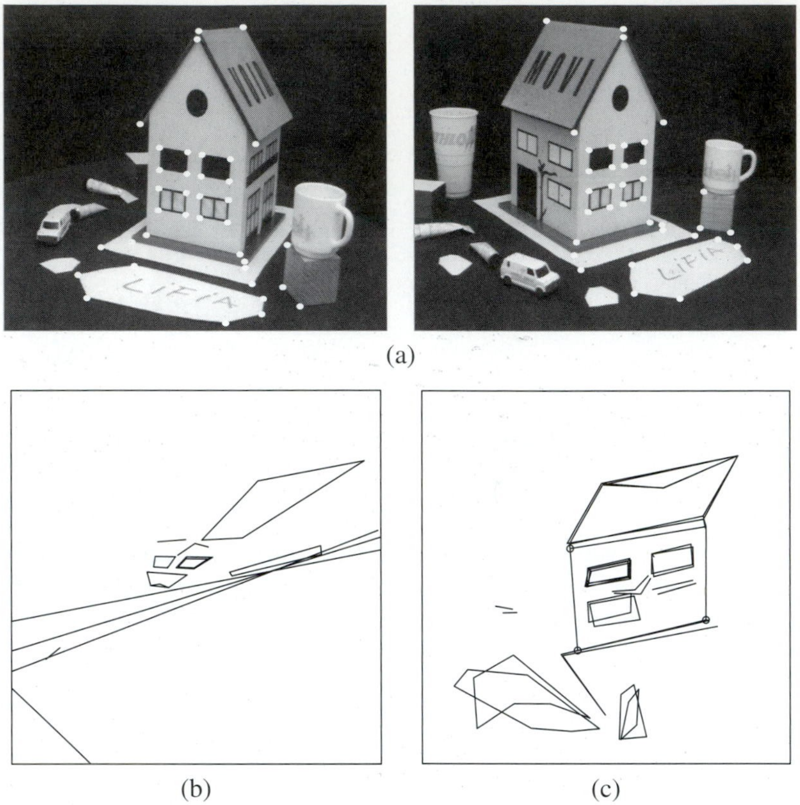

Figure 13.4 Geometric point reconstruction: (a) input data, (b) raw projective coordinates, (c) corrected projective coordinates. *Reprinted from "Relative Stereo and Motion Reconstruction," by J. Ponce, T.A. Cass, and D.H. Marimont, Tech. Report UIUC-BI-AI-RCV-93-07, Beckman Institute, Univ. of Illinois (1993). Data courtesy of Boubakeur Boufama and Roger Mohr.*

In particular, let us consider five points A_0, A_1, A_2, A_3, A_4 and choose them as a basis for $\tilde{\mathbb{E}}^3$, with A_4 playing the role of the unit point. We consider a camera observing these points, with projection matrix $\mathcal{M}$, and denote by a_0, a_1, a_2, a_3, a_4 the images of these points, choosing the points a_0 to a_3 as a projective basis of the image plane, a_3 being this time the unit point. We also denote by α, β, and γ the coordinates of a_4 in this basis.

Writing that $z_i \boldsymbol{a}_i = \mathcal{M}\boldsymbol{A}_i$ for $i = 0, 1, 2, 3, 4$ yields immediately

$$\mathcal{M} = \begin{pmatrix} z_0 & 0 & 0 & z_3 \\ 0 & z_1 & 0 & z_3 \\ 0 & 0 & z_2 & z_3 \end{pmatrix} \quad \text{and} \quad \begin{cases} z_4\alpha = z_0 + z_3, \\ z_4\beta = z_1 + z_3, \\ z_4\gamma = z_2 + z_3. \end{cases}$$

Since a perspective projection matrix is only defined up to scale, we can divide its coefficients by z_3, and defining $\lambda = z_4/z_3$ yields

$$\mathcal{M} = \begin{pmatrix} \lambda\alpha - 1 & 0 & 0 & 1 \\ 0 & \lambda\beta - 1 & 0 & 1 \\ 0 & 0 & \lambda\gamma - 1 & 1 \end{pmatrix}.$$

Let us now suppose we have a second image of the same scene, with projection matrix $\mathcal{M}'$ and image points $a'_0, a'_1, a'_2, a'_3, a'_4$. The same construction applies in this case, and we obtain

$$\mathcal{M}' = \begin{pmatrix} \lambda'\alpha' - 1 & 0 & 0 & 1 \\ 0 & \lambda'\beta' - 1 & 0 & 1 \\ 0 & 0 & \lambda'\gamma' - 1 & 1 \end{pmatrix}.$$

The stereo configuration of our two cameras is thus completely determined by the two parameters λ and λ'. The epipolar geometry of the rig can now be used to compute these parameters. Let us denote by C the optical center of the first camera and by e' the associated epipole in the image plane of the second camera, with coordinate vectors $\boldsymbol{C}$ and $\boldsymbol{e}'$ in the corresponding projective bases. We have $\mathcal{M}\boldsymbol{C} = \boldsymbol{0}$, and thus

$$\boldsymbol{C} = \left(\frac{1}{1-\lambda\alpha}, \frac{1}{1-\lambda\beta}, \frac{1}{1-\lambda\gamma}, 1 \right)^T.$$

Substituting in the equation $\mathcal{M}'\boldsymbol{C} = \boldsymbol{e}'$ then yields

$$\boldsymbol{e}' = \left(1 - \frac{\lambda'\alpha' - 1}{\lambda\alpha - 1}, 1 - \frac{\lambda'\beta' - 1}{\lambda\beta - 1}, 1 - \frac{\lambda'\gamma' - 1}{\lambda\gamma - 1} \right)^T.$$

Now if μ' and ν' denote this time the *nonhomogeneous* coordinates of e' in the projective basis formed by the points a'_i, we finally obtain

$$\begin{cases} \mu'(\lambda\gamma - \lambda'\gamma')(\lambda\alpha - 1) = (\lambda\alpha - \lambda'\alpha')(\lambda\gamma - 1), \\ \nu'(\lambda\gamma - \lambda'\gamma')(\lambda\beta - 1) = (\lambda\beta - \lambda'\beta')(\lambda\gamma - 1). \end{cases} \tag{13.5}$$

A system of two quadratic equations in two unknowns λ and λ' such as Eq. (13.5) admits in general four solutions, that can be thought of as the four intersections of the conic sections defined by the two equations in the (λ, λ') plane. Inspection of Eq. (13.5) reveals immediately that $(\lambda, \lambda') = (0, 0)$ and $(\lambda, \lambda') = (1/\gamma, 1/\gamma')$ are always solutions of these equations. It is easy (if a bit tedious) to show that the two remaining solutions are identical (geometrically, the two conics are tangent to each other at their point of intersection) and the corresponding values of the parameters λ and λ' are given by

$$\lambda = \frac{\text{Det}\begin{pmatrix} \mu' & \alpha & \alpha' \\ \nu' & \beta & \beta' \\ 1 & \gamma & \gamma' \end{pmatrix}}{\text{Det}\begin{pmatrix} \mu'\alpha & \alpha & \alpha' \\ \nu'\beta & \beta & \beta' \\ \gamma & \gamma & \gamma' \end{pmatrix}} \quad \text{and} \quad \lambda' = \frac{\text{Det}\begin{pmatrix} \mu & \alpha & \alpha' \\ \nu & \beta & \beta' \\ 1 & \gamma & \gamma' \end{pmatrix}}{\text{Det}\begin{pmatrix} \mu\alpha' & \alpha & \alpha' \\ \nu\beta' & \beta & \beta' \\ \gamma' & \gamma & \gamma' \end{pmatrix}}.$$

These values uniquely determine the projection matrices $\mathcal{M}$ and $\mathcal{M}'$. Note that taking into account the equations defining the second epipole would not add independent constraints because of the epipolar constraint that relates matching epipolar lines. Once the projection matrices are known, it is a simple matter to reconstruct the scene points.

13.3 PROJECTIVE MOTION ESTIMATION FROM MULTILINEAR CONSTRAINTS

The methods given in the previous two sections reconstruct the scene relative to five of its points, thus the quality of the reconstruction strongly depends on the accuracy of the localization of these points in the two images. In contrast, the approach presented in this section takes all points into account in a uniform manner and uses the multilinear constraints introduced in chapter 10 to reconstruct the camera motion in the form of the associated projection matrices.

13.3.1 Motion Estimation from Fundamental Matrices

Let us assume that the fundamental matrix $\mathcal{F}$ associated with two pictures has been estimated from binocular correspondences. As in the affine case, the projection matrices can in fact be estimated from a parameterization of $\mathcal{F}$ that exploits the inherent ambiguity of projective structure from motion. Since in the projective setting the scene structure and camera motion are only defined up to an arbitrary projective transformation, we can reduce the two matrices to canonical forms $\tilde{\mathcal{M}} = \mathcal{M}\mathcal{Q}$ and $\tilde{\mathcal{M}}' = \mathcal{M}'\mathcal{Q}'$ by postmultiplying them by an appropriate 4×4 matrix $\mathcal{Q}$. This time we take $\tilde{\mathcal{M}}'$ to be proportional to $(\text{Id} \quad \mathbf{0})$ and leave $\tilde{\mathcal{M}}$ in the general form $(\mathcal{A} \quad \boldsymbol{b})$. This reduction process determines 11 of the entries of $\mathcal{Q}$, and we refrain from using the 4 remaining degrees of freedom of $\mathcal{Q}$ to reduce $\tilde{\mathcal{M}}$ to a simpler form.

Let us now derive a new expression for the fundamental matrix using the canonical form of $\tilde{\mathcal{M}}'$. If $\tilde{\boldsymbol{P}} = (x, y, z, 1)^T$ denotes the homogeneous coordinate vector of the point P in the corresponding world coordinate system, we can write the projection equations associated with the two cameras as $z\boldsymbol{p} = (\mathcal{A} \quad \boldsymbol{b})\tilde{\boldsymbol{P}}$ and $z'\boldsymbol{p}' = (\text{Id} \quad \mathbf{0})\tilde{\boldsymbol{P}}$ or, equivalently,

$$z\boldsymbol{p} = \mathcal{A}(\text{Id} \quad \mathbf{0})\tilde{\boldsymbol{P}} + \boldsymbol{b} = z'\mathcal{A}\boldsymbol{p}' + \boldsymbol{b}.$$

It follows that $z\boldsymbol{b} \times \boldsymbol{p} = z'\boldsymbol{b} \times \mathcal{A}\boldsymbol{p}'$, and forming the dot product of this expression with $\boldsymbol{p}$ finally yields

$$\boldsymbol{p}^T \mathcal{F} \boldsymbol{p}' = 0 \quad \text{where} \quad \mathcal{F} = [\boldsymbol{b}_\times]\mathcal{A}.$$

Note the similarity with the expression for the essential matrix derived in chapter 10.

In particular, we have $\mathcal{F}^T\boldsymbol{b} = 0$, so (as could have been expected) $\boldsymbol{b}$ is the homogeneous coordinate vector of the first epipole in the corresponding image coordinate system. This new parameterization of the matrix $\mathcal{F}$ provides a simple method for computing the projection matrix $\tilde{\mathcal{M}}$. First note that, since the overall scale of $\tilde{\mathcal{M}}$ is irrelevant, we can always take $|\boldsymbol{b}| = 1$. This allows us to first compute $\boldsymbol{b}$ as the linear least-squares solution of $\mathcal{F}^T\boldsymbol{b} = 0$ with unit norm, then pick $\mathcal{A}_0 = -[\boldsymbol{b}_\times]\mathcal{F}$ as the value of $\mathcal{A}$. It is easy to show that, for any vector $\boldsymbol{a}$, $[\boldsymbol{a}_\times]^2 = \boldsymbol{a}\boldsymbol{a}^T - |\boldsymbol{a}|^2\text{Id}$; thus,

$$[\boldsymbol{b}_\times]\mathcal{A}_0 = -[\boldsymbol{b}_\times]^2\mathcal{F} = -\boldsymbol{b}\boldsymbol{b}^T\mathcal{F} + |\boldsymbol{b}|^2\mathcal{F} = \mathcal{F},$$

since $\mathcal{F}^T\boldsymbol{b} = \mathbf{0}$ and $|\boldsymbol{b}|^2 = 1$. This shows that $\tilde{\mathcal{M}} = \begin{pmatrix}\mathcal{A}_0 & \boldsymbol{b}\end{pmatrix}$ is *a* solution of our problem. As shown in the exercises, there is in fact a 4-parameter family of solutions whose general form is

$$\tilde{\mathcal{M}} = \begin{pmatrix}\mathcal{A} & \boldsymbol{b}\end{pmatrix} \quad \text{with} \quad \mathcal{A} = \lambda\mathcal{A}_0 + (\ \mu\boldsymbol{b} \mid \nu\boldsymbol{b} \mid \tau\boldsymbol{b}\ ).$$

The four parameters correspond, as could have been expected, to the remaining degrees of freedom of the projective transformation $\mathcal{Q}$. Once the matrix $\tilde{\mathcal{M}}$ is known, we can compute the position of any point P by solving in the least-squares sense the nonhomogeneous linear system of equations in z and z' defined by $z\boldsymbol{p} = z'\mathcal{A}\boldsymbol{p}' + \boldsymbol{b}$.

13.3.2 Motion Estimation from Trifocal Tensors

We now rewrite in a projective setting the trilinear constraints associated with the trifocal tensor first introduced in chapter 10. As in the previous section, we can postmultiply the projection matrices by an appropriate 4×4 matrix so they take the form

$$\tilde{\mathcal{M}}_1 = (\text{Id} \quad \mathbf{0}), \quad \tilde{\mathcal{M}}_2 = (\mathcal{A}_2 \quad \boldsymbol{b}_2), \quad \text{and} \quad \tilde{\mathcal{M}}_3 = (\mathcal{A}_3 \quad \boldsymbol{b}_3).$$

Under this transformation, $\boldsymbol{b}_2$ and $\boldsymbol{b}_3$ can still be interpreted as the homogeneous image coordinates of the epipoles e_{12} and e_{13}, and the trilinear constraints in Eqs. (10.14) and (10.15) still hold, with the trifocal tensor defined this time by the three matrices[2]

$$\mathcal{G}_1^i = \boldsymbol{b}_2 \boldsymbol{A}_3^{iT} - \boldsymbol{A}_2^i \boldsymbol{b}_3^T, \tag{13.6}$$

where $\boldsymbol{A}_2^i$ and $\boldsymbol{A}_3^i$ ($i = 1, 2, 3$) denote the columns of $\mathcal{A}_2$ and $\mathcal{A}_3$.

Assuming that the trifocal tensor has been estimated from point or line correspondences as described in chapter 10, our goal in this section is to recover the projection matrices $\tilde{\mathcal{M}}_2$ and $\tilde{\mathcal{M}}_3$. Let us first observe that

$$(\boldsymbol{b}_2 \times \boldsymbol{A}_2^i)^T \mathcal{G}_1^i = \left[(\boldsymbol{b}_2 \times \boldsymbol{A}_2^i)^T \boldsymbol{b}_2\right] \boldsymbol{A}_3^{iT} - \left[(\boldsymbol{b}_2 \times \boldsymbol{A}_2^i)^T \boldsymbol{A}_2^i\right] \boldsymbol{b}_3^T = \boldsymbol{0},$$

and, likewise,

$$\mathcal{G}_1^i (\boldsymbol{b}_3 \times \boldsymbol{A}_3^i) = \left[\boldsymbol{A}_3^{iT} (\boldsymbol{b}_3 \times \boldsymbol{A}_3^i)\right] \boldsymbol{b}_2 - \left[\boldsymbol{b}_3^T (\boldsymbol{b}_3 \times \boldsymbol{A}_3^i)\right] \boldsymbol{A}_2^i = \boldsymbol{0}.$$

It follows that the matrix $\mathcal{G}_1^i$ is singular (a fact already mentioned in chapter 10) and the vectors $\boldsymbol{b}_2 \times \boldsymbol{A}_2^i$ and $\boldsymbol{b}_3 \times \boldsymbol{A}_3^i$ lie, respectively, in its left and right nullspaces. In turn, this means that, once the trifocal tensor is known, we can compute the epipole $\boldsymbol{b}_2$ (resp. $\boldsymbol{b}_3$) as the common normal to the left (resp. right) nullspaces of the matrices $\mathcal{G}_1^i$ ($i = 1, 2, 3$).

Once the epipoles are known, writing Eq. (13.6) for $i = 1, 2, 3$ provides 27 homogeneous linear equations in the 18 unknown entries of the matrices $\mathcal{A}_j$ ($j = 2, 3$). These equations can be solved up to scale using linear least squares. Alternatively, it is possible to estimate the matrices $\mathcal{A}_j$ directly from the trilinear constraints associated with pairs of matching points or lines by writing the trifocal tensor coefficients as functions of these matrices, which leads once again to a linear estimation process.

Once the projection matrices have been recovered, the projective structure of the scene can be recovered as well by using the perspective projection equations as linear constraints on the homogeneous coordinate vectors of the observed points and lines.

13.4 PROJECTIVE STRUCTURE AND MOTION FROM MULTIPLE IMAGES

Section 13.3 used the epipolar and trifocal constraints to reconstruct the camera motion and the corresponding scene structure from a pair or triple of images. Likewise, the quadrifocal tensor introduced in chapter 10 can in principle be used to estimate the projection matrices associated with four cameras and the corresponding projective scene structure. However, multilinear constraints do not provide a direct method for handling $m > 4$ views in a uniform manner. Instead, the structure and motion parameters estimated from pairs, triples, or quadruples of successive views must be stitched together iteratively. We now present an alternative where all images are taken into account at once in a nonlinear optimization scheme.

13.4.1 A Factorization Approach to Projective Structure from Motion

In this section, we present a factorization algorithm for motion analysis that generalizes the Tomasi–Kanade algorithm presented in chapter 12 to the projective case. Given m images of n

[2]Formally, postmultiplying the three projection matrices by $\mathcal{Q}$ has the same effect as taking the calibration matrix $\mathcal{K}_1$ equal to the identity in the equations defining the (uncalibrated) trifocal tensor in chapter 10. Note, however, that we do not assume here that the calibration parameters are known. Instead, we use an appropriate change of projective coordinates to simplify the form of the projection matrices.

points, we can rewrite Eq. (13.1) as

$$\mathcal{D} = \mathcal{M}\mathcal{P}, \tag{13.7}$$

where

$$\mathcal{D} \stackrel{\text{def}}{=} \begin{pmatrix} z_{11}\boldsymbol{p}_{11} & z_{12}\boldsymbol{p}_{12} & \dots & z_{1n}\boldsymbol{p}_{1n} \\ z_{21}\boldsymbol{p}_{21} & z_{22}\boldsymbol{p}_{22} & \dots & z_{2n}\boldsymbol{p}_{2n} \\ \dots & \dots & \dots & \dots \\ z_{m1}\boldsymbol{p}_{m1} & z_{m2}\boldsymbol{p}_{m2} & \dots & z_{mn}\boldsymbol{p}_{mn} \end{pmatrix}, \quad \mathcal{M} \stackrel{\text{def}}{=} \begin{pmatrix} \mathcal{M}_1 \\ \mathcal{M}_2 \\ \dots \\ \mathcal{M}_m \end{pmatrix} \quad \text{and } \mathcal{P} \stackrel{\text{def}}{=} (\boldsymbol{P}_1 \ \boldsymbol{P}_2 \ \dots \ \boldsymbol{P}_n).$$

As the product of $3m \times 4$ and $4 \times n$ matrices, the $3m \times n$ matrix $\mathcal{D}$ has (at most) rank 4; thus, if the projective depths z_{ij} were known, we could compute $\mathcal{M}$ and $\mathcal{P}$, just as in the affine case, by using singular value decomposition to factor $\mathcal{D}$. On the other hand, if $\mathcal{M}$ and $\mathcal{P}$ were known, we could read out the values of the projective depths z_{ij} from Eq. (13.7). This suggests an iterative scheme for estimating the unknowns z_{ij}, $\mathcal{M}$ and $\mathcal{P}$ by alternating steps where some of these unknowns are held constant while others are estimated.

We minimize the squared Frobenius norm of $\mathcal{D} - \mathcal{M}\mathcal{P}$—that is,

$$E \stackrel{\text{def}}{=} |\mathcal{D} - \mathcal{M}\mathcal{P}|^2 = \sum_{i,j} |z_{ij}\boldsymbol{p}_j - \mathcal{M}_i\boldsymbol{P}_j|^2$$

with respect to the unknowns $\mathcal{M}_i$, $\boldsymbol{P}_j$ and z_{ij}. Note that the minimization of E is ill-posed unless some constraints are imposed on the parameters $\mathcal{M}_i$, $\boldsymbol{P}_j$, and z_{ij}. Indeed, as mentioned earlier, these unknowns are not independent: The matrices $\mathcal{M}_i$ and the vectors $\boldsymbol{P}_j$ are only defined up to scale. If $\mathcal{M}_i$, $\boldsymbol{P}_j$, and z_{ij} are solutions of Eq. (13.1), so are $\alpha_i\mathcal{M}_i$, $\beta_j\boldsymbol{P}_j$, and $\alpha_i\beta_j z_{ij}$ for arbitrary values of the scalars α_i and β_j. In particular, Eq. (13.1) always admits the trivial solution $\mathcal{M}_i = 0, \boldsymbol{P}_j = 0, z_{ij} = 0$. In fact, this equation admits a much wider class of trivial, nonphysical solutions—for example, $z_{ij} = 0$, $\mathcal{M}_i = \mathcal{M}_0$, and $\boldsymbol{P}_j = \boldsymbol{P}_0$, where $\mathcal{M}_0$ is an arbitrary rank-3 3×4 matrix and $\boldsymbol{P}_0$ is a unit vector in its kernel. Here we impose the constraint that the columns $\boldsymbol{d}_j$ of the matrix $\mathcal{D}$ have unit norm, which eliminates these trivial solutions.

Let us assume that we are at some stage of the minimization process, fix the value of $\mathcal{M}$ to its current estimate and compute, for $j = 1, \dots, n$, the values of $\boldsymbol{z}_j \stackrel{\text{def}}{=} (z_{1j}, \dots, z_{mj})^T$ and $\boldsymbol{P}_j$ that minimize

$$E_j \stackrel{\text{def}}{=} \sum_{i=1}^{m} |z_{ij}\boldsymbol{p}_j - \mathcal{M}_i\boldsymbol{P}_j|^2.$$

These values minimize E as well. Writing that the gradient of E_j with respect to the vector $\boldsymbol{P}_j$ should be zero at a minimum yields

$$0 = \frac{\partial E_j}{\partial \boldsymbol{P}_j} = 2\sum_{i=1}^{m} \mathcal{M}_i^T (z_{ij}\boldsymbol{p}_{ij} - \mathcal{M}_i\boldsymbol{P}_j),$$

or

$$\mathcal{M}^T\boldsymbol{d}_j = \mathcal{M}^T\mathcal{M}\boldsymbol{P}_j \Longleftrightarrow \boldsymbol{P}_j = \mathcal{M}^\dagger\boldsymbol{d}_j,$$

where $\mathcal{M}^\dagger \stackrel{\text{def}}{=} (\mathcal{M}^T\mathcal{M})^{-1}\mathcal{M}^T$ is the pseudoinverse of $\mathcal{M}$. In turn, substituting this value in the definition of E_j yields $E_j = |(\text{Id} - \mathcal{M}\mathcal{M}^\dagger)\boldsymbol{d}_j|^2$.

Now $\mathcal{M}$ is a $3m \times 4$ matrix of rank 4 whose singular value decomposition $\mathcal{U}\mathcal{W}\mathcal{V}^T$ is formed by the product of a column-orthogonal $3m \times 4$ matrix $\mathcal{U}$, a 4×4 nonsingular diagonal matrix $\mathcal{W}$, and a 4×4 orthogonal matrix $\mathcal{V}^T$. The pseudoinverse of $\mathcal{M}$ is $\mathcal{M}^\dagger = \mathcal{V}\mathcal{W}^{-1}\mathcal{U}^T$; substituting this value in the expression of E_j and taking into account the fact that $|\boldsymbol{d}_j|^2 = 1$ immediately yields

$$E_j = \left|[\text{Id} - \mathcal{U}\mathcal{U}^T]\boldsymbol{d}_j\right|^2 = 1 - |\mathcal{U}\boldsymbol{d}_j|^2.$$

In turn, this means that minimizing E_j with respect to z_j and $\boldsymbol{P}_j$ is equivalent to maximizing $|\mathcal{U}\boldsymbol{d}_j|^2$ under the constraint $|\boldsymbol{d}_j|^2 = 1$. Finally, observing that

$$\boldsymbol{d}_j = \mathcal{Q}_j z_j, \quad \text{where} \quad \mathcal{Q}_j \stackrel{\text{def}}{=} \begin{pmatrix} \boldsymbol{p}_{1j} & \mathbf{0} & \dots & \mathbf{0} \\ \mathbf{0} & \boldsymbol{p}_{2j} & \dots & \mathbf{0} \\ \dots & \dots & \dots & \dots \\ \mathbf{0} & \mathbf{0} & \dots & \boldsymbol{p}_{mj} \end{pmatrix},$$

shows that minimizing E_j is equivalent to maximizing $|\mathcal{R}_j z_j|^2$ with respect to z_j under the constraint $|\mathcal{Q}_j z_j|^2 = 1$, where $\mathcal{R}_j \stackrel{\text{def}}{=} \mathcal{U}^T \mathcal{Q}_j$. This is a generalized eigenvalue problem, whose solution is the unit vector z_j corresponding to the largest scalar λ such that $\mathcal{R}_j^T \mathcal{R}_j z_j = \lambda \mathcal{Q}_j^T \mathcal{Q}_j z_j$.

The minimization step where the projective depths are held constant and $\mathcal{M}$ and $\mathcal{P}$ are updated is the same as in the Tomasi–Kanade approach to affine structure from motion. The overall process is summarized in Algorithm 13.1. The initial projective depth values are set to 1 or they can be computed as before from estimates of the epipolar geometry.

It is easy to show that the error E eventually converges to some value E^*. Indeed, let E_0 be the current error value at the beginning of each iteration; the first two steps of the algorithm do not change the vectors z_j, but minimize E with respect to the unknowns $\mathcal{M}$ and $\boldsymbol{P}_j$. If E_2 is the value of the error at the end of Step 2, we have $E_2 \leq E_0$. Now Step 3 does not change the matrix $\mathcal{M}$, but minimizes each error term E_j with respect to both the vectors z_j and $\boldsymbol{P}_j$. Therefore, the error E_3 at the end of this step is smaller than or equal to E_2. This shows that the error decreases in a monotone manner at each iteration. Since it is bounded below by zero, we conclude that the error converges to some value $E^* \geq 0$. The convergence of its error is not sufficient to guarantee the convergence of an optimization algorithm to a local minimum. However, a convergence proof for Algorithm 13.13.1, based on the *Global Convergence Theorem* (Luenberger, 1985) from numerical analysis and far too complex to be included here, can be found in Mahamud *et al.* (2001). Whether this local minimum turns out to be the global one depends, of course, on the choice of initial values chosen for the various unknown parameters. A

Algorithm 13.1: A Factorization Algorithm for Projective Shape from Motion.

1. Compute an initial estimate of the projective depths z_{ij}, with $i = 1, \dots, m$ and $j = 1, \dots, n$.
2. Normalize each column of the data matrix $\mathcal{D}$.
3. Repeat:
   - **(a)** use singular value decomposition to compute the $2m \times 4$ matrix $\mathcal{M}$ and the $4 \times n$ matrix $\mathcal{P}$ that minimize $|\mathcal{D} - \mathcal{M}\mathcal{P}|^2$;
   - **(b)** for $j = 1$ to n, compute the matrices $\mathcal{R}_j$ and $\mathcal{Q}_j$ and find the value of z_j that maximize $|\mathcal{R}_j z_j|^2$ under the constraint $|\mathcal{Q}_j z_j|^2 = 1$ as the solution of a generalized eigenvalue problem;
   - **(c)** update the value of $\mathcal{D}$ accordingly;

   until convergence.

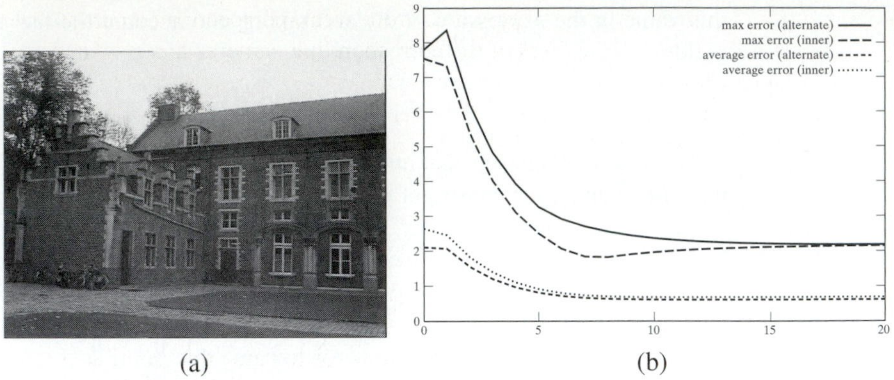

Figure 13.5 Iterative projective estimation of camera motion and scene structure: (a) a sample image in the sequence; (b) plot of the average and maximum reprojection error as a function of iteration number. Two experiments were conducted: In the first one (alternate), alternate images in the sequence are used as training and testing datasets; in the second experiment (inner), the first five and last five pictures were used as training set, and the remaining images were used for testing. In both cases, the average error falls below 1 pixel after 15 iterations. *Reprinted from "Iterative Projective Reconstruction from Multiple Views," by S. Mahamud and M. Hebert, Proc. IEEE Conference on Computer Vision and Pattern Recognition, (2000).*

possible choice, used in the experiments presented in Mahamud and Hebert (2000), is to initialize the projective depths z_{ij} to 1, which effectively amounts to starting with a weak-perspective projection model.

Figure 13.5(a) shows the first image in a sequence of 20 pictures of an outdoor scene. Thirty points were tracked manually across the sequence, with a localization error of ∓ 1 pixel. Figure 13.5(b) plots the evolution of the average and maximum errors between the observed and predicted image point positions when various subsets of the image sequence are used for training and testing.

13.4.2 Bundle Adjustment

Given initial estimates for the matrices $\mathcal{M}_i$ $(i = 1, \dots, m)$ and vectors $\boldsymbol{P}_j$ $(j = 1, \dots, n)$, we can refine these estimates by using nonlinear least squares to minimize the global error measure

$$E = \frac{1}{mn} \sum_{i,j} \left[\left(u_{ij} - \frac{\boldsymbol{m}_{i1} \cdot \boldsymbol{P}_j}{\boldsymbol{m}_{i3} \cdot \boldsymbol{P}_j} \right)^2 + \left(v_{ij} - \frac{\boldsymbol{m}_{i2} \cdot \boldsymbol{P}_j}{\boldsymbol{m}_{i3} \cdot \boldsymbol{P}_j} \right)^2 \right].$$

This is the method of *bundle adjustment*, whose name originates from the field of photogrammetry. Although it may be expensive, it offers the advantage of combining all measurements to minimize a physically significant error measure—namely, the mean-squared error between the actual image point positions and those predicted using the estimated scene structure and camera motion.

13.5 FROM PROJECTIVE TO EUCLIDEAN IMAGES

Although projective structure is useful by itself, in most cases it is the Euclidean structure of the scene that is the true object of interest. We saw in chapter 12 that the absolute scale of a scene cannot be recovered from weak-perspective or paraperspective images even when the intrinsic parameters of the corresponding cameras are known. The same ambiguity holds in the perspective case: Given a camera with known intrinsic parameters, we can take the calibration matrix to be the identity and write the perspective projection Eq. (2.15) in some Euclidean world coordinate system (W) as

$$\boldsymbol{p} = \frac{1}{z}(\mathcal{R} \quad \boldsymbol{t})\begin{pmatrix}\boldsymbol{P}\\1\end{pmatrix} = \frac{1}{\lambda z}(\mathcal{R} \quad \lambda\boldsymbol{t})\begin{pmatrix}\lambda\boldsymbol{P}\\1\end{pmatrix}$$

for any nonzero scale factor λ. This ambiguity is not surprising given the fact that $\boldsymbol{t}$ is defined in Eq. (2.15) as the position of the origin of (W) relative to the camera: Moving both the scene and the camera observing it away from (or toward) this point at constant speed alters the apparent depth of the scene, but does not change its image. Adding more cameras does not help, and the best we can hope for is to estimate the Euclidean shape of the scene, defined, as in chapter 2, up to an arbitrary similarity transformation.

Let us assume from now on that one of the techniques presented in Section 13.4 has been used to estimate the projection matrices $\mathcal{M}_i$ $(i = 1, \ldots, m)$ and the point positions $\boldsymbol{P}_j$ $(j = 1, \ldots, n)$ from m images of these points. We know that any other reconstruction *and in particular a Euclidean one* is separated from this one by a projective transformation. In other words, if $\hat{\mathcal{M}}_i$ and $\hat{\boldsymbol{P}}_j$ denote the shape and motion parameters measured in some Euclidean coordinate system, there must exist a 4×4 matrix $\mathcal{Q}$ such that $\hat{\mathcal{M}}_i = \mathcal{M}_i\mathcal{Q}$ and $\hat{\boldsymbol{P}}_j = \mathcal{Q}^{-1}\boldsymbol{P}_j$. The rest of this section presents a method for computing the *Euclidean upgrade* matrix $\mathcal{Q}$ and thus recovering the Euclidean shape and motion from the projective ones when (some of) the intrinsic parameters of the camera are known.

Let us first note that, since the individual matrices $\mathcal{M}_i$ are only defined up to scale, so are the matrices $\hat{\mathcal{M}}_i$ that can be written (in the most general case where some of the intrinsic parameters are unknown) as

$$\hat{\mathcal{M}}_i = \rho_i\mathcal{K}_i(\mathcal{R}_i \quad \boldsymbol{t}_i),$$

where ρ_i accounts for the unknown scale of $\mathcal{M}_i$, and $\mathcal{K}_i$ is a calibration matrix as defined by Eq. (2.13). In particular, if we write the Euclidean upgrade matrix as $\mathcal{Q} = (\mathcal{Q}_3 \quad \boldsymbol{q}_4)$, where $\mathcal{Q}_3$ is a 4×3 matrix and $\boldsymbol{q}_4$ is a vector in $\mathbb{R}^4$, we obtain immediately

$$\mathcal{M}_i\mathcal{Q}_3 = \rho_i\mathcal{K}_i\mathcal{R}_i. \tag{13.8}$$

Using this equation, it is a simple matter to adapt the affine methods introduced in chapter 12 to the projective setting when the intrinsic parameters of all cameras are known so the matrices $\mathcal{K}_i$ can be taken equal to the identity: According to Eq. (13.8), the 3×3 matrices $\mathcal{M}_i\mathcal{Q}_3$ are in this case scaled rotation matrices. Writing that their rows $\boldsymbol{m}_{ij}^T$ $(j = 1, 2, 3)$ are perpendicular to each other and have the same norm yields

$$\begin{cases} \boldsymbol{m}_{i1}^T\mathcal{Q}_3\mathcal{Q}_3^T\boldsymbol{m}_{i2} = 0,\\ \boldsymbol{m}_{i2}^T\mathcal{Q}_3\mathcal{Q}_3^T\boldsymbol{m}_{i3} = 0,\\ \boldsymbol{m}_{i3}^T\mathcal{Q}_3\mathcal{Q}_3^T\boldsymbol{m}_{i1} = 0,\\ \boldsymbol{m}_{i1}^T\mathcal{Q}_3\mathcal{Q}_3^T\boldsymbol{m}_{i1} - \boldsymbol{m}_{i2}^T\mathcal{Q}_3\mathcal{Q}_3^T\boldsymbol{m}_{i2} = 0,\\ \boldsymbol{m}_{i2}^T\mathcal{Q}_3\mathcal{Q}_3^T\boldsymbol{m}_{i2} - \boldsymbol{m}_{i3}^T\mathcal{Q}_3\mathcal{Q}_3^T\boldsymbol{m}_{i3} = 0. \end{cases} \tag{13.9}$$

The upgrade matrix $\mathcal{Q}$ is of course only defined up to an arbitrary similarity. To determine it uniquely, we can assume that the world coordinate system and the first camera's frame coincide. Given m images, we obtain 12 linear equations and $5(m-1)$ quadratic ones in the coefficients of $\mathcal{Q}$. These equations can be solved using nonlinear least squares.

Alternatively, the constraints in Eq. (13.9) are linear in the coefficients of the symmetric matrix $\mathcal{A} \stackrel{\text{def}}{=} \mathcal{Q}_3\mathcal{Q}_3^T$, allowing its estimation from at least two images via linear least squares. Note that $\mathcal{A}$ has rank 3—a constraint not enforced by our construction. To recover $\mathcal{Q}_3$, let us also note that, since $\mathcal{A}$ is symmetric, it can be diagonalized in an orthonormal basis as $\mathcal{A} = \mathcal{U}\mathcal{D}\mathcal{U}^T$, where $\mathcal{D}$ is the diagonal matrix formed by the eigenvalues of $\mathcal{A}$ and $\mathcal{U}$ is the orthogonal matrix formed by its eigenvectors. In the absence of noise, $\mathcal{A}$ is positive semidefinite with three positive and one zero eigenvalues, and $\mathcal{Q}_3$ can be computed as $\mathcal{U}_3\sqrt{\mathcal{D}_3}$, where $\mathcal{U}_3$ is the matrix formed by the columns of $\mathcal{U}$ associated with the positive eigenvalues of $\mathcal{A}$, and $\mathcal{D}_3$ is the corresponding submatrix of $\mathcal{D}$. Because of noise, however, $\mathcal{A}$ usually has maximal rank, and its smallest eigenvalue may even be negative. As shown in Ponce (2000), if we take this time $\mathcal{U}_3$ and $\mathcal{D}_3$ to be the submatrices of $\mathcal{U}$ and $\mathcal{D}$ associated with the three largest (positive) eigenvalues of $\mathcal{A}$, then $\mathcal{U}_3\mathcal{D}_3\mathcal{U}_3^T$ provides the best positive semidefinite rank-3 approximation of $\mathcal{A}$ in the sense of the Frobenius norm,[3] and we can take as before $\mathcal{Q}_3 = \mathcal{U}_3\sqrt{\mathcal{D}_3}$. At this point, the last column vector $\boldsymbol{q}_4$ of $\mathcal{Q}$ can be determined by (arbitrarily) picking the origin of the frame attached to the first camera as the origin of the world coordinate system.

This method can easily be adapted to the case where only some of the intrinsic camera parameters are known: Using the fact that $\mathcal{R}_i$ is an orthogonal matrix allows us to write

$$\mathcal{M}_i\mathcal{A}\mathcal{M}_i^T = \rho_i^2\mathcal{K}_i\mathcal{K}_i^T. \tag{13.10}$$

Thus, every image provides a set of constraints between the entries of $\mathcal{K}_i$ and $\mathcal{A}$. Assuming, for example, that the center of the image is known for each camera, we can take $u_0 = v_0 = 0$ and write the square of the matrix $\mathcal{K}_i$ as

$$\mathcal{K}_i\mathcal{K}_i^T = \begin{pmatrix} \alpha_i^2\dfrac{1}{\sin^2\theta_i} & -\alpha_i\beta_i\dfrac{\cos\theta_i}{\sin^2\theta_i} & 0 \\ -\alpha_i\beta_i\dfrac{\cos\theta_i}{\sin^2\theta_i} & \beta_i^2\dfrac{1}{\sin^2\theta_i} & 0 \\ 0 & 0 & 1 \end{pmatrix}.$$

In particular, the part of Eq. (13.10) corresponding to the zero entries of $\mathcal{K}_i\mathcal{K}_i^T$ provides two independent linear equations in the 10 coefficients of the 4×4 symmetric matrix $\mathcal{A}$. With $m \geq 5$ images, these parameters can be estimated via linear least squares. Once $\mathcal{A}$ is known, $\mathcal{Q}$ can be estimated as before. Figure 13.6 shows a texture-mapped picture of the 3D model of a castle obtained by a variant of this method (Pollefeys *et al.* (1999)).

13.6 NOTES

The short introduction to projective geometry given at the beginning of this chapter focuses on the analytical side of things. See, for example, Todd (1946), Berger (1987), and Samuel (1988) for thorough introductions to analytical projective geometry, and Coxeter (1974) for an axiomatic presentation. Projective structure from motion is covered in detail in the books of Hartley and Zisserman (2000) and Faugeras, Luong, and Papadopoulo (2001).

As mentioned by Faugeras (1993), the problem of calculating the epipoles and the epipolar transformations compatible with seven point correspondences was first posed by Chasles (1855) and solved by Hesse (1863). The problem of estimating the epipolar geometry from five point correspondences for internally calibrated cameras was solved by Kruppa (1913). An excellent

[3] Note the obvious similarity between this result and Theorem 4.

Figure 13.6 A synthetic texture-mapped image of a castle constructed via projective motion analysis followed by a Euclidean upgrade. The principal point is assumed to be known. *Reprinted from "Self-Calibration and Metric 3D Reconstruction from Uncalibrated Image Sequences," by M. Pollefeys, PhD Thesis, Katholieke Universiteit, Leuven, (1999).*

modern account of Hesse's and Kruppa's techniques can be found in Faugeras and Maybank (1990), where the *absolute conic*, an imaginary conic section invariant through similarities, is used to derive two tangency constraints that make up for the missing point correspondences. These methods are of course mostly of theoretical interest since their reliance on a minimal number of correspondences limits their ability to deal with noise. The weak-calibration methods of Luong *et al.* (1993, 1996) and Hartley (1995) described in chapter 10 provide reliable and accurate alternatives.

Faugeras (1992) and Hartley *et al.* (1992) introduced independently the idea of using a pair of uncalibrated cameras to recover the projective structure of a scene. Other notable work in this area includes, for example, Mohr *et al.* (1992) and Shashua (1993). Section 13.2.2 presents Faugeras' original method, and its geometric variant presented in Section 13.2.1 is taken from Ponce *et al.* (1993). The two- and three-view motion analysis techniques also presented in this chapter are variants of the methods proposed by Hartley (1992, 1994*b*, 1997) and Beardsley *et al.* (1997). When the cameras are calibrated, it is also possible, as shown in the exercises and (Longuet–Higgins, 1981), to recover (up to a two-fold ambiguity) the similitude associated with the corresponding *essential* matrix. An iterative algorithm for perspective motion and structure recovery using calibrated cameras is given in Christy and Horaud (1996). The extension of factorization approaches to structure and motion recovery was first proposed by Sturm and Triggs (1996). The variant presented in Section 13.4.1 is due to Mahamud and Hebert (2000) and has the advantage of being provably convergent (Mahamud *et al.*, 2001). Algorithms for stitching together pairs, triples or quadruples of successive views can be found in Beardsley *et al.* (1997) and Pollefeys *et al.* (1999) for example.

The problem of computing Euclidean upgrades of projective reconstructions when some of the intrinsic parameters are known has been addressed by a number of authors (e.g., Heyden and Åström, 1996, Triggs, 1997, Pollefeys, 1999). The matrix $\mathcal{A} = \mathcal{Q}_3\mathcal{Q}_3^T$ introduced in Section 13.5 can be interpreted geometrically as the projective representation of the dual of the absolute conic, the *absolute dual quadric* (Triggs, 1997). Like the absolute conic, this quadric surface is invariant through similarities, and the (dual) conic section associated with $\mathcal{K}_i\mathcal{K}_i^T$ is simply the projection of this quadric surface into the corresponding image. Self-calibration is the process of computing the intrinsic parameters of a camera from point correspondences with unknown Euclidean positions. Work in this area was pioneered by Faugeras and Maybank (1992) for cameras with fixed intrinsic parameters. A number of reliable self-calibration methods are now available (Hartley, 1994*a*, Fitzgibbon and Zisserman, 1998, Pollefeys *et al.*, 1999), and they can also be used to upgrade projective reconstructions to Euclidean ones. The problem of computing Euclidean upgrades of projective reconstructions under minimal camera constraints such as zero skew is addressed in Heyden and Åström (1998, 1999), Pollefeys *et al.* (1999), and Ponce (2000).

PROBLEMS

13.1. Use a simple counting argument to determine the minimum number of point correspondences required to solve the projective structure-from-motion problem in the trinocular case.

13.2. Show that the change of coordinates between two projective frames (A) and (B) can be represented by Eq. (13.2).

13.3. Show that any two distinct lines in a projective plane intersect in exactly one point and that two parallel lines Δ and Δ' in an affine plane intersect at the point at infinity associated with their common direction $\boldsymbol{v}$ in the projective completion of this plane.

Hint: Use J_A to embed the affine plane in its projective closure, and write the vector of $\Pi \times \mathbb{R}$ associated with any point in $J_A(\Delta)$ (resp. $J_A(\Delta')$) as a linear combination of the vectors $(\overrightarrow{AB}, 1)$ and $(\overrightarrow{AB} + \boldsymbol{v}, 1)$ (resp. $(\overrightarrow{AB'}, 1)$ and $(\overrightarrow{AB'} + \boldsymbol{v}, 1)$), where B and B' are arbitrary points on Δ and Δ'.

13.4. Show that a perspective projection between two planes of $\mathbb{P}^3$ is a projective transformation.

13.5. Given an affine space X and an affine frame $(A_0, \ldots, A_n)$ for that space, what is the projective basis of $\tilde{X}$ associated with the vectors $\boldsymbol{e}_i \stackrel{\text{def}}{=} (\overrightarrow{A_0A_i}, 0)$ $(i = 1, \ldots, n)$ and the vector $\boldsymbol{e}_{n+1} = (\boldsymbol{0}, 1)$? Are the points $J_{A_0}(A_i)$ part of that basis?

13.6. In this exercise, you will show that the cross-ratio of four collinear points A, B, C, and D is equal to

$$\{A, B; C, D\} = \frac{\sin(\alpha + \beta)\sin(\beta + \gamma)}{\sin(\alpha + \beta + \gamma)\sin\beta},$$

where the angles α, β, and γ are defined as in Figure 13.2.

(a) Show that the area of a triangle PQR is

$$A(P, Q, R) = \frac{1}{2} PQ \times RH = \frac{1}{2} PQ \times PR \sin\theta,$$

where PQ denotes the distance between the two points P and Q, H is the projection of R onto the line passing through P and Q, and θ is the angle between the lines joining the point P to the points Q and R.

(b) Define the ratio of three collinear points A, B, C as

$$R(A, B, C) = \frac{\overline{AB}}{\overline{BC}}$$

for some orientation of the line supporting the three points. Show that

$$R(A, B, C) = A(A, B, O)/A(B, C, O),$$

where O is some point not lying on this line.

(c) Conclude that the cross-ratio $\{A, B; C, D\}$ is indeed given by the formula above.

13.7. Show that the homography between two epipolar pencils of lines can be written as

$$\tau \to \tau' = \frac{a\tau + b}{c\tau + d},$$

where τ and τ' are the slopes of the lines.

13.8. Here we revisit the three-point reconstruction problem in the context of the *homogeneous* coordinates of the point D in the projective basis formed by the tetrahedron (A, B, C, O') and the unit point O''. Note that the ordering of the reference points, and thus the ordering of the coordinates, is different from the one used earlier: This new choice is, like the previous one, made to facilitate the reconstruction.

We denote the (unknown) coordinates of the point D by (x, y, z, w), equip the first (resp. second) image plane with the triangle of reference a', b', c' (resp. a'', b'', c'') and the unit point e' (resp. e''), and denote by (x', y', z') (resp. (x'', y'', z'')) the coordinates of the point d' (resp. d'').

Hint: Drawing a diagram similar to Figure 13.3 helps.

(a) What are the homogeneous projective coordinates of the points D', D'', and E where the lines $O'D$, $O''D$, and $O'O''$ intersect the plane of the triangle?

(b) Write the coordinates of D as a function of the coordinates of O' and D' (resp. O'' and D'') and some unknown parameters.

Hint: Use the fact that the points D, O', and D' are collinear.

(c) Give a method for computing these unknown parameters and the coordinates of D.

13.9. Show that if $\tilde{\mathcal{M}} = (\mathcal{A} \quad \boldsymbol{b})$ and $\tilde{\mathcal{M}}' = (\text{Id} \quad \mathbf{0})$ are two projection matrices, and if $\mathcal{F}$ denotes the corresponding fundamental matrix, then $[\boldsymbol{b}_\times]\mathcal{A}$ is proportional to $\mathcal{F}$ whenever $\mathcal{F}^T\boldsymbol{b} = 0$ and

$$\mathcal{A} = -\lambda[\boldsymbol{b}_\times]\mathcal{F} + (\ \mu\boldsymbol{b} \mid \nu\boldsymbol{b} \mid \tau\boldsymbol{b}\ ).$$

13.10. We derive in this exercise a method for computing a minimal parameterization of the fundamental matrix and estimating the corresponding projection matrices. This is similar in spirit to the technique presented in Section 12.2.2 of Chapter 12 in the affine case.

(a) Show that two projection matrices $\mathcal{M}$ and $\mathcal{M}'$ can always be reduced to the following canonical forms by an appropriate projective transformation:

$$\tilde{\mathcal{M}} = \begin{pmatrix} 1 & 0 & 0 & 0 \\ 0 & 1 & 0 & 0 \\ 0 & 0 & 1 & 0 \end{pmatrix} \quad \text{and} \quad \tilde{\mathcal{M}}' = \begin{pmatrix} \boldsymbol{a}_1^T & b_1 \\ \boldsymbol{a}_2^T & b_2 \\ \mathbf{0}^T & 1 \end{pmatrix}.$$

Note: For simplicity, you can assume that all the matrices involved in your solution are nonsingular.

(b) Note that applying this transformation to the projection matrices amounts to applying the inverse transformation to every scene point P. Let us denote by $\tilde{\boldsymbol{P}} = (x, y, z)^T$ the position of the transformed point $\tilde{P}$ in the world coordinate system and by $\boldsymbol{p} = (u, v, 1)^T$ and $\boldsymbol{p}' = (u', v', 1)^T$ the homogeneous coordinate vectors of its images. Show that

$$(u' - b_1)(\boldsymbol{a}_2 \cdot \boldsymbol{p}) = (v' - b_2)(\boldsymbol{a}_1 \cdot \boldsymbol{p}).$$

(c) Derive from this equation an eight-parameter parameterization of the fundamental matrix, and use the fact that $\mathcal{F}$ is only defined up to a scale factor to construct a minimal seven-parameter parameterization.

(d) Use this parameterization to derive an algorithm for estimating $\mathcal{F}$ from at least seven point correspondences and for estimating the projective shape of the scene.

13.11. Here we address the problem of recovering the rotation $\mathcal{R}$ and translation $\boldsymbol{t}$ associated with an essential matrix $\mathcal{E} = [\boldsymbol{t}_\times]\mathcal{R}$ (this exercise is courtesy of Andrew Zisserman). The translation part is easy since $\boldsymbol{t}$ can be recovered (up to scale since we know that the structure of a scene can only be determined up to a similitude) as the unit vector satisfying $\mathcal{E}^T\boldsymbol{t}$.

(a) Show that the SVD of the essential matrix can be written as

$$\mathcal{E} = \mathcal{U}\ \text{diag}(1, 1, 0)\mathcal{V}^\top,$$

and conclude that $\boldsymbol{t}$ is the third column vector of $\mathcal{U}$.

(b) Show that the two matrices

$$\mathcal{R}_1 = \mathcal{U}\mathcal{W}\mathcal{V}^\top \quad \mathcal{R}_2 = \mathcal{U}\mathcal{W}^\top\mathcal{V}^\top$$

satisfy $\mathcal{E} = [\boldsymbol{t}]_\times \mathcal{R}$, where

$$\mathcal{W} = \begin{pmatrix} 0 & -1 & 0 \\ 1 & 0 & 0 \\ 0 & 0 & 1 \end{pmatrix}.$$

Programming Assignments

13.12. Implement the geometric approach to projective scene estimation introduced in Section 13.2.1.

13.13. Implement the algebraic approach to projective scene estimation introduced in Section 13.2.2.

13.14. Implement the factorization approach to projective scene estimation introduced in Section 13.4.1.

PART IV
Mid-Level Vision

14

Segmentation by Clustering

An attractive broad view of vision is that it is an inference problem: We have some measurements and a model, and we wish to determine what caused the measurement. There are crucial features that distinguish vision from many other inference problems: First, there is an awful lot of data; and second, we don't know which of these data items may help solve the inference problem and which may not. For example, one huge difficulty in building good object-recognition programs is knowing *which* pixels to recognize and which to ignore. It is difficult to tell whether a pixel lies on the surfaces in Figure 14.1 simply by looking at the pixel. This problem can be addressed by working with a compact representation of the interesting image data that emphasizes the properties that make it interesting. Obtaining such representation is known variously as *segmentation*, *grouping*, *perceptual organization*, or *fitting*. We use the term *segmentation* for a wide range of activities because, although techniques may differ, the motivation for all these activities is the same: Obtain a compact representation of what is helpful in the image. It's hard to see that there could be a comprehensive theory of segmentation, not least because what is interesting and what is not depends on the application. There is certainly no comprehensive theory of segmentation at time of writing, and the term is used in different ways in different quarters. In this chapter, we describe segmentation processes that currently have no probabilistic interpretation. In the following chapters, we deal with more complex probabilistic algorithms.

14.1 WHAT IS SEGMENTATION?

Assume that we would like to recognize objects in an image. There are too many pixels to handle each individually. Instead, we should like some form of compact, summary representation. The details of what that representation should be depend on the task, but there are a number of quite

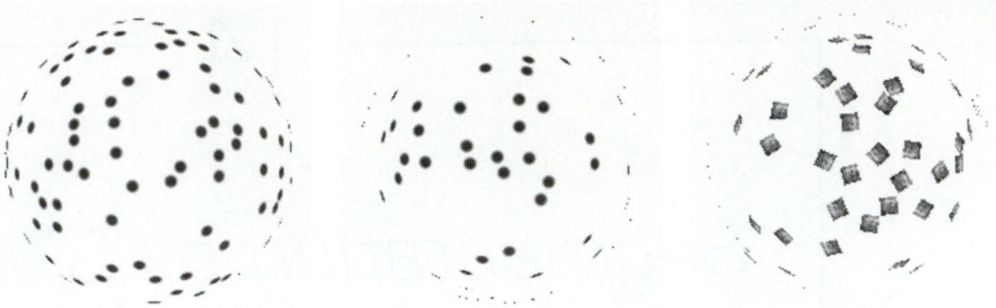

Figure 14.1 As these images suggest, an important component of vision involves organizing image information into meaningful assemblies. The human vision system seems to do so surprisingly well. In each of these three images, blobs are organized together to form textured surfaces that appear to bulge out of the page (you may feel that they are hemispheres). The blobs appear to be assembled "because they form surfaces," hardly a satisfactory explanation and one that begs difficult computational questions. Notice that saying that they are assembled because together they form the same texture also begs questions (How do we know?). In the case of the surface on the **left**, it might be quite difficult to write programs that can recognize a single coherent texture. This process of organization can be applied to many different kinds of input. *Reprinted from "Shape from texture and integrability," by D. A. Forsyth, Proc. Int. Conf. Computer Vision, 2001, © 2001, IEEE*

general desirable features. First, there should be relatively few (= not more than later algorithms can cope with) components in the representation computed for typical pictures. Second, these components should be suggestive. It should be pretty obvious from these components whether the objects we are looking for are present, again for typical pictures.

Methods deal with different kinds of data set: Some are intended for images, some are intended for video sequences, and some are intended to be applied to *tokens* —placeholders that indicate the presence of an interesting pattern in an image, say a spot, dot, or edge point. Tokens can, in fact, occur in video, too; an example might be a spot moving according to some parametric rule.

Although superficially these methods may seem quite different, there is a strong similarity among them (which is why they appear together!). Each method attempts to obtain a compact representation of its data set using some form of model of similarity (in some cases, one has to look quite hard to spot the model). These general features manifest themselves in different problems. We review a few examples:

- **Summarizing video:** Users may wish to browse large collections of video sequences. We need to supply a representation that encapsulates "what's in a sequence." One way to do this is to break each sequence into shots—subsequences that look similar—and then represent it with a montage of frames, one for each shot. This suggests segmenting the sequences into shots.
- **Finding machined parts:** Assume we wish to find a machined part in an image (a circumstance that arises far less often than one might think). Machined objects tend to contain lines—where plane faces meet—and circles—where holes have been drilled. This suggests segmenting the image into sets of lines and circles; typically, one would find edges first and then fit lines and circles to them.

- **Finding people:** Assume we wish to find people in images. This problem remains open as of writing, but the general outlines of the solution are clear. One should look for body segments first, then assemble them. These segments appear in the image as extended regions; if the people are wearing clothing that isn't textured, then they are extended regions of a single color—we must look for bars of a constant color.
- **Finding buildings in satellite images:** The vast majority of buildings are polyhedral, particularly at the scale at which they appear in satellite images. This suggests representing the image in terms of polygonal regions on some background. Typically, this is done by looking for collections of edge points that can be assembled into line segments and then assembling line segments into polygons.
- **Searching a collection of images:** For users to search a collection of images, the images must be represented in a way that both makes sense to the user and is related to the content of the picture. Since the content typically involves the objects that are present, and objects tend to have coherent color and texture, it is natural to try and break the images into regions of coherent color and texture, and use these regions as a representation.

14.1.1 Model Problems

Segmentation is a big topic. There are a variety of examples of segmentation problem, some of which don't obviously look like a segmentation problem. Natural model problems are segmentation problems, where it is valuable to know more than one method of solving them and that appear commonly in applications. We generally refer our discussion to one or another of the following problems when we can:

- **Forming image segments:** We should like to decompose an image into superpixels, image regions that have roughly coherent color and texture. Typically, the shape of these regions isn't particularly important, but the coherence is important. This process is quite widely studied—it is often referred to as the exclusive meaning of the term segmentation—and usually thought of as a first step in recognition.
- **Fitting lines to edge points:** As we saw before, there are a number of reasons it might be useful to fit a set of lines to a set of points. This problem goes from being quite simple (in the case where we know how many lines there are, and which point belongs to which line) to being remarkably subtle (in most other cases). It is a segmentation problem because we are organizing some tokens that belong together because they fit on a line. If we try and fit a line to a set of points, some of which do not lie close to any line, the resulting line can be meaningless unless we are careful. This illustrates an important, quite general, principle: Ignorance of correspondence can behave like noise. Typically, we need to estimate both the parameters of the lines and the correspondence between points and lines simultaneously.
- **Fitting a fundamental matrix to a set of feature points:** Assume we have two views of a set of feature points. It is typically difficult to be sure which points correspond to which, although we may have some cues. One important cue is that, if the correspondence is right, there is a fundamental matrix connecting the points. We should like to determine this fundamental matrix without knowing the correspondence in advance. There are several reasons this problem is worth solving. First, it is impossible to construct sensible shape representations from multiple views without a solution to this problem. Second, solutions to this problem can be used as a cue to whether a set of points is moving rigidly. If a sequence shows two moving objects, they'll have different fundamental matrices. Again, correspondence errors look like noise in this problem.

We use these problems to illustrate various segmentation algorithms, but you should keep in mind that not every technique offers a plausible solution to each of these model problems.

14.1.2 Segmentation as Clustering

One natural view of segmentation is that we are attempting to determine which components of a data set naturally belong together. This is a problem known as *clustering*; there is a wide literature. Generally, we can cluster in two ways:

- **Partitioning:** Here we have a large data set, and carve it up according to some notion of the association between items inside the set. We would like to decompose it into pieces that are good according to our model. For example, we might
  - decompose an image into regions that have coherent color and texture inside them;
  - decompose an image into extended blobs, consisting of regions that have coherent color, texture, and motion and look like limb segments;
  - take a video sequence and decompose it into *shots*—segments of video showing about the same stuff from about the same viewpoint.
- **Grouping:** Here we have a set of distinct data items, and we wish to collect sets of data items that make sense together according to our model. Effects like occlusion mean that image components that belong to the same object are often separated. Examples of grouping include the following:
  - collecting together tokens that, taken together, form a line;
  - collecting together tokens that seem to share a fundamental matrix.

Of course, the key issue here is to determine what representation is suitable for the problem at hand. Occasionally this is pretty obvious; more often, the question is subtle. Typically, one has to know what various methods do and make an informed choice. Even so, we need to know by what criteria a segmentation method should decide which pixels (or tokens) belong together and which do not. A fruitful source of insight is the human vision system, which has to solve a completely general form of this problem and, remarkably, displays strong and easily evoked preferences for how tokens are grouped.

14.2 HUMAN VISION: GROUPING AND GESTALT

A key feature of the human vision system is that context affects how things are perceived (e.g., see the illusion of Figure 14.2). This observation led the Gestalt school of psychologists to reject the study of responses to stimuli and to emphasize grouping as the key to understanding visual perception. To them, grouping meant the tendency of the visual system to assemble some components of a picture together and to perceive them together (this supplies a rather rough meaning to the word context used above). Grouping, for example, is what causes the Müller–Lyer illusion of Figure 14.2—the vision system assembles the components of the two arrows, and the horizontal lines look different from one another because they are peceived as components of a whole, rather than as lines. Furthermore, many grouping effects can't be disrupted by cognitive input; for example, you can't make the lines in Figure 14.2 look equal in length by deciding not to group the arrows.

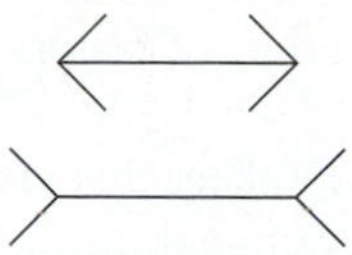

Figure 14.2 The famous Muller–Lyer illusion; the horizontal lines are in fact the same length, although that belonging to the upper figure looks longer. Clearly, this effect arises from some property of the relationships that form the whole (the *gestaltqualität*), rather than from properties of each separate segment.

A common experience of segmentation is the way that an image can resolve itself into a **figure**—typically, the significant, important object—and a **ground**—the background on which the figure lies. However, as Figure 14.3 illustrates, what is figure and what is ground can be profoundly ambiguous, meaning that a richer theory is required.

The Gestalt school used the notion of a *gestalt*—a whole or a group—and of its **gestalt-qualität**—the set of internal relationships that makes it a whole (e.g., Figure 14.2) as central components in their ideas. Their work was characterized by attempts to write down a series of rules by which image elements would be associated together and interpreted as a group. There were also attempts to construct algorithms, which are of purely historical interest (see Gordon, 1997 for an introductory account that places their work in a broad context).

The Gestalt psychologists identified a series of factors, which they felt predisposed a set of elements to be grouped. These factors are important because it is quite clear that the human vision system uses them in some way. Furthermore, it is reasonable to expect that they represent a set of preferences about when tokens belong together that lead to a useful intermediate representation.

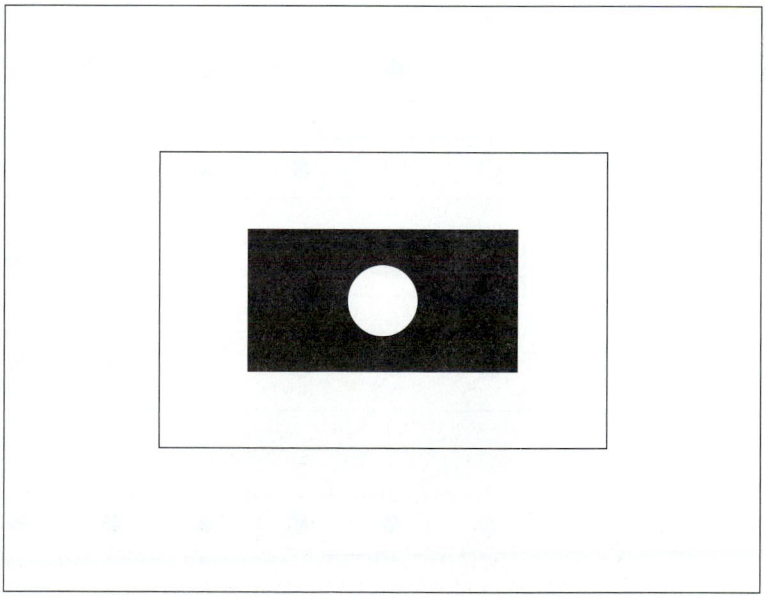

Figure 14.3 One view of segmentation is that it determines which component of the image forms the figure and which the ground. The figure illustrates one form of ambiguity that results from this view; the white circle can be seen as figure on the black rectangular ground, or as ground where the figure is a black rectangle with a circular hole in it—the ground is then a white square.

There are a variety of factors, some of which postdate the main Gestalt movement:

- **Proximity:** Tokens that are nearby tend to be grouped.
- **Similarity:** Similar tokens tend to be grouped together.
- **Common fate:** Tokens that have coherent motion tend to be grouped together.
- **Common region:** Tokens that lie inside the same closed region tend to be grouped together.
- **Parallelism:** Parallel curves or tokens tend to be grouped together.
- **Closure:** Tokens or curves that tend to lead to closed curves tend to be grouped together.
- **Symmetry:** Curves that lead to symmetric groups are grouped together.
- **Continuity:** Tokens that lead to continuous—as in joining up nicely, rather than in the formal sense—curves tend to be grouped.
- **Familiar configuration:** Tokens that, when grouped, lead to a familiar object tend to be grouped together.

These laws are illustrated in Figures 14.4, 14.5, 14.7, and 14.1.

These rules can function fairly well as explanations, but they are insufficiently crisp to be regarded as forming an algorithm. The Gestalt psychologists had serious difficulty with the details, such as when one rule applied and when another. It is difficult to supply a satisfactory algorithm for using these rules—the Gestalt movement attempted to use an extremality principle.

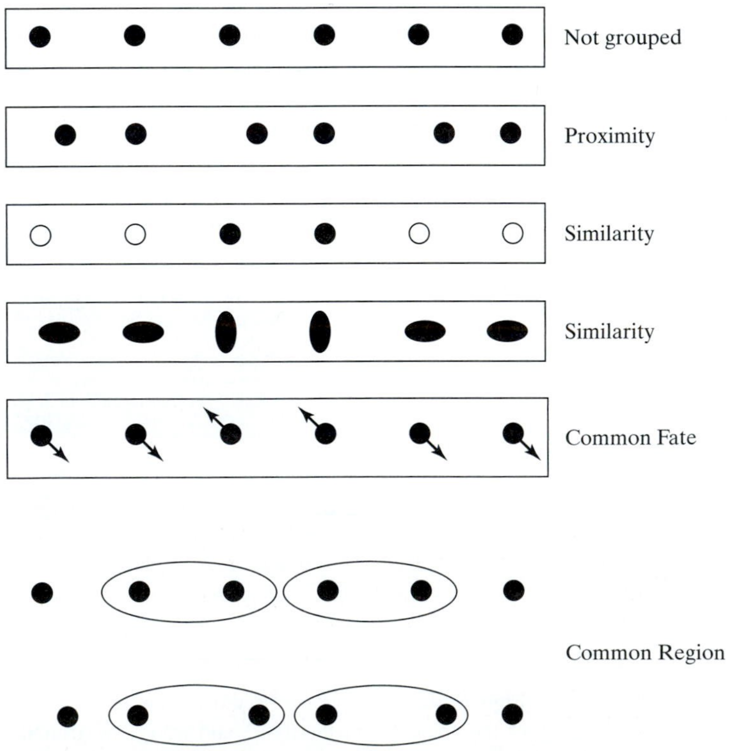

Figure 14.4 Examples of Gestalt factors that lead to grouping (which are described in greater detail in the text).

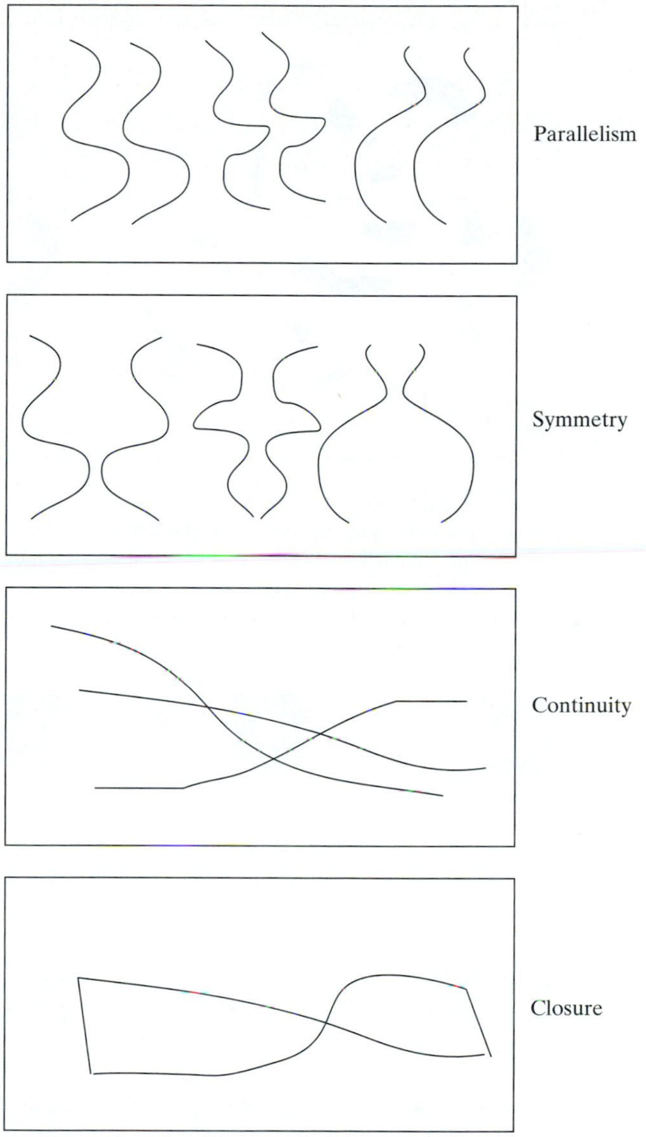

Figure 14.5 Examples of Gestalt factors that lead to grouping (which are described in greater detail in the text).

Familiar configuration is a particular problem. The key issue is to understand just *what* familiar configuration applies in a problem and how it is selected. For example, look at Figure 14.1. One might argue that the blobs are grouped because they yield a sphere. The difficulty with this view is explaining how this occurred—where did the hypothesis that a sphere is present come from? A search through all views of all objects is one explanation, but one must then explain how this search is organized. Do we check *every view* of *every* sphere with *every* pattern of spots? How can this be done efficiently?

The Gestalt rules do offer some insight because they explain what happens in various examples. These explanations seem to be sensible because they suggest that the rules help solve problems posed by visual effects that arise commonly in the real world—that is, they are *ecolog-*

Figure 14.6 Occlusion appears to be an important cue in grouping. It may be possible to see the pattern on the **left** as a collection of digits; that on the **right** is quite clearly some occluded digits. The black regions on the left and right are the same. The visual system appears to be helped by evidence that separated tokens are separated for a reason, rather than just scattered.

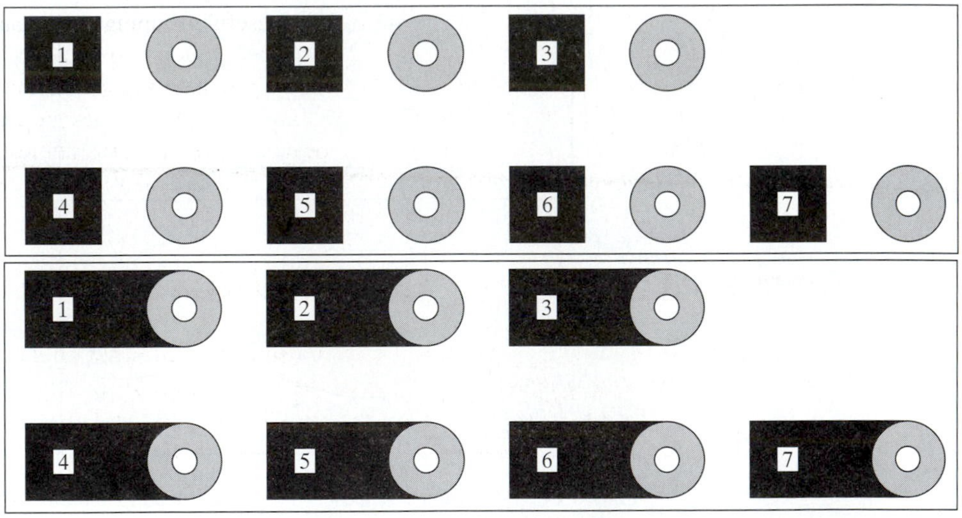

Figure 14.7 An example of grouping phenomena in real life. The buttons on an elevator in the computer science building at U.C. Berkeley used to be laid out as in the **top** figure. It was common to arrive at the wrong floor and discover that this was because you'd pressed the wrong button—the buttons are difficult to group unambiguously with the correct label, and it is easy to get the wrong grouping at a quick glance. A public-spirited individual filled in the gap between the numbers and the buttons, as in the **bottom** figure, and the confusion stopped because the proximity cue had been disambiguated.

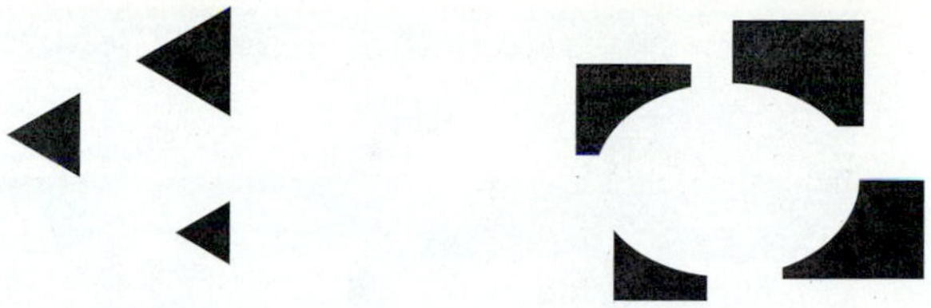

Figure 14.8 The tokens in these images suggest the presence of occluding objects whose boundaries don't contrast with much of the image. Notice that one has a clear impression of the position of the entire contour of the occluding figures. These contours are known as *illusory contours*.

ically valid. For example, continuity may represent a solution to problems posed by occlusion—sections of the contour of an occluded object could be joined up by continuity (see Figure 14.6).

This tendency to prefer interpretations that are explained by occlusion leads to interesting effects. One is the *illusory contour*, illustrated in Figure 14.8. Here a set of tokens suggests the presence of an object most of whose contour has no contrast. The tokens appear to be grouped together because they provide a cue to the presence of an occluding object, which is so strongly suggested by these tokens that one could fill in the no-contrast regions of contour.

This ecological argument has some force because it is possible to interpret most grouping factors using it. Common fate can be seen as a consequence of the fact that components of objects tend to move together. Equally, symmetry is a useful grouping cue because there are a lot of real objects that have symmetric or close to symmetric contours. Essentially, the ecological argument says that tokens are grouped because doing so produces representations that are helpful for the visual world that people encounter. The ecological argument has an appealing, although vague, statistical flavor. From our perspective, Gestalt factors provide interesting hints, but should be seen as the *consequences* of a larger grouping process, rather than the process itself.

14.3 APPLICATIONS: SHOT BOUNDARY DETECTION AND BACKGROUND SUBTRACTION

Simple segmentation algorithms are often useful in significant applications. Generally, simple algorithms work best when it is easy to tell what a useful decomposition is. Two important cases are *background subtraction*—where anything that doesn't look like a known background is interesting—and *shot boundary detection*—where substantial changes in a video are interesting.

14.3.1 Background Subtraction

In many applications, objects appear on a largely stable background. The standard example is detecting parts on a conveyor belt. Another example is counting motor cars in an overhead view of a road—the road is pretty stable in appearance. Another, less obvious, example is in human–computer interaction. Quite commonly, a camera is fixed (say, on top of a monitor) and views a room. Pretty much anything in the view that doesn't look like the room is interesting.

In these kinds of applications, a useful segmentation can often be obtained by subtracting an estimate of the appearance of the background from the image and looking for large absolute values in the result. The main issue is obtaining a good estimate of the background. One method is simply to take a picture. This approach works rather poorly because the background typically changes slowly over time. For example, the road may get more shiny as it rains and less when the weather dries up; people may move books and furniture around in the room, and so on.

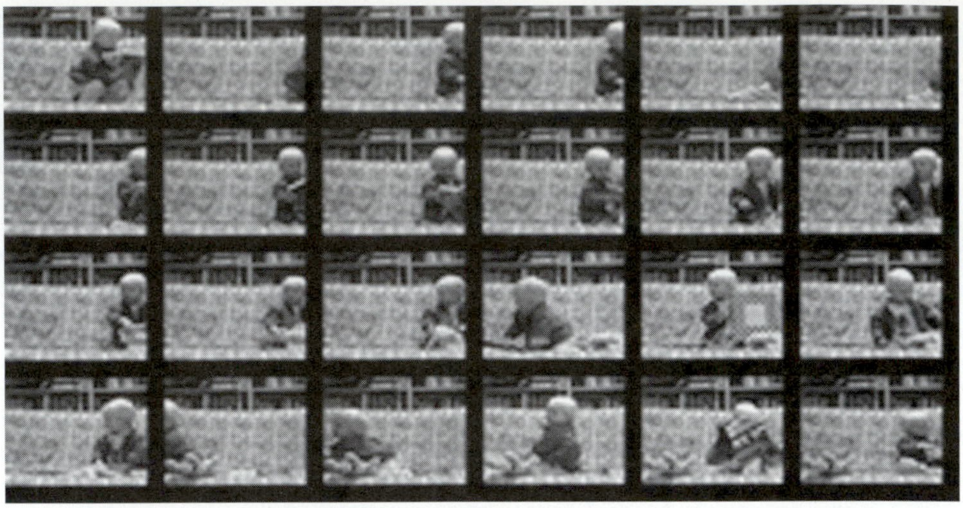

Figure 14.9 The figure shows every fifth frame from a sequence of 120 frames of a child playing on a patterned sofa. The frames are used at an 80×60 resolution, for reasons we discuss in Figure 14.11. Notice that the child moves from one side of the frame to the other during the sequence.

Algorithm 14.1: Background Subtraction

Form a background estimate $\mathcal{B}^{(0)}$. At each frame $\mathcal{F}$

Update the background estimate, typically by forming $\mathcal{B}^{(n+1)} = \frac{w_a \mathcal{F} + \sum_i w_i \mathcal{B}^{(n-i)}}{w_c}$ for a choice of weights w_a, w_i and w_c.

Subtract the background estimate from the frame, and report the value of each pixel where the magnitude of the difference is greater than some threshold.

end

An alternative that usually works quite well is to estimate the value of background pixels using a **moving average**. In this approach, we estimate the value of a particular background pixel as a weighted average of the previous values. Typically, pixels in the distant past should be weighted at zero, and the weights increase smoothly. Ideally, the moving average should track the changes in the background, meaning that if the weather changes quickly (or the book mover is frenetic) relatively few pixels should have nonzero weights, and if changes are slow the number of past pixels with nonzero weights should increase. This yields Algorithm 14.1. For those who have read the filters chapter, this is a filter that smooths a function of time, and we would like it to suppress frequencies that are larger than the typical frequency of change in the background and pass those that are at or below that frequency. The approach can be quite successful, but needs to be used on quite coarse scale images as Figures 14.10 and 14.11 illustrate.

14.3.2 Shot Boundary Detection

Long sequences of video are composed of *shots*—much shorter subsequences that show largely the same objects. These shots are typically the product of the editing process. There is seldom

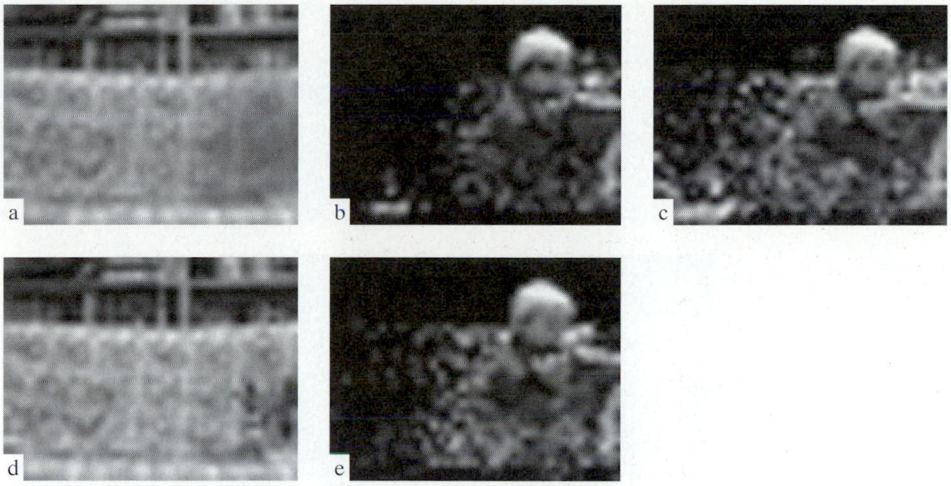

Figure 14.10 Background subtraction results for the sequence of Figure 14.9 using 80×60 frames. We compare two methods of computing the background: (**a**) The average of all 120 frames—notice that the child spent more time on one side of the sofa than the other, leading to the faint blur in the average there. (**b**) Pixels whose difference from the average exceeds a threshold. (**c**) Those whose difference from the average exceeds a somewhat smaller threshold. Notice that, in each case, there are some excess pixels and some missing pixels. (**d**) A background computed using a somewhat more sophisticated method (described briefly in Section 16.2.5). (**e**) Pixels that this method believes are different from the background. Again, notice the missing pixels.

any record of where the boundaries between shots fall. It is helpful to represent a video as a collection of shots; each shot can then be represented with a *key frame*. This representation can be used to search for videos or to encapsulate their content for a user to browse a video or a set of videos.

Finding the boundaries of these shots automatically—*shot boundary detection*—is an important practical application of simple segmentation algorithms. A shot boundary detection algorithm must find frames in the video that are significantly different from the previous frame. Our test of significance must take account of the fact that, within a given shot, both objects and the background can move around in the field of view. Typically, this test takes the form of a distance; if the distance is larger than a threshold, a shot boundary is declared (Algorithm 14.2).

Algorithm 14.2: Shot Boundary Detection Using Interframe Differences

```
For each frame in an image sequence
    Compute a distance between this frame and the previous frame
    If the distance is larger than some threshold,
      classify the frame as a shot boundary.
end
```

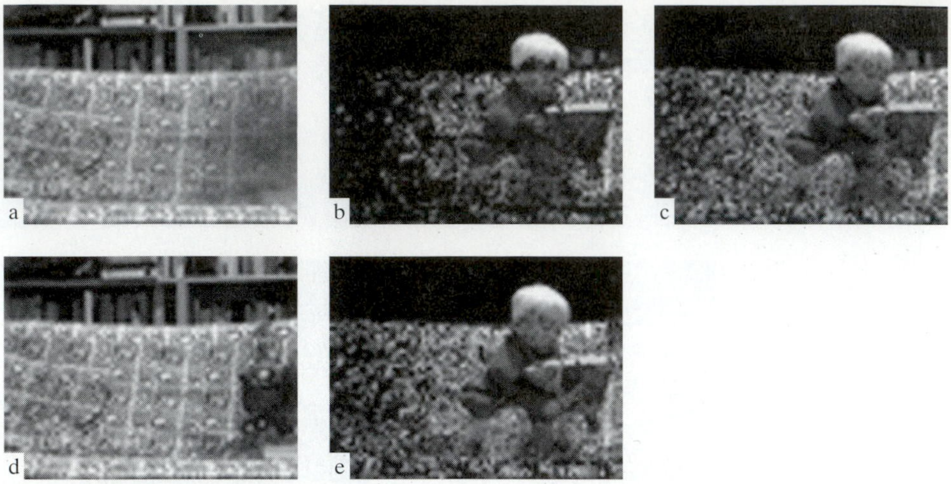

Figure 14.11 Registration can be a significant nuisance in background subtraction, particularly for textures. These figures show results for the sequence of Figure 14.9, using 160 × 120 frames. We compare two methods of computing the background: (**a**) The average of all 120 frames—notice that the child spent more time on one side of the sofa than the other, leading the a faint blur in the average there. (**b**) Pixels whose difference from the average exceeds a threshold. (**c**) Those whose difference from the average exceeds a somewhat smaller threshold. (**d**) A background computed using a somewhat more sophisticated method (described briefly in Section 16.2.5). (**e**) Pixels that this method believes are different from the background. Notice that the number of problem pixels—where the pattern on the sofa has been mistaken for the child—has markedly increased. This is because small movements can cause the high spatial frequency pattern on the sofa to be misaligned, leading to large differences.

There are a variety of standard techniques for computing a distance:

- **Frame differencing** algorithms take pixel-by-pixei differences between each two frames in a sequence and sum the squares of the differences. These algorithms are unpopular, because they are slow—there are many differences—and because they tend to find many shots when the camera is shaking.
- **Histogram-based** algorithms compute color histograms for each frame and compute a distance between the histograms. A difference in color histograms is a sensible measure to use because it is insensitive to the spatial arrangement of colors in the frame (e.g., small camera jitters will not affect the histogram).
- **Block comparison** algorithms compare frames by cutting them into a grid of boxes and comparing the boxes. This is to avoid the difficulty with color histograms, where a red object disappearing off-screen in the bottom left corner is equivalent to a red object appearing on screen from the top edge. Typically, these block comparison algorithms compute an interframe distance that is a composite—taking the maximum is one natural strategy—of interblock distances, each computed using methods like those used for interframe distances.
- **Edge differencing** algorithms compute edge maps for each frame, and then compare these edge maps. Typically, the comparison is obtained by counting the number of potentially corresponding edges (nearby, similar orientation, etc.) in the next frame. If there

are few potentially corresponding edges, there is a shot boundary. A distance can be obtained by transforming the number of corresponding edges.

These are relatively *ad hoc* methods, but are often sufficient to solve the problem at hand.

14.4 IMAGE SEGMENTATION BY CLUSTERING PIXELS

Clustering is a process whereby a data set is replaced by **clusters**, which are collections of data points that belong together. It is natural to think of image segmentation as clustering; we would like to represent an image in terms of clusters of pixels that belong together. The specific criterion to be used depends on the application. Pixels may belong together because they have the same color, they have the same texture, they are nearby, and so on.

14.4.1 Segmentation Using Simple Clustering Methods

It is relatively easy to take a clustering method and build an image segmenter from it. Much of the literature on image segmentation consists of papers that are, in essence, papers about clustering (although this isn't always acknowledged).

Simple Clustering Methods There are two natural algorithms for clustering. In **divisive clustering**, the entire data set is regarded as a cluster, and then clusters are recursively split to yield a good clustering (Algorithm 14.4). In **agglomerative clustering**, each data item is regarded as a cluster, and clusters are recursively merged to yield a good clustering (Algorithm 14.3).

There are two major issues in thinking about clustering:

- *What is a good inter-cluster distance?* Agglomerative clustering uses an intercluster distance to fuse nearby clusters; divisive clustering uses it to split insufficiently coherent clusters. Even if a natural distance between data points is available (which may not be the case for vision problems), there is no canonical intercluster distance. Generally,

Algorithm 14.3: Agglomerative Clustering or Clustering by Merging

```
Make each point a separate cluster
Until the clustering is satisfactory
    Merge the two clusters with the smallest inter-cluster distance
end
```

Algorithm 14.4: Divisive Clustering, or Clustering by Splitting

```
Construct a single cluster containing all points
Until the clustering is satisfactory
    Split the cluster that yields the two components with the largest inter-cluster distance
end
```

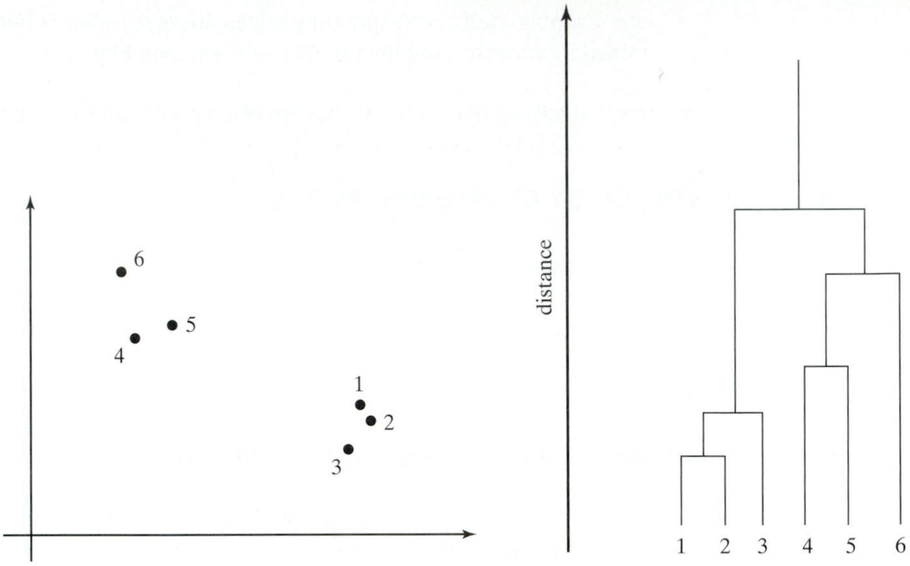

Figure 14.12 **Left**, a data set; **right**, a dendrogram obtained by agglomerative clustering using single-link clustering. If one selects a particular value of distance, then a horizontal line at that distance splits the dendrogram into clusters. This representation makes it possible to guess how many clusters there are and to get some insight into how good the clusters are.

one chooses a distance that seems appropriate for the data set. For example, one might choose the distance between the closest elements as the intercluster distance—this tends to yield extended clusters (statisticians call this method *single-link clustering*). Another natural choice is the maximum distance between an element of the first cluster and one of the second—this tends to yield rounded clusters (statisticians call this method *complete-link clustering*). Finally, one could use an average of distances between elements in the clusters—this also tends to yield "rounded" clusters (statisticians call this method *group average clustering*).

- *How many clusters are there?* This is an intrinsically difficult task if there is no model for the process that generated the clusters. The algorithms we have described generate a hierarchy of clusters. Usually, this hierarchy is displayed to a user in the form of a *dendrogram*—a representation of the structure of the hierarchy of clusters that displays intercluster distances—and an appropriate choice of clusters is made from the dendrogram (see the example in Figure 14.12).

Building Segmenters Using Clustering Methods The distance used depends entirely on the application, but measures of color difference and of texture difference are commonly used as clustering distances. It is often desirable to have clusters that are blobby; this can be achieved by using difference in position in the clustering distance.

The main difficulty in using either agglomerative or divisive clustering methods directly is that there are an awful lot of pixels in an image. There is no reasonable prospect of examining a dendrogram because the quantity of data means that it will be too big. In practice, this means that the segmenters decide when to stop splitting or merging by using a set of threshold tests—for example, an agglomerative segmenter may stop merging when the distance between clusters is sufficiently low or when the number of clusters reaches some value.

Another difficulty created by the number of pixels is that it is impractical to look for the best split of a cluster (for a divisive method) or the best merge (for an agglomerative method). **Divisive methods** are usually modified by using some form of summary of a cluster to suggest a good split. A natural summary to use is a histogram of pixel colors (or gray levels).

Agglomerative methods also need to be modified. First, the number of pixels means that one needs to be careful about the intercluster distance (the distance between cluster centers of gravity is often used). Second, it is usual to merge only clusters with shared boundaries (we probably don't wish to represent the U.S. flag as three clusters—one red, one white, and one blue). Finally, it can be useful to merge regions simply by scanning the image and merging all pairs whose distance falls below a threshold, rather than searching for the closest pair.

14.4.2 Clustering and Segmentation by K-means

Simple clustering methods use greedy interactions with existing clusters to come up with a good overall representation. For example, in agglomerative clustering, we repeatedly make the best available merge. However, the methods are not explicit about the objective function that the methods are attempting to optimize. An alternative approach is to write down an objective function that expresses how good a representation is and then build an algorithm for obtaining the best representation.

A natural objective function can be obtained by assuming that we know there are k clusters, where k is known. Each cluster is assumed to have a center; we write the center of the ith cluster as $\boldsymbol{c}_i$. The jth element to be clustered is described by a feature vector $\boldsymbol{x}_j$. For example, if we were segmenting scattered points, then $\boldsymbol{x}$ would be the coordinates of the points; if we were segmenting an intensity image, $\boldsymbol{x}$ might be the intensity at a pixel.

We now assume that elements are close to the center of their cluster, yielding the objective function

$$\Phi(\text{clusters, data}) = \sum_{i \in \text{clusters}} \left\{ \sum_{j \in i\text{thcluster}} (\boldsymbol{x}_j - \boldsymbol{c}_i)^T (\boldsymbol{x}_j - \boldsymbol{c}_i) \right\}.$$

Notice that if the allocation of points to clusters is known, it is easy to compute the best center for each cluster. However, there are far too many possible allocations of points to clusters to search this space for a minimum. Instead, we define an algorithm that iterates through two activities:

- Assume the cluster centers are known and allocate each point to the closest cluster center.
- Assume the allocation is known and choose a new set of cluster centers. Each center is the mean of the points allocated to that cluster.

We then choose a start point by randomly choosing cluster centers and then iterate these stages alternately. This process eventually converges to a local minimum of the objective function (the value either goes down or is fixed at each step and it is bounded below). It is not guaranteed to converge to the global minimum of the objective function, however. It is also not guaranteed to produce k clusters unless we modify the allocation phase to ensure that each cluster has some nonzero number of points. This algorithm is usually referred to as *k-means*. It is possible to search for an appropriate number of clusters by applying k-means for different values of k and comparing the results; we defer a discussion of this issue until Section 16.3.

One difficulty with using this approach for segmenting images is that segments are not connected and can be very widely scattered (Figures 14.13 and 14.14). This effect can be reduced

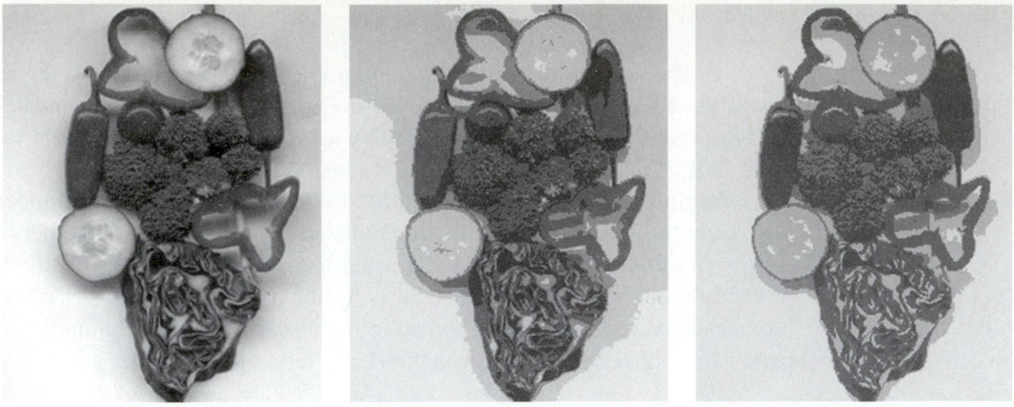

Figure 14.13 On the **left**, an image of mixed vegetables, which is segmented using k-means to produce the images at **center** and on the **right**. We have replaced each pixel with the mean value of its cluster; the result is somewhat like an adaptive requantization as one would expect. In the center, a segmentation obtained using only the intensity information. At the right, a segmentation obtained using color information. Each segmentation assumes five clusters.

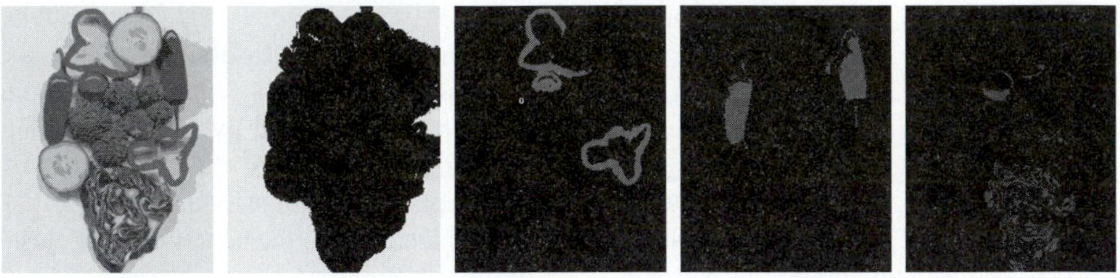

Figure 14.14 Here we show the image of vegetables segmented with k-means, assuming a set of 11 components. The **left** figure shows all segments shown together, with the mean value in place of the original image values. The other figures show four of the segments. Note that this approach leads to a set of segments that are not necessarily connected. For this image, some segments are actually quite closely associated with objects, but one segment may represent many objects (the peppers); others are largely meaningless. The absence of a texture measure creates serious difficulties as the many different segments resulting from the slice of red cabbage indicate.

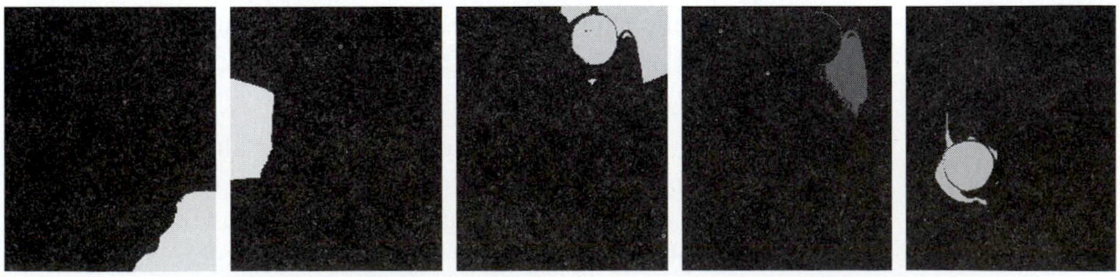

Figure 14.15 Five of the segments obtained by segmenting the image of vegetables with a k-means segmenter that uses position as part of the feature vector describing a pixel, now using 20 segments rather than 11. Note that the large background regions that should be coherent have been broken up because points got too far from the center. The individual peppers are now better separated, but the red cabbage is still broken up because there is no texture measure.

Algorithm 14.5: Clustering by K-Means

Choose k data points to act as cluster centers
Until the cluster centers are unchanged
  Allocate each data point to cluster whose center is nearest
  Now ensure that every cluster has at least one data point; possible techniques for doing this include supplying empty clusters with a point chosen at random from points far from their cluster center.
  Replace the cluster centers with the mean of the elements in their clusters.
end

by using pixel coordinates as features—an approach that results in large regions being broken up (Figure 14.15).

14.5 SEGMENTATION BY GRAPH-THEORETIC CLUSTERING

Clustering can be seen as a problem of cutting graphs into good pieces. In effect, we associate each data item with a vertex in a weighted graph, where the weights on the edges between elements are large if the elements are similar and small if they are not. We then attempt to cut the graph into connected components with relatively large interior weights (which components correspond to clusters) by cutting edges with relatively low weights. This view leads to a series of different, quite successful segmentation algorithms.

14.5.1 Terminology for Graphs

We review terminology here very briefly, as it's quite easy to forget.

- A *graph* is a set of vertices V and edges E that connect various pairs of vertices. A graph can be written $G = \{V, E\}$. Each edge can be represented by a pair of vertices—that is, $E \subset V \times V$. Graphs are often drawn as a set of points with curves connecting the points.
- A *directed graph* is one in which edges (a, b) and (b, a) are distinct; such a graph is drawn with arrowheads indicating which direction is intended.
- An *undirected graph* is one in which no distinction is drawn between edges (a, b) and (b, a).
- A *weighted graph* is one in which a weight is associated with each edge.
- A *self-loop* is an edge that has the same vertex at each end; self-loops don't occur in practice in our applications.
- Two vertices are said to be *connected* if there is a sequence of edges starting at the one and ending at the other; if the graph is directed, then the arrows in this sequence must point the right way.
- A *connected graph* is one where every pair of vertices is connected.
- Every graph consists of a disjoint set of *connected components*—that is, $G = \{V_1 \cup V_2 \ldots V_n, E_1 \cup E_2 \ldots E_n\}$, where $\{V_i, E_i\}$ are all connected graphs and there is no edge in E that connects an element of V_i with one of V_j for $i \neq j$.

14.5.2 The Overall Approach

It is useful to understand that a weighted graph can be represented by a square matrix (Figure 14.16). There is a row and a column for each vertex. The i, jth element of the matrix represents the weight on the edge from vertex i to vertex j; for an undirected graph, we use a symmetric matrix and place half the weight in each of the i, jth and j, ith elements.

The application of graphs to clustering is this: Take each element of the collection to be clustered and associate it with a vertex on a graph. Now construct an edge from every element to every other, and associate with this edge a weight representing the extent to which the elements are similar. Now cut edges in the graph to form a good set of connected components—ideally, the within-component edges are large compared with the across-component edges. Each component is a cluster. For example, Figure 14.17 shows a set of well separated points and the weight matrix (i.e., undirected weighted graph, just drawn differently) that results from a particular similarity measure; a desirable algorithm would notice that this matrix looks a lot like a block diagonal matrix—because intercluster similarities are strong and intracluster similarities are weak—and

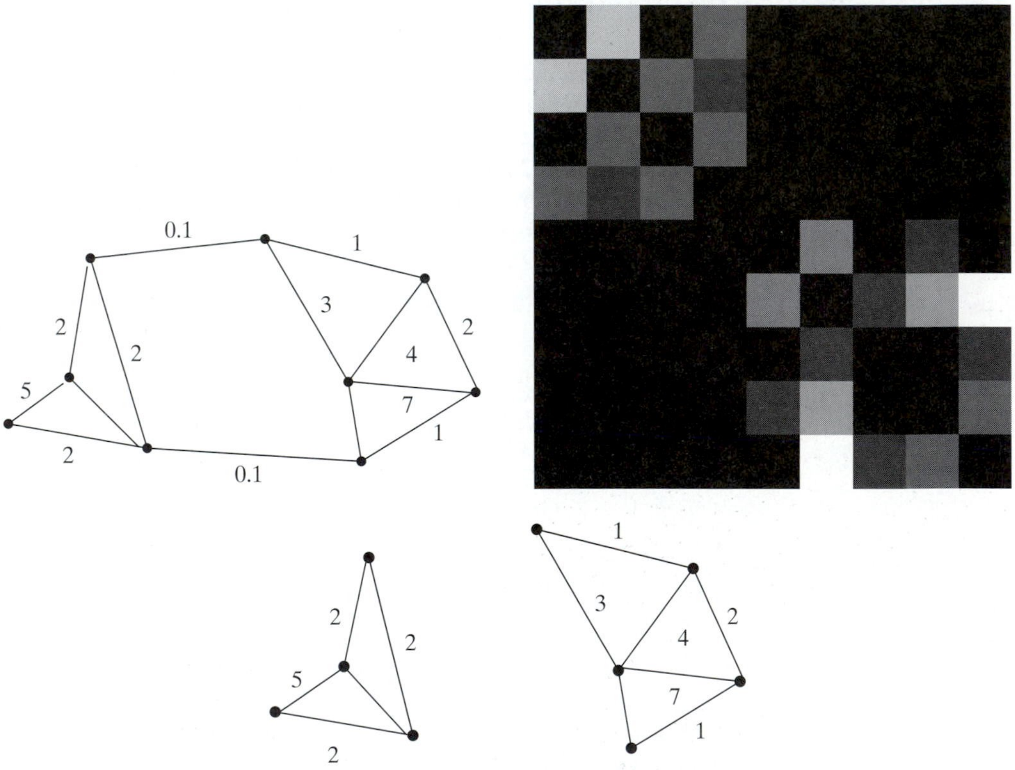

Figure 14.16 On the **top left**, a drawing of an undirected weighted graph; on the **top right**, the weight matrix associated with that graph. Larger values are lighter. By associating the vertices with rows (and columns) in a different order, the matrix can be shuffled. We have chosen the ordering to show the matrix in a form that emphasizes it it is largely block-diagonal. The figure on the **bottom** shows a cut of that graph that decomposes the graph into two tightly linked components. This cut decomposes the graph's matrix into the two main blocks on the diagonal.

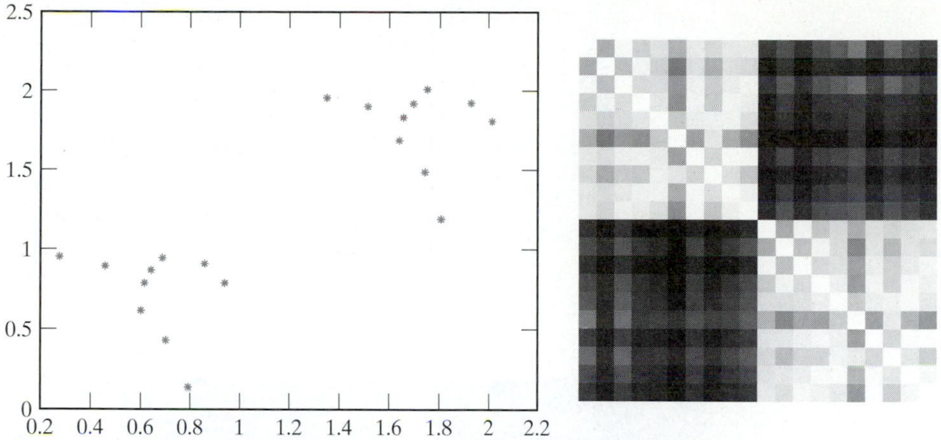

Figure 14.17 On the left, a set of points on the plane. On the right, the affinity matrix for these points computed using a decaying exponential in distance (Section 14.5.3), where large values are light and small values are dark. Notice the near block diagonal structure of this matrix; there are two off-diagonal blocks that contain terms that are close to zero. The blocks correspond to links internal to the two obvious clusters, and the off-diagonal blocks correspond to links between these clusters.

split it into two matrices, each of which is a block. The issues to study are the criteria that lead to good connected components and the algorithms for forming these connected components.

14.5.3 Affinity Measures

When we viewed segmentation as simple clustering, we needed to supply some measure of how similar clusters were. The current model of segmentation simply requires us a weight to place on each edge of the graph; these weights are usually called *affinity measures* in the literature. Clearly, the affinity measure depends on the problem at hand. The weight of an arc connecting similar nodes should be large, and the weight on an arc connecting different nodes should be small.

Affinity by Distance Affinity should go down quite sharply with distance once the distance is over some threshold. One appropriate expression has the form

$$\text{aff}(\boldsymbol{x}, \boldsymbol{y}) = \exp\left\{-\left((\boldsymbol{x}-\boldsymbol{y})^t(\boldsymbol{x}-\boldsymbol{y})/2\sigma_d^2\right)\right\}$$

where σ_d is a parameter that is large if quite distant points should be grouped and small if only nearby points should be grouped (this is the expression used for Figure 14.17; notice that the choice of scale has significant effects, illustrated in Figure 14.18).

Affinity by Intensity Affinity should be large for similar intensities and smaller as the difference increases. Again, an exponential form suggests itself, and we can use

$$\text{aff}(\boldsymbol{x}, \boldsymbol{y}) = \exp\left\{-\left((I(\boldsymbol{x})-I(\boldsymbol{y}))^t(I(\boldsymbol{x})-I(\boldsymbol{y}))/2\sigma_I^2\right)\right\}.$$

Affinity by Color We need a color metric to construct a meaningful color affinity function. It's a good idea to use a uniform color space, and it's a bad idea to use RGB space (for

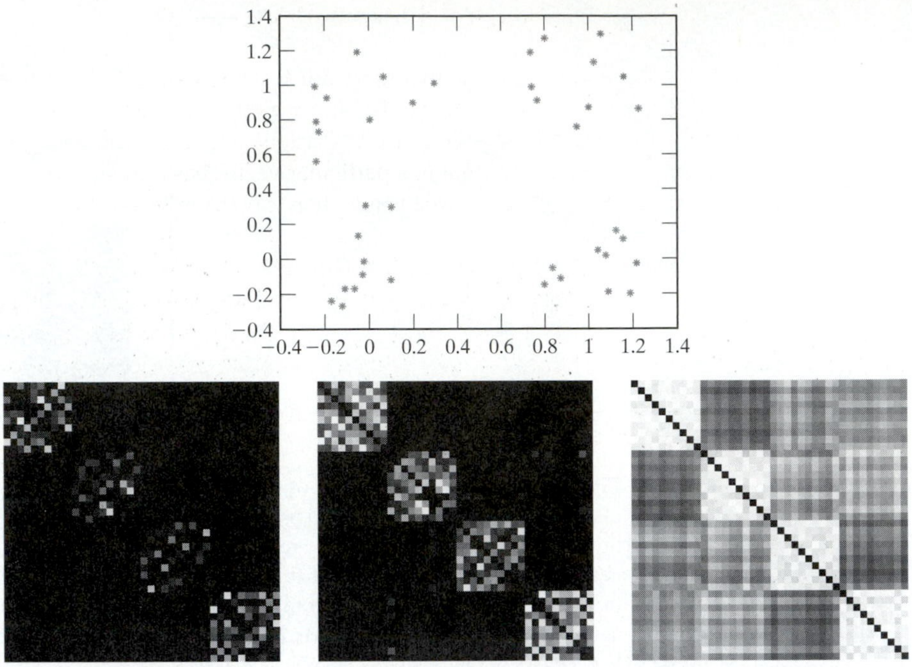

Figure 14.18 The choice of scale for the affinity affects the affinity matrix. The top row shows a dataset, which consists of four groups of 10 points drawn from a rotationally symmetric normal distribution with four different means. The standard deviation in each direction for these points is 0.2. In the second row, affinity matrices computed for this dataset using different values of σ_d. On the **left**, $\sigma_d = 0.1$, in the **center** $\sigma_d = 0.2$, and on the **right**, $\sigma_d = 1$. For the finest scale, the affinity between all points is rather small; for the next scale, there are four clear blocks in the affinity matrix; and for the coarsest scale, the number of blocks is less obvious.

reasons that should be obvious; otherwise reread Section 6.3.2). An appropriate expression has the form

$$\text{aff}(\boldsymbol{x}, \boldsymbol{y}) = \exp\left\{-\left(\text{dist}(\boldsymbol{c}(\boldsymbol{x}), \boldsymbol{c}(\boldsymbol{y}))^2/2\sigma_c^2\right)\right\},$$

where $\boldsymbol{c}_i$ is the color at pixel i.

Affinity by Texture The affinity should be large for similar textures and smaller as the difference increases. We adopt a collection of filters $f_1, \ldots, f_n$ and describe textures by the outputs of these filters, which should span a range of scales and orientations. Now for most textures, the filter outputs are not the same at each point in the texture—think of a chessboard—but a histogram of the filter outputs constructed over a reasonably sized neighborhood will be well behaved. This suggests a process where we first establish a local scale at each point—perhaps by looking at energy in coarse scale filters—and then compute a histogram of filter outputs over a region determined by that scale—perhaps a circular region centered on the point in question. We then write $\boldsymbol{h}$ for this histogram and use an exponential form:

$$\text{aff}(\boldsymbol{x}, \boldsymbol{y}) = \exp\left\{-\left((\boldsymbol{f}(\boldsymbol{x}) - \boldsymbol{f}(\boldsymbol{y}))^t(\boldsymbol{f}(\boldsymbol{x}) - \boldsymbol{f}(\boldsymbol{y}))/2\sigma_I^2\right)\right\}$$

14.5.4 Eigenvectors and Segmentation

In the first instance, assume that there are k elements and c clusters. We can represent a cluster by a vector with k components. We allow elements to be associated with clusters using some continuous weight—we need to be a bit vague about the semantics of these weights, but the intention is that if a component in a particular vector has a small value, then it is weakly associated with the cluster; if it has a large value, then it is strongly associated with a cluster.

Extracting a Single Good Cluster A good cluster is one where elements that are strongly associated with the cluster also have large values connecting one another in the affinity matrix. Write the matrix representing the element affinities as $\mathcal{A}$ and the vector of weights linking elements to the nth cluster as $\mathbf{w}_n$. In particular, we can construct an objective function

$$\mathbf{w}_n^T \mathcal{A} \mathbf{w}_n.$$

This is a sum of terms of the form

$$\{\text{association of element } i \text{ with cluster } n\} \times \{\text{affinity between } i \text{ and } j\} \times \{\text{association of element } j \text{ with cluster } n\}.$$

We can obtain a cluster by choosing a set of association weights that maximize this objective function. The objective function is useless on its own because scaling $\mathbf{w}_n$ by λ scales the total association by λ^2. However, we can normalise the weights by requiring that $\mathbf{w}_n^T \mathbf{w}_n = 1$.

This suggests maximizing $\mathbf{w}_n^T \mathcal{A} \mathbf{w}_n$ subject to $\mathbf{w}_n^T \mathbf{w}_n = 1$. The Lagrangian is

$$\mathbf{w}_n^T \mathcal{A} \mathbf{w}_n + \lambda \left(\mathbf{w}_n^T \mathbf{w}_n - 1 \right)$$

(where λ is a Lagrange multiplier). Differentiation and dropping a factor of two yields

$$\mathcal{A} \mathbf{w}_n = \lambda \mathbf{w}_n,$$

meaning that $\mathbf{w}_n$ is an eigenvector of $\mathcal{A}$. This means that we could form a cluster by obtaining the eigenvector with the largest eigenvalue—the cluster weights are the elements of the eigenvector. For problems where reasonable clusters are apparent, we expect that these cluster weights are large for some elements, which belong to the cluster, and nearly zero for others, which do not (Figure 14.19). In fact, we can get the weights for other clusters from other eigenvectors of $\mathcal{A}$ as well.

Extracting Weights for a Set of Clusters In typical vision problems, there are strong association weights between relatively few pairs of elements. We can reasonably expect to be dealing with clusters that are quite tight and distinct.

These properties lead to a fairly characteristic structure in the affinity matrix. In particular, if we relabel the nodes of the graph, then the rows and columns of the matrix $\mathcal{A}$ are shuffled. We expect to be dealing with relatively few collections of nodes with large association weights. Furthermore, we expect these collections actually form a series of relatively coherent, largely disjoint clusters. This means that we could shuffle the rows and columns of M to form a matrix that is roughly block-diagonal (the blocks being the clusters). Shuffling M simply shuffles the elements of its eigenvectors so that we can reason about the eigenvectors by thinking about a shuffled version of M (i.e., Figure 14.16 is a fair source of insight).

The eigenvectors of block-diagonal matrices consist of eigenvectors of the blocks padded out with zeros. We expect that each block has an eigenvector corresponding to a rather large eigenvalue—corresponding to the cluster—and then a series of small eigenvalues of no particular

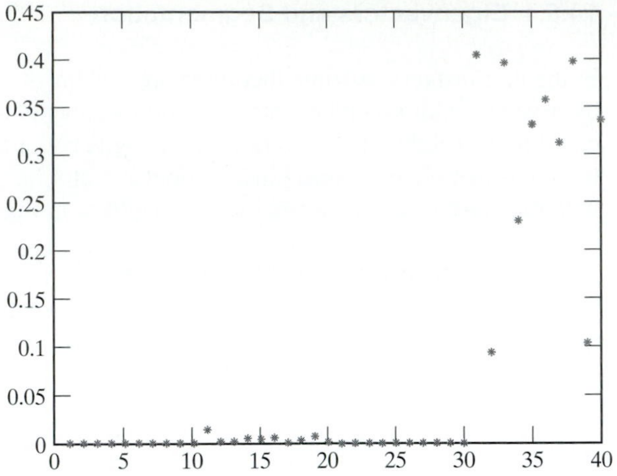

Figure 14.19 The eigenvector corresponding to the largest eigenvalue of the affinity matrix for the dataset of Figure 14.18 using $\sigma_d = 0.2$. Notice that most values are small, but some—corresponding to the elements of the main cluster—are large. The sign of the association is not significant, because a scaled eigenvector is still an eigenvector.

significance. From this, we expect that, if there are c significant clusters (where $c < k$), the eigenvectors corresponding to the c largest eigenvalues each represent a cluster.

This means that each of these eigenvectors is an eigenvector of a block padded with zeros. In particular, a typical eigenvector has a small set of large values—corresponding to its block—and a set of near-zero values. We expect that only one of these eigenvectors has a large value for any given component; all the others will be small (Figure 14.20). Thus, we can interpret eigenvectors corresponding to the c largest magnitude eigenvalues as cluster weights for the first c clusters. One can usually quantize the cluster weights to zero or one to obtain discrete clusters; this is what has happened in the figures.

This is a qualitative argument, and there are graphs for which the argument is decidedly suspect. Furthermore, we have been vague about how to determine c, although our

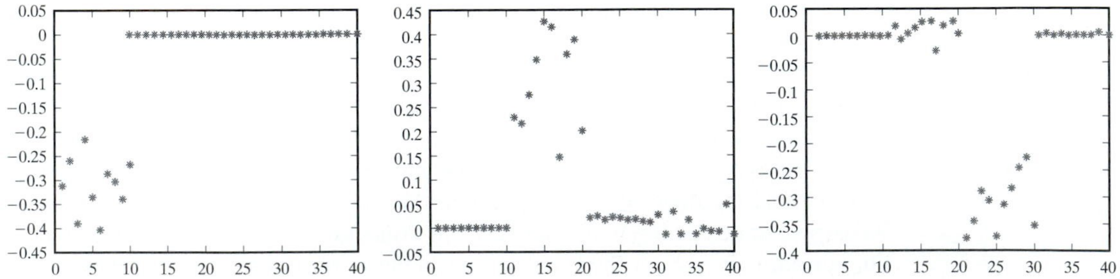

Figure 14.20 The three eigenvectors corresponding to the next three largest eigenvalues of the affinity matrix for the dataset of Figure 14.18 using $\sigma_d = 0.2$ (the eigenvector corresponding to the largest eigenvalue is given in Figure 14.19). Notice that most values are small, but for (disjoint) sets of elements, the corresponding values are large. This follows from the block structure of the affinity matrix. The sign of the association is not significant because a scaled eigenvector is still an eigenvector.

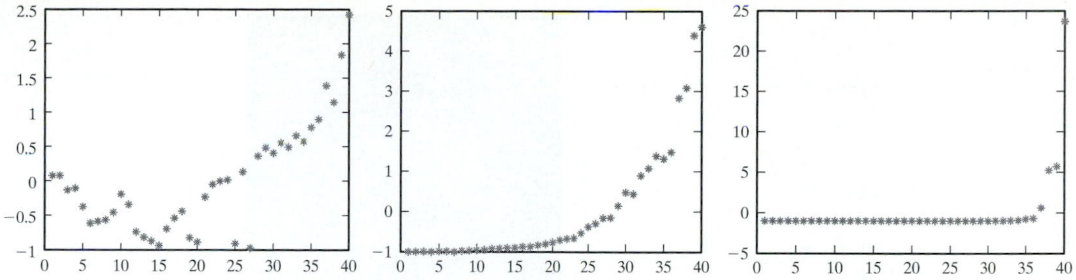

Figure 14.21 The number of clusters is reflected in the eigenvalues of the affinity matrix. The figure shows eigenvalues of the affinity matrices for each of the cases in Figure 14.18. On the **left**, $\sigma_d = 0.1$; in the **center**, $\sigma_d = 0.2$; and on the **right**, $\sigma_d = 1$. For the finest scale, there are many rather large eigenvalues because the affinity between all points is rather small; for the next scale, there are four eigenvalues rather larger than the rest; and for the coarsest scale, there are only two eigenvalues rather larger than the rest.

Algorithm 14.6: Clustering by Graph Eigenvectors

Construct an affinity matrix
Compute the eigenvalues and eigenvectors of the affinity matrix
Until there are sufficient clusters
    Take the eigenvector corresponding to the largest unprocessed eigenvalue; zero all components corresponding to elements that have already been clustered, and threshold the remaining components to determine which element belongs to this cluster, choosing a threshold by clustering the components, or using a threshold fixed in advance.
    If all elements have been accounted for, there are sufficient clusters
end

argument suggests that poking around in the spectrum of $\mathcal{A}$ might be rewarding—one would hope to find a small set of large eigenvalues and a large set of small eigenvalues (Figure 14.21).

14.5.5 Normalized Cuts

The qualitative argument of the previous section is somewhat soft. For example, if the eigenvalues of the blocks are similar, we could end up with eigenvectors that do not split clusters because any linear combination of eigenvectors with the same eigenvalue is also an eigenvector (Figure 14.22).

An alternative approach is to cut the graph into two connected components such that the cost of the cut is a small fraction of the total affinity within each group. We can formalize this as decomposing a weighted graph V into two components A and B and scoring the decomposition with

$$\frac{cut(A, B)}{assoc(A, V)} + \frac{cut(A, B)}{assoc(B, V)}$$

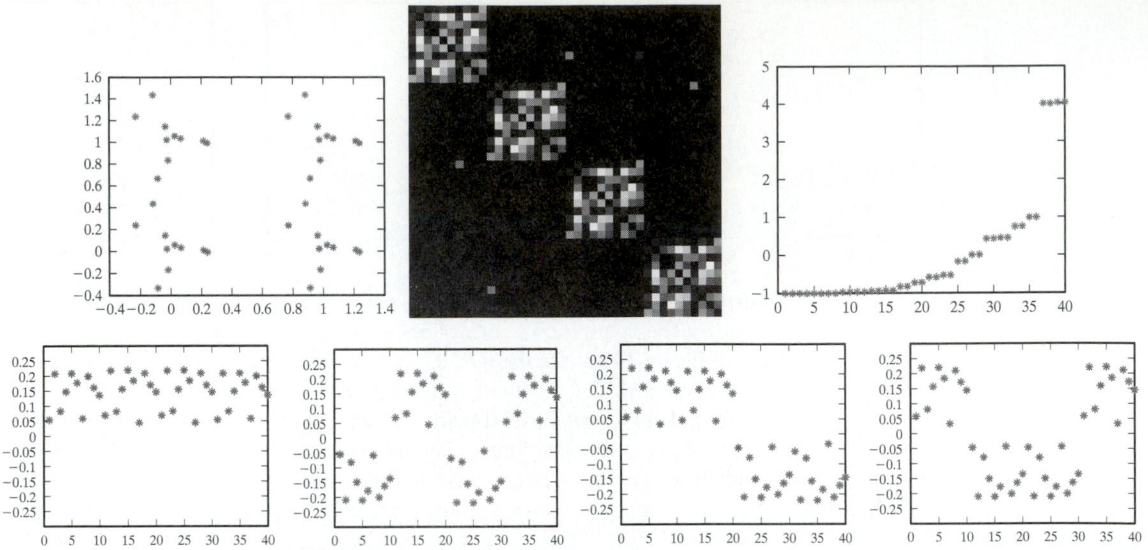

Figure 14.22 Eigenvectors of an affinity matrix can be a misleading guide to clusters. The dataset on the **top left** consists of four copies of the same set of points; this leads to a repeated block structure in the affinity matrix shown in the **top center**. Each block has the same spectrum, and this results in a spectrum for the affinity matrix that has (roughly) four copies of the same eigenvalue (**top right**). The bottom row shows the eigenvectors corresponding to the four largest eigenvalues; notice (a) that the values don't suggest clusters, and (b) a linear combination of the eigenvectors might lead to a quite good clustering.

(where $cut(A, B)$ is the sum of weights of all edges in V that have one end in A and the other in B, and $assoc(A, V)$ is the sum of weights of all edges that have one end in A). This score is small if the cut separates two components that have few edges of low weight between them and many internal edges of high weight. We would like to find the cut with the minimum value of this criterion, called a *normalized cut*. The criterion is successful in practice (Figures 14.23 and 14.24).

We write $\mathbf{y}$ is a vector of elements, one for each graph node, *whose values are either* 1 *or* $-b$. The values of $\mathbf{y}$ are used to distinguish between the components of the graph: If the ith component of $\mathbf{y}$ is 1, then the corresponding node in the graph belongs to one component; if it is $-b$, the node belongs to the other. We write the affinity matrix as $\mathcal{A}$ and write $\mathcal{D}$ for the *degree matrix*, each diagonal element of which matrix is the sum of weights coming into the corresponding node; that is,

$$D_{ii} = \sum_j A_{ij}$$

and the off-diagonal elements of $\mathcal{D}$ are zero. In this notation and with a little manipulation, our criterion can be rewritten as

$$\frac{\mathbf{y}^T(\mathcal{D} - \mathcal{A})\mathbf{y}}{\mathbf{y}^T\mathcal{D}\mathbf{y}}.$$

We now wish to find a vector $\mathbf{y}$ that minimizes this criterion. The problem we have set up is an *integer programming* problem. Because it is exactly equivalent to the graph cut problem, it isn't

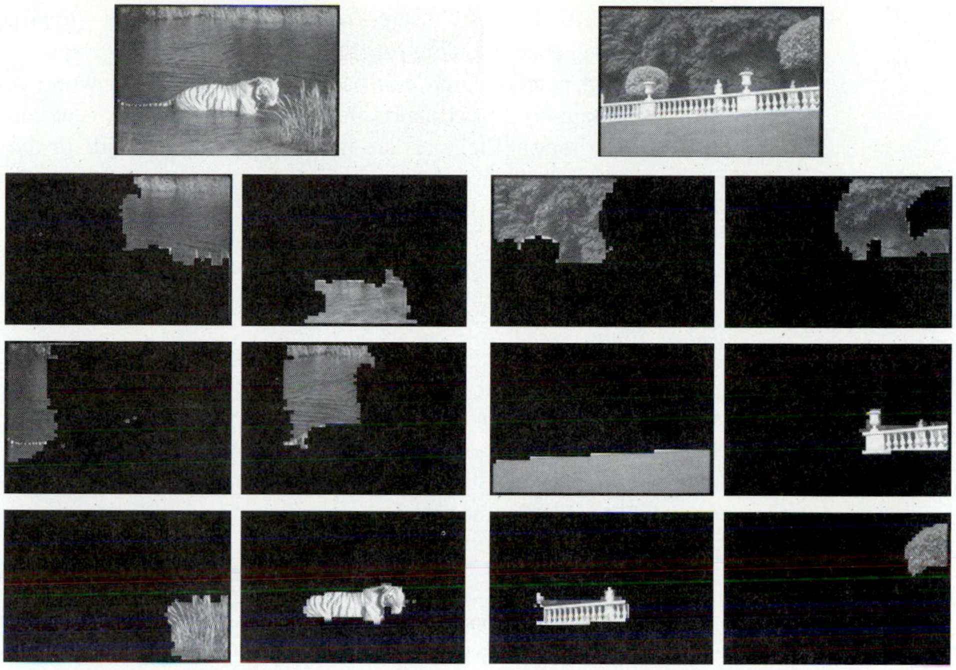

Figure 14.23 The images on top are segmented using the normalized cuts framework, described in the text, into the components shown. The affinity measures used involved intensity and texture as in Section 14.5.3. The image of the swimming tiger yields one segment that is essentially tiger, one that is grass, and four components corresponding to the lake. Similarly, the railing shows as three reasonably coherent segments. Note the improvement over k-means segmentation obtained by having a texture measure. *Reprinted from "Image and video segmentation: the normalized cut framework," by J. Shi et al., Proc. IEEE Int. Conf. Image Processing, 1998 © 1998 IEEE*

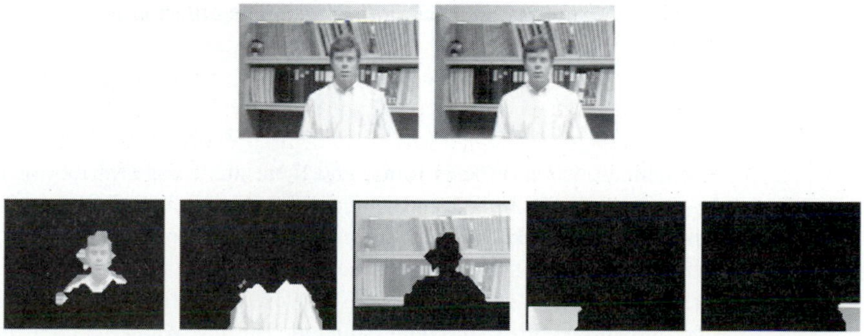

Figure 14.24 **Top:** two frames from a motion sequence, that shows a moving view of a person. **Bottom:** spatiotemporal segments established using normalized cuts and a spatiotemporal affinity function (Section 14.5.3). *Reprinted from "Normalized cuts and image segmentation," by J. Shi and J. Malik, IEEE Trans. PAMI 2000 © 2000 IEEE*

any easier. The difficulty is the discrete values for elements of $\mathbf{y}$—in principle, we could solve the problem by testing every possible $\mathbf{y}$, but this involves searching a space whose size is exponential in the number of pixels, which will be slow (as in probably won't finish before the universe burns out). A common approximate solution to such problems is to compute a *real* vector $\mathbf{y}$ that minimizes the criterion. Elements are then assigned to one side or the other by testing against a threshold. There are then two issues: First, we must obtain the real vector; second, we must choose a threshold.

Obtaining a Real Vector The real vector is easily obtained. It is an exercise to show that a solution to

$$(\mathcal{D} - \mathcal{A})\mathbf{y} = \lambda \mathcal{D}\mathbf{y}$$

is a solution to our problem *with real values*. The only question is which generalized eigenvector to use. It turns out that the smallest eigenvalue is guaranteed to be zero, so the eigenvector corresponding to the second smallest eigenvalue is appropriate. The easiest way to determine this eigenvector is to perform the transformation $z = \mathcal{D}^{1/2}\mathbf{y}$ and get

$$\mathcal{D}^{-1/2}(\mathcal{D} - \mathcal{A})\mathcal{D}^{-1/2}z = \lambda z,$$

whereupon $\mathbf{y}$ follows easily. Note that solutions to this problem are also solutions to

$$\mathcal{N}z = \mathcal{D}^{-1/2}\mathcal{A}\mathcal{D}^{-1/2}z = \mu z,$$

and $\mathcal{N}$ is sometimes called the *normalized affinity matrix*.

Choosing a Threshold Finding the appropriate threshold value is not particularly difficult; assume there are N nodes in the graph, so that there are N elements in $\mathbf{y}$ and at most N different values. Now if we write $ncut(v)$ for the value of the normalized cut criterion at a particular threshold value v, there are at most $N + 1$ values of $ncut(v)$. We can form each of these values and choose a threshold that leads to the smallest. Notice also that this formalism lends itself to recursion, in that each component of the result is a graph, and these new graphs can be split, too. A simpler criterion, which appears to work in practice, is to walk down the eigenvalues and use eigenvectors corresponding to smaller eigenvalues to obtain new clusters.

14.6 NOTES

Segmentation is a difficult topic, and there are a huge variety of methods. Surveys of mainly historical interest are Fu and Mui (1981), Haralick and Shapiro (1985), Nevatia (1986), and Riseman and Arbib (1977). More recent surveys are rare, but there is Pal and Pal (1993). One reason is that it is typically quite hard to assess the performance of a segmenter at a level more useful than that of showing some examples. Evaluation is easier in the context of a specific task; papers dealing with assorted tasks include Hartley, Wang, Kitchen and Rosenfeld (1982), Ranade and Prewitt (1980), Yasnoff, Mui and Bacus (1977), and Zhang (1996).

The original clustering segmenter is Ohlander, Price and Reddy (1978). Clustering methods tend to be rather arbitrary—remember, this doesn't mean they're not useful—because there really isn't much theory available to predict what should be clustered and how. It is clear that what we should be doing is forming clusters that are helpful to a particular application, but this criterion hasn't been formalized in any useful way. In this chapter, we have attempted to give the big picture while ignoring detail, because a detailed record of what has been done would be unenlightening.

A variety of graph theoretical clustering methods have been used in vision (see Sarkar and Boyer, 1998, and Wu and Leahy, 1993; there is a summary in Weiss, 1999). The normalized cuts formalism is due to Shi and Malik (1997) and (2000). Variants include applications to motion segmentation (Shi and Malik, 1998*a*) and methods for deducing similarity metrics from outputs (Shi and Malik, 1998*b*). There are numerous alternate criteria (e.g., Cox, Zhong and Rao, 1996, Perona and Freeman, 1998). We have stressed the graph theoretical clustering methods because their ability to deal with any affinity function one cares to name is an attractive feature.

Segmentation is also a key open problem in vision, which is why a detailed record of what has been done would be huge. Until quite recently, it was usual to talk about recognition and segmentation as if they were distinct activities. This view is going out of fashion—as it should—because there isn't much point in creating a segmented representation that doesn't help with some application. Furthermore, if we can be crisp about what should be recognized, that should make it possible to be crisp about what a segmented representation should look like.

Segmentation and Grouping in People

There is a large literature on the role of grouping in human visual perception. Standard Gestalt handbooks include Kanizsa (1979) and Koffka (1935). Subjective contours were first described by Kanisza; there is a broad summary discussion in Kanizsa (1976). The authoritative book by Palmer (1999) gives a much broader picture than we can supply here. There is a great deal of information about the development of different theories of vision and the origins of Gestalt thinking in Gordon (1997). Some groups appear to be formed remarkably early in the visual process, a phenomenon known as *pop out* (Triesman, 1982).

Perceptual Grouping

On occasion, a distinction is drawn between perceptual organization, which is seen as clustering image tokens into useful groups, and segmentation, which is seen as decomposing images into regions. We don't accept this distinction; the advantage of seeing these problems as manifestations of the same activity is that one can convert algorithmic advances from one problem to another freely. We haven't discussed some aspects of perceptual organization in great detail mainly because our emphasis is on exposition rather than historical accuracy, and these methods follow from the unified view. For example, there is a long thread of literature on clustering image edge points or line segments into configurations that are unlikely to have arisen by accident. We cover some of these ideas in the following chapter, but also draw the readers attention to Amir and Lindenbaum (1996), Huttenlocher and Wayner (1992), Lowe (1985), Mohan and Nevatia (1992), Sarkar and Boyer (1993) and (1994). In building user interfaces, it can (as we hinted before) be helpful to know what is perceptually salient (e.g., Saund and Moran, 1995).

PROBLEMS

14.1. We wish to cluster a set of pixels using color and texture differences. The objective function

$$\Phi(\text{clusters, data}) = \sum_{i \in \text{clusters}} \left\{ \sum_{j \in i\text{thcluster}} (\boldsymbol{x}_j - \boldsymbol{c}_i)^T (\boldsymbol{x}_j - \boldsymbol{c}_i) \right\}$$

used in Section 14.4.2 may be inappropriate—for example, color differences could be too strongly weighted if color and texture are measured on different scales.

(a) Extend the description of the k-means algorithm to deal with the case of an objective function of the form

$$\Phi(\text{clusters, data}) = \sum_{i \in \text{clusters}} \left\{ \sum_{j \in i\text{'thcluster}} (\boldsymbol{x}_j - \boldsymbol{c}_i)^T \mathcal{S} (\boldsymbol{x}_j - \boldsymbol{c}_i) \right\},$$

where $\mathcal{S}$ is an a symmetric, positive definite matrix.

(b) For the simpler objective function, we had to ensure that each cluster contained at least one element (otherwise we can't compute the cluster center). How many elements must a cluster contain for the more complicated objective function?

(c) As we remarked in Section 14.4.2, there is no guarantee that k-means gets to a global minimum of the objective function; show that it must always get to a local minimum.

(d) Sketch two possible local minima for a k-means clustering method clustering data points described by a two-dimensional feature vector. Use an example with only two clusters for simplicity. You shouldn't need many data points. You should do this exercise for both objective functions.

14.2. Read Shi and Malik (2000) and follow the proof that the normalized cut criterion leads to the integer programming problem given in the text.

14.3. This exercise explores using normalized cuts to obtain more than two clusters. One strategy is to construct a new graph for each component separately and call the algorithm recursively. You should notice a strong similarity between this approach and classical divisive clustering algorithms. The other strategy is to look at eigenvectors corresponding to smaller eigenvalues.

(a) Explain why these strategies are not equivalent.

(b) Now assume that we have a graph that has two connected components. Describe the eigenvector corresponding to the largest eigenvalue.

(c) Now describe the eigenvector corresponding to the second largest eigenvalue.

(d) Turn this information into an argument that the two strategies for generating more clusters should yield quite similar results under appropriate conditions; what are appropriate conditions?

Programming Assignments

14.4. Build a background subtraction algorithm using a moving average and experiment with the filter.

14.5. Build a shot boundary detection system using any two techniques that appeal, and compare performance on different runs of video.

14.6. Implement a segmenter that uses k-means to form segments based on color and position. Describe the effect of different choices of the number of segments and investigate the effects of different local minima.

15 Segmentation by Fitting a Model

One view of segmentation is to assert that pixels (tokens, etc.) belong together because together they conform to some model. This view is rather similar to the clustering view; the main difference is that the model is now explicit and may involve relations at a larger scale than from token to token. For example, imagine a program that attempts to assemble tokens into groups that "look like" a line (in some sense we don't need to make precise at this point). It isn't possible to do this by looking only at pairwise relations between tokens—instead we must look at some properties of the collective of tokens, identified by choosing a model, and then declaring some criterion for a good fit. An alternative view is that we are clustering tokens because, together, they form a familiar geometric configuration—for example, they all lie on a line or on a circle. However one looks at it, this activity is usually called *fitting*.

15.1 THE HOUGH TRANSFORM

There are three problems in line fitting. First, given the points that belong to a line, what is the line? Second, which points belong to which line? Finally, how many lines are there? The Hough transform is a method that promises a solution to all three (although in practice rarely delivers it). It is something worth understanding because the underlying method is quite general and appears in a number of applications.

One way to cluster points that could lie on the same structure is to record all the structures on which each point lies and then look for structures that get many votes. This (quite general) technique is known as the *Hough transform*. We take each image token and determine all structures *that could pass through that token*. We make a record of this set—you should think of this as voting—and repeat the process for each token. We decide on what is present by looking at the votes. For example, if we are grouping points that lie on lines, we take each point and vote for all lines that could go through it; we now do this for each point. The line (or lines) that are present should make themselves obvious because they pass through many points and so have many votes.

15.1.1 Fitting Lines with the Hough Transform

Hough transforms tend to be most successfully applied to line finding. We do this example to illustrate the method and its drawbacks. A line is easily parametrized as a collection of points (x, y) such that

$$x \cos\theta + y \sin\theta + r = 0.$$

Now any pair of (θ, r) represents a unique line, where $r \geq 0$ is the perpendicular distance from the line to the origin and $0 \leq \theta < 2\pi$. We call the set of pairs (θ, r) *line space*; the space can be visualized as a half-infinite cylinder. There is a family of lines that passes through any point token. In particular, the lines that lie on the curve *in line space* given by $r = -x_0 \cos\theta + y_0 \sin\theta$ all pass through the point token at (x_0, y_0).

Because the image has a known size, there is some R such that we are not interested in lines for $r > R$—these lines are too far away from the origin for us to see them. This means that the lines we are interested in form a bounded subset of the plane, and we discretize this with some convenient grid (which we'll discuss later). The grid elements can be thought of as buckets into which we place votes. This grid of buckets is referred to as the *accumulator array*. For each point token, we add a vote to the total formed for every grid element on the curve corresponding to the point token. If there are many point tokens that are collinear, we expect there to be many votes in the grid element corresponding to the line they lie on.

15.1.2 Practical Problems with the Hough Transform

Unfortunately, the Hough transform comes with a number of important practical problems:

- **Quantization errors:** An appropriate grid size is difficult to pick. Too coarse a grid can lead to large values of the vote being obtained falsely because many quite different lines correspond to a bucket. Too fine a grid can lead to lines not being found because votes resulting from tokens that are not exactly collinear end up in different buckets, and no bucket has a large vote (Figure 15.1).
- **Difficulties with noise:** The attraction of the Hough transform is that it connects widely separated tokens that lie close to some form of parametric curve. This is also a weakness; it is usually possible to find many quite good phantom lines in a large set of reasonably uniformly distributed tokens (Figure 15.2). This means that regions of texture can generate peaks in the voting array that are larger than those associated with the lines sought (Figures 15.3 and 15.4).

The Hough transform is worth talking about because, despite these difficulties, it can often be implemented in a way that is quite useful for well-adapted problems. In practice, it is almost always used to find lines in sets of edge points. The following are useful implementation guidelines:

- **Ensure the minimum of irrelevant tokens:** This can often be done by tuning the edge detector to smooth out texture, setting the illumination to produce high-contrast edges, and so on.
- **Choose the grid carefully:** This is usually done by trial and error. It can be helpful to vote for all neighbors of a grid element at the same time one votes for the element.

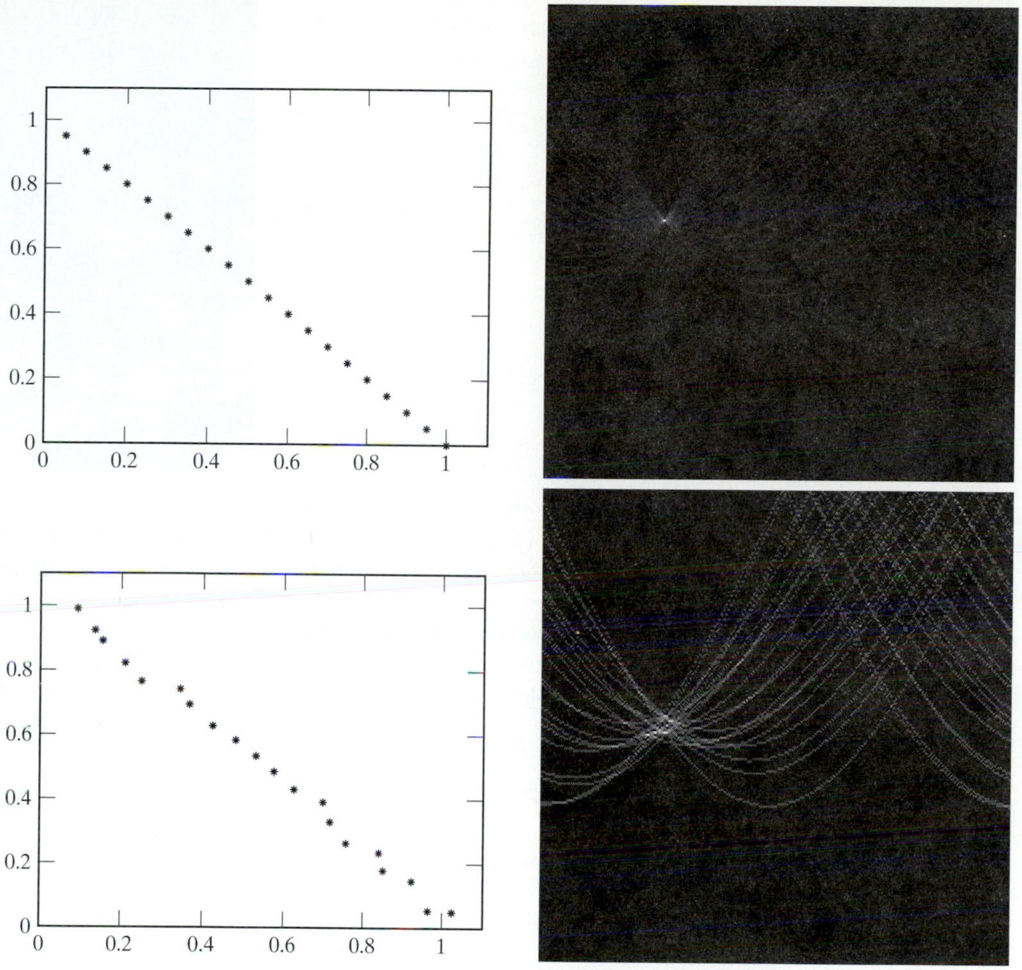

Figure 15.1 The Hough transform maps each point like token to a curve of possible lines (or other parametric curves) through that point. These figures illustrate the Hough transform for lines. The **left-hand column** shows points, and the **right-hand column** shows the corresponding accumulator arrays (the number of votes is indicated by the grey level, with a large number of votes being indicated by bright points). The **top** row shows what happens using a set of 20 points drawn from a line. On the **top right**, the accumulator array for the Hough transform of these points. Corresponding to each point is a curve of votes in the accumulator array; the largest set of votes is 20 (which corresponds to the brightest point). The horizontal variable in the accumulator array is θ and the vertical variable is r; there are 200 steps in each direction, and r lies in the range [0, 1.55]. On the **bottom**, these points have been offset by a random vector each element of which is uniform in the range [0, 0.05]; note that this offsets the curves in the accumulator array shown next to the points; the maximum vote is now 6 (which corresponds to the brightest value in this image—this value would be difficult to see on the same scale as the top image).

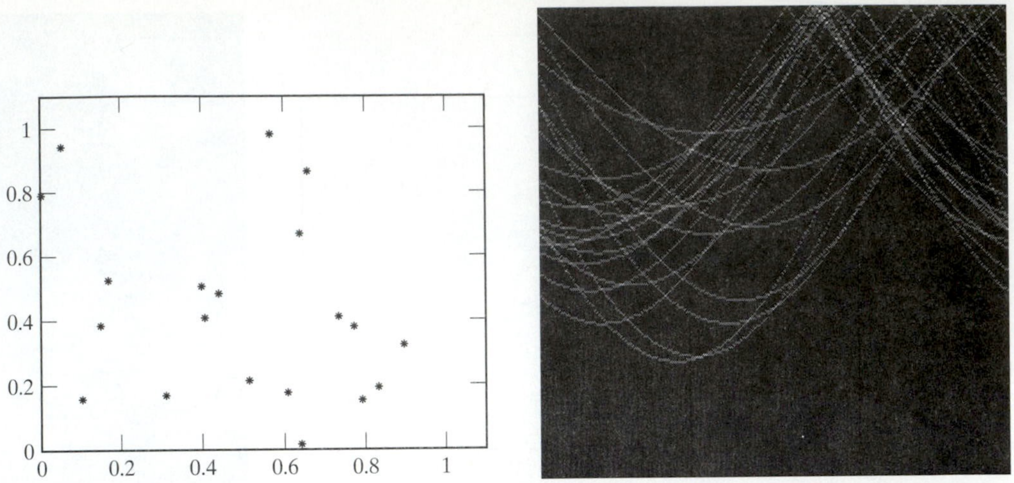

Figure 15.2 The Hough transform for a set of random points can lead to quite large sets of votes in the accumulator array. As in Figure 15.1, the **left-hand column** shows points, and the **right-hand column** shows the corresponding accumulator arrays (the number of votes is indicated by the grey level, with a large number of votes being indicated by bright points). In this case, the data points are noise points (both coordinates are uniform random numbers in the range [0, 1]); the accumulator array in this case contains many points of overlap, and the maximum vote is now 4 (compared with 6 in Figure 15.1). Figures 15.3 and 15.4 explore noise issues somewhat further.

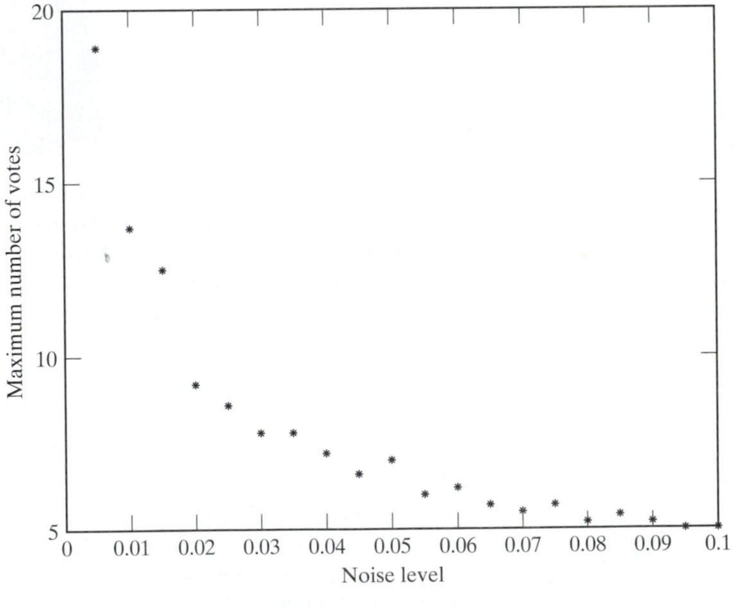

Figure 15.3 The effects of noise make it difficult to use a Hough transform robustly. The plot shows the maximum number of votes in the accumulator array for a Hough transform of 20 points on a line perturbed by uniform noise plotted against the magnitude of the noise. The noise displaces the curves from each other and quickly leads to a collapse in the number of votes. The plot has been averaged over 10 trials. The accumulator array had the same quantization for each case shown here.

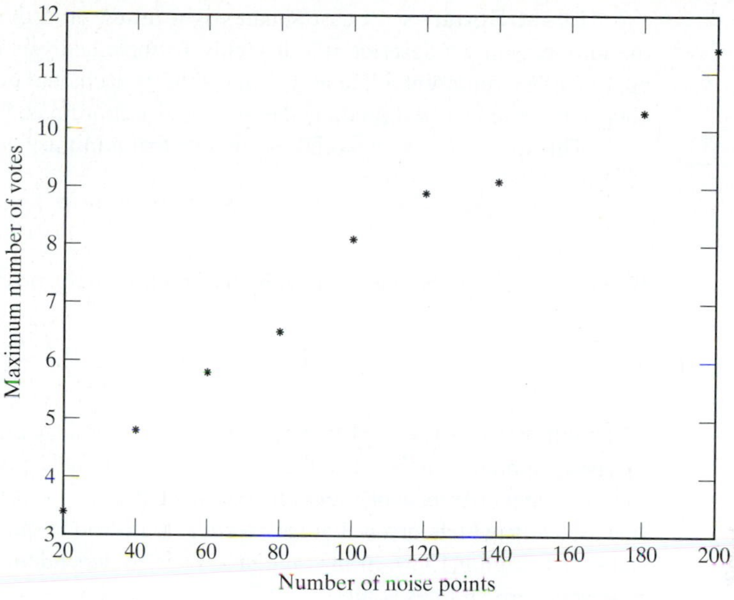

Figure 15.4 A plot of the maximum number of votes in the accumulator array for a Hough transform of a set of points whose coordinates are uniform random numbers in the range [0, 1] plotted against the number of points. As the level of noise goes up, the number of votes in the right bucket goes down, and the prospect of obtaining a large spurious vote in the accumulator array goes up. The plots have again been averaged over 10 trials. Compare this figure with Figure 15.3, but notice the slightly different scales; the comparison suggests that it can be quite difficult to pull a line out of noise with a Hough transform (because the number of votes for the line might be comparable with the number of votes for a line due to noise). These figures illustrate the importance of ruling out as many noise tokens as possible before performing a Hough transform.

15.2 FITTING LINES

Line fitting is extremely useful. In many applications, objects are characterized by the presence of straight lines. For example, we might wish to build models of buildings using pictures of the buildings (as in the application in chapter 26). This application uses polyhedral models of buildings, meaning that straight lines in the image are important. Similarly, many industrial parts have straight edges of one form or another; if we wish to recognise industrial parts in an image, straight lines could be helpful. In either case, a report of all straight lines in the image is an extremely useful segmentation. We review the special case of lines here for people who have not read the more general treatment of section 3.2.

15.2.1 Line Fitting with Least Squares

We first assume that all the points that belong to a particular line are known, and the parameters of the line must be found. We adopt the notation

$$\overline{u} = \frac{\sum u_i}{k}$$

to simplify the presentation.

Least Squares Least squares is a fitting procedure with a long tradition (which is the only reason we describe it!). It yields a simple analysis but has a substantial bias. For this approach, we represent a line as $y = ax + b$. At each data point, we have (x_i, y_i); we decide to choose the line that best predicts the measured y coordinate for each measured x coordinate.

This means we want to choose the line that minimises

$$\sum_i (y_i - ax_i - b)^2.$$

By differentiation, the line is given by the solution to the problem

$$\begin{pmatrix} \overline{y^2} \\ \overline{y} \end{pmatrix} = \begin{pmatrix} \overline{x^2} & \overline{x} \\ \overline{x} & 1 \end{pmatrix} \begin{pmatrix} a \\ b \end{pmatrix}.$$

Although this is a standard linear solution to a classical problem, it's actually not much help in vision applications because the model is an extremely poor model. The difficulty is that the measurement error is dependent on coordinate frame—we are counting vertical offsets from the line as errors, which means that near vertical lines lead to quite large values of the error and quite funny fits (Figure 15.5). In fact, the process is so dependent on coordinate frame that it doesn't represent vertical lines at all.

Total Least Squares We could work with the actual distance between the point and the line (rather than the vertical distance). This leads to a problem known as *total least squares*. We can represent a line as the collection of points where $ax + by + c = 0$. Every line can be represented in this way, and we can think of a line as a triple of values (a, b, c). Notice that for $\lambda \neq 0$, the line given by $\lambda(a, b, c)$ is the same as the line represented by (a, b, c). In the exercises, you are asked to prove the simple, but extremely useful, result that the perpendicular distance from a point (u, v) to a line (a, b, c) is given by

$$\text{abs}(au + bv + c) \text{ if } a^2 + b^2 = 1.$$

In our experience, this fact is useful enough to be worth memorizing.

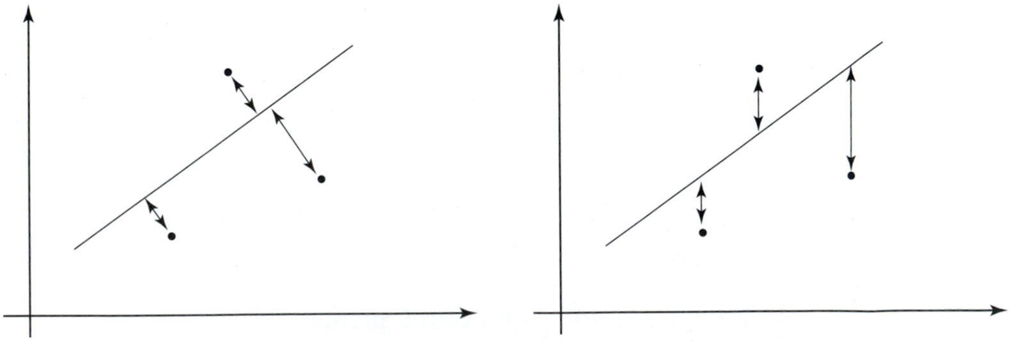

Figure 15.5 **Left:** Total least-squares models data points as being generated by an abstract point along the line to which is added a vector perpendicular to the line. We wish to choose a line that minimizes the sum of distances to tokens measured (as distance usually is!) perpendicular to the line. **Right:** Least squares follows the same general outline, but assumes that the error appears only in the y coordinate. This yields a (very slightly) simpler mathematical problem at the cost of a poor fit.

To minimize the sum of perpendicular distances between points and lines, we need to minimize

$$\sum_i (ax_i + by_i + c)^2,$$

where $a^2 + b^2 = 1$ and C is some normalizing constant of no interest. Thus, a maximum-likelihood solution is obtained by maximizing this expression. Now using a Lagrange multiplier λ, we have a solution if

$$\begin{pmatrix} \overline{x^2} & \overline{xy} & \overline{x} \\ \overline{xy} & \overline{y^2} & \overline{y} \\ \overline{x} & \overline{y} & 1 \end{pmatrix} \begin{pmatrix} a \\ b \\ c \end{pmatrix} = \lambda \begin{pmatrix} 2a \\ 2b \\ 0 \end{pmatrix}$$

This means that

$$c = -a\overline{x} - b\overline{y}$$

and we can substitute this back to get the eigenvalue problem

$$\begin{pmatrix} \overline{x^2} - \overline{x}\,\overline{x} & \overline{xy} - \overline{x}\,\overline{y} \\ \overline{xy} - \overline{x}\,\overline{y} & \overline{y^2} - \overline{y}\,\overline{y} \end{pmatrix} \begin{pmatrix} a \\ b \end{pmatrix} = \mu \begin{pmatrix} a \\ b \end{pmatrix}$$

Because this is a 2D eigenvalue problem, two solutions up to scale can be obtained in closed form (for those who care—it's usually done numerically!). The scale is obtained from the constraint that $a^2 + b^2 = 1$. The two solutions to this problem are lines at right angles, one of which minimises it.

15.2.2 Which Point Is on Which Line?

This problem can be difficult because it can involve search over a large combinatorial space. One approach is to notice that we seldom encounter isolated points; instead, we are fitting lines to edge points. We can use the orientation of an edge point as a hint to the position of the next point on the line. If we are stuck with isolated points, then k-means can be applied.

Incremental Fitting **Incremental line fitting** algorithms take connected curves of edge points and fit lines to runs of points along the curve. Connected curves of edge points are fairly easily obtained from an edge detector whose output gives orientation (see Exercises). An incremental fitter then starts at one end of a curve of edge points and walks along the curve, cutting off runs of pixels that fit a line well (the structure of the algorithm is shown in Algorithm 15.1). Incremental line fitting can work well, despite the lack of an underlying statistical model. One feature is that it reports groups of lines that form closed curves. This is attractive when the lines one is interested in can reasonably be expected to form a closed curve (e.g., in some object recognition applications) because it means that the algorithm reports natural groups without further fuss. This strategy often leads to occluded edges resulting in more than one fitted line. This difficulty can be addressed by postprocessing the lines to find pairs that (roughly) coincide, but the process is somewhat unattractive because it is hard to give a sensible criterion by which to decide when two lines do coincide.

Allocating points to lines with K-means Assume that points carry no hints about which line they lie on (i.e., there is no color, etc., information to help, and, crucially, the points

Algorithm 15.1: Incremental line fitting by walking along a curve, fitting a line to runs of pixels along the curve, and breaking the curve when the residual is too large

```
Put all points on curve list, in order along the curve
Empty the line point list
Empty the line list
Until there are too few points on the curve
   Transfer first few points on the curve to the line point list
   Fit line to line point list
   While fitted line is good enough
      Transfer the next point on the curve
         to the line point list and refit the line
   end
   Transfer last point(s) back to curve
   Refit line
   Attach line to line list
end
```

are not linked). We can attempt to determine which point lies on which line is to use a modified version of k-means. In this case, the model is that there are k lines, each of which generates some subset of the data points; the best solution for lines and data points is obtained by minimizing

$$\sum_{l_i \in \text{lines}} \sum_{\substack{x_j \in \text{data due} \\ \text{to } i\text{th line}}} \text{dist}(l_i, x_j)^2$$

over both correspondences and lines. Again, there are too many correspondences to search this space.

It is easy to modify k-means to deal with this problem. The two phases are as follows:

- allocate each point to the closest line;
- fit the best line to the points allocated to each line.

Algorithm 15.2: K-means line fitting by allocating points to the closest line and then refitting.

```
Hypothesize k lines (perhaps uniformly at random)
or
Hypothesize an assignment of lines to points
   and then fit lines using this assignment

Until convergence
   Allocate each point to the closest line
   Refit lines
end
```

This results in Algorithm 15.2. Convergence can be tested by looking at the size of the change in the lines, at whether labels have been flipped (probably the best test), or at the sum of perpendicular distances of points from their lines (which operates as a log likelihood).

15.3 FITTING CURVES

In principle, fitting curves is similar to fitting lines. We minimize the sum of squared distances between the points and the curve. This generates quite difficult practical problems: It is usually very hard to tell the distance between a point and a curve. We can either solve this problem or apply various approximations (which are usually chosen because they are computationally simple, not because they result from clean models). We sketch some solutions for the distance problem for the two main representations of curves.

15.3.1 Implicit Curves

The coordinates of *implicit curves* satisfy some parametric equation; if this equation is a polynomial, then the curve is said to be *algebraic*, and this case is by far the most common. Some common cases are given in Table 15.1.

The Distance from a Point to an Implicit Curve Now we would like to know the distance from a data point to the closest point on the implicit curve. Assume that the curve has the form $\phi(x, y) = 0$. The vector from the closest point on the implicit curve to the data point is normal to the curve, so the closest point is given by finding all the (u, v) with the following properties:

TABLE 15.1 Some implicit curves used in vision applications. Note that not all of these curves are guaranteed to have any real points on them—e.g., $x^2 + y^2 + 1 = 0$ doesn't. Higher degree curves are seldom used because it can be difficult to get stable fits to these curves.

Curve	Equation
Line	$ax + by + c = 0$
Circle, center (a, b), and radius r	$x^2 + y^2 - 2ax - 2by + a^2 + b^2 - r^2 = 0$
Ellipses (including circles)	$ax^2 + bxy + cy^2 + dx + ey + f = 0$ where $b^2 - 4ac < 0$
Hyperbolae	$ax^2 + bxy + cy^2 + dx + ey + f = 0$ where $b^2 - 4ac > 0$
Parabolae	$ax^2 + bxy + cy^2 + dx + ey + f = 0$ where $b^2 - 4ac = 0$
General conic sections	$ax^2 + bxy + cy^2 + dx + ey + f = 0$

1. (u, v) is a point on the curve—this means that $\phi(u, v) = 0$;
2. $s = (d_x, d_y) - (u, v)$ is normal to the curve.

Given all such s, the length of the shortest is the distance from the data point to the curve.

The second criterion requires a little work to determine the normal. The normal to an implicit curve is the direction in which we leave the curve fastest; along this direction, the value of ϕ must change fastest, too. This means that the normal at a point (u, v) is

$$\left(\frac{\partial \phi}{\partial x}, \frac{\partial \phi}{\partial y}\right),$$

evaluated at (u, v). If the tangent to the curve is $\boldsymbol{T}$, then we must have $\boldsymbol{T}.\boldsymbol{s} = 0$. Because we are working in 2D, we can determine the tangent from the normal, so that we must have

$$\psi(u, v; d_x, d_y) = \frac{\partial \phi}{\partial y}(u, v)\{d_x - u\} - \frac{\partial \phi}{\partial x}(u, v)\left\{d_y - v\right\} = 0$$

at the point (u, v). We now have two equations in two unknowns and, *in principle* can solve them. However, this is very seldom as easy as it looks, as Example 15.1 indicates.

Example 15.1 The distance between a point and a conic

A conic section is given by $ax^2 + bxy + cy^2 + dx + ey + f = 0$. Given a data point (d_x, d_y), the nearest point on the conic satisfies two equations:

$$au^2 + buv + cv^2 + du + ev + f = 0$$

and

$$2(a - c)uv - (2ad_y + e)u + (2cd_x + d)v + (ed_x - dd_y) = 0.$$

There can be up to four real solutions of this pair of equations (in the exercises, you are asked to demonstrate this, given an algorithm for obtaining the solutions, and asked to sketch various cases). As an example, choose the ellipse $2x^2 + y^2 - 1 = 0$, which yields the equations

$$2u^2 + v^2 - 1 = 0 \quad \text{and} \quad 2uv - 4d_y u + 2d_x v = 0.$$

Let us consider a family of data points $(d_x, d_y) = (0, \lambda)$; then we can rearrange these equations to get

$$2u^2 + v^2 - 1 = 0 \quad \text{and} \quad 2uv - 4\lambda u = 2u(v - 2\lambda) = 0.$$

The second equation helps: Either $u = 0$ or $v = 2\lambda$. Two of our solutions will be $(0, 1)$, $(0, -1)$. The other two are obtained by solving $2u^2 + 4\lambda^2 - 1 = 0$, which has solutions only if $-1/2 \leq \lambda \leq 1/2$. The situation is illustrated in Figure 15.6.

Approximations to the Distance Notice that for a relatively simple curve, we already have an unpleasant problem to solve. A curve with a slightly more complicated geometry—obtained by choosing ϕ to be a polynomial of higher degree, say d—leads to openly nasty problems. This is because the closest point on the curve would be obtained by solving two simultaneous polynomial equations, *both* of degree d. It can be shown that this can lead to as many as d^2 solutions, which are usually hard to obtain in practice. Various approximations to the distance between a point and an implicit algebraic curve have come into practice.

The best known is *algebraic distance*: In this case, we measure the distance between a curve and a point by evaluating the polynomial equation at that point, that is, we make the approximation

$$\text{distance between } (d_x, d_y) \text{ and } \phi(x, y) = 0 = \phi(d_x, d_y).$$

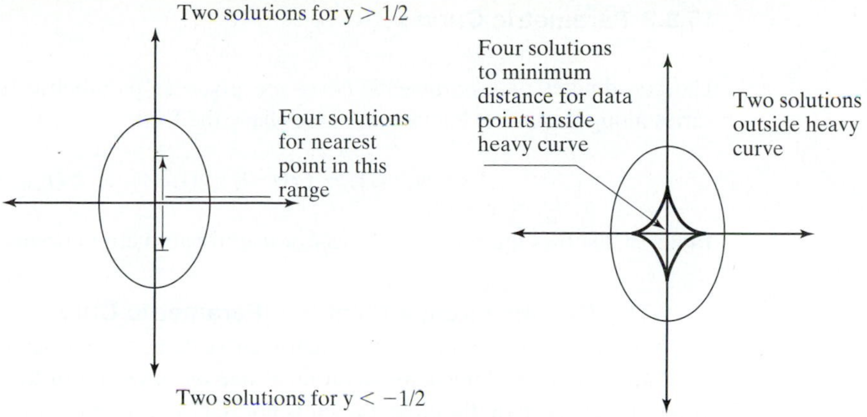

Figure 15.6 On the **left**, the example worked in the text, where we study the number of possible solutions for the distance between a point and an ellipse for data points lying on the vertical axis. The figure on the **right** indicates the general case for this ellipse.

This approximation can be (rather roughly!) justified when the data points are quite close to the curve. For a point sufficiently close to the curve *and to first order*, $\phi(d_x, d_y)$ increases as (d_x, d_y) moves normal to the curve—because the normal to the curve is given by the gradient of ϕ—and does not increase as (d_x, d_y) moves tangent to the curve. One significant difficulty is that, as it stands, algebraic distance is ill defined because many polynomials correspond to the same curve. In particular, the curve given by $\mu\phi(x, y) = 0$ is the same as the curve given by $\phi(x, y) = 0$. This problem can be solved by normalizing the coefficients of the polynomial in some way.

We have already seen one example of this process in Section 15.2, where we fitted a line ($\phi(x, y) = ax + by + c = 0$) to a set of points by minimizing the algebraic distance subject to the constraint that $a^2 + b^2 = 1$. In this case, the algebraic distance is the same as the actual distance. The choice of normalization is important. For example, if we try to fit conics ($ax^2 + bxy + cy^2 + dx + ey + f = 0$) using the constraint $b = 1$, we cannot fit circles. An alternative approximation is to use

$$\frac{\phi(d_x, d_y)}{\left|\nabla\phi(d_x, d_y)\right|},$$

which has the advantage of not requiring a normalizing constant; in the case of a line, this approximation is exact. Notice that this approximation has the same properties as algebraic distance—it goes up as one moves along the normal, and so on. The advantage of the approximation is that it is somewhat more accurate than algebraic distance because it is normalised by the length of the normal. This means that it can be read—roughly!—as giving the percentage distance along the normal from the curve to the point. In practice, this approximation is seldom used mainly because the use of algebraic distance yields simpler numerical problems.

Both of these approximations are dangerous because their behavior for data points that are far from the curve is strange and not well understood. As a result, the relationship between a fitted curve and a set of data points becomes a bit mysterious if the data points don't lie close to a curve of that class. Algebraic distance is used quite widely in practice because it yields easy numerical problems and can be used for higher dimensional problems like approximating the distance between points and implicit surfaces. The exact distance is often difficult to compute for such problems.

15.3.2 Parametric Curves

The coordinates of a *parametric curve* are given as parametric functions of a parameter that varies along the curve. Parametric curves have the form

$$(x(t), y(t)) = (x(t;\theta), y(t;\theta)) \quad t \in [t_{\min}, t_{\max}].$$

Table 15.2 shows the form of a variety of useful parametric curves.

The Distance from a Point to a Parametric Curve Assume we have a data point (d_x, d_y). The closest point on a parametric curve can be identified by its parameter value, which we shall write as τ. This point could lie at one or other end of the curve. Otherwise, the vector from our data point to the closest point is normal to the curve. This means that $\boldsymbol{s}(\tau) = (d_x, d_y) - (x(\tau), y(\tau))$ is normal to the tangent vector, so that $\boldsymbol{s}(\tau).\boldsymbol{T} = 0$. The tangent vector is

$$\left(\frac{dx}{dt}(\tau), \frac{dy}{dt}(\tau)\right),$$

which means that τ must satisfy the equation

$$\frac{dx}{dt}(\tau)\{d_x - x(\tau)\} + \frac{dy}{dt}(\tau)\left\{d_y - y(\tau)\right\} = 0.$$

Now this is only one equation, rather than two, but the situation is not much better than that for parametric curves. It is almost always the case that $x(t)$ and $y(t)$ are polynomials because it is usually easier to do root finding for polynomials. If $x(t)$ and $y(t)$ are ratios of polynomials we can rearrange the left-hand side of our equation to come up with a polynomial in this case, too. However, we are still faced with a possibly large number of roots.

There is a second difficulty that makes fitting to parametric curves unpopular. Parametric curves with different coefficients may represent the same curve—for example, the curve $(x(t), y(t))$ for $t \in [0, 1]$ is the same as the curve $(x(2t), y(2t))$ for $t \in [0, 1/2]$. This situation can be very bad depending on the class of parametric curves that we use.

TABLE 15.2 A selection of parametric curves often used in vision applications. It is quite common to put together a set of cubic curves, with constraints on their coefficients such that they form a single continuous differentiable curve; the result is known as a *cubic spline*.

Curves	Parametric Form	Parameters
Circles centered at the origin	$(r\sin(t), r\cos(t))$	$\theta = r$ $t \in [0, 2\pi)$
Circles	$(r\sin(t) + a, r\cos(t) + b)$	$\theta = (r, a, b)$ $t \in [0, 2\pi)$
Axis aligned ellipses	$(r_1\sin(t) + a, r_2\cos(t) + b)$	$\theta = (r_1, r_2, a, b)$ $t \in [0, 2\pi)$
Ellipses	$(\cos\phi\,(r_1\sin(t) + a) - \sin\phi\,(r_2\cos(t) + b)$, $\sin\phi\,(r_1\sin(t) + a) + \cos\phi\,(r_2\cos(t) + b))$	$\theta = (r_1, r_2, a, b, \phi)$ $t \in [0, 2\pi)$
cubic segments	$(at^3 + bt^2 + ct + d, et^3 + ft^2 + gt + h)$	$\theta = (a, b, c, d, e, f, g, h)$ $t \in [0, 1]$

15.4 FITTING AS A PROBABILISTIC INFERENCE PROBLEM

Up to this point, our criteria for fitting to a model have been arbitrary. Total least squares seems like a reasonable criterion, but (of course!) the criterion should depend on the kind of error model that we expect—how did the tokens come to not lie on a line in the first place? We return to the problem of fitting a line to a set of points that are known to have come from the line. It turns out that total least squares is, quite naturally, a probabilistic criterion. We start with a model that indicates how image measurements arise.

Generative Model We assume that our measurements are generated by choosing a point along the line and then perturbing it perpendicular to the line using Gaussian noise. We assume that the process that chooses points along the line is uniform—in principle, it can't be because the line is infinitely long, but in practice we can assume that any difference from uniformity is too small to bother with. This means we have a sequence of k measurements (x_i, y_i), that are obtained from the model

$$\begin{pmatrix} x \\ y \end{pmatrix} = \begin{pmatrix} u \\ v \end{pmatrix} + n \begin{pmatrix} a \\ b \end{pmatrix},$$

where $n \sim N(0, \sigma)$, $au+bv+c = 0$, and $a^2+b^2 = 1$. This model yields the likelihood function:

$$P(\text{measurements} \mid a, b, c) = \prod_i P(x_i, y_i \mid a, b, c).$$

Now we could choose either the maximum likelihood or the maximum *a posteriori* line, given this model. Typically, we have no particular reason to prefer one line over another, and maximum likelihood is fine. Now the log-likelihood is

$$-\frac{1}{2\sigma^2} \sum_i (ax_i + by_i + c)^2$$

given that $a^2 + b^2 = 1$. Maximizing the likelihood boils down to minimizing the sum of perpendicular distances between points and lines as in Section 15.2.1. There are two significant phenomena that we must deal with in using this criterion:

- **Robustness:** The total least-squares criterion places huge weight on large errors. This could become a serious problem. For example, if one data point lay a long way from a line that fits all others well (we discuss some mechanisms by which this can happen later), the resulting fitted line will be heavily biased by that data point. This phenomenon can become a serious problem. For example, if we are fitting a fundamental matrix to a data set, we need correspondences between data points in left and right images; but if we get one correspondence wrong, we have a potentially huge error in our data set. We discuss this difficulty in detail in the following two sections.
- **Missing data:** We assumed that we knew which points belonged to the line; it is usually the case that we do not. For example, we might have a set of measured points, some of which come from a line and others of which are noise. If we knew which points came from a line, it would be easy to determine what the line was. Similarly, if we knew what line generated the points, it would be easy to determine which points had come from the line. The missing data—which point is noise and which is not—is a crucial component of the problem. Most segmentation problems can be seen as missing data problems; we devote most of chapter 16 to this view.

15.5 ROBUSTNESS

All of the line fitting methods described involve squared error terms. This can lead to poor fits in practice because a single wildly inappropriate data point can give errors that dominate those due to many good data points; these errors can result in a substantial bias in the fitting process (Figure 15.7). It is difficult to avoid such data points—usually called *outliers*—in practice. Errors in collecting or transcribing data points is one important source of outliers. Another common source is a problem with the model—perhaps some rare but important effect has been ignored or the magnitude of an effect has been badly underestimated. Finally, errors in correspondence are particularly prone to generating outliers. Practical vision problems usually involve outliers.

One approach to this problem puts the model at fault: The model predicts these outliers occurring perhaps once in the lifetime in the universe, and they clearly occur much more often than that. The natural response is to improve the model either by giving the noise "heavier tails" (Section 15.5.1) or by allowing an explicit outlier model. The second strategy requires a study of missing data problems—we don't know which point is an outlier and which isn't—and we defer

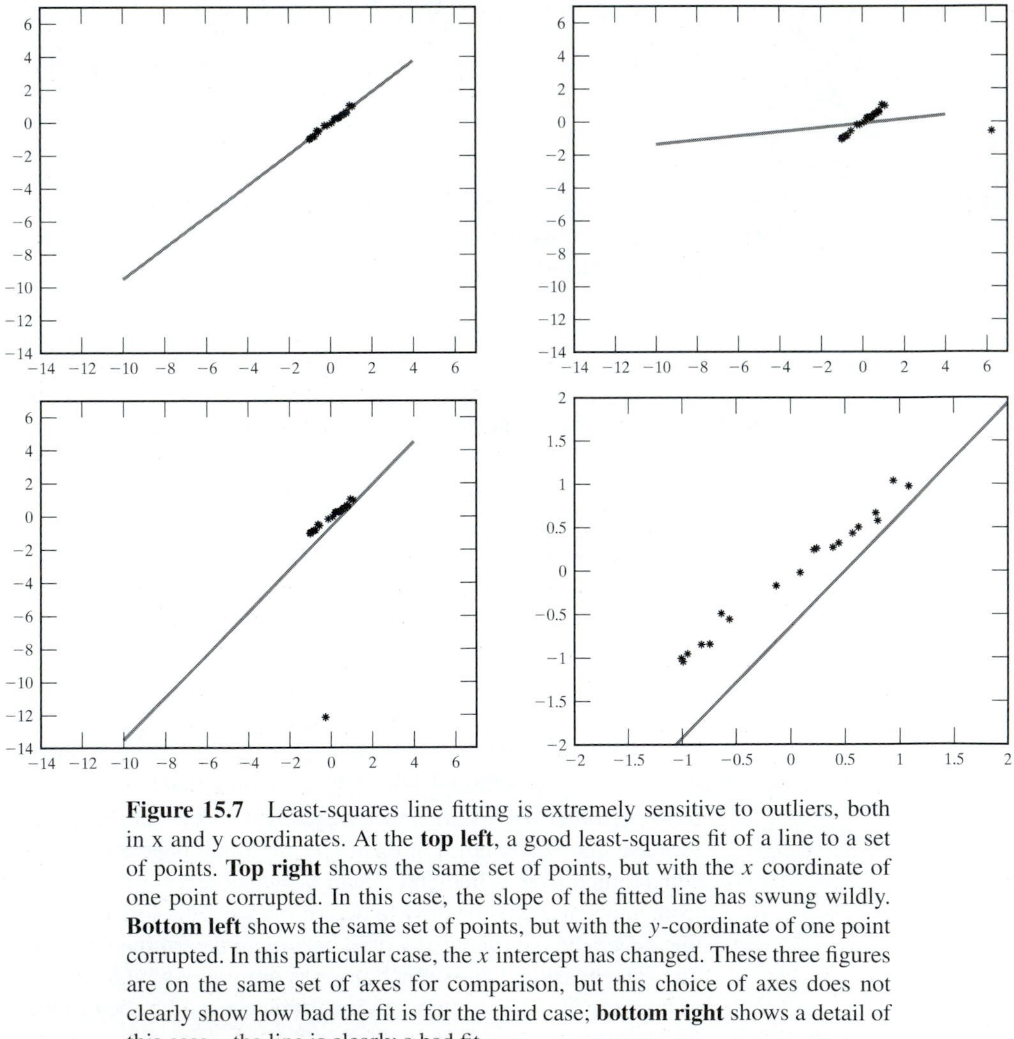

Figure 15.7 Least-squares line fitting is extremely sensitive to outliers, both in x and y coordinates. At the **top left**, a good least-squares fit of a line to a set of points. **Top right** shows the same set of points, but with the x coordinate of one point corrupted. In this case, the slope of the fitted line has swung wildly. **Bottom left** shows the same set of points, but with the y-coordinate of one point corrupted. In this particular case, the x intercept has changed. These three figures are on the same set of axes for comparison, but this choice of axes does not clearly show how bad the fit is for the third case; **bottom right** shows a detail of this case—the line is clearly a bad fit.

discussion until Section 16.2.4 in the following chapter. An alternative approach is to search for points that appear to be good (Section 15.5.2).

15.5.1 M-estimators

The difficulty with modeling the source of outliers is that the model might be wrong. Generally, the best we can hope for from a probabilistic model of a process is that it is quite close to the right model. Assume that we are guaranteed that our model of a process is close to the right model—say, the distance between the density functions in some appropriate sense is less than ϵ. We can use this guarantee to reason about the design of estimation procedures for the parameters of the model. In particular, we can choose an estimation procedure by assuming that nature is malicious and well informed about statistics. These are generally sound assumptions for any enterprise; the world is full of opportunities for painful and expensive lessons in practical statistics. In this line of reasoning, we assess the goodness of an estimator by assuming that somewhere in the collection of processes close to our model is the real process, and it just happens to be the one that makes the estimator produce the worst possible estimates. The best estimator is the one that behaves best on the worst distribution close to the parametric model. This is a criterion that can be used to produce a wide variety of estimators.

An *M-estimator* estimates parameters by minimizing an expression of the form

$$\sum_i \rho(r_i(\boldsymbol{x}_i, \theta); \sigma),$$

where θ are the parameters of the model being fitted and $r_i(\boldsymbol{x}_i, \theta)$ is the residual error of the model on the ith data point. Generally, $\rho(u; \sigma)$ looks like u^2 for part of its range and then flattens out. A common choice is

$$\rho(u; \sigma) = \frac{u^2}{\sigma^2 + u^2}.$$

The parameter σ controls the point at which the function flattens out; we have plotted a variety of examples in Figure 15.8. There are many other M-estimators available. Typically, they are discussed in terms of their *influence function*, which is defined as

$$\frac{\partial \rho}{\partial \theta}.$$

This is natural because our criterion is

$$\sum_i \rho(r_i(\boldsymbol{x}_i, \theta); \sigma)\frac{\partial \rho}{\partial \theta} = 0.$$

For the kind of problems we consider, we would expect a good influence function to be antisymmetric—there is no difference between a slight overprediction and a slight underprediction—and to tail off with large values—because we want to limit the influence of the outliers.

There are two tricky issues with using M-estimators. First, the extremization problem is non-linear and must be solved iteratively. The standard difficulties apply: There may be more than one local minimum, the method may diverge, and the behavior of the method is likely to be quite dependent on the start point. A common strategy for dealing with this problem is to draw a subsample of the data set, fit to that subsample using least squares, and use this as a start point for the fitting process. We do this for a large number of different subsamples—enough to ensure that there is a high probability that in that set there is at least one that consists entirely of good data points.

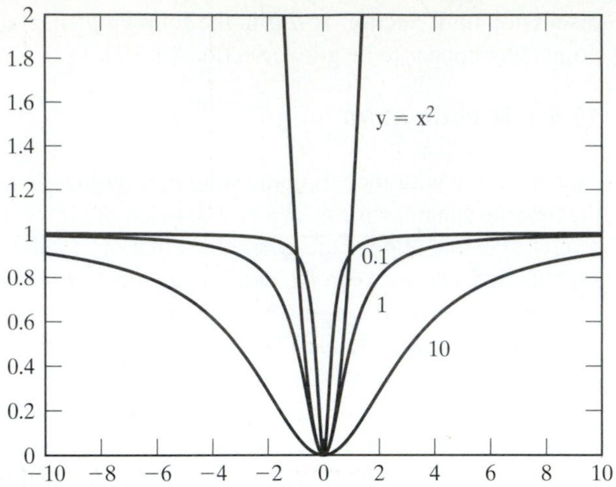

Figure 15.8 The function $\rho(x;\sigma) = x^2/(\sigma^2 + x^2)$, plotted for $\sigma^2 = 0.1$, 1, and 10, with a plot of $y = x^2$ for comparison. Replacing quadratic terms with ρ reduces the influence of outliers on a fit—a point that is several multiples of σ away from the fitted curve is going to have almost no effect on the coefficients of the fitted curve because the value of ρ will be close to 1 and will change extremely slowly with the distance from the fitted curve.

Second, as Figures 15.9 and 15.10 indicate, the estimators require a sensible estimate of σ, which is often referred to as *scale*. Typically, the scale estimate is supplied at each iteration of the solution method; a popular estimate of scale is

$$\sigma^{(n)} = 1.4826 \text{ median}_i \, |r_i^{(n)}(x_i; \theta^{(n-1)})|.$$

Algorithm 15.3: Using an M-estimator to Fit a Probabilistic Model

For $s = 1$ to $s = k$
  Draw a subset of r distinct points, chosen uniformly at random
  Fit to this set of points using maximum likelihood
    (usually least squares) to obtain θ_s^0
  Estimate σ_s^0 using θ_s^0
  Until convergence (usually $|\theta_s^n - \theta_s^{n-1}|$ is small):
    Take a minimizing step using θ_s^{n-1}, σ_s^{n-1} to get θ_s^n
    Now compute σ_s^n
  end
end
Report the best fit of this set, using the median of the
  residuals as a criterion

An M-estimator can be thought of as a trick for ensuring that there is more probability in the tails than would otherwise occur with a quadratic error. The function that is minimized looks like distance for small values of $\boldsymbol{x}$—thus, for valid data points, the behavior of the M-estimator

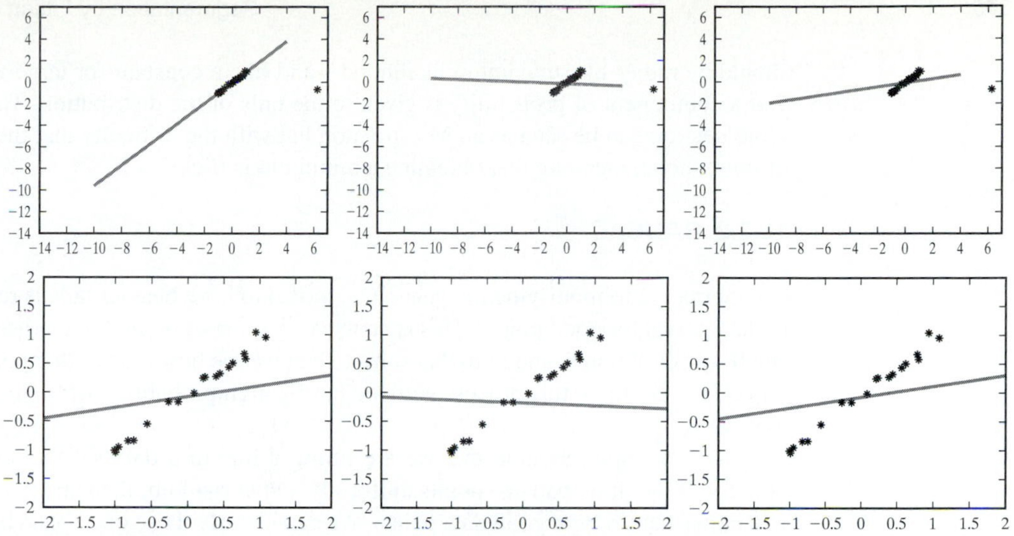

Figure 15.9 The **top row** shows lines fitted to the second dataset of Figure 15.7 using a weighting function that deemphasizes the contribution of distant points (the function ϕ of Figure 15.8). On the **left**, σ has about the right value; the contribution of the outlier has been down-weighted, and the fit is good. In the **center**, the value of σ is too small so that the fit is insensitive to the position of all the data points, meaning that its relationship to the data is obscure. On the **right**, the value of σ is too large, meaning that the outlier makes about the same contribution that it does in least-squares. The **bottom row** shows closeups of the fitted line and the non-outlying data points for the same cases.

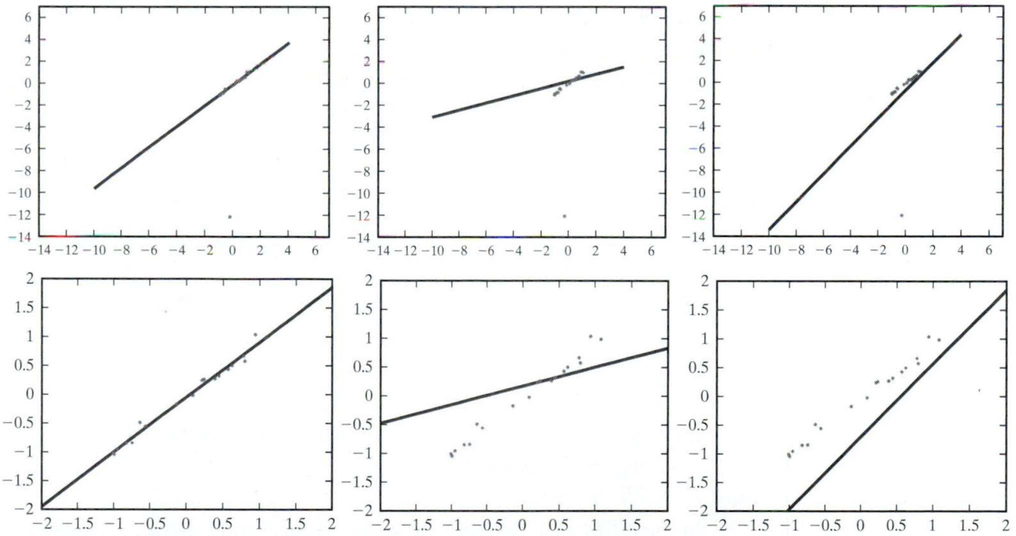

Figure 15.10 The **top row** shows lines fitted to the third dataset of Figure 15.7 using a weighting function that deemphasizes the contribution of distant points (the function ϕ of Figure 15.8). On the **left**, σ has about the right value; the contribution of the outlier has been down-weighted, and the fit is good. In the **center**, the value of σ is too small, so that the fit is insensitive to the position of all the data points, meaning that its relationship to the data is obscure. On the **right**, the value of σ is too large, meaning that the outlier makes about the same contribution that it does in least-squares. The **bottom row** shows closeups of the fitted line and the non-outlying data points, for the same cases.

should be rather like maximum likelihood—and like a constant for large values of $\boldsymbol{x}$—meaning that a component of probability is given to the tails of the distribution. The strategy of the previous section can be seen as an M-estimator, but with the difficulty that the influence function is discontinuous, meaning that obtaining a minimum is tricky.

15.5.2 RANSAC

An alternative to modifying the generative model to have heavier tails is to search the collection of data points for good points. This is quite easily done by an iterative process: First, we choose a small subset of points and fit to that subset, then we see how many other points fit to the resulting object. We continue this process until we have a high probability of finding the structure we are looking for.

For example, assume that we are fitting a line to a data set that consists of about 50% outliers. If we draw pairs of points uniformly and at random, then about a quarter of these pairs will consist entirely of good data points. We can identify these good pairs by noticing that a large collection of other points lie close to the line fitted to such a pair. Of course, a better estimate of the line could then be obtained by fitting a line to the points that lie close to our current line.

This approach leads to an algorithm—search for a random sample that leads to a fit on which many of the data points agree. The algorithm is usually called `RANSAC`, for RANdom SAmple Consensus, and is displayed in Algorithm 15.4. To make this algorithm practical, we need to choose three parameters.

Algorithm 15.4: RANSAC: Fitting Lines Using Random Sample Consensus

Determine:
  n—the smallest number of points required
  k—the number of iterations required
  t—the threshold used to identify a point that fits well
  d—the number of nearby points required
    to assert a model fits well
Until k iterations have occurred
  Draw a sample of n points from the data uniformly and at random
  Fit to that set of n points
  For each data point outside the sample
    Test the distance from the point to the line against t; if the distance from the point
      to the line is less than t, the point is close
  end
  If there are d or more points close to the line then there is a good fit. Refit the line
    using all these points.
end
Use the best fit from this collection, using the fitting error as a criterion

The Number of Samples Required Our samples consist of sets of points drawn uniformly and at random from the data set. Each sample contains the minimum number of points required to fit the abstraction we wish to fit. For example, if we wish to fit lines, we draw pairs of points; if we wish to fit circles, we draw triples of points, and so on. We assume that we need to draw n data points, and that w is the fraction of these points that are good (we need only a reasonable estimate of this number). Now the expected value of the number of draws k required to get one point is given by

$$\begin{aligned} E[k] &= 1P(\text{one good sample in one draw}) + 2P(\text{one good sample in two draws}) + \cdots \\ &= w^n + 2(1-w^n)w^n + 3(1-w^n)^2 w^n + \cdots \\ &= w^{-n} \end{aligned}$$

(where the last step takes a little manipulation of algebraic series). We would like to be fairly confident that we have seen a good sample, so we wish to draw more than w^{-n} samples; a natural thing to do is to add a few standard deviations to this number. The standard deviation of k can be obtained as

$$SD(k) = \frac{\sqrt{1-w^n}}{w^n}.$$

An alternative approach to this problem is to look at a number of samples that guarantees a low probability z of seeing only bad samples. In this case, we have

$$(1-w^n)^k = z,$$

which means that

$$k = \frac{\log(z)}{\log(1-w^n)}.$$

It is common to have to deal with data where w is unknown. However, each fitting attempt contains information about w. In particular, if n data points are required, then we can assume that the probability of a successful fit is w^n. If we observe a long sequence of fitting attempts, we can estimate w from this sequence. This suggests that we start with a relatively low estimate of w, generate a sequence of attempted fits, and then improve our estimate of w. If we have more fitting attempts than we need for the new, the process can stop. The problem of updating the estimate of w reduces to estimating the probability that a coin comes up heads or tails given a sequence of fits.

Telling Whether a Point Is Close We need to determine whether a point lies close to a line fitted to a sample. We do this by determining the distance between the point and the fitted line, and testing that distance against a threshold d; if the distance is below the threshold, the point lies close. In general, specifying this parameter is part of the modeling process. For example, when we fitted lines using maximum likelihood, there was a term σ in the model (which disappeared in the manipulations to find an maximum). This term gives the average size of deviations from the model being fitted.

In general, obtaining a value for this parameter is relatively simple. We generally need only an order of magnitude estimate, and the same value applies to many different experiments. The parameter is often determined by trying a few values and seeing what happens; another approach is to look at a few characteristic data sets, fitting a line by eye, and estimating the average size of the deviations.

The Number of Points That Must Agree Assume that we have fitted a line to some random sample of two data points. We need to know whether that line is good. We do this by counting the number of points that lie within some distance of the line (the distance was determined in the previous section). In particular, assume that we know the probability that an outlier lies in this collection of points; write this probability as y. We should like to choose some number of points t such that y^t is small (say less than 0.05).

There are two ways to proceed. One is to notice that $y \leq (1 - w)$ and to choose t such that $(1 - w)^t$ is small. Another is to get an estimate of y from some model of outliers—for example, if the points lie in a unit square, the outliers are uniform, and the distance threshold is d, then $y \leq 2\sqrt{2}d$.

15.6 EXAMPLE: USING RANSAC TO FIT FUNDAMENTAL MATRICES

A point in 3D generates two measurements—one in the left view and one in the right. We write the actual coordinates of the 3D point as $\boldsymbol{X}_i$, the coordinates in the left (resp. right) image as $\boldsymbol{x}_{li}$ (resp. $\boldsymbol{x}_{ri}$), the measured coordinates in the left (resp. right) image as $\boldsymbol{m}_{ri}$ (resp. $\boldsymbol{m}_{ri}$). The fundamental matrix is an expression of the epipolar constraint. In particular, using the hat to indicate that we are employing homogeneous coordinates, we have that $\hat{\boldsymbol{x}}_{ri}^T \mathcal{F} \hat{\boldsymbol{x}}_{li} = 0$ for every point. Here $\mathcal{F}$ is the fundamental matrix.

15.6.1 An Expression for Fitting Error

The constraint generally does not hold for the measured values. We assume that measurements are subject to additive Gaussian noise of uniform rotationally symmetric covariance. We write $\boldsymbol{m}_{li}$ and $\boldsymbol{m}_{ri}$ in affine (conventional, or non-homogeneous!) coordinates, too. This means that

$$P(\boldsymbol{m}_{li}, \boldsymbol{m}_{ri} | \boldsymbol{x}_{ri}, \boldsymbol{x}_{li}, \mathcal{F}) \propto \exp -\frac{1}{2\sigma^2} \left\{ \begin{array}{c} (\boldsymbol{x}_{ri} - \boldsymbol{m}_{ri})^T (\boldsymbol{x}_{ri} - \boldsymbol{m}_{ri}) + \\ (\boldsymbol{x}_{li} - \boldsymbol{m}_{li})^T (\boldsymbol{x}_{li} - \boldsymbol{m}_{li}) \end{array} \right\}.$$

Now this is a complicated function of the data, with parameters $\mathcal{F}$, x_{ri}, y_{ri} and x_{li}. However, with sufficient data points we could, in principle, obtain an extremum. We can't do this currently, because we don't know which points correspond between left and right images. Furthermore, we should suspect that the sum-of-squares form of the log-likelihood will lead to robustness problems.

15.6.2 Correspondence as Noise

One way to deal with the correspondence problem is to assume that there is only a small camera motion between the two views. In turn, we can assert that feature points whose position in the second view is "close" to their position in the first view correspond. This is a fairly dangerous assumption; it can lead to fits that are bad, but look quite good. The difficulty is that *a correspondence error behaves like an outlier*. An alternative strategy is to search the correspondences for a set that is consistent with a good fundamental matrix.

This search can be simplified if we attach some representation of the local image neighborhood to each point. This means that we have a more detailed description of each point. For example, we might have a set of filter outputs computed at that point over a range of scales. We would expect a matching point to have a local image neighborhood that wasn't all that different from the original point, so the set of filter outputs shouldn't change all that much. This criterion will yield a set of possible correspondences, which may contain errors—these errors will behave like outliers. We now apply RANSAC to this set of putative correspondences.

15.6.3 Applying RANSAC

As we see next, seven point correspondences yield a fundamental matrix. With this information and one further trick, applying RANSAC to the set of possible correspondences is relatively straighforward. Although we may not know what percentage of correspondences is good, it is possible to estimate this using fitting attempts (as before). The distance threshold to determine whether a point is an inlier is usually of the order of a pixel or so (this must depend on the quality of the cameras, etc.).

Obtaining a Fundamental Matrix from Seven Points Assume that we have seven hypothesized correspondences. We assume that at each point the measured value is the same as the actual value of the point. Each constraint $\hat{\boldsymbol{x}}_{ri}^T \mathcal{F} \hat{\boldsymbol{x}}_{li} = 0$ yields a single linear equation in the coefficients of the fundamental matrix for *known* $\boldsymbol{x}_{ri}, \boldsymbol{x}_{li}$. Furthemore, the equation $\hat{\boldsymbol{x}}_{ri}^T \mathcal{F} \hat{\boldsymbol{x}}_{li} = 0$ is homogeneous in the elements of the fundamental matrix—that is, if $\mathcal{U}$ satisfies these constraints, then so does $\lambda\mathcal{U}$. Finally, recall from chapter 10 that the fundamental matrix has rank 2, and so $\det(\mathcal{F}) = 0$.

This means that we need only seven point correspondences to estimate $\mathcal{F}$. Each point yields a single homogeneous equation in the elements of $\mathcal{F}$. The solution of seven of these homogeneous equations is a two-dimensional linear space, which we can write as $\lambda\mathcal{F}_0 + \mu\mathcal{F}_1$ for known $\mathcal{F}_0$ and $\mathcal{F}_1$, and arbitrary λ, μ. But we need an element of this space with zero determinant, and the equation $\det(\lambda\mathcal{F}_0 + \mu\mathcal{F}_1) = 0$ is a homogeneous cubic in λ, μ. We can divide both sides by μ and solve for λ/μ. There is either one real root—and so only one solution—or three.

Comparing Other Points with This Fundamental Matrix Now assume that we have an estimate of the fundamental matrix. We need to know whether a particular hypothesized correspondence agrees with that estimate. For this correspondence, we know measured points $\boldsymbol{m}_{ri}$, $\boldsymbol{m}_{li}$ and some estimate $\mathcal{F}_0$ of the fundamental matrix. We do not know the actual points $\boldsymbol{x}_{ri}$, $\boldsymbol{x}_{li}$ that generated our measurements. However, we have

$$P(\boldsymbol{m}_{li}, \boldsymbol{m}_{ri} | \boldsymbol{x}_{ri}, \boldsymbol{x}_{li}, \mathcal{F}) \propto \exp -\frac{1}{2\sigma^2} \left\{ \begin{array}{c} (\boldsymbol{x}_{ri} - \boldsymbol{m}_{ri})^T (\boldsymbol{x}_{ri} - \boldsymbol{m}_{ri}) + \\ (\boldsymbol{x}_{li} - \boldsymbol{m}_{li})^T (\boldsymbol{x}_{li} - \boldsymbol{m}_{li}) \end{array} \right\}.$$

If our estimate of the fundamental matrix is good *and* if the correspondence is good, then there are some actual points that are close to our measurements and are consistent with our fundamental matrix estimate. This means we need to know the distance between the closest such points and the measurements—if this distance is small, then the correspondence is good. Assume that we can recover this distance (a nice trick for doing so is given later; it would distract us to do it here). Then obtaining the fundamental matrix using RANSAC is simple. We put all the pieces together in Algorithm 15.5.

15.6.4 Finding the Distance

We have a pair of measured points $\boldsymbol{m}_{li}$ and $\boldsymbol{m}_{ri}$, and we wish to find the distance between these points and the closest pair of points *consistent with our fundamental matrix estimate*. This means that the closest pair of points must have the property that $\boldsymbol{x}_{li}^T \mathcal{F}_0 \boldsymbol{x}_{ri} = 0$. We sketch a clever solution to this problem due to Hartley and Sturm (1997).

Notice that, first, rotating and translating the image coordinate system in both the left and the right images does not affect the distance. We ignore the possibility that the best pair of points is at the epipole because it never happens (is non-generic, in polite language). We could translate both coordinate systems to put the epipole at the origin. Now the right epipole has the property that $\mathcal{F}\hat{\boldsymbol{e}}_r = \boldsymbol{0}$ and the left epipole that $\hat{\boldsymbol{e}}_l^T \mathcal{F} = \boldsymbol{0}^T$. This means that, in a coordinate system where the epipole in both the left image and the right image is at the origin (and rotation and translation will put it there unless it is at infinity, a case we ignore),

$$\mathcal{F} = \begin{pmatrix} \mathcal{M} & \boldsymbol{0} \\ \boldsymbol{0}^T & 0 \end{pmatrix}.$$

Algorithm 15.5: RANSAC: Fitting a Fundamental Matrix Using Random Sample Consensus

Determine:
  the smallest number of corresponding pairs of points required is seven
  k—the number of iterations required
  t—the threshold used to identify a point that fits well
  d—the number of nearby points required
    to assert a model fits well
Until k iterations have occurred
  Draw a sample of seven correspondences from the data uniformly and at random
  Use the seven point algorithm to obtain an estimate of the fundamental matrix $\mathcal{F}_0$.
  For each putative correspondence outside the sample
    Test the distance from the closest points consistent with $\mathcal{F}_0$ to the measured points
      against t;
      if the distance is less than t, the correspondence is consistent
  end
  If there are d or more consistent correspondences then there is a good fit. Refit the
    estimate of the fundamental matrix using all these points.
end
Use the best fit from this collection, using the fitting error as a criterion

This is a three-by-three matrix because $\mathcal{M}$ is some two-by-two block. Now what happens if we rotate the coordinate system about the origin? Write $\boldsymbol{u}_{ri} = \mathcal{C}_{ri}\boldsymbol{x}_{ri}$, where

$$\mathcal{C}_{ri} = \begin{pmatrix} \mathcal{R}_{ri} & \mathbf{0} \\ \mathbf{0}^T & 0 \end{pmatrix}$$

and $\mathcal{R}_{ri}$ is a plane rotation (when we wish to talk about the left image, we'll use the same notation, but with the subscript "li"). Now if we work in $\boldsymbol{u}$ (rotated) coordinates, the new fundamental matrix has the form

$$\mathcal{F}_0' = \mathcal{C}_{ri}^T \mathcal{F}_0 \mathcal{C}_{li} = \begin{pmatrix} \mathcal{R}_{ri}^T \mathcal{M} \mathcal{R}_{li} & \mathbf{0} \\ \mathbf{0}^T & 0 \end{pmatrix},$$

which means that, by a suitable choice of rotations, we can diagonalize the fundamental matrix.

Now let us consider the question of distance. The distance from the measured point to the actual point in the left image is the same as the distance from the measured point and some epipolar line in the left image. This epipolar line is transformed into some other epipolar line in the right image, *and the fundamental matrix tells us which one*. The distance from the measured point to the actual point in the right image is the distance from the measured point *to this epipolar line*. What we need to do is minimize the sum of distances.

The epipoles are at the origin in each image. In the left image, the epipolar lines are the family of lines $(s, t, 0)$ (meaning $sx+ty = 0$—all these lines pass through the origin, and varying s and t changes the line). In the right image because we have rotated the coordinate systems to diagonalize the fundamental matrix, the corresponding lines have the form $(s, \lambda t, 0)$ (meaning $sx + \lambda ty = 0$) for some λ depending on the fundamental matrix. Write m_{xri} for the x coordinate of $\boldsymbol{m}_{ri}$, and so on. Now the distance is

$$\frac{(sm_{xli} + tm_{yli})^2}{(s^2 + t^2)} + \frac{(sm_{xri} + t\lambda m_{yri})^2}{(s^2 + \lambda^2 t^2)}$$

and we have to minimize this as a function of s and t. In fact, it is homogeneous in s and t, so we need to consider it only as a function of $u = s/t$. We maximize this function *and then* consider it as a function of $v = t/s$ just in case there was a maximum at $t = 0$. Maximization is easy, and we consider only the case of u. We have a rational function of u, whose numerator and denominator are of degree four. This must be equal to a rational function of u, whose numerator is of degree three and whose denominator is of degree four, plus a constant (this works because we can do long division to get rid of the degree four term in the numerator). Now the numerator of the derivative of this expression must vanish, and this is a polynomial in u of degree six. We find its roots, and we are done.

15.6.5 Fitting a Fundamental Matrix to Known Correspondences

Now we have a system of known correspondences and wish to fit a fundamental matrix. We wish to obtain a maximum likelihood solution for $\mathcal{F}$; we already have an expression for the log likelihood, but this incorporates a set of points $\boldsymbol{x}_{ri}, \boldsymbol{x}_{li}$, which are unknown. We can either attempt to marginalize with respect to these points—which looks hard—or solve for them, too.

The appropriate strategy is to solve for them *as points in 3D*. We can fix one camera at the origin, and the second camera then takes a standard form as a function of the fundamental matrix alone. We now minimize the negative log likelihood as a function of the fundamental matrix and the 3D configuration of the points. We have an estimated start point for this latter because we know the correspondences and the fundamental matrix.

15.7 NOTES

We have covered a few important points from a large body of technique here. Fitting is a problem that occurs in any number of contexts; it is almost always possible and often helpful to see a problem as a fitting problem. This means it is difficult to supply a useful guide to the literature.

Usually the main difficulties one encounters in practice are: (a) determining distances (which can be very hard indeed); (b) ensuring that outliers do not overwhelm good data; and (c) deciding what to fit in the first place.

Approximations to the distance from a point to a curve or to a surface are discussed in numerous papers: We particularly recommend Agin (1981), Bookstein (1979), Cabrera and Meer (1996), Porrill (1990), Sabin (1994), Sullivan, Sandford and Ponce (1994*a,b*), Sullivan and Ponce (1998), and Taubin, Cukierman, Sullivan, Ponce and Kriegman (1994*a,b*).

Robustness hasn't had as much impact on practices within the vision community as it probably should have. Good starting points for reading include Huber (1981), Meer, Mintz, Kim and Rosenfeld (1991), Rousseeuw (1987), and Stewart (1999). The ideas are genuinely useful despite the subject's tendency to inspire zealotry. We haven't gone deeply into the topic mainly because a superficial acquaintance with the topic is sufficient to deal with any issues likely to arise in practice.

We used the example of computing a fundamental matrix using RANSAC because this is currently the best method of doing so. Anyone interested in applications of RANSAC should be looking at structure from motion, and particularly the account in Hartley and Zisserman (2000), which shows the tremendous impact the algorithm has had. We recommend reading also Fischler and Bolles (1981) and Torr and Murray (1997). We expect that in future it will be used to start EM-like methods; more on that later once we have discussed EM.

We haven't really stressed fitting as inference. The advantage of thinking about fitting as inference is that anything one knows about fitting can be exchanged into knowledge about fitting, too, and the exchange rate seems to be favorable. The next chapter shows some of that, but you should also be aware that our comments of the probability chapter on the website apply to pretty much everything in this chapter, too. In particular, one might use MAP inference—all this would require is attaching a prior term to the fitting error. In years gone by, it was fashionable to construct a prior term that penalized models that wiggled too quickly (a phenomenon often confused with a failure to be smooth). This practice is useful but seems to have diminished in the last few years; samples of this literature include Horn and Schunck (1981), Bertero, Poggio and Torre (1988), and Poggio, Torre and Koch (1985).

Fitting can also be used for image reconstruction. One fits a surface to image data—which is interpreted as a height map—tearing the surface at edges. It is then important to account correctly for the cost of tearing versus the cost of a poor fit. This general strategy can also be applied to reconstruction of depth maps, stereo maps, and so on. Various constraints are available. In Grimson (1981*b*), the absence of features is taken to imply smoothness ("No news is good news"). More recent work accounts for discontinuities in various forms, including tears and creases (Blake and Zisserman, 1987, Mumford and Shah, 1985 and 1988).

PROBLEMS

15.1. Prove the simple, but extremely useful, result that the perpendicular distance from a point (u, v) to a line (a, b, c) is given by $\text{abs}(au + bv + c)$ *if* $a^2 + b^2 = 1$.

15.2. Derive the eigenvalue problem

$$\begin{pmatrix} \overline{x^2} - \overline{x}\,\overline{x} & \overline{xy} - \overline{x}\,\overline{y} \\ \overline{xy} - \overline{x}\,\overline{y} & \overline{y^2} - \overline{y}\,\overline{y} \end{pmatrix} \begin{pmatrix} a \\ b \end{pmatrix} = \mu \begin{pmatrix} a \\ b \end{pmatrix}$$

from the generative model for total least squares. This is a simple exercise—maximum likelihood and a little manipulation will do it—but worth doing right and remembering; the technique is extremely useful.

15.3. How do we get a curve of edge points from an edge detector that returns orientation? Give a recursive algorithm.

15.4. A slightly more stable variation of incremental fitting cuts the first few pixels and the last few pixels from the line point list when fitting the line because these pixels may have come from a corner

(a) Why would this lead to an improvement?

(b) How should one decide how many pixels to omit?

15.5. A conic section is given by $ax^2 + bxy + cy^2 + dx + ey + f = 0$.

(a) Given a data point (d_x, d_y), show that the nearest point on the conic (u, v) satisfies two equations:

$$au^2 + buv + cv^2 + du + ev + f = 0$$

and

$$b(u^2 + v^2) + 2(a - c)\,uv - (2a\,d_y + e)u + (2c\,d_x + d)v + (e\,d_x - d\,d_y) = 0.$$

(b) These are two quadratic equations. Write $\boldsymbol{u}$ for the vector $(u, v, 1)$. Now show that we can write these equations as $\boldsymbol{u}^T \mathcal{M}_1 \boldsymbol{u} = 0$ and $\boldsymbol{u}^T \mathcal{M}_2 \boldsymbol{u} = 0$, for $\mathcal{M}_1$ and $\mathcal{M}_2$ symmetric matrices.

(c) Show that there is a transformation $\mathcal{T}$, such that $\mathcal{T}^T \mathcal{M}_1 \mathcal{T} = Id$ and $\mathcal{T}^T \mathcal{M}_2 \mathcal{T}$ is diagonal.

(d) Now show how to use this transformation to obtain a set of solutions to the equations; in particular, show that there can be up to four real solutions.

(e) Show that there are four, two, or zero real solutions to these equations.

(f) Sketch an ellipse and indicate the points for which there are four or two solutions.

15.6. Show that the curve

$$\left(\frac{1-t^2}{1+t^2}, \frac{2t}{1+t^2}\right)$$

is a circular arc (the length of the arc depending on the interval for which the parameter is defined).

(a) Write out the equation in t for the closest point on this arc to some data point (d_x, d_y). What is the degree of this equation? How many solutions in t could there be?

(b) Now substitute $s^3 = t$ in the parametric equation, and write out the equation for the closest point on this arc to the same data point. What is the degree of the equation? Why is it so high? What conclusions can you draw?

15.7. Show that the viewing cone for a cone is a family of planes, all of which pass through the focal point and the vertex of the cone. Now show the outline of a cone consists of a set of lines passing through a vertex. You should be able to do this by a simple argument without any need for calculations.

Programming Assignments

15.8. Implement an incremental line fitter. Determine how significant a difference results if you leave out the first few pixels and the last few pixels from the line point list (put some care into building this, as it's a useful piece of software to have lying around in our experience).

15.9. Implement a hough transform line finder.

15.10. Count lines with an HT line finder—how well does it work?

16

Segmentation and Fitting Using Probabilistic Methods

All the segmentation algorithms described in the previous chapter involve essentially local models of similarity. Although some algorithms attempt to build clusters that are good globally, the underlying *model* of similarity compares individual pixels. Furthermore, none of these algorithms involved an explicit probabilistic model of how measurements differed from the underlying abstraction that we are seeking.

We now look at explicitly probabilistic methods for segmentation. These methods attempt to explain data using global models. These models attempt to explain a large collection of data with a small number of parameters. For example, we might take a set of tokens and fit a line to them, or we might take a pair of images and attempt to fit a parametric set of motion vectors that explain how pixels move from one to the other.

16.1 MISSING DATA PROBLEMS, FITTING, AND SEGMENTATION

A number of important vision problems can be phrased as problems that happen to be missing useful elements of the data. For example, we can think of segmentation as the problem of determining from which of a number of sources a measurement came. This is a general view: Segmenting an image into regions involves determining which source of color and texture pixels generated the image pixels; segmenting a set of tokens into collinear groups involves determining which tokens lie on which line; and segmenting a motion sequence into moving regions involves allocating moving pixels to motion models. Each of these problems would be easy if we happened to possess some data that is currently missing (respectively, which region a pixel comes from, which line a token comes from, and which motion model a pixel comes from).

16.1.1 Missing Data Problems

A *missing data problem* is a statistical problem where some data is missing. There are two natural contexts in which missing data are important: In the first, some terms in a data vector are missing for some instances and present for others (perhaps someone responding to a survey was embarrassed by a question). In the second, which is far more common in our applications, an inference problem can be made much simpler by rewriting it using some variables whose values are unknown. Fortunately, there is an effective algorithm for dealing with missing data problems—in essence, we take an expectation over the missing data. We demonstrate this method and appropriate algorithms with two examples.

Example 16.1 Image Segmentation.

At each pixel in an image, we compute a d-dimensional feature vector $\boldsymbol{x}$, which encapsulates position, color, and texture information. This feature vector could contain various color representations, and the output of a series of filters centered at a particular pixel. Our image model is that each pixel is produced by a density associated with one of g image segments. Thus, to produce a pixel, we choose an image segment and then generate the pixel from the density associated with that segment.

We assume that the lth segment is chosen with probability π_l, and we model the density associated with the lth segment as a Gaussian, with parameters $\theta_l = (\boldsymbol{\mu}_l, \Sigma_l)$ that depend on the particular segment. This means that we can write the probability of generating a pixel vector as

$$p(\boldsymbol{x}) = \sum_i p(\boldsymbol{x} \mid \theta_l)\pi_l.$$

This form of model is known as a *mixture model* (because it is a weighted sum or *mixture* of probability models; the π_l are usually called *mixing weights*). We encounter this form often.

One way to interpret such a mixture model is to think of it as a *generative model*. In this view, each pixel in the image is obtained by (a) selecting the lth component of the model with probability π_l, and then (b) drawing a sample from $p(\boldsymbol{x} \mid \theta_l)$. One can visualize this model as a density in feature vector space that consists of a set of g "blobs", each of which is associated with an image segment. We should like to determine: (a) the parameters of each of these blobs, (b) the mixing weights, and (c) from which component each pixel came (thereby segmenting the image).

We encapsulate these parameters into a parameter vector, writing the mixing weights as α_l and the parameters of each blob as $\theta_l = (\boldsymbol{\mu}_l, \Sigma_l)$ to get $\Theta = (\alpha_1, \ldots, \alpha_g, \theta_1, \ldots, \theta_g)$. The mixture model then has the form

$$p(\boldsymbol{x} \mid \Theta) = \sum_{l=1}^{g} \alpha_l p_l(\boldsymbol{x} \mid \theta_l).$$

Each component density is the usual Gaussian:

$$p_l(\boldsymbol{x} \mid \theta_l) = \frac{1}{(2\pi)^{d/2} \det(\Sigma_i)^{1/2}} \exp\left\{-\frac{1}{2}(\boldsymbol{x} - \boldsymbol{\mu}_i)^T \Sigma_i^{-1} (\boldsymbol{x} - \boldsymbol{\mu}_i)\right\}.$$

The likelihood function for an image is

$$\prod_{j \in \text{observations}} \left(\sum_{l=1}^{g} \alpha_l p_l(\boldsymbol{x}_j \mid \theta_l) \right).$$

Each component is associated with a segment, and Θ is unknown.

The important point is this: If we knew the component from which each pixel came, it would be simple to determine Θ. We could use maximum likelihood estimates for each θ_l, and then the fraction of the image in each component would give the α_l. Similarly, if we knew Θ, then for each pixel we could determine the component that is most likely to have produced that pixel—this yields an image segmentation. The difficulty is that we know neither.

Example 16.2 Fitting Lines to Point Sets.

There are g different lines in the plane. The lth line is parametrized by $\boldsymbol{a}_l$ and generates tokens with probability π_l. Each token results in a measurement vector $\boldsymbol{W}$, and the value of the jth measurement is $\boldsymbol{W}_j$. For the lth line, there is a probability density function describing how it emits tokens, which we write as $p(\boldsymbol{W} \mid \boldsymbol{a}_l)$. This means that the probability density function for a set of measurements of a token is

$$p(\boldsymbol{W}) = \sum_l \pi_l p(\boldsymbol{W} \mid \boldsymbol{a}_l).$$

This is another mixture model. Under this model, the likelihood of a set of observations is

$$\prod_{j \in \text{observations}} \left(\sum_{l=1}^{g} \pi_l p(\boldsymbol{W}_j \mid \boldsymbol{a}_l) \right).$$

We would like to infer $\boldsymbol{a}_l$ and π_l. As in the case of segmentation, if we knew which point was generated by which line, the problem would be easy. We could estimate $\boldsymbol{a}_l$ by line fitting and obtain π_l by counting the number of tokens generated by the lth line and dividing by the total number of lines. The difficulty is that we have only the measurements of the tokens, not the association between tokens and lines.

A Formal Statement of Missing Data Problems Assume we have two spaces—the *complete data space* $\mathcal{X}$ and the *incomplete data space* $\mathcal{Y}$. There is a map f, that takes $\mathcal{X}$ to $\mathcal{Y}$. This map "loses" the missing data; for example, it could be a projection. For the example of image segmentation, the complete data consists of the measurements at each pixel *and* a set of variables indicating from which component of the mixture the measurements came; the incomplete data is obtained by dropping this second set of variables. For the example of lines and tokens, the complete data space consists of the measurements of the tokens (position, certainly, but color and shape could come into this) *and* a set of variables indicating from which line the token came; the incomplete data is obtained by dropping this second set of variables.

There is a parameter space $\mathcal{U}$. For the image segmentation example, the parameter space consists of the mixing weights and of the parameters of each mixture component; for the lines and tokens, the parameter space consists of the mixing weights and the parameters of each line. We wish to obtain a maximum likelihood estimate for these parameters given only incomplete data. If we had complete data, we could use the probability density function for the complete data space, written $p_c(\boldsymbol{x}; \boldsymbol{u})$. The complete data log-likelihood is

$$\begin{aligned} L_c(\boldsymbol{x}; \boldsymbol{u}) &= \log \left\{ \prod_j p_c(\boldsymbol{x}_j; \boldsymbol{u}) \right\} \\ &= \sum_j \log \left(p_c(\boldsymbol{x}_j; \boldsymbol{u}) \right). \end{aligned}$$

In either of our examples, this log-likelihood would be relatively easy to work with. In the case of image segmentation, the problem would be to estimate the parameters for each image segment given the segment from which each pixel came. In the case of the lines and tokens, the problem would be to estimate the mixing weights and parameters given the line from which each token came.

The problem is that we don't have the complete data. The probability density function for the *incomplete* data space is $p_i(\boldsymbol{y}; \boldsymbol{u})$. The probability density function for the incomplete data space is obtained by integrating the probability density function for the complete data space over all values that give the same $\boldsymbol{y}$. That is,

$$p_i(\boldsymbol{y}; \boldsymbol{u}) = \int_{\{\boldsymbol{x} | f(\boldsymbol{x}) = \boldsymbol{y}\}} p_c(\boldsymbol{x}; \boldsymbol{u})\, d\eta$$

(where η measures volume on the space of $\boldsymbol{x}$ such that $f(\boldsymbol{x}) = \boldsymbol{y}$). The incomplete data likelihood is

$$\prod_{j \in \text{observations}} p_i(\boldsymbol{y}_j; \boldsymbol{u}).$$

We could form a maximum likelihood estimate for $\boldsymbol{u}$, given $\boldsymbol{y}$ by writing out the likelihood and maximizing it. This isn't easy because both the integral and the maximization can be quite difficult to do. The usual strategy of taking logs doesn't make things easier because of the integral *inside* the log. We have

$$\begin{aligned} L_i(\boldsymbol{y}; \boldsymbol{u}) &= \log \left\{ \prod_j p_i(\boldsymbol{y}_j; \boldsymbol{u}) \right\} \\ &= \sum_j \log \left(p_i(\boldsymbol{y}_j; \boldsymbol{u}) \right) \\ &= \sum_j \log \left(\int_{\{\boldsymbol{x} | f(\boldsymbol{x}) = \boldsymbol{y}_j\}} p_c(\boldsymbol{x}; \boldsymbol{u})\, d\eta \right) \end{aligned}$$

This form of expression is difficult to deal with. The reason we are stuck with the incomplete data likelihood is that we don't know which of the many possible $\boldsymbol{x}$s that could correspond to the $\boldsymbol{y}$s that we observe actually does correspond. Forming the incomplete data likelihood involves averaging over all such $\boldsymbol{x}$s.

Strategy For each of our examples, if we knew the missing data, we could estimate the parameters effectively. Similarly, if we knew the parameters, the missing data would follow. This suggests an iterative algorithm:

1. Obtain some estimate of the missing data using a guess at the parameters.
2. Form a maximum likelihood estimate of the free parameters using the estimate of the missing data.

We would iterate this procedure until (hopefully!) it converged. For image segmentation, this would look like the following:

1. Obtain some estimate of the component from which each pixel's feature vector came using an estimate of the θ_l.
2. Update the θ_l and the mixing weights, using this estimate.

In the case of the tokens and the lines, the algorithm would look like this:

1. Obtain some estimate of the correspondence between tokens and lines, using a guess at $\boldsymbol{a}_l$.
2. Form a revised estimate of $\boldsymbol{a}_l$ using the estimated correspondence.

16.1.2 The EM Algorithm

Although it would be nice if the procedures given for missing data converged, there is no particular reason to believe that they do. In fact, given appropriate choices in each stage, they do. This is most easily shown by showing that they are examples of a general algorithm—the *expectation-maximization* algorithm.

EM for Mixture Models Now we assume that the complete data log-likelihood is *linear* in the missing variables. This case is common, because it is associated with mixture models; all the examples we show have this property.

In a mixture model, the missing data consists of variables that indicate the mixture component from which a data item is drawn (e.g., which line a point came from or whether it came from noise). We represent this information by associating with each data point a vector z of g elements (recall that each of our examples had g components in the mixture). If the jth data point comes from the lth mixture component, then the lth component of $\boldsymbol{z}_j$ (which we write z_{jl}) is one, otherwise it is zero. Then $\boldsymbol{x}_j = [\boldsymbol{y}_j, \boldsymbol{z}_j]$. Now if we write the mixture model as

$$p(\boldsymbol{y}) = \sum_l \pi_l p(\boldsymbol{y} \mid \boldsymbol{a}_l),$$

the complete data log-likelihood is

$$\sum_{j \in \text{observations}} \left(\sum_{l=1}^{g} z_{lj} \log p(\boldsymbol{y}_j \mid \boldsymbol{a}_l) \right)$$

(which is linear in the missing variables).

The key idea in EM is to obtain a set of working values for the missing data (and so for $\boldsymbol{x}$) by substituting an expectation for each missing value. In particular, we fix the parameters at some value, and then compute the expected value of each z_j, given the value of $\boldsymbol{y}_j$ and the parameter values. We then plug the expected value of z_j into the complete data log-likelihood, which is much easier to work with, and obtain a value of the parameters by maximizing that. At this point, the expected values of z_j may have changed. We obtain an algorithm by alternating the expectation step with the maximization step and iterate until convergence.

More formally, given $\boldsymbol{u}^s$, we form $\boldsymbol{u}^{s+1}$ by:

1. Computing an expected value for the *complete* data using the incomplete data and the current value of the parameters. We know the expected value of $\boldsymbol{y}_j$ and so need compute only the expected values of z_j for each j. We write these values $\overline{z}_j^{(s)}$. We use a superscript to indicate that the expectation depends on the current value of the parameters. This is referred to as the *E-step*.
2. Maximizing the *complete* data log likelihood with respect to $\boldsymbol{u}$ using the expected value of the complete data computed in the E-step. That is, we compute

$$\begin{aligned} \boldsymbol{u}^{s+1} &= \arg\max_{\boldsymbol{u}} L_c(\overline{\boldsymbol{x}}^s; \boldsymbol{u}) \\ &= \arg\max_{\boldsymbol{u}} L_c([\boldsymbol{y}, \overline{\boldsymbol{z}}^s]; \boldsymbol{u}). \end{aligned}$$

This is known as the *M-step*.

It can be shown that the incomplete data log-likelihood is increased at each step, meaning that the sequence $\boldsymbol{u}^s$ converges to a (local) maximum of the incomplete data log-likelihood (e.g., Dempster, Laird and Rubin (1977) or McLachlan and Krishnan (1996)). Of course, there is no guarantee that this algorithm converges to the *right* local maximum, and in some of the following examples we will show that finding the right local maximum can be a nuisance.

16.1.3 The EM Algorithm in the General Case

If the complete data log-likelihood is not linear in the missing data, then we cannot simply substitute expectations of these variables. We must deal with the missing variables by taking an expectation of the complete data log-likelihood with respect to the missing variables conditioned on the current value of the parameter. Assume that we know an estimate of the parameters $\boldsymbol{u}^{(s)}$. Now we average the complete data log-likelihood over all values of the complete data, weighting by the probability of each case given our estimate of the parameters and our knowledge of the incomplete data. This yields a function

$$Q(\boldsymbol{u}; \boldsymbol{u}^{(s)}) = \int L_c(\boldsymbol{x}; \boldsymbol{u}) p(\boldsymbol{x} \mid \boldsymbol{u}^{(s)}, \boldsymbol{y})\, d\boldsymbol{x},$$

which is a function of the previous estimate of the parameters. We now maximize this with respect to $\boldsymbol{u}$ to get $\boldsymbol{u}^{(s+1)} = \arg\max_{\boldsymbol{u}} Q(\boldsymbol{u}; \boldsymbol{u}^{(s)})$. It is a straightforward exercise (mainly in notation!) to show that this reduces to the algorithm described for the linear case.

16.2 THE EM ALGORITHM IN PRACTICE

Missing data problems turn up all over computer vision. We have collected a variety of examples here to illustrate the general story. The calculations are usually straightforward; once we have shown a few, we pass over the rest in silence.

16.2.1 Example: Image Segmentation, Revisited

Assume there are n pixels. The missing data forms an n by g array of indicator variables $\mathcal{I}$. In each row, there is a single one, and all other values are zero—this indicates the blob from which each pixel's feature vector came. We note the notation of example 16.1 above.

The E-step: The l, mth element of $\mathcal{I}$ is one if the lth pixel comes from the mth blob, and zero otherwise. This means that

$$\begin{aligned} \mathrm{E}(I_{lm}) &= 1P(l\text{th pixel comes from the } m\text{th blob}) \\ &\quad + 0.P(l\text{th pixel does not come from the } m\text{th blob}) \\ &= P(l\text{th pixel comes from the } m\text{th blob}). \end{aligned}$$

Assuming that the parameters are for the sth iteration are $\Theta^{(s)}$, we have

$$\overline{I}_{lm} = \frac{\alpha_m^{(s)} p_m(\boldsymbol{x}_l \mid \theta_l^{(s)})}{\sum_{k=1}^{K} \alpha_k^{(s)} p_k(\boldsymbol{x}_l \mid \theta_l^{(s)})}$$

(keeping in mind that $\alpha_m^{(s)}$ means the value of α_m on the sth iteration!).

M-step: Once we have an expected value of $\mathcal{I}$, the rest is easy. We are essentially forming maximum likelihood estimates of Θ^{s+1}. Again, the expected value of the indicator variables is not in general going to be zero or one; instead, they take some value in that range. This should be interpreted as an observation of that particular case that occurs with that frequency, meaning that the term in the likelihood corresponding to a particular indicator variable is raised to the power of the expected value. The calculation yields expressions for a weighted mean and weighted standard deviation that should be familiar:

$$\alpha_m^{(s+1)} = \frac{1}{r}\sum_{l=1}^{r} p\left(m \mid \boldsymbol{x}_l, \Theta^{(s)}\right)$$

$$\mu_m^{(s+1)} = \frac{\sum_{l=1}^{r} \boldsymbol{x}_l\, p\left(m \mid \boldsymbol{x}_l, \Theta^{(s)}\right)}{\sum_{l=1}^{r} p\left(m \mid \boldsymbol{x}_l, \Theta^{(s)}\right)}$$

$$\Sigma_m^{s+1} = \frac{\sum_{l=1}^{r} p\left(m \mid \boldsymbol{x}_l, \Theta^{(s)}\right) \left\{ (\boldsymbol{x}_l - \mu_m^{(s)}) (\boldsymbol{x}_l - \mu_m^{(s)})^T \right\}}{\sum_{l=1}^{r} p(m \mid \boldsymbol{x}_l, \Theta^{(s)})}$$

(again keeping in mind that $\alpha_m^{(s)}$ means the value of α_m on the sth iteration!).

It remains to specify appropriate feature vectors and discuss such matters as starting the EM algorithm. The results shown in Figures 16.1 and16.2 use three color features—the coordinates of the pixel in L*a*b* after the image has been smoothed—and three texture features—which use filter outputs to estimate local scale, anisotropy and contrast (Figure 16.1). Other features may well be more effective—and the position of the pixel.

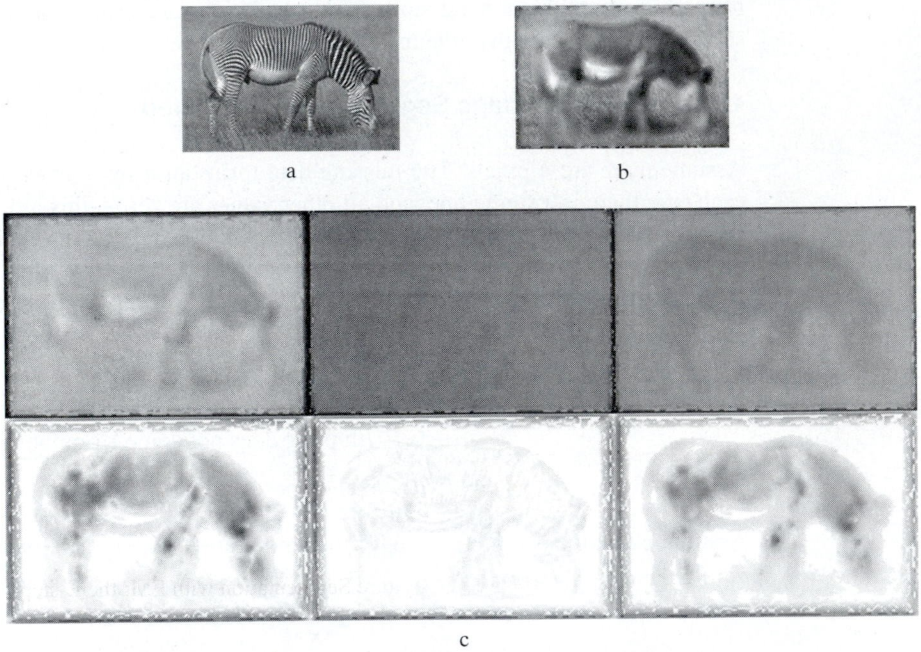

Figure 16.1 The image of the zebra in (a) is smoothed at varying scales to yield (b). This smoothing is done using local estimates of scale. These scale measurements essentially measure the scale of the change around a pixel; at edges, the scale is narrow; in stripey regions, it is broad. The features that result are shown in (c); the top three images show the smoothed color coordinates, and the bottom three show the texture features. *Reprinted from "Color and Texture Based Image Segmentation Using EM and Its Application to Content Based Image Retrieval," by S.J. Belongie et al., Proc. Int. Conf. Computer Vision, 1998 © 1998 IEEE*

Figure 16.2 Each pixel of the zebra image (which is the same as that in Figure 16.1) is labeled with the value of m for which $p(m|\boldsymbol{x}_l, \Theta^s)$ is a maximum to yield a segmentation. The images show the result of this process for $K = 2, 3, 4, 5$. Each image has K gray-level values corresponding to the segment indexes. *Reprinted from "Color and Texture Based Image Segmentation Using EM and Its Application to Content Based Image Retrieval," by S.J. Belongie et al., Proc. Int. Conf. Computer Vision, 1998 © 1998 IEEE*

Algorithm 16.1: Color and Texture Segmentation with EM

Choose a number of segments
Construct a set of support maps, one per segment, containing one element per pixel. These support maps will contain the weight associating a pixel with a segment
Initialize the support maps by either:
    Estimating segment parameters from small blocks of pixels, and then computing weights using the E-step;
    or
    Randomly allocating values to the support maps.
Until convergence
    Update the support maps with an E-Step
    Update the segment parameters with an M-Step
end

Algorithm 16.2: Color and Texture Segmentation with EM: the E-step

For each pixel location l
    For each segment m
        Insert $\alpha_m^{(s)} p_m(\boldsymbol{x}_l \mid \theta_l^{(s)})$ in pixel location l in the support map m
    end
    Add the support map values to obtain $\sum_{k=1}^{K} \alpha_k^{(s)} p_k(\boldsymbol{x}_l \mid \theta_l^{(s)})$ and divide the value in location l in each support map by this term
end

Algorithm 16.3: Color and Texture Segmentation with EM: the M-step

For each segment m

Form new values of the segment parameters using the expressions:

$$\alpha_m^{(s+1)} = \frac{1}{r}\sum_{l=1}^{r} p(m \mid \boldsymbol{x}_l, \Theta^{(s)})$$

$$\mu_m^{(s+1)} = \frac{\sum_{l=1}^{r} \boldsymbol{x}_l p(m \mid \boldsymbol{x}_l, \Theta^{(s)})}{\sum_{l=1}^{r} p(m \mid \boldsymbol{x}_l, \Theta^{(s)})}$$

$$\Sigma_m^{s+1} = \frac{\sum_{l=1}^{r} p(m \mid \boldsymbol{x}_l, \Theta^{(s)}) \left\{(\boldsymbol{x}_l - \mu_m^{(s)})(\boldsymbol{x}_l - \mu_m^{(s)})^T\right\}}{\sum_{l=1}^{r} p(m \mid \boldsymbol{x}_l, \Theta^{(s)})}$$

Where $p(m \mid \boldsymbol{x}_l, \Theta_{(s)})$ is the value in the mth support map for pixel location l

end

Algorithm 16.4: EM Line Fitting by Weighting the Allocation of Points to Each Line, with the Closest Line Getting the Highest Weight

Choose k lines (perhaps uniformly at random)

or

Choose $\overline{\mathcal{L}}$

Until convergence

E-step:

Recompute $\overline{\mathcal{L}}$, from perpendicular distances

M-step:

Refit lines using weights in $\overline{\mathcal{L}}$

end

What should the segmenter report? One option is to choose for each pixel the value of m for which $p(m \mid \boldsymbol{x}_l, \Theta^s)$ is a maximum. Another is to report these probabilities and build an inference process on top of them.

16.2.2 Example: Line Fitting with EM

An EM line fitting algorithm follows the lines of the segmentation example; the missing data is an array of indicator variables $\mathcal{M}$ whose k, lth element m_{kl} is one if point k is drawn from line l, zero otherwise. As in that example, the expected value is given by determining $P(m_{kl} = 1 \mid$ point k, line l's parameters), and this probability is proportional to

$$\exp\left(-\frac{\text{distance from point } k \text{ to line } l^2}{2\sigma^2}\right)$$

for σ as before. The constant of proportionality is most easily determined from the fact that

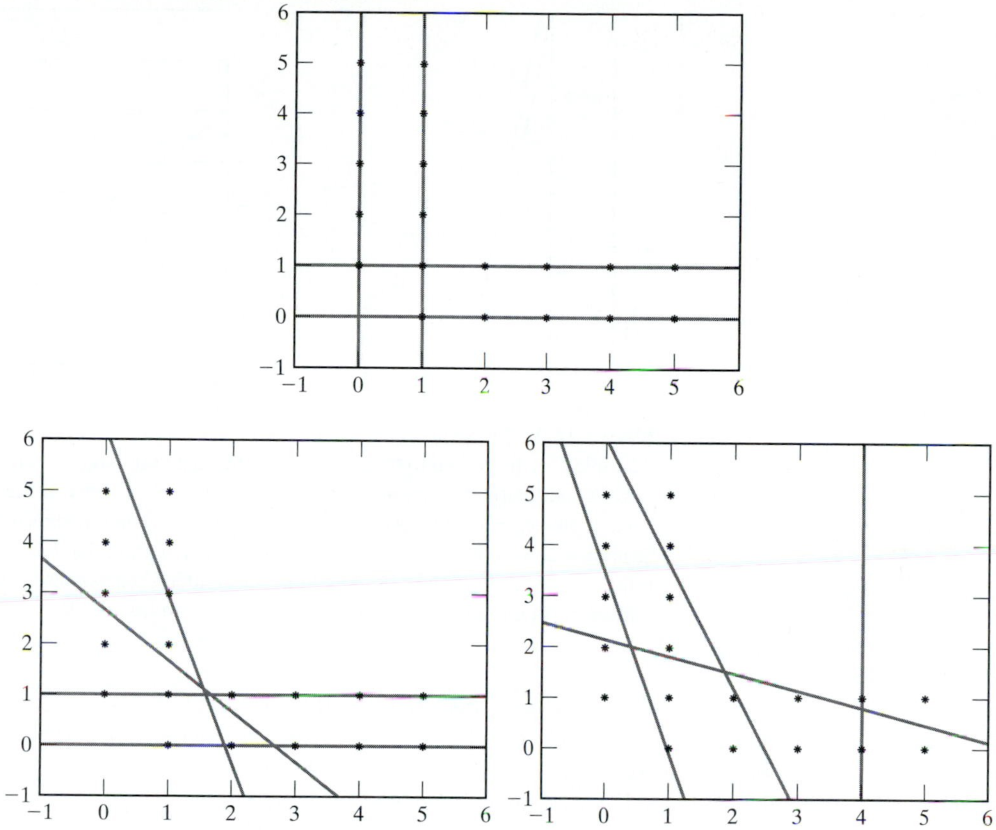

Figure 16.3 The **top figure** shows a good fit obtained using EM line fitting. The two bad examples in the **bottom** row were run with the right number of lines, but have converged to poor fits, which can be fairly good interpretations of the data, and are definitely local minima. This implementation adds a term to the mixture model that models the data point as arising uniformly and at random on the domain; a point that has a high probability of coming from this component has been identified as noise. Further examples of poor fits appear in Figure 16.4.

$$\sum_k P(m_{kl} = 1 \mid \text{point } k, \text{line } l\text{'s parameters})$$
$$= \sum_l P(m_{kl} = 1 \mid \text{point } k, \text{line } l\text{'s parameters})$$
$$= 1.$$

The maximization follows the form of that for fitting a single line to a set of points, only now it must be done g times, and the point coordinates are weighted by the value of $\overline{l_{kl}}$. Convergence can be tested by looking at the size of the change in the lines or by looking at the sum of perpendicular distances of points from their lines (which operates as a log likelihood, see Exercises).

16.2.3 Example: Motion Segmentation and EM

For example, motion sequences often consist of large regions that have similar motion internally. Let us assume for the moment that we have a very short sequence—two frames—and wish to

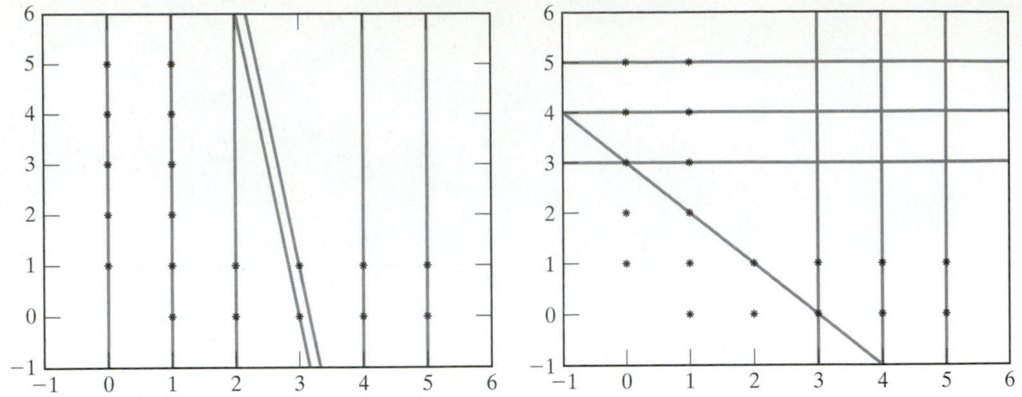

Figure 16.4 More poor fits to the data shown in Figure 16.3. For these examples, we have tried to fit seven lines to this data set. Notice that these fits are fairly good interpretations of the data; they are local extrema of the likelihood. This implementation adds a term to the mixture model that models the data point as arising uniformly and at random on the domain; a point that has a high probability of coming from this component has been identified as noise. The fit on the bottom **right** has allocated some points to noise and fits the others very well.

determine the motion field at each point on the first frame. We assume that the motion field comes from a mixture model. Recall that a general mixture model is a weighted sum of densities—the components do not have to have the Gaussian form used in Section 16.1 (missing data, the EM algorithm, and general mixture models turn up rather naturally together in vision applications).

A generative model for a motion sequence would have the following form:

- At each pixel in each image, there is a motion vector connecting it to a pixel in the next image;
- there are a set of different parametric motion fields, each of which is given by a different probabilistic model;
- the overall motion is given by a mixture model, meaning that to determine the image motion at a pixel, we determine which component the motion comes from and then draw a sample from this component.

This model encapsulates a set of distinct, internally consistent motion fields—which might come from, say, a set of rigid objects at different depths and a moving camera (Figure 16.5)—rather well. The separate motion fields are often referred to as *layers* and the model as a *layered motion* model.

Now assume that the motion fields have a parametric form and that there are g different motion fields. Given a pair of images, we wish to determine (a) which motion field a pixel belongs to and (b) the parameter values for each field. All this should look a great deal like the first two examples, in that if we knew the first, the second would be easy, and if we knew the second, the first would be easy. This is again a missing data problem: The missing data is the motion field to which a pixel belongs, and the parameters are the parameters of each field and the mixing weights.

Assume that the pixel at (u, v) in the first image belongs to the lth motion field, with parameters θ_l. This means that this pixel has moved to $(u, v) + \boldsymbol{m}(u, v; \theta_l)$ in the second frame, and so that the intensity at these two pixels is the same up to measurement noise. We write

Figure 16.5 Frames 1, 15, and 30 of the MPEG flower garden sequence, which is often used to demonstrate motion segmentation algorithms. This sequence appears to be taken from a translating camera, with the tree much closer to the camera than the house and a flower garden on the ground plane. As a result, the tree appears to be translating quickly across the frame, and the house slowly; the plane generates an affine motion field. *Reprinted from "Representing moving images with layers," by J. Wang and E.H. Adelson, IEEE Transactions on Image Processing, 1994, © 1994, IEEE*

$I_1(u, v)$ for the image intensity of the first image at the u, vth pixel, and so on. The missing data is the motion field to which the pixel belongs. We can represent this by an indicator variable $V_{uv,l}$, where

$$V_{uv,l} = \left\{ \begin{array}{c} 1, \text{if the } u, v\text{th pixel belongs to the } l\text{th motion field} \\ 0, \text{otherwise} \end{array} \right\}.$$

We assume Gaussian noise with standard deviation σ in the image intensity values, so the complete data log-likelihood is

$$L(V, \Theta) = -\sum_{ij,l} V_{uv,l} \frac{(I_1(u, v) - I_2(u + m_1(u, v; \theta_l), v + m_2(u, v; \theta_l)))^2}{2\sigma^2} + C,$$

where $\Theta = (\theta_1, \ldots, \theta_g)$. Setting up the EM algorithm from here on is straightforward. As before, the crucial issue is determining

$$P\left\{V_{uv,l} = 1 \mid I_1, I_2, \Theta\right\}.$$

These probabilities are often represented as *support maps*—maps assigning a gray-level representing the maximum probability layer to each pixel (Figure 16.6). The more interesting question is the appropriate choice of parametric motion model. A common choice is an *affine motion model*, where

$$\left\{ \begin{array}{c} m_1 \\ m_2 \end{array} \right\}(i, j; \theta_l) = \left\{ \begin{array}{cc} a_{11} & a_{12} \\ a_{21} & a_{22} \end{array} \right\}\left\{ \begin{array}{c} i \\ j \end{array} \right\} + \left\{ \begin{array}{c} a_{13} \\ a_{23} \end{array} \right\}$$

and $\theta_l = (a_{11}, \ldots, a_{23})$. Layered motion representations are useful for several reasons: First, they cluster together points moving "in the same way". Second, they expose motion boundaries. Finally, new sequences can be reconstructed from the layers in interesting ways (Figure 16.7).

16.2.4 Example: Using EM to Identify Outliers

The line fitters we described have difficulty with outliers because they encounter outliers with a frequency wildly underpredicted by the model. Outliers are often referred to as being "in the *tails*" of a probability distribution. In probability distributions like the normal distribution, there is a large collection of values with small probability; these values are the tails of the distribution (probably because these values are where the distribution *tails off*). A natural mechanism for

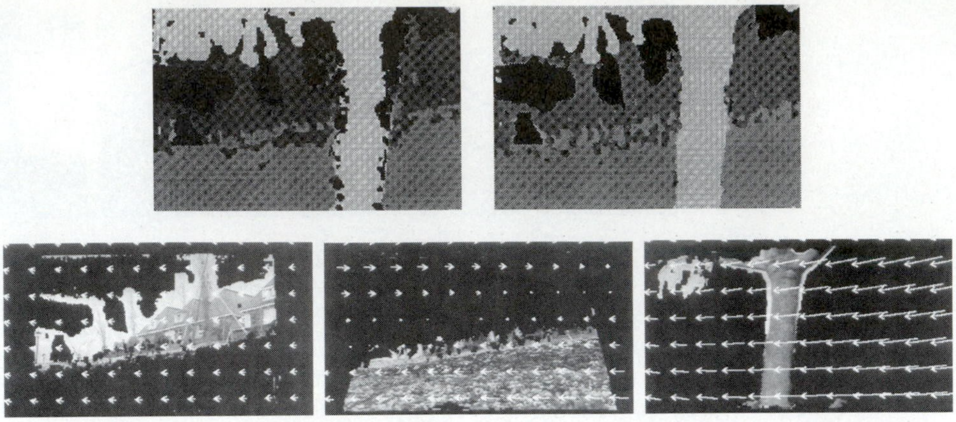

Figure 16.6 On the **top left**, a map indicating to which layer pixels in a frame of the flower garden sequence belong, obtained by clustering local estimates of image motion. Each gray level corresponds to a layer, and each layer is moving with a different affine motion model. This map can be refined by checking the extent to which the motion of pixel neighborhoods is consistent with neighborhoods in future and past frames, resulting in the map on the **top right**. Three of the layers and their motion models are shown on the **bottom**. *Reprinted from "Representing moving images with layers," by J. Wang and E.H. Adelson, IEEE Transactions on Image Processing, 1994, © 1994, IEEE*

Figure 16.7 One feature of representing motion in terms of layers is that one can reconstruct a motion sequence *without* some of the layers. In this example, the MPEG garden sequence has been reconstructed with the tree layer omitted. The figure on the **left** shows frame 1, that in the **center** shows frame 15, and that on the **right** shows frame 30. *Reprinted from "Representing moving images with layers," by J. Wang and E.H. Adelson, IEEE Transactions on Image Processing, 1994, © 1994, IEEE*

dealing with outliers is to modify the model so that the distribution has *heavier tails* (i.e., so there is more probability in the tails).

One way to do this is to construct an explicit model of outliers, which is usually quite easy. We form a weighted sum of the likelihood $P(\text{measurements} \mid \text{model})$ and a term for outliers $P(\text{outliers})$ to obtain

$$(1-\lambda)P(\text{measurements} \mid \text{model}) + \lambda P(\text{outliers}).$$

Here $\lambda \in [0, 1]$ models the frequency with which outliers occur, and $P(\text{outliers})$ is some probability model for outliers. Failing anything better, it could be uniform over the possible range of the data.

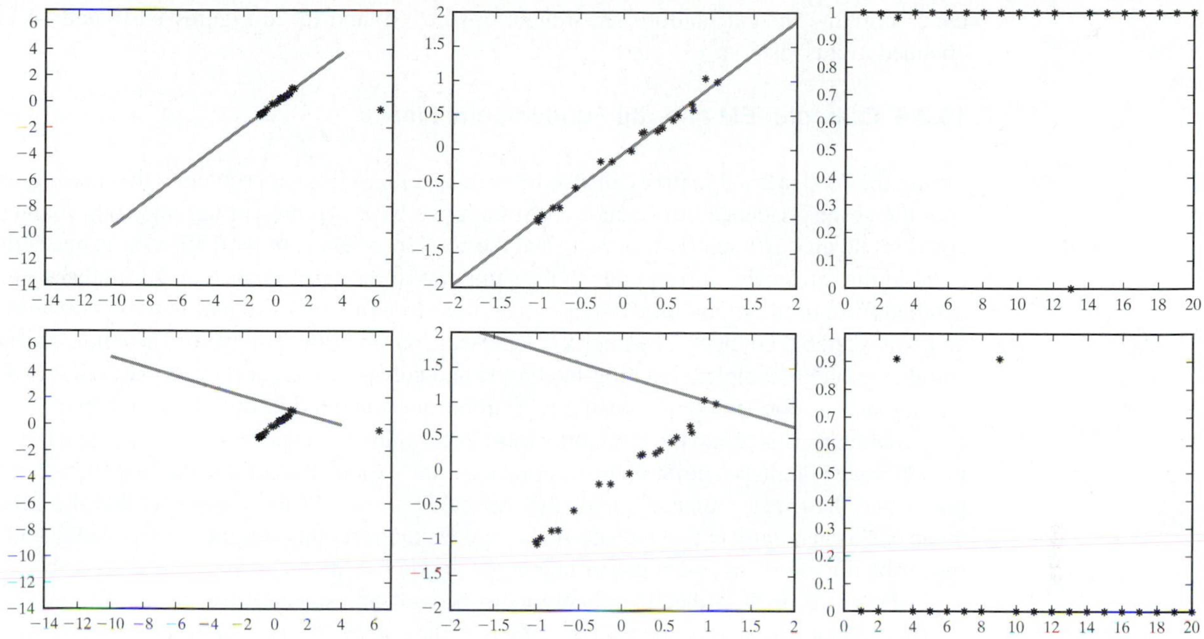

Figure 16.8 EM can be used to reject outliers. Here we demonstrate a line fit to the second data set of figure 15.7. The **top row** shows the correct local minimum, and the **bottom row** shows another local minimum. The **first column** shows the line superimposed on the data points using the same axes as Figure 15.7; the **second column** shows a detailed view of the line, indicating the region around the data points; and the **third column** shows a plot of the probability that a point comes from the line, rather than from the noise model, plotted against the index of the point. Notice that at the correct local minimum, all but one point is associated with the line, whereas at the incorrect local minimum, there are two points associated with the line and the others are allocated to noise.

The natural way to deal with this model is to construct a variable that indicates which component generated each point. With this variable, we have a complete data likelihood function with an easy form. Of course, we don't *know* this variable, but this is a missing data problem, and we know how to proceed here using EM (you provide the details in the exercises!). The usual difficulties with EM occur here, too (Figure 16.8). In particular, it is easy to get trapped in local minima, and we may need to be careful about the numerical representation adopted for small probabilities.

16.2.5 Example: Background Subtraction Using EM

As we saw in Section 14.3.1, estimating the background can be difficult. Simply averaging over frames has this difficulty: An object that spends a lot of time in one place can bias the average quite seriously. We could see this as a missing variable problem: The image in each frame of video is the same (multiplied by some constant to take account of various adjustments made by automatic gain control) with noise added. We model the noise as coming from some uniform source. This has the added benefit that every pixel that belongs to noise is not background; we can obtain the "subtraction" by simply looking at the expected values of the missing variables at the extremum. The calculations are straightforward; (d) and (e) in Figures 14.10 and 14.11 are obtained in this fashion.

16.2.6 Example: EM and the Fundamental Matrix

Fitting the fundamental matrix can also be seen as a missing data problem—the missing data is now the correspondence information. Assuming we have n points in left and right images, we could set up an $n \times n$ matrix $\mathcal{C}$ to represent the correspondence. In particular, c_{ij} is one if the ith point in the left image corresponds to the jth point in the right image and zero otherwise. The detailed form of the E- and the M-steps are tedious to write out—we leave them to the exercises and only sketch them here. The matrix $\mathcal{C}$ represents our missing data; if we knew this matrix, we could compute a complete data log-likelihood and solve for the parameters—which are $\mathcal{A}$, and x_{ri}, y_{ri}, and x_{li} for each point—using an extremisation method. Similarly, given an estimate of the parameters, it is relatively straightforward to compute the expected value of $\mathcal{C}$ (each point in the left image predicts a point in the right image—the exponential of the distances between these points and their measurements yields the expected values). Again, we expect that the expected value of $\mathcal{C}$ is generally not an integer. As before, we interpret this weight as a frequency and raise the probability to the relevant power in computing the likelihood.

This is a method that is unlikely to work well unless carefully started The problem is simple: EM may be able to alleviate the effects of the combinatorial search component of missing variable problems, but cannot make them disappear. The space of correspondences represented by $\mathcal{C}$ is huge ($n!$), and a substantial fraction of this space contains few or no good correspondences. Because the algorithm looks for the best available fundamental matrix given a set of correspondences, a poor initialization is likely to lead to serious trouble. This is because the original set of correspondences may be wrong, leading to a poor estimate of the fundamental matrix, leading to a prediction of more incorrect correspondences, and so on. One possibility is to start EM with RANSAC. The advantage of this approach over conventional RANSAC is that, by using EM in the final phase, we can weight the contribution of conforming correspondences appropriately.

16.2.7 Difficulties with the EM Algorithm

EM is inclined to get stuck in local minima. These local minima are typically associated with combinatorial aspects of the problem being studied, as in the line fitting example or in the fundamental matrix examples. These difficulties follow from the assumption that the points are interchangeable (see Figure 16.3 and Figure 16.4). This can be dodged by noticing that the final configuration of either fitter is a deterministic function of its start point and using carefully chosen start points. One strategy is to start in many different (randomly chosen) configurations and sift through the results looking for the best fit. Another is to preprocess the data using something like a Hough transform to guess good initial line fits. Neither is guaranteed. A cleaner approach is to notice that we are seldom, if ever, faced with a cloud of indistinguishable points and required to infer some structure on that cloud. Usually this is the result of posing a problem poorly. If points are not indistinguishable and have some form of linking structure, then a good start point should be much easier to choose.

A second difficulty is that some points will have extremely small expected weights. This presents us with a numerical problem; it isn't clear what happens if we regard small weights as being equivalent to zero (this isn't usually a wise thing to do). In turn, we may need to adopt a numerical representation that allows us to add many very small numbers and come up with a non-zero result. This issue is rather outside the scope of this book; you should not underestimate its nuisance value because we don't treat it in detail.

16.3 MODEL SELECTION: WHICH MODEL IS THE BEST FIT?

At each stage of our discussion of missing variable problems, we have assumed that the number of components in the mixture model is known. This is generally not the case in practice. Finding the number of components is, in essence, a model selection problem—we search through the collection of models (where different models have different numbers of components) to determine which fits the data best. Generally, the value of the negative log-likelihood is a poor guide to the number of components because, in general, a model with more parameters will fit a dataset better than a model with fewer parameters. This means that simply minimizing the negative log-likelihood as a function of the number of components tends to lead to too many components. For example, we can fit a set of lines extremely accurately by passing a line through each pair of points—there may be a lot of lines, but the fitting error is zero. We resolve this difficulty by adding a term that *increases* with the number of components—this penalty compensates for the decrease in negative log-likelihood caused by the increasing number of parameters.

It is important to understand that there is no *canonical* model selection process. Instead, we can choose from a variety of techniques, each of which uses a different discount corresponding to a different extremality principle (and different approximations to these criteria!).

16.3.1 Basic Ideas

Model selection is a general problem in fitting parametric models. The problem can be set up as follows: There is a data set, which is a sample from a parametric model that is a member of a family of models. We wish to determine (a) which model the data set was drawn from, and (b) what the parameters of that model were. A proper choice of the parameters predicts future samples from the model—a *test set—as well as* the data set (which is often called the *training set*). Unfortunately, these future samples are not available. Furthermore, the estimate of the model's parameters obtained using the data set is likely to be biased because the parameters chosen ensure that the model fits the *training set*—rather than the entire set of possible data—optimally. The effect is known as *selection bias*. The training set is a subset of the entire set of data that could have been drawn from the model; it represents the model exactly only if it is infinitely large. This is why the negative log-likelihood is a poor guide to the choice of model: The fit looks better because it is increasingly biased.

The correct penalty to use comes from the *deviance* given by

twice (log-likelihood of the best model minus log-likelihood of the current model)

(from Ripley (1996), p. 348). The best model should be the true model. Ideally, the deviance would be zero; the argument above suggests that the deviance on a training set is larger than the deviance on a test set. A natural penalty to use is the difference between these deviances averaged over both test and training sets. This penalty is applied to twice the log-likelihood of the fit—the factor of two appears for reasons we cannot explain, but has no effect in practice. Let us write the best choice of parameters as Θ^* and the log-likelihood of the fit to the data set as $L(\boldsymbol{x}; \Theta^*)$.

16.3.2 AIC—An Information Criterion

Akaike proposed a penalty, widely called *AIC* (for "An information criterion," *not* "Akaike information criterion") that leads to minimizing

$$-2L(\boldsymbol{x}; \Theta^*) + 2p$$

where p is the number of free parameters. There is a collection of statistical debate about the AIC. The first main point is that it lacks a term in the *number* of data points. This is suspicious because the deviance between a fitted model and the real model should go down as the number of data points goes up. Second, there is a body of experience that the AIC tends to *overfit*—that is, to choose a model with too many parameters that fits the training set well but doesn't perform as well on test sets.

16.3.3 Bayesian Methods and Schwartz' BIC

For simplicity, let us write $\mathcal{D}$ for the data, $\mathcal{M}$ for the model, and θ for the parameters. Bayes' rule then yields:

$$\begin{aligned} P(\mathcal{M} \mid \mathcal{D}) &= \frac{P(\mathcal{D} \mid \mathcal{M})}{P(\mathcal{M})} P(\mathcal{D}) \\ &= \frac{\int P(\mathcal{D} \mid \mathcal{M}_i, \theta) P(\theta)\, d\theta\, P(\mathcal{M})}{P(\mathcal{D})}. \end{aligned}$$

Now we could choose the model for which the posterior is large. Computing this posterior can be difficult; however, by a series of approximations, we can obtain a criterion

$$-L(\mathcal{D}; \theta^*) + \frac{p}{2} \log N$$

called the *Bayes information criterion* or BIC.

16.3.4 Description Length

Models can be selected by criteria not intrinsically statistical. After all, we are selecting the model and we can say why we want to select it. A criterion that is somewhat natural is to choose the model that encodes the data set most crisply. This *minimum description length* criterion chooses the model that allows the most efficient transmission of the data set. To transmit the data set, one codes and transmits the model parameters and then codes and transmits the data given the model parameters. If the data fits the model poorly, then this latter term is large because one has to code a noise-like signal.

A derivation of the criterion used in practice is rather beyond our needs. The details appear in Rissanen (1983) and (1987), or in Wallace and Freeman (1987); there are similar ideas rooted in information theory, due to Kolmogorov, and expounded in Cover and Thomas (1991). Surprisingly, the BIC emerges from this analysis, yielding

$$-L(\mathcal{D}; \theta^*) + \frac{p}{2} \log N.$$

16.3.5 Other Methods for Estimating Deviance

The key difficulty in model selection is that we should be using a quantity we can't measure—the model's ability to predict data not in the training set. Given a sufficiently large training set, we could split the training set into two components, use one to fit the model and the other the test the fit. This approach is known as *cross-validation*.

We can use cross-validation to determine the number of components in a model by splitting the data set, fitting a variety of different models to one side of the split, and then choosing the model that performs best on the other side. We expect this process to estimate the number of components because a model that has too many parameters will fit the one data set well, but fit the other badly.

Using a single choice of a split into two components introduces a different form of selection bias, and the safest thing to do is average the estimate over all such splits. This becomes unwieldy if the test set is large because the number of splits is huge. The most usual version is *leave-one-out cross-validation*. In this approach, we fit a model to each set of $N - 1$ of the training set, compute the error on the remaining data point, and sum these errors to obtain an estimate of the model error. The model that minimizes this estimate is then chosen.

To our knowledge, this approach, which is standard for model selection in other kinds of problems, has not been used in fitting applications. It is certainly appropriate for estimating the number of components. First, assume that we compute the model error for a model with too few components to describe the image accurately. In this case, the model error is large, because for many pixels the model is insufficiently flexible to describe the pixel that was left out. Similarly, if we use too many components, the model predicts the left out pixel rather poorly.

16.4 NOTES

It should be obvious that we think missing variable models are important. EM was first formally described in the statistical literature by Dempster et al. (1977). A very good summary reference is McLachlan and Krishnan (1996), which describes numerous variants. For example, it isn't necessary to find the maximum of $Q(\boldsymbol{u}; \boldsymbol{u}^{(s)})$; all that is required is to obtain a better value. As another example, the expectation can be estimated using stochastic integration methods.

EM and Missing Variable Models

Missing variable models seem to crop up in all sorts of places. All the models we are aware of in computer vision arise from mixture models (and so have complete data log-likelihood that is linear in the missing variables) and so we have concentrated on this case. It is natural to use a missing variable model for segmentation (Adelson and Weiss 1996, Belongie, Carson, Greenspan and Malik 1998, Feng and Perona 1998, Vasconcelos and Lippman 1997, Wells, Grimson, Kikinis and Jolesz 1996). The model is in the process of reforming how we think about multiple images (i.e., both motion and stereo). The general idea is that the set is decomposed into different layers, where the elements of a layer share the same motion model (Adelson and Weiss, 1996, Dellaert, Seitz, Thorpe and Thrun, 2000, Tao, Sawhney and Kumar, 2000, Wang and Adelson, 1994, and Weiss, 1997) or lie at the same depth (Baker, Szeliski and Anandan 1998, Brostow and Essa 1999, Torr, Szeliski and Anandan 1999*b*), or have some other common property.Other interesting cases include motions resulting from transparency, specularities, and so on (Black and Anandan 1996, Darrell and Simoncelli 1993, Hsu, Anandan and Peleg 1994, Jepson and Black 1993, Szeliski, Avidan and Anandan 2000). The resulting representation can be used for quite efficient image-based rendering (Shade, Gortler, Li-wei and Szeliski 1998). This is a mixture model. Although the problem isn't always seen as a hidden variable problem (the hidden variable is the layer to which a pixel belongs or, equivalently, that generated it), it should probably be. We expect great things fairly shortly.

EM is an extremely successful inference algorithm, but it isn't magical. The primary source of difficulty for the kinds of problem described is local maxima. It is common for problems that have large numbers of missing variables to have large numbers of local maxima. This could be dealt with by starting the optimization close to the right answer, which rather misses the point. In practice, many vision problems that can be attacked with EM seem to be easier than they should be given the number of missing variables, meaning that it is possible to obtain solutions relatively easily. It would be attractive to be able to talk about how hard a missing variable problem is.

Model Selection

Model selection is a topic that hasn't received as much attention as it deserves. There is significant work in motion, the question being which camera model (orthographic, perspective, etc.) to apply (Kinoshita and Lindenbaum, 2000, Maybank and Sturm, 1999, and Torr, 1997, 1999).

Similarly, there is work in segmentation of range data, where the question is to what set of parametric surfaces the data should be fitted (i.e., are there two planes or three, etc.; Bubna and Stewart (2000)). In reconstruction problems, one must sometimes decide whether a degenerate camera motion sequence is present (Torr, Fitzgibbon and Zisserman 1999*a*). The standard problem in segmentation is how many segments are present (Adelson and Weiss 1996, Belongie et al. 1998, Raja, McKenna and Gong 1998). If one is using models predictively, it is sometimes better to compute a weighted average over model predictions (real Bayesians don't do model selection) (Ripley 1996, Torr and Zisserman 1998). We have described only some of the available methods; one important omission is Kanatani's geometric information criterion (Kanatani 1998).

PROBLEMS

16.1. Derive the expressions of Section 16.1 for segmentation. One possible modification is to use the new mean in the estimate of the covariance matrices. Perform an experiment to determine whether this makes any difference in practice.

16.2. Supply the details for the case of using EM for background subtraction. Would it help to have a more sophisticated foreground model than uniform random noise?

16.3. Describe using leave-one-out cross-validation for selecting the number of segments.

Programming Assignments

16.4. Build an EM background subtraction program. Is it practical to insert a dither term to overcome the difficulty with high spatial frequencies illustrated in Figure 14.11?

16.5. Build an EM segmenter that uses color and position (ideally, use texture too) to segment images; use a model selection term to determine how many segments there should be. How significant a phenomenon is the effect of local minima?

16.6. Build an EM line fitter that works for a fixed number of lines. Investigate the effects of local minima. One way to avoid being distracted by local minima is to start from many different start points and then look at the best fit obtained from that set. How successful is this? How many local minima do you have to search to obtain a good fit for a typical data set? Can you improve things using a Hough transform?

16.7. Expand your EM line fitter to incorporate a model selection term so that the fitter can determine how many lines fit a dataset. Compare the choice of AIC and BIC.

16.8. Insert a noise term in your EM line fitter, so that it is able to perform robust fits. What is the effect on the number of local minima? Notice that, if there is a low probability of a point arising from noise, most points will be allocated to lines, but the fits will often be quite poor If there is a high probability of a point arising from noise, points will be allocated to lines only if they fit well. What is the effect of this parameter on the number of local minima?

16.9. Construct a RANSAC fitter that can fit an arbitrary (but known) number of lines to a given data set. What is involved in extending your fitter to determine the best number of lines?

17

Tracking with Linear Dynamic Models

Tracking is the problem of generating an inference about the motion of an object given a sequence of images. Good solutions to this problem have a variety of applications:

- **Motion capture:** If we can track a moving person accurately, then we can make an accurate record of their motions. Once we have this record, we can use it to drive a rendering process; for example, we might control a cartoon character, thousands of virtual extras in a crowd scene, or a virtual stunt avatar. Furthermore, we could modify the motion record to obtain slightly different motions. This means that a single performer can produce sequences they wouldn't want to do in person.
- **Recognition from motion:** The motion of objects is quite characteristic. We may be able to determine the identity of the object from its motion; we should be able to tell what it's doing.
- **Surveillance:** Knowing what objects are doing can be very useful. For example, different kinds of trucks should move in different, fixed patterns in an airport; if they do not, then something is going wrong. Similarly, there are combinations of places and patterns of motions that should never occur (e.g., no truck should ever stop on an active runway). It could be helpful to have a computer system that can monitor activities and give a warning if it detects a problem case.
- **Targeting:** A significant fraction of the tracking literature is oriented toward (a) deciding what to shoot, and (b) hitting it. Typically, this literature describes tracking using radar or infrared signals (rather than vision), but the basic issues are the same—what do we

infer about an object's future position from a sequence of measurements? Where should we aim?

In typical tracking problems, we have a model for the object's motion and some set of measurements from a sequence of images. These measurements could be the position of some image points, the position and moments of some image regions, or pretty much anything else. They are not guaranteed to be relevant, in the sense that some could come from the object of interest and some might come from other objects or from noise.

Tracking is properly thought of as an inference problem. The moving object has some form of internal state, which is measured at each frame. We need to combine our measurements as effectively as possible to estimate the object's state. There are two important cases. Either both **dynamics and measurement are linear**, in which case the inference problem is straightforward and has a standard solution. If we are faced with **nonlinear dynamics**, even slight nonlinearities in system dynamics have tremendous effects. As a result, inference can be difficult and appears to be impossible in general. If the dimension of the state space is low, there is a useful algorithm that often works. As tracking through nonlinear dynamics is a somewhat technical activity, we have confined it to its own chapter which appears on the book website In this chapter, we concentrate on the formulation of tracking through linear dynamics. Section 17.1 sketches the overall view of tracking as an inference problem. In Section 17.2, we deal with linear dynamics and the Kalman filter. We then sketch out some examples of tracking applications in Section 17.5. The most interesting current tracking application—tracking people—requires a discussion of nonlinear dynamics and is discussed in the chapter on the book website.

17.1 TRACKING AS AN ABSTRACT INFERENCE PROBLEM

Much of this chapter deals with the algorithmics of tracking. In particular, we see tracking as a probabilistic inference problem. The key technical difficulty is maintaining an accurate representation of the posterior on object position given measurements and doing so efficiently. We model the object as having some internal state; the state of the object at the ith frame is typically written as $\boldsymbol{X}_i$. The capital letters indicate that this is a random variable—when we want to talk about a particular value that this variable takes, we use small letters. The measurements obtained in the ith frame are values of a random variable $\boldsymbol{Y}_i$; we write $\boldsymbol{y}_i$ for the value of a measurement, and, on occasion, we write $\boldsymbol{Y}_i = \boldsymbol{y}_i$ for emphasis. There are three main problems:

- **Prediction:** We have seen $\boldsymbol{y}_0, \ldots, \boldsymbol{y}_{i-1}$—what state does this set of measurements predict for the ith frame? To solve this problem, we need to obtain a representation of $P(\boldsymbol{X}_i \mid \boldsymbol{Y}_0 = \boldsymbol{y}_0, \ldots, \boldsymbol{Y}_{i-1} = \boldsymbol{y}_{i-1})$.
- **Data association:** Some of the measurements obtained from the i-th frame may tell us about the object's state. Typically, we use $P(\boldsymbol{X}_i \mid \boldsymbol{Y}_0 = \boldsymbol{y}_0, \ldots, \boldsymbol{Y}_{i-1} = \boldsymbol{y}_{i-1})$ to identify these measurements.
- **Correction:** Now that we have $\boldsymbol{y}_i$—the relevant measurements—we need to compute a representation of $P(\boldsymbol{X}_i \mid \boldsymbol{Y}_0 = \boldsymbol{y}_0, \ldots, \boldsymbol{Y}_i = \boldsymbol{y}_i)$.

17.1.1 Independence Assumptions

Tracking is difficult without the following assumptions:

- **Only the immediate past matters:** Formally, we require that

$$P(\boldsymbol{X}_i \mid \boldsymbol{X}_1, \ldots, \boldsymbol{X}_{i-1}) = P(\boldsymbol{X}_i \mid \boldsymbol{X}_{i-1}).$$

This assumption hugely simplifies the design of algorithms as we shall see. Furthermore, it isn't terribly restrictive if we're clever about interpreting $\boldsymbol{X}_i$, as we shall show in the next section.

- **Measurements depend only on the current state:** We assume that $\boldsymbol{Y}_i$ is conditionally independent of all other measurements given $\boldsymbol{X}_i$. This means that

$$P(\boldsymbol{Y}_i, \boldsymbol{Y}_j, \ldots \boldsymbol{Y}_k \mid \boldsymbol{X}_i) = P(\boldsymbol{Y}_i \mid \boldsymbol{X}_i) P(\boldsymbol{Y}_j, \ldots, \boldsymbol{Y}_k \mid \boldsymbol{X}_i).$$

Again, this isn't a particularly restrictive or controversial assumption, but it yields important simplifications.

These assumptions mean that a tracking problem has the structure of inference on a hidden Markov model (where both state and measurements may be on a continuous domain). You should compare this chapter with Section 23.4, which describes the use of hidden Markov models in recognition.

17.1.2 Tracking as Inference

We proceed inductively. First, we assume that we have $P(\boldsymbol{X}_0)$, which is our prediction in the absence of any evidence. Now correcting this is easy: When we obtain the value of $\boldsymbol{Y}_0$—which is $\boldsymbol{y}_0$—we have

$$\begin{aligned}
P(\boldsymbol{X}_0 \mid \boldsymbol{Y}_0 = \boldsymbol{y}_0) &= \frac{P(\boldsymbol{y}_0 \mid \boldsymbol{X}_0) P(\boldsymbol{X}_0)}{P(\boldsymbol{y}_0)} \\
&= \frac{P(\boldsymbol{y}_0 \mid \boldsymbol{X}_0) P(\boldsymbol{X}_0)}{\int P(\boldsymbol{y}_0 \mid \boldsymbol{X}_0) P(\boldsymbol{X}_0) d\boldsymbol{X}_0} \\
&\propto P(\boldsymbol{y}_0 \mid \boldsymbol{X}_0) P(\boldsymbol{X}_0).
\end{aligned}$$

All this is just Bayes rule, and we either compute or ignore the constant of proportionality depending on what we need. Now assume we have a representation of $P(\boldsymbol{X}_{i-1} \mid \boldsymbol{y}_0, \ldots, \boldsymbol{y}_{i-1})$.

Prediction Prediction involves representing

$$P(\boldsymbol{X}_i \mid \boldsymbol{y}_0, \ldots, \boldsymbol{y}_{i-1}).$$

Our independence assumptions make it possible to write

$$\begin{aligned}
P(\boldsymbol{X}_i \mid \boldsymbol{y}_0, \ldots, \boldsymbol{y}_{i-1}) &= \int P(\boldsymbol{X}_i, \boldsymbol{X}_{i-1} \mid \boldsymbol{y}_0, \ldots, \boldsymbol{y}_{i-1}) d\boldsymbol{X}_{i-1} \\
&= \int P(\boldsymbol{X}_i \mid \boldsymbol{X}_{i-1}, \boldsymbol{y}_0, \ldots, \boldsymbol{y}_{i-1}) P(\boldsymbol{X}_{i-1} \mid \boldsymbol{y}_0, \ldots, \boldsymbol{y}_{i-1}) d\boldsymbol{X}_{i-1} \\
&= \int P(\boldsymbol{X}_i \mid \boldsymbol{X}_{i-1}) P(\boldsymbol{X}_{i-1} \mid \boldsymbol{y}_0, \ldots, \boldsymbol{y}_{i-1}) d\boldsymbol{X}_{i-1}.
\end{aligned}$$

Correction Correction involves obtaining a representation of

$$P(\boldsymbol{X}_i \mid \boldsymbol{y}_0, \ldots, \boldsymbol{y}_i).$$

Our independence assumptions make it possible to write

$$\begin{aligned}
P(X_i \mid y_0, \ldots, y_i) &= \frac{P(X_i, y_0, \ldots, y_i)}{P(y_0, \ldots, y_i)} \\
&= \frac{P(y_i \mid X_i, y_0, \ldots, y_{i-1}) P(X_i \mid y_0, \ldots, y_{i-1}) P(y_0, \ldots, y_{i-1})}{P(y_0, \ldots, y_i)} \\
&= P(y_i \mid X_i) P(X_i \mid y_0, \ldots, y_{i-1}) \frac{P(y_0, \ldots, y_{i-1})}{P(y_0, \ldots, y_i)} \\
&= \frac{P(y_i \mid X_i) P(X_i \mid y_0, \ldots, y_{i-1})}{\int P(y_i \mid X_i) P(X_i \mid y_0, \ldots, y_{i-1}) dX_i}.
\end{aligned}$$

17.1.3 Overview

The key algorithmic issue involves finding a representation of the relevant probability densities that (a) is sufficiently accurate for our purposes, and (b) allows these two crucial sums to be done quickly and easily. The simplest case occurs when the dynamics are linear, the measurement model is linear, and the noise models are Gaussian (Section 17.2). We discuss data association in Section 17.4, and show some examples of tracking systems in action in Section 17.5. Nonlinearities introduce a host of unpleasant problems; we discuss some current methods for handling them in a chapter that appears on the website.

17.2 LINEAR DYNAMIC MODELS

There are good relations between linear transformations and Gaussian probability densities. The practical consequence is that if we restrict attention to linear dynamic models and linear measurement models, both with additive Gaussian noise, all the densities we are interested in will be Gaussians. Furthermore, the question of solving the various integrals we encounter can usually be avoided by tricks that allow us to determine directly *which* Gaussian we are dealing with.

In the simplest possible dynamic model, the state is advanced by multiplying it by some known matrix (which may depend on the frame) and then adding a normal random variable of zero mean and known covariance. Similarly, the measurement is obtained by multiplying the state by some matrix (which may depend on the frame) and then adding a normal random variable of zero mean and known covariance. We use the notation

$$\boldsymbol{x} \sim N(\boldsymbol{\mu}, \Sigma)$$

to mean that $\boldsymbol{x}$ is the value of a random variable with a normal probability distribution with mean $\boldsymbol{\mu}$ and covariance Σ; notice that this means that, if x is one-dimensional—we'd write $x \sim N(\mu, v)$—that its standard deviation is $\sqrt{v}$. We can write our dynamic model as

$$\begin{aligned}
x_i &\sim N(\mathcal{D}_i \boldsymbol{x}_{i-1}; \Sigma_{d_i}); \\
y_i &\sim N(\mathcal{M}_i \boldsymbol{x}_i; \Sigma_{m_i}).
\end{aligned}$$

Notice that the covariances could be different from frame to frame as could the matrices. Although this model appears limited, it is in fact extremely powerful; we show how to model some common situations next.

17.2.1 Drifting Points

Let us assume that $\boldsymbol{x}$ encodes the position of a point. If $\boldsymbol{D}_i = Id$, then the point is moving under random walk—its new position is its old position plus some Gaussian noise term. This form of dynamics isn't obviously useful because it appears that we are tracking stationary objects. It is quite commonly used for objects for which no better dynamic model is known—we assume that the random component is quite large and hope we can get away with it.

This model also illustrates aspects of the **measurement matrix** $\mathcal{M}$. The most important thing to keep in mind is that we don't need to measure every aspect of the state of the point at every step. For example, assume that the point is in 3D: Now if $\mathcal{M}_{3k} = (0, 0, 1)$, $\mathcal{M}_{3k+1} = (0, 1, 0)$, and $\mathcal{M}_{3k+2} = (1, 0, 0)$, then at every third frame we measure, respectively, the z, y, or x position of the point. Notice that we can still expect to track the point, even though we measure only one component of its position at a given frame. If we have sufficient measurements, we can reconstruct the state—the state is **observable**. We explore observability in the exercises.

17.2.2 Constant Velocity

Assume that the vector $\boldsymbol{p}$ gives the position and $\boldsymbol{v}$ the velocity of a point moving with constant velocity. In this case, $\boldsymbol{p}_i = \boldsymbol{p}_{i-1} + (\Delta t)\boldsymbol{v}_{i-1}$ and $\boldsymbol{v}_i = \boldsymbol{v}_{i-1}$. This means that we can stack the position and velocity into a single state vector, and our model applies (Figure 17.1). In particular,

$$\boldsymbol{x} = \left\{ \begin{array}{c} \boldsymbol{p} \\ \boldsymbol{v} \end{array} \right\}$$

and

$$\mathcal{D}_i = \left\{ \begin{array}{cc} Id & (\Delta t)Id \\ 0 & Id \end{array} \right\}.$$

Notice that, again, we don't have to observe the whole state vector to make a useful measurement. For example, in many cases, we would expect that

$$\mathcal{M}_i = \left\{ \begin{array}{cc} Id & 0 \end{array} \right\}$$

(i.e., that we see only the position of the point). Because we know that it's moving with constant velocity—that's the model—we expect that we could use these measurements to estimate the whole state vector rather well.

17.2.3 Constant Acceleration

Assume that the vector $\boldsymbol{p}$ gives the position, vector $\boldsymbol{v}$ the velocity, and vector $\boldsymbol{a}$ the acceleration of a point moving with constant acceleration. In this case, $\boldsymbol{p}_i = \boldsymbol{p}_{i-1} + (\Delta t)\boldsymbol{v}_{i-1}$, $\boldsymbol{v}_i = \boldsymbol{v}_{i-1} + (\Delta t)\boldsymbol{a}_{i-1}$, and $\boldsymbol{a}_i = \boldsymbol{a}_{i-1}$. Again, we can stack the position, velocity and acceleration into a single state vector, and our model applies (Figure 17.2). In particular,

$$\boldsymbol{x} = \left\{ \begin{array}{c} \boldsymbol{p} \\ \boldsymbol{v} \\ \boldsymbol{a} \end{array} \right\}$$

and

$$\mathcal{D}_i = \left\{ \begin{array}{ccc} Id & (\Delta t)Id & 0 \\ 0 & Id & (\Delta t)Id \\ 0 & 0 & Id \end{array} \right\}.$$

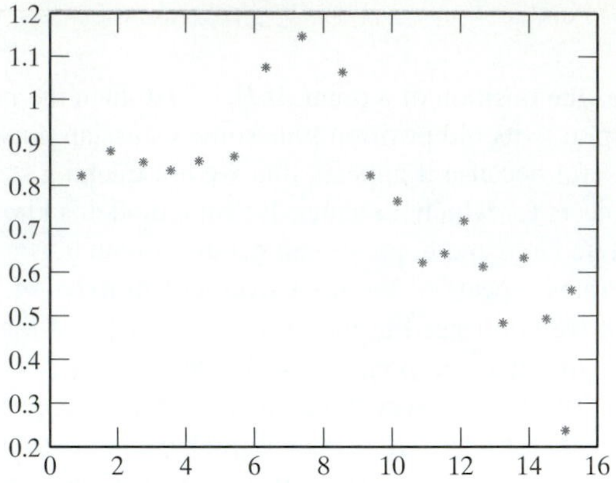

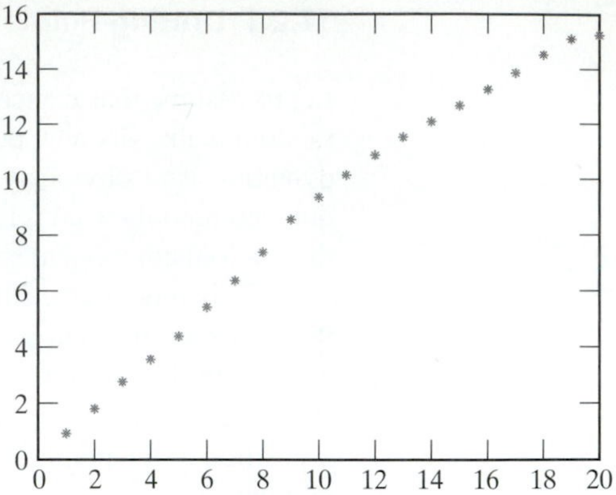

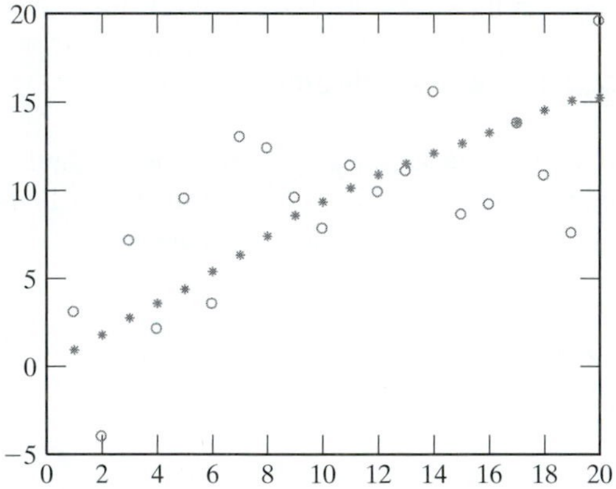

Figure 17.1 A constant velocity dynamic model for a point on the line. In this case, the state space is two dimensional—one coordinate for position, one for velocity. The figure on the **top left** shows a plot of the state; each asterisk is a different state. Notice that the vertical axis (velocity) shows some small change compared with the horizontal axis. This small change is generated only by the random component of the model, so the velocity is constant up to a random change. The figure on the **top right** shows the first component of state (which is position) plotted against the time axis. Notice we have something that is moving with roughly constant velocity. The figure on the **bottom** overlays the measurements (the circles) on this plot. We are assuming that the measurements are of position only, and are quite poor; as we see, this doesn't significantly affect our ability to track.

Notice that, again, we don't have to observe the whole state vector to make a useful measurement. For example, in many cases, we would expect that

$$\mathcal{M}_i = \left\{ \begin{array}{ccc} Id & 0 & 0 \end{array} \right\}$$

(i.e., that we see only the position of the point). Because we know that it's moving with constant acceleration—that's the model—we expect that we could use these measurements to estimate the whole state vector rather well.

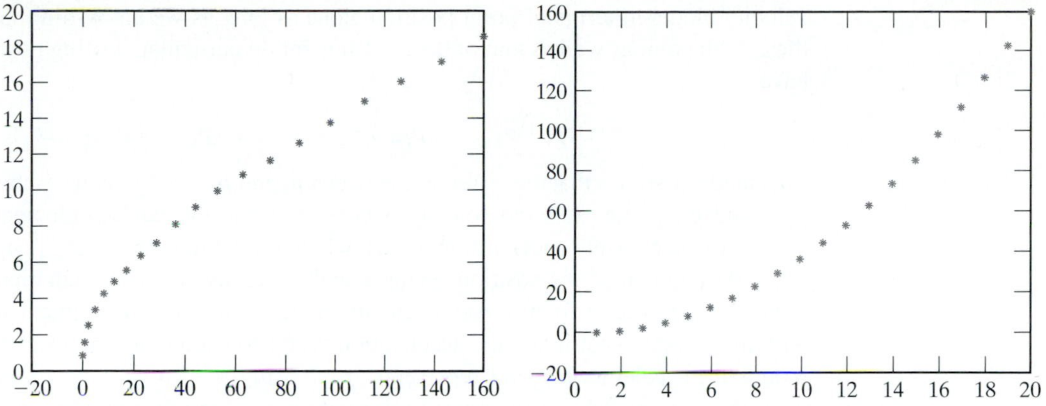

Figure 17.2 This figure illustrates a constant acceleration model for a point moving on the line. On the **left**, we show a plot of the first two components of state—the position on the x-axis and the velocity on the y-axis. In this case, we expect the plot to look like (t^2, t), which it does. On the **right**, we show a plot of the position against time—note that the point is moving away from its start position increasingly quickly.

17.2.4 Periodic Motion

Assume we have a point moving on a line with a periodic movement. Typically, its position p satisfies a differential equation like

$$\frac{d^2 p}{dt^2} = -p.$$

This can be turned into a first order linear differential equation by writing the velocity as v and stacking position and velocity into a vector $\boldsymbol{u} = (p, v)$; we then have

$$\frac{d\boldsymbol{u}}{dt} = \begin{pmatrix} 0 & 1 \\ -1 & 0 \end{pmatrix} \boldsymbol{u} = \mathcal{S}\boldsymbol{u}.$$

Now assume we are integrating this equation with a forward Euler method, where the steplength is Δt; we have

$$\begin{aligned} \boldsymbol{u}_i &= \boldsymbol{u}_{i-1} + \Delta t \frac{d\boldsymbol{u}}{dt} \\ &= \boldsymbol{u}_{i-1} + \Delta t \mathcal{S} \boldsymbol{u}_{i-1} \\ &= \begin{pmatrix} 1 & \Delta t \\ -\Delta t & 1 \end{pmatrix} \boldsymbol{u}_{i-1}. \end{aligned}$$

We can either use this as a state equation, or we can use a different integrator. If we used a different integrator, we might have some expression in $\boldsymbol{u}_{i-1}, \ldots, \boldsymbol{u}_{i-n}$—we would need to stack $\boldsymbol{u}_{i-1}, \ldots, \boldsymbol{u}_{i-n}$ into a state vector and arrange the matrix appropriately (see Exercises). This method works for points on the plane, in 3D, and so on, as well (again, see Exercises).

17.2.5 Higher Order Models

Another way to look at a constant velocity model is that we have augmented the state vector to get around the requirement that $P(\boldsymbol{x}_i \mid \boldsymbol{x}_1, \ldots, \boldsymbol{x}_{i-1}) = P(\boldsymbol{x}_i \mid \boldsymbol{x}_{i-1})$. We could write a constant velocity model in terms of point position alone as long as we are willing to use the position of the $i-2$th point as well as that of the $i-1$th point. In particular, writing position as $\boldsymbol{p}$, we would have

$$P(\boldsymbol{p}_i \mid \boldsymbol{p}_1, \ldots, \boldsymbol{p}_{i-1}) = N(\boldsymbol{p}_{i-1} + (\boldsymbol{p}_{i-1} - \boldsymbol{p}_{i-2}), \Sigma_{d_i}).$$

This model assumes that the difference between $\boldsymbol{p}_i$ and $\boldsymbol{p}_{i-1}$ is the same as the difference between $\boldsymbol{p}_{i-1}$ and $\boldsymbol{p}_{i-2}$—i.e., that the velocity is constant up to the random element. A similar remark applies to the constant acceleration model, which is now in terms of $\boldsymbol{p}_{i-1}$, $\boldsymbol{p}_{i-2}$, and $\boldsymbol{p}_{i-3}$.

We augmented the position vector with the velocity vector (which represents $\boldsymbol{p}_{i-1} - \boldsymbol{p}_{i-2}$) to get the state vector for a constant velocity model. Similarly, we augmented the position vector with the velocity vector and the acceleration vector to get a constant acceleration model. In this model, the acceleration vector represents $(\boldsymbol{p}_{i-1} - \boldsymbol{p}_{i-2}) - (\boldsymbol{p}_{i-2} - \boldsymbol{p}_{i-3})$. We might reasonably want the new position of the point to depend on $\boldsymbol{p}_{i-4}$ or other points even further back in the history of the point's track. To represent dynamics like this, all we need to do is augment the state vector to a suitable size. Notice that it can be somewhat difficult to visualize how the model will behave. There are two approaches to determining what $\mathcal{D}_i$ needs to be; in the first, we know something about the dynamics and can write it down, as we have done here; in the second, we need to learn it from data.

17.3 KALMAN FILTERING

An important feature of linear dynamic models is that all the conditional probability distributions we need to deal with are normal distributions. In particular, $P(\boldsymbol{X}_i \mid \boldsymbol{y}_1, \ldots, \boldsymbol{y}_{i-1})$ is normal, as is $P(\boldsymbol{X}_i \mid \boldsymbol{y}_1, \ldots, \boldsymbol{y}_i)$. This means that they are relatively easy to represent—all we need to do is maintain representations of the mean and the covariance for the prediction and correction phase. In particular, our model admits a relatively simple process where the representation of the mean and covariance for the prediction and estimation phase are updated.

17.3.1 The Kalman Filter for a 1D State Vector

The dynamic model is now

$$x_i \sim N(d_i x_{i-1}, \sigma_{d_i}^2);$$
$$y_i \sim N(m_i x_i, \sigma_{m_i}^2).$$

We need to maintain a representation of $P(X_i \mid y_0, \ldots, y_{i-1})$ and of $P(X_i \mid y_0, \ldots, y_i)$. In each case, we need only represent the mean and the standard deviation because the distributions are normal.

Notation We represent the mean of $P(X_i \mid y_0, \ldots, y_{i-1})$ as $\overline{X_i^-}$ and the mean of $P(X_i \mid y_0, \ldots, y_i)$ as $\overline{X_i^+}$—the superscripts suggest that they represent our belief about X_i immediately before and immediately after the ith measurement arrives. Similarly, we represent the standard deviation of $P(X_i \mid y_0, \ldots, y_{i-1})$ as σ_i^- and of $P(X_i \mid y_0, \ldots, y_i)$ as σ_i^+. In each case, we assume that we know $P(X_{i-1} \mid y_0, \ldots, y_{i-1})$, meaning that we know $\overline{X_{i-1}^+}$ and σ_{i-1}^+.

Tricks with Integrals The main reason we work with normal distributions is that their integrals are quite well behaved. We are going to obtain values for various parameters as integrals usually by change of variable. Our current notation can make appropriate changes a bit difficult to spot, so we write

$$g(x;\mu,v)=\exp\left(-\frac{(x-\mu)^2}{2v}\right).$$

We have dropped the constant and, for convenience, are representing the *variance* (as v), rather than the standard deviation. This expression allows some convenient transformations; in particular, we have

$$g(x;\mu,v)=g(x-\mu;0,v);$$
$$g(m;n,v)=g(n;m,v);$$
$$g(ax;\mu,v)=g(x;\mu/a,v/a^2).$$

We also need the following fact:

$$\int_{-\infty}^{\infty} g(x-u;\mu,v_a)g(u;0,v_b)\,du \propto g(x;\mu,v_a^2+v_b^2).$$

There are several ways to confirm that this is true: The easiest is to look it up in tables; more subtle is to think about convolution directly; more subtle still is to think about the sum of two independent random variables. We need a further identity. We have

$$g(x;a,b)g(x;c,d)=g\left(x;\frac{ad+cb}{b+d},\frac{bd}{b+d}\right)f(a,b,c,d),$$

where the form of f is not significant, but the fact that it is not a function of x is. The exercises show you how to prove this identity.

Prediction We have

$$P(X_i \mid y_0,\dots,y_{i-1})=\int P(X_i \mid X_{i-1})P(X_{i-1}\mid y_0,\dots,y_{i-1})\,dX_{i-1}.$$

Now

$$\begin{aligned}
P(X_i \mid y_0,\dots,y_{i-1}) &= \int P(X_i \mid X_{i-1})P(X_{i-1}\mid y_0,\dots,y_{i-1})dX_{i-1})\\
&\propto \int_{-\infty}^{\infty} g(X_i;d_iX_{i-1},\sigma_{d_i}^2)g(X_{i-1};\overline{X}_{i-1}^{+},(\sigma_{i-1}^{+})^2)dX_{i-1}\\
&\propto \int_{-\infty}^{\infty} g((X_i-d_iX_{i-1});0,\sigma_{d_i}^2)g((X_{i-1}-\overline{X}_{i-1}^{+});0,(\sigma_{i-1}^{+})^2)dX_{i-1}\\
&\propto \int_{-\infty}^{\infty} g((X_i-d_i(u+\overline{X}_{i-1}^{+}));0,(\sigma_{d_i})^2)g(u;0,(\sigma_{i-1}^{+})^2)du\\
&\propto \int_{-\infty}^{\infty} g((X_i-d_iu);d_i\overline{X}_{i-1}^{+},\sigma_{d_i}^2)g(u;0,(\sigma_{i-1}^{+})^2)du\\
&\propto \int_{-\infty}^{\infty} g((X_i-v);d_i\overline{X}_{i-1}^{+},\sigma_{d_i}^2)g(v;0,(d_i\sigma_{i-1}^{+})^2)dv\\
&\propto g(X_i;d_i\overline{X}_0^{+},\sigma_{d_i}^2+(d_i\sigma_{i-1}^{+})^2),
\end{aligned}$$

where we have applied various of the transformations given earlier, and changed variables twice. All this means that

$$\overline{X}_i^- = d_i \overline{X}_{i-1}^+;$$
$$(\sigma_i^-)^2 = \sigma_{d_i}^2 + (d_i \sigma_{i-1}^+)^2.$$

Correction We have

$$P(X_i \mid y_0, \ldots, y_i) = \frac{P(y_i \mid X_i) P(X_i \mid y_0, \ldots, y_{i-1})}{\int P(y_i \mid X_i) P(X_i \mid y_0, \ldots, y_{i-1}) dX_i}$$
$$\propto P(y_i \mid X_i) P(X_i \mid y_0, \ldots, y_{i-1}).$$

We know $\overline{X}_i^-$ and σ_i^-, which represent $P(X_i \mid y_0, \ldots, y_{i-1})$.

Using the notation for Gaussians given earlier, we have

$$P(X_i \mid y_0, \ldots, y_i) \propto g(y_i; m_i X_i, \sigma_{m_i}^2) g(X_i; \overline{X}_i^-, (\sigma_i^-)^2)$$
$$= g(m_i X_i; y_i, \sigma_{m_i}^2) g(X_i; \overline{X}_i^-, (\sigma_i^-)^2)$$
$$= g\left(X_i; \frac{y_i}{m_i}, \frac{\sigma_{m_i}^2}{m_i^2}\right) g(X_i; \overline{X}_i^-, (\sigma_i^-)^2),$$

Algorithm 17.1: The 1D Kalman filter updates estimates of the mean and covariance of the various distributions encountered while tracking a one-dimensional state variable using the given dynamic model.

Dynamic Model:

$$x_i \sim N(d_i x_{i-1}, \sigma_{d_i})$$
$$y_i \sim N(m_i x_i, \sigma_{m_i})$$

Start Assumptions: $\overline{x}_0^-$ and σ_0^- are known

Update Equations: Prediction

$$\overline{x}_i^- = d_i \overline{x}_{i-1}^+$$
$$\sigma_i^- = \sqrt{\sigma_{d_i}^2 + (d_i \sigma_{i-1}^+)^2}$$

Update Equations: Correction

$$x_i^+ = \left(\frac{\overline{x}_i^- \sigma_{m_i}^2 + m_i y_i (\sigma_i^-)^2}{\sigma_{m_i}^2 + m_i^2 (\sigma_i^-)^2}\right)$$
$$\sigma_i^+ = \sqrt{\left(\frac{\sigma_{m_i}^2 (\sigma_i^-)^2}{(\sigma_{m_i}^2 + m_i^2 (\sigma_i^-)^2)}\right)}$$

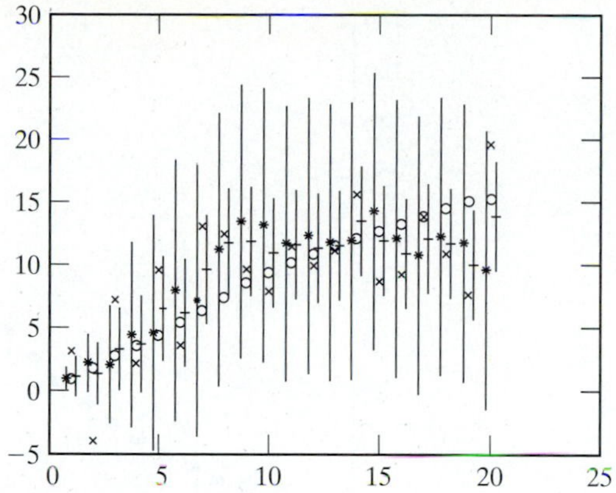

Figure 17.3 The Kalman filter for a point moving on the line under our model of constant velocity (compare with Figure 17.1). The state is plotted with open circles as a function of the step i. The *-s give $\overline{x}_i^-$, which is plotted slightly to the left of the state to indicate that the estimate is made before the measurement. The x-s give the measurements, and the +-s give $\overline{x}_i^+$, which is plotted slightly to the right of the state. The vertical bars around the *-s and the +-s are three standard deviation bars using the estimate of variance obtained before and after the measurement, respectively. When the measurement is noisy, the bars don't contract all that much when a measurement is obtained (compare with Figure 17.4).

and by pattern matching to our identities, we have

$$X_i^+ = \left(\frac{\overline{X}_i^- \sigma_{m_i}^2 + m_i y_i (\sigma_i^-)^2}{\sigma_{m_i}^2 + m_i^2 (\sigma_i^-)^2} \right);$$

$$\sigma_i^+ = \sqrt{\left(\frac{\sigma_{m_i}^2 (\sigma_i^-)^2}{(\sigma_{m_i}^2 + m_i^2 (\sigma_i^-)^2)} \right)}.$$

Figure 17.3 shows a Kalman filter tracking a constant velocity model, and Figure 17.4 shows a Kalman filter tracking a constant acceleration model.

17.3.2 The Kalman Update Equations for a General State Vector

We obtained a 1D tracker without having to do any integration using special properties of normal distributions. This approach works for a state vector of arbitrary dimension, but the process of guessing integrals, and so on, is a good deal more elaborate than that shown in Section 17.3.1. We omit the necessary orgy of notation—it's a tough but straightforward exercise for those who really care (you should figure out the identities first and the rest follows)—and simply give the result in Algorithm 17.2.

17.3.3 Forward–Backward Smoothing

It is important to notice that $P(\boldsymbol{X}_i \mid \boldsymbol{y}_0, \ldots, \boldsymbol{y}_i)$ is not the best available representation of $\boldsymbol{X}_i$; this is because it doesn't take into account the future behavior of the point. In particular, all the measurements *after* $\boldsymbol{y}_i$ could affect our representation of $\boldsymbol{X}_i$. This is because these future

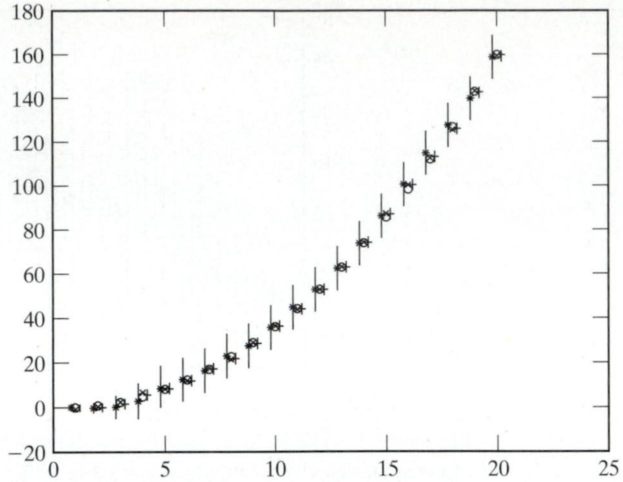

Figure 17.4 The Kalman filter for a point moving on the line under our model of constant acceleration (compare with Figure 17.2). The state is plotted with open circles as a function of the step i. The *-s give $\overline{x}_i^-$, which is plotted slightly to the left of the state to indicate that the estimate is made before the measurement. The x-s give the measurements, and the +-s give $\overline{x}_i^+$, which is plotted slightly to the right of the state. The vertical bars around the *-s and the +-s are three standard deviation bars using the estimate of variance obtained before and after the measurement, respectively. When the measurement is not noisy, the bars contract when a measurement is obtained.

Algorithm 17.2: The Kalman filter updates estimates of the mean and covariance of the various distributions encountered while tracking a state variable of some fixed dimension using the given dynamic model.

Dynamic Model:

$$\boldsymbol{x}_i \sim N(\mathcal{D}_i \boldsymbol{x}_{i-1}, \Sigma_{d_i})$$
$$\boldsymbol{y}_i \sim N(\mathcal{M}_i \boldsymbol{x}_i, \Sigma_{m_i})$$

Start Assumptions: $\overline{\boldsymbol{x}}_0^-$ and Σ_0^- are known

Update Equations: Prediction

$$\overline{\boldsymbol{x}}_i^- = \mathcal{D}_i \overline{\boldsymbol{x}}_{i-1}^+$$
$$\Sigma_i^- = \Sigma_{d_i} + \mathcal{D}_i \sigma_{i-1}^+ \mathcal{D}_i$$

Update Equations: Correction

$$\mathcal{K}_i = \Sigma_i^- \mathcal{M}_i^T \left[\mathcal{M}_i \Sigma_i^- \mathcal{M}_i^T + \Sigma_{m_i}\right]^{-1}$$
$$\overline{\boldsymbol{x}}_i^+ = \overline{\boldsymbol{x}}_i^- + \mathcal{K}_i \left[\boldsymbol{y}_i - \mathcal{M}_i \overline{\boldsymbol{x}}_i^-\right]$$
$$\Sigma_i^+ = [Id - \mathcal{K}_i \mathcal{M}_i]\, \Sigma_i^-$$

measurements might contradict the estimates obtained to date—perhaps the future movements of the point are more in agreement with a slightly different estimate of the position of the point. However, $P(\boldsymbol{X}_i \mid \boldsymbol{y}_0, \ldots, \boldsymbol{y}_i)$ *is* the best estimate available at step i.

What we do with this observation depends on the circumstances. If our application requires an immediate estimate of position—perhaps we are tracking a car in the opposite lane—there isn't much we can do. If we are tracking off-line—perhaps for forensic purposes, we need the best estimate of what an object was doing given a videotape—then we can use all data points, and so we want to represent $P(\boldsymbol{X}_i \mid \boldsymbol{y}_0, \ldots, \boldsymbol{y}_N)$. A common alternative is that we need a rough estimate immediately, and can use an improved estimate that has been time-delayed by a number of steps. This means we want to represent $P(\boldsymbol{X}_i \mid \boldsymbol{y}_0, \ldots, \boldsymbol{y}_{i+k})$—we have to wait until time $i+k$ for this representation, but it should be an improvement on $P(\boldsymbol{X}_i \mid \boldsymbol{y}_0, \ldots, \boldsymbol{y}_i)$.

Introducing a Backward Filter Now we have

$$\begin{aligned}
P(\boldsymbol{X}_i \mid \boldsymbol{y}_0, \ldots, \boldsymbol{y}_N) &= \frac{P(\boldsymbol{X}_i, \boldsymbol{y}_{i+1}, \ldots, \boldsymbol{y}_N \mid \boldsymbol{y}_0, \ldots, \boldsymbol{y}_i) P(\boldsymbol{y}_0, \ldots, \boldsymbol{y}_i)}{P(\boldsymbol{y}_0, \ldots, \boldsymbol{y}_N)} \\
&= \frac{P(\boldsymbol{y}_{i+1}, \ldots, \boldsymbol{y}_N \mid \boldsymbol{X}_i, \boldsymbol{y}_0, \ldots, \boldsymbol{y}_i) P(\boldsymbol{X}_i \mid \boldsymbol{y}_0, \ldots, \boldsymbol{y}_i) P(\boldsymbol{y}_0, \ldots, \boldsymbol{y}_i)}{P(\boldsymbol{y}_0, \ldots, \boldsymbol{y}_N)} \\
&= \frac{P(\boldsymbol{y}_{i+1}, \ldots, \boldsymbol{y}_N \mid \boldsymbol{X}_i) P(\boldsymbol{X}_i \mid \boldsymbol{y}_0, \ldots, \boldsymbol{y}_i) P(\boldsymbol{y}_0, \ldots, \boldsymbol{y}_i)}{P(\boldsymbol{y}_0, \ldots, \boldsymbol{y}_N)} \\
&= P(\boldsymbol{X}_i \mid \boldsymbol{y}_{i+1}, \ldots, \boldsymbol{y}_N) P(\boldsymbol{X}_i \mid \boldsymbol{y}_0, \ldots, \boldsymbol{y}_i) \alpha
\end{aligned}$$

where

$$\alpha = \left(\frac{P(\boldsymbol{y}_{i+1}, \ldots, \boldsymbol{y}_N) P(\boldsymbol{y}_0, \ldots, \boldsymbol{y}_i)}{P(\boldsymbol{X}_i) P(\boldsymbol{y}_0, \ldots, \boldsymbol{y}_N)} \right).$$

This term should look like a potential source of problems to you; in fact, we can avoid tangling with it by a clever trick. What is important about this form is that we are combining $P(\boldsymbol{X}_i \mid \boldsymbol{y}_0, \ldots, \boldsymbol{y}_i)$—which we know how to obtain—with $P(\boldsymbol{X}_i \mid \boldsymbol{y}_{i+1}, \ldots, \boldsymbol{y}_N)$. We actually know how to obtain a representation of $P(\boldsymbol{X}_i \mid \boldsymbol{y}_{i+1}, \ldots, \boldsymbol{y}_N)$, too. We could simply run the Kalman filter *backward* in time, using *backward dynamics* and take the predicted representation of $\boldsymbol{X}_i$ (we leave the details of relabeling the sequence, etc., to the exercises).

Combining Representations Now we have two representations of $\boldsymbol{X}_i$: one obtained by running a forward filter and incorporating all measurements up to $\boldsymbol{y}_i$; and one obtained by running a backward filter and incorporating all measurements after $\boldsymbol{y}_i$. We need to combine these representations. Instead of explicitly determining the value of α (which should look hard), we can get the answer by noting that *this is like having another measurement*. In particular, we have a new measurement generated by $\boldsymbol{X}_i$—that is, the result of the backward filter—to combine with our estimate from the forward filter. We know how to combine estimates with measurements because that's what the Kalman filter equations are for.

All we need is a little notation. We attach the superscript f to the estimate from the forward filter and the superscript b to the estimate from the backward filter. We write the mean of $P(\boldsymbol{X}_i \mid \boldsymbol{y}_0, \ldots, \boldsymbol{y}_N)$ as $\overline{\boldsymbol{X}}_i^*$ and the covariance of $P(\boldsymbol{X}_i \mid \boldsymbol{y}_0, \ldots, \boldsymbol{y}_N)$ as Σ_i^*. We regard the representation of $\boldsymbol{X}_i^b$ as a measurement of $\boldsymbol{X}_i$ with mean $\overline{\boldsymbol{X}}_i^{b,-}$ and covariance $\Sigma_i^{b,-}$—the minus sign is because the ith measurement cannot be used twice, meaning the backward filter predicts $\boldsymbol{X}_i$ using $\boldsymbol{y}_N \ldots \boldsymbol{y}_{i+1}$. This measurement needs to be combined with $P(\boldsymbol{X}_i \mid \boldsymbol{y}_0, \ldots, \boldsymbol{y}_i)$, which has mean

$\overline{X}_i^{f,+}$ and covariance $\Sigma_i^{f,+}$ (when we substitute into the Kalman equations, these take the role of the representation *before* a measurement because the value of the measurement is now $\overline{X}_i^{b,-}$).

Substituting into the Kalman equations, we find that

$$\mathcal{K}_i^* = \Sigma_i^{f,+}\left[\Sigma_i^{f,+} + \Sigma_i^{b,-}\right]^{-1};$$

$$\Sigma_i^* = [I - \mathcal{K}_i]\,\Sigma_i^{+,f};$$

$$\overline{X}_i^* = \overline{X}_i^{f,+} + \mathcal{K}_i^*\left[\overline{X}_i^{b,-} - \overline{X}_i^{f,+}\right].$$

It turns out that a little manipulation (exercise!) yields a simpler form, which we give in Algorithm 17.3. Forward–backward estimates can make a substantial difference as Figure 17.5 illustrates.

Algorithm 17.3: The forward–backward algorithm combines forward and backward estimates of state to come up with an improved estimate.

Forward filter: Obtain the mean and variance of $P(X_i \mid y_0, \ldots, y_i)$ using the Kalman filter. These are $\overline{X}_i^{f,+}$ and $\Sigma_i^{f,+}$.

Backward filter: Obtain the mean and variance of $P(X_i \mid y_{i+1}, \ldots, y_N)$ using the Kalman filter running backward in time. These are $\overline{X}_i^{b,-}$ and $\Sigma_i^{b,-}$.

Combining forward and backward estimates: Regard the backward estimate as a new measurement for X_i, and insert into the Kalman filter equations to obtain

$$\Sigma_i^* = \left[(\Sigma_i^{f,+})^{-1} + (\Sigma_i^{b,-})^{-1}\right]^{-1};$$

$$\overline{X}_i^* = \Sigma_i^*\left[(\Sigma_i^{f,+})^{-1}\overline{X}_i^{f,+} + (\Sigma_i^{b,-})^{-1}\overline{X}_i^{b,-}\right].$$

Priors In typical vision applications, we are tracking forward in time. This leads to an inconvenient asymmetry: We may have a good idea of where the object started, but only a poor one of where it stopped (i.e., we are likely to have a fair prior for $P(\boldsymbol{x}_0)$, but may have difficulty supplying a prior for $P(\boldsymbol{x}_N)$ for the forward–backward filter). One option is to use $P(\boldsymbol{x}_N \mid \boldsymbol{y}_0, \ldots, \boldsymbol{y}_N)$ as a prior. This is a dubious act as this probability distribution does not in fact reflect our prior belief about $P(\boldsymbol{x}_N)$—we've used all the measurements to obtain it. The consequences can be that this distribution understates our uncertainty in $\boldsymbol{x}_N$ and so leads to a forward–backward estimate that significantly underestimates the covariance for the later states. An alternative is to use the mean supplied by the forward filter, but enlarge the covariance substantially; the consequences are a forward–backward estimate that overestimates the covariance for the later states (compare Figure 17.5 with Figure 17.6).

Not all applications have this asymmetry. For example, if we are engaged in a forensic study of a videotape, we might be able to start both the forward tracker and the backward tracker by hand and provide a good estimate of the prior in each case. If this is possible, then we have a good deal more information which may be able to help choose correspondences, and so on—the forward tracker should finish rather close to where the backward tracker starts.

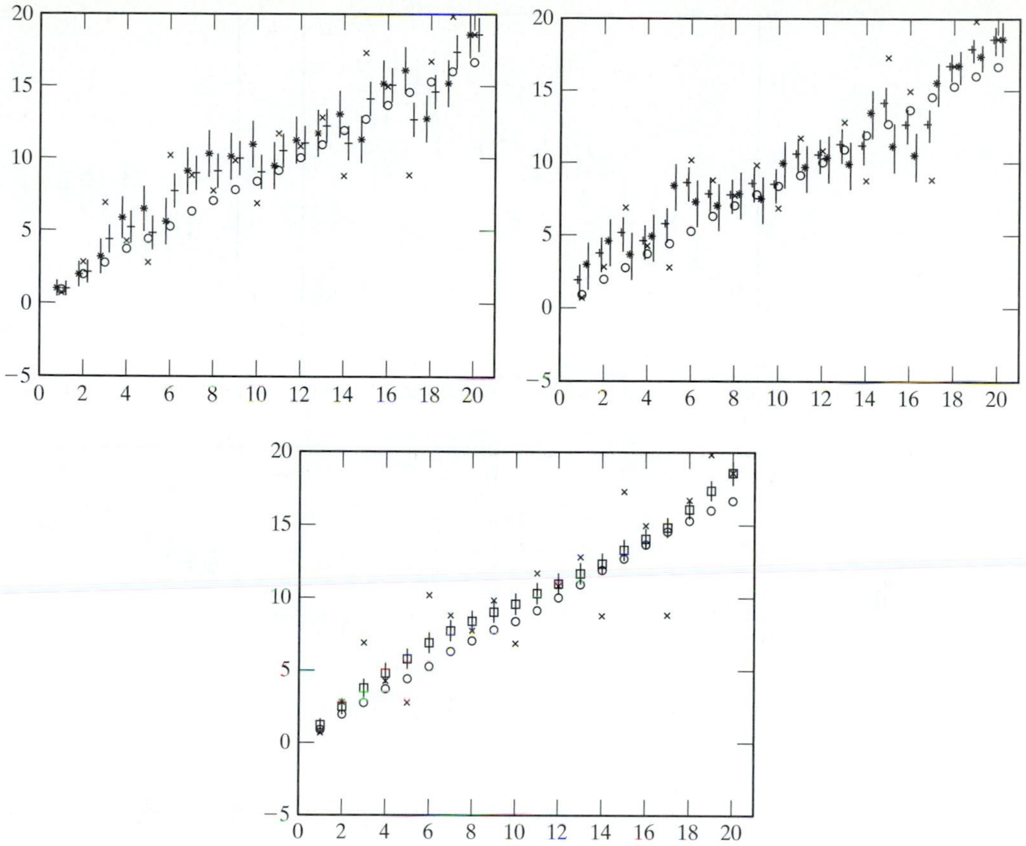

Figure 17.5 Forward–backward estimation for a dynamic model of a point moving on the line with constant velocity. We are plotting the position component of state against time. On the **top left**, we show the forward estimates, again using the convention that the state is shown with circles, the data is shown with an x, the prediction is shown with a *, and the corrected estimate is shown with a +; the bars give one standard deviation in the estimate. The predicted estimate is shown slightly behind the state and the corrected estimate is shown slightly ahead of the state. You should notice that the measurements are noisy. On the **top right** we show the backward estimates. Now time is running backward (although we have plotted both curves on the same axis) so that the prediction is slightly ahead of the measurement and the corrected estimate is slightly behind. We have used the final corrected estimate of the forward filter as a prior. Again, the bars give one standard deviation in each variable. On the **bottom**, we show the combined forward–backward estimate. The squares give the estimates of state. Notice the significant improvement in the estimate.

Smoothing over an Interval Although our formulation of forward–backward smoothing assumed that the backward filter started at the last data point, it is easy to start this filter a fixed number of steps ahead of the forward filter. If we do this, we obtain an estimate of state in real time (essentially immediately after the measurement) and an improved estimate some fixed numbers of measurements later. This is sometimes useful. Furthermore, it is an efficient way to obtain most of the improvement available from a backward filter if we can assume that the effect of the distant future on our estimate is relatively small compared with the effect of the immediate

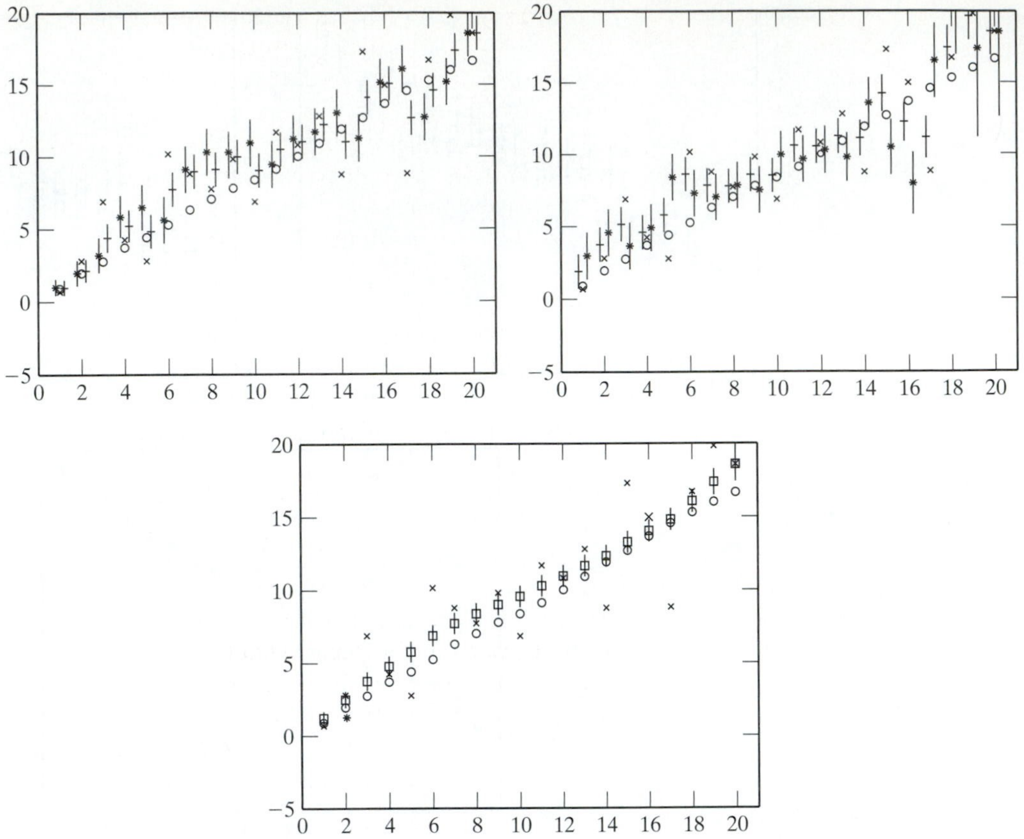

Figure 17.6 We now show the effects of using a diffuse prior for the position of the final point in forward–backward estimation for a dynamic model of a point moving on the line with constant velocity. We are plotting the position component of state against time. On the **top left**, we show the forward estimates, again using the convention that the state is shown with circles, the data is shown with an x, the prediction is shown with a *, and the corrected estimate is shown with a +; the bars give one stanlard deviation in the estimate. The predicted estimate is shown slightly behind the state, and the corrected estimate is shown slightly ahead of the state. You should notice that the measurements are noisy. On the **top right** we show the backward estimates. Now time is running backward (although we have plotted both curves on the same axis) so that the prediction is slightly ahead of the measurement and the corrected estimate is slightly behind. Again, the bars give one standard deviation in each variable. On the **bottom**, we show the combined forward–backward estimate. The squares give the estimates of state. Notice the significant improvement in the estimate.

future. Notice that we need to be careful about priors for the backward filter here; we might take the forward estimate and enlarge its covariance somewhat.

17.4 DATA ASSOCIATION

Not every aspect of every measurement conveys information about the state of the object being tracked. In fact, we have been somewhat disingenuous up to this point and have not really talked about what is in $\boldsymbol{y}_i$ at all. Usually, there are measurements that are informative and measurements that are not informative (usually referred to as **clutter**).

Determining which measurements are informative is usually referred to as **data association**. Typically, one wishes to map a series of measurements to a series of tracks, possibly ignoring some—or almost all—of them. The main work in this problem relates to tracking moving objects (aeroplanes, missiles, etc., all conveniently belonging to the bad guys) with radar returns. Typically, there may be many radar returns at any given timestep—we should like to update our representations of the motion of the objects being tracked without necessarily knowing which returns come from which object. As we have seen, tracking algorithms are complicated, but not particularly difficult. Data association is probably the biggest source of difficulties in vision applications, and is not often discussed in the literature. We expect this to change under the impact of practical applications. We will confine our discussion to the case where there is a single moving object. The problem here is that some pixels in the image are very informative about that object, and some are not—which should we use to guide our tracking process?

17.4.1 Choosing the Nearest—Global Nearest Neighbours

In the easiest case, we need to track a single object moving in clutter. For example, we might be tracking a ball moving on a fixed or very slowly varying background. We segment the image into regions, with the reasonable expectation that the ball tends to produce one region, and that the segmentation of the background might change with time. Intuitively, it would be very difficult to confuse the ball with a background region, because we have a strong model of how the ball is moving. This means we would have to be unlucky if there was a new background region that (a) was easily confused with the ball region and (b) confused the dynamic model. This suggests one fairly popular strategy for data association: the rth region offers a measurement $\boldsymbol{y}_i^r$, and we choose the region with the best value of

$$P(\boldsymbol{Y}_i = \boldsymbol{y}_i^r \mid \boldsymbol{y}_0, \ldots, \boldsymbol{y}_{i-1}) = \int P(\boldsymbol{Y}_i = \boldsymbol{y}_i^r \mid \boldsymbol{X}_i, \boldsymbol{y}_0, \ldots, \boldsymbol{y}_{i-1}) P(\boldsymbol{X}_i \mid \boldsymbol{y}_0, \ldots, \boldsymbol{y}_{i-1})\, d\boldsymbol{X}_i$$

$$= \int P(\boldsymbol{Y}_i = \boldsymbol{y}_i^r \mid \boldsymbol{X}_i) P(\boldsymbol{X}_i \mid \boldsymbol{y}_0, \ldots, \boldsymbol{y}_{i-1})\, d\boldsymbol{x}_i$$

Determining $P(\boldsymbol{Y}_i = \boldsymbol{y}_i \mid \boldsymbol{y}_0, \ldots, \boldsymbol{y}_{i-1})$ is a particularly easy calculation with the Kalman filter. We know how $\boldsymbol{Y}_i$ is obtained from $\boldsymbol{X}_i$—we take a normal random variable with mean $\overline{\boldsymbol{X}}_i^-$, and covariance Σ_i^-, apply the linear operator $\mathcal{D}_i$ to it, and add some other random variable. The output of the linear operator must have mean $\mathcal{D}_i\overline{\boldsymbol{X}}_i^-$ and covariance $\mathcal{D}_i\Sigma_i^-\mathcal{D}^T$. To this we are going to add a random variable with zero mean and covariance Σ_{m_i}; the result must have mean

$$\mathcal{D}_i\overline{\boldsymbol{X}}_i^-$$

and covariance

$$\mathcal{D}_i\Sigma_i^-\mathcal{D}^T + \Sigma_{m_i}.$$

In Figure 17.7, we have plotted bounds on the position of an expected measurement for a Kalman filter following various dynamic models.

Notice that this strategy can be relatively robust depending on the accuracy of the dynamic model. If we are able to use a tight dynamic model, anything that is easily confused with the object being tracked must be more similar to the predicted measurement than the real object does. This means that an occasional misidentification may not create major problems because one is unlikely to find a region that is both similar to the predicted measurement and able to throw off the dynamic model badly. In Figure 17.8, we show a Kalman filter tracking the state of

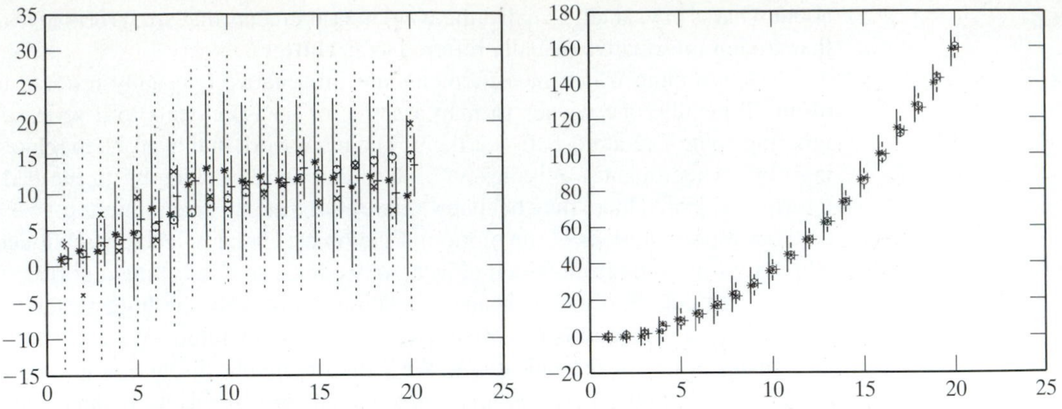

Figure 17.7 Data association for a Kalman filter for a point moving on the line under our model of constant velocity on the **left** and constant acceleration on the **right**. Compare with Figure 17.1 for the constant velocity model and with Figure 17.2 for the constant acceleration model. We have used the conventions of Figure 17.3. We have now overlaid three standard deviation bars for the measurement (the dashed bars passing through the state). These are obtained using the estimate of state before a measurement and our knowledge of the variance of the measurement process. Notice that the measurements lie within these windows.

a point by choosing the best measurement at each step; it does not always correctly identify the point, but its estimate of state is always good.

Notice that what we are doing here is using only measurements that are consistent with our predictions. This may or may not be dangerous: It can be easy to track nonexistent objects this way or to claim to be tracking an object without ever obtaining a measurement from it. If the dynamic model can give only weak predictions (i.e., the object doesn't really behave like that or Σ_{d_i} is consistently large) we may have serious problems because we need to rely on the measurements. These problems occur because the error can accumulate—it is now relatively easy to continue tracking the wrong point for a long time, and the longer we do this the less chance there is of recovering the right point. Figure 17.9 shows a Kalman filter becoming hopelessly confused in this manner.

17.4.2 Gating and Probabilistic Data Association

Again, we assume that we are tracking a single object in clutter and use the example of tracking a ball moving on a fixed or slowly varying background. Instead of choosing the region most like the predicted measurement, we could exclude all regions that are too different and then use all others, weighting them according to their similarity to the prediction.

The first step is called *gating*. We exclude all measurements that are too different from the predicted measurement. What "too different" means rather depends on the application: If we are too aggressive in excluding measurements, we may find nothing left. It is usual to exclude measurements that lie more than some number—commonly, three—of standard deviations from the predicted mean. A more sophisticated strategy is required if the object being tracked has more than one dynamic behavior; for example, military aircraft often engage in high-speed maneuvers. In cases like this, it is common to have several gates and to take all measurements that lie in the tightest gate that contains any measurements.

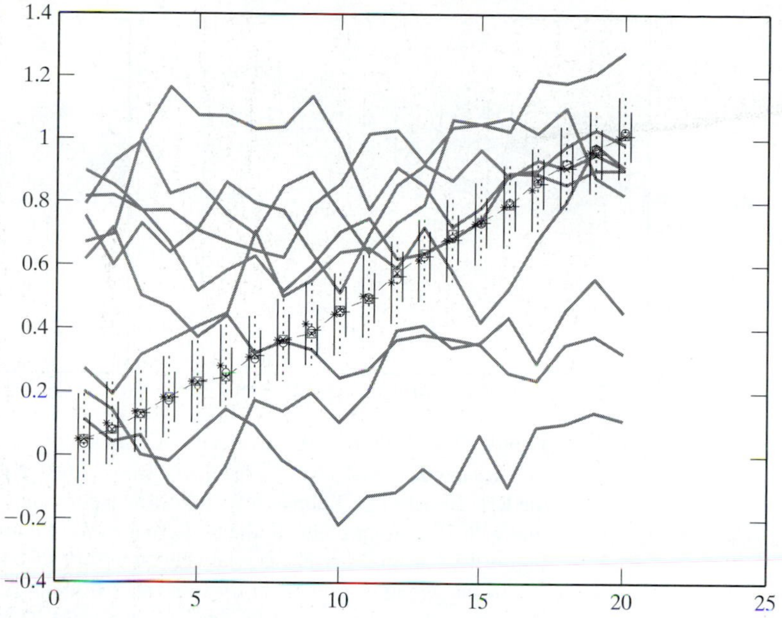

Figure 17.8 Predictions of the point position can identify "good" measurements for a Kalman filter. We are using a Kalman filter to identify a point, moving with constant velocity on a line, and with a small Σ_{d_i} at every stage. There are also 10 drifting points. This plot shows position plotted against time for the drifting points—which are shown with a solid line—and for the point that is being tracked. The trajectory of the point that should be tracked is shown in a dashed line, and each measurement on this trajectory is shown with a square. We have used the conventions of Figure 17.3 (i.e., the state is plotted with open circles, as a function of the step i; * gives $\overline{x}_i^-$, which is plotted slightly to the left of the state to indicate that the estimate is made before the measurement; x gives the measurements, and + gives $\overline{x}_i^+$, which is plotted slightly to the right of the state; the vertical bars around the * and the + are three standard deviation bars using the estimate of variance obtained before and after the measurement, respectively; we have overlaid one standard deviation bars in each case). This filter chooses the measurement at each step by choosing the measurement that maximizes $P(y_i^r | y_0, \ldots, y_{i-1})$; notice that it doesn't choose the right measurement at every step (i.e., the x is not always in the square), but it maintains a good estimate of the state (i.e., the +'s are close to the circles).

The next step is called *probabilistic data association* (PDA). Assume that, in the gate, we have a set of N regions each producing a vector of measurements $\boldsymbol{y}_i^k$, where the superscript indicates the region. We have a set of possible hypotheses to deal with: Either no region comes from the object, which we call h_0, or region k comes from the object, which we call h_k. The measurement we report is

$$E_h[\boldsymbol{y}_i] = \sum_j P(h_j \mid \boldsymbol{y}_0, \ldots, \boldsymbol{y}_{i-1})\boldsymbol{y}_i^j,$$

where the expectation is taken over the space of hypotheses (which is why we have given it the subscript h). Now the probability that none of the measurements comes from the object depends on the details of our detection process. For some detection processes, this parameter can be calculated; for example, in chapter 4 of Blackman and Popoli (1999), there is a worked example

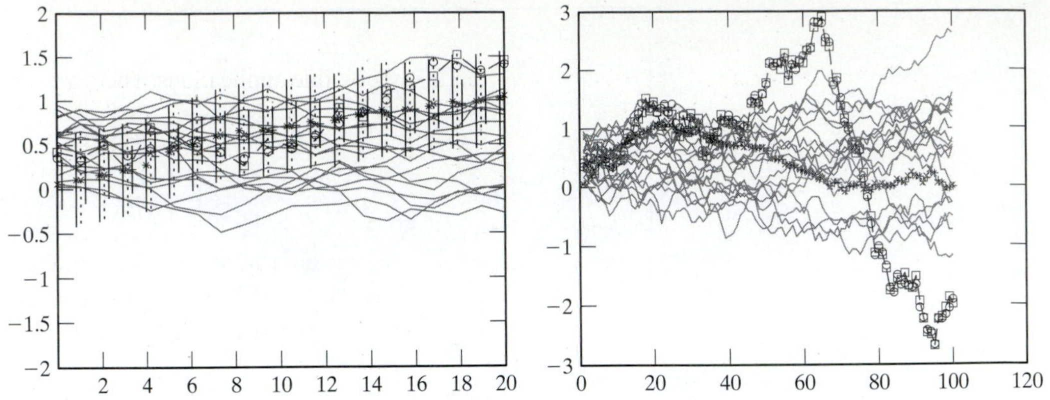

Figure 17.9 If the dynamic model is not sufficiently constrained, then choosing the measurement that gives the best $P(y_i^r | y_0, \ldots, y_{i-1})$ can lead to disaster. On the **left**, 20 steps of a Kalman filter following a point moving periodically on the line with 20 drifting points in the background. We are using the conventions of Figure 17.3 again. Now Σ_{d_i} is relatively large for each step, and so it is easy to follow the wrong measurement for some way. It looks as if the filter is tracking the state well, but in fact as the figure on the **right**—which gives 100 steps—shows, it quickly becomes hopelessly lost.

for a radar system. In other cases, we need to search for a value of the parameter that results in good behavior on a set of training examples. Assume that we have calculated or learned this parameter, which we can write as β. We must also assume that either the object is not detected or only one measurement comes from the object. Now

$$\begin{aligned} P(h_j \mid \boldsymbol{y}_0, \ldots, \boldsymbol{y}_{i-1}) &= \int P(h_j \mid \boldsymbol{X}_i) P(\boldsymbol{X}_i \mid \boldsymbol{y}_0, \ldots, \boldsymbol{y}_{i-1}) d\boldsymbol{X}_i \\ &= P(\boldsymbol{Y}_i = \boldsymbol{y}_i^j \mid \boldsymbol{y}_0, \ldots, \boldsymbol{y}_{i-1}) P(\text{object detected}) \\ &= P(\boldsymbol{Y}_i = \boldsymbol{y}_i^j \mid \boldsymbol{y}_0, \ldots, \boldsymbol{y}_{i-1})(1 - \beta). \end{aligned}$$

In what follows, we write $P(h_j \mid \boldsymbol{y}_0, \ldots, \boldsymbol{y}_{i-1})$ as p_j. In practice, this method is usually used with a Kalman filter. To do so, we report the measurement

$$\boldsymbol{y}_i' = \sum_j p_j \boldsymbol{y}_i^j$$

to the Kalman update equations. Note that the term for not having a measurement appears here as the factor $(1 - \beta)$ in the expressions for the p_j, but our uncertainty about which measurement should contribute to the update should also appear in the covariance update. We modify the covariance update equations to take the form

$$\begin{aligned} \Sigma_i^+ = &(1 - \beta)\left[Id - \mathcal{K}_i \mathcal{M}_i\right] \Sigma_i^- + \beta \Sigma_i^- \\ &+ \mathcal{K}_i \left[\sum_j p_j (\mathcal{H}_i \overline{\boldsymbol{x}}_i^- - \boldsymbol{y}_i^j)(\mathcal{H}_i \overline{\boldsymbol{x}}_i^- - \boldsymbol{y}_i^j)^T - \boldsymbol{y}_i' (\boldsymbol{y}_i')^T \right]. \end{aligned}$$

Here the first term is the update for the standard Kalman filter weighted by the probability that one observation is good, the second term deals with the prospect that all observations are bad, and the third term contributes uncertainty due to the correspondence uncertainty.

17.5 APPLICATIONS AND EXAMPLES

Tracking is a technology with a number of possible applications. There are three dominant topics.

- **Vehicle tracking** systems could report traffic congestion, accidents, and dangerous or illegal behavior by road users. Traffic congestion reports are useful for potential road users—who might change their travel plans—and to authorities—who might arrange to remove immobilised vehicles blocking lanes, etc. Accident reports can be used to alert emergency services; if the tracking system can read vehicle number plates, it might use reports of dangerous or illegal behavior to send a summons to the vehicle owner.
- **Surveillance** systems report what people are doing, usually with the aim of catching people who are doing things they shouldn't. The police might wish to know which member of a sports audience threw a bottle onto the field, for example, or if the same person visited several different banks just before they were robbed. Customs might wish to know exactly who is loading and unloading aircraft flying to foreign ports.
- **Human–computer interaction** systems use people's actions to drive various devices. For example, the living room might decide for itself, by watching what people are doing, when low lights and soft music are appropriate. The television set might change channels when you wave at it. Your computer might watch what you write on your whiteboard and make a record of the contents when you tell it to.

Currently, the most convincing applications are in vehicle tracking. These systems work reliably under a large range of circumstances. We survey vehicle tracking systems briefly here, and then we discuss human trackers in a chapter that appears on the website.

17.5.1 Vehicle Tracking

Systems that can track cars using video from fixed cameras can be used to predict traffic volume and flow; the ideal is to report on and act to prevent traffic problems as quickly as possible. A number of systems can track vehicles successfully. The crucial issue is initiating a track automatically. In the two systems we describe here, the problem is attacked quite differently. Sullivan, Baker, Worrall, Attwood and Remagnino (1997) construct a set of regions of interest (ROIs) in each frame. Because the camera is fixed, these regions of interest can be chosen to span each lane (Figure 17.10); this means that almost all vehicles must pass directly through a region of interest in a known direction (there are mild issues if a vehicle chooses to change lanes while in the ROI, but these can be ignored). Their system then watches for characteristic edge signatures in the ROI that indicate the presence of a vehicle (Figure 17.10). These signatures can alias slightly—typically, a track is initiated when the front of the vehicle enters the ROI, another is initiated when the vehicle lies in the ROI, and a third is initiated close to the vehicle's leaving—because some of the vehicle's edges are easily mistaken for others.

Each initiated track is tracked for a sequence of frames, during which time it accumulates a quality score—essentially, an estimate of the extent to which predictions of future position were accurate. If this quality score is sufficiently high, the track is accepted as an hypothesis. An exclusion region in space and time is constructed around each hypothesis, such that there can be only one track in this region; if the regions overlap, the track with the highest quality is chosen. The requirement that the exclusion regions do not overlap derives from the fact that two cars can't occupy the same region of space at the same time. Once a track has passed these tests, the position in which and the time at which it will pass through another ROI can be predicted. The track is finally confirmed or rejected by comparing this ROI at the appropriate time with a template that predicts the car's appearance. Typically, relatively few tracks that are initiated reach this stage (Figure 17.11).

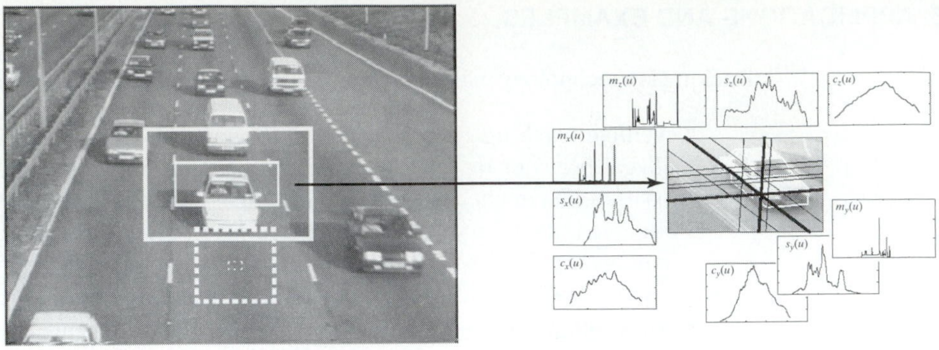

Figure 17.10 Sullivan *et al.* track vehicles in views of the road from a stationary camera. Their tracker uses a series of regions of interest registered to the road, which are shown on the **left**. They initiate tracks by looking for characteristic edge signatures in a particular ROI; these signatures are projected onto three distinct coordinate axes—if the edges projected on these axes have a high enough correlation with the expected form, then a track is initiated (**right**). *Reprinted from "Model-based Vehicle Detection and Classification using Orthographic Approximations," by G D Sullivan, et al., Proc. British Machine Vision Association Conference, 1996, by permission of the K.D. Baker*

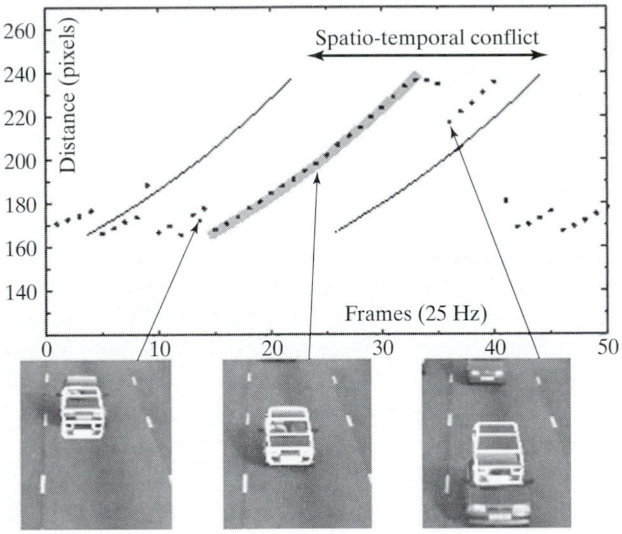

Figure 17.11 In the system of Sullivan *et al.*, tracks are continued if they are of sufficient quality measured by comparing the prediction of the track with measurements. Tracks exclude other tracks: By the time a car reaches the bottom of the view, the system must decide which track to accept. It does so by comparing the track prediction with another ROI. This figure plots a series of tracks (position on the vertical axis and time on the horizontal axis). Notice that the typical alias tracks (that arise because the front of a car and the back of a car both tend to look rather like a registered car to the track initiation process) tend to die out quite quickly; the real track (and its exclusion regions) is indicated. If two tracks attempt to exclude one another, the winner is the track of the highest quality. *Reprinted from "Model-based Vehicle Detection and Classification using Orthographic Approximations," by G D Sullivan, et al., Proc. British Machine Vision Association Conference, 1996, by permission of the K.D. Baker*

An alternative method for initiating car tracks is to track individual features and then group those tracks into possible cars. Beymer, McLauchlan, Coifman and Malik (1997) used this strategy rather successfully. Because the road is plane and the camera is fixed, the homography connecting the road plane and the camera can be determined. This homography can be used to determine the distance between points; and points can lie together on a car only if this distance doesn't change with time. Their system tracks corner points, identified using a second moment matrix (Section 8.3.3), using a Kalman filter. Points are grouped using a simple algorithm using a graph abstraction: Each feature track is a vertex, and edges represent a grouping relationship between the tracks. When a new feature comes into view—and a track is thereby initiated—it is given an edge joining it to every feature track that appears nearby in that frame. If, at some future time, the distance between points in a track changes by too much, the edge is discarded. An exit region is defined near where vehicles leave the frame. When tracks reach this exit region, connected components are defined to be vehicles. This grouper is successful, both in example images (Figure 17.12) and in estimating traffic parameters over long sequences (Figure 17.13).

The ground plane to camera transformation can provide a great deal of information. We have already used this to determine whether points are on rigid objects (by figuring out velocity on the ground plane and comparing velocities). This allowed us to assemble features into objects. Now once an object has been tracked, we can use this transformation to reason about spatial layout and occlusion. Furthermore, we can track cars from moving vehicles. In this case, there are two issues to manage: First, the motion of the camera platform (so-called *ego-motion*); and second, the motion of other vehicles. Ferryman, Maybank and Worrall (2000) estimate the ego-motion by matching views of the road to one another from frame to frame (Figure 17.14). With an estimate of the homography and of the ego-motion, we can now refer tracks of other moving vehicles into the road coordinate system to come up with reconstructions of all vehicles visible on the road from a moving vehicle (Figure 17.14).

Figure 17.12 The figure on the **left** shows individual tracks for the system of Beymer et al. These tracks are obtained by tracking corner points with a Kalman filter. Because the camera position with respect to the road plane is known, the camera transformation can be inverted for points lying on a plane parallel to the road plane. This means that we can determine pairs of points that remain at a constant distance from one another. The figure on the **right** shows groups of such points. These groups are assumed to represent vehicles. *Reprinted from "A Real-Time Computer Vision System for Measuring Traffic Parameters," by D. Beymer et al., Proc. IEEE Conf. on Computer Vision and Pattern Recognition, 1997 © 1997, IEEE*

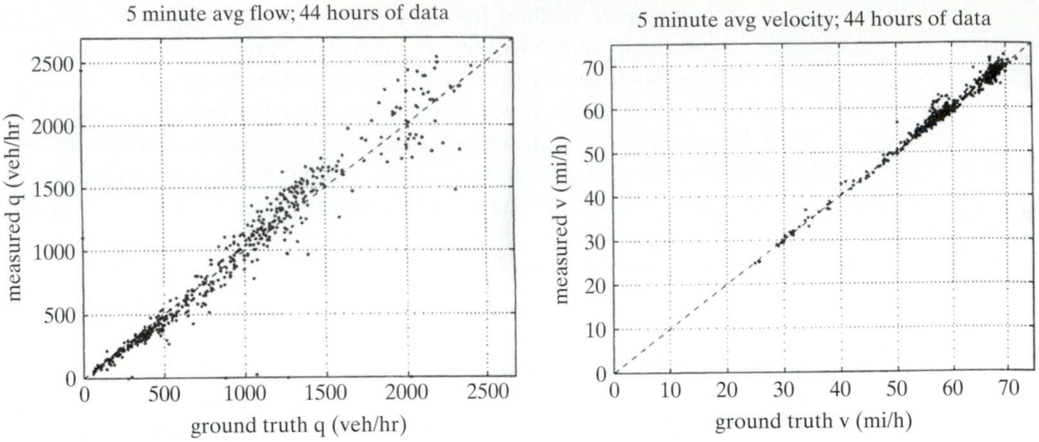

Figure 17.13 The system of Beymer et al. can produce rather accurate estimates of traffic flow and traffic velocity. On the **left**, a scatter plot of estimates of flow versus ground truth; on the **right**, a scatter plot of estimates of velocity vs. ground truth. *Reprinted from "A Real-Time Computer Vision System for Measuring Traffic Parameters," by D. Beymer et al., Proc. IEEE Conf. on Computer Vision and Pattern Recognition, 1997 © 1997, IEEE*

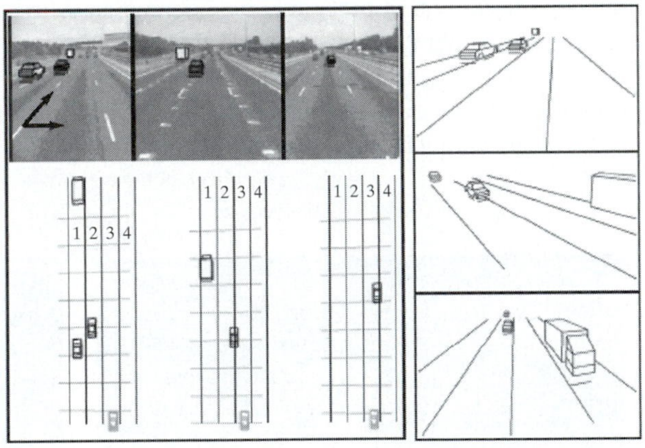

Figure 17.14 Once we know the homography to a ground plane, tracks of other vehicles obtained using the moving camera platform can be referred to the coordinate system relative to the camera platform on this ground plane. This allows detailed reconstructions of traffic geometry illustrated on the **left**. Furthermore, we can use the movement of fixed objects on the ground plane (such as the white marks) to estimate the movement of the camera platform. All this means that we can (a) interpret traffic geometry, for example, predicting impending collision between the camera platform and some other vehicle, and (b) render views of the traffic from some other platform (shown on the **right**). *Reprinted from "Visual Surveillance for Moving Vehicles," by J.M. Ferryman, S.J. Maybank and A.D. Worrall, Proc. 1998 IEEE Workshop on Visual Surveillance, © 1998, IEEE*

17.6 NOTES

The Kalman filter is an extremely useful trick. It is regularly rediscovered, and appears in different guises in different fields. Often dynamics that are not linear can be represented as linear dynamics well enough to fit a Kalman filter. We refer interested readers to Chui (1991), Staff of the Analytical Sciences Corporation (1974) and West and Harrison (1997).

We have not discussed the process of fitting a linear dynamic model. The matter is relatively straightforward *if* one knows the order of the model, a natural state space to use, and a reasonable measurement model. Otherwise, things get tricky—there is an entire field of control theory dedicated to the topic in this case known as *system identification*. We recommend, in the first instance, Ljung (1995).

PROBLEMS

17.1. Assume we have a model $\boldsymbol{x}_i = \mathcal{D}_i \boldsymbol{x}_{i-1}$ and $y_i = \boldsymbol{M}_i^T \boldsymbol{x}_i$. Here the measurement y_i is a one-dimensional vector (i.e., a single number) for each i and $\boldsymbol{x}_i$ is a k-dimensional vector. We say model is *observable* if the state can be reconstructed from any sequence of k measurements.

(a) Show that this requirement is equivalent to the requirement that the matrix

$$\left[\boldsymbol{M}_i \mathcal{D}_i^T \boldsymbol{M}_{i+1} \mathcal{D}_i^T \mathcal{D}_{i+1}^T \boldsymbol{M}_{i+2} \ldots \mathcal{D}_i^T \ldots \mathcal{D}_{i+k-2}^T \boldsymbol{M}_{i+k-1}\right]$$

has full rank.

(b) The point drifting in 3D, where $\mathcal{M}_{3k} = (0, 0, 1)$, $\mathcal{M}_{3k+1} = (0, 1, 0)$, and $\mathcal{M}_{3k+2} = (1, 0, 0)$ is observable.

(c) A point moving with constant velocity in any dimension, with the observation matrix reporting position only, is observable.

(d) A point moving with constant acceleration in any dimension, with the observation matrix reporting position only, is observable.

17.2. A point on the line is moving under the drift dynamic model. In particular, we have $x_i \sim N(x_{i-1}, 1)$. It starts at $x_0 = 0$.

(a) What is its average velocity? (Remember, velocity is *signed*.)

(b) What is its average speed? (Remember, speed is *unsigned*.)

(c) How many steps, on average, before its distance from the start point is greater than two (i.e., what is the expected number of steps, etc.?)

(d) How many steps, on average, before its distance from the start point is greater than ten (i.e., what is the expected number of steps, etc.)?

(e) (This one requires some thought.) Assume we have two nonintersecting intervals, one of length 1 and one of length 2; what is the limit of the ratio (average percentage of time spent in interval one)/ (average percentage of time spent in interval two) as the number of steps becomes infinite?

(f) You probably guessed the ratio in the previous question; now run a simulation and see how long it takes for this ratio to look like the right answer.

17.3. We said that

$$g(x; a, b)g(x; c, d) = g\left(x; \frac{ad + cb}{b + d}, \frac{bd}{b + d}\right) f(a, b, c, d).$$

Show that this is true. The easiest way to do this is to take logs and rearrange the fractions.

17.4. Assume that we have the dynamics

$$x_i \sim N(d_i x_{i-1}, \sigma_{d_i}^2);$$

$$y_i \sim N(m_i x_i, \sigma_{m_i}^2).$$

(a) $P(x_i \mid x_{i-1})$ is a normal density with mean $d_i x_{i-1}$ and variance $\sigma_{d_i}^2$. What is $P(x_{i-1} \mid x_i)$?

(b) Now show how we can obtain a representation of $P(\boldsymbol{x}_i \mid \boldsymbol{y}_{i+1}, \ldots, \boldsymbol{y}_N)$ using a Kalman filter.

Programming Assignments

17.5. Implement a 2D Kalman filter tracker to track something in a simple video sequence. We suggest that you use a background subtraction process and track the foreground blob. The state space should probably involve the position of the blob, its velocity, its orientation—which you can get by computing the matrix of second moments—and its angular velocity.

17.6. If one has an estimate of the background, a Kalman filter can improve background subtraction by tracking illumination variations and camera gain changes. Implement a Kalman filter that does this; how substantial an improvement does this offer? Notice that a reasonable model of illumination variation has the background multiplied by a noise term that is near one—you can turn this into linear dynamics by taking logs.

PART V
High-Level Vision: Geometric Methods

18

Model-Based Vision

This chapter poses object recognition as a correspondence problem—which image feature corresponds to which feature on which object? This simple view of recognition, which naturally focuses on the relationship among object features, image features, and camera models, is useful.

We discuss a variety of different algorithms that use this correspondence approach. The key observation underlying these algorithms is that objects do not scatter features in the image; if we know correspondences for a small set of features, it is fairly easy to obtain correspondences for a much larger set. This is because cameras are fairly orderly and have relatively few degrees of freedom.

There are a number of practical reasons to understand the relationship between the position of image features, and the position and orientation of an object. In Section 18.6, we describe one application, that uses the techniques from the rest of the chapter to register medical images with actual patients so that a surgeon can see where features in the image lie on the patient.

18.1 INITIAL ASSUMPTIONS

All the algorithms we discuss assume that there is a collection of geometric models of the objects that should be recognized. This collection is usually referred to as the *modelbase*. We assume that, if information about an object turns out to be useful in an algorithm, we can ensure it is in the modelbase.

All the algorithms we describe in this chapter are of a single type, usually known as *hypothesize and test*. Each algorithm will

- Hypothesize a correspondence between a collection of image features and a collection of object features, and then use this to generate a hypothesis about the projection from the object coordinate frame to the image frame. There are a variety of different ways of generating hypotheses. When camera intrinsic parameters are known, the hypothesis is equivalent to a hypothetical position and orientation—*pose* —for the object.

- Use this projection hypothesis to generate a rendering of the object. This step is usually known as *backprojection* (for no reason we know; "projection", "forward projection" or "rendering" all seem more reasonable names).
- Compare the rendering to the image, and, if the two are sufficiently similar, accept the hypothesis.

For this approach (illustrated in Figure 18.1) to be effective, we must generate relatively few hypotheses, relatively quickly, and have good methods for comparing renderings and images. The process of comparison is usually called *verification* and can be quite unreliable; we describe verification techniques in Section 18.5. Generally, these methods compute some score of the hypothesis that an object is present at a particular pose, which we call the *verification score*.

This approach works for point features and for curved surfaces, although the details are much more difficult for curved surfaces. The vast majority of the literature deals with object models that consist of geometric features *that project like points*. This means that different views of the object give different views of the same set of features—although some may be occluded—rather than of different sets of features. We deal mainly with this case (which is by far the most useful in practice). However, in Section 18.7, we describe some methods for obtaining and verifying hypotheses for images of curved surfaces.

We generally avoid detailed discussion of the question of what features should be matched. Most of the algorithms that we describe involve a certain amount of search among features—clearly, if we can describe features well, this search is going to be reduced. For example, if our features are simply image points, perhaps obtained by intersecting edge curves, all points are equivalent, and there may be a fair amount of search. If, instead, our points are described using a local representation of the image (say, a vector of filter outputs), then the number of available correspondences goes down and so does the amount of search required.

18.1.1 Obtaining Hypotheses

The main difference between algorithms is the mechanism by which hypotheses are obtained. The most obvious approach is to take all M geometric features in the image and all N geometric features on each of the L objects and enumerate all the possible correspondences between object and image features (i.e., image feature 3 corresponds to feature 5 on object 7, and so on). This is a terrible algorithm because the number of possible correspondences is enormous—$O(LM^N)$.

Geometric constraints between object points limit the size of this space. For example, if we were matching 3D models to 3D data, we would expect pairs of points on the model to be the same distance apart as corresponding pairs of points on the data. Any correspondence for which this constraint is violated can be ignored, whatever the other components. This reasoning is equivalent to pruning a search tree—the approach of searching an aggressively pruned search tree is known as an *interpretation tree* algorithm, after work of Grimson and Lozano-Perez (1984).

Geometric constraints also apply when 3D models must be matched to 2D data. This is because the parameters of the projection model can usually be determined from a fairly small number of point correspondences. Once these parameters are known, the position of all other projected features is known too—this is a constraint because we can't choose the position of these features arbitrarily. We can exploit these constraints by determining the projection parameters explicitly from a small number of correspondences and then using the projection model to predict other correspondences (Section 18.2 describes this well-established strategy whose origins are uncertain).

In fact, it isn't necessary to determine the projection parameters by the following argument. Once we have established a correspondence between a small number of object features and a small number of image features—the base set—the camera constraints could be used to predict the position of other image features. This means that, in an appropriate sense, the position of the other image features is fixed *relative to the base set*. By an appropriate interpretation of this relative position, we can obtain measurements that are independent of the projection parameters and use these to identify the object (Section 18.4).

18.2 OBTAINING HYPOTHESES BY POSE CONSISTENCY

Assume we have an image of some object obtained using a camera model of known type, but with unknown parameters (e.g., we might be viewing an object in a calibrated perspective camera with unknown extrinsic parameters with respect to the object frame). If we hypothesize a match between a sufficiently large group of image features and a sufficiently large group of object features, we can recover the missing camera parameters from this hypothesis (and so render the rest of the object). Methods of this form (Algorithm 18.1) are known as *pose consistency methods*. We describe this family of methods—whose general form is shown in Figure 18.1—with a set of examples. This form of algorithm is increasingly being called *alignment*; the term refers to the idea that the object is being aligned with its image. The name was coined for a relatively recent version of the algorithm, which appears in the literature in many forms. It is only relatively recently that the similarity between these forms has become apparent.

We are really dealing with a family of methods here because the details depend on what is known about the camera and whether the objects are two or three-dimensional. We call a group that can be used to yield a camera hypothesis a *frame group* (there can be both object and image frame groups).

Algorithm 18.1: Alignment: Matching Object and Image Groups to Infer a Camera Model

```
For all object frame groups O
   For all image frame groups F
      For all correspondences C between
         elements of F and elements of O
         Use F, C, and O to infer the missing parameters in a camera model
         Use the camera model estimate to render the object
         If the rendering conforms to the image, the object is present
      end
   end
end
```

18.2.1 Pose Consistency for Perspective Cameras

Assume we have a perspective camera for which the intrinsic parameters are known. This camera is viewing an object in the modelbase. Let us work in the object's coordinate system—the extrinsic parameters now boil down to the position and orientation of the camera in the object's frame. Now if we use Algorithm 18.1, we have correspondences between a group of image features in an image coordinate system and a group of object features in the object coordinate system. From this information, we can determine the extrinsic parameters of the camera. Once the extrinsic

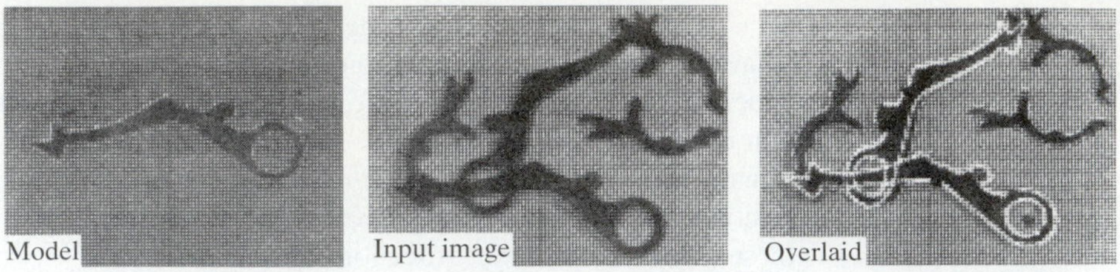

Figure 18.1 The alignment method applied to a plane object. On the **left**, an image of an object; in the **center**, an image containing two instances of this object, along with some other stuff (the popular term is *clutter*). Feature points are detected, and then correspondences between groups—in this case, triples of points—are searched; each correspondence gives rise to an affine transformation from the model to the image. Satisfactory correspondences align many model edge points with image edge points, as in the figure on the **right**. The images in this figure come from one of the earliest papers on the subject and are affected by the poor reproduction techniques of the time. *Reprinted from "Object recognition using alignment," by D.P. Huttenlocher and S. Ullman, Proc. Int. Conf. Computer Vision, 1986 © 1986, IEEE*

parameters are known, the entire camera is known, and we can use this to render the rest of the object.

There is a variety of frame groups available for this problem. Typically, good frame groups contain few features of several different types (to reduce the number of correspondences to be searched). Groups that have been popular include

- three points;
- three directions—often known as a *trihedral vertex*—and a point (which is necessary to establish scale);
- and a *dihedral vertex* (two directions emanating from a shared origin) and a point.

Usually, directions are obtained by using line segments. This is attractive because it is quite often the case that it is quite likely that part of a line segment will appear, but it is often difficult to localize the endpoints exactly.

The Intrinsic Parameters It is quite common in the literature to assume that the intrinsic parameters of the camera are unknown, too. This doesn't change the problem all that much, although we might need to use more complicated frame groups, but it does offer some opportunities for more aggressive consistency reasoning. In images *with more than one object*, we can require that the *intrinsic* camera parameters are the same for different objects.

The line of reasoning is quite simple. First, we use Algorithm 18.1 to recognise individual objects. Associated with each object is a camera solution. Now for each *pair* of recognized objects, we compare the intrinsic parameters of the camera solution: If they are (sufficiently) different, then the two hypotheses are incompatible.

18.2.2 Affine and Projective Camera Models

Calibrating perspective cameras is complicated because the extrinsic parameters involve a rotation. It is often possible to use a camera model that allows simpler calibration at the possible cost of greater ambiguity in the model identity. The two important simplifications are:

- **Affine cameras**, which model a perspective view as an affine transformation followed by an orthographic projection.
- **Projective cameras**, which model a perspective view as a projective transformation followed by perspective projection.

We deal with each case in some detail. Remember, the only real issue here is how to obtain a camera model from an hypothesized correspondence between an object frame group and an image frame group; the rest is supplied by Algorithm 18.1.

Affine Cameras In homogeneous coordinates, we can write an affine camera as $\Pi\mathcal{A}$, where $\mathcal{A}$ is a general affine transformation and Π is an orthographic camera transformation. For reference, this means that

$$\Pi = \begin{pmatrix} 1 & 0 & 0 & 0 \\ 0 & 1 & 0 & 0 \\ 0 & 0 & 0 & 1 \end{pmatrix}$$

and

$$\mathcal{A} = \begin{pmatrix} a_{00} & a_{01} & a_{02} & a_{03} \\ a_{10} & a_{11} & a_{12} & a_{13} \\ a_{20} & a_{21} & a_{22} & a_{23} \\ 0 & 0 & 0 & 1 \end{pmatrix}.$$

We use capital letters for points on the model, small letters for points in the image, and subscripts to denote correspondence (so that $\boldsymbol{p}_1 = \Pi\mathcal{A}\boldsymbol{P}_1$).

One possible frame group consists of four points. In this case, we must determine the camera—essentially, $\mathcal{A}$—from a correspondence between four image points and four object points. Now assume we have a (hypothesized) correspondence between four image points ($\boldsymbol{p}_i$) and four object points ($\boldsymbol{P}_i$). We can interpret $\boldsymbol{p}_i = \Pi\mathcal{A}\boldsymbol{P}_i$ as two linear equations in the first two rows of $\mathcal{A}$—that is

$$\begin{pmatrix} p_{i0} \\ p_{i1} \end{pmatrix} = \begin{pmatrix} a_{00}P_{i0} + a_{01}P_{i1} + a_{02}P_{i2} + a_{03} \\ a_{10}P_{i0} + a_{11}P_{i1} + a_{12}P_{i2} + a_{13} \end{pmatrix}.$$

There are eight elements in the first two rows of $\mathcal{A}$; with four points in general position, we can solve these equations to obtain a unique solution for the first two rows of $\mathcal{A}$. Notice that the rest of $\mathcal{A}$ doesn't contribute to the projection and doesn't need to be known to compute the projection of all other points. This means that knowing the first two rows of $\mathcal{A}$ is all we need to know to generate a backprojection.

Some models that are distinct under rotations and translations are ambiguous under affine cameras. Assume that one model is given by a set of points $\boldsymbol{P}_j$, a second is given by $\boldsymbol{Q}_j$ *and* there is some affine transformation $\mathcal{B}$ such that for each j, $\boldsymbol{P}_j = \mathcal{B}\boldsymbol{Q}_j$. These models can't be distinguished under an affine camera. A view of the first model in an affine camera is a set of image points $\boldsymbol{p}_j = \Pi\mathcal{A}_1\boldsymbol{P}_j$, and a view of the second model in some other affine camera is given by a set of image points $\boldsymbol{q}_j = \Pi\mathcal{A}_2\boldsymbol{Q}_j$. If $\mathcal{A}_2 = \mathcal{A}_1\mathcal{B}$, then we have that

$$\boldsymbol{q}_j = \Pi\mathcal{A}_2\boldsymbol{Q}_j = \Pi\mathcal{A}_1\mathcal{B}\boldsymbol{Q}_j = \Pi\mathcal{A}_1\boldsymbol{P}_j = \boldsymbol{p}_j$$

that is, there is an affine camera that makes the second model look exactly like a view of the first model in some other affine camera—so they are indistinguishable.

Projective Cameras In homogeneous coordinates, we can write a projective camera as $\Pi\mathcal{A}$, where $\mathcal{A}$ is a general projective transformation and Π is a perspective camera transformation. For reference, this means that

$$\Pi = \begin{pmatrix} 1 & 0 & 0 & 0 \\ 0 & 1 & 0 & 0 \\ 0 & 0 & 1 & 0 \end{pmatrix}$$

and

$$\mathcal{A} = \begin{pmatrix} a_{00} & a_{01} & a_{02} & a_{03} \\ a_{10} & a_{11} & a_{12} & a_{13} \\ a_{20} & a_{21} & a_{22} & a_{23} \\ a_{30} & a_{31} & a_{32} & a_{33} \end{pmatrix}.$$

Again, we use capital letters for points on the model, small letters for points in the image, and subscripts to denote correspondence (so that $\boldsymbol{p}_1 = \Pi\mathcal{A}\boldsymbol{P}_1$). Notice that because we are working in homogeneous coordinates, $\mathcal{A}$ and $\lambda\mathcal{A}$ represent the same transformation if $\lambda \neq 0$.

One possible frame group consists of five points. In this case, we must determine the camera—essentially, $\mathcal{A}$—from a correspondence between five image points and five object points Now assume we have a (hypothesized) correspondence between five image points ($\boldsymbol{p}_i$) and five object points ($\boldsymbol{P}_i$). We can interpret $\boldsymbol{p}_i = \Pi\mathcal{A}\boldsymbol{P}_i$ as two *nonlinear* equations in the first two rows of $\mathcal{A}$; that is,

$$\begin{pmatrix} p_{i0} \\ p_{i1} \end{pmatrix} = \frac{1}{a_{30}P_{i0} + a_{31}P_{i1} + a_{32}P_{i2} + a_{33}} \begin{pmatrix} a_{00}P_{i0} + a_{01}P_{i1} + a_{02}P_{i2} + a_{03} \\ a_{10}P_{i0} + a_{11}P_{i1} + a_{12}P_{i2} + a_{13} \end{pmatrix}.$$

There are 12 elements in the first three rows of $\mathcal{A}$; with five points in general position, we can solve these equations to obtain a unique solution for the first three rows of $\mathcal{A}$. Notice that the rest of $\mathcal{A}$ doesn't contribute to the projection and doesn't need to be known to compute the projection of all other points. This means that knowing the first three rows of $\mathcal{A}$ is all we need to know to generate a backprojection. Notice also that all we have done here is to repeat, in significantly less detail, the activities of Section 18.2.2.

Some models that are distinct under affine cameras (and so under rotations and translations) are ambiguous under projective cameras. Assume that one model is given by a set of points $\boldsymbol{P}_j$, a second is given by $\boldsymbol{Q}_j$, *and* there is some projective transformation $\mathcal{B}$ such that for each j $\boldsymbol{P}_j = \mathcal{B}\boldsymbol{Q}_j$. These models can't be distinguished under a projective camera. A view of the first model in a projective camera is a set of image points $\boldsymbol{p}_j = \Pi\mathcal{A}_1\boldsymbol{P}_j$, and a view of the second model in some other projective camera is given by a set of image points $\boldsymbol{q}_j = \Pi\mathcal{A}_2\boldsymbol{Q}_j$. If $\mathcal{A}_2 = \mathcal{A}_1\mathcal{B}$, then we have

$$\boldsymbol{q}_j = \Pi\mathcal{A}_2\boldsymbol{Q}_j = \Pi\mathcal{A}_1\mathcal{B}\boldsymbol{Q}_j = \Pi\mathcal{A}_1\boldsymbol{P}_j = \boldsymbol{p}_j$$

that is, there is a projective camera that makes the second model look exactly like a view of the first model in some other projective camera, so they are indistinguishable.

18.2.3 Linear Combinations of Models

The case of an affine camera used the correspondences to perform explicit camera calibration. We can hide the camera calibration process with a little linear algebra. We use homogeneous coordinates and can write a general uncalibrated affine camera as $\Pi\mathcal{A}$, where $\mathcal{A}$ is a general affine transformation and

$$\Pi = \begin{pmatrix} 1 & 0 & 0 & 0 \\ 0 & 1 & 0 & 0 \\ 0 & 0 & 0 & 1 \end{pmatrix}.$$

We use capital letters for points on the model, small letters for points in the image, and subscripts to denote correspondence (so that $\boldsymbol{p}_1 = \Pi \mathcal{A} \boldsymbol{P}_1$).

Let us identify one point on the model as an origin and consider offsets from that point (this means that translations can be ignored). Use the notation $\boldsymbol{v}_i = \boldsymbol{p}_i - \boldsymbol{p}_0$ and $\boldsymbol{V}_i = \boldsymbol{P}_i - \boldsymbol{P}_0$. Now obtain three views of the object in different general affine cameras with affine transformations $\mathcal{A}$, $\mathcal{B}$, and $\mathcal{C}$ so that for the ith point on the object we have

$$\begin{aligned} \boldsymbol{v}_i^A &= \Pi \mathcal{A} \boldsymbol{V}_i; \\ \boldsymbol{v}_i^B &= \Pi \mathcal{B} \boldsymbol{V}_i; \\ \boldsymbol{v}_i^C &= \Pi \mathcal{C} \boldsymbol{V}_i. \end{aligned}$$

Because Π contains a lot of zeros and the fourth row of $\boldsymbol{V}_i$ is zero, we can simplify things considerably with these three views.

Write the jth **row** of $\mathcal{A}$ as $\boldsymbol{a}^T{}_j$. We now have

$$\begin{aligned} \boldsymbol{v}_i^A &= (\boldsymbol{a}_0^T.\boldsymbol{V}_i, \boldsymbol{a}_1^T.\boldsymbol{V}_i, 0)^T; \\ \boldsymbol{v}_i^B &= (\boldsymbol{b}_0^T.\boldsymbol{V}_i, \boldsymbol{b}_1^T.\boldsymbol{V}_i, 0)^T; \\ \boldsymbol{v}_i^C &= (\boldsymbol{c}_0^T.\boldsymbol{V}_i, \boldsymbol{c}_1^T.\boldsymbol{V}_i, 0)^T. \end{aligned}$$

We would like to generate some arbitrary new view of the object, which could be obtained by applying $\Pi\mathcal{D}$ to the points, where $\mathcal{D}$ is some new affine transformation. To obtain this view, we must first decide where $\boldsymbol{p}_0$ lies; having done so, we need the $\boldsymbol{v}_i^D$ for the ith point.

Now $\boldsymbol{v}_i^D = (\boldsymbol{d}_0^T.\boldsymbol{V}_i, \boldsymbol{d}_1^T.\boldsymbol{V}_i, 0)$. Assuming that $\mathcal{A}$, $\mathcal{B}$, and $\mathcal{C}$ are general, we have that $\boldsymbol{d}_j$ must be a fixed linear combination of $\boldsymbol{a}_j$, $\boldsymbol{b}_j$, and $\boldsymbol{c}_j$, say,

$$\boldsymbol{d}_j = \lambda_{(\boldsymbol{a}_j)} \boldsymbol{a}_j + \lambda_{(\boldsymbol{b}_j)} \boldsymbol{b}_j + \lambda_{(\boldsymbol{c}_j)} \boldsymbol{c}_j.$$

Then we have

$$\boldsymbol{v}_i^D = (\lambda_{(\boldsymbol{a}_0)} \boldsymbol{a}_0^T.\boldsymbol{V}_i + \lambda_{(\boldsymbol{b}_0)} \boldsymbol{b}_0^T.\boldsymbol{V}_i + \lambda_{(\boldsymbol{c}_0)} \boldsymbol{c}_0^T.\boldsymbol{V}_i, \lambda_{(\boldsymbol{a}_1)} \boldsymbol{a}_1^T.\boldsymbol{V}_i + \lambda_{(\boldsymbol{b}_1)} \boldsymbol{b}_1^T.\boldsymbol{V}_i + \lambda_{(\boldsymbol{c}_1)} \boldsymbol{c}_1^T.\boldsymbol{V}_i, 0),$$

which means that, given three unknown affine views of the object, we can reconstruct a fourth by determining the values of these λs.

This strategy has become known as *linear combinations of models*. Generating hypotheses with this method requires searching correspondences, too; we select some image points to be $\boldsymbol{p}_0$, $\boldsymbol{p}_1$, and so on. and then solve for the λ values. Once these are known, we can render the object, although additional ingenuity is required for hidden line removal. Notice that the approach is simply an alternative version of affine camera calibration. It has the attractive feature that the object model is constructed from three views of the object in a fairly simple way. It turns out that an object model is also easily constructed from three views for the approach of Section 18.4, too.

18.3 OBTAINING HYPOTHESES BY POSE CLUSTERING

Most objects have many frame groups. This means that there should be many correspondences between object and image frame groups that verify satisfactorily. Each of these correspondences should yield approximately the same estimate of position and orientation for the object with respect to the camera (or the camera with respect to the object—it doesn't matter which we work with). However, image frame groups that come from noise (or *clutter*, a term used for objects that are not of interest and not in the modelbase) are likely to yield estimates of pose that are uncorrelated. This motivates the use of some form of clustering method to filter hypotheses before verification.

For each object, we set up an accumulator array that represents pose space—each element in the accumulator array corresponds to a bucket in pose space. Now we take each image frame group, and hypothesize a correspondence between it and every frame group on every object. For each of these correspondences, we determine pose parameters and make an entry in the accumulator array for the current object at the pose value. If there are large numbers of votes in any object's accumulator array, this can be interpreted as evidence for the presence of that object at that pose; this evidence can be checked using a verification method. It is important to note the similarity between this method (which is given in Algorithm 18.2) and the Hough transform (Section 15.1).

Algorithm 18.2: Pose Clustering: Voting on Pose, Correspondence, and Identity

```
For all objects O
   For all object frame groups F(O)
      For all image frame groups F(I)
         For all correspondences C between
            elements of F(I) and elements of F(O)
            Use F(I), F(O), and C to infer object pose P(O)
            Add a vote to O's pose space at the bucket
            corresponding to P(O).
         end
      end
   end
end
For all objects O
   For all elements P(O) of O's pose space that have enough votes
      Use the P(O) and the camera model estimate to render the object
      If the rendering conforms to the image, the object is present
   end
end
```

There are two difficulties with these methods (which mirror the difficulties in using the Hough transform in practice):

1. In an image containing noise or texture that generates many spurious frame groups, the number of votes in the pose arrays corresponding to real objects may be smaller than the number of spurious votes (the details are in Grimson and Huttenlocher, 1990*b*).
2. Choosing the size of the buckets in the pose arrays is difficult; buckets that are too small mean there is no accumulation of votes (because it is hard to compute pose accurately); buckets that are too large mean too many buckets will have enough votes to trigger a verification attempt.

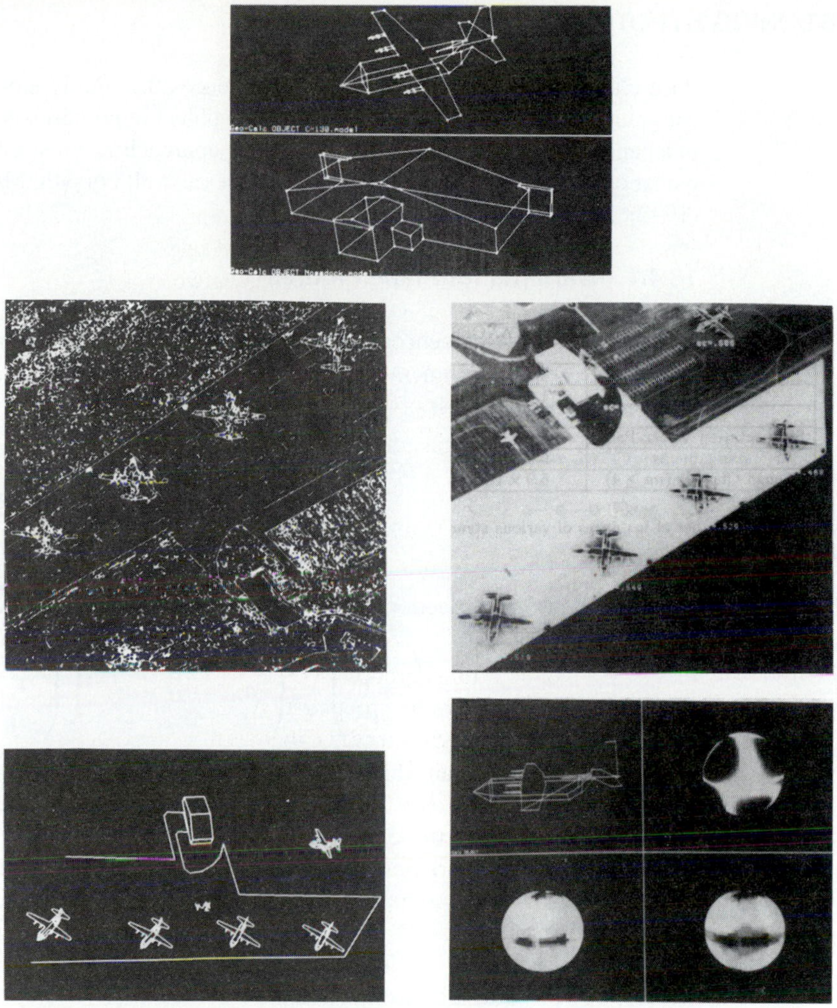

Figure 18.2 Pose clustering methods use frame-bearing groups to generate pose estimates and then cluster these estimates. **Top:** two models used in an early pose clustering system. **Center left:** edge points marked for an image used in testing. **Center right:** edges of models that are found overlaid on the image. **Bottom left:** a new view of the layout of the models in space to indicate their pose; notice the curious pose of the aircraft off the runway. **Bottom right:** for each frame-bearing group, some views are better than others because the estimate of pose is more stable; next to the model of the aircraft, we see a sphere representing different viewpoints, with light regions corresponding to high error views of the pair of features marked on the model. *Reprinted from "The evolution and testing of a model-based object recognition system," by J.L. Mundy and A. Heller, Proc. Int. Conf. Computer Vision, 1990 © 1990 IEEE*

We can improve the noise resistance of the method by not counting votes for objects at poses where the vote is obviously unreliable—for example, in cases where, if the object were at that pose, the object frame group would be invisible. These improvements are sufficient to yield working systems (Figure 18.2).

18.4 OBTAINING HYPOTHESES USING INVARIANTS

Pose clustering methods collect correspondences that imply similar hypotheses about camera calibration and pose. Another way to obtain object hypotheses is to use measurements that are independent of the camera properties. This approach is most easily developed for images of planar objects, but can be applied to other cases as well Forsyth, Mundy, Zisserman and Rothwell (1992), Forsyth (1996), Huang (1981).

18.4.1 Invariants for Plane Figures

Recall that an affine camera can be written as $\Pi\mathcal{A}$, where $\mathcal{A}$ is a general affine transformation and Π is an orthographic camera transformation. Assume we have a set of model points $\boldsymbol{P}_j$, which are coplanar; without loss of generality, we can assume they lie on the $z = 0$ plane. Now we have

$$\begin{pmatrix} p_{i0} \\ p_{i1} \\ 1 \end{pmatrix} = \Pi\mathcal{A} \begin{pmatrix} P_{i0} \\ P_{i1} \\ 0 \\ 1 \end{pmatrix}$$

(using the notation of Section 18.2.2). We can substitute for Π and $\mathcal{A}$ to obtain

$$\begin{pmatrix} p_{i0} \\ p_{i1} \\ 1 \end{pmatrix} = \begin{pmatrix} a_{00} & a_{01} & a_{03} \\ a_{10} & a_{11} & a_{13} \\ 0 & 0 & 1 \end{pmatrix} \begin{pmatrix} P_{i0} \\ P_{i1} \\ 1 \end{pmatrix}.$$

This is important; it means that views of a set of coplanar points in an affine camera are generated by *plane affine transformations* —this means that we can abstract away the camera and reason only about the effect of these transformations on the model.

A similar result applies to views of plane points in a projective camera—in this case, the transformation is a *plane projective transformation*. To get this result, recall that a projective camera can be written as $\Pi\mathcal{A}$, where $\mathcal{A}$ is a general projective transformation and Π is a perspective camera transformation. Assume we have a set of model points $\boldsymbol{P}_j$, which are coplanar; without loss of generality, we can assume they lie on the $z = 0$ plane. Now we have

$$\begin{pmatrix} p_{i0} \\ p_{i1} \\ 1 \end{pmatrix} = \Pi\mathcal{A} \begin{pmatrix} P_{i0} \\ P_{i1} \\ 0 \\ 1 \end{pmatrix}$$

(using the notation of Section 18.2.2). We can substitute for Π and $\mathcal{A}$ to obtain

$$\begin{pmatrix} p_{i0} \\ p_{i1} \end{pmatrix} = \frac{1}{a_{20}P_{i0} + a_{21}P_{i1} + a_{23}} \begin{pmatrix} a_{00} & a_{01} & a_{03} \\ a_{10} & a_{11} & a_{13} \end{pmatrix} \begin{pmatrix} P_{i0} \\ P_{i1} \\ 1 \end{pmatrix}.$$

Recalling that we are working in homogeneous coordinates, a more convenient form is

$$\begin{pmatrix} p_{i0} \\ p_{i1} \\ p_{i2} \end{pmatrix} = \begin{pmatrix} a_{00} & a_{01} & a_{03} \\ a_{10} & a_{11} & a_{13} \\ a_{20} & a_{21} & a_{23} \end{pmatrix} \begin{pmatrix} P_{i0} \\ P_{i1} \\ P_{i3} \end{pmatrix}.$$

Again, this means that views of a set of coplanar points in a projective camera are generated by these plane projective transformations. This means that we can abstract away the camera and reason only about the effect of these transformations on the model.

Affine Invariants for Coplanar Points Assume we have a model that is a set of coplanar points. Choose three of these points $\boldsymbol{P}_0$, $\boldsymbol{P}_1$, and $\boldsymbol{P}_2$. This gives a coordinate frame, and any other point $\boldsymbol{P}_i$ in the model can be expressed as $\boldsymbol{P}_0 + \mu_{i1}(\boldsymbol{P}_1 - \boldsymbol{P}_0) + \mu_{i2}(\boldsymbol{P}_2 - \boldsymbol{P}_0)$. It takes only a little linear algebra to compute the μ values associated with each point.

Now the camera takes model points $\boldsymbol{P}_i$ to image points $\boldsymbol{p}_i$. For an uncalibrated affine camera viewing a set of plane points, the effect of the camera can be written as an (unknown) plane affine transformation. We can write the camera as $\mathcal{C}$. Now

$$\begin{aligned}
\boldsymbol{p}_i &= \mathcal{C}\boldsymbol{P}_i \\
&= \mathcal{C}(\boldsymbol{P}_0 + \mu_{i1}(\boldsymbol{P}_1 - \boldsymbol{P}_0) + \mu_{i2}(\boldsymbol{P}_2 - \boldsymbol{P}_0)) \\
&= (1 - \mu_{i1} - \mu_{i2})(\mathcal{C}\boldsymbol{P}_0) + \mu_{i1}(\mathcal{C}\boldsymbol{P}_1) + \mu_{i2}(\mathcal{C}\boldsymbol{P}_2) \\
&= (1 - \mu_{i1} - \mu_{i2})\boldsymbol{p}_0 + \mu_{i1}\boldsymbol{p}_1 + \mu_{i2}\boldsymbol{p}_2 \\
&= \boldsymbol{p}_0 + \mu_{i1}(\boldsymbol{p}_1 - \boldsymbol{p}_0) + \mu_{i2}(\boldsymbol{p}_2 - \boldsymbol{p}_1).
\end{aligned}$$

This means that the μ_{ij} describe the geometry of the object and *are independent of the view* (i.e., if we compute the μ_{ij} in the model plane or in some affine view, we obtain the same values). Measurements with this property are often referred to as *affine invariants* (other constructions for affine invariants are given in the exercises).

Projective Invariants for Coplanar Points and Lines In homogeneous coordinates, we can write the relationship between image points and (plane) model points as $\boldsymbol{p}_i = \mathcal{A}\boldsymbol{P}_i$, where $\mathcal{A}$ is a general 3×3 matrix. Now we have

$$\frac{\det\left[\boldsymbol{p}_i\boldsymbol{p}_l\boldsymbol{m}_k\right]}{\det\left[\boldsymbol{p}_i\boldsymbol{p}_j\boldsymbol{p}_l\right]}\frac{\det\left[\boldsymbol{p}_i\boldsymbol{p}_l\boldsymbol{p}_m\right]}{\det\left[\boldsymbol{p}_i\boldsymbol{p}_k\boldsymbol{p}_m\right]}$$

is equal to

$$\frac{\det\left[(\mathcal{A}\boldsymbol{P}_i)(\mathcal{A}\boldsymbol{P}_j)(\mathcal{A}\boldsymbol{P}_k)\right]}{\det\left[(\mathcal{A}\boldsymbol{P}_i)(\mathcal{A}\boldsymbol{P}_j)(\mathcal{A}\boldsymbol{P}_l)\right]}\frac{\det\left[(\mathcal{A}\boldsymbol{P}_i)(\mathcal{A}\boldsymbol{P}_l)(\mathcal{A}\boldsymbol{P}_m)\right]}{\det\left[(\mathcal{A}\boldsymbol{P}_i)(\mathcal{A}\boldsymbol{P}_k)(\mathcal{A}\boldsymbol{P}_m)\right]},$$

and this is equal to

$$\frac{\det\mathcal{A}\det\left[\boldsymbol{P}_i\boldsymbol{P}_j\boldsymbol{P}_k\right]}{\det\mathcal{A}\det\left[\boldsymbol{P}_i\boldsymbol{P}_j\boldsymbol{P}_l\right]}\frac{\det\mathcal{A}\det\left[\boldsymbol{P}_i\boldsymbol{P}_l\boldsymbol{P}_m\right]}{\det\mathcal{A}\det\left[\boldsymbol{P}_i\boldsymbol{P}_k\boldsymbol{P}_m\right]},$$

and, in turn, this is equal to

$$\frac{\det\left[\boldsymbol{P}_i\boldsymbol{P}_j\boldsymbol{P}_k\right]}{\det\left[\boldsymbol{P}_i\boldsymbol{P}_j\boldsymbol{P}_l\right]}\frac{\det\left[\boldsymbol{P}_i\boldsymbol{P}_l\boldsymbol{P}_m\right]}{\det\left[\boldsymbol{P}_i\boldsymbol{P}_k\boldsymbol{P}_m\right]}.$$

(as long as no two of i, j, k, l, and m are the same and no three of the points are collinear). There are other arrangements of determinants that are invariant as well (see Exercises).

Plane Algebraic Curves and Projective Transformations Algebraic curves consist of all points on the plane where a polynomial vanishes. A line is an algebraic curve. If we write points in homogeneous coordinates as $\boldsymbol{p}_i = [p_{i0}, p_{i1}, p_{i2}]^T$, a line is the locus of points $\boldsymbol{p}$ for which $l_0 p_0 + l_1 p_1 + l_2 p_2 = 0$. We can write this as $\boldsymbol{l}^T\boldsymbol{p} = 0$, where the line is represented by $\boldsymbol{l}$. Now if our points transform by $\boldsymbol{p} = \mathcal{A}\boldsymbol{P}$, then the lines transform by $\boldsymbol{l} = \mathcal{A}^{-T}\boldsymbol{L}$. This is easiest to see by observing that

$$l^T p = l^T \mathcal{A}P = L^T \mathcal{A}^{-1}\mathcal{A}P$$

so that if $\boldsymbol{p}$ lies on line $\boldsymbol{l}$, then $\boldsymbol{P}$ lies on line $\boldsymbol{L}$.

Because lines transform (basically) like points, we have

$$\begin{aligned}\frac{\det\left[l_i l_j l_k\right]\det\left[l_i l_l l_m\right]}{\det\left[l_i l_j l_l\right]\det\left[l_i l_k l_m\right]} &= \frac{\det\left[(\mathcal{A}^{-T}L_i)(\mathcal{A}^{-T}L_j)(\mathcal{A}^{-T}L_k)\right]\det\left[(\mathcal{A}^{-T}L_i)(\mathcal{A}^{-T}L_l)(\mathcal{A}^{-T}L_m)\right]}{\det\left[(\mathcal{A}^{-T}L_i)(\mathcal{A}^{-T}L_j)(\mathcal{A}^{-T}L_l)\right]\det\left[(\mathcal{A}^{-T}L_i)(\mathcal{A}^{-T}L_k)(\mathcal{A}^{-T}L_m)\right]}\\ &= \frac{\det\mathcal{A}\det\left[L_i L_j L_k\right]\det\mathcal{A}\det\left[L_i L_l L_m\right]}{\det\mathcal{A}\det\left[L_i L_j L_l\right]\det\mathcal{A}\det\left[L_i L_k L_m\right]}\\ &= \frac{\det\left[L_i L_j L_k\right]\det\left[L_i L_l L_m\right]}{\det\left[L_i L_j L_l\right]\det\left[L_i L_k L_m\right]}\end{aligned}$$

(as long as no two of i, j, k, l, and m are the same and no three of the lines pass through a single point).

In fact, algebraic invariants abound in the projective case. Useful examples occur in particular for plane conics. A plane conic is the locus of points $\boldsymbol{x}$ such that $\boldsymbol{x}^t\mathcal{M}\boldsymbol{x} = 0$, where $\boldsymbol{x}$ is the vector of homogeneous coordinates describing a point, and the matrix $\mathcal{M}$ contains the coefficients of the conic. Now if we transform the coordinate system somehow (say, by observing the points in a camera), then we have $\boldsymbol{x}' = \mathcal{P}\boldsymbol{x}$ for some plane projective transformation $\mathcal{P}$. The equation of the conic in the new coordinate system can be obtained by noticing that the new equation must vanish for every point that used to lie on the old conic (i.e., for every point for which the old conic's equation vanished in the old coordinate frame). In particular, if we invert the transformation and plug the resulting point into the old conic's equation, we should get a zero. This line of reasoning means that $\mathcal{M}' = \mathcal{P}^{-t}\mathcal{M}\mathcal{P}^{-1}$ is the equation of the conic in the new frame.

Now assume that we have two conics, $\mathcal{M}$ and $\mathcal{N}$. Each transforms in this fashion, meaning that $\mathcal{A}_{MN} = \mathcal{M}^{-1}\mathcal{N}$ transforms to $\mathcal{A}'_{MN} = \mathcal{P}\mathcal{M}^{-1}\mathcal{N}\mathcal{P}^{-1}$, which we observe. This means, in turn, that the eigenvalues of $\mathcal{A}'_{MN}$ are the same as $\mathcal{A}_{MN}$. We can observe both $\mathcal{A}_{MN}$—by looking at the model—and $\mathcal{A}'_{MN}$—by looking at the image—but only up to a constant scale factor. This means that the eigenvalues we observe may have been scaled by a constant but unknown factor; however, appropriate ratios of eigenvalues are invariant. A useful example is $\text{trace}(\mathcal{A}_{MN})^3/\det(\mathcal{A}_{MN})$.

It is quite easy to construct invariants for mixed sets of points and lines, too. For example, assume we have a set of points $\boldsymbol{p}_i$ and a set of lines $\boldsymbol{l}_j$. Notice that

$$\begin{aligned}\frac{(l_i^T p_k)(l_j^T p_l)}{(l_i^T p_l)(l_j^T p_k)} &= \frac{(L_i^T \mathcal{A}^{-1}\mathcal{A}P_k)(L_j^T \mathcal{A}^{-1}\mathcal{A}P_l)}{(L_i^T \mathcal{A}^{-1}\mathcal{A}P_l)(L_j^T \mathcal{A}^{-1}\mathcal{A}P_k)}\\ &= \frac{(L_i^T P_k)(L_j^T P_l)}{(L_i^T P_l)(L_j^T P_k)},\end{aligned}$$

which means that this expression is invariant, too (as long as i and j are not equal and k and l are not equal).

Projective invariants for various mixtures of points, lines, and conics are known and have been used successfully in object recognition (some examples are explored in the exercises). Projective invariants are known for plane algebraic curves of higher degree, but are of little practical significance because such curves are seldom encountered in practice and are hard to fit accurately.

18.4.2 Geometric Hashing

Geometric hashing is an algorithm that uses geometric invariants to vote for object hypotheses. You should keep pose clustering in mind as an analogy—we will be voting again—but we are now voting on geometry rather than on pose. The idea was originally developed for uncalibrated affine views of plane models and is easiest to explain in this context.

For any set of three points on the model, we can use the techniques of Section 18.4.1 to compute values of μ_1 and μ_2 for every other point in the model. We now set up a table indexed by the values of μ_1 and μ_2. For every model in the modelbase and for every group of three points on that model, we compute the μs for every other point. Using these μs as an index, insert an entry recording the name of the model and the three points on the model that gave rise to the values obtained. Thus, a pair of μs acts as an hypothesis about the identity of the model *and* three points on that model.

Now that we have the table, we can find the model by searching correspondences. We take any triple of *image* points and compute the μs for every other point in the image. We recover the contents of the table indexed by all of these μs. If the triple corresponds to a triple on the object, we are going to obtain many votes for the combination of the object *and* the three points. Hopefully, noise votes are uncorrelated, meaning that there is a lot of uncoordinated votes for various triples on various objects and many votes for an object triple combination. This implies that, if we have many votes for the same object *and* the same three points on that object, the object may well be present. Notice that this set of three points can act as a frame group for verification purposes. Voting on the μ values is sketched in Algorithm 18.3.

Algorithm 18.3: Geometric Hashing: Voting on Identity and Point Labels

```
For all groups of three image points T(I)
  For every other image point p
    Compute the μs from p and T(I)
    Obtain the table entry at these values
      if there is one, it will label the three points in T(I)
      with the name of the object
      and the names of these particular points.
    Cluster these labels;
      if there are enough labels, backproject and verify
    end
  end
end
```

This algorithm can be generalized to work for other geometric groups than points (see the assignments). If we have uncalibrated affine views of 3D objects, then there are three μs for each point, and we cannot determine them uniquely for each point, but the method extends (see the assignments, again). As with pose clustering and the hough transform (which is really what this method is), it is difficult to choose the size of the buckets; it is hard to be sure what enough means; and there is some danger that the table will get clogged (Grimson and Huttenlocher, 1990*a*).

18.4.3 Invariants and Indexing

Geometric hashing searches correspondences, but does so extracting acceptable labels from a hash table. The main feature of geometric hashing is that we do not need to search over models at recognition time—the hash table has been preloaded in a way that avoids this. This is a desirable feature, usually called **indexing**. One version of indexing applies in alignment when different models have different types of frame group. Clearly, an image frame group of a particular kind need only be checked against the models for which that type of frame group applies.

The trick in geometric hashing is to look for image groups that contained information that was independent of the object pose and changed from object to object (the μs). These μs then generate object information. Geometric hashing explores all possible groups of points. The motivating trick can be extended to all sorts of other geometric features as well. We call groups of features that carry information independent of object pose and changes from object to object *invariant bearing groups*; we saw some examples in Section 18.4.1. Assume that we know the different types of invariant-bearing group that are available. We can now modify the alignment algorithm to come up with Algorithm 18.4.

Algorithm 18.4: Invariant Indexing Using Invariant Bearing Groups

```
For each type T of invariant-bearing group
  For each image group G of type T
    Determine the values V of the invariants of G
    For each model feature group M of type T whose invariants have the values V
      Determine the transformation that takes M to G
      Render the model using this transformation
      Compare the result with the image, and accept if similar
    end
  end
end
```

This approach could be extremely efficient if invariant bearing groups were distinctive, in the sense that there are few model feature groups with the same values of the invariants. We also need to measure the values of the invariants accurately. Notice that we have to look at every model feature group with the same values as the image feature group because we don't know which group we might have. This is again a search over correspondences, with a reasonable hope that the correspondences to be searched could be few.

Indexing in Uncalibrated Perspective Views with Lines and Conics There are numerous invariant bearing groups that can be plugged into the general algorithm given before. The process works best with plane objects viewed in unknown perspective cameras, where invariants of the imaging transformations are quite freely available. Generally, the most useful cases appear to be the invariants of three kinds of groups: five lines; two conics; and a conic and two lines.

Given these functions, a typical system (illustrated in Figure 18.3) works like this:

- **Extract primitive groups:** Images are passed through an edge detector, and conics and lines are fitted to groups of edge points. The edge points are discarded and the fitted curves retained. It is excessively onerous to look at all groups of five line segments to form invariants. It is also unnecessary because objects do not consist of scattered lines, so that open curves of line segments are all that are required. These are groups where, at one end point of each line segment, there is an end point of at least another segment nearby.

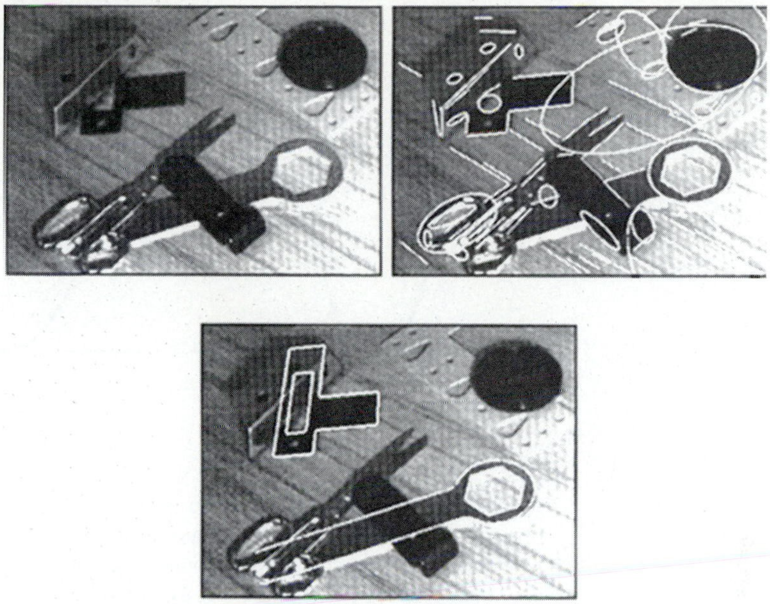

Figure 18.3 These figures illustrate the overall structure of a system that recognizes plane objects using invariants. At the top left is an image; top right shows edge points that have been fitted with lines or conics; and at the bottom, outline points from the objects that have been recognized and verified have been overlaid. *Reprinted from "Efficient model library access by projectively invariant indexing functions," by C.A. Rothwell et al., Proc. Computer Vision and Pattern Recognition, 1992, © 1992, IEEE*

- **Index using invariants:** Relevant assemblies of lines and conics are used to obtain object hypotheses usually by indexing in arrays using quantized values of the invariants. Typically there are one or two invariants available for each type of group and relatively small numbers of models, so that using an array is not particularly wasteful. The size of the quantization buckets is usually determined by trial and error; a wise implementer searches the neighbors of a selected bucket, too. Again, the fact that there are only one or two invariants for each type of group means this is not too wasteful.
- **Backproject and verify:** For each object hypothesis, the transformation that takes the model group to the image group is determined. This transformation is used to backproject the model, and verification proceeds.

Invariant Indexing for Plane Curves One source of geometric invariants is to use *covariant constructions*—constructions that commute with the transformation in mind, meaning that if you perform the construction and transform the result, you get the same result as if you transform the geometry and then perform the construction. In this approach, the construction yields a coordinate frame that is then transformed into some convenient universal coordinate frame, and measurements are made in that frame (usually called a *canonical frame*). Since the measurements are made in a fixed coordinate frame, any property measured in a canonical frame is an invariant. You should notice the similarity of this idea to that used in geometric hashing—the μs there are measured in a canonical frame.

For example, take a curve and construct a line that is tangent to it at two distinct points (closed plane curves that are not convex have such lines—convex closed curves do not, and for

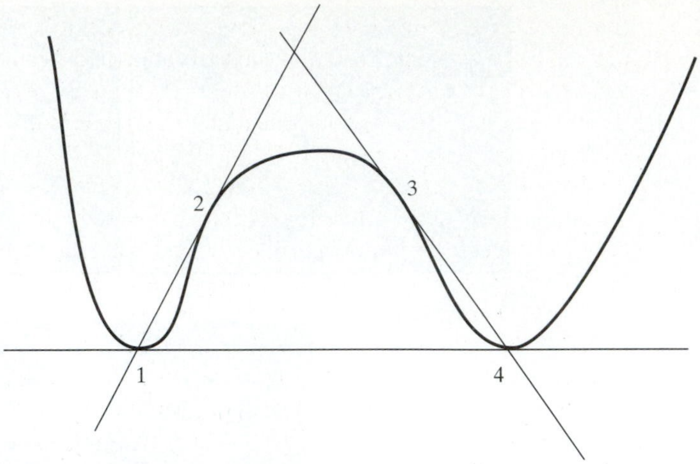

Figure 18.4 Because tangency and incidence are covariant (i.e., a tangent in one coordinate system will, when transformed, be a tangent in the other coordinate system), constructions based around tangency and incidence yield canonical coordinate frames. This figure shows a construction called the *M-curve construction*. For a curve shaped like the letter M (upside-down by convention), a bitangent yields two points (1 and 4) from which tangents can be produced to meet the curve at 2 and 3; this yields four points. Since any set of four points with no three collinear can be mapped to any other such set, we can take these points to the unit square on the plane and look at the curve in this coordinate system the *canonical frame*. Any measurement in this system is invariant.

open curves there are no guarantees). Now apply a transformation so that one of the points of tangency lies at the origin and the tangent line lies on the x-axis. In this coordinate system, any measurement you care to take is an invariant *if* you are careful about the fact that you may not know which point to place at the origin (Figure 18.4). This freedom brings with it a troubling question: Given that we can make many different invariant measurements, which should we make?

18.5 VERIFICATION

Accurate verification requires good tests for whether a rendering of an object model is similar to an image. The choice of test depends on the amount of information about the world available to generate the rendering. For example, if the lighting in the world and the camera response to illumination are both known precisely—for a system that viewed parts on a conveyor belt, this might be possible—we could reasonably expect the rendering to predict image pixel values accurately. In this case, comparing pixel values would be a sensible test.

Usually, all we know about the illumination is that it is bright enough to have generated an hypothesis. This means that comparisons should be robust to changes in illumination. The only test used in practice is to render the silhouette of the object and then compare it to edge points in an image. We describe some other possible tests, too.

18.5.1 Edge Proximity

A natural test is to overlay object silhouette edges on the image using the camera model, and then score the hypothesis by comparing these points with actual image edge points. The usual score is the fraction of the length of predicted silhouette edges that lie nearby actual image edge points. This is invariant to rotation and translation in the camera frame, which is a good thing, but changes with scale, which may not be a bad thing. It is usual to allow edge points to contribute to a verification score only if their orientation is similar to the orientation of the silhouette edge to which they are being compared. The principle here is that the more detailed the description of the edge point, the more likely one is to know whether it came from the object.

It is a bad idea to include invisible silhouette components in the score, so the rendering should be capable of removing hidden lines. The silhouette is used because edges internal to a silhouette may have low contrast under a bad choice of illumination. This means that their absence may be evidence about the illumination rather than the presence or absence of the object.

Edge proximity tests can be quite unreliable. Even orientation information doesn't really overcome these difficulties. When we project a set of model boundaries into an image, the *absence* of edges lying near these boundaries could well be a quite reliable sign that the model isn't there, but the *presence* of edges lying near the boundaries is *not* a particularly reliable sign that the object is there. This is because there are a lot of different sources of edges, and we have no guarantee that the edges being used in the scoring process are the right ones.

A poor pose estimate can lead to silhouette edges on the backprojected object lying a long way from the actual image edges. For example, if we have an object that projects to a long thin region, like a spanner, and we estimate its image plane orientation using one end, the backprojected edges at the other end may lie a long way from the image edge points we are interested in (Figure 18.5). This is an example of error propagation—essentially, the camera estimate is good for features nearby those used to obtain it, but increasingly bad for those a long way away. Several fixes are possible:

- **Maximize over the pose estimate:** The situation can sometimes be improved by maximizing the verification score with respect to the pose. This doesn't always work; if the

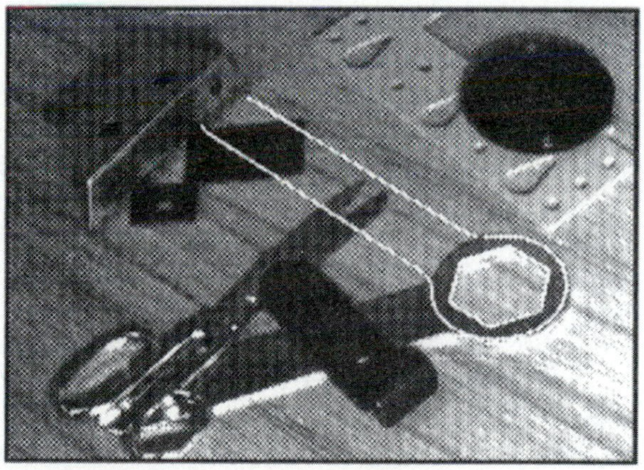

Figure 18.5 Edge orientation can be a deceptive cue for verification as this figure illustrates. The edge points marked on the image come from a model of a spanner, recognized and verified with 52% of its outline points matching image edge points with corresponding orientations. Unfortunately, the image edge points come from the oriented texture on the table, not from an instance of the spanner. As the text suggests, this difficulty could be avoided with a much better description of the spanner's interior as untextured, which would be a poor match to the oriented texture of the table. *Reprinted from "Efficient model library access by projectively invariant indexing functions," by C.A. Rothwell et al., Proc. Computer Vision and Pattern Recognition, 1992, © 1992, IEEE*

original pose estimate is poor, the object may indeed lie close to edges, but to the wrong edges—think about an attempt to verify the presence of an object on a textured background. Furthermore, the optimization can be hard, particularly if the `nearby` test is a threshold on distance.

- **Count only edges for which the camera estimate is reliable:** To our knowledge, this has never been tried in practice. The advantage of the approach is that it should deal with issues like the example of Figure 18.5 relatively easily; the disadvantage is that one may not use a large portion of the object in the verification process, meaning that there is a good prospect of false positives.

Counting the wrong edges in the score is another major source of difficulties. In textured regions, there are many edge points in small collections. The whole point of verification is to use the model to link up image evidence that is too hard to gather together in the hypothesis formation stage so we can't exclude small groups of edge points from the score. This means that, in highly textured regions, it is possible to get high verification scores for almost any model at almost any pose (e.g., see Figure 18.5). Notice that counting similarity in edge orientation in the verification score hasn't made any difference here.

We can tune the edge detector to smooth texture heavily, in the hope that textured regions will disappear. This is a dodge, and a dangerous one, because it usually affects the contrast sensitivity so that the objects disappear, too. It can be made to work acceptably and is widely used.

18.5.2 Similarity in Texture, Pattern, and Intensity

If we score edge matches, then texture is not much more than a nuisance. However, some objects have quite distinctive textures—for example, camouflage paint—and this should probably be used. We could describe model regions using texture descriptors (like the statistics of filter outputs in a region in chapter 9) and then compare those descriptors with the image. The comparison would need to estimate the significance of the difference between the image and the backprojected object regions. The most promising approach appears to be comparing the probability of obtaining the image region covered by the object region by drawing a texture from the object family, with the probability of obtaining this texture when an object is absent.

Comparing silhouette edges ignores a great deal of useful information. If objects are patterned, meaning there are large-scale colored regions like the markings on a soda can, we could compare backprojected pattern edges as well. A more sophisticated approach would compare backprojected pattern regions with the image regions by using texture descriptors (which should agree on the absence of texture) and perhaps descriptions of hue and saturation.

We do not usually have enough information about illumination to predict object intensities. As a result, intensities tend to be ignored in practice in verification. This is a mistake. Many differences in intensity patterns seldom, if ever, come from light sources—for example, only strange light sources like movie projectors generated textured intensity patterns. This suggests that one could probably obtain a verification score by comparing differences in absolute intensity with differences that have arisen in practice and differences that could not arise in practice.

18.6 APPLICATION: REGISTRATION IN MEDICAL IMAGING SYSTEMS

There are numerous problems where pose is far more important than recognition. Many recognition algorithms were designed in the expectation that a selection of industrial parts would be scattered in a bin or on a table; it turns out that production engineers are quite careful to ensure that their parts do not get mixed up, but would often like accurate measurements of pose. Medical applications are similar in that it is usually known *what* is being looked at, but there is a crucial need for an accurate measurement of *where* it is.

18.6.1 Imaging Modes

There are a variety of imaging technologies available, including *magnetic resonance imaging* (MRI), which uses magnetic fields to measure the density of protons and is typically used for descriptions of organs and soft tissue; *computed tomography imaging* (CTI or CT), which measures the density of X-ray absorbtion and is typically used for descriptions of bones; *nuclear medical imaging* (NMI), which measures the density of various injected radioactive molecules and is typically used for functional imaging; and *ultra-sound imaging*, which measures variations in the speed of ultrasound propagation and is often used to obtain information about moving organs (Figure 18.6 illustrates these modes). All of these techniques can be used to obtain slices of data,

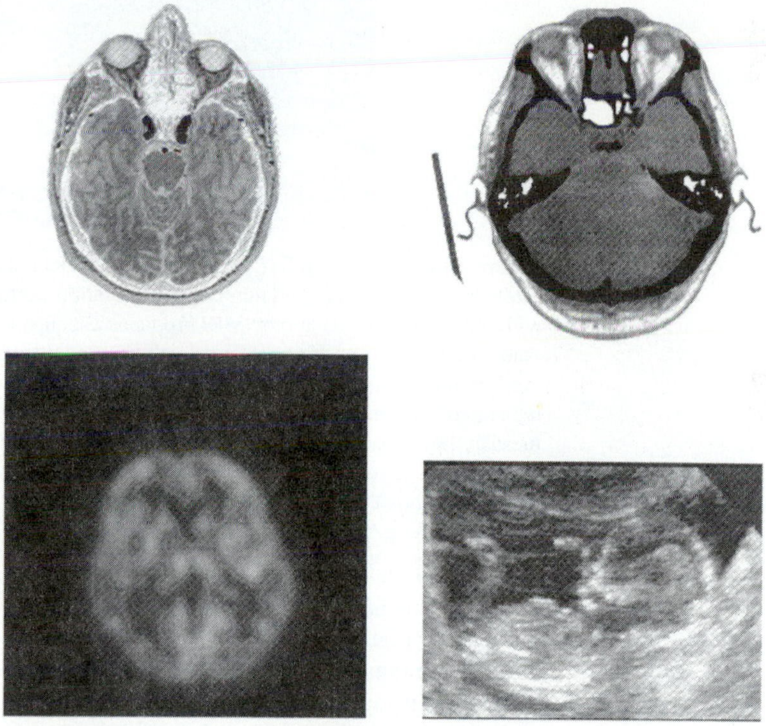

Figure 18.6 Images obtained with four different imaging modes. On the top left, an MRI image of a cross-section of the skull; on the top right, a CTI image of a cross-section of the skull; on the bottom left, an NMI image of a brain; and on the bottom right, a USI image of a foetus in a womb. Notice how each modality shows different detail in different ways; there is high-resolution detail of the brain in the MRI image and of the skull in the CTI image. The NMI image is at low resolution, but (in fact) reflects function because regions that respond strongly have taken up some reagent. Finally, the USI image has a significant noise component, but shows details of soft tissue—you should be able to see a leg, the body, the head and a hand of the fetus. Reprinted from Image and Vision Computing, v. 13, N. Ayache, "Medical computer vision, virtual reality and robotics," Page 296, ©, (1995), with permission from Elsevier Science

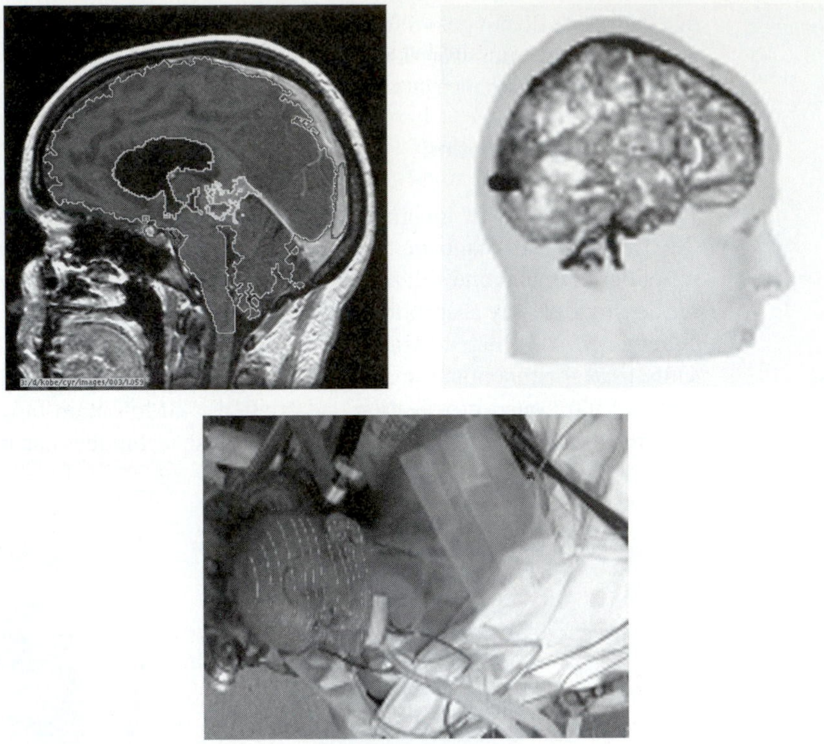

Figure 18.7 On the **top left**, a single slice of MRI data with an automatically acquired segmentation overlaid. The segmentation outlines the brain, vacuoles within the brain, and the tumor. MRI produces a sequence of slices, which yield a volume model; a view of a segmented volume model, with different colors showing different regions, is shown at the **top right**. Once this data is obtained, it is registered to a patient lying on a table. Registration is obtained using depth data measured by a laser ranger; the **bottom** figure shows a camera view of a patient with laser ranger data overlaid. *Figures by kind permission of Eric Grimson; further information can be obtained from his web site* `http://www.ai.mit.edu/people/~welg/welg.html`.

which allow a 3D volume to be reconstructed. A standard problem is to segment these volumes into various structures. Figure 18.7 shows an MRI image with the brain, brain ventricles and tumor segmented. Since tumours are essentially fixed with respect to the skull and skin, this data gives us the position of the tumor with respect to the head.

Registration in medical imaging is almost always a 3D from 3D problem, and the only transformations to care about are 3D rotations and translations. Geometric hashing is the dominant mechanism because it can be used to search correspondences efficiently. The literature differs largely in the matter of what data is used.

18.6.2 Applications of Registration

In **brain surgery** applications, surgeons are attempting to remove tumors while doing the minimum damage to a patient's faculties. We show examples due to Grimson and colleagues. The general approach is to obtain images of the patients brain, segment these images to show the tumor, and then display the images to the surgeon. The display is overlaid on pictures of the patient on the table, obtained using a camera near the surgeon's view, to cue the surgeon to the exact position of the tumor. Various methods exist for attaching functional tags to the image of the brain—usually one stimulates a region of the brain and watches to see what happens—and this information can also be displayed to the surgeon so that the impact of any damage done can be minimized. The problem here is pure pose estimation; we need to know the pose of the brain image and the brain measurements with respect to the person on the table.

Reconstructive surgery offers similar applications. For example, in facial reconstruction in a planning phase, surgeons can be allowed to work out a sequence of activities on a visualization of a patient's skull. The results of this visualisation need to be displayed to the surgeons when they operate, again registered to a view of the patient.

Diagnostic applications include creating 3D visualisations to display results from many different imaging modes. For example, a surgeon may have images from MRI—which quite often has relatively high resolution—and PET—which are linked to functional properties. It is natural to wish to fuse these images, and so they need to be registered.

18.6.3 Geometric Hashing Techniques in Medical Imaging

The main differences between algorithms in applications is the type of measurement used for geometric hashing. We discuss a few cases next.

Point Correspondences We have seen how to search point correspondences. For example, head MRI data can be registered to a patient's head on a table by obtaining 3D measurements of the head with a laser ranger on the operating table and using these to register with the skin points on the MRI data. We can hash pairs of skin points from the MRI on their distance apart and the angle between their normals and then query with pairs of points from the laser data. Pairs with similar distances apart and similar angles could correspond. With a corresponding pair, we can estimate pose and then check the total error of this pose estimate. Although there is no true correspondence, neither the skin points in the MRI image nor the laser ranger data consists of isolated points; they are samples from surfaces, the sampling is sufficiently dense that it is possible to obtain good initial hypotheses about pose. If the error in the registration is then measured sensibly, counting only error components normal to the surface to avoid being put off by small localization errors, an excellent pose estimate can be obtained by minimizing the error. Grimson is the main proponent of such systems, one of which is illustrated in Figure 18.8.

Curves Curves can be used to drive geometric hashing, too. In this case, we fit a surface to the dataset and then mark significant curves on this surface. Using parabolic curves is impractical because some datasets have many flat regions. Ayache (1995*a,b*) successfully used curves where the maximum normal curvature is an extremal along the curves of maximun normal curvature on the surface. Whatever curve is used, at any point on a curve we have a complete 3D frame (the Serret–Frenet frame), and we can use this frame as a canonical frame to obtain measurements for geometric hashing. Natural choices include the curvature and torsion of the curve, and the angle between the curve normal and the surface normal.

Frame Pairs Given two coordinate frames, the transformation from one to the other is invariant to a shared transformation, and so can be used to supply indexes for hashing. We call this cue a *frame pair*. A natural way to obtain these pairs is to fit a surface to the dataset, identify significant curves or points, and then use local frames. For a significant point on a surface (e.g., an umbilic point) a frame can be obtained from the normal and the two directions of extremal curvature. For any point on a significant curve, we can use either the Serret–Frenet frame or the frame on the surface (or compare the two frames). Ayache and his group have emphasized the use of curves and of frame pairs; a broad review appears in Ayache (1995*a,b*).

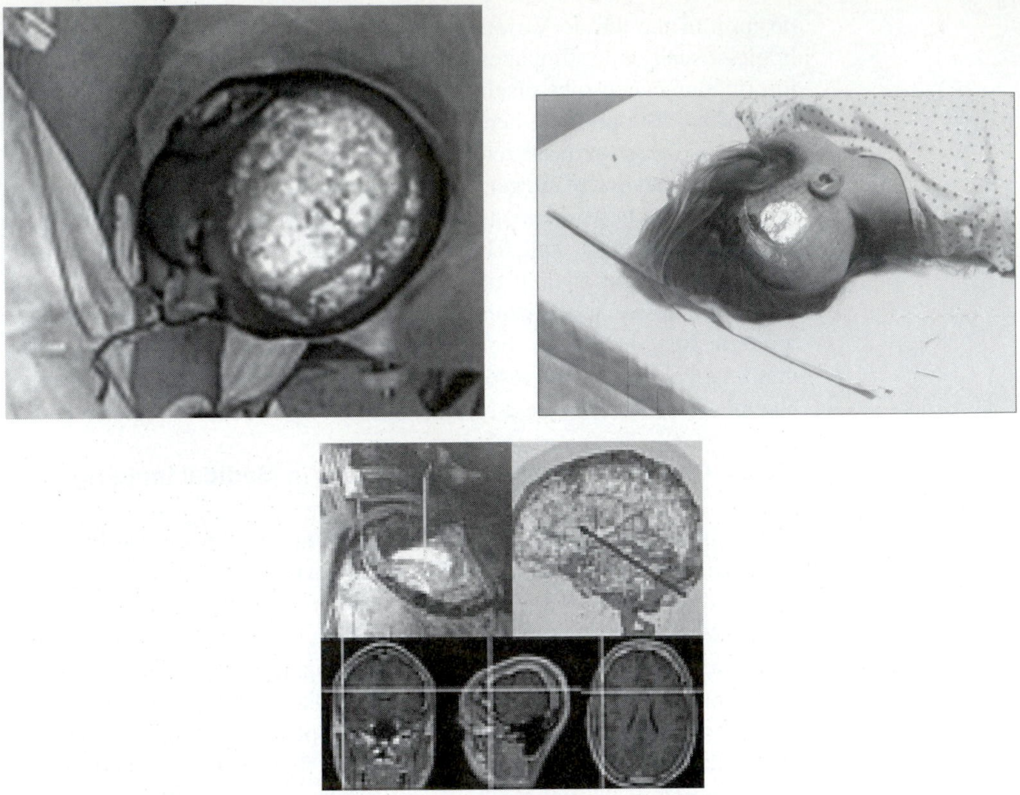

Figure 18.8 The figure on the **top left** shows skin data overlaid on a view of a patient to indicate the success of the registration process. Once data is registered to a patient, a number of uses are available. At the **top right**, we see a view of a patient on an operating table with MRI imagery showing part of the brain and a tumor overlaid for the surgeon's information. At the **bottom**, we see imagery obtained by registering the position of a surgical instrument with the MRI data set. This means that the position of the surgical instrument can be displayed to the surgeon, however deep the instrument is within tissue. *Figures by kind permission of Eric Grimson; further information can be obtained from his web site* `http://www.ai.mit.edu/people/~welg/welg.html`.

18.7 CURVED SURFACES AND ALIGNMENT

Curved surfaces can be aligned, too. The hypothesis generating process can be more complicated, but rendering and verification are straightforward generalizations.

The natural strategy is to find frame-bearing groups that behave like points. For example, if a curved surface has points painted on it or if three surfaces meet discontinuously at points, the hypothesis generating process behaves like those described before. A geometric model of the surface can then be projected into the image, and used to verify as below.

A more difficult approach sees alignment as minimization, rather like the linear combinations of models approach discussed before. In this approach, the outline of the surface is predicted

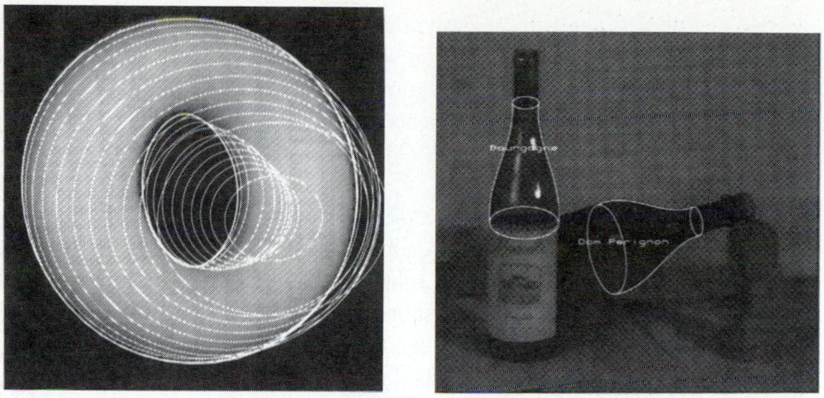

Figure 18.9 An algebraic surface is viewed in a calibrated perspective camera. The contour generator is clearly an algebraic curve (because we can write down a set of polynomial equations that it satisfies), and the outline is also an algebraic curve. This curve is a function of the pose of the surface. The figure on the left shows a family of curves obtained by overlaying the outline of a surface on an image and obtaining pose by minimizing the sum of the minimum distances between selected edge points and the outline. The curves come from different points in the minimization processs. The figure on the right shows two different algebraic surfaces aligned successfully with image contours; the surfaces used identify two different bottles. *Reprinted from "On Recognizing and positioning curved 3d objects from image contours," by D.J. Kriegman and J. Ponce, IEEE Trans. Pattern Analysis and Machine Intelligence, 1990, © IEEE 1990*

as a function of the pose of the surface. We can adopt as an objective function the sum of the minimum distances of selected edge points from this outline and minimize the objective function over pose, as Figure 18.9 illustrates. The mechanics of predicting outline curves is simplified for algebraic surfaces; all examples in the literature use algebraic surfaces as an example for this reason. The details appear in Kriegman and Ponce (1990*b*). Algebraic surfaces are so rigid that a single outline completely determines the surface geometry; enthusiasts can look up the details in Forsyth (1996).

18.8 NOTES

Typically, systems built around the algorithms described in this chapter can recognize small numbers of objects in quite cluttered scenes. They are important because pose recovery and registration is useful in applications, and because their weaknesses point out important issues in recognition.

The term alignment is due to Huttenlocher and Ullman (1990). It is a convenient term for a general class of algorithm that reasons about pose consistency. It is hard to determine who used the approach first, though it is quite likely Roberts (1965); other possibilities include Faugeras, Hebert, Pauchon and Ponce (1984). A contemporary survey is Chin and Dyer (1986). Alignment is **quite general** because, for most conceivable images of most conceivable objects, some form of camera consistency constraint applies and can be exploited. It is also quite **noise resistant**—meaning that we can find objects even in heavily cluttered images—because relatively little image evidence needs to be collected to construct an object hypothesis. Testing a large hypothesis robustly tends to be easier than assembling a large hypothesis because the hypothesis under test gives strong constraints on what image evidence can contribute to a decision. The noise behavior of some alignment algorithms has been studied in detail (Grimson, Huttenlocher and Alter, 1992, Grimson, Huttenlocher and Jacobs, 1994, Grimson, Lozano-Perez and Huttenlocher, 1990, Sarachik and Grimson, 1993). As a result, alignment algorithms are widely used and there are numerous variants.

However, alignment **scales poorly** with increasing numbers of models. Linear growth in the number of models occurs because the modelbase is *flat*. There is no hierarchy, and every model is treated in the same way. Furthermore, while constrained search for a model that is present can be efficient, showing that a model is absent is expensive (Grimson, 1992).

Pose clustering is due to Thompson and Mundy (1987). The analogy to the Hough transform means that the method can behave quite badly in the presence of noise (Grimson and Huttenlocher, 1990*b*). Geometric hashing in various forms is due to Lamdan, Schwartz and Wolfson (1990), Wolfson (1990) and Wolfson and Lamdan (1988). The use of various invariants for indexing recognition hypotheses is described in Forsyth, Mundy, Zisserman, Coelho, Heller and Rothwell (1991); in collections and books (Mundy and Zisserman, 1992, Mundy, Zisserman and Forsyth, 1993, Rothwell, 1995); and in various papers (Barrett, Payton, Haag and Brill, 1991, Forsyth et al., 1992, Rothwell, Zisserman, Forsyth and Mundy, 1995, Rothwell, Zisserman, Marinos, Forsyth and Mundy, 1992, Zisserman, Forsyth, Mundy, Rothwell, Liu and Pillow, 1995*a*). One lively area of research we have not described is the measurement of invariant properties of 3D objects from multiple, uncalibrated views (Barrett, Brill, Haag and Payton, 1992); another is the process of computing invariant values conditioned on some model knowledge (Shashua, 1995, Weinshall, 1993). There has been some work in vision circles on methods for deriving invariants, too (Csurka and Faugeras, 1998, 1999).

Pose consistency can be used in a variety of forms. For example, recognition hypotheses yield estimates of camera intrinsic parameters. This means that if there are several objects in an image, all must give consistent estimates of camera intrinsic parameters (Forsyth, Mundy, Zisserman and Rothwell, 1994).

Each algorithm attempts to get enough information to perform verification with as little trouble as possible while trying to reduce the number of spurious verification attempts. This dependency on *image level* verification is inconsistent with model abstraction; we could not verify that a picture contained a fish by looking at pixel values or edges if we did not have exact details of its species, configuration, and the like. Most applications of recognition desperately need abstraction. For example, if we want to search the Internet for pictures of the Pope, we shouldn't have to know his exact geometry. Similarly, if we want to deploy our automatic motor car on real roads, it has to be able to decide what it should swerve to avoid and what it may run down. Instead of matching features (like points and lines), we might match *tokens* (where a token might be a hairy patch, an eye, or something of the sort) to models containing tokens. Here, the abstraction is in the token (Ettinger, 1988, Ullman, 1996).

The main role of verification is to find evidence for an hypothesis that could not be collected in other ways. Since collecting evidence is poorly understood, current recognition systems work well when verification works well and badly otherwise. Verification works well when sufficient evidence is used and scored appropriately. Unfortunately, it is difficult to translate these platitudes into algorithms. For a topic so central to the performance of recognition systems, verification has been extremely poorly studied (but see Grimson and Huttenlocher, 1991). Verification based on generic evidence—say, edge points—has the difficulty that we cannot tell which evidence should be counted. Similarly, if we use specific evidence—say, a particular camouflage pattern—we have problems with abstraction. Template-matching and appearance-based vision, which we discuss in greater detail in chapter 22, can be seen as mechanisms to involve more kinds of evidence in the verification process.

Medical Applications

This is not a topic on which we speak with any authority. Valuable surveys are Ayache (1995*a,b*), Duncan and Ayache (2000), and Gerig, Pun and Ratib (1994).

The three main topics appear to be: **segmentation**, which is used to identify regions of (often 3D) images that correspond to particular organs; **registration**, which is used to construct correspondences between images of different modalities and between images and patients; and **analysis** of both morphology—how big is this? has it grown?—and function. McInerney and Terzopolous (1996) survey the use of deformable models. There are surveys of registration methods and issues in Lavallee (1996) and in Maintz and Viergever (1998), and a comparison between registration output and "ground truth" in West, Fitzpatrick, Wang, Dawant, Maurer, Kessler, Maciunas, Barillot, Lemoine, Collignon, Maes, Suetens, Vandermeulen, van den Elsen, Napel, Sumanaweera, Harkness, Hemler, Hill, Hawkes, Studholme, Antoine Maintz, Viergever, Malandain, Pennec, Noz, Maguire, Pollack, Pelizzari, Robb, Hanson and Woods (1997).

PROBLEMS

18.1. Assume that we are viewing objects in a calibrated perspective camera and wish to use a pose consistency algorithm for recognition.

(a) Show that three points is a frame group.

(b) Show that a line and a point is *not* a frame group.

(c) Explain why it is a good idea to have frame groups composed of different types of feature.

(d) Is a circle and a point not on its axis a frame group?

18.2. We have a set of plane points $\boldsymbol{P}_j$; these are subject to a plane affine transformation. Show that

$$\frac{\det\left[\boldsymbol{P}_i\boldsymbol{P}_j\boldsymbol{P}_k\right]}{\det\left[\boldsymbol{P}_i\boldsymbol{P}_j\boldsymbol{P}_l\right]}$$

is an affine invariant (as long as no two of i, j, k, and l are the same and no three of these points are collinear).

18.3. Use the result of the previous exercise to construct an affine invariant for:

(a) four lines,

(b) three coplanar points,

(c) a line and two points (these last two will take some thought).

18.4. In chamfer matching at any step, a pixel can be updated if the distances from some or all of its neighbors to an edge are known. Borgefors counts the distance from a pixel to a vertical or horizontal neighbor as 3 and to a diagonal neighbor as 4 to ensure the pixel values are integers. Why does this mean $\sqrt{2}$ is approximated as $4/3$? Would a better approximation be a good idea?

18.5. One way to improve pose estimates is to take a verification score and then optimize it as a function of pose. We said that this optimization could be hard particularly if the test to tell whether a backprojected curve was close to an edge point was a threshold on distance. Why would this lead to a hard optimization problem?

18.6. We said that for an uncalibrated affine camera viewing a set of plane points, the effect of the camera can be written as an unknown plane affine transformation. Prove this. What if the camera is an uncalibrated perspective camera viewing a set of plane points?

18.7. Prepare a summary of methods for registration in medical imaging other than the geometric hashing idea we discussed. You should keep practical constraints in mind, and you should indicate which methods you favor, and why.

18.8. Prepare a summary of nonmedical applications of registration and pose consistency.

Programming Assignments

18.9. Representing an object as a linear combination of models is often represented as abstraction because we can regard adjusting the coefficients as obtaining the same view of different models. Furthermore, we could get a parametric family of models by adding a basis element to the space. Explore these ideas by building a system for matching rectangular buildings where the width, height, and depth of the building are unknown parameters. You should extend the linear combinations idea to handle orthographic cameras; this involves constraining the coefficients to represent rotations.

19

Smooth Surfaces and Their Outlines

Several chapters of this book have explored the *quantitative* relationship between simple geometric figures such as points, lines, and planes and the parameters of their image projections. In this one, we investigate instead the *qualitative* relationship between three-dimensional shapes and their pictures, focusing on the outlines of solids bounded by smooth surfaces. Specifically, consider such a solid and assume that its surface's reflectance function is also smooth. Ignoring shadows (e.g., considering a point light source and a camera colocated far from the scene), the image is also smooth except along the solid's *outline*, also called object *silhouette* or *image contour* in the rest of this chapter (Figure 19.1).

The outline is formed by intersecting the retina with a viewing cone (or cylinder in the case of orthographic projection) whose apex coincides with the pinhole and whose surface grazes the object along a surface curve called the *occluding contour* or *rim*. It can be shown that the occluding contour is in general a smooth curve, formed by *fold points* where the viewing ray is tangent to the surface and a discrete set of *cusp points* where the ray is tangent to the occluding contour as well. The image contour is piecewise smooth, and its only singularities are a discrete set of *cusps* formed by the projection of cusp points and *T-junctions* formed by the transversal superposition of pairs of fold points (Figure 19.2). The intuitive meaning of these exotic terms should be pretty clear: A fold is a point where the surface folds away from its viewer, and a contour cusps at a point where it suddenly decides to turn back, following a different path along the same tangent (this is for transparent objects only: contours of opaque objects terminate at cusps; see Figure 19.2). Likewise, two smooth pieces of contour cross at a T-junction (unless the object is opaque and one of the branches terminates at the junction).

Interestingly, attached shadows are delineated by the occluding contours associated with the light sources, and cast shadows are bounded by the corresponding object outlines (Figure 19.3). Thus, we also know what they look like. (Caveat: The objects onto which shadows are

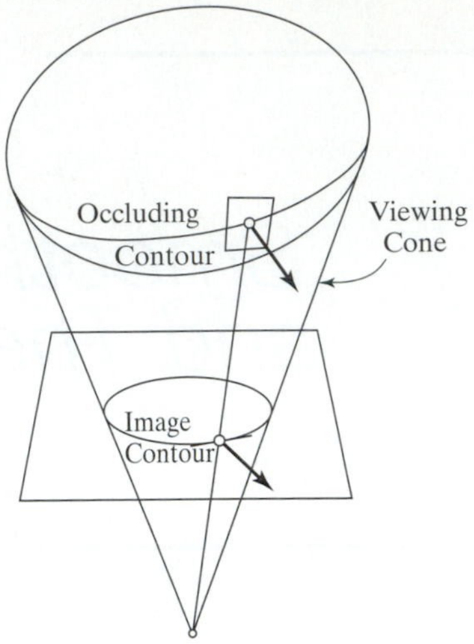

Figure 19.1 The occlusion boundaries of a smooth surface.

cast may have curved surfaces. Even in this case, however, the boundaries of attached shadows are just occluding contours. Of course, light sources are rarely punctual, which further complicates things.)

The image contour of a solid shape constrains it to lie within the associated viewing cone, but does not reveal the depth of its occluding contour. In the case of solids bounded by smooth surfaces, it provides additional information. In particular, the plane defined by the eye and the tangent to the image contour is tangent to the surface. Thus, the contour orientation determines the surface orientation along the occluding contour. More generally, the rest of this chapter focuses on the geometric relationship between curved surfaces and their projections and on the type of information about surface geometry that can be inferred from contour geometry. For example, we show that the contour curvature also reveals information about the surface curvature. In the meantime, let us start by introducing elementary notions of differential geometry that provide a natural mathematical setting for our study. Differential geometry proves useful again in the next chapter, when we study the changes in object appearance that stem from viewpoint changes.

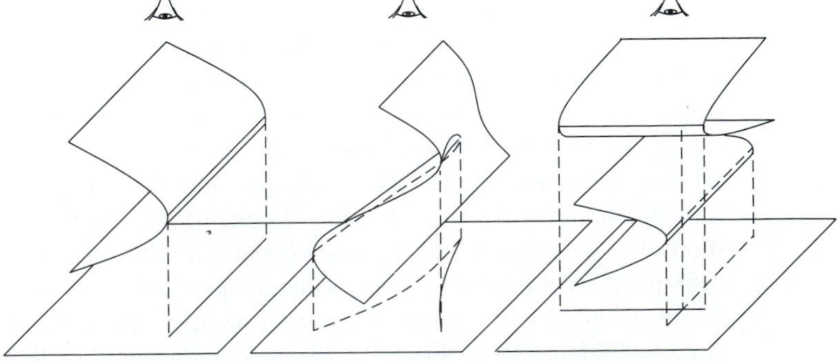

Figure 19.2 Contour components: Folds, cusps, and T-junctions. *Reprinted from "Computing Exact Aspect Graphs of Curved Objects: Algebraic Surfaces," by S. Petitjean, J. Ponce and D.J. Kriegman, International Journal of Computer Vision, 9(3):231–255, (1992). © 1992 Kluwer Academic Publishers.*

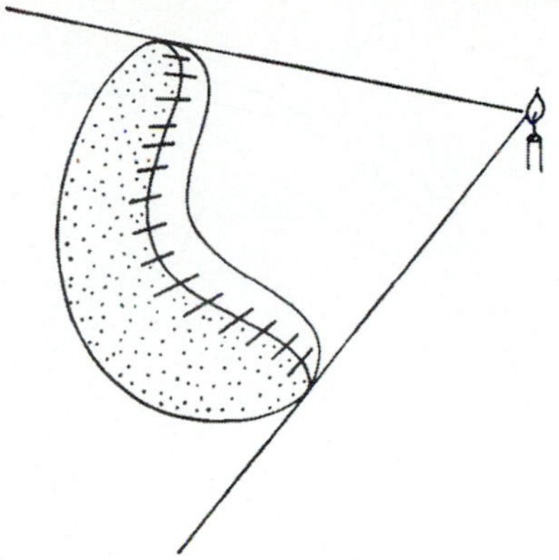

Figure 19.3 Shadow boundaries and occluding contours. *Reprinted from "Solid Shape," by J.J. Koenderink, MIT Press, (1990). © 1990 by The Massachusetts Institute of Technology.*

19.1 ELEMENTS OF DIFFERENTIAL GEOMETRY

This section presents the rudiments of Euclidean differential geometry necessary to understand the local relationship between light rays and solid objects. We limit our discussion of surfaces to those bounding compact solids in $\mathbb{E}^3$. The topic of our discussion is of course technical, but we attempt to stay at a fairly informal level, emphasizing descriptive over analytical geometry. In particular, we refrain from picking a *global* coordinate system for $\mathbb{E}^3$, although *local* coordinate systems attached to a curve or surface in the vicinity of one of its points are used on several occasions. This is appropriate for the type of qualitative geometric reasoning that is the focus of this chapter. Analytical differential geometry is discussed in chapter 21 in the (quantitative) context of range data analysis.

19.1.1 Curves

Let us start with the study of curves that lie in a plane. We examine a curve γ in the immediate vicinity of some point P and assume that γ does not intersect itself or, for that matter, terminate in P. If we draw a straight line L through P, it generally intersects γ in some other point Q, defining a *secant* of this curve (Figure 19.4). As Q moves closer to P, the secant L rotates about P and approaches a limit position T called the *tangent line* to γ in P.

By construction, the tangent T has more intimate contact with γ than any other line passing through P. Let us now draw a second line N through P and perpendicular to T and call it the *normal* to γ in P. Given an (arbitrary) choice for a unit *tangent vector* $\boldsymbol{t}$ along T, we can construct a right-handed coordinate frame whose origin is P and whose axes are $\boldsymbol{t}$ and a unit *normal vector* $\boldsymbol{n}$ along N. This *local* coordinate system is particularly well adapted to the study of the curve in the neighborhood of P: Its axes divide the plane into four quadrants that can be numbered in counterclockwise order as shown in Figure 19.5, the first quadrant being chosen so it contains a particle traveling along the curve toward (and close to) the origin. In which quadrant will this particle end up just after passing P?

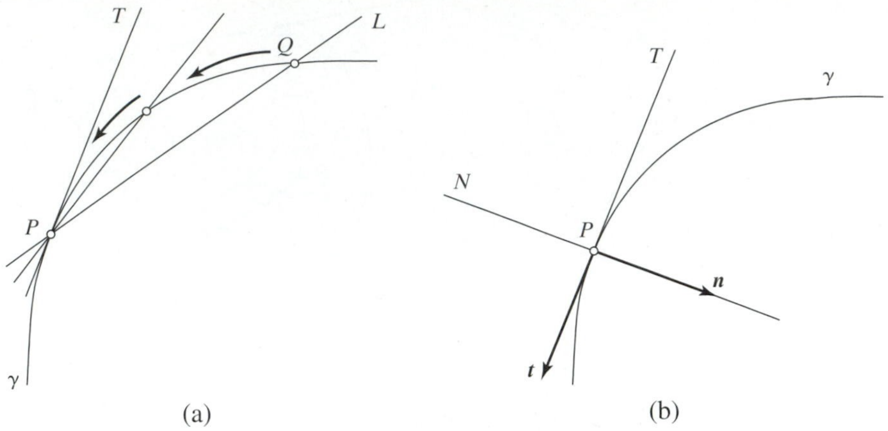

Figure 19.4 Tangents and normals: (a) definition of the tangent as the limit of secants, (b) the coordinate system defined by the (oriented) tangent and normal.

As shown by the figure, there are four possible answers to this question, and they characterize the shape of the curve near P. We say that P is *regular* when the moving point ends up in the second quadrant and *singular* otherwise. When the particle traverses the tangent and ends up in the third quadrant, P is called an *inflection* of the curve, and we say that P is a *cusp* of the *first* or *second kind* in the two remaining cases, respectively. This classification is independent of the orientation chosen for γ, and it turns out that almost all points of almost all curves are regular, with singularities only occurring at isolated points.

As noted before, the tangent to a curve γ in P is the closest linear approximation of γ passing through this point. In turn, constructing the closest *circular* approximation now allows us to define the *curvature* in P—another fundamental characteristic of the curve shape. Consider a point P' as it approaches P along the curve, and let M denote the intersection of the normal lines N and N' in P and P' (Figure 19.6). As P' moves closer to P, M approaches a limit position C along the normal N, called the *center of curvature* of γ in P.

At the same time, if $\delta\theta$ denotes the (small) angle between the normals N and N' and δs denotes the length of the (short) curve arc joining P and P', the ratio $\delta\theta/\delta s$ also approaches a definite limit κ, called the *curvature* of the curve in P, as δs nears zero. It turns out that κ is just the inverse of the distance r between C and P (this follows easily from the fact that $\sin u \approx u$ for small angles; see Exercises). The circle centered in C with radius r is called the *circle of curvature* in P, and r is the *radius of curvature*.

It can also be shown that a circle drawn through P and two close-by points P' and P'' approaches the circle of curvature as P' and P'' move closer to P. This circle is indeed the

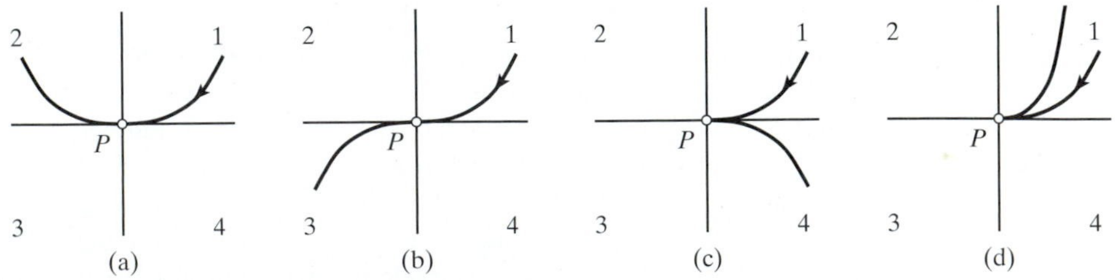

Figure 19.5 A classification of curve points: (a) a regular point, (b) an inflection, (c) a cusp of the first kind, (d) a cusp of the second kind. Note that the curve stays on the same side of the tangent at regular points.

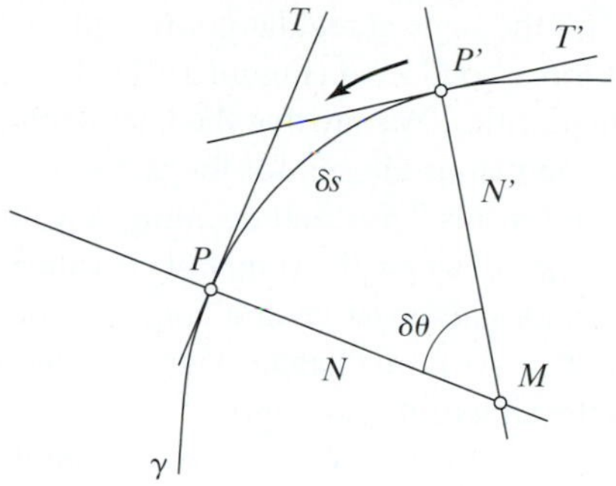

Figure 19.6 Definition of the center of curvature as the limit of the intersection of normal lines through neighbours of P.

closest circular approximation to γ passing through P. The curvature is zero at inflections, and the circle of curvature degenerates to a straight line (the tangent) there: Inflections are the flattest points along a curve.

Let us now introduce a device that proves to be extremely important in the study of both curves and surfaces—the *Gauss map*. Let us pick an orientation for the curve γ and associate with every point P on γ the point Q on the unit circle where the tip of the associated normal vector meets the circle (Figure 19.7). This mapping from γ to the unit circle is the Gauss map associated with γ.[1]

Let us have another look at the limiting process used to define the curvature. As P' approaches P on the curve, the Gaussian image Q' of P' approaches the image Q of P. The (small) angle between N and N' is equal to the length of the arc joining Q and Q' on the unit circle. The curvature is therefore the limit of the ratio between the lengths of corresponding arcs of the Gaussian image and of the curve as both approach zero.

The Gauss map also provides an interpretation of the classification of curve points introduced earlier: Consider a particle traveling along a curve and the motion of its Gaussian image. The direction of traversal of γ stays the same at regular points and inflections, but reverses for both types of cusps (Figure 19.5). On the other hand, the direction of traversal of the Gaussian

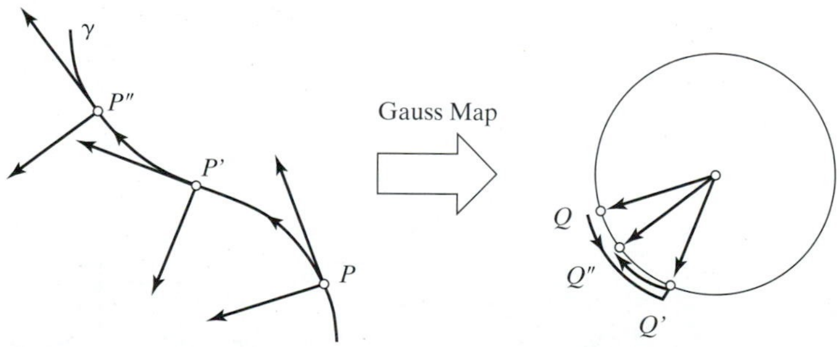

Figure 19.7 The Gaussian image of a plane curve. Observe how the direction of traversal of the Gaussian image reverses at the inflection P' of the curve. Also note that there are close-by points with parallel tangents/normals on either side of P'. The Gauss map folds at the corresponding point Q'.

[1]The Gauss map could have been defined just as well by associating with each curve point the tip of its unit tangent on the unit circle. The two representations are equivalent in the case of planar curves. The situation will be different when we generalize the Gauss map to twisted curves and surfaces.

image stays the same at regular points and cusps of the first kind, but it reverses at inflections and cusps of the second kind (Figure 19.7). This indicates a double covering of the unit circle near these singularities: We say that the Gauss map *folds* at these points.

A sign can be chosen for the curvature at every point of a plane curve γ by picking some orientation for this curve and deciding, say, that the curvature can (arbitrarily) be taken positive at a *convex point* where the center of curvature lies on the same side of γ as the tip of the oriented normal vector and negative at a *concave point* where these two points lie on opposite sides of γ. Thus, the curvature changes sign at inflections, and reversing the orientation of a curve also reverses the sign of its curvature.

Twisted space curves are more complicated animals that their planar counterparts. Although the tangent can be defined as before as a limit of secants, there is now an infinity of lines perpendicular to the tangent at a point P, forming a *normal plane* to the curve at this point (Figure 19.8).

In general, a twisted curve does not lie in a plane in the vicinity of one of its points, but there exists a unique plane that lies closest to it. This is the *osculating plane*, defined as the limit of the plane containing the tangent line in P and some close-by curve point Q as the latter approaches P. We finish the construction of a local coordinate frame in P by drawing a *rectifying plane* through P perpendicular to both the normal and osculating planes. The axes of this coordinate system, called *moving trihedron*, or *Frénet frame*, are the tangent, the *principal normal* formed by the intersection of the normal and osculating planes, and the *binormal* defined by the intersection of the normal and rectifying planes (Figure 19.8).

As in the planar case, the curvature of a twisted curve can be defined in a number of ways, as the inverse of the radius of the limit circle defined by three curve points as they approach each other (this circle of curvature lies in the osculating plane), as the limit ratio of the angle between the tangents at two close-by points and the distance separating these points as it approaches zero, and so on. Likewise, the concept of Gauss map can be extended to space curves, but this time the tips of the tangents, principal normals, and binormals draw curves on a unit *sphere*. Note that it

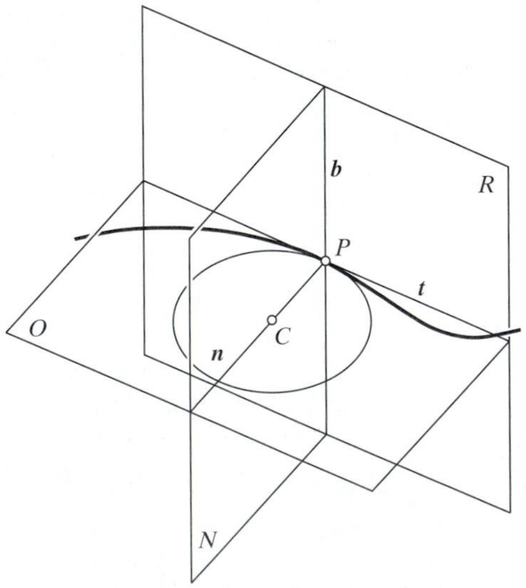

Figure 19.8 The local geometry of a space curve: N, O, and R are, respectively, the normal, osculating, and rectifying plane; t, n, and b are, respectively, the tangent, (principal) normal and binormal lines, and C is the center of curvature.

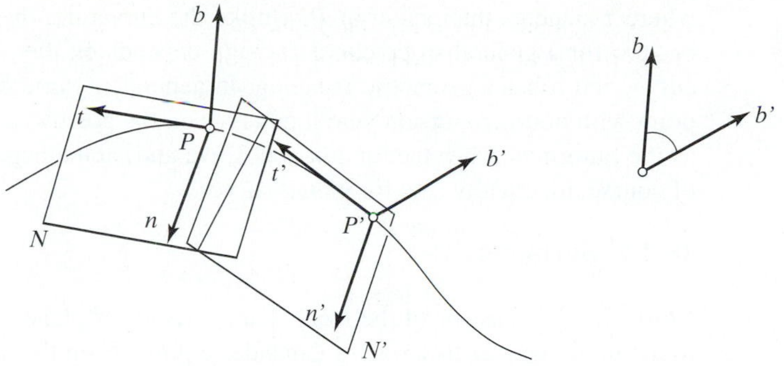

Figure 19.9 Geometric definition of the torsion as the limit, as both quantities approach zero, of the ratio obtained by dividing the angle between the binormals by the distance between the associated surface points.

is not possible to give a meaningful sign to the curvature of a twisted curve. In general, such a curve does not have inflections, and its curvature is positive everywhere.

The curvature can be thought of as a measure of the rate of change of the tangent direction along a curve. It is also possible to define the rate of change of the osculating plane direction along a twisted curve: Consider two close-by points P and P' on the curve; we can measure the angle between their osculating planes or, equivalently, between the associated binormals and divide this angle by the distance between the two points. The limit of this ratio as P' approaches P is called the *torsion* of the curve in P. Not surprisingly, its inverse is the limit of the ratio between the lengths of corresponding arcs on the curve and the spherical image of the binormals.

A space curve can be oriented by considering it as the trajectory of a moving particle and picking a direction of travel for this particle. Furthermore, we can pick an arbitrary reference point P_0 on the curve and define the *arc length* s associated with any other point P as the (signed) length of the curve arc separating P_0 and P. Although the arc length depends on the choice of P_0, its differential does not (moving P_0 along the curve amounts to adding a constant to the arc length), and it is often convenient to parameterize a curve by its arc length, where some unspecified choice of origin P_0 is assumed. In particular, the tangent vector $\boldsymbol{t}$ at the point P is the unit *velocity* $\frac{d}{ds}P$ (defined as the limit of the vector $\frac{1}{\delta s}\overrightarrow{PP'}$ as the curve point P' approaches P and the [signed] distance δs between them approaches zero). Reversing s also reverses $\boldsymbol{t}$. It can be shown that the *acceleration* $\frac{d^2}{ds^2}P$, the curvature κ, and the (principal) normal $\boldsymbol{n}$ are related by

$$\frac{d^2}{ds^2}P = \frac{d}{ds}\boldsymbol{t} = \kappa\boldsymbol{n}.$$

Note that κ and $\boldsymbol{n}$ are both independent of the curve orientation (the negative signs introduced by reversing the direction of traversal of the curve cancel during differentiation), and the curvature is the magnitude of the acceleration. The binormal vector can be defined as $\boldsymbol{b} = \boldsymbol{t} \times \boldsymbol{n}$; like $\boldsymbol{t}$, it depends on the orientation chosen for the curve. In general, it is easy to show that

$$\begin{cases} \frac{d}{ds}\boldsymbol{t} = & \kappa\boldsymbol{n} \\ \frac{d}{ds}\boldsymbol{n} = -\kappa\boldsymbol{t} & \quad + \tau\boldsymbol{b}\,, \\ \frac{d}{ds}\boldsymbol{b} = & -\tau\boldsymbol{n} \end{cases}$$

where τ denotes the torsion in P. Unlike the curvature, the torsion may be positive, negative, or zero for a general space curve. Its sign depends on the direction of traversal chosen for the curve, and it has a geometric meaning: In general, a curve crosses its osculating plane at every point with non-zero torsion, and it emerges on the positive side of that plane (i.e., the same side as the binormal) when the torsion is positive and on the negative side otherwise. The torsion is, of course, identically zero for planar curves.

19.1.2 Surfaces

Most of the discussion of the local characteristics of plane and space curves can be generalized in a simple manner to surfaces. Consider a point P on the surface S and all the curves passing through P and lying on S. It can be shown that the tangents to these curves lie in the same plane Π, appropriately called the *tangent plane* in P (Figure 19.10a). The line N passing through P and perpendicular to Π is called the *normal line* to P in S, and the surface can be oriented (locally) picking a sense for a unit *normal vector* along N (unlike curves, surfaces admit a single normal but an infinity of tangents at every point). The surface bounding a solid admits a canonical orientation defined by letting the normal vectors locally point toward the outside of the solid.[2]

Intersecting a surface with the planes that contain the normal in P yields a one-parameter family of planar curves called *normal sections* (Figure 19.10b). These curves are, in general, regular in P or may exhibit an inflection there. The curvature of a normal section is called the *normal curvature* of the surface in the associated tangent direction. By convention, we choose a positive sign for the normal curvature when the normal section lies (locally) on the same side of the tangent plane as the inward-pointing surface normal and a negative sign when it lies on the other side. The normal curvature is, of course, zero when P is an inflection of the corresponding normal section.

With this convention, we can record the normal curvature as the sectioning plane rotates about the surface normal. It generally assumes its maximum value κ_1 in a definite direction of the tangent plane, and reaches its minimum value κ_2 in a second definite direction. These

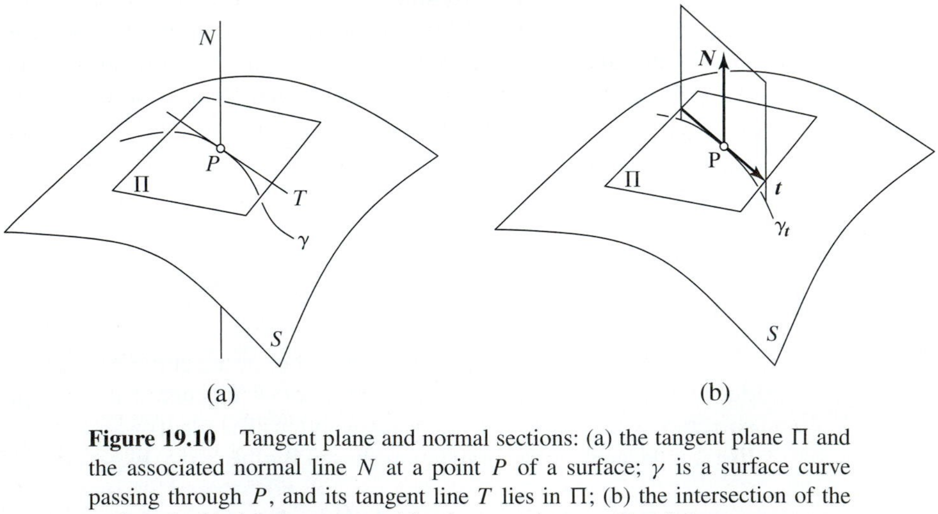

Figure 19.10 Tangent plane and normal sections: (a) the tangent plane Π and the associated normal line N at a point P of a surface; γ is a surface curve passing through P, and its tangent line T lies in Π; (b) the intersection of the surface S with the plane spanned by the normal vector $\boldsymbol{N}$ and the tangent vector $\boldsymbol{t}$ forms a normal section γ_t of S.

[2]Of course, the reverse orientation, where, as Koenderink (1990, p. 137) puts it, "the normal vector points into the 'material' of the blob like the arrows in General Custer's hat," is just as valid. The main point is that either choice yields a coherent global orientation of the surface. Certain surfaces (e.g., Möbius strips) do not admit a global orientation, but they do not bound solids.

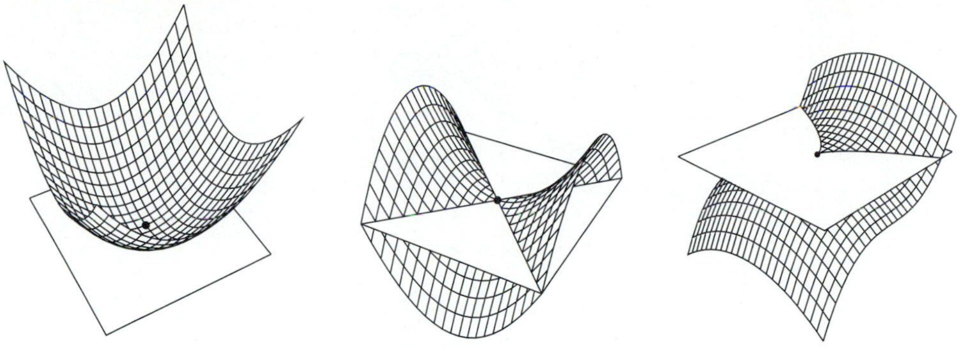

Figure 19.11 Local shape of a surface: (a) an elliptic point, (b) a hyperbolic point, and (c) a parabolic point (there are actually two distinct kinds of parabolic points; we come back to those in chapter 20). *Reprinted from "On Computing Structural Changes in Evolving Surfaces and their Appearance," by S. Pae and J. Ponce, International Journal of Computer Vision, 43(2):113–131, (2001).* © *2001 Kluwer Academic Publishers.*

two directions are called the *principal directions* in P, and it can be shown that, unless the normal curvature is constant over all possible orientations, they are orthogonal to each other (see Exercises). The *principal curvatures* κ_1 and κ_2 and the associated directions define the best local quadratic approximation of the surface. In particular, we can set up a coordinate system in P with x and y axes along the principal directions and z axis along the outward-pointing normal; the surface can be described (up to second order) in this frame by the *paraboloid* $z = -1/2(\kappa_1 x^2 + \kappa_2 y^2))$.

The neighborhood of a surface point can locally take three different shapes depending on the sign of the principal curvatures (Figure 19.11). A point P where both curvatures have the same sign is said to be *elliptic*, and the surface in its vicinity is egg-shaped (Figure 19.11a). It does not cross its tangent plane and looks like the outside shell of an egg (positive curvatures) or the inside of its broken shell (negative curvatures). We say that P is *convex* in the former case and *concave* in the latter one. When the principal curvatures have opposite signs, we have a *hyperbolic* point. The surface is locally saddle-shaped and crosses its tangent plane along two curves (Figure 19.11b). The corresponding normal sections have an inflection in P, and their tangents are called the *asymptotic directions* of the surface in P. They are bisected by the principal directions. The elliptic and hyperbolic points form patches on a surface. These areas are in general separated by curves formed by *parabolic* points where one of the principal curvatures vanishes. The corresponding principal direction is also an asymptotic direction, and the intersection of the surface and its tangent plane has (in general) a cusp in that direction (Figure 19.11c).

Naturally, we can define the Gaussian image of a surface by mapping every point onto the place where the associated unit normal pierces the unit sphere (that is sometimes referred to as the *Gauss sphere* in the sequel). In the case of plane curves, the Gauss map is one-to-one in the neighborhood of regular points, but the direction of traversal of the Gaussian image reverses in the vicinity of certain singularities. Likewise, it can be shown that the Gauss map is one-to-one in the neighborhood of elliptic or hyperbolic points. The orientation of a small, closed curve centered at an elliptic point is preserved by the Gauss map, but the orientation of a curve centered at a hyperbolic point is reversed (Figure 19.12).

The situation is a bit more complicated at a parabolic point. In this case, any small neighborhood contains points with parallel normals, indicating a double covering of the sphere near

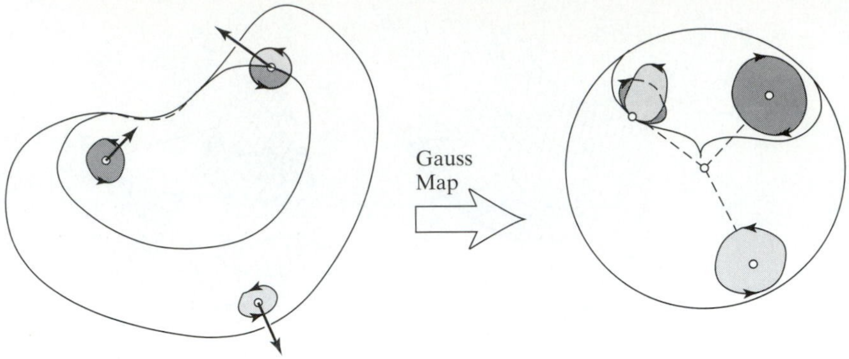

Figure 19.12 Left: A surface in the shape of a kidney bean. It is formed of a convex area, a hyperbolic region, and the parabolic curve separating them. Right: The corresponding Gaussian image. Darkly shaded areas indicate hyperbolic areas, lightly shaded ones indicate elliptic ones. Note that the bean is not convex, but does not have any concavity. *Reprinted from "On Computing Structural Changes in Evolving Surfaces and their Appearance," by S. Pae and J. Ponce, International Journal of Computer Vision, 43(2):113–131, (2001). © 2001 Kluwer Academic Publishers.*

the parabolic point (Figure 19.12). We say that the Gaussian map *folds* along the parabolic curve. Note the similarity with inflections of planar curves.

Let us now consider a surface curve γ passing through P and parameterized by its arc length s in the neighborhood of P. Since the restriction of the surface normal to γ has constant (unit) length, its derivative with respect to s lies in the tangent plane in P. It is easy to show that the value of this derivative only depends on the unit tangent $\boldsymbol{t}$ to γ and not on γ itself. Thus, we can define a mapping $d\boldsymbol{N}$ that associates with each unit vector $\boldsymbol{t}$ in the tangent plane in P the corresponding derivative of the surface normal (Figure 19.13). Using the convention $d\boldsymbol{N}(\lambda\boldsymbol{t}) \stackrel{\text{def}}{=} \lambda d\boldsymbol{N}(\boldsymbol{t})$ when $\lambda \neq 1$, we can extend $d\boldsymbol{N}$ to a linear mapping defined over the whole tangent plane and called the *differential of the Gauss map* in P.

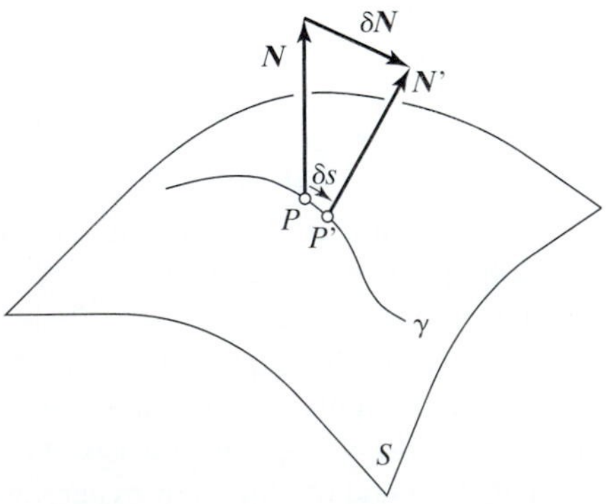

Figure 19.13 The directional derivative of the surface normal: If P and P' are nearby points on the curve γ, and $\boldsymbol{N}$ and $\boldsymbol{N}'$ denote the associated surface normals, with $\delta\boldsymbol{N} = \boldsymbol{N}' - \boldsymbol{N}$, the derivative is defined as the limit of $\frac{1}{\delta s}\delta\boldsymbol{N}$ as the length δs of the curve arc separating P and P' tends toward zero.

The *second fundamental form* in P is the bilinear form that associates with any two vectors $\boldsymbol{u}$ and $\boldsymbol{v}$ lying in the tangent plane the quantity[3]

$$\mathrm{II}(\boldsymbol{u}, \boldsymbol{v}) \stackrel{\text{def}}{=} \boldsymbol{u} \cdot dN(\boldsymbol{v}).$$

Since II is easily shown to be symmetric—that is, $\mathrm{II}(\boldsymbol{u}, \boldsymbol{v}) = \mathrm{II}(\boldsymbol{v}, \boldsymbol{u})$, the mapping that associates with any tangent vector $\boldsymbol{u}$ the quantity $\mathrm{II}(\boldsymbol{u}, \boldsymbol{u})$ is a quadratic form. In turn, this quadratic form is intimately related to the curvature of the surface curves passing through P. Indeed, note that the tangent $\boldsymbol{t}$ to a surface curve is everywhere orthogonal to the surface normal $\boldsymbol{N}$. Differentiating the dot product of these two vectors with respect to the curve arc length yields

$$\kappa \boldsymbol{n} \cdot \boldsymbol{N} + \boldsymbol{t} \cdot dN(\boldsymbol{t}) = 0,$$

where $\boldsymbol{n}$ denotes the principal normal to the curve and κ denotes its curvature. This can be rewritten as

$$\mathrm{II}(\boldsymbol{t}, \boldsymbol{t}) = -\kappa \cos \phi, \tag{19.1}$$

where ϕ is the angle between the surface and curve normals. For normal sections, we have $\boldsymbol{n} = \mp \boldsymbol{N}$, and it follows that the normal curvature in some direction $\boldsymbol{t}$ is

$$\kappa_t = \mathrm{II}(\boldsymbol{t}, \boldsymbol{t}),$$

where, as before, we use the convention that the normal curvature is positive when the principal normal to the curve and the surface normal point in opposite directions. In addition, Eq. (19.1) shows that the curvature κ of a surface curve whose principal normal makes an angle ϕ with the surface normal is related to the normal curvature κ_t in the direction of its tangent $\boldsymbol{t}$ by $\kappa \cos \phi = -\kappa_t$. This is known as *Meusnier's theorem* (Figure 19.14).

It turns out that the principal directions are the eigenvectors of the linear map dN, and the principal curvatures are the associated eigenvalues. The determinant K of this map is called the Gaussian curvature, and it is equal to the product of the principal curvatures. Thus, the sign of the Gaussian curvature determines the local shape of the surface: A point is elliptic when $K > 0$, hyperbolic when $K < 0$, and parabolic when $K = 0$. If δA is the area of a small patch centered

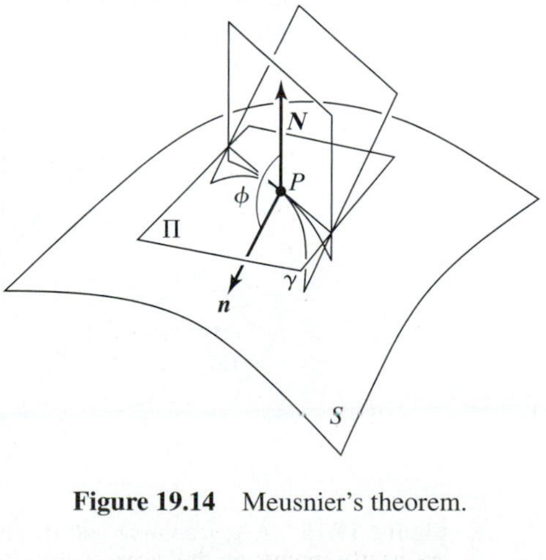

Figure 19.14 Meusnier's theorem.

[3]The second fundamental form is sometimes defined as $\mathrm{II}(\boldsymbol{u}, \boldsymbol{v}) = -\boldsymbol{u} \cdot dN(\boldsymbol{v})$ (see, e.g., do Carmo 1976, Struik 1988). Our definition allows us to assign positive normal curvatures to the surfaces bounding convex solids with outward-pointing normals.

in P on a surface S and $\delta A'$ is the area of the corresponding patch of the Gaussian image of S, it can also be shown that the Gaussian curvature is the limit of the (signed) ratio $\delta A'/\delta A$ as both areas approach zero (by convention, the ratio is chosen to be positive when the boundaries of both small patches have the same orientation and negative otherwise; see Figure 19.12). Note again the strong similarity with the corresponding concepts (Gaussian image and plain curvature) in the context of planar curves.

19.2 CONTOUR GEOMETRY

Before studying the geometry of surface outlines, let us pose for a minute and examine the relationship between the local shape of a space curve Γ and that of its orthographic projection γ onto some plane Π (Figure 19.15). Let us denote by α the angle between the plane Π and the tangent $\boldsymbol{t}$ to Γ and by β the angle between Π and the osculating plane of Γ (or equivalently between the normal to Π and the binormal $\boldsymbol{b}$ to Γ). These two angles completely define the local orientation of the curve relative to the image plane.

If κ denotes the curvature at some point on Γ and κ_a denotes its *apparent curvature* (i.e., the curvature of γ at the corresponding image point), it is easy to show analytically (see Exercises) that

$$\kappa_a = \kappa \frac{\cos\beta}{\cos^3\alpha}. \tag{19.2}$$

In particular, when the viewing direction is in the osculating plane ($\cos\beta = 0$), the apparent curvature κ_a vanishes, and the image of the curve acquires an inflection. When, in addition, the viewing direction is tangent to the curve ($\cos\alpha = \cos\beta = 0$), κ_a is not well defined anymore and the projection acquires a cusp.

These two properties of the projections of space curves are well known and mentioned in all differential geometry textbooks. Is it possible to relate in a similar fashion the local shape of the surface bounding a solid object to the shape of its image contour? The answer is a resounding "yes!", as shown by Koenderink (1984) in his delightful paper "What Does the Occluding Contour Tell Us About Solid Shape?" We present in this section a few elementary properties of image contours before stating and proving the main theorem of Koenderink's paper, and concluding by discussing some of its implications.

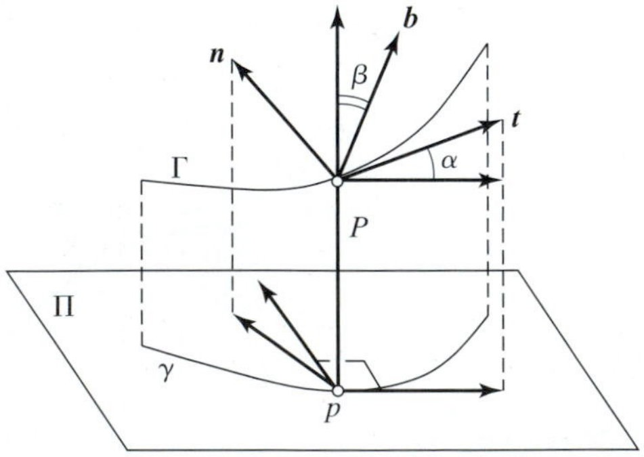

Figure 19.15 A space curve and its projection. Note that the tangent to γ is the projection of the tangent to Γ (e.g., think of the tangent as the velocity of a particle traveling along the curve). The normal to γ is *not*, in general, the projection of the normal $\boldsymbol{n}$ to Γ.

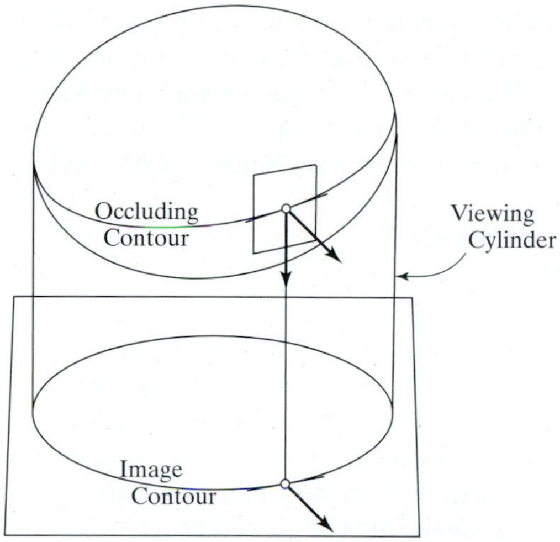

Figure 19.16 Occlusion boundaries under orthographic projection.

19.2.1 The Occluding Contour and the Image Contour

As noted earlier, the image of a solid bounded by a smooth surface is bounded by an image curve, called the *contour*, *silhouette*, or *outline* of this solid. This curve is the intersection of the retina with a viewing cone whose apex coincides with the pinhole and whose surface grazes the object along a second curve, called the *occluding contour* or *rim* (Figure 19.1).

We assume orthographic projection in the rest of this section. In this case, the pinhole moves to infinity and the viewing cone becomes a cylinder whose generators are parallel to the (fixed) viewing direction. The surface normal is constant along each one of these generators, and it is parallel to the image plane (Figure 19.16). The tangent plane at a point on the occluding contour projects onto the tangent to the image contour, and it follows that the normal to this contour is equal to the surface normal at the corresponding point of the occluding contour.

It is important to note that the viewing direction $\boldsymbol{v}$ is *not*, in general, perpendicular to the occluding contour tangent $\boldsymbol{t}$ (as noted by Nalwa, 1988, for example, the occluding contour of a tilted cylinder is parallel to its axis and not to the image plane). As shown in the next section, these two directions are *conjugated*—an extremely important property of the occluding contour.

19.2.2 The Cusps and Inflections of the Image Contour

Two directions $\boldsymbol{u}$ and $\boldsymbol{v}$ in the tangent plane are said to be conjugated when $\mathrm{II}(\boldsymbol{u}, \boldsymbol{v}) = 0$. For example, the principal directions are conjugated since they are orthogonal eigenvectors of dN, and asymptotic directions are self-conjugated.

It is easy to show that the tangent $\boldsymbol{t}$ to the occluding contour is always conjugated to the corresponding projection direction $\boldsymbol{v}$. Indeed, $\boldsymbol{v}$ is tangent to the surface at every point of the occluding contour, and differentiating the identity $\boldsymbol{N} \cdot \boldsymbol{v} = 0$ with respect to the arc length of this curve yields

$$0 = \left(\frac{d}{ds}\boldsymbol{N}\right) \cdot \boldsymbol{v} = d\boldsymbol{N}(\boldsymbol{t}) \cdot \boldsymbol{v} = \mathrm{II}(\boldsymbol{t}, \boldsymbol{v}).$$

Let us now consider a hyperbolic point P_0 and project the surface onto a plane perpendicular to one of its asymptotic directions. Since asymptotic directions are self-conjugated, the occluding contour in P_0 must run along this direction. As shown by Eq. (19.2), the curvature of the contour must be infinite in that case, and the contour acquires a cusp of the first kind.

We state in a moment a theorem by Koenderink (1984) that provides a quantitative relationship between the curvature of the image contour and the Gaussian curvature of the surface. In the meantime, we prove (informally) a weaker, but still remarkable result.

Theorem 5. *Under orthographic projection, the inflections of the contour are images of parabolic points (Figure 19.17).*

To see why this theorem holds, first note that, under orthographic projection, the surface normal at a point on the occluding contour is the same as the normal at the corresponding point of the image contour. Since the Gauss map folds at a parabolic point, the Gaussian image of the image contour must reverse direction at such a point. As shown earlier, the Gaussian image of a planar curve reverses at its inflections and cusps of the second kind. It is possible to show that the latter singularity does not occur for a general viewpoint, which proves the result.

In summary, the occluding contour is formed by points where the viewing direction $\boldsymbol{v}$ is tangent to the surface (the fold points mentioned in the introduction). Occasionally, it becomes tangent to $\boldsymbol{v}$ at a hyperbolic cusp point or crosses a parabolic line, and cusps (of the first kind) or inflections appear on the contour accordingly. Unlike the curves mentioned so far, the image contour may also cross itself (transversally) when two distinct branches of the occluding contour project onto the same image point, forming a T-junction (Figure 19.1). For general viewpoints, these are the only possibilities: There is no cusp of the second kind, nor any tangential self-intersection, for example. We come back to the study of exceptional viewpoints and the corresponding contour singularities in the next chapter.

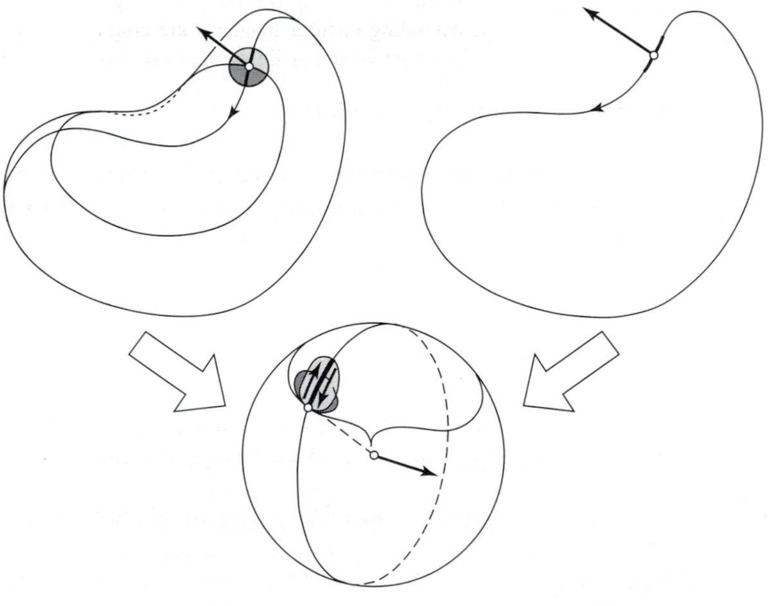

Figure 19.17 The inflections of the contour are images of parabolic points: The top-left side of this diagram shows the bean-shaped surface with an occluding contour overlaid, and its top-right side shows the corresponding image contour. As shown in the bottom part of the drawing, the Gauss map folds at the parabolic point, and so does its restriction to the great circle formed by the images of the occluding and image contours.

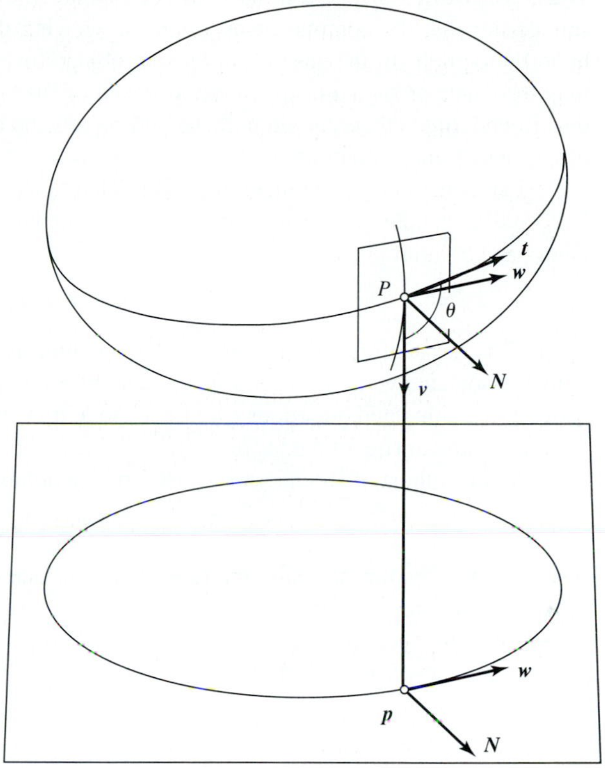

Figure 19.18 Occluding contour and image contour: The viewing direction $\boldsymbol{v}$ and the occluding contour tangent $\boldsymbol{t}$ are conjugated, and the radial curvature is always nonnegative at a visible point of the contour for opaque solids.

19.2.3 Koenderink's Theorem

Let us now state the theorem by Koenderink (1984) that has already been mentioned several times. We assume as before orthographic projection, consider a point P on the occluding contour of a surface S, and denote by p its image on the contour.

Theorem 6. *The Gaussian curvature K of S in P and the contour curvature κ_c in p are related by*

$$K = \kappa_c \kappa_r,$$

where κ_r denotes the curvature of the radial curve *formed by the intersection of S with the plane defined by the normal to S in P and the projection direction (Figure 19.18).*

This remarkably simple relation has several important corollaries (starting with Theorem 5, of course): Note that the *radial curvature* κ_r remains positive (or zero) along the occluding contour since the projection ray locally lies inside the imaged object at any point where $\kappa_r < 0$. It follows that κ_c is positive when the Gaussian curvature is positive and negative otherwise. In particular, the theorem shows that convexities of the contour corresponds to elliptic points of the surface, whereas contour concavities correspond to hyperbolic points and contour inflections correspond to parabolic points.

Among elliptic surface points, it is clear that concave points never appear on the occluding contour of an opaque solid since their tangent plane lies (locally) completely inside this solid.

Thus, convexities of the contour also correspond to convexities of the surface. Likewise, we saw earlier that the contour cusps when the viewing direction is an asymptotic direction at a hyperbolic point. In the case of an opaque object, this means that concave arcs of the contour may terminate at such a cusp, where a branch of the contour becomes occluded. Thus, we see that Koenderink's theorem strengthens and refines the earlier characterization of the geometric properties of image contours.

Let us now prove the theorem. It is related to a general property of conjugated directions: If κ_u and κ_v denote the normal curvatures in conjugated directions $\boldsymbol{u}$ and $\boldsymbol{v}$, and K denotes the Gaussian curvature, then

$$K \sin^2 \theta = \kappa_u \kappa_v, \tag{19.3}$$

where θ is the angle between $\boldsymbol{u}$ and $\boldsymbol{v}$. This relation is easy to prove by using the fact that the matrix associated with the second fundamental form is diagonal in the basis of the tangent plane formed by conjugated directions (see Exercises). It is obviously satisfied for principal directions ($\theta = \pi/2$) and asymptotic ones ($\theta = 0$).

In the context of Koenderink's theorem, we obtain

$$K \sin^2 \theta = \kappa_r \kappa_t,$$

where κ_t denotes the normal curvature of the surface along the occluding contour direction $\boldsymbol{t}$ (which is, of course, different from the actual curvature of the occluding contour). To complete the proof of the theorem, we use another general property of surfaces: The apparent curvature of any surface curve with tangent $\boldsymbol{t}$ is

$$\kappa_a = \frac{\kappa_t}{\cos^2 \alpha}, \tag{19.4}$$

where α denotes as before the angle between $\boldsymbol{t}$ and the image plane. As shown in the exercises, this property easily follows from Eq. (19.2) and Meusnier's theorem.

In other words, the apparent curvature of any surface curve is obtained by dividing the associated normal curvature by the square of the cosine of the angle between its tangent and the image plane. Noting that κ_c is just the apparent curvature of the occluding contour now allows us to write

$$\kappa_c = \frac{\kappa_t}{\sin^2 \theta} \tag{19.5}$$

since $\alpha = \theta - \pi/2$. Substituting Eq. (19.5) into Eq. (19.3) concludes the proof of the theorem.

19.3 NOTES

There are many excellent textbooks on differential geometry, including the accessible presentations found in do Carmo (1976) and Struik (1988). Our presentation is closer in spirit (if not in elegance) to the descriptive introduction to differential geometry found in Hilbert and Cohn-Vossen's (1952) wonderful book "Geometry and the Imagination."

It was not always recognized that the image contour carries vital information about surface shape (see Marr, 1977, Horn, 1986, for arguments to the contrary). The theorem proved in this chapter clarified the situation and appeared first in Koenderink (1984). Our proof is different from the original one, but it is close in spirit to the proof given by Koenderink (1990) in his book "Solid Shape" (which, like Hilbert and Cohn-Vossen's book, should be required reading for anybody seriously interested in the geometric aspects of computer vision). Our choice here was motivated by our reluctance to use any formulas that require setting a particular coordinate system. Alternate proofs for various kinds of projection geometries can be found in Brady *et al.* (1985*a*), Arbogast and Mohr (1991), Cipolla and Blake (1992), Vaillant and Faugeras (1992), and Boyer (1996).

PROBLEMS

19.1. What is (in general) the shape of the silhouette of a sphere observed by a perspective camera?

19.2. What is (in general) the shape of the silhouette of a sphere observed by an orthographic camera?

19.3. Prove that the curvature κ of a planar curve in a point P is the inverse of the radius of curvature r at this point.

Hint: Use the fact that $\sin u \approx u$ for small angles.

19.4. Given a fixed coordinate system, let us identify points of $\mathbb{E}^3$ with their coordinate vectors and consider a parametric curve $\boldsymbol{x} : I \subset \mathbb{R} \to \mathbb{R}^3$ not necessarily parameterized by arc length. Show that its curvature is given by

$$\kappa = \frac{|\boldsymbol{x}' \times \boldsymbol{x}''|}{|\boldsymbol{x}'|^3}, \tag{19.6}$$

where $\boldsymbol{x}'$ and $\boldsymbol{x}''$ denote, respectively, the first and second derivatives of $\boldsymbol{x}$ with respect to the parameter t defining it.

Hint: Reparameterize $\boldsymbol{x}$ by its arc length and reflect the change of parameters in the differentiation.

19.5. Prove that, unless the normal curvature is constant over all possible directions, the principal directions are orthogonal to each other.

19.6. Prove that the second fundamental form is bilinear and symmetric.

19.7. Let us denote by α the angle between the plane Π and the tangent to a curve Γ and by β the angle between the normal to Π and the binormal to Γ, and by κ the curvature at some point on Γ. Prove that if κ_a denotes the apparent curvature of the image of Γ at the corresponding point, then

$$\kappa_a = \kappa \frac{\cos\beta}{\cos^3\alpha}.$$

(Note: This result can be found in Koenderink, 1990, p. 191.)

Hint: Write the coordinates of the vectors $\boldsymbol{t}$, $\boldsymbol{n}$, and $\boldsymbol{b}$ in a coordinate system whose z-axis is orthogonal to the image plane, and use Eq. (19.6) to compute κ_a.

19.8. Let κ_u and κ_v denote the normal curvatures in conjugated directions $\boldsymbol{u}$ and $\boldsymbol{v}$ at a point P, and let K denote the Gaussian curvature; prove that

$$K \sin^2\theta = \kappa_u \kappa_v,$$

where θ is the angle between the $\boldsymbol{u}$ and $\boldsymbol{v}$.

Hint: Relate the expressions obtained for the second fundamental form in the bases of the tangent plane respectively formed by the conjugated directions and the principal directions.

19.9. Show that the occluding contour is a smooth curve that does not intersect itself.

Hint: Use the Gauss map.

19.10. Show that the apparent curvature of any surface curve with tangent $\boldsymbol{t}$ is

$$\kappa_a = \frac{\kappa_t}{\cos^2\alpha},$$

where α is the angle between the image plane and $\boldsymbol{t}$.

Hint: Write the coordinates of the vectors $\boldsymbol{t}$, $\boldsymbol{n}$, and $\boldsymbol{b}$ in a coordinate system whose z axis is orthogonal to the image plane, and use Eq. (19.2) and Meusnier's theorem.

20

Aspect Graphs

This chapter continues our investigation of the qualitative relationship between solid shapes and their pictures. This time, however, we focus on the characterization of the set of *all* views of a solid using tools from differential geometry and singularity theory to group images that share the same topological structure into equivalence classes. To illustrate this idea, let us first pause in *Flatland*, a world where tiny one-eyed gnats roam a two-dimensional landscape, and consider a planar object bounded by a smooth closed curve and observed by one of these critters (Figure 20.1). We model the gnat's eye by an *omnidirectional circular camera* (i.e., a pinhole perspective camera with a circular imaging surface and a 360° field of view), and assume that

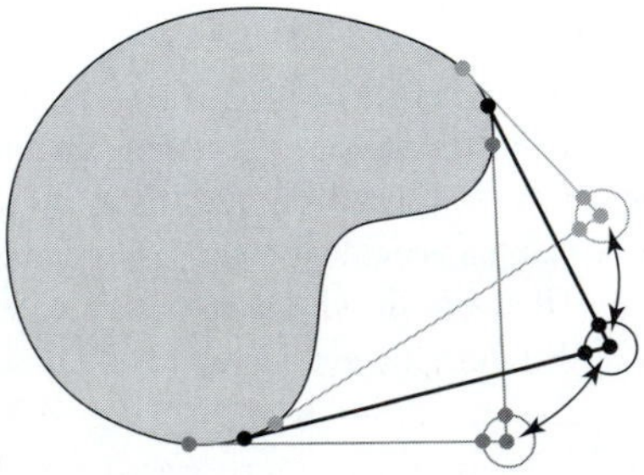

Figure 20.1 Imaging in Flatland: Three omnidirectional circular cameras observe a two-dimensional solid from close-by viewpoints. In typical views like these, the order and number of contour points (or equivalently, of the associated projection rays) are the same for all cameras, indicating a smooth mapping from viewpoint to image structure.

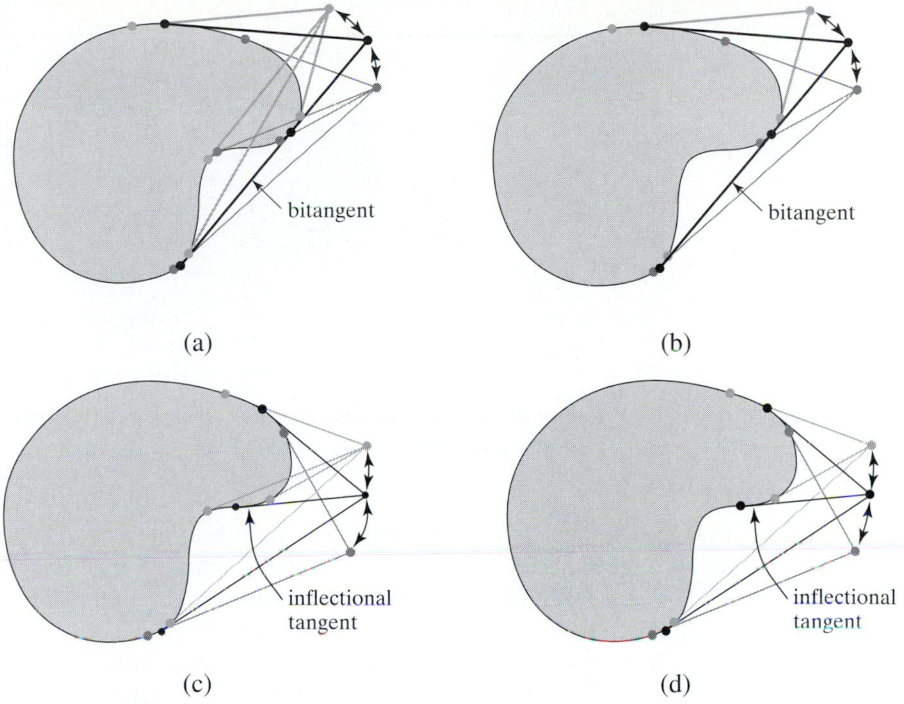

Figure 20.2 Exceptional views induce qualitative image changes—namely: (a) a change of the ordering of contour points along the retina, or (b)–(d) a change in the number of these points. The retinas are omitted in this figure to limit clutter. See text for details.

the retina has been oriented (say, counterclockwise) so we can speak of the ordering of points along it.

The "occluding contour" consists in this case of the (discrete) points where rays of light issued from the optical center graze the object boundary, and the "image contour" is formed by the intersections of these tangent rays with the circular retina of the camera. Objects in Flatland may be transparent or opaque; in the latter case, tangent rays that intersect transversally some part of the object boundary are blocked before reaching the retina. Either way, small camera motions alter the position of the contour points, but do not change (in general) their number or their ordering along the retina.

For certain viewpoints, however, a small camera motion may cause a dramatic change in image structure: In the case of transparent objects, close-by contour points merge when the eye crosses a *bitangent* (i.e., a line that grazes the object boundary in two different places, then separate again as their order reverses; Figure 20.2a). For opaque objects, a new contour point is revealed (or an old one vanishes) as the eye traverses the bitangent, so the view also changes radically there, but in a different way: The order reversal is replaced by the appearance (or disappearance) of a contour point (Figure 20.2b). Likewise, a pair of contour points appears or disappears when the eye crosses the tangent line at a curve inflection of a transparent object (Figure 20.2c). For opaque objects, only one of these points is visible, the second one being hidden by the object (Figure 20.2d).

These structural image changes are called *visual events*. Are the events associated with bitangents or inflectional tangents the only possible ones? Two bitangents may of course intersect, an inflectional tangent and a bitangent may intersect, and so on, yielding more complex

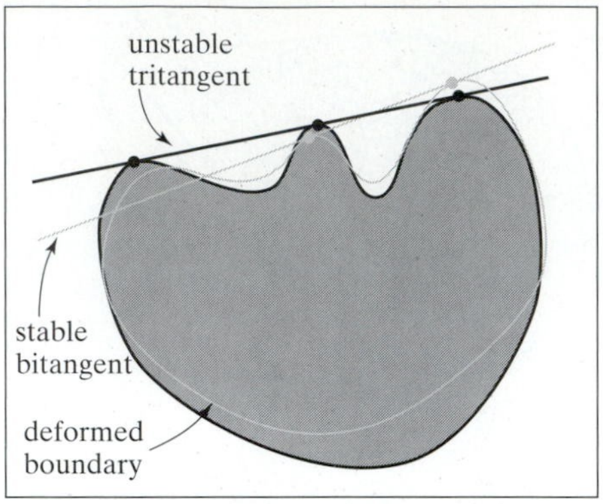

Figure 20.3 Exceptional and generic curves: Unlike bitangents, tritangents are not stable under small curve deformations.

changes in object appearance. Tritangent lines and inflectional bitangents are a priori possible too. It is clear, however, that a tritangent would not survive any small deformation of the associated curve (Figure 20.3). In contrast, a bitangent or the intersection of two inflectional tangents would merely follow the deformation with a motion of its own.

We restrict our attention in the rest of this chapter to *generic curves and surfaces* whose features are not affected by small deformations. Genericity is a mathematical concept related to the intuitive notion of general position and the topological notions of openness, density, and transversality. Its formal definition is rather technical and would be out of place here. Let us just note that, although the genericity assumption rules out certain simple geometric figures (lines, planes, etc.), the boundaries of all *real* objects are generic (it is impossible to draw a true straight line in the real world). What may be more remarkable, at least in the context of this chapter, is that restricting one's attention to generic curves and surfaces also limits the number and type of possible visual events: Indeed, structural changes in contour structure can be shown to occur in the orderly fashion predicted by the branch of mathematics called *singularity theory* or *catastrophe theory*. In Flatland, this translates to restricting the visual events of generic curves to those associated with bitangents, inflectional tangents, and their intersections.

For omnidirectional circular cameras, the *aspect* of an object (i.e., the number and order of its contour points) only depends on the eye position (Figure 20.2). Visual events correspond to certain types of contacts between curves and projection rays and have nothing to do with the location of the points where these rays intersect the retina. Thus, the set of viewpoints for a Flatland camera can be modeled by the plane where its optical center lies. This plane can be organized into a cellular structure, the *aspect graph*, where visual event rays and their intersections delineate maximal cells where the object's aspect does not change (Figure 20.4, left).

For distant observers, the omnidirectional circular cameras considered so far can be replaced by conventional orthographic cameras since the viewing direction determines the direction of the linear retina in this case and, conversely, the orientation of the image line can always be chosen so the object of interest entirely lies *in front* of that line. In this setting, the view space becomes a unit circle of (oriented) projection vectors partitioned by the directions of the bitangents and inflectional tangents into a finite set of aspects (Figure 20.4, right).

The same principles apply in the three-dimensional world: As in the Flatland case, choosing a projection model (omnidirectional spherical perspective, planar perspective, or orthographic camera) and a viewpoint determines the aspect of an object (in this case, a graphical representation of its image contour, with nodes corresponding to T-junctions and cusps, and arcs to the smooth contour pieces between them). The range of possible viewpoints is again parti-

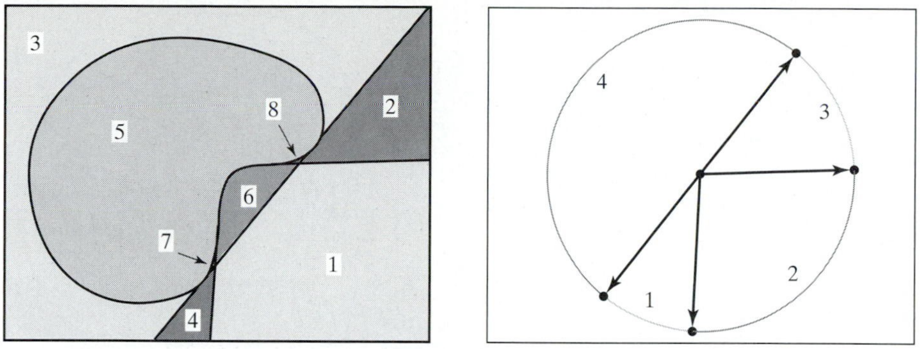

Figure 20.4 The aspect graph of an (opaque) object in Flatland. Left: A two-dimensional object bounded by a smooth closed curve and the cells of its (omnidirectional circular) perspective aspect graph; note the two small regions numbered 7 and 8. Right: The corresponding orthographic aspect graph. The arcs joining pairs of adjacent aspects and the aspects are not shown. Easy exercise: Draw the aspects. How would the aspect graph change for a transparent object?

tioned by visual event boundaries into maximal cells where the aspect does not change and these cells, labeled by representative aspects, form the nodes of a three-dimensional aspect graph, with arcs attached to the visual event boundaries separating adjacent cells. The aspect graph was first introduced by Koenderink and Van Doorn (1979) under the name of *visual potential*. The next section introduces the additional elements of differential geometry necessary to understand its construction. The following sections present algorithms for constructing exact and approximate aspect graphs of solid objects and using the latter in bin-picking tasks.

20.1 VISUAL EVENTS: MORE DIFFERENTIAL GEOMETRY

From now on, we assume orthographic projection (the perspective case is briefly discussed in Section 20.4), and model the set of all viewpoints by a unit sphere of oriented projection directions. Inflections, cusps, and T-junctions are stable features of the image contour that generally survive small eye movements: Let us consider, for example, a contour inflection; as shown in chapter 19, it is the projection of a point where the occluding contour and a parabolic curve of the associated surface intersect (normally at a nonzero angle). Any small change in viewpoint deforms the occluding contour a bit, but the two curves still intersect transversally at a close-by point projecting onto a contour inflection.

It is natural to ask: What are the (peculiar) eye motions that make a stable contour feature appear or disappear? To answer this question, we take another look at the Gauss map and introduce the *asymptotic spherical map*, showing in the process that the boundaries of the images of a surface through these mappings determine the appearance and disappearance of inflections and cusps of its contour. This provides us with a characterization of *local visual events* (i.e., the changes in contour structure associated with the differential geometry of the surface at these boundaries). We also consider multiple contacts between visual rays and a surface. This leads us to the concept of *bitangent ray manifold*, and the characterization of its boundaries allows us to understand the genesis and annihilation of T-junctions and introduce the associated *multilocal visual events*. Together, the local and multilocal events capture the totality of the structural contour changes that determine the aspect graph.

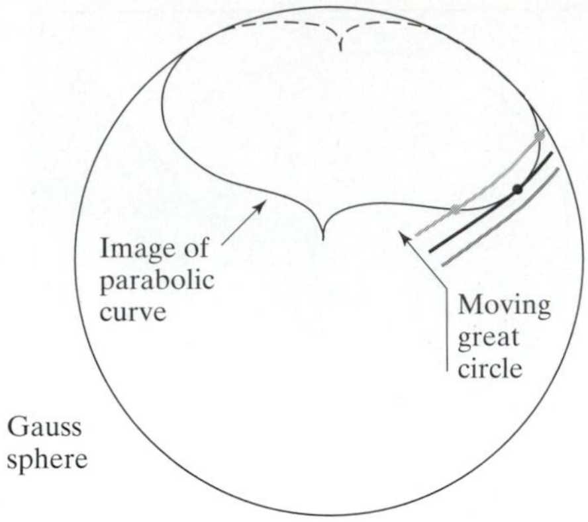

Figure 20.5 As the viewpoint changes, the great circle of the Gauss sphere associated with the (orthographic) occluding contour may become tangent to the spherical image of the parabolic curve. Afterward, the circle intersects this curve in two close-by points corresponding to two contour inflections.

20.1.1 The Geometry of the Gauss Map

The Gauss map provides a natural setting for the study of the image contour and its inflections. Indeed, under orthographic projection, we saw in chapter 19 that the occluding contour maps onto a great circle of the unit sphere, and that the intersections of this circle with the spherical image of the parabolic curves yield inflections of the contour. Therefore, it is clear that the contour gains (or loses) two inflections when a camera movement causes the corresponding great circle to cross the image of the parabolic curve (Figure 20.5).

A finer understanding of the creation of inflection pairs may be gained by taking a closer look at the geometry of the Gauss map. As shown in chapter 19, the image of a surface on the Gauss sphere folds along the image of its parabolic curves. Figure 20.6 shows an example, with a single covering of the sphere on one side of the parabolic curve and a triple covering on the other side. The easiest way to think about the creation of such a fold is to grab (in your mind) a bit of the rubber skin of a deflated balloon, pinch it, and fold it over. As illustrated by the figure, this process generally introduces not only a fold of the spherical image, but two cusps as well (whose preimages are aptly named *cusps of Gauss* in differential geometry). Cusps and inflections of the image of the parabolic curve always come in pairs (two inflection pairs and one cusp pair here, but of course there may be no cusp or inflection at all). The inflections split the fold of the Gauss map into convex and concave parts, and their preimages are called *gutterpoints* (Figure 20.6).

What happens to the occluding contour as the associated great circle crosses the spherical image of the parabolic curve depends on *where* the crossing happens. As shown by Figure 20.6, there are several cases: When the crossing occurs along a convex fold of the Gauss map, an isolated point appears on the spherical image of the occluding contour before exploding into a small closed loop on the unit sphere (Figure 20.6, bottom right). In contrast, if the crossing occurs along a concave fold, two separate loops merge and then separate with a different connectivity (Figure 20.6, top right). These changes are, of course, reflected on the image contour in a way that is detailed in the next couple of sections.

The great circle associated with the occluding contour may also cross the image of the parabolic curve at a cusp. Unlike crossings that occur at regular fold points, this one is in general transversal and does not impose a tangency condition on the orientation of the great circle. There is no change in the topology of the intersection, but two inflections appear or disappear on the image contour. Finally, the great circle may cross the Gaussian image of a parabolic curve at

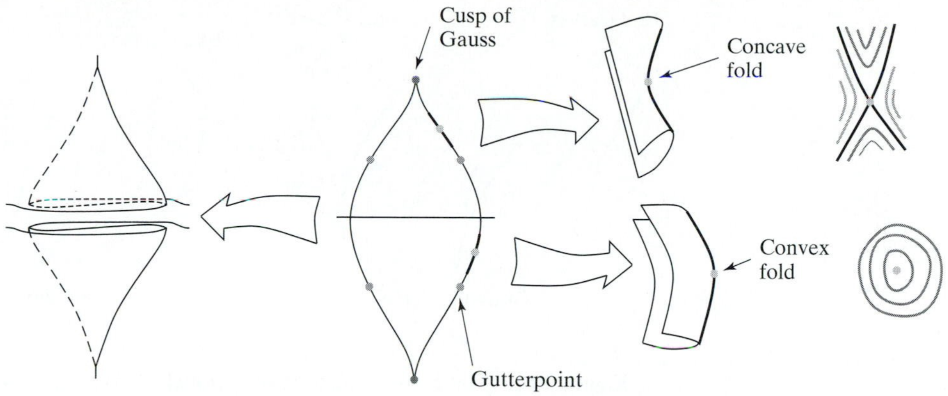

Figure 20.6 Folds and cusps of the Gauss map. The gutterpoints are the preimages of the inflections of the spherical image of the parabolic curve. To clarify the structure of the fold, it is drawn in the left and right sides of the figure as a surface folding in space. The changes in topology of the intersection between a great circle and the Gaussian image of the surface as the circle crosses the fold are illustrated in the far right portion of the figure. *Reprinted from "On Computing Structural Changes in Evolving Surfaces and their Appearance," by S. Pae and J. Ponce, International Journal of Computer Vision, 43(2):113–131, (2001).*

one of its inflections. The change in topology in this case is too complicated to be described here. The good news is that there is only a finite number of viewpoints for which this situation may occur (since there is only a finite number of gutterpoints on a generic surface). In contrast, the other types of fold crossings occur for infinite one-parameter families of viewpoints: This is because the tangential crossings associated with convex or concave portions of the fold may occur anywhere along an extended curve arc drawn on the unit sphere, whereas the transversal crossings associated with cusps occur at isolated points, but for arbitrary orientations of the great circle. We identify the associated families of singular viewpoints in the next section.

20.1.2 Asymptotic Curves

We saw in chapter 19 that ordinary hyperbolic points admit two distinct asymptotic tangents. More generally, the set of all asymptotic tangents on a hyperbolic patch can neatly be divided into two families such that each family admits a smooth field of integral curves, called *asymptotic curves*. Following Koenderink (1990), we give a color to each family and talk about the associated *red* and *blue* asymptotic curves. These curves only cover the hyperbolic part of a surface and must therefore be singular in the neighborhood of its parabolic boundary: Indeed, a red asymptotic curve merges with a blue one at an ordinary parabolic point forming a cusp and intersecting the parabolic curve at a nonzero angle (Figure 20.7a).[1]

Let us now examine the behavior of the asymptotic curves under the Gauss map. Remember from chapter 19 that asymptotic directions are self-conjugated. This (literally) means that the derivative of the surface normal along an asymptotic curve is orthogonal to the tangent to this curve: The asymptotic curve and its spherical image have perpendicular tangents. On the

[1]The situation is different at cusps of Gauss, where the asymptotic curves meet the parabolic curve tangentially. This unusual behavior also occurs for planar parabolic curves of nongeneric objects (e.g., the two circular parabolic curves at the top and bottom of a torus lying on its side or, more generally, the parabolic lines of a solid of revolution that are associated with local extrema of the cross-section height along its axis).

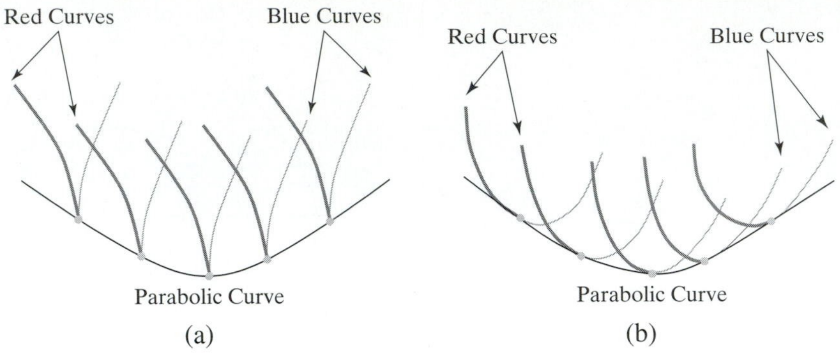

Figure 20.7 Contact of asymptotic and parabolic curves on (a) the surface, and (b) the Gauss sphere.

other hand, all the directions of the tangent plane are conjugated to the asymptotic direction at a parabolic point, so the Gaussian image of *any* surface curve passing through a parabolic point is perpendicular to the corresponding asymptotic direction. In particular, the Gaussian image of a parabolic curve is the *envelope* of the images of the asymptotic curves intersecting it (i.e., it is tangent to these curves everywhere; see Figure 20.7b).

We can now characterize the viewpoints for which a pair of inflections appears (or disappears). Since the great circle associated with the occluding contour becomes tangent to the image of the parabolic curve on the Gauss sphere as they cross, the viewing direction normal to this great circle is along the corresponding asymptotic direction of the parabolic curve. A pair of inflections may of course also appear when the great circle crosses the image of a cusp of Gauss or, equivalently, when the line of sight crosses the tangent plane at such a point. As noted earlier, the topology of the image contour not change in this case; it simply gains (or loses) an *undulation* (i.e., a small concave dent in one of its convex parts or a convex bump in one of its concave ones). The next section shows how the contour structure changes at the other types of singularities.

20.1.3 The Asymptotic Spherical Map

The Gauss map associates with every surface point the place where the tip of the corresponding normal pierces the unit sphere. We now define the *asymptotic spherical map*, which associates with every (hyperbolic) point the corresponding asymptotic directions. Let us make a few remarks before proceeding. First, there is really one asymptotic spherical image for each family of asymptotic curves, and the two images may or may not overlap on the sphere. Second, elliptic points obviously have no asymptotic spherical image at all, and the unit sphere may not be fully covered by the images of the hyperbolic points. However, it may also be fully covered, and, at least locally, it may be covered several times by members of a single family of asymptotic directions.

Since an image contour cusps when the line of sight is along an asymptotic direction, a cusp pair appears (or disappears) when the line of sight crosses a fold of the asymptotic spherical map (note the close analogy with contour inflections and folds of the Gauss map). As could be expected, the asymptotic spherical image of an asymptotic curve is singular at the fold boundary. There are again two possibilities (the image may join the boundary tangentially or cusp there), and they occur at two types of fold points: Those associated with asymptotic directions along parabolic curves (since there is no asymptotic direction at all on their elliptic side), and those associated with asymptotic directions at *flecnodal points*. These points are inflections of

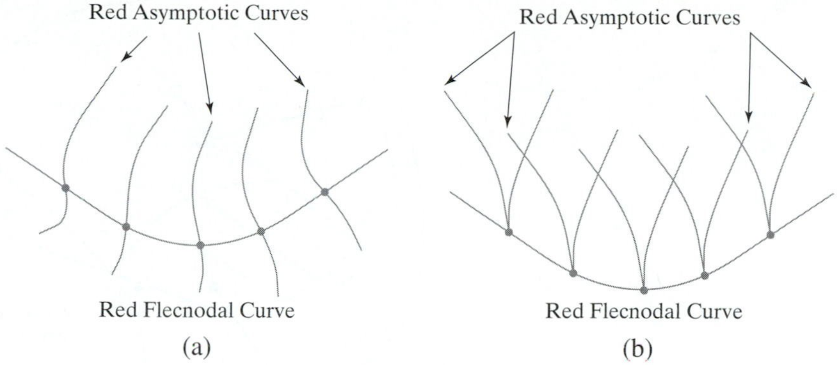

Figure 20.8 Contact of asymptotic and flecnodal curves on (a) the surface, and (b) the asymptotic spherical image.

the asymptotic curves' projections into their tangent plane (Figure 20.8a), and they form curves intersecting transversally the corresponding asymptotic curves. Like those, they come in two colors depending on which asymptotic family has an inflection. The asymptotic spherical image of an asymptotic curve cusps at a flecnodal point (Figure 20.8b). It should also be noted that flecnodal curves intersect parabolic ones tangentially at cusps of Gauss.

It is clear that the contour structure changes when the line of sight crosses the parabolic or flecnodal boundaries of the asymptotic spherical image. Such a change is called a *visual event*, and the associated boundaries are called *visual event curves*. Before examining in more detail the various visual events, let us note a different, but equivalent way of thinking of the associated boundaries. If we draw the singular asymptotic tangent line at each point along a parabolic or flecnodal surface we obtain *ruled surfaces* swept by the tangents. A visual event occurs whenever the line of sight crosses one of these ruled surfaces, whose intersections with the sphere at infinity are exactly the visual event curves when this sphere is identified with the unit sphere. Thinking of contour evolution in terms of these ruled surfaces has the advantages of pointing toward a generalization of visual events to perspective projection (the view changes whenever the optical center crosses them) and allowing a clear visualization of the relationship between singular viewpoints and surface shape.

20.1.4 Local Visual Events

We are now in a position to understand how the contour structure changes at visual event boundaries. There are three *local* visual events that are completely characterized by the local differential surface geometry: *lip*, *beak-to-beak*, and *swallowtail*. Their colorful names are inherited from Thom's (1972) catalogue of elementary catastrophes and are closely related to the shape of the contour near the associated events.

Let us first examine the *lip* event, which occurs when the line of sight crosses the asymptotic spherical image of a convex parabolic point or, equivalently, the ruled surface defined by the associated asymptotic tangents (Figure 20.9, top). We have shown earlier that the intersection between the great circle associated with the occluding contour and the Gaussian image of the surface acquires a loop during the event (Figure 20.6, bottom right) with the creation of two inflections and two cusps on the contour. More precisely, there is no image contour before the event, with an isolated contour point appearing out of nowhere at the singularity before exploding into a closed contour loop consisting of a pair of branches meeting at two cusps (Figure 20.9, bottom). One of the branches is formed by the projection of both elliptic and hyperbolic points,

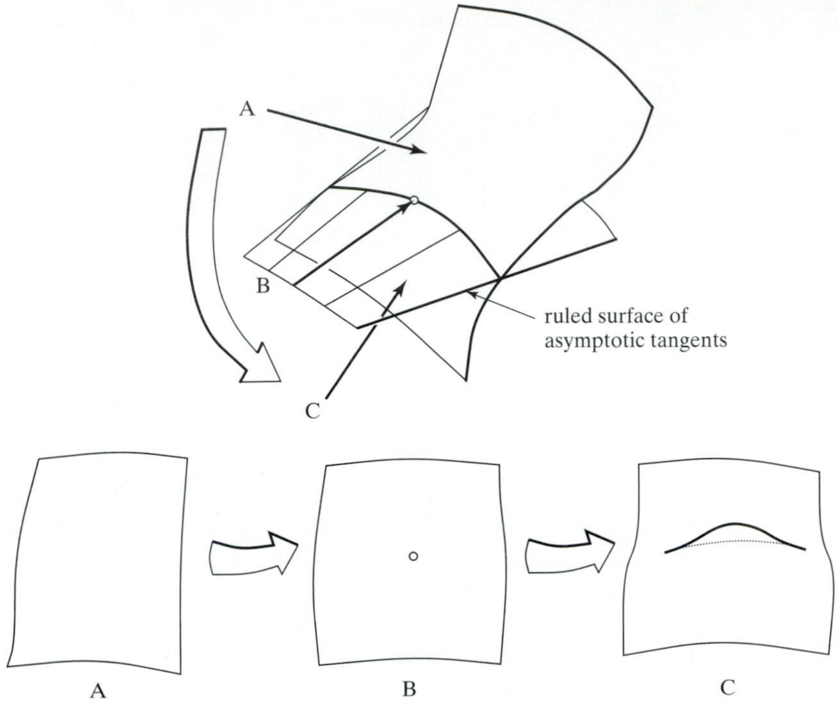

Figure 20.9 A lip event. The name is related to the shape of the contour on the right of the figure. Here, as in latter figures, the dashed part of the contour would be invisible due to occlusion for an opaque object. In this example, the two inflections are on the visible part of the contour, the hidden part being all hyperbolic, but the situation would be reversed by taking a viewpoint along the opposite direction. *Reprinted from "On Computing Structural Changes in Evolving Surfaces and their Appearance," by S. Pae and J. Ponce, International Journal of Computer Vision, 43(2):113–131, (2001). © 2001 Kluwer Academic Publishers.*

with two inflections, whereas the other one is formed by the projection of hyperbolic points only. For opaque objects, one of the branches is always occluded by the object.

The *beak-to-beak* event occurs when the line of sight crosses the asymptotic spherical image of a concave parabolic point or, once again, the ruled surface defined by the associated asymptotic tangents (Figure 20.10, top). As shown earlier, the topology of the intersection between the great circle associated with the occluding contour and the Gaussian image of the surface changes during this event, with two loops merging and then splitting again with a different connectivity (Figure 20.6, top right). In the image, two distinct portions of the contour, each having a cusp and an inflection, meet at a point in the image. Before the event, each of the branches is divided by the associated cusp into a purely hyperbolic portion and a mixed elliptic-hyperbolic arc, one of which is always occluded. After the event, two contour cusps and two inflections disappear as the contour splits into two smooth arcs with a different connectivity. One of these is purely elliptic, whereas the other is purely hyperbolic, with one of the two always being occluded for opaque objects (Figure 20.10, bottom). The reverse transition is, of course, also possible, as for all other visual events.

Finally, the *swallowtail* event occurs when the eye crosses the surface ruled by the asymptotic tangents along a flecnodal curve of the same color. We know that two cusps appear (or disappear) in this event. As shown in Figure 20.11(a)–(b), the intersection of the surface and

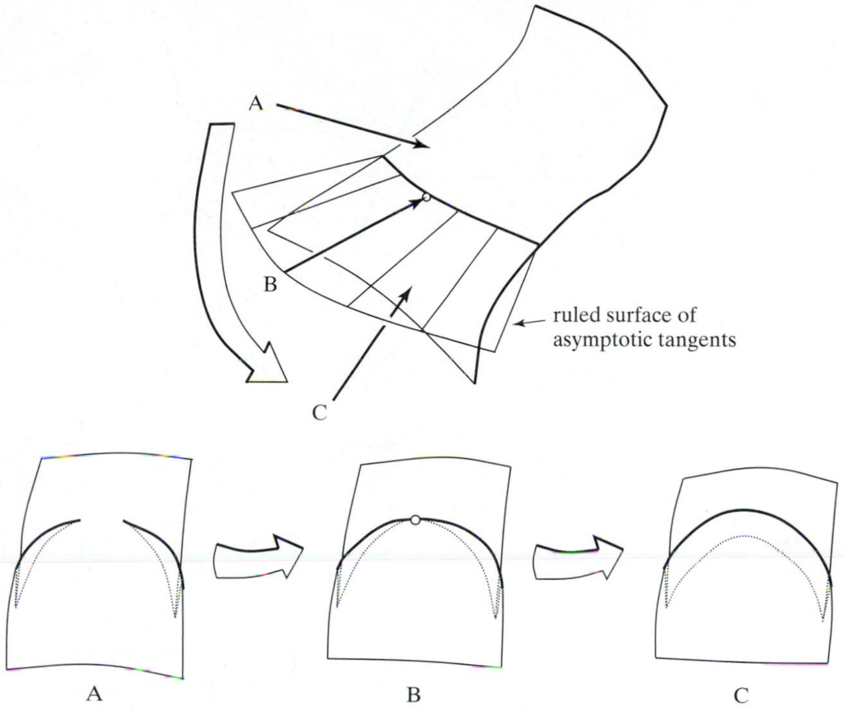

Figure 20.10 A beak-to-beak event (the name is related to the shape of the contour on the left of the figure). *Reprinted from "On Computing Structural Changes in Evolving Surfaces and their Appearance," by S. Pae and J. Ponce, International Journal of Computer Vision, 43(2):113-131, (2001). © 2001 Kluwer Academic Publishers.*

its tangent plane at a flecnodal point consists of two curves, one of which has an inflection. The corresponding asymptotic tangent is, of course, associated with the family of asymptotic curves having an inflection there, too. Unlike ordinary asymptotic rays (Figure 20.11c), which are blocked by the observed solid, this one grazes the solid's surface (Koenderink, 1990), causing a sharp V on the image contour at the singularity. The contour is smooth before the transition, but it acquires two cusps and a T-junction after it (Figure 20.11, bottom). All surface points involved in the event are hyperbolic. For opaque objects, one branch of the contour ends at the T-junction and the other one ends at a cusp.

20.1.5 The Bitangent Ray Manifold

Now remember from chapter 19 that cusps and inflections are not the only kinds of stable contour features: T-junctions also occur over open sets of viewpoints. They form when two distinct pieces of the occluding contour project onto the same image location. The corresponding surface normals must be orthogonal to the *bitangent* line of sight joining the two points, but they are not (in general) parallel. That T-junctions are stable over small eye movements is intuitively clear: Consider a convex point P and its tangent plane (Figure 20.12, left). This plane intersects (in general) the surface along a closed (but possibly empty) curve, and there is an even number of points (P' and P'' in the figure) such that the rays drawn from P through these points are tangent to the curve. Each such tangency yields a bitangent ray and an associated T-junction. A small

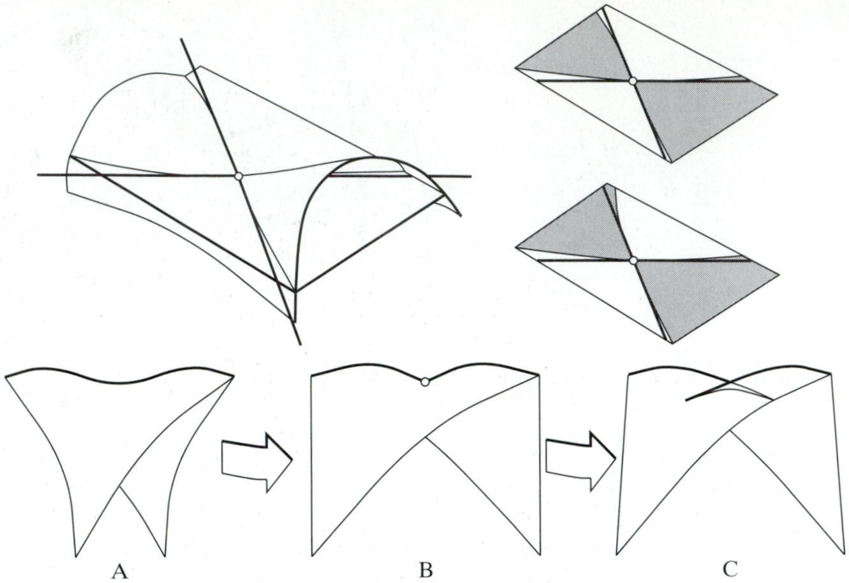

Figure 20.11 A swallowtail event. Top: Surface shape in the neighborhood of a flecnodal point (a), and comparison of the intersection of the associated solid and its tangent plane near such a point (b) and an ordinary hyperbolic point (c). Bottom: The event itself. *Reprinted from "On Computing Structural Changes in Evolving Surfaces and their Appearance," by S. Pae and J. Ponce, International Journal of Computer Vision, 43(2):113-131, (2001). © 2001 Kluwer Academic Publishers.*

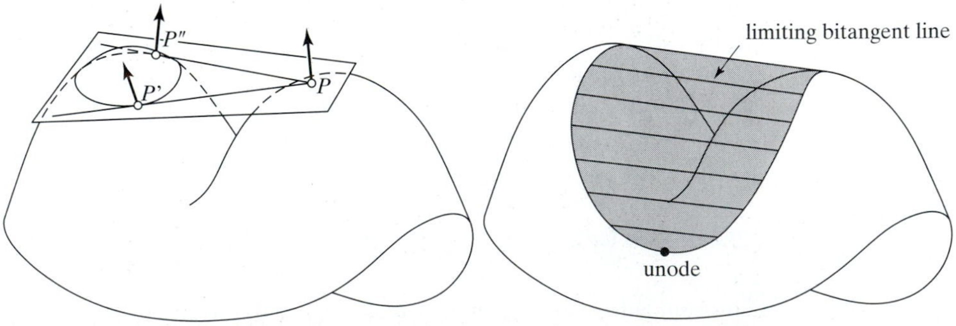

Figure 20.12 Bitangent rays. Left: The tangent plane to the surface in P intersects it along a closed curve with two bitangent rays PP' and PP'' grazing the surface along this curve. Right: The limiting bitangent developable surface ruled by the lines where a plane bitangent to a surface grazes it. Here the two curves where it touches the observed surface merge tangentially at a *unode*, a type of cusp of Gauss. *Reprinted from "Toward a Scale-Space Aspect Graph: Solids of Revolution," by S. Pae and J. Ponce, Proc. IEEE Conference on Computer Vision and Pattern Recognition, (1999). © 1999 IEEE.*

motion of the eye induces a small deformation of the intersection curve, but does not change (in general) the number of tangent points. Thus, T-junctions are indeed stable.

The bitangent rays form a two-dimensional *bitangent ray manifold*[2] in the four-dimensional space formed by all straight lines. Since bitangents map onto T-junctions in the projection process, it is clear that these contour features are created or destroyed at boundaries of the manifold. Since a T-junction appears or disappears during a swallowtail transition, it is also obvious that the singular asymptotic tangents along a flecnodal curve form one of these boundaries. What is not as clear is what the remaining boundaries are made of. This is the topic of the next section.

20.1.6 Multilocal Visual Events

A pair of T-junctions appear or disappear when the line of sight crosses the boundary of the bitangent ray manifold. The corresponding change in contour structure is called a *multilocal* visual event. This section shows that there are three types of multilocal events—namely, the *tangent crossing*, *cusp crossing*, and *triple point*, besides the singularity associated with the crossing of a flecnodal curve that was mentioned in the previous section.

Let us first have a look at the *tangent crossing* event. An obvious boundary of the bitangent ray manifold is formed by the *limiting bitangents* (Figure 20.12, right), that occur when the curve formed by the intersection between the tangent plane at some point and the rest of the surface shrinks to a single point, and the plane becomes bitangent to the surface. The limiting bitangents sweep a ruled surface called the *limiting bitangent developable*. A tangent crossing occurs when the line of sight crosses this surface (Figure 20.13, top), with two separate pieces of contour becoming tangent to each other at the event before crossing transversally at two T-junctions (Fig-

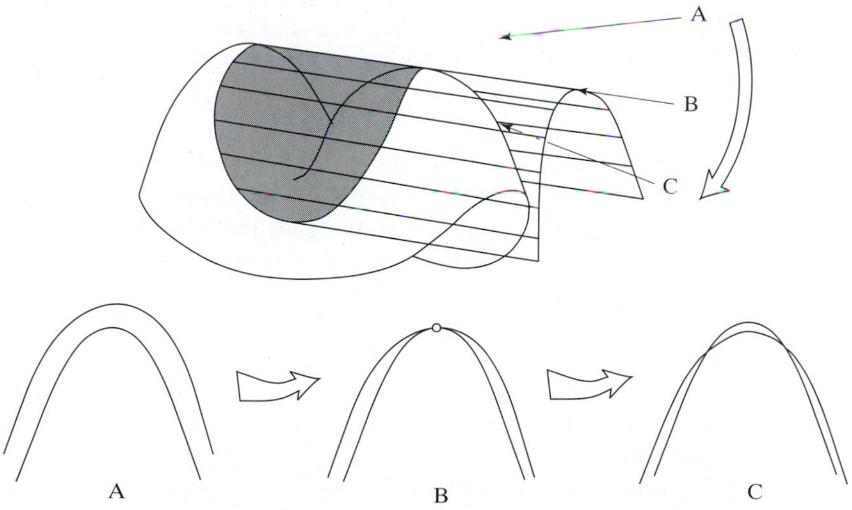

Figure 20.13 A tangent crossing event. The occlusion relationship between spatially distinct parts of the occluding contour changes when the viewpoint crosses the limiting bitangent developable surface in B. *Reprinted from "On Computing Structural Changes in Evolving Surfaces and their Appearance," by S. Pae and J. Ponce, International Journal of Computer Vision, 43(2):113–131, (2001). © 2001 Kluwer Academic Publishers.*

[2]A manifold is a topological concept generalizing surfaces defined in Euclidean space to more abstract settings; its formal definition is omitted here. It is intuitively clear that the bitangent ray manifold is two-dimensional since there is a finite number of bitangent rays for each point of the (two-dimensional) surface being observed.

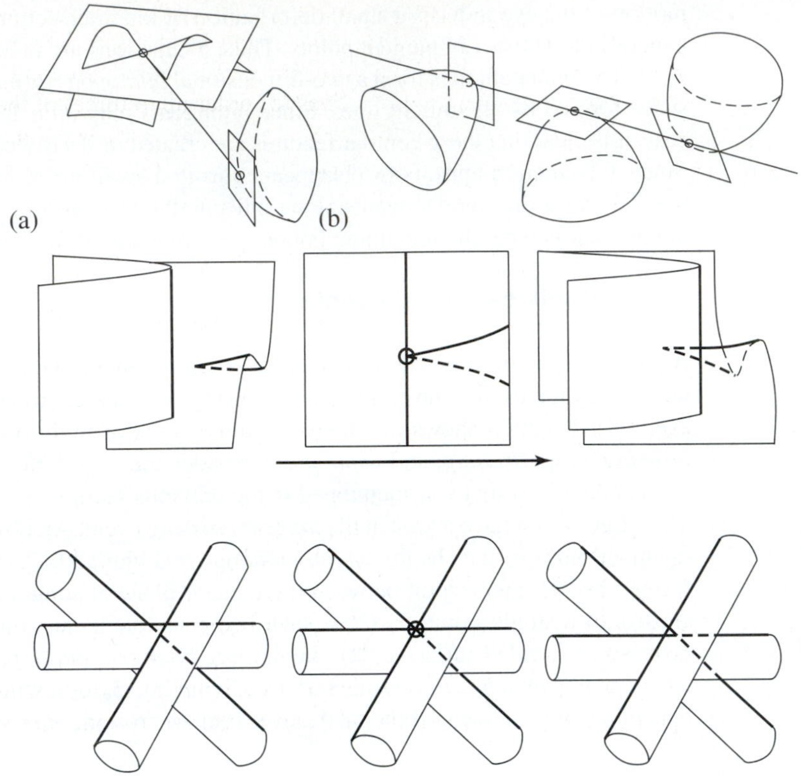

Figure 20.14 Multilocal events: (a) an asymptotic bitangent ray, (b) a tritangent ray, (c) a cusp crossing, and (d) a triple point. After Petitjean *et al.* (1992, Figure 6).

ure 20.13, bottom). For opaque objects, either a previously hidden part of the contour becomes visible after the transition or, (as in the figure) another contour arc disappears due to occlusion.

The bitangent ray manifold admits two other kinds of boundaries also associated with bitangents that touch the surface along a set of curves and sweep developable surfaces: the *asymptotic bitangents*, which intersect the surface along an asymptotic direction at one of their endpoints (Figure 20.14a), and the *tritangents*, which graze the surface in three distinct points (Figure 20.14b). The corresponding visual events occur when the line of sight crosses one of the associated developable surfaces, and they also involve the appearance or disappearance of a pair of T-junctions: a *cusp crossing* occurs when a smooth piece of the image contour crosses another part of the contour at a cusp (or end point for an opaque object) of the latter (Figure 20.14c). Two T-junctions are created (or destroyed) in the process, only one of which is visible for opaque objects. A *triple point* is formed when three separate pieces of the contour momentarily join at nonzero angles (Figure 20.14d). For transparent objects, three T-junctions merge at the singularity before separating again. For opaque objects, a contour branch and two T-junctions disappear (or appear), while another T-junction appears (or disappears).

20.2 COMPUTING THE ASPECT GRAPH

Given some model of a solid's boundary (in the form of a polyhedron, the zero set of a volumetric density, or whatever your fancy may be), the next question one may ask is how to actually construct the corresponding aspect graph. We assume in this section that a fixed Euclidean coordinate system has been chosen so that points of $\mathbb{E}^3$ can be identified with their coordinate vectors in $\mathbb{R}^3$. It is easy to rewrite the geometric definitions of parabolic points, limiting bitangent rays, and so on in terms of the derivatives (of order up to three) of the surface parameterization associated with the given model at one, two, or three surface points (Petitjean *et al.*, 1992). In each case, the surface curves associated with the ruled surfaces defined earlier can be characterized in $\mathbb{R}^{n+1}$ by a system of n equations in $n+1$ unknowns:

$$\begin{cases} P_1(x_0, x_1, \dots, x_n) = 0, \\ \dots \\ P_n(x_0, x_1, \dots, x_n) = 0, \end{cases} \tag{20.1}$$

with $1 \le n \le 8$ depending on the type of event and whether the surface is defined parametrically or implicitly. For example, in the case of a surface defined implicitly as the zero set of some density function $F(x, y, z) = 0$, we have $n = 2$ for a local event and $n = 8$ for a triple point involving three separate surface points (see Exercises).

Given these equations, a general approach to constructing the aspect graph of a given object involves the following steps: (a) tracing the visual event curves of the transparent object, (b) constructing the regions of the view sphere delineated by these curves, (c) eliminating the occluded events and merging the incident regions, and (d) constructing the corresponding aspects. The aspect graph of a transparent solid can be constructed by using the same procedure but omitting step (c).

Here we address the problem of constructing the aspect graph of a solid bounded by an *algebraic surface* (i.e., the zero set of a volumetric *polynomial* density)

$$S = \left\{ (x, y, z) \in \mathbb{R}^3 | \sum_{i+j+k \le d} a_{ijk} x^i y^j z^k = 0 \right\}.$$

Examples of algebraic surfaces include planes and quadric surfaces (i.e., ellipsoids, hyperboloids and paraboloids) as well as the zero set of higher degree polynomial densities. Most important, the constraints that define the visual events in Eq. (20.1) are all polynomials in the unknowns of interest in this case. This is the key to successfully implementing the various steps of the general algorithm outlined earlier since both numerical and symbolic computational tools are available for characterizing explicitly the solutions of multivariate polynomial equations.

20.2.1 Step 1: Tracing Visual Events

A visual event is associated with two curves: The first one, call it Γ, is either drawn on the object surface or (in the case of multilocal events) in a higher dimensional space. The second one, call it Δ, is drawn on the view sphere as the set of corresponding viewpoints. The curve Γ is defined by the n equations of Eq. (20.1) in $\mathbb{R}^{n+1}$.

This section addresses the problem of *tracing* the curve Γ (i.e., identifying its smooth arcs and its singularities, then establishing the connectivity of the various curve components). A simple solution to this problem is provided by Algorithm 20.1, which is illustrated by Figure 20.15. This algorithm takes as input the n equations in $n+1$ unknowns defining Γ and outputs a graph G whose nodes are the extremal points (including singularities) of Γ in the x_0 direction and whose arcs are the smooth curve branches joining them.

Step (1.1) requires the computation of the extrema of Γ in the x_0 direction. These points are characterized by a system of $n+1$ polynomial equations in $n+1$ unknowns obtained by adding the equation $\text{Det}(\mathcal{J}) = 0$ to Eq. (20.1), where $\mathcal{J}$ denotes the Jacobian matrix $(\partial P_i / \partial x_j)$, with $i, j = 1, \dots, n$. They can be found using *homotopy continuation* (Morgan, 1987), a numerical

Algorithm 20.1: A Curve-Tracing Algorithm

(1.1) Compute all extremal points (including singular points) of Γ in some direction, say x_0. These points form the nodes of the graph G.

(1.2) Compute all intersections of Γ with the hyperplanes orthogonal to the x_0 axis at the extremal points.

(1.3) For each interval of the x_0 axis delimited by these hyperplanes, intersect Γ and the hyperplane passing through the midpoint of the interval to obtain one sample for each real branch.

(1.4) March numerically from the sample points found in Step (1.3) to the intersection points found in Step (1.2) by predicting new points through Taylor expansion and correcting them through Newton iterations.

(1.5) Merge the smooth arcs of Γ meeting at an intersection that is not an extremal point, and add an arc to G for each pair of extremal (or singular) points joined by a curve branch.

method for computing *all* the roots of a square system of polynomial equations. Steps (1.2) and (1.3) require computing the intersections of a curve with a hyperplane. Again, these points are the solutions of polynomial equations, and they are found using homotopy continuation. The curve is actually traced (in the classical sense of the term) in step (1.4) using a Newton prediction/correction approach based on a Taylor expansion of the P_is. This involves inverting the matrix $\mathcal{J}$, which is guaranteed to be nonsingular on extrema-free intervals.

Once G has been constructed, a similar graphical description D for the curve Δ is easily obtained by mapping the points of Γ onto the corresponding asymptotic or bitangent directions.

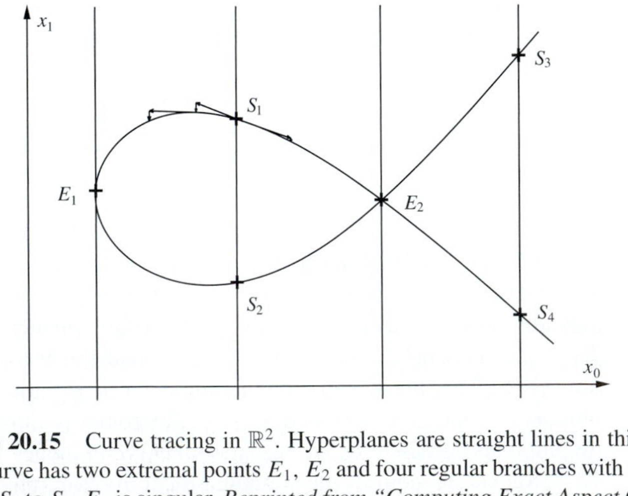

Figure 20.15 Curve tracing in $\mathbb{R}^2$. Hyperplanes are straight lines in this case. This curve has two extremal points E_1, E_2 and four regular branches with sample points S_1 to S_4. E_2 is singular. *Reprinted from "Computing Exact Aspect Graphs of Curved Objects: Algebraic Surfaces," by S. Petitjean, J. Ponce and D.J. Kriegman, International Journal of Computer Vision, 9(3):231–255, (1992).*

20.2.2 Step 2: Constructing the Regions

Let us assume now that we have constructed the graphs D_i associated with all visual event curves Δ_i ($i = 1, \ldots, p$). To construct the aspect graph regions delineated by the curves Δ_i on the view sphere, we map these curves onto the plane using spherical coordinates, for example, and refine the curve-tracing algorithm into a cell-decomposition procedure (see Algorithm 20.2 and Figure 20.16), whose output is a description of the regions, their boundary curves, and their adjacency relationships. Note that this refinement is only possible for planar curves. In practice, this algorithm takes as input the polygonal curves obtained by mapping the discrete representation of the curves Δ_i associated with the graphs D_i onto the plane, and the extremal points and intersections of the corresponding planar curves are easily found from these polygons. The visual event curves and the cells found by the algorithm are easily mapped back onto the view sphere if necessary.

Algorithm 20.2: A cell-decomposition algorithm. This algorithm takes as input a set of planar curves and outputs a description of the regions bounded by these curves and their adjacency relationships (two regions are said to be adjacent when they share a common boundary—i.e., a vertical line segment or curve branch).

(2.1) Compute all extremal points of the curves in the x_0 direction.
(2.2) Compute all the intersection points between these curves.
(2.3) Compute all intersections of the curves with the vertical lines orthogonal to the x_0 axis at the extremal and intersection points.
(2.4) For each interval of the x_0 axis delimited by these lines, do the following:
- **(2.4.1)** Intersect the curves and the vertical line passing through the midpoint of the interval to obtain a sample point on each curve branch.
- **(2.4.2)** Sort the sample points in increasing x_1 order.
- **(2.4.3)** March from these samples to the intersections found in Step (2.3).
- **(2.4.4)** Two consecutive branches within an interval of x_0 and the vertical segments joining their extremities bound a region.

(2.5) For each region, construct a sample point as the midpoint of consecutive samples found in step (2.4.1).
(2.6) Merge regions adjacent along vertical line segments into maximal regions.

20.2.3 Remaining Steps of the Algorithm

Steps (3) and (4) of the aspect-graph construction algorithm are concerned with the elimination of the occluded visual events and the construction of a sample aspect for each region. Both steps are conceptually simple. Note that all visual events of the transparent object are found in step (1) of the algorithm, and their intersections are found in step (2). For an opaque object, some of the events are occluded and should be eliminated. Since the visibility of a visual event curve only changes at its singularities and intersections with other visual events, occluded events can be eliminated through ray tracing at the sample point of each branch found in step (3). Regions adjacent to occluded events are readily merged. The final step of the algorithm involves determining the contour structure of a single view for each region. Given the sample viewpoint of a region, the curve-tracing algorithm described in Section 20.2.1 is applied to the image contour to construct an *image structure graph* whose nodes are the contour extrema and singularities (cusps

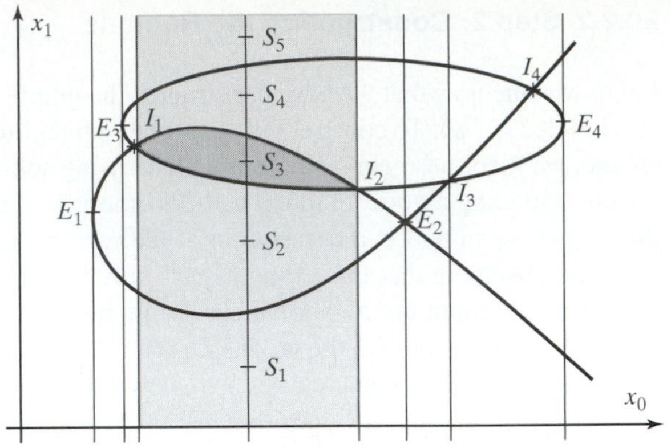

Figure 20.16 An example of cell decomposition. Two curves are shown, with their extremal points $\mathbf{E}_i$ and their intersection points I_j. The shaded rectangle delimited by I_1 and I_2 is divided into five regions with sample points S_1 to S_5. The region corresponding to S_3 is shown in a darker shade. *Reprinted from "Computing Exact Aspect Graphs of Curved Objects: Algebraic Surfaces," by S. Petitjean, J. Ponce and D.J. Kriegman, International Journal of Computer Vision, 9(3):231–255, (1992). © 1992 Kluwer Academic Publishers.*

and T-junctions) and whose arcs are the smooth branches joining them. As before, ray tracing is then used to determine whether the sample point associated with each branch is occluded.

20.2.4 An Example

Figure 20.17(top) shows two line drawings of a squash-shaped solid whose surface is defined as the zero set of a polynomial density function. The two curves running roughly parallel to each other in Figure 20.17(top left) are the parabolic curves of the squash, and they split its surface into two convex blobs separated by a saddle-shaped region. The self-intersecting curve also shown in the figure is the flecnodal curve. Figure 20.17(top right) shows the limiting bitangent developable surface associated with the squash, whose rulings are the lines joining pairs of points on the squash surface that admit the same bitangent plane. The parabolic and flecnodal curves and the limiting bitangent developable have been found using the curve-tracing algorithm presented in this section. There are no asymptotic bitangents or tritangents in this case. Figure 20.17(bottom) shows the orthographic aspect graph of the opaque squash, computed using the cell-decomposition algorithm presented earlier.

20.3 ASPECT GRAPHS AND OBJECT LOCALIZATION

Aspect graphs are intuitively appealing and mathematically elegant. In practice, however, it is probably fair to say that *exact* aspect graphs, as described in the previous sections, have not (yet) fulfilled their promise in visual tasks such as object recognition, partly because the reliable extraction of contour features such as terminations and T-junctions from real images is extremely difficult and partly because even relative simple objects may have extremely complicated aspect graphs. *Approximate* aspect graphs, on the other hand, have been successfully applied to a number of practical problems from part localization to object recognition. Specifically, we consider in this section the *bin-picking* problem addressed by Ikeuchi and Kanade (1987*b*, 1988), where a number of instances of the same object (usually a mechanical part in the automation context,

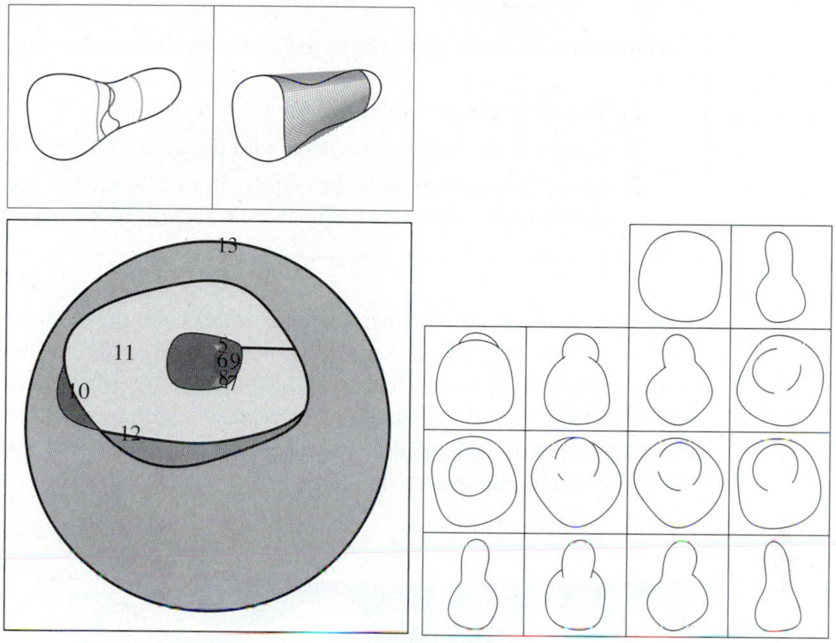

Figure 20.17 Top: A squash-shaped solid with the corresponding parabolic and flecnodal curves (left) and limiting bitangent developable (right). Bottom: Its (opaque) orthographic aspect graph with (left) the view sphere cells and (right) the corresponding aspects. Note that only 9 of the actual 14 cells are visible on the hemisphere shown here, and that some of these (e.g., region 7) are quite small.

but a plastic toy in the experiments described in this section) are piled up in random orientations waiting to be picked up by a robot, and their three-dimensional position and orientation (or *pose*) must be determined from sensory data. We suppose that polyhedral models of these parts are available, and that the data consist of *needle maps* constructed from overhead images using the photometric stereo techniques introduced in chapter 5.

Let us focus on the part located at the top of the bin and assume that it is observed from afar (orthographic projection) without partial occlusion due to other objects in the pile. This is reasonable in bin-picking tasks where the size of the parts is normally small compared with the height of the overhead camera observing them. Under these assumptions, the visibility of any point on the surface of the top part can be determined from the viewing direction (defined here by the orientation of the vertical relative to the local coordinate frame attached to the object) alone. The corresponding aspect is defined as the list of visible faces ordered in descending surface area. For a polyhedral part, it is easy to identify the aspect associated with a given object orientation using simple hidden-surface elimination techniques such as z-buffering. Objects such as the plastic toy used in our example, which are bounded by piecewise-smooth surfaces and approximated by polyhedral meshes, do not pose particular problems either. Their (curved) faces are deemed visible when any one of the planar facets approximating them is itself visible. In this setting, an *approximate* aspect graph is easily constructed using Algorithm 20.3.

This algorithm assumes that visibility does not change within the range of viewpoints associated with each cell of the discretized viewing sphere, and thus only constructs an approximation of the true aspect graph. Figure 20.18 illustrates the construction of the aspect graph of the plastic toy. At this point, it is possible to introduce constraints associated with the data-acquisition process. In particular, a face can only be found by photometric stereo methods when the angle

Algorithm 20.3: An Approximate Aspect-Graph Construction Algorithm.

1. Tessellate the unit sphere of viewing directions.
2. Compute the aspect associated with the center of each cell in the tessellation.
3. Group adjacent aspects into equivalence classes labeled by binary strings where a one indicates that a face is visible and a zero indicates it is not.

between its normal and the line of sight is small enough: In the example shown in Figure 20.18, none of the visible faces in Aspect 7 is actually detectable according to this criterion, yielding an all-zero label. Although the front "face" of the toy (numbered 4 in Figure 20.18) should be detectable in Aspect 6, it actually corresponds to an unmodeled recess in its surface, and no representative view is generated for this aspect either. A representative view is computed for each one of the remaining aspects by finding the pose that maximizes the area of the object's picture

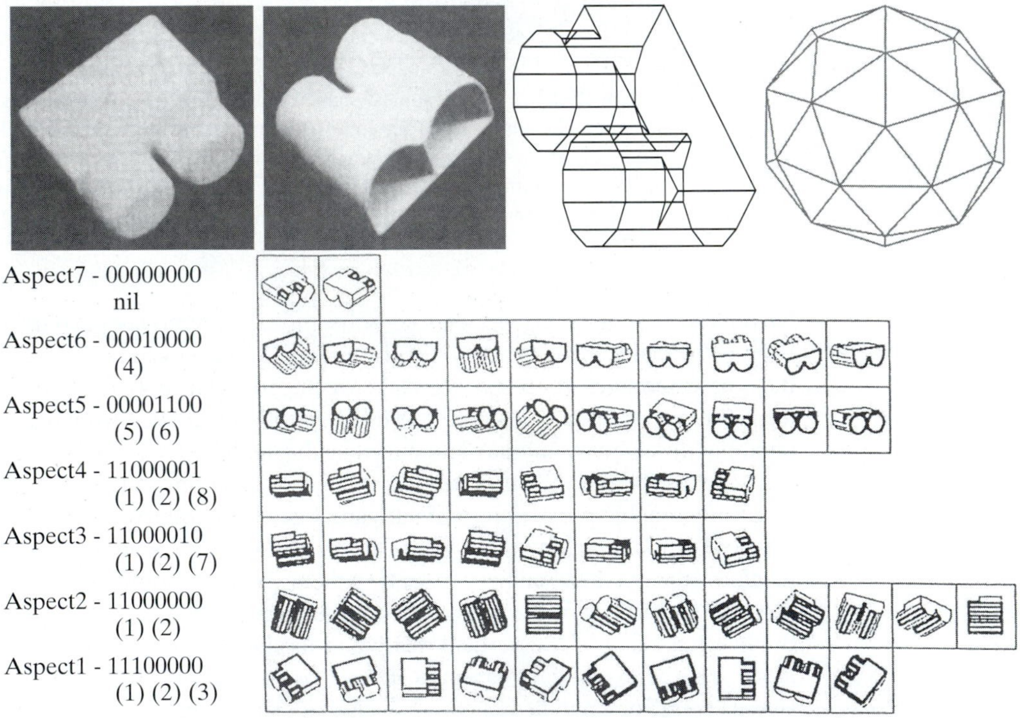

Figure 20.18 Approximate aspect graphs. Top, from left to right: Two photographs of a plastic toy, its polyhedral model (note that each cylindrical surface is approximated by several planar faces, but it is considered as a single object face during the aspect graph construction) and a semi-regular tessellation of the unit sphere with 60 triangular faces. *Reprinted from "Automatic Generation of Object Recognition Programs," by K. Ikeuchi and T. Kanade, Proceedings of the IEEE, 76(8):1016–1035, (1988).* © *1988 IEEE.* Bottom: The corresponding seven aspects. *Reprinted from "Precompiling a Geometrical Model into an Interpretation Tree for Object Recognition in Bin-Picking Tasks," by K. Ikeuchi, Proc. DARPA Image Understanding Workshop, 1987.*

over the corresponding range of viewpoints and aligning its axis of maximum inertia with the horizontal image direction.

To facilitate the localization process, it is convenient to organize the aspects into a decision tree called the *interpretation tree*. Each node in this tree is associated with a range of viewpoints, the list of the corresponding aspects, and an object face. The tree is constructed iteratively: At each stage, the face F associated with a node N is used to split the corresponding viewpoints into a subset V' where F is visible and a subset V'' where it is not. The left child of N is assigned the range of viewpoints V', the corresponding subset of aspects, and the next largest face visible for some viewpoint in V'. Its right child is constructed in a similar manner from V'' and the corresponding subset of aspects. When two faces have the same area, they are used together in the subdivision process, and V' becomes the set of views where one of them may be visible, whereas V'' is the set of views where neither of them is visible. The tree construction starts with a root node associated with the largest face of the object, the whole viewing sphere, and the list of all possible aspects.

The raw interpretation tree can be refined by adding to it a set of classification and pose-determination rules. These rules are constructed by hand from geometric and topological features associated with each face in the geometric model, including the direction of its axes of inertia and the corresponding moments; a classification of its overall shape as planar, cylindrical, elliptic or hyperbolic; its *extended Gaussian image* (*EGI*; i.e., a histogram of the surface normals within the face, computed over a discretized Gaussian sphere); contour information extracted from one of the input images by an edge detector; and adjacency relationships between the face and its neighbors. A typical rule may use a comparison between, say, the moments of an observed face and the moments available from the model to decide which branch to follow during an interpretation task. Pose-determination rules use the geometric information associated with matched object faces to compute the object pose. For example, the viewing direction can be calculated by aligning the predicted center of mass of a face's EGI with the observed one, and the object's orientation in the plane perpendicular to the line of sight can be recovered from the predicted and observed axes of inertia of that face.

Figure 20.19 shows the results of a localization experiment. There are three sources of sensory data in this case: The main one is a *needle map* extracted from several images taken from the same viewpoint under different illumination patterns. A rough *depth map* is also extracted by matching the needle maps associated with two cameras (*dual photometric stereo*), and a *contour map* is finally found by running an edge detector in one of the input images. The

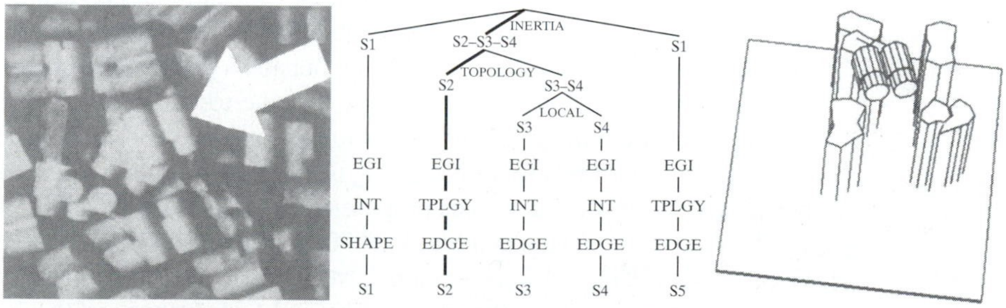

Figure 20.19 Left: a sample bin image; the top instance of the toy is indicated by the arrow. Middle: the path (in bold lines) followed by applying the interpretation tree to the corresponding image region. Right: the toy has been found by the localization program and is shown in its estimated pose. *Reprinted from "Precompiling a Geometrical Model into an Interpretation Tree for Object Recognition in Bin-Picking Tasks," by K. Ikeuchi, Proc. DARPA Image Understanding Workshop, 1987.*

three maps are registered in the same coordinate system, and the top region (Figure 20.19, left) is identified from the depth map and passed to the interpretation tree. The localization proceeds as a tree traversal, the test associated with each node being used to select the correct branches and successively identify the observed aspect (Figure 20.19, middle), the corresponding viewing direction, and finally the rotation about the line of sight that aligns the predicted and observed aspects (Figure 20.19, right).

20.4 NOTES

The material in this chapter is largely based on the work of Koenderink and Van Doorn, including the seminal papers that introduced the idea of an aspect graph, albeit under a different name (Koenderink and Van Doorn, 1976*b*, 1979), the article that presents in a very accessible manner the geometric foundations of this shape representation (Koenderink, 1986, see also Platonova, 1981, Kergosien, 1981), and, of course, the somewhat more demanding (but so rewarding for the patient student) book by Koenderink (1990). See also Callahan and Weiss (1985) for an excellent informal introduction to aspect graphs. Genericity, singularity and catastrophe theories are discussed in many books, including Whitney (1955), Arnol'd (1984), and Demazure (2000). See also Koenderink (1990) for a discussion of why chairs wobble and Thom (1972) for an in-depth discussion of this argument. The algorithm presented in Section 20.2 is due to Petitjean *et al.* (1992), and the material in that section is largely based on the corresponding article, which relies heavily on the numerical *homotopy continuation* method proposed by Morgan (1987) for finding all the roots (including the complex ones as well as those at infinity) of a square system of multivariate polynomial equations. Symbolic methods such as multivariate resultants (Macaulay, 1916, Collins, 1971, Canny, 1988, Manocha, 1992) and cylindrical algebraic decomposition (Collins, 1975, Arnon *et al.*, 1984) exist as well, and they have been used by Rieger (1987, 1990, 1992) in a different algorithm for constructing the aspect graph of an algebraic surface.

We have focused on orthographic aspect graphs in this chapter, but the singular tangents and bitangents discussed in Section 20.1 also form visual event boundaries under perspective projection, and the corresponding ruled surfaces carve the three-dimensional view space into cells that form the nodes of the perspective aspect graph. Strictly speaking, the imaging model used in perspective aspect graphs is only valid for cameras with a $360° \times 360°$ field of view, such as the omnidirectional spherical cameras discussed in the introduction.[3] For observers lying outside the convex hull of an object, a more conventional perspective projection model can be used with the orientation of the retina chosen so the object lies in front of the pinhole. This is always possible because (a) a point lying outside a convex object can always be separated from it by a plane, and (b) it is easy to show that the topological structure of the image contour is independent of the image plane orientation as long as the object remains in front of the pinhole.

Many algorithms for constructing the aspect graph of a polyhedron under orthographic or perspective projection have been proposed (see, e.g., Castore, 1984, Stewman and Bowyer, 1987, 1988, Watts, 1987, Gigus and Malik, 1990, Plantinga and Dyer, 1990, Wang and Freeman, 1990, Gigus *et al.*, 1991), and some of them have been implemented. There are only two kinds of visual event curves in this case—the so-called *EV* and *EEE* events, and they occur when the line of sight touches a polyhedron at both an edge and a vertex or grazes its surface at three different edge points. Algorithms for constructing the exact aspect graph of simple curved objects such as solids bounded by quadric surfaces (Chen and Freeman, 1991) and solids of revolution (Eggert and Bowyer, 1989, 1991, Kriegman and Ponce, 1990*a*) have also been proposed and implemented.

[3] Of course, true omnidirectional cameras are unusual, to say the least, but commercial models based on the catadioptric omnidirectional camera technology developed by Nayar (1997) are commercially available.

The algorithms for solids bounded by algebraic surfaces that we presented in this chapter and the approaches based on cylindrical algebraic decomposition (Rieger, 1987, 1990, 1992) are probably the most general to date, but they cannot handle typical CAD models comprised of hundreds of bicubic patches.

Aspect graphs are unfortunately very large: a polyhedron with n faces has an orthographic aspect graph with $O(n^6)$ cells (Gigus *et al.*, 1991). The size increases to $O(n^6 d^{12})$ for a piecewise-smooth surface made of n polynomial patches of degree d (Petitjean, 1995). The situation is, of course, even worse in the perspective case. Should we blame the huge size of aspect graphs on the representation itself? Or is it an artifact of the (combinatorial and/or algebraic) complexity of the underlying surface model (e.g., the number of polyhedral faces or the degree of the patches used to approximate relatively simple free-form surfaces)? Noble *et al.* (1997) address the problem of constructing the aspect graph of a solid defined by the zero set of a volumetric density function (e.g., the boundary of an organ in a CT image) and present preliminary results. Another part of the size problem lies in modeling appropriately the optical blur introduced by real lenses together with the finite spatial resolution of cameras. Preliminary efforts at tackling this question can be found in Eggert *et al.* (1993), Shimshoni and Ponce (1997), and Pae and Ponce (2001), partly based on recent results about the effect of one-parameter families of deformations on generic surfaces (Bruce *et al.*, 1996*a*, 1996*b*).

Approximate aspect graphs of polyhedra have been successfully used in object localization tasks. Section 20.3 presented the approach proposed by Ikeuchi and Kanade (1987*b*, 1988). Variants include Chakravarty (1982) and Hebert and Kanade (1985). The extended Gaussian image was introduced by Horn (1984), and a method for dual photometric stereo is described in Ikeuchi (1987*a*).

Descriptions of the ray-tracing and z-buffer algorithms for hidden-surface elimination mentioned in this chapter can be found in classical computer graphics texts (e.g., Foley *et al.*, 1990).

PROBLEMS

20.1. Draw the orthographic and spherical perspective aspect graphs of the transparent Flatland object below along with the corresponding aspects.

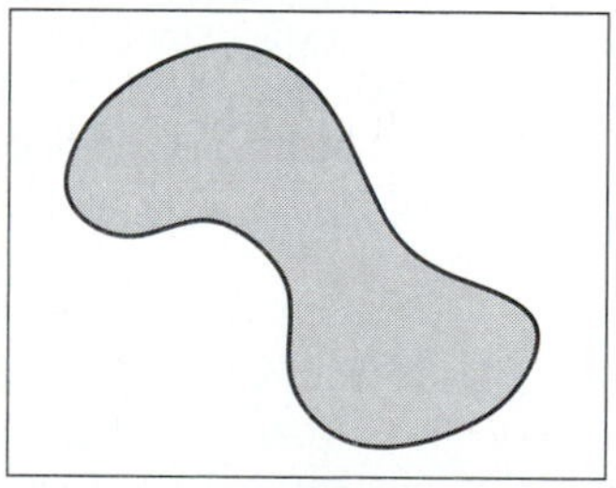

20.2. Draw the orthographic and spherical perspective aspect graphs of the opaque object along with the corresponding aspects.

20.3. Is it possible for an object with a single parabolic curve (such as a banana) to have no cusp of Gauss at all? Why (or why not)?

20.4. Use an equation-counting argument to justify the fact that contact of order six or greater between lines and surfaces does not occur for generic surfaces. (Hint: Count the parameters that define contact.)

20.5. We saw that an asymptotic curve and its spherical image have perpendicular tangents. Lines of curvature are the integral curves of the field of principal directions. Show that these curves and their Gaussian image have parallel tangents.

20.6. Use the fact that the Gaussian image of a parabolic curve is the envelope of the asymptotic curves intersecting it to give an alternate proof that a pair of cusps is created (or destroyed) in a lip or beak-to-beak event.

20.7. Lip and beak-to-beak events of implicit surfaces. It can be shown (Pae and Ponce, 2001) that the parabolic curves of a surface defined implicitly as the zero set of some density function $F(x, y, z) = 0$ are characterized by this equation and $\nabla F^T \mathcal{A} \nabla F = 0$, where ∇F is the gradient of F and $\mathcal{A}$ is the symmetric matrix

$$\mathcal{A} \stackrel{\text{def}}{=} \begin{pmatrix} F_{yy}F_{zz} - F_{yz}^2 & F_{xz}F_{yz} - F_{zz}F_{xy} & F_{xy}F_{yz} - F_{yy}F_{xz} \\ F_{xz}F_{yz} - F_{zz}F_{xy} & F_{zz}F_{xx} - F_{xz}^2 & F_{xy}F_{xz} - F_{xx}F_{yz} \\ F_{xy}F_{yz} - F_{yy}F_{xz} & F_{xy}F_{xz} - F_{xx}F_{yz} & F_{xx}F_{yy} - F_{xy}^2 \end{pmatrix}.$$

It can also be shown that the asymptotic direction at a parabolic point is $\mathcal{A}\nabla F$.

(a) Show that $\mathcal{A}\mathcal{H} = \text{Det}(\mathcal{H})\text{Id}$, where $\mathcal{H}$ denotes the Hessian of F.

(b) Show that cusps of Gauss are parabolic points that satisfy the equation $\nabla P^T \mathcal{A} \nabla F = 0$. Hint: Use the fact that the asymptotic direction at a cusp of Gauss is tangent to the parabolic curve, and that the vector ∇F is normal to the tangent plane of the surface defined by $F = 0$.

(c) Sketch an algorithm for tracing the lip and beak-to-beak events of an implicit surface.

20.8. Swallowtail events of implicit surfaces. It can be shown that the asymptotic directions $\boldsymbol{a}$ at a hyperbolic point satisfy the two equations $\nabla F \cdot \boldsymbol{a} = 0$ and $\boldsymbol{a}^T \mathcal{H} \boldsymbol{a} = 0$, where $\mathcal{H}$ denotes the Hessian of F. These two equations simply indicate that the order of contact between a surface and its asymptotic tangents is at least equal to three. Asymptotic tangents along flecnodal curves have order-four contact with the surface, and this is characterized by a third equation, namely

$$\begin{pmatrix} \boldsymbol{a}^T \mathcal{H}_x \boldsymbol{a} \\ \boldsymbol{a}^T \mathcal{H}_y \boldsymbol{a} \\ \boldsymbol{a}^T \mathcal{H}_z \boldsymbol{a} \end{pmatrix} \cdot \boldsymbol{a} = 0.$$

Sketch an algorithm for tracing the swallowtail events of an implicit surface.

20.9. Derive the equations characterizing the multilocal events of implicit surfaces. You can use the fact that, as mentioned in the previous exercise, the asymptotic directions $\boldsymbol{a}$ at a hyperbolic point satisfy the two equations $\nabla F \cdot \boldsymbol{a} = 0$ and $\boldsymbol{a}^T \mathcal{H} \boldsymbol{a} = 0$.

Programming Assignments

20.10. Write a program to explore multilocal visual events: Consider two spheres with different radii and assume orthographic projection. The program should allow you to change viewpoint interactively as well as explore the tangent crossings associated with the limiting bitangent developable.

20.11. Write a similar program to explore cusp points and their projections. You have to trace a plane curve.

21

Range Data

This chapter discusses *range images* (or *depth maps*) that store, instead of brightness or color information, the depth at which the ray associated with each pixel first intersects the scene observed by a camera. In a sense, a range image is exactly the desired output of stereo, motion, or other *shape-from-X* vision modules. In this chapter, however, we focus our attention on range images acquired by *active sensors* that project some sort of light pattern on the scene, using it to avoid the difficult and costly problem of establishing correspondences and construct dense and accurate depth pictures. After a brief review of range-sensing technology, we discuss image segmentation, multiple-image registration, three-dimensional model construction, and object recognition, focusing on the aspects of these problems that are specific to the range data domain.

21.1 ACTIVE RANGE SENSORS

Triangulation-based active range finders date back to the early 1970s (e.g., Agin, 1972, Shirai, 1972). They function along the same principles as passive stereo vision systems, one of the cameras being replaced by a source of controlled illumination (*structured light*) that avoids the correspondence problem mentioned in chapter 11. For example, a laser and a pair of rotating mirrors may be used to sequentially scan a surface. In this case, as in conventional stereo, the position of the bright spot where the laser beam strikes the surface of interest is found as the intersection of the beam with the projection ray joining the spot to its image. Contrary to the stereo case, however, the laser spot can normally be identified without difficulty since it is in general much brighter than the other scene points (especially when a filter tuned to the laser wavelength is placed in front of the camera), altogether avoiding the correspondence problem. Alternatively,

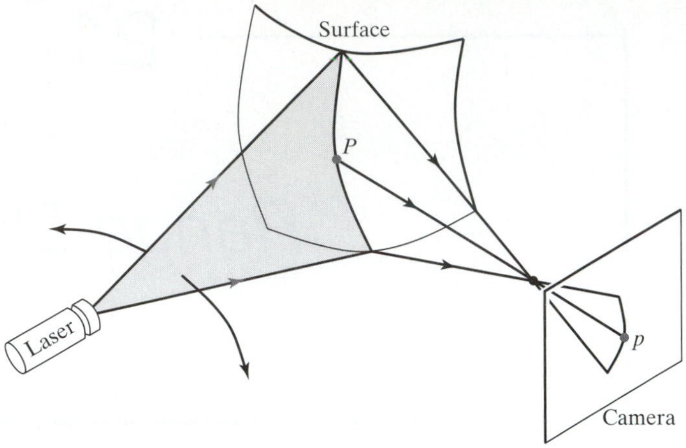

Figure 21.1 A range sensor using a plane of light to scan the surface of an object.

the laser beam can be transformed by a cylindrical lens into a plane of light (Figure 21.1). This simplifies the mechanical design of the range finder since it only requires one rotating mirror. More important, perhaps, it shortens the time required to acquire a range image since a laser stripe—the equivalent of a whole image column—can be acquired at each frame. It should be noted that this setup does not introduce matching ambiguities since the laser spot associated with an image pixel can be retrieved as the (unique) intersection of the corresponding projection ray with the plane of light.

Variants of these two techniques include using multiple cameras to improve measurement accuracy and exploiting (possibly time-coded) two-dimensional light patterns to improve data-acquisition speed. The main drawbacks of the active triangulation technology are relatively low speed, missing data at points where the laser spot is hidden from the camera by the object, and missing or erroneous data due to specularities. The latter difficulty is actually common to all active ranging techniques: A purely specular surface does not reflect any light in the direction of the camera unless it happens to lie in the corresponding mirror direction. Worse, the reflected beam may induce secondary reflections giving false depth measurements. Additional difficulties include keeping the laser stripe in focus during the entire scanning procedure and the loss of accuracy inherent in all triangulation techniques as depth increases (see the exercises in chapter 11; intuitively this is due to the fact that depth is inversely proportional to disparity). Several triangulation-based scanners are commercially available today. Figure 21.2 shows an example obtained using the Minolta VIVID range finder, which can acquire a 200×200 range image together with a registered 400×400 color image in 0.6 s within an operating range of 0.6 to 2.5 m.

The second main approach to active ranging involves a signal transmitter, a receiver, and electronics for measuring the time of flight of the signal during its round trip from the range sensor to the surface of interest. Time-of-flight range finders are normally equipped with a scanning mechanism, and the transmitter and receiver are often coaxial, eliminating the problem of missing data common in triangulation approaches. There are three main classes of time-of-flight laser range sensors: *pulse time delay* technology directly measures the time of flight of a laser pulse; *AM phase-shift* range finders measure the phase difference between the beam emitted by an amplitude-modulated laser and the reflected beam, a quantity proportional to the time of flight; and *FM beat* sensors measure the frequency shift (or *beat frequency*) between a frequency-modulated laser beam and its reflection, another quantity proportional to the round-trip flight

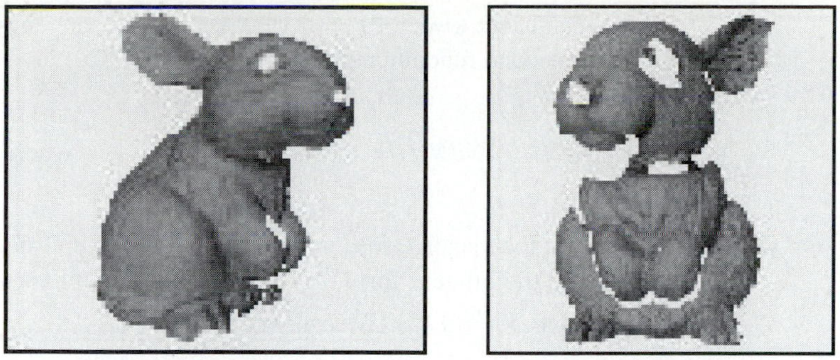

Figure 21.2 Two range images captured by the Minolta VIVID scanner. As in several other figures in this chapter, the range data are displayed as a shaded mesh of $(x, y, z(x, y))$ points viewed in perspective. *Courtesy of D. Huber and M. Hebert.*

time. Compared to triangulation-based systems, time-of-flight sensors offer a greater operating range (up to tens of meters), which is valuable in outdoor robotic navigation tasks.

New technologies continue to emerge, including range sensors equipped with acoustico-optical scanning systems and capable of extremely high image-acquisition rates, and range cameras that eliminate scanning altogether, using instead a large array of receivers to analyze a laser pulse covering the entire field of view.

21.2 RANGE DATA SEGMENTATION

This section adapts some of the edge detection and segmentation methods introduced in chapters 8 and 14 to the specific case of range images. As shown in the rest of this section, the fact that surface geometry is readily available greatly simplifies the segmentation process because this provides objective, physically meaningful criteria for finding surface discontinuities and merging contiguous patches with a similar shape. Let us start by introducing some elementary notions of *analytical* differential geometry, which form the basis for the approach to edge detection in range images discussed in this section.

21.2.1 Elements of Analytical Differential Geometry

Here we revisit the notions of differential geometry introduced in chapter 19 in an analytical setting. Specifically, we assume that $\mathbb{E}^3$ has been equipped with a fixed coordinate system and identify this space with $\mathbb{R}^3$ and each point with its coordinate vector. We consider a *parametric surface* defined as the smooth (i.e., indefinitely differentiable) mapping $\boldsymbol{x} : U \subset \mathbb{R}^2 \rightarrow \mathbb{R}^3$ that associates with any couple (u, v) in the open subset U of $\mathbb{R}^2$ the point $\boldsymbol{x}(u, v)$ in $\mathbb{R}^3$. To ensure that the tangent plane is well defined everywhere, we assume that the partial derivatives $\boldsymbol{x}_u \stackrel{\text{def}}{=} \partial\boldsymbol{x}/\partial u$ and $\boldsymbol{x}_v \stackrel{\text{def}}{=} \partial\boldsymbol{x}/\partial v$ are linearly independent. Indeed, let $\boldsymbol{\alpha} : I \subset \mathbb{R} \rightarrow U$ denote a smooth planar curve, with $\boldsymbol{\alpha}(t) = (u(t), v(t))$, then $\boldsymbol{\beta} \stackrel{\text{def}}{=} \boldsymbol{x} \circ \boldsymbol{\alpha}$ is a parameterized space curve lying on the surface. According to the chain rule, a tangent vector to $\boldsymbol{\beta}$ at the point $\boldsymbol{\beta}(t)$ is $u'(t)\boldsymbol{x}_u + v'(t)\boldsymbol{x}_v$, and it follows that the plane tangent to the surface in $\boldsymbol{x}(u, v)$ is parallel to the vector plane spanned by the vectors $\boldsymbol{x}_u$ and $\boldsymbol{x}_v$. The (unit) surface normal is thus

$$\boldsymbol{N} = \frac{1}{|\boldsymbol{x}_u \times \boldsymbol{x}_v|}(\boldsymbol{x}_u \times \boldsymbol{x}_v).$$

Let us consider a vector $\boldsymbol{t} = u'\boldsymbol{x}_u + v'\boldsymbol{x}_v$ in the tangent plane at the point $\boldsymbol{x}$. It is easy to show that the second fundamental form is given by[1]

$$\mathrm{II}(\boldsymbol{t}, \boldsymbol{t}) = \boldsymbol{t} \cdot d\boldsymbol{N}(\boldsymbol{t}) = eu'^2 + 2fu'v' + gv'^2, \quad \text{where} \quad \begin{cases} e = -\boldsymbol{N} \cdot \boldsymbol{x}_{uu}, \\ f = -\boldsymbol{N} \cdot \boldsymbol{x}_{uv}, \\ g = -\boldsymbol{N} \cdot \boldsymbol{x}_{vv}. \end{cases}$$

Note that the vector $\boldsymbol{t}$ does not (in general) have unit norm. Let us define the *first fundamental form* as the bilinear form that associates with two vectors in the tangent plane their dot product (i.e., $\mathrm{I}(\boldsymbol{u}, \boldsymbol{v}) \stackrel{\text{def}}{=} \boldsymbol{u} \cdot \boldsymbol{v}$). We can write

$$\mathrm{I}(\boldsymbol{t}, \boldsymbol{t}) = |\boldsymbol{t}|^2 = Eu'^2 + 2Du'v' + Gv'^2, \quad \text{where} \quad \begin{cases} E = \boldsymbol{x}_u \cdot \boldsymbol{x}_u, \\ F = \boldsymbol{x}_u \cdot \boldsymbol{x}_v, \\ G = \boldsymbol{x}_v \cdot \boldsymbol{x}_v, \end{cases}$$

and it follows immediately that the normal curvature in the direction $\boldsymbol{t}$ is given by

$$\kappa_t = \frac{\mathrm{II}(\boldsymbol{t}, \boldsymbol{t})}{\mathrm{I}(\boldsymbol{t}, \boldsymbol{t})} = \frac{eu'^2 + 2fu'v' + gv'^2}{Eu'^2 + 2Du'v' + Gv'^2}.$$

Likewise, it is easily shown that the matrix associated with the differential of the Gauss map *in the basis* $(\boldsymbol{x}_u, \boldsymbol{x}_v)$ of the tangent plane is

$$d\boldsymbol{N}(\boldsymbol{t}) = \begin{pmatrix} e & f \\ f & g \end{pmatrix} \begin{pmatrix} E & F \\ F & G \end{pmatrix}^{-1}.$$

Thus, since the Gaussian curvature is equal to the determinant of the operator $d\boldsymbol{N}$, it is given by

$$K = \frac{eg - f^2}{EG - F^2}.$$

Asymptotic and principal directions are also easily found by using this parameterization: Since an asymptotic direction verifies $\mathrm{II}(\boldsymbol{t}, \boldsymbol{t}) = 0$, the corresponding values of u' and v' are the (homogeneous) solutions of $eu'^2 + 2fu'v' + gv'^2 = 0$. The principal directions can be shown to verify

$$\begin{vmatrix} v'^2 & -u'v' & u'^2 \\ E & F & G \\ e & f & g \end{vmatrix} = 0. \tag{21.1}$$

Example 21.1 Monge patches.

An important example of parametric surface is provided by *Monge patches*: consider the surface $\boldsymbol{x}(u, v) = (u, v, h(u, v))$. In this case, we have

$$\begin{cases} \boldsymbol{N} = \dfrac{1}{(1 + h_u^2 + h_v^2)^{1/2}}(-h_u, -h_v, 1)^T, \\ E = 1 + h_u^2,\ F = h_u h_v,\ G = 1 + h_v^2, \\ e = -\dfrac{h_{uu}}{(1 + h_u^2 + h_v^2)^{1/2}},\ f = -\dfrac{h_{uv}}{(1 + h_u^2 + h_v^2)^{1/2}},\ g = -\dfrac{h_{vv}}{(1 + h_u^2 + h_v^2)^{1/2}}, \end{cases}$$

[1]This definition is in keeping with the orientation conventions defined in chapter 19. The coefficients e, f, g are often defined with opposite signs (e.g., do Carmo, 1976, Struik, 1988).

and the Gaussian curvature has a simple form:

$$K = \frac{h_{uu}h_{vv} - h_{uv}^2}{(1 + h_u^2 + h_v^2)^2}.$$

Example 21.2 Local surface parameterization.

Another fundamental example is provided by the local parameterization of a surface in the coordinate system formed by its principal directions. This is a special case of a Monge patch. Writing that the origin of the coordinate system lies in the tangent plane immediately yields $h(0,0) = h_u(0,0) = h_v(0,0) = 0$. As expected, the normal is simply $\boldsymbol{N} = (0, 0, 1)^T$ at the origin, and the first fundamental form is the identity there. As shown in the exercises, it follows easily from Eq. (21.1) that a necessary and sufficient condition for the coordinate curves of a parameterized surface to be principal directions is that $f = F = 0$ (e.g., this implies that the lines of curvature of a surface of revolution are its meridians and parallels). In our context, we already know that $F = 0$, and this condition reduces to $h_{uv}(0,0) = 0$. The principal curvatures in this case are simply $\kappa_1 = e/E = -h_{uu}(0,0)$ and $\kappa_2 = g/G = -h_{vv}(0,0)$. In particular, we can write a Taylor expansion of the height function in the neighborhood of $(0, 0)$ as

$$h(u,v) = h(0,0) + (h_u, h_v)\begin{pmatrix} u \\ v \end{pmatrix} + \frac{1}{2}(u,v)\begin{pmatrix} h_{uu} & h_{uv} \\ h_{uv} & h_{vv} \end{pmatrix}\begin{pmatrix} u \\ v \end{pmatrix} + \varepsilon(u^2 + v^2)^{3/2},$$

where the argument $(0, 0)$ for the derivatives of h has been omitted for conciseness. This shows that the best second-order approximation to the surface in this neighborhood is the paraboloid defined by

$$h(u,v) = -\frac{1}{2}(\kappa_1 u^2 + \kappa_2 v^2)$$

(i.e., the expression already encountered in chapter 19).

21.2.2 Finding Step and Roof Edges in Range Images

This section presents a method for finding various types of edges in range images (Ponce and Brady, 1987). This technique combines tools from analytical differential geometry and scale-space image analysis to detect and locate depth and orientation discontinuities in range data.

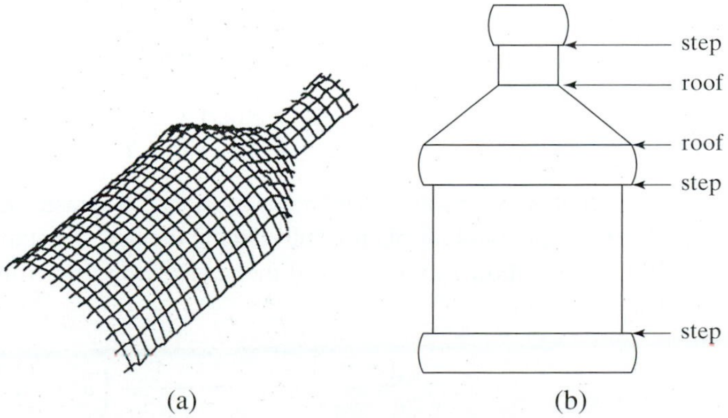

Figure 21.3 An oil bottle: (a) a range image of the bottle (the background has been thresholded away), and (b) a sketch of its depth and orientation discontinuities. This 128 × 128 picture was acquired using the INRIA range finder (Boissonnat and Germain, 1981), with a depth accuracy of about 0.5 mm.

Figure 21.3 shows a range image of a bottle of motor oil that serves to illustrate the concepts introduced in this section.

The surface of the oil bottle can be modeled as a Monge patch $z(x, y)$ in the coordinate system attached to the sensor, and it presents two types of discontinuities: *steps*, where the actual depth is discontinuous; and *roofs*, where the depth is continuous but the orientation changes abruptly. As shown in the next section, it is possible to characterize the behavior of analytical models of step and roof edges under Gaussian smoothing and to show that, respectively, they give rise to parabolic points and extrema of the dominant principal curvature in the corresponding principal direction. This is the basis for the multiscale edge-detection scheme outlined in Algorithm 21.1.

Algorithm 21.1: The model-based edge-detection algorithm of Ponce and Brady (1987).

1. Smooth the range image with Gaussian distributions at a set of scales σ_i ($i = 1, \dots, 4$). Compute the principal directions and curvatures at each point of the smoothed images $z_{\sigma_i}(x, y)$.
2. Mark in each smoothed image $z_{\sigma_i}(x, y)$ the zero crossings of the Gaussian curvature and the extrema of the dominant principal curvature in the corresponding principal direction.
3. Use the analytical step and roof models to match the features found across scales and output the points lying on these surface discontinuities.

Edge Models In the neighborhood of a discontinuity, the shape of a surface changes much faster in the direction of the discontinuity than in the orthogonal direction. Accordingly, we assume in the rest of this section that the direction of the discontinuity is one of the principal directions, with the corresponding (dominant) principal curvature changing rapidly in this direction, while the other one remains roughly equal to zero. This allows us to limit our attention to *cylindrical* models of surface discontinuities (i.e., models of the form $z(x, y) = h(x)$). These models are only intended to be valid in the neighborhood of an edge, with the direction of the $x - z$ plane being aligned with the corresponding dominant principal direction.

In particular, a step edge can be modeled by two sloped half-planes separated by a vertical gap, with normals in the $x - z$ plane. This model is cylindrical, and it is sufficient to study its univariate formulation (Figure 21.4, left), whose equation is

$$z = \begin{cases} k_1 x + c & \text{when} \quad x < 0, \\ k_2 x + c + h & \text{when} \quad x > 0. \end{cases} \tag{21.2}$$

In this expression, c and h are constants, h measuring the size of the gap and k_1 and k_2 the slopes of the two half-planes. Introducing the new constants $k = (k_1 + k_2)/2$ and $\delta = k_2 - k_1$, it is easy to show (see Exercises) that convolving the z function with the second derivative of a Gaussian yields

$$z''_\sigma \stackrel{\text{def}}{=} \frac{\partial^2}{\partial \sigma^2} G_\sigma * z = \frac{1}{\sigma\sqrt{2\pi}} \left(\delta - \frac{hx}{\sigma^2}\right) \exp\left(-\frac{x^2}{2\sigma^2}\right). \tag{21.3}$$

As shown in the exercises of chapter 19, the curvature of a twisted parametric curve is $\kappa = |\boldsymbol{x}' \times \boldsymbol{x}''|/|\boldsymbol{x}'|^3$. In the case of plane curves, the curvature can be given a meaningful sign, and this formula becomes $\kappa = (\boldsymbol{x}' \times \boldsymbol{x}'')/|\boldsymbol{x}'|^3$, where "×" denotes this time the operator associating

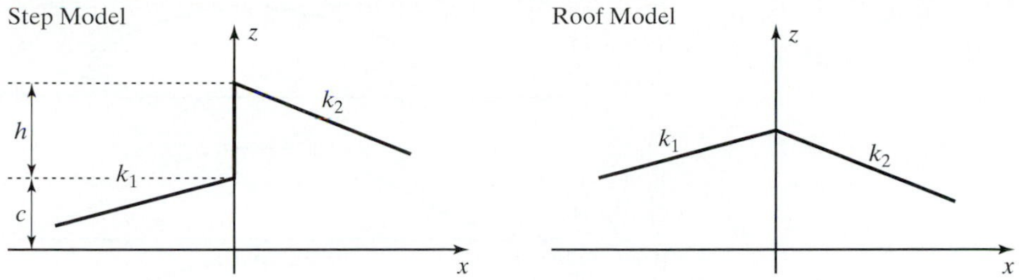

Figure 21.4 Edge models: A step consists of two half-planes separated by a distance h at the origin, and a roof consists of two half-planes meeting at the origin with different slopes. After Ponce and Brady (1987, Figure 4).

with two vectors in $\mathbb{R}^2$ the determinant of their coordinates. It follows that the corresponding curvature κ_σ vanishes in $x_\sigma = \sigma^2\delta/h$. This point is only at the origin when $k_1 = k_2$, and its position is a quadratic function of σ otherwise. This suggests identifying step edges with zero crossings of one of the principal curvatures (or equivalently of the Gaussian curvature), whose position changes with scale. To characterize qualitatively the behavior of these features as a function of σ, let us also note that, since $z''_\sigma = 0$ in x_σ, we have

$$\frac{\kappa''_\sigma}{\kappa'_\sigma}(x_\sigma) = \frac{z''''_\sigma}{z''_\sigma}(x_\sigma) = -2\frac{\delta}{\sigma}.$$

In other words, the ratio of the second and first derivatives of the curvature is independent of σ.

An analytical model for roof edges is obtained by taking $h = 0$ and $\delta \neq 0$ in the step model (Figure 21.4, right). In this case, it is easy to show (see Exercises) that

$$\kappa_\sigma = \frac{1}{\sigma\sqrt{2\pi}} \frac{\delta \exp(-\frac{x^2}{2\sigma^2})}{\left[1 + \left(k + \frac{\delta}{\sqrt{2\pi}} \int_0^{x/\sigma} \exp(-\frac{u^2}{2})du\right)^2\right]^{3/2}}. \quad (21.4)$$

It follows that, when $x_2 = \lambda x_1$ and $\sigma_2 = \lambda\sigma_1$, we must have $\kappa_{\sigma_2}(x_2) = \kappa_{\sigma_1}(x_1)/\lambda$. In turn, the maximum value of $|\kappa_\sigma|$ must be inversely proportional to σ, and it is reached at a point whose distance from the origin is proportional to σ. This maximum tends toward infinity as σ tends toward zero, indicating that roofs can be found as local curvature extrema. In actual range images, these extrema should be sought in the direction of the dominant principal direction, in keeping with our assumptions about local shape changes in the vicinity of surface edges.

Computing the Principal Curvatures and Directions According to the models derived in the previous section, instances of step and roof edges can be found as zero crossings of the Gaussian curvature and extrema of the dominant principal curvature in the corresponding direction. Computing these differential quantities requires estimating the first and second partial derivatives of the depth function at each point of a range image. This can be done, as in chapter 8, by convolving the images with the derivatives of a Gaussian distribution. However, range images are different from usual pictures: For example, the pixel values in a photograph are usually assumed to be piecewise constant in the neighborhood of step edges,[2] which is justified for Lambertian objects since the shape of a surface is, to first order, piecewise-constant near an edge,

[2]This corresponds to taking $k_1 = k_2 = 0$ in the model given in the previous section; note that in that case zero crossings do not move as scale changes.

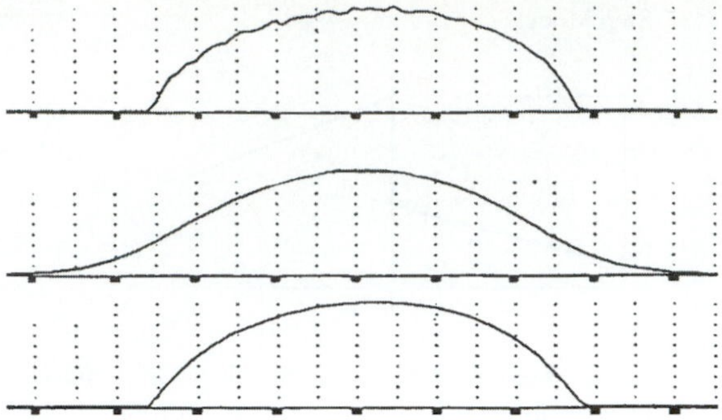

Figure 21.5 Smoothing a range image. Top: A slice of the range image shown in Figure 21.3. The background has been thresholded away. Middle: Result of Gaussian smoothing. Bottom: smoothing using computational molecules. *Reprinted from "Describing Surfaces," by J.M. Brady, J. Ponce, A. Yuille, and H. Asada, Proc. International Symposium on Robotics Research, H. Hanafusa and H. Inoue (eds.), pp. 5–16, MIT Press, (1985*b*). ©*1985 by The Massachusetts Institute of Technology.*

with a piecewise-planar intensity in that case. In contrast, piecewise-constant (local) models of range images are unsatisfactory. Likewise, the maximum values of contrast along the significant edges of a photograph are usually assumed to have roughly the same magnitude. In range images, however, there are two different types of step edges: The large depth discontinuities that separate solid objects from each other and from their background, and the much smaller gaps that usually separate patches of the same surface. The edge-detection scheme discussed in this section is aimed at the latter class of discontinuities.

Blindly applying Gaussian smoothing across object boundaries introduces radical shape changes that may overwhelm the surface details we are interested in (Figure 21.5, top and middle). This suggests finding the major depth discontinuities first (thresholding suffices in many cases), and then somehow restricting the smoothing process to the surface patches enclosed by these boundaries. This can be achieved by convolving the range image with *computational molecules* (Terzopoulos, 1984), which are linear templates that, added together, form a 3×3 averaging mask; e.g.,

$$\begin{array}{|c|c|c|}\hline 1 & & \\ \hline & 2 & \\ \hline & & 1 \\ \hline\end{array} + \begin{array}{|c|c|c|}\hline 2 & 4 & 2 \\ \hline\end{array} + \begin{array}{|c|}\hline 2 \\ \hline 4 \\ \hline 2 \\ \hline\end{array} + \begin{array}{|c|c|c|}\hline & & 1 \\ \hline & 2 & \\ \hline 1 & & \\ \hline\end{array} = \begin{array}{|c|c|c|}\hline 1 & 2 & 1 \\ \hline 2 & 12 & 2 \\ \hline 1 & 2 & 1 \\ \hline\end{array}.$$

Repeatedly convolving the image with the 3×3 mask (normalized so its weights add to one) yields, according to the central limit theorem, a good approximation of Gaussian smoothing with a mask whose σ value is proportional to $\sqrt{n}$ after n iterations. To avoid smoothing across discontinuities, the molecules crossing these discontinuities are not used and the remaining ones are once again normalized so the total sum of the weights is equal to one. The effect is shown in Figure 21.5(bottom).

After the surface has been smoothed, the derivatives of the height function can be computed via finite differences. The gradient of the height function is computed by convolving the smoothed image with the masks:

$$\frac{\partial}{\partial x} = \frac{1}{6} \begin{array}{|c|c|c|}\hline -1 & 0 & 1 \\ \hline -1 & 0 & 1 \\ \hline -1 & 0 & 1 \\ \hline\end{array} \quad \text{and} \quad \frac{\partial}{\partial y} = \frac{1}{6} \begin{array}{|c|c|c|}\hline 1 & 1 & 1 \\ \hline 0 & 0 & 0 \\ \hline -1 & -1 & -1 \\ \hline\end{array}.$$

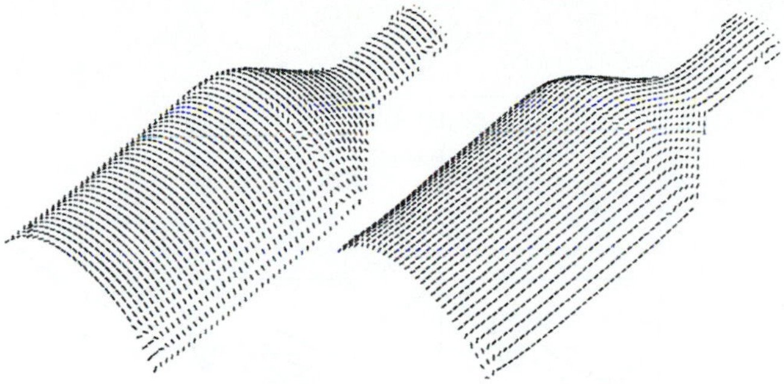

Figure 21.6 The two principal direction fields for the oil bottle. *Reprinted from "Describing Surfaces," by J.M. Brady, J. Ponce, A. Yuille, and H. Asada, H. Asada, Proc. International Symposium on Robotics Research, H. Hanafusa and H. Inoue (eds.), pp. 5–16, MIT Press, (1985b). ©1985 by The Massachusetts Institute of Technology.*

The Hessian is computed by convolving the smoothed image with the masks

$$\frac{\partial^2}{\partial x^2} = \frac{1}{3}\begin{array}{|c|c|c|}\hline 1 & -2 & 1 \\ \hline 1 & -2 & 1 \\ \hline 1 & -2 & 1 \\ \hline \end{array}, \quad \frac{\partial^2}{\partial x \partial y} = \frac{1}{4}\begin{array}{|c|c|c|}\hline -1 & 0 & 1 \\ \hline 0 & 0 & 0 \\ \hline 1 & 0 & -1 \\ \hline \end{array} \quad \text{and} \quad \frac{\partial^2}{\partial y^2} = \frac{1}{3}\begin{array}{|c|c|c|}\hline 1 & 1 & 1 \\ \hline -2 & -2 & -2 \\ \hline 1 & 1 & 1 \\ \hline \end{array}.$$

Once the derivatives are known, the principal directions and curvatures are easily computed. Figure 21.6 shows the two sets of principal directions found for the oil bottle after 20 iterations of the molecules. As expected, they lie along the meridians and parallels of this surface of revolution.

Matching Features Across Scales Given the principal curvatures and directions, parabolic points can be detected as (nondirectional) zero crossings of the Gaussian curvature, whereas local extrema of the dominant curvature along the corresponding principal direction can be found using the nonmaximum suppression techniques discussed in chapter 8. Although there may be a considerable amount of noise at fine resolutions (i.e., after a few iterations only), the situation improves as smoothing proceeds. Features due to noise can also be eliminated, at least in part, via thresholding of the zero crossing slope for parabolic points and of the curvature magnitude for extrema of principal curvatures. Nonetheless, experiments show that smoothing and thresholding are not sufficient to eliminate all irrelevant features. In particular, as illustrated by Figure 21.7(left), curvature extrema parallel to the axis of the oil bottle show up more and more clearly as smoothing proceeds. These are due to the fact that points near the occluding boundary of the bottle do not get smoothed as much by the computational molecules as points closer to its center.

A multiscale approach to edge detection solves this problem. Features are tracked from coarse to fine scales, all features at a given scale not having an ancestor at a coarser one being eliminated. The evolution of the principal curvatures and their derivatives is also monitored. Surviving parabolic features such that the ratio $\kappa''_\sigma/\kappa'_\sigma$ remains (roughly) constant across scales are output as step edge points, whereas directional extrema of the dominant curvature such that $\sigma\kappa_\sigma$ remains (roughly) constant are output as roof points. Finally, since, for both our models, the distance between the true discontinuity and the corresponding zero crossing or extremum increases with scale, the finest scale is used for edge localization. Figure 21.7(right) shows the results of applying this strategy to the oil bottle.

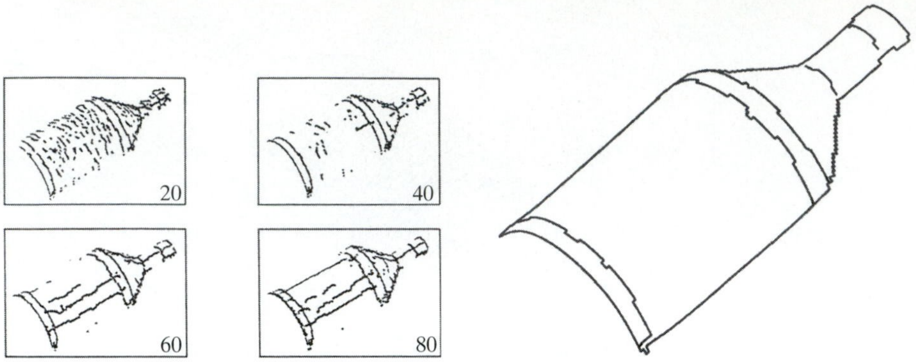

Figure 21.7 Finding step and roof edges on the oil bottle. Left: The features found after 20, 40, 60 and 80 smoothing iterations and thresholding. The thresholds have been chosen empirically to eliminate most false features while retaining those corresponding to true surface discontinuities. Still, artifacts such as the extrema of curvature parallel to the axis of the bottle subsist. Right: The output of model-based edge detection: The three step edges and two roof discontinuities of the oil bottle have been correctly identified. *Reprinted from "Towards a Surface Primal Sketch," by J. Ponce and J.M. Brady, in THREE-DIMENSIONAL MACHINE VISION, T. Kanade (ed.), pp. 195–240, Kluwer Academic Publishers, (1987). ©1987 Kluwer Academic Publishers.*

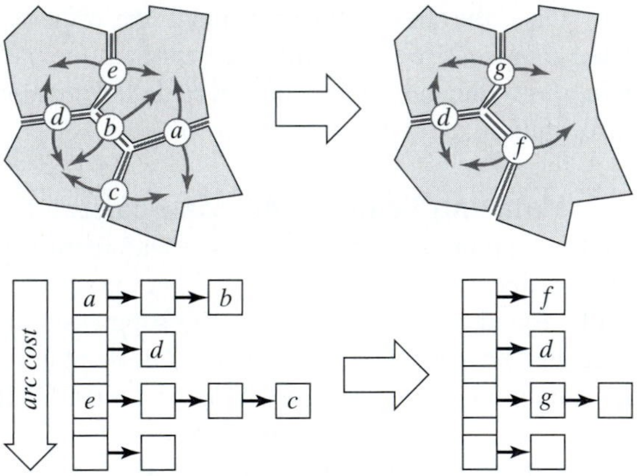

Figure 21.8 This diagram illustrates one iteration of the region-growing process during which the two patches incident to the minimum-cost arc labeled a are merged. The heap shown in the bottom part of the figure is updated as well: The arcs a, b, c, and e are deleted, and two new arcs f and g are created and inserted in the heap.

21.2.3 Segmenting Range Images into Planar Regions

We saw in the last section that edge detection is implemented by quite different processes in photographs and depth maps. The situation is similar for image segmentation into regions. In particular, meaningful segmentation criteria are elusive in the intensity domain because pixel brightness (or color) is only a cue to physical properties such as shape or reflectance. In the range domain, however, geometric information is directly available, making it possible to use,

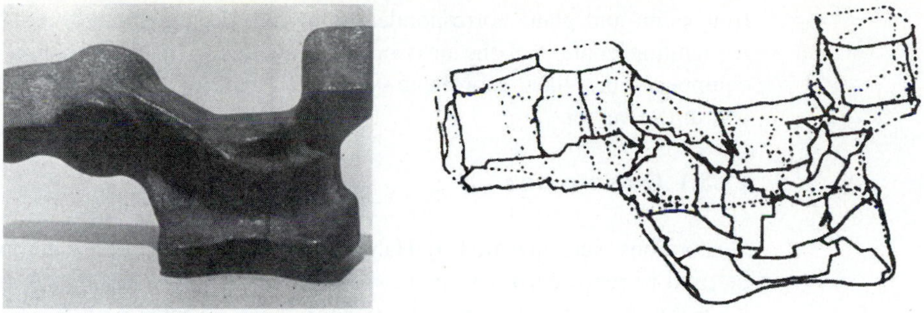

(a) (b)

Figure 21.9 The Renault part: (a) photo of the part, and (b) its model. *Reprinted from "The Representation, Recognition, and Locating of 3D Objects," by O.D. Faugeras and M. Hebert, International Journal of Robotics Research, 5(3):27–52, (1986). © 1986 Sage Publications. Reprinted by permission of Sage Publications.*

say, the average distance between a set of surface points and the plane best fitting them as an effective segmentation criterion. The region-growing technique of Faugeras and Hebert (1986) is a good example of this approach. This algorithm iteratively merges planar patches by maintaining a graph whose nodes are the patches and arcs associated with their common boundary link adjacent patches. Each arc is assigned a cost equal to the average error between the points of the two patches and the plane best fitting these points. The best arc is always selected, and the corresponding patches are merged. Note that the remaining arcs associated with these patches must be deleted while new arcs linking the new patch to its neighbors are introduced. The situation is illustrated by Figure 21.8.

The graph structure is initialized by using a triangulation of the range data, and it is efficiently updated by maintaining a heap of active arcs. The triangulation can either be constructed directly from a range image (by splitting the quadrilaterals associated with the pixels along one of their diagonals), or from a global surface model constructed from multiple images as described in the next section. The heap storing the active arcs can be represented, for example, by an array of buckets indexed by increasing costs, which supports fast insertion and deletion (Figure 21.8, bottom). Figure 21.9 shows an example, where the complex shape of a Renault automobile part is approximated by 60 planar patches.

21.3 RANGE IMAGE REGISTRATION AND MODEL ACQUISITION

Geometric models of real objects are useful in manufacturing (e.g., for process and assembly planning or inspection). Closer to the theme of this book, they are also key components of many object recognition systems, and are more and more in demand in the entertainment industry, as synthetic pictures of real objects now routinely appear in feature films and video games (we come back to this issue in much greater detail in chapter 26). Range images are an excellent source of data for constructing accurate geometric models of real objects, but a single picture shows, at best, half of the surface of a given solid, and the construction of complete object models requires the integration of multiple range images. This section addresses the dual problems of registering multiple images in the same coordinate system and fusing the three-dimensional data provided by these pictures into a single integrated surface model. Before attacking these two problems, let us introduce quaternions, which afford linear methods for estimating rigid transformations from point and plane correspondences in both the registration context of this section and the recognition context of the next one. We assume in the rest of this chapter that $\mathbb{E}^3$ has been equipped with a fixed coordinate system and identify this space with $\mathbb{R}^3$ and each point with its coordinate vector.

21.3.1 Quaternions

Quaternions were invented by Hamilton (1844). Like complex numbers in the plane, they can be used to represent rotations in space in a convenient manner. A quaternion $\mathtt{q}$ is defined by its *real part*, a scalar a, and its *imaginary part*, a vector $\boldsymbol{\alpha}$ in $\mathbb{R}^3$, and it is usually denoted by $\mathtt{q} = a + \boldsymbol{\alpha}$. This is justified by the fact that real numbers can be identified with quaternions with a zero imaginary part, and vectors can be identified with quaternions with a zero real part, while addition between quaternions is defined by

$$(a + \boldsymbol{\alpha}) + (b + \boldsymbol{\beta}) \stackrel{\text{def}}{=} (a + b) + (\boldsymbol{\alpha} + \boldsymbol{\beta}).$$

The multiplication of a quaternion by a scalar is defined naturally by $\lambda(a+\boldsymbol{\alpha}) \stackrel{\text{def}}{=} \lambda a + \lambda\boldsymbol{\alpha}$, and these two operations give the set of all quaternions the structure of a four-dimensional vector space. It is also possible to define a multiplication operation that associates with two quaternions the quaternion

$$(a + \boldsymbol{\alpha})(b + \boldsymbol{\beta}) \stackrel{\text{def}}{=} (ab - \boldsymbol{\alpha} \cdot \boldsymbol{\beta}) + (a\boldsymbol{\beta} + b\boldsymbol{\alpha} + \boldsymbol{\alpha} \times \boldsymbol{\beta}).$$

Quaternions, equipped with the operations of addition and multiplication as defined before, form a noncommutative field, whose zero and unit elements are the scalars 0 and 1, respectively. The *conjugate* of the quaternion $\mathtt{q} = a + \boldsymbol{\alpha}$ is the quaternion $\bar{\mathtt{q}} \stackrel{\text{def}}{=} a - \boldsymbol{\alpha}$ with opposite imaginary part. The squared norm of a quaternion is defined by

$$|\mathtt{q}|^2 \stackrel{\text{def}}{=} \mathtt{q}\bar{\mathtt{q}} = \bar{\mathtt{q}}\mathtt{q} = a^2 + |\boldsymbol{\alpha}|^2,$$

and it is easily verified that $|\mathtt{q}\mathtt{q}'| = |\mathtt{q}||\mathtt{q}'|$ for any pair of quaternions $\mathtt{q}$ and $\mathtt{q}'$.

Now it can be shown that the quaternion

$$\mathtt{q} = \cos\frac{\theta}{2} + \sin\frac{\theta}{2}\boldsymbol{u}$$

represents the rotation $\mathcal{R}$ of angle θ about the *unit* vector $\boldsymbol{u}$ in the following sense: If $\boldsymbol{\alpha}$ is some vector in $\mathbb{R}^3$, then

$$\mathcal{R}\boldsymbol{\alpha} = \mathtt{q}\boldsymbol{\alpha}\bar{\mathtt{q}}. \tag{21.5}$$

Note that $|\mathtt{q}| = 1$ and that $-\mathtt{q}$ also represents the rotation $\mathcal{R}$. Reciprocally, the rotation matrix $\mathcal{R}$ associated with a given unit quaternion $\mathtt{q} = a + \boldsymbol{\alpha}$ with $\boldsymbol{\alpha} = (b, c, d)^T$ is

$$\mathcal{R} = \begin{pmatrix} a^2 + b^2 - c^2 - d^2 & 2(bc - ad) & 2(bd + ac) \\ 2(bc + ad) & a^2 - b^2 + c^2 - d^2 & 2(cd - ab) \\ 2(bd - ac) & 2(cd + ab) & a^2 - b^2 - c^2 + d^2 \end{pmatrix}, \tag{21.6}$$

a fact easily deduced from Eq. (21.5). (Note that the four parameters a, b, c, d are not independent since they satisfy the constraint $a^2 + b^2 + c^2 + d^2 = 1$.)

Finally, if $\mathtt{q}_1$ and $\mathtt{q}_2$ are unit quaternions and $\mathcal{R}_1$ and $\mathcal{R}_2$ are the corresponding rotation matrices, the quaternions $\mathtt{q}_1\mathtt{q}_2$ and $-\mathtt{q}_1\mathtt{q}_2$ are both representations of the rotation matrix $\mathcal{R}_1\mathcal{R}_2$.

21.3.2 Registering Range Images Using the Iterative Closest-Point Method

Besl and McKay (1992) proposed an algorithm capable of *registering* two sets of three-dimensional points (i.e., of computing the rigid transformation that maps the first point set onto the second one). Their algorithm simply minimizes the average distance between the two point sets by iterating over the following steps: First establish correspondences between scene and model features by matching every scene point to the model point closest to it, estimate the rigid transformation mapping the scene points onto their matches, and finally apply the computed displacement to the scene. The iterations stop when the change in mean distance between the matched points falls below some preset threshold. Pseudocode for this *iterated closest-point (ICP)* algorithm is given next.

Algorithm 21.2: The iterative closest-point algorithm of Besl and McKay (1992). The auxiliary function Initialize-Registration uses some global registration method based on moments, for example, to compute a rough initial estimate of the rigid transformation mapping the scene onto the model. The function Return-Closest-Pairs returns the indexes (i, j) of the points in the registered scene and the model such that point number j is the closest to point number i. The function Update-Registration estimates the rigid transformation between selected pairs of points in the scene and the model, and the function Apply-Registration applies a rigid transformation to all the points in the scene.

```
      Function ICP(Model, Scene);
begin
E' ← +∞;
(Rot, Trans) ← Initialize-Registration(Scene, Model);
repeat
  E ← E';
  Registered-Scene ← Apply-Registration(Scene, Rot, Trans);
  Pairs ← Return-Closest-Pairs(Registered-Scene, Model);
  (Rot, Trans, E') ← Update-Registration(Scene, Model, Pairs, Rot, Trans);
  until |E' − E| < τ;
return (Rot, Trans);
end
```

It is easy to show that Algorithm 21.2 forces the error E to decrease monotonically with each iteration: indeed, the average error decreases during the registration stage, and the individual errors decrease as well during the determination of the closest point pairs. By itself, this does not guarantee convergence to a global (or even local) minimum, and a reasonable guess for the rigid transformation sought by the algorithm must be provided. A variety of methods are available for that purpose, including roughly sampling the set of all possible transformations, and using the moments of both the scene and model point sets to estimate the transformation.

Finding the Closest-Point Pairs At every iteration of the algorithm, finding the closest point M in the model to a given (registered) scene point S takes (naively) $O(n)$ time, where n is the number of model points. In fact, various algorithms can be used to answer such a nearest-neighbor query in $\mathbb{R}^3$ in $O(\log n)$ time at the cost of additional preprocessing of the model using, for example, *k-d trees* (Friedman et al., 1977, for which the logarithmic query time only holds on average) or more complex data structures. For example, Clarkson's (1988) general randomized algorithm takes preprocessing time $O(n^{2+\varepsilon})$, where ε is an arbitrarily small positive number, and query time $O(\log n)$. The efficiency of repeated queries can also be improved by *caching* the results of previous computations. For example, Simon et al. (1994) store at each iteration of the ICP algorithm the k closest model points to each scene point (a typical value for k is 5). Since the incremental update of the rigid transformation is normally small, it is likely that the closest neighbor of a point after an iteration is among its k closest neighbors from the previous one. It is in fact possible to determine efficiently and conclusively whether the closest point is in the cached set (see Simon et al., 1994, for details).

Estimating the Rigid Transformation Under the rigid transformation defined by the rotation matrix $\mathcal{R}$ and the translation vector $\boldsymbol{t}$, a point $\boldsymbol{x}$ maps onto the point $\boldsymbol{x}' = \mathcal{R}\boldsymbol{x} + \boldsymbol{t}$. Thus, given n pairs of matching points $\boldsymbol{x}_i$ and $\boldsymbol{x}'_i$, with $i = 1, \dots, n$, we seek the rotation matrix $\mathcal{R}$ and translation vector $\boldsymbol{t}$ minimizing the error

$$E = \sum_{i=1}^{n} |\boldsymbol{x}'_i - \mathcal{R}\boldsymbol{x}_i - \boldsymbol{t}|^2.$$

Let us first note that the value of $\boldsymbol{t}$ minimizing E must satisfy

$$0 = \frac{\partial E}{\partial \boldsymbol{t}} = -2\sum_{i=1}^{n}(\boldsymbol{x}'_i - \mathcal{R}\boldsymbol{x}_i - \boldsymbol{t})$$

or

$$\boldsymbol{t} = \bar{\boldsymbol{x}}' - \mathcal{R}\bar{\boldsymbol{x}}, \quad \text{where} \quad \bar{\boldsymbol{x}} \stackrel{\text{def}}{=} \frac{1}{n}\sum_{i=1}^{n}\boldsymbol{x}_i \quad \text{and} \quad \bar{\boldsymbol{x}}' \stackrel{\text{def}}{=} \frac{1}{n}\sum_{i=1}^{n}\boldsymbol{x}'_i \tag{21.7}$$

denote, respectively, the centroids of the two sets of points $\boldsymbol{x}_i$ and $\boldsymbol{x}'_i$.

Introducing the centered points $\boldsymbol{y}_i = \boldsymbol{x}_i - \bar{\boldsymbol{x}}$ and $\boldsymbol{y}'_i = \boldsymbol{x}'_i - \bar{\boldsymbol{x}}$ $(i = 1, \dots, n)$ yields

$$E = \sum_{i=1}^{n} |\boldsymbol{y}'_i - \mathcal{R}\boldsymbol{y}_i|^2.$$

Quaternions can now be used to minimize E as follows: Let q denote the quaternion associated with the matrix $\mathcal{R}$. We use the fact that $|\mathsf{q}|^2 = 1$ and the multiplicativity properties of the quaternion norm to write

$$E = \sum_{i=1}^{n} |\boldsymbol{y}'_i - \mathsf{q}\boldsymbol{y}_i\bar{\mathsf{q}}|^2|\mathsf{q}|^2 = \sum_{i=1}^{n} |\boldsymbol{y}'_i\mathsf{q} - \mathsf{q}\boldsymbol{y}_i|^2.$$

As shown in the exercises, this allows us to rewrite the rotational error as $E = \mathsf{q}^T\mathcal{B}\mathsf{q}$, where $\mathcal{B} = \sum_{i=1}^{n} \mathcal{A}_i^T\mathcal{A}_i$ and

$$\mathcal{A}_i = \begin{pmatrix} 0 & \boldsymbol{y}_i^T - \boldsymbol{y}_i'^T \\ \boldsymbol{y}'_i - \boldsymbol{y}_i & [\boldsymbol{y}_i + \boldsymbol{y}'_i]_\times \end{pmatrix}.$$

Note that the matrix $\mathcal{A}_i$ is antisymmetric with (in general) rank 3, but that the matrix $\mathcal{B}$ has, in the presence of noise, rank 4. As shown in chapter 3, minimizing E under the constraint $|\mathsf{q}|^2 = 1$ is a (homogeneous) linear least-squares problem whose solution is the eigenvector of $\mathcal{B}$ associated with the smallest eigenvalue of this matrix. Once $\mathcal{R}$ is known, $\boldsymbol{t}$ is obtained from Eq. (21.7).

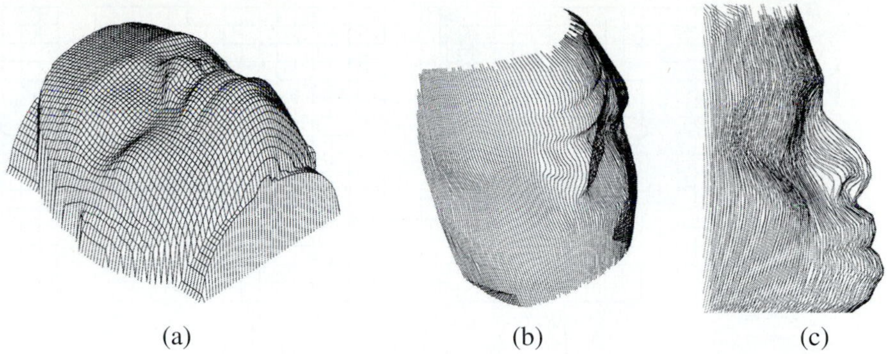

Figure 21.10 Registration results: (a) a range image serving as model for an African mask; (b) a (decimated) view of the model, serving as scene data; (c) a view of the two datasets after registration. *Reprinted from "A Method for Registration of 3D Shapes," by P.J. Besl and N.D. McKay, IEEE Transactions on Pattern Analysis and Machine Intelligence, 14(2):238–256, (1992). © 1992 IEEE.*

Results Figure 21.10 shows an example, where two range images of an African mask are matched by the algorithm. The average distance between matches is 0.59 mm for this 9 cm object.

21.3.3 Fusing Multiple Range Images

Given a set of registered range images of a solid object, it is possible to construct an integrated surface model of this object. In the approach proposed by Curless and Levoy (1996), this model is constructed as the zero set S of a volumetric density function $D : \mathbb{R}^3 \rightarrow \mathbb{R}$ (i.e., as the set of points (x, y, z) such that $D(x, y, z) = 0$). Like any other level set of a continuous density function, S is by construction guaranteed to be a closed, watertight surface, although it may have several connected components (Figure 21.11).

The difficulty, of course, is to construct an appropriate density function from registered range measurements. Curless and Levoy embed the corresponding surface fragments into a cubic grid and assign to each cell of this grid, or *voxel*, a weighted sum of the signed distances between its center and the closest surface point intersecting it (Figure 21.12, left). This averaged signed distance is the desired density function, and its zero set can be found using classical tech-

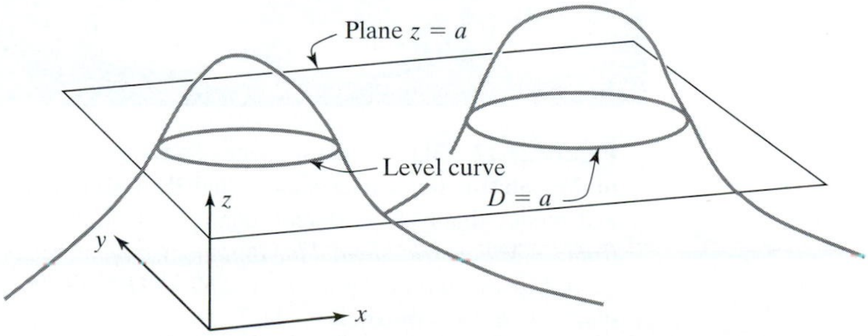

Figure 21.11 A 2D illustration of volumetric density functions and their level sets. In this case, the "volume" is, of course, the (x, y) plane, and the "surface" is a curve in this plane, with two connected components in the example shown here.

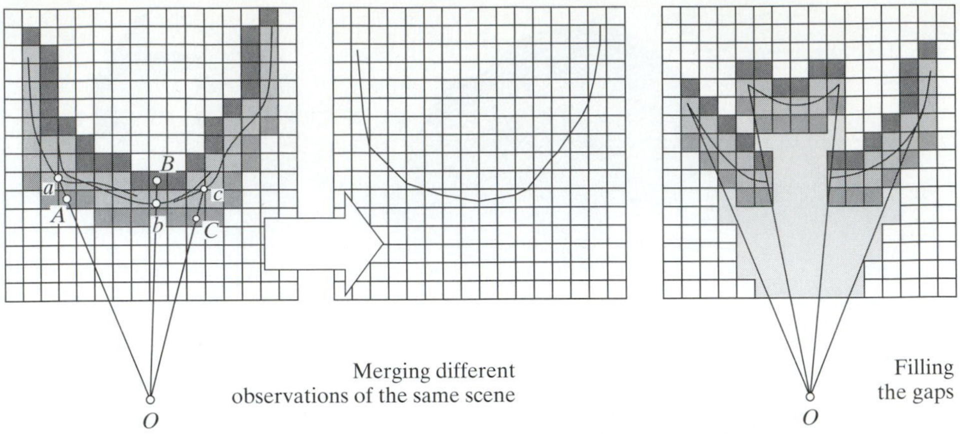

Figure 21.12 A 2D illustration of the Curless–Levoy method for fusing multiple range images. In the left part of the figure, three views observed by the same sensor located at the point O are merged by computing the zero set of a weighted average of the signed distances between voxel centers (e.g., points A, B, and C) and surface points (e.g., a, b, and c) along viewing rays. In general, distances to different sensors would be used instead. The light gray area in the right part of the figure is the set of voxels marked as empty in the gap-filling part of the procedure.

Figure 21.13 3D Fax of a statuette of a Buddha. From left to right: photograph of the statuette; range image; integrated 3D model; model after hole filling; physical model obtained via stereolithography. Courtesy of Marc Levoy. *Reprinted from "A Volumetric Method for Building Complex Models from Range Images," by B. Curless and M. Levoy, Proc. SIGGRAPH, (1996). © 1996 ACM, Inc. Included here by permission.*

niques, such as the *marching cubes* algorithm developed by Lorensen and Cline (1987) to extract isodensity surfaces from volumetric medical data.

Missing surface fragments corresponding to unobserved parts of the scene are handled by initially marking all voxels as *unseen* or, equivalently, assigning them a depth equal to some large positive value (standing for $+\infty$), then assigning as before to all voxels close to the measured surface patches the corresponding signed distance, and finally carving out (i.e., marking as *empty* or having a large negative depth standing for $-\infty$) the voxels that lie between the observed surface patches and the sensor (Figure 21.12, right).

Figure 21.13 shows an example of a model built from multiple range images of a Buddha statuette acquired with a Cyberware 3030 MS optical triangulation scanner, as well as a physical model constructed from the geometric one via stereolithography (Curless and Levoy, 1996).

21.4 OBJECT RECOGNITION

We now turn to actual object recognition from range images. The registration techniques introduced in the previous section play a crucial role in the two algorithms discussed in this one.

21.4.1 Matching Piecewise-Planar Surfaces Using Interpretation Trees

The recognition algorithm proposed by Faugeras and Hebert (1986) is a recursive procedure exploiting rigidity constraints to efficiently search an interpretation tree for the path(s) corresponding to the best sequence(s) of plane matches. The basic procedure is given in pseudocode in Algorithm 21.3. To correctly handle occlusions (and the fact that, as noted earlier, a range finder sees, at best, one half of the object facing it), the algorithm must consider, at every stage of the search, the possibility that a model plane may not match any scene plane. This is done by always incorporating in the list of potential matches of a given plane a token *null* plane.

Algorithm 21.3: The plane-matching algorithm of Faugeras and Hebert (1986). The recursive function Match returns the best set of matching plane pairs found by recursively visiting the interpretation tree. It is initially called with an empty list of pairs and nil values for the rotation and translation arguments rot and trans. The auxiliary function Potential-Matches returns the subset of the planes in the scene that are compatible with the model plane Π and the current estimate of the rigid transformation mapping the model planes onto their scene matches (see text for details). The auxiliary function Update-Registration-2 uses the matched plane pairs to update the current estimate of the rigid transformation.

```
    Function Match(model, scene, pairs, rot, trans);
begin
bestpairs ← nil; bestscore ← 0;
for Π in model do
  for Π′ in Potential-Matches(scene, pairs, Π, rot, trans) do
    (rot,trans) ← Update-Registration-2(pairs, Π, Π′, rot, trans);
    (score, newpairs) ← Match(model−Π, scene−Π′, pairs+(Π, Π′), rot, trans);
    if score> bestscore then bestscore ← score; bestpairs ← newpairs endif;
    endfor;
  endfor;
return (bestscore,bestpairs);
end
```

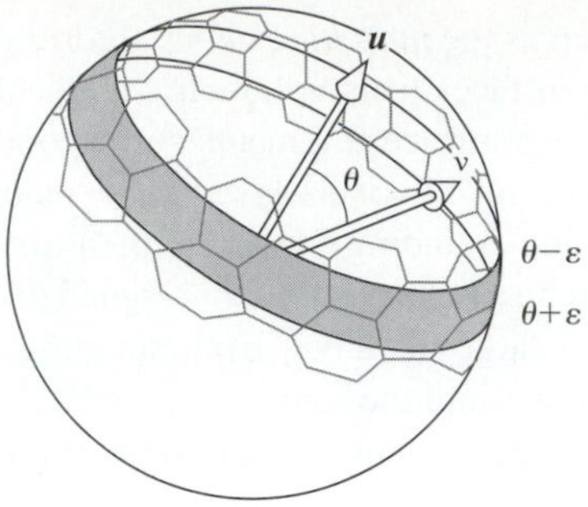

Figure 21.14 Finding all vectors $\boldsymbol{v}$ that make an angle in the $[\theta - \varepsilon, \theta + \varepsilon]$ range with a given vector $\boldsymbol{u}$. It should be noted that the unit sphere does not admit tesselations with an arbitrary level of detail by regular (spherical) polygons. The tesselation shown in the diagram is made of hexagons with unequal edge lengths (see, e.g., Horn 1986, chap. 16 for a discussion of this problem and various tesselation schemes).

Selecting Potential Matches The selection of potential matches for a given model plane is based on various criteria depending on the number of correspondences already established, with each new correspondence providing new geometric constraints and more stringent criteria. At the beginning of the search, we only know that a model plane with area A should only be matched to scene planes with a compatible area (i.e., in the range $[\alpha A, \beta A]$). Reasonable values for the two thresholds might be 0.5 and 1.1, which allows for some discrepancy between the unoccluded areas and also affords a degree of occlusion up to 50%.

After the first correspondence has been established, it is still too early to estimate the rigid transformation mapping the model onto the scene, but it is clear that the angle between the normals to any matching planes should be (roughly) equal to the angle θ between the normals to the first pair of planes—say in the interval $[\theta - \varepsilon, \theta + \varepsilon]$. The normals to the corresponding planes lie in a band of the Gauss sphere, and they can be efficiently retrieved by discretizing this sphere and associating to each cell a bucket that stores the scene planes whose normal falls into it (Figure 21.14).

A second pairing is sufficient to completely determine the rotation separating the model from its instance in the scene: This is geometrically clear (and is confirmed analytically in the next section) since a pair of matching vectors constrains the rotation axis to lie in the plane bisecting these vectors. Two pairs of matching planes determine the axis of rotation as the intersection of the corresponding bisecting planes, and the rotation angle is readily computed from either of the matches. Given the rotation and a third model plane, one can predict the orientation of the normal to its possible matches in the scene, which can be efficiently recovered using once again the discrete Gauss sphere mentioned before. After three pairings have been found, the translation can also be estimated and used to predict the distance between the origin and any scene plane matching a fourth scene plane. The same is true for any further pairing.

Estimating the Rigid Transformation Let us consider a plane Π defined by the equation $\boldsymbol{n} \cdot \boldsymbol{x} - d = 0$ in some fixed coordinate system. Here $\boldsymbol{n}$ denotes the unit normal to the plane and d its (signed) distance from the origin. Under the rigid transformation defined by the rotation matrix $\mathcal{R}$ and the translation vector $\boldsymbol{t}$, a point $\boldsymbol{x}$ maps onto the point $\boldsymbol{x}' = \mathcal{R}\boldsymbol{x} + \boldsymbol{t}$, and Π maps onto the plane Π' whose equation is $\boldsymbol{n}' \cdot \boldsymbol{x}' - d' = 0$, with

$$\begin{cases} \boldsymbol{n}' = \mathcal{R}\boldsymbol{n}, \\ d' = \boldsymbol{n}' \cdot \boldsymbol{t} + d. \end{cases}$$

Thus, estimating the rigid transformation that maps n planes Π_i onto the matching planes Π'_i $(i = 1, \dots, n)$ amounts to finding the rotation matrix $\mathcal{R}$ that minimizes the error

$$E_r = \sum_{i=1}^{n} |\boldsymbol{n}'_i - \mathcal{R}\boldsymbol{n}_i|^2$$

and the translation vector $\boldsymbol{t}$ that minimizes

$$E_t = \sum_{i=1}^{n} (d'_i - d_i - \boldsymbol{n}'_i \cdot \boldsymbol{t})^2.$$

The rotation $\mathcal{R}$ minimizing E_r can be computed, exactly as in Section 21.4.1, by using the quaternion representation of matrices and solving an eigenvector problem. The translation vector $\boldsymbol{t}$ minimizing E_t is the solution of a (nonhomogeneous) linear least-squares problem, whose solution can be found using the techniques presented in chapter 3.

Results Figure 21.15 shows recognition results obtained using a bin of Renault parts such as the one shown in Figure 21.9. The range image of the bin has been segmented into planar patches using the technique presented in Section 21.2.3. The matching algorithm is run

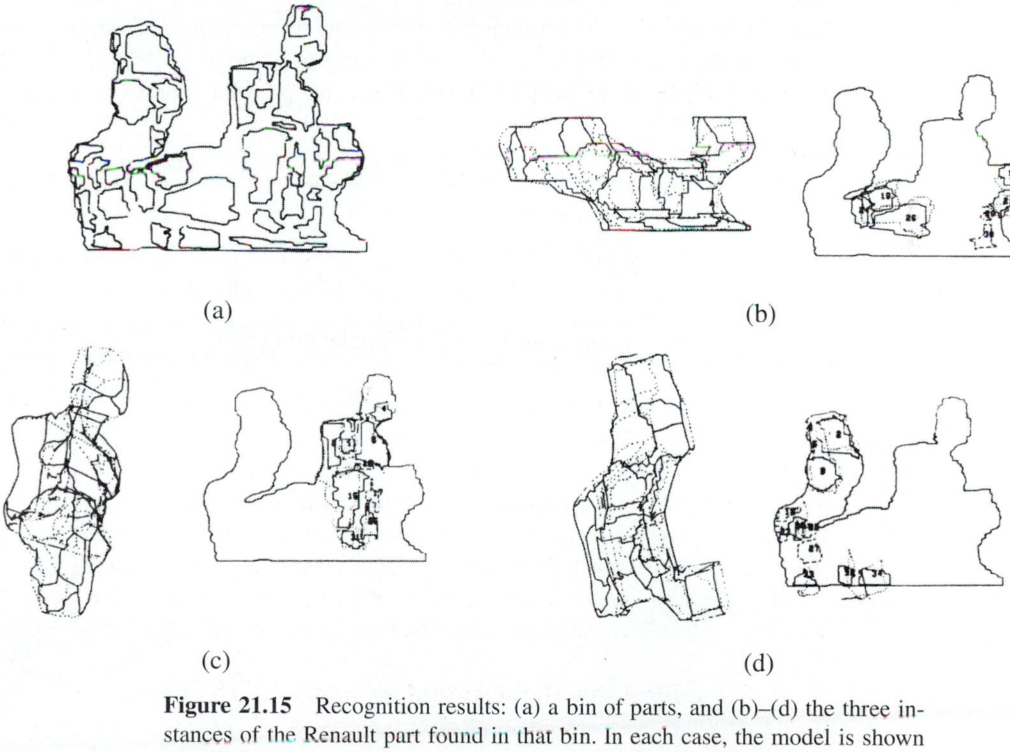

Figure 21.15 Recognition results: (a) a bin of parts, and (b)–(d) the three instances of the Renault part found in that bin. In each case, the model is shown both by itself in the position and orientation estimated by the algorithm, as well as superimposed (dotted lines) in this pose over the corresponding planes of the range image. *Reprinted from "The Representation, Recognition, and Locating of 3D Objects," by O.D. Faugeras and M. Hebert, International Journal of Robotics Research, 5(3):27–52, (1986). © 1986 Sage Publications. Reprinted by permission of Sage Publications.*

three times on the scene, with patches matched during each run removed from the scene before the next iteration. As shown by the figure, the three instances of the part present in the bin are correctly identified, and the accuracy of the pose estimation process is attested by the reprojection into the range image of the model in the computed pose.

21.4.2 Matching Free-Form Surfaces Using Spin Images

As demonstrated in Section 21.2.2, differential geometry provides a powerful language for describing the shape of a surface *locally* (i.e., in a small neighborhood of each one of its points). On the other hand, the region-growing algorithm discussed in Section 21.2.3 is aimed at constructing a *globally* consistent surface description in terms of planar patches. We introduce in this section a *semilocal* surface representation—the *spin image* of Johnson and Hebert (1998, 1999)—that captures the shape of a surface in a relatively large neighborhood of each one of its points. As shown in the rest of this section, the spin image is invariant under rigid transformations, and it affords an efficient algorithm for pointwise surface matching, thus completely bypassing segmentation in the recognition process.

Spin Image Definition Let us assume, as in Section 21.2.3, that the surface Σ of interest is given in the form of a triangular mesh. The (outward-pointing) surface normal at each vertex can be estimated by fitting a plane to this vertex and its neighbors, turning the triangulation into a net of *oriented points*. Given an oriented point P, the *spin coordinates* of any other point Q can now be defined as the (nonnegative) distance α separating Q from the (oriented) normal line in P and the (signed) distance β from the tangent plane to Q (Figure 21.16). Accordingly, the *spin map* $s_P : \Sigma \to \mathbb{R}^2$ associated with P is defined for any point Q on Σ as

$$s_P(Q) \stackrel{\text{def}}{=} (\underbrace{|\overrightarrow{PQ} \times \boldsymbol{n}|}_{\alpha}, \underbrace{\overrightarrow{PQ} \cdot \boldsymbol{n}}_{\beta}).$$

As shown by Figure 21.16, this mapping is not injective. This is not surprising since the spin map only provides a partial specification of a cylindrical coordinate system: The third coordinate that would normally record the angle between some reference vector in the tangent plane and the projection of $\overrightarrow{PQ}$ into this plane is missing. The principal directions are obvious choices

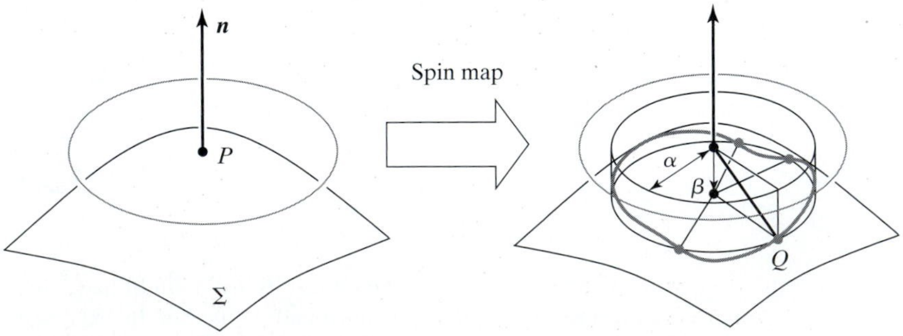

Figure 21.16 Definition of the spin map associated with a surface point P: The spin coordinates (α, β) of the point Q are, respectively, defined by the lengths of the projections of $\overrightarrow{PQ}$ onto the tangent plane and its surface normal. Note that there are three other points with the same (α, β) coordinates as Q in this example.

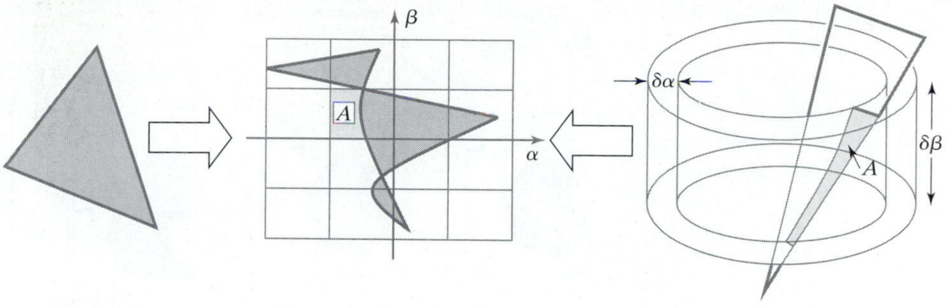

Figure 21.17 Spin image construction: The triangle shown in the left of the diagram maps onto a region with hyperbolic boundaries in the spin image; the value of each bin intersected by this region is incremented by the area of the portion of the triangle that intersects the annulus associated with the bin. After Carmichael et al. (1999, Figure 3).

for such a reference vector, but focusing on the spin coordinates avoids their computation—a process that is susceptible to noise since it involves second derivatives and may be ambiguous for (almost) planar or spherical patches.

The *spin image* associated with an oriented point is a histogram of the α, β coordinates in a neighborhood of this point. Concretely, the α, β plane is divided into a rectangular array of $\delta\alpha \times \delta\beta$ bins that accumulate the total surface area spanned by points with α, β values in that range.[3] As shown in Carmichael et al. (1999) and the exercises, each triangle in the surface mesh maps onto a region of the α, β plane whose boundaries are hyperbola arcs. Its contribution to the spin image can thus be computed by assigning to each bin that this region traverses the area of the patch where the triangle intersects the annular region of $\mathbb{R}^3$ associated with the bin (Figure 21.17). The bins can be found efficiently using *scan conversion* (Foley et al., 1990)—a process routinely used in computer graphics to find in optimal time the pixels traversed by a generalized polygon with straight or curved edges.

Spin images are defined by several key parameters (Johnson and Hebert, 1999): The first one is the support distance d that limits to a sphere of radius d centered in P the range of the *support points* used to construct the image. This sphere must be large enough to provide good descriptive power, but small enough to support recognition in the presence of clutter and occlusion. In practice, an appropriate choice for d might be a tenth of the object's diameter: Thus, as noted earlier, the spin image is indeed a semilocal description of the shape of a surface in an *extended* neighborhood of one of its points. Robustness to clutter can be improved by limiting the range of surface normals at the support points to a cone of half-angle θ centered in $\boldsymbol{n}$. As in the support distance case, choosing the right value for θ involves a trade-off between descriptive power and insensitivity to clutter; a value of 60° has empirically been shown to be satisfactory. The last parameter defining a spin image is its size (in pixels) or, equivalently, given the support distance, its bin size (in meters). It can be shown that an appropriate choice for the bin size is the average distance between mesh vertices in the model. Figure 21.18 shows the spin images associated with three oriented points on the surface of a rubber duck.

[3]The corresponding point sets may actually be divided into several connected components. For example, for small enough values of $\delta\alpha$ and $\delta\beta$, there are four connected components in the example shown in Figure 21.16, corresponding to small patches centered at the points having the same α, β coordinates as Q.

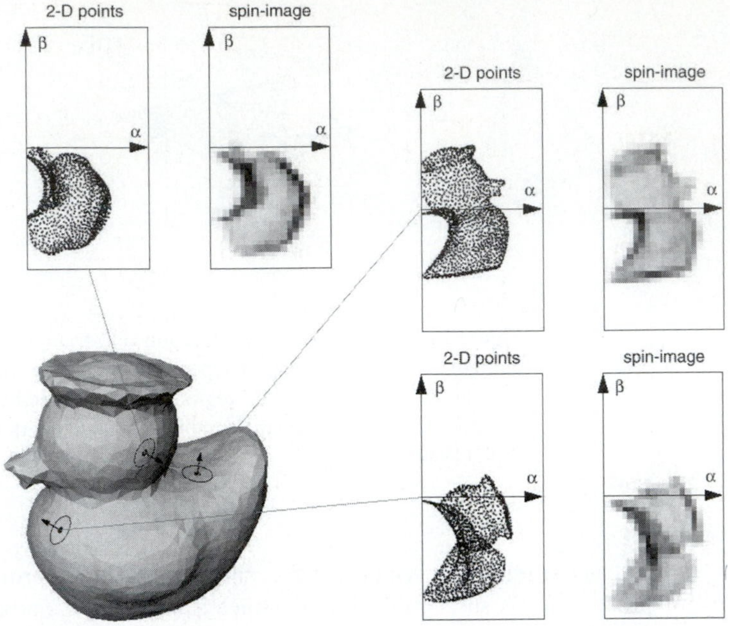

Figure 21.18 Three oriented points on the surface of a rubber duck and the corresponding spin images. The α, β coordinates of the mesh vertices are shown besides the actual spin images. *Reprinted from "Using Spin Images for Efficient Object Recognition in Cluttered 3D Scenes," by A.E. Johnson and M. Hebert, IEEE Transactions on Pattern Analysis and Machine Intelligence, 21(5):433–449, (1999). © 1999 IEEE.*

Matching Spin Images One of the most important features of spin images is that they are (obviously) invariant under rigid transformations. Thus, an image comparison technique such as correlation can in principle be used to match the spin images associated with oriented points in the scene and object model. Things are not that simple, however. We already noted that the spin map is not injective; in general, it is not surjective either, and empty bins (or, equivalently, zero-valued pixels) may occur for values of α and β that do not correspond to physical surface points (e.g., see the blank areas in Figure 21.18). Occlusion may cause the appearance of zero pixels in the scene image, whereas clutter may introduce irrelevant nonempty bins. Therefore, it is reasonable to restrict the comparison of two spin images to their common nonzero pixels. In this context, Johnson and Hebert (1998) have shown that

$$S(\boldsymbol{I}, \boldsymbol{J}) \stackrel{\text{def}}{=} [\text{Arctanh}(C(\boldsymbol{I}, \boldsymbol{J}))]^2 - \frac{3}{N-3}$$

is an appropriate similarity measure for two spin images whose overlap regions contain N pixels and are represented by the vectors $\boldsymbol{I}$ and $\boldsymbol{J}$ of $\mathbb{R}^N$. In this formula, $C(\boldsymbol{I}, \boldsymbol{J})$ denotes the normalized correlation of the vectors $\boldsymbol{I}$ and $\boldsymbol{J}$, and Arctanh denotes the hyperbolic arc tangent function. Armed with this similarity measure, we can now outline a recognition procedure (Algorithm 21.4) that uses spin images to establish pointwise correspondences.

The various stages of this algorithm are mostly straightforward. Let us note, however, that the filtering/grouping step relies on comparing the spin coordinates of model points relative to the other mesh vertices in their group with the spin coordinates of the corresponding scene points relative to their own group. Once consistent groups have been identified, an initial estimate of

Algorithm 21.4: The algorithm of Johnson and Hebert (1998, 1999) for pointwise matching of free-form surfaces using spin images.

Off-line:
Compute the spin images associated with the oriented points of a surface model and store them into a table.

On-line:

1. Form correspondences between a set of spin images randomly selected in the scene and their best matches in the model table using the similarity measure S to rank order the matches.
2. Filter and group correspondences using geometric consistency constraints, and compute the rigid transformations best aligning the matched scene and model features.
3. Verify the matches using the ICP algorithm.

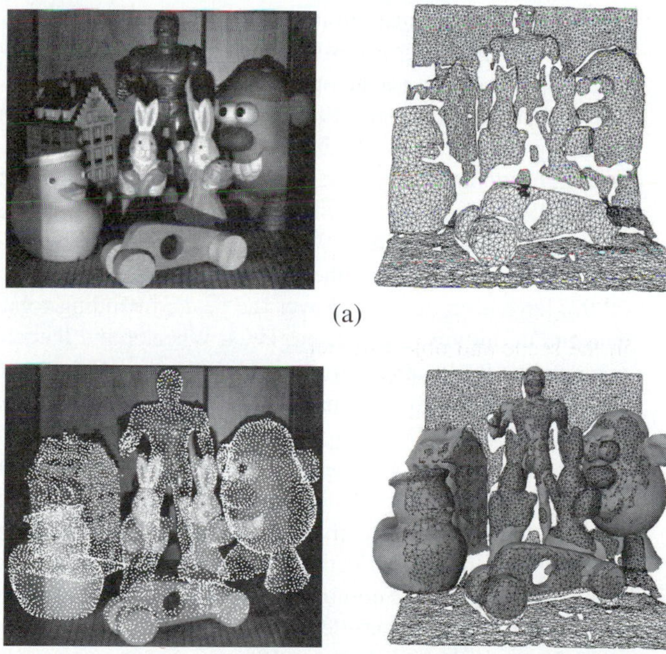

(b)

Figure 21.19 Spin image recognition results: (a) a cluttered image of toys and the mesh constructed from the corresponding range image; (b) recognized objects overlaid on the original pictures. *Reprinted from "Using Spin Images for Efficient Object Recognition in Cluttered 3D Scenes," by A.E. Johnson and M. Hebert, IEEE Transactions on Pattern Analysis and Machine Intelligence, 21(5):433–449, (1999). © 1999 IEEE.*

the rigid transformation aligning the scene and model is computed from (oriented) point matches using the quaternion-based registration technique described in Section 21.3.2. Finally, consistent sets of correspondences are verified by iteratively spreading the matching process to their neighbors, updating along the way the rigid transformation that aligns the scene and model.

Results The matching algorithm presented in the previous section has been extensively tested in recognition tasks with cluttered indoor scenes that contain both industrial parts and various toys (Johnson and Hebert, 1998, 1999). It has also been used in outdoor navigation/mapping tasks with large datasets covering thousands of squared meters of terrain (Carmichael et al., 1999). Figure 21.19 shows sample recognition results in the toy domain.

21.5 NOTES

Excellent surveys of active range-finding techniques can be found in Jarvis (1983), Nitzan (1988), Besl (1989), and Hebert (2000). The model-based approach to edge detection presented in Section 21.2.2 is only one of the many techniques that have been proposed for segmenting range pictures using notions from differential geometry (see, e.g., Fan et al., 1987, Besl and Jain, 1988). An alternative to the computational molecules used to smooth a range image in that section is provided by anisotropic diffusion, where the amount of smoothing at each point depends on the value of the gradient (Perona and Malik, 1990*c*). The method for segmenting surfaces into (almost) planar patches presented in Section 21.2.3 is easily extended to quadric patches (see Faugeras and Hebert, 1986, and Exercises). Extensions to higher order surface primitives is more problematic, in part because surface fitting is more difficult in that case. There is a vast amount of literature on the latter problem using superquadrics (e.g., Pentland, 1986, Bajcsy and Solina, 1987, Gross and Boult, 1988) and algebraic surfaces (e.g., Taubin et al., 1994*a,b*, Keren et al., 1994, Sullivan et al., 1994*a,b*).

Different variants of the ICP algorithm presented in Section 21.3.2 and Besl and McKay (1992) have been developed over the years, including robust ones capable of handling missing data and/or outliers (e.g., Zhang, 1994, Wheeler and Ikeuchi, 1995), and they have been applied to a number of global registration problems (e.g., Shum et al., 1995, Curless and Levoy, 1996).

Alternatives to the Curless and Levoy (1996) approach to the fusion of multiple range images include the Delaunay triangulation algorithm of Boissonnat (1984), the zippered polygonal meshes of Turk and Levoy (1994), and the crust technique of Amenta et al. (1998). The quaternion-based approach to the estimation of rigid transformations described in this chapter was developed independently by Faugeras and Hebert (1986) and Horn (1987). The recognition technique discussed in Section 21.4.1 is closely related to other algorithms using interpretation trees to control the combinatorial cost of feature matching in the two- and three-dimensional cases (Gaston and Lozano-Pérez, 1984, Ayache and Faugeras, 1986, Grimson and Lozano-Pérez, 1987, Huttenlocher and Ullman, 1987).

The spin images discussed in Section 21.4.2 have been used to establish pointwise correspondences between range images and surface models. Related approaches to this problem include the structural indexing method of Stein and Medioni (1992) and the point signatures proposed by Chua and Jarvis (1996). A (local) variant of the same idea is discussed in chapter 23 in the context of object recognition from photographs (Schmid and Mohr, 1997*a,b*). To conclude, let us note that the original algorithm described in Section 21.4.2 has been extended in various directions: A scene can now be matched simultaneously to several models using principal component analysis (Johnson and Hebert, 1999), while learning techniques are used to prune false matches in cluttered scenes (Carmichael and Hebert, 1999).

PROBLEMS

21.1. Use Eq. (21.1) to show that a necessary and sufficient condition for the coordinate curves of a parameterized surface to be principal directions is that $f = F = 0$.

21.2. Show that the lines of curvature of a surface of revolution are its meridians and parallels.

21.3. Step model: Compute $z_\sigma(x) = G_\sigma * z(x)$, where $z(x)$ is given by Eq. (21.2). Show that z''_σ is given by Eq. (21.3). Conclude that $\kappa''_\sigma/\kappa'_\sigma = -2\delta/h$ in the point x_σ where z''_σ and κ_σ vanish.

21.4. Roof model: Show that κ_σ is given by Eq. (21.4).

21.5. Show that the quaternion $\mathsf{q} = \cos\frac{\theta}{2} + \sin\frac{\theta}{2}\boldsymbol{u}$ represents the rotation $\mathcal{R}$ of angle θ about the unit vector $\boldsymbol{u}$ in the sense of Eq. (21.5).
Hint: Use the Rodrigues formula derived in the exercises of Chapter 3.

21.6. Show that the rotation matrix $\mathcal{R}$ associated with a given unit quaternion $\mathsf{q} = a + \boldsymbol{\alpha}$ with $\boldsymbol{\alpha} = (b, c, d)^T$ is given by Eq. (21.6).

21.7. Show that the matrix $\mathcal{A}_i$ constructed in Section 21.3.2 is equal to

$$\mathcal{A}_i = \begin{pmatrix} 0 & \boldsymbol{y}_i^T - \boldsymbol{y}_i'^T \\ \boldsymbol{y}_i' - \boldsymbol{y}_i & [\boldsymbol{y}_i + \boldsymbol{y}_i']_\times \end{pmatrix}.$$

21.8. As mentioned earlier, the ICP method can be extended to various types of geometric models. We consider here the case of polyhedral models and piecewise parametric patches.
- **(a)** Sketch a method for computing the point Q in a polygon that is closest to some point P.
- **(b)** Sketch a method for computing the point Q in the parametric patch $\boldsymbol{x} : I \times J \to \mathbb{R}^3$ that is closest to some point P. Hint: Use Newton iterations.

21.9. Develop a linear least-squares method for fitting a quadric surface to a set of points under the constraint that the quadratic form has unit Frobenius form.

21.10. Show that a surface triangle maps onto a patch with hyperbolic edges in α, β space for spin images.

Programming Assignments

21.11. Implement molecule-based smoothing and the computation of principal directions and curvatures.

21.12. Implement the region-growing approach to plane segmentation described in this chapter.

21.13. Implement an algorithm for computing the lines of curvature of a surface from its range image. Hint: Use a curve-growing algorithm analogous to the region-growing algorithm for plane segmentation.

21.14. Implement the Besl–McKay ICP registration algorithm.

21.15. Marching squares in the plane: Develop and implement an algorithm for finding the zero set of a planar density function. Hint: Work out the possible ways a curve may intersect the edges of a pixel, and use linear interpolation along these edges to identify the zero set.

21.16. Implement the registration part of the Faugeras–Hebert algorithm.

PART VI

High-Level Vision: Probabilistic and Inferential Methods

22

Finding Templates Using Classifiers

There are a number of important object recognition problems that involve looking for image windows that have a simple shape and stylized content. For example, frontal faces appear as oval windows, and (at a coarse scale) all faces look pretty much the same—a dark horizontal bar at the eyes and mouth, a light vertical bar along the nose, and not much texture on the cheeks and forehead. As another example, a camera mounted on the front of a car always sees relevant stop signs as having about the same shape and appearance.

This suggests a view of object recognition where we take all image windows of a particular shape and test them to tell whether the relevant object is present. If we don't know how big the object will be, we can search over scale, too; if we don't know its orientation, we might search over orientation as well; and so on. Generally, this approach is referred to as *template matching*. There are some objects that can be found effectively with a template matcher. Faces and road signs are important examples. Second, although many objects are hard to find with simple template matchers (it would be hard to find a person this way because the collection of possible image windows that represent a person is immense), there is some evidence that reasoning about *relations* among many different kinds of templates can be an effective way to find objects. In chapter 23, we explore this line of reasoning further.

The main issue to study in template matching is how one builds a test that can tell whether an oval represents a face. Ideally, this test is obtained using a large set of examples. The test is known as a *classifier*—a classifier is anything that takes a feature set as an input and produces a class label. In this chapter, we describe a variety of techniques for building classifiers, with examples of their use in vision applications. We first present the key ideas and terminology used (Section 22.1); we then show two successful classifiers built using histograms (Section 22.2); for more complex classifiers, we need to choose the features a classifier should use; and we discuss two methods in (Section 22.3). Finally, we describe two different methods for building classifiers with current applications in vision. Section 22.4 is an introduction to the use of neural nets in classification, and Section 22.5 describes a useful classifier known as a support vector machine.

22.1 CLASSIFIERS

Classifiers are built by taking a set of labeled examples and using them to come up with a rule that assigns a label to any new example. In the general problem, we have a training dataset $(\mathbf{x}_i, y_i)$; each of the $\boldsymbol{x}_i$ consists of measurements of the properties of different types of object, and the y_i are labels giving the type of the object that generated the example. We know the relative costs of mislabeling each class and must come up with a rule that can take any plausible $\boldsymbol{x}$ and assign a class to it.

The cost of an error significantly affects the decision that is made. In Section 22.1.1, we study this question. It emerges that the probability of a class label given a measurement is the key matter. In Section 22.1.2, we discuss methods for building appropriate models in a general way. Finally, we discuss how to estimate the performance of a given classifier (Section 22.1.5).

22.1.1 Using Loss to Determine Decisions

The choice of classification rule must depend on the cost of making a mistake. For example, doctors engage in classification all the time—given a patient, they produce the name of a disease. A doctor who decided that a patient suffering from a dangerous and easily treated disease is well is going to have problems. It would be better to err on the side of misclassifying healthy patients as sick even if doing so involves treating some healthy patients unnecessarily.

The cost depends on what is misclassified to what. Generally, we write outcomes as $(i \rightarrow j)$, meaning that an item of type i is classified as an item of type j. Each outcome has its own cost, which is known as a *loss*. Hence, we have a loss function that we write as $L(i \rightarrow j)$, meaning the loss incurred when an object of type i is classified as having type j. Since losses associated with correct classification should not affect the design of the classifier, $L(i \rightarrow i)$ must be zero, but the other losses could be any positive numbers.

The *risk function* of a particular classification strategy is the expected loss when using it, as a function of the kind of item. The *total risk* is the total expected loss when using the classifier. Thus, if there were two classes, the total risk of using strategy s would be

$$R(s) = Pr\,\{1 \rightarrow 2 \mid \text{using } s\}\, L(1 \rightarrow 2) + Pr\,\{2 \rightarrow 1 \mid \text{using } s\}\, L(2 \rightarrow 1).$$

The desirable strategy is one that minimizes this total risk.

Building a Two-Class Classifier That Minimizes Total Risk Assume that the classifier can choose between two classes and we have a known loss function. There is some boundary in the feature space, which we call the *decision boundary*, such that points on one side belong to class one and points on the other side to class two.

We can resort to a trick to determine where the decision boundary is. It must be the case that, *for points on the decision boundary of the optimal classifier*, either choice of class has the same expected loss—if this weren't so, we could obtain a better classifier by always choosing one class (and so moving the boundary). This means that, for measurements on the decision boundary, choosing class one yields the same expected loss as choosing class two.

A choice of class one for a point $\boldsymbol{x}$ at the decision boundary yields an expected loss

$$\begin{aligned} P\,\{\text{class is } 2 \mid \boldsymbol{x}\}\, L(2 \rightarrow 1) + P\,\{\text{class is } 1 \mid \boldsymbol{x}\}\, L(1 \rightarrow 1) &= P\,\{\text{class is } 2 \mid \boldsymbol{x}\}\, L(2 \rightarrow 1) + 0 \\ &= p(2 \mid \boldsymbol{x}) L(2 \rightarrow 1). \end{aligned}$$

You should watch the one's and two's closely here. Similarly, a choice of class two for this point yields an expected loss

$$P\{\text{class is } 1 \mid \boldsymbol{x}\} L(1 \to 2) = p(1 \mid \boldsymbol{x}) L(1 \to 2),$$

and these two terms must be equal. This means our decision boundary consists of the points $\boldsymbol{x}$, where

$$p(1 \mid \boldsymbol{x}) L(1 \to 2) = p(2 \mid \boldsymbol{x}) L(2 \to 1).$$

We can come up with an expression that is often slightly more practical by using Bayes' rule. Rewrite our expression as

$$\frac{p(\boldsymbol{x} \mid 1) p(1)}{p(\boldsymbol{x})} L(1 \to 2) = \frac{p(\boldsymbol{x} \mid 2) p(2)}{p(\boldsymbol{x})} L(2 \to 1)$$

and clear denominators to get

$$p(\boldsymbol{x} \mid 1) p(1) L(1 \to 2) = p(\boldsymbol{x} \mid 2) p(2) L(2 \to 1).$$

This expression identifies points $\boldsymbol{x}$ on a class boundary; we now need to know how to classify points off a boundary.

At points off the boundary, we must choose the class with the *lowest* expected loss. Recall that if we choose class two for a point $\boldsymbol{x}$, the expected loss is

$$p(1 \mid \boldsymbol{x}) L(1 \to 2),$$

etc. This means that we should choose class one if

$$p(1 \mid \boldsymbol{x}) L(1 \to 2) > p(2 \mid \boldsymbol{x}) L(2 \to 1)$$

and class two if

$$p(1 \mid \boldsymbol{x}) L(1 \to 2) < p(2 \mid \boldsymbol{x}) L(2 \to 1).$$

A Classifier for Multiple Classes From now on, we assume that $L(i \to j)$ is zero for $i = j$ and one otherwise – that is, that each outcome has the same loss. In some problems, there is another option, which is to refuse to decide which class an object belongs to. This option involves some loss, too, which we assume to be $d < 1$ (if the loss involved in refusing to decide is greater than the loss involved in any decision, then we'd never refuse to decide).

For our loss function, the best strategy, known as the *Bayes classifier*, is given in Algorithm 22.1. The total risk associated with this rule is known as the *Bayes risk*; this is the smallest possible risk that we can have in using a classifier. It is usually rather difficult to know what the Bayes classifier—and hence the Bayes risk—is because the probabilities involved are not known exactly. In a few cases, it is possible to write the rule out explicitly. One way to tell the effectiveness of a technique for building classifiers is to study the behavior of the risk as the number of examples increases (e.g., one might want the risk to converge to the Bayes risk in probability if the number of examples is large). The Bayes risk is seldom zero as Figure 22.1 illustrates.

22.1.2 Overview: Methods for Building Classifiers

Usually, we do not know $Pr\{\mathbf{x} \mid k\}$ exactly—which are often called *class-conditional densities*—or $Pr\{k\}$, and we must determine a classifier from an example dataset. There are two rather general strategies:

Algorithm 22.1: The Bayes classifier classifies points using the posterior probability that an object belongs to a class, the loss function, and the prospect of refusing to decide.

For a loss function

$$L(i \to j) = \begin{cases} 1 & i \neq j \\ 0 & i = j \\ d < 1 & \text{no decision} \end{cases}$$

the best strategy is

- *if* $Pr\{k \mid \mathbf{x}\} > Pr\{i \mid \mathbf{x}\}$ *for all i not equal to k, and if this probability is greater than* $1 - d$*, choose type k*
- *if there are several classes* $k_1 \dots k_j$ *for which* $Pr\{k_1 \mid \mathbf{x}\} = Pr\{k_2 \mid \mathbf{x}\} = \cdots = Pr\{k_j \mid \mathbf{x}\} > Pr\{i \mid \mathbf{x}\}$ *for all i not in* $k_1, \dots k_j$*, choose uniformly and at random between* $k_1, \dots k_j$
- *if for all k we have* $Pr\{k \mid \mathbf{x}\} \leq 1 - d$*, refuse to decide.*

- **Explicit probability models:** We can use the example data set to build a probability model (of either the likelihood or the posterior, depending on taste). There is a wide variety of ways of doing this, some of which we see in the following sections. In the simplest case, we know that the class-conditional densities come from some known parametric form of distribution. In this case, we can compute estimates of the parameters from the dataset and plug these estimates into the Bayes rule. This strategy is often known as a *plug-in classifier* (Section 22.1.3). This approach covers other parametric density models and other methods of estimating parameters. One subtlety is that the best estimate of a parameter may not give the best classifier because the parametric model may not

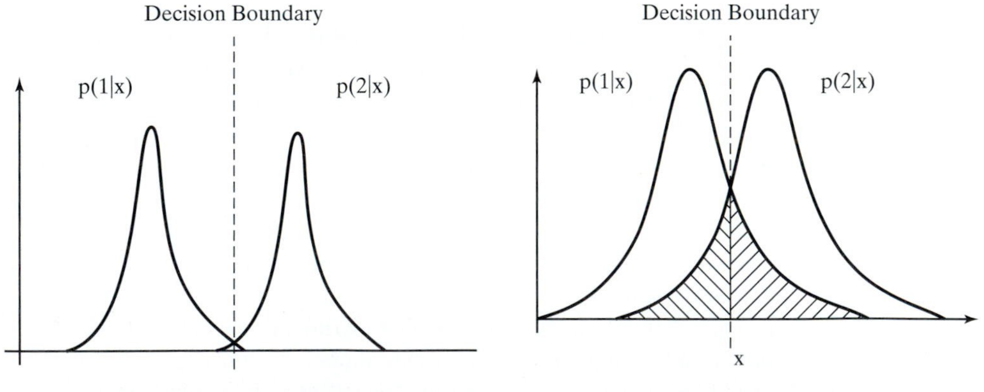

Figure 22.1 This figure shows typical elements of a two-class classification problem. We have plotted $p(\text{class}|x)$ as a function of the feature x. Assuming that $L(1 \to 2) = L(2 \to 1)$, we have marked the classifier boundaries. In this case, the Bayes risk is the sum of the amount of the posterior for class one in the class two region and the amount of the posterior for class two in the class one region (the hatched area in the figures). For the case on the **left**, the classes are well separated, which means that the Bayes risk is small; for the case on the **right**, the Bayes risk is rather large.

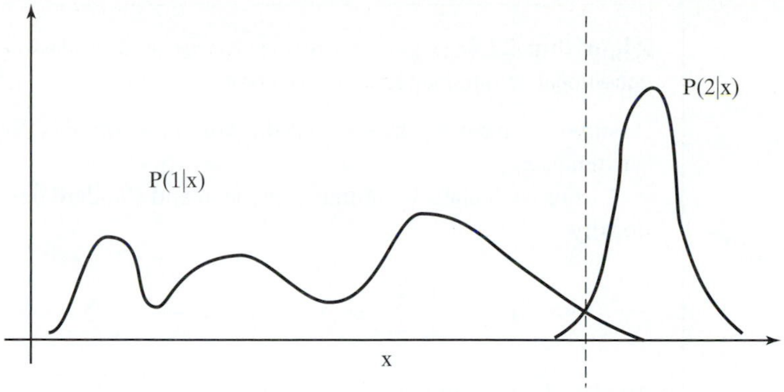

Figure 22.2 The figure shows posterior densities for two classes. The optimal decision boundary is shown as a dashed line. Notice that, although a normal density may provide rather a poor fit *to the posteriors*, the quality of the classifier it provides depends only on *how well it predicts the position of the boundaries*. In this case, assuming that the posteriors are normal may provide a fairly good classifier because $P(2|x)$ looks normal, and the mean and covariance of $P(1|x)$ look as if they would predict the boundary in the right place.

be correct. Another subtlety is that a good classifier may be obtained using a parametric density model that is not an accurate description of the data (see Figure 22.2). In many cases, it is hard to obtain a satisfactory model with a small number of parameters. More sophisticated modeling tools (such as neural nets, which we deal with in some detail in Section 22.4) provides flexible density models that can be fitted using data.

- **Determining decision boundaries directly:** Quite bad probability models can produce good classifiers, as Figure 22.2 indicates. This is because the decision boundaries are what determine the performance of a classifier, not the details of the probability model (the main role of the probability model in the Bayes classifier is to identify the decision boundaries). This suggests that we might ignore the probability model and attempt to construct good decision boundaries directly. This approach is often extremely successful; it is particularly attractive when there is no reasonable prospect of modeling the data source. One strategy assumes that the decision boundary comes from one or another class and constructs an extremization problem to choose the best element of that class. A particularly important case comes when the data is *linearly separable*—which means that there exists a hyperplane with all the positive points on one side and all the negative points on the other—and thus that a hyperplane is all that is needed to separate the data (Section 22.5).

22.1.3 Example: A Plug-in Classifier for Normal Class-conditional Densities

An important plug-in classifier occurs when the class-conditional densities are known to be normal. We can either assume that the priors are known or estimate the priors by counting the number of data items from each class. Now we need to provide the parameters for the class-conditional densities. We do this as an estimation problem using the data items to estimate the mean $\boldsymbol{\mu}_k$ and covariance Σ_k for each class. Now since $\log a > \log b$ implies $a > b$, we can work with the logarithm of the posterior. This yields a classifier of the form in Algorithm 22.2.

The term $\delta(\mathbf{x}; \boldsymbol{\mu}_k, \Sigma_k)$ in this algorithm is known as the *Mahalanobis distance* (e.g., see Ripley, 1996). The algorithm can be interpreted geometrically as saying that the correct

Algorithm 22.2: A plug-in classifier can be used to classify objects into classes if the class-conditional densities are known to be normal

Assume we have N classes, and the kth class contains N_k examples, of which the ith is written as $x_{k,i}$.

For each class k, estimate the mean and standard deviation for that class-conditional density.

$$\mu_k = \frac{1}{N_k}\sum_{i=1}^{N_k} x_{k,i}; \qquad \Sigma_k = \frac{1}{N_k - 1}\sum_{i=1}^{N_k}(x_{k,i} - \mu_k)(x_{k,i} - \mu_k)^T;$$

To classify an example x,

Choose the class k with the smallest value of $\delta(\mathbf{x}; \mu_k, \Sigma_k)^2 - Pr\{k\} + \frac{1}{2}\log|\Sigma_k|$

where

$$\delta(\mathbf{x}; \mu_k, \Sigma_k) = \frac{1}{2}\left((\mathbf{x} - \mu_k)^T \Sigma_k^{-1}(\mathbf{x} - \mu_k)\right)^{(1/2)}.$$

class is the one whose mean is closest to the data item *taking into account the variance*. In particular, distance from a mean along a direction where there is little variance has a large weight and distance from the mean along a direction where there is a large variance has little weight. This classifier can be simplified by assuming that each class has the same covariance (with the advantage that we have fewer parameters to estimate). In this case, because the term $x^T\Sigma^{-1}x$ is common to all expressions, the classifier actually involves comparing expressions that are *linear in* x (see Exercises). If there are only two classes, the process boils down to determining whether a linear expression in $\mathbf{x}$ is greater than or less than zero (Exercises).

22.1.4 Example: A Nonparametric Classifier Using Nearest Neighbors

It is reasonable to assume that example points near an unclassified point should indicate the class of that point. *Nearest neighbors* methods build classifiers using this heuristic. We could classify a point by using the class of the nearest example whose class is known, or use several example points and make them vote. It is reasonable to require that some minimum number of points vote for the class we choose.

A (k, l) nearest neighbor classifier finds the k example points closest to the point being considered, and classifies this point with the class that has the highest number of votes, as long as this class has more than l votes (otherwise the point is classified as unknown). A $(k, 0)$-nearest neighbor classifier is usually known as a *k-nearest neighbor classifier*, and a $(1, 0)$-nearest neighbor classifier is usually known as a *nearest neighbor classifier*.

Nearest neighbor classifiers are known to be good, in the sense that the risk of using a nearest neighbor classifier with a sufficiently large number of examples lies within quite good bounds of the Bayes risk. As k grows, the difference between the Bayes risk and the risk of using a k-nearest neighbor classifier goes down as $1/\sqrt{k}$. In practice, one seldom uses more than three nearest neighbors. Furthermore, if the Bayes risk is zero, the expected risk of using a k-nearest neighbor classifier is also zero (see Devroye, Gyorfi and Lugosi, 1996 for more detail on all these points).

Nearest neighbor classifiers come with some computational subtleties, however. The first is the question of finding the k nearest points, which is no mean task in a high-dimensional space (surprisingly, checking the distance to each separate example one by one is, at present, a fairly competitive algorithm). This task can be simplified by noticing that some of the example points may be superfluous. If, when we remove a point from the example set, the set still classifies every point in the space in the same way (the decision boundaries have not moved), that point is redundant and can be removed. However, it is hard to know *which* points to remove. The decision regions for (k, l)-nearest neighbor classifiers are convex polytopes; this makes familiar algorithms available in 2D (where Voronoi diagrams implement the nearest neighbor classifier) but leads to complications in high dimensions.

Algorithm 22.3: A (k, l) nearest neighbor classifier uses the type of the nearest training examples to classify a feature vector

Given an feature vector $\boldsymbol{x}$

1. determine the k training examples that are nearest, $\boldsymbol{x}_1, \ldots, \boldsymbol{x}_k$;
2. determine the class c that has the largest number of representatives n in this set;
3. if $n > l$, classify $\boldsymbol{x}$ as c, otherwise refuse to classify it.

A second difficulty in building such classifiers is the choice of distance. For features that are obviously of the same type, such as lengths, the usual metric may be good enough. But what if one feature is a length, one is a color, and one is an angle? One possibility is to use a covariance estimate to compute a Mahalanobis-like distance.

22.1.5 Estimating and Improving Performance

Typically, classifiers are chosen to work well on the training set, and this can mean that the performance of the classifier on the training set is a poor guide to its overall performance. One example of this problem is the (silly) classifier that takes any data point and, if it is the same as a point in the training set, emits the class of that point and otherwise chooses randomly between the classes. This classifier has been learned from data, and has a zero error rate on the training dataset; it is likely to be unhelpful on any other dataset, however.

The difficulty occurs because classifiers are subject to *overfitting* effects. The phenomenon, which is known by a variety of names (*selection bias* is quite widely used), has to do with the fact that the classifier is chosen to perform well *on the training dataset*. The training data is a (possibly representative) subset of the available possibilities. The term *overfitting* is descriptive of the source of the problem, which is that the classifier's performance on the training dataset may have to do with quirks of that dataset that don't occur in other sets of examples. If the classifier does this, it is quite possible that it will perform well on the training data and badly on any other dataset (this phenomenon is often referred to as *generalizing badly*).

Generally, we expect classifiers to perform somewhat better on the training set than on the test set (e.g., see Figure 22.18, which shows training set and test set errors for a classifier that is known to work well). Overfitting can result in a substantial difference between performance on the training set and performance on the test set. This leaves us with the problem of predicting performance. There are two possible approaches: We can hold back some training data to check the performance of the classifier (an approach we describe later), or we can use theoretical methods to bound the future error rate of the classifier (see, e.g., Vapnik, 1996 or 1998).

Estimating Total Risk with Cross-Validation We can make direct estimates of the expected risk of using a classifier if we split the dataset into two subsets, train the classifier on one subset, and test it on the other. This is a waste of data, particularly if we have few data items for a particular class, and may lead to an inferior classifier. However, if the size of the test subset is small, the difficulty may not be significant. In particular, we could then estimate total risk by averaging over all possible splits. This technique, known as *cross-validation* allows an estimate of the likely future performance of a classifier, at the expense of substantial computation.

Algorithm 22.4: Cross-Validation

Choose some class of subsets of the training set, for example, singletons.

For each element of that class, construct a classifier by omitting that element in training, and compute the classification errors (or risk) on the omitted subset.

Average these errors over the class of subsets to estimate the risk of using the classifier trained on the entire training dataset.

The most usual form of this algorithm involves omitting single items from the dataset and is known as *leave-one-out cross-validation*. Errors are usually estimated by simply averaging over the class, but more sophisticated estimates are available (see, e.g., Ripley, 1996). We do not justify this tool mathematically; however, it is worth noticing that leave-one-out cross-validation, in some sense, looks at the sensitivity of the classifier to a small change in the training set. If a classifier performs well under this test, then large subsets of the dataset look similar to one another, which suggests that a representation of the relevant probabilities derived from the dataset might be quite good.

Using Bootstrapping to Improve Performance Generally, more training data leads to a better classifier. However, training classifiers with large datasets can be difficult, and there are diminishing returns. Typically, only a relatively small number of example items are really important in determining the behavior of a classifier (we see this phenomenon in greater detail in Section 22.5). The really important examples tend to be rare cases that are quite hard to discriminate. This is because these cases affect the position of the decision boundary most significantly. We need a large dataset to ensure that these cases are present, but it appears inefficient to go to great effort to train on a large dataset, most of whose elements aren't particularly important.

There is a useful trick that avoids much redundant work. We train on a subset of the examples, run the resulting classifier on the rest of the examples, and then insert the false positives and false negatives into the training set to retrain the classifier. This is because the false positives and false negatives are the cases that give the most information about errors in the configuration of the decision boundaries. This strategy is known as *bootstrapping* (the name is potentially confusing because there is an unrelated statistical procedure known as bootstrapping; nonetheless, we're stuck with it at this point).

22.2 BUILDING CLASSIFIERS FROM CLASS HISTOGRAMS

One simple way to build a probability model for a classifier is to use a histogram. If a histogram is divided by the total number of pixels, we get a representation of the class-conditional probability density function. It is a fact that, as the dataset gets larger and the histogram bins get smaller, the histogram divided by the total number of data items will almost certainly converge to the probability density function (e.g., Devroye, Gyorfi and Lugosi, 1996, Vapnik, 1996, 1998). In low-dimensional problems, this approach can work quite well (Section 22.2.1). It isn't practical for high-dimensional data because the number of histogram bins required quickly becomes intractable unless we use strong independence assumptions to control the complexity (Section 22.2.2).

22.2.1 Finding Skin Pixels Using a Classifier

Skin finding is useful for activities like building gesture-based interfaces. Skin has a quite characteristic range of colors, suggesting that we can build a skin finder by classifying pixels on their color. Jones and Rehg (1999) construct a histogram of RGB values due to skin pixels and a second histogram of RGB values due to non-skin pixels. These histograms serve as models of the class-conditional densities.

We write $\boldsymbol{x}$ for a vector containing the color values at a pixel. We subdivide this color space into boxes and count the percentage of skin pixels that fall into each box—this histogram supplies $p(\boldsymbol{x} \mid \text{skin pixel})$, which we can evaluate by determining the box corresponding to $\boldsymbol{x}$ and then reporting the percentage of skin pixels in this box. Similarly, a count of the percentage of non skin pixels that fall into each box supplies $p(\boldsymbol{x} \mid \text{not skin pixel})$. We need $p(\text{skin pixel})$ and $p(\text{not skin pixel})$—or rather, we need only one of the two as they sum to one. Assume for the moment that the prior is known. We can now build a classifier using Bayes' rule to obtain the posterior (keep in mind that $p(\boldsymbol{x})$ is easily computed as $p(\boldsymbol{x} \mid \text{skin pixel}) + p(\boldsymbol{x} \mid \text{not skin pixel})$).

One way to estimate the prior is to model $p(\text{skin pixel})$ as the fraction of skin pixels in some (ideally large) training set. Notice that our classifier compares

$$\frac{p(\boldsymbol{x} \mid \text{skin})p(\text{skin})}{p(\boldsymbol{x})} L(\text{skin} \to \text{not skin})$$

with

$$\frac{p(\boldsymbol{x} \mid \text{not skin})p(\text{not skin})}{p(\boldsymbol{x})} L(\text{not skin} \to \text{skin}).$$

Now by rearranging terms and noticing that $p(\text{skin} \mid \boldsymbol{x}) = 1 - p(\text{not skin} \mid \boldsymbol{x})$, our classifier becomes

- if $p(\text{skin} \mid \boldsymbol{x}) > \theta$, classify as skin
- if $p(\text{skin} \mid \boldsymbol{x}) < \theta$, classify as not skin
- if $p(\text{skin} \mid \boldsymbol{x}) = \theta$, choose classes uniformly and at random

where θ is an expression that doesn't depend on $\boldsymbol{x}$ and encapsulates the relative loss. This yields a family of classifiers, one for each choice of θ. For an appropriate choice of θ, the classifier can be good (Figure 22.3).

Each classifier in this family has a different false-positive and false-negative rate. These rates are functions of θ, so we can plot a parametric curve that captures the performance of the family of classifiers. This curve is known as a *receiver operating curve* (*ROC*). Figure 22.4 shows the ROC for a skin finder built using this approach. The ROC is invariant to choice of prior (Exercises)—this means that if we change the value of $p(\text{skin})$, we can choose some new value of θ to get a classifier with the same performance. This yields another approach to estimating a prior. We choose some value rather arbitrarily, plot the loss on the training set as a function of θ, and then select the value of θ that minimizes this loss.

Figure 22.3 The figure shows a variety of images together with the output of the skin detector of Jones and Rehg applied to the image. Pixels marked black are skin pixels and white are background. Notice that this process is relatively effective and could certainly be used to focus attention on, say, faces and hands. *Reprinted from "Statistical color models with application to skin detection," by M.J. Jones and J. Rehg, Proc. Computer Vision and Pattern Recognition, 1999 © 1999, IEEE*

22.2.2 Face Finding Assuming Independent Template Responses

Histogram models become impractical in high dimensions because the number of boxes required goes up as a power of the dimension. We can dodge this phenomenon. Recall that independence assumptions reduce the number of parameters that must be learned in a probabilistic model (or see the chapter on probability on the book website); by assuming that terms are independent, we can reduce the dimension sufficiently to use histograms. Although this appears to be an aggressive oversimplification—it is known by the pejorative name of *naive Bayes*— it can result in useful systems. In one such system, due to Schneiderman and Kanade (1998), this model is used to find faces. Assume that the face occurs at a fixed, known scale (we could search smoothed and resampled versions of the image to find larger faces) and occupies a region of known shape. In the case of frontal faces, this might be an oval or a square; for a lateral face, this might be some more complicated polygon. We now need to model the image pattern generated by the face. This is a likelihood model—we want a model giving $P(\text{image pattern} \mid \text{face})$. As usual, it is helpful to think in terms of generative models; the process by which a face gives rise to an image patch. The set of possible image patches is somewhat difficult to deal with because it is big, but we can

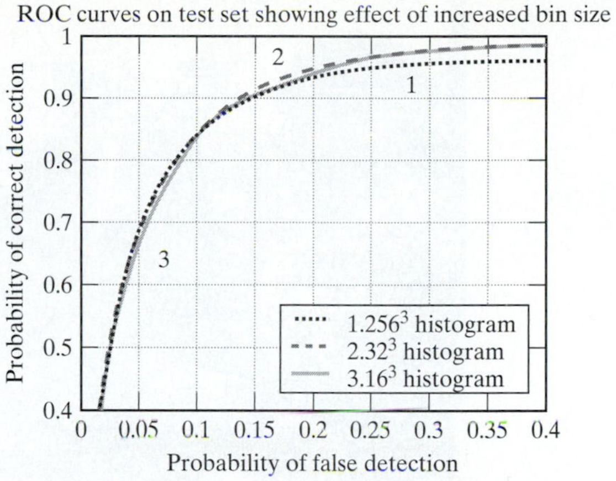

Figure 22.4 The receiver operating curve for the skin detector of Jones and Rehg. This plots the detection rate against the false-negative rate for a variety of values of the parameter θ. A perfect classifier has an ROC that, on these axes, is a horizontal line at 100% detection. Notice that the ROC varies slightly with the number of boxes in the histogram. *Reprinted from "Statistical color models with application to skin detection," by M.J. Jones and J. Rehg, Proc. Computer Vision and Pattern Recognition, 1999 © 1999, IEEE*

avoid this by dividing the image patch into a set of subregions and then labeling the subregions using a small set of labels.

An appropriate labeling can be obtained using a clustering algorithm and a large number of example images. For example, we might cluster the subregions in a large number of example images using k-means; now each cluster center represents a typical form of subregion. The subregions in our image patch can then be labeled with the cluster center to which they are closest. This approach has the advantage that minor variations in the image pattern—caused perhaps by noise, skin irregularities, and so on—are suppressed.

At this point, a number of models are available. The simplest practical model is to assume that the probability of encountering each pattern is independent of the configuration of the other patterns (but not of position) given that a face is present. This means that our model is

$$\begin{aligned} P(\text{image} \mid \text{face}) &= P(\text{label 1 at } (x_1, y_1), \ldots, \text{label } k \text{ at } (x_k, y_k) \mid \text{face}) \\ &= P(\text{label 1 at } (x_1, y_1) \mid \text{face}) \ldots P(\text{label } k \text{ at } (x_k, y_k) \mid \text{face}). \end{aligned}$$

In this case, each term of the form $P(\text{label } k \text{ at } (x_k, y_k) \mid \text{face})$ can be learned fairly easily by labeling a large number of example images and then forming a histogram. Because the histograms are now two dimensional, the number of boxes is no longer problematic. A similar line of reasoning leads to a model of $P(\text{image} \mid \text{no face})$. A classifier follows from the line of reasoning given earlier. This approach has been used successfully by Schneiderman and Kanade to build detectors for faces and cars (Figure 22.5).

22.3 FEATURE SELECTION

Assume we have a set of pixels that we believe belong together and should be classified. What features should we present to a classifier? One approach is to present all the pixel values: This

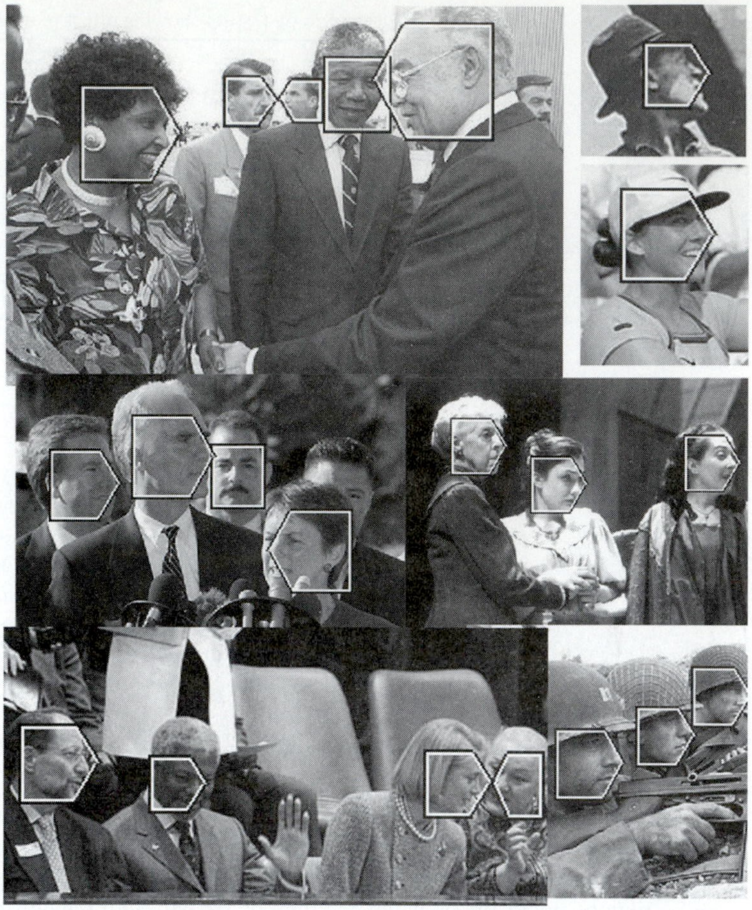

Figure 22.5 Faces found using the method of Section 22.2.2. Image windows at various scales are classified as frontal face, lateral face, or non-face using a likelihood model learned from data. Subregions in the image window are classified into a set of classes learned from data; the face model assumes that labels from these classes are emitted independently of one another at different positions. This likelihood model yields a posterior value for each class and for each window, and the posterior value is used to identify the window. *Reprinted from A Statistical Method for 3D Object Detection Applied to Faces and Cars, H. Schneiderman and T. Kanade, Proc. Computer Vision and Pattern Recognition, 2000, © 2000, IEEE*

gives the classifier the maximum possible amount of information about the set of pixels, but creates a variety of problems.

First, high-dimensional spaces are big in the sense that large numbers of examples can be required to represent the available possibilities fairly. For example, a face at low resolution has a fairly simple structure: It consists (rather roughly) of some dark bars (the eyebrows and eyes) and light bars (the specular reflections from the nose and forehead) on a textureless background. However, if we are working with high-resolution faces, it might be difficult to supply enough examples to determine that this structure is significant and that minor variations in skin texture are irrelevant. Instead, we would like to choose a feature space that would make these properties obvious, typically by imposing some form of structure on the examples.

Second, we may know some properties of the patterns in advance. For example, we have models of the behavior of illumination. Forcing a classifier to use examples to, in essence, come up with a model that we already know is a waste of examples. We would like to use features that are consistent with our knowledge of the patterns. This might involve preprocessing regions (e.g., to remove the effects of illumination changes) or choosing features that are invariant to some kinds of transformation (e.g., scaling an image region to a standard size).

You should notice a similarity between feature selection and model selection (as described in Section 16.3). In model selection, we attempt to obtain a model that best explains a dataset; here we are attempting to find a set of features that best classifies a dataset. The two are basically the same activity in slightly distinct forms (you can view a set of features as a model and classification as explanation). Here we describe methods that are used mainly for feature selection. We concentrate on two standard methods for obtaining *linear features*—features that are a linear function of the initial feature set.

22.3.1 Principal Component Analysis

The core goal in feature selection is to obtain a smaller set of features that accurately represents the original set. What this means rather depends on the application. However, one important possibility is that the new set of features should capture as much of the old set's variance as possible. The easiest way to see this is to consider an extreme example. If the value of one feature can be predicted precisely from the value of the others, it is clearly redundant and can be dropped. By this argument, if we are going to drop a feature, the best one to drop is the one whose value is most accurately predicted by the others. We can do more than drop features: We can make new features as functions of the old features.

In *principal component analysis*, the new features are linear functions of the old features. In principal component analysis, we take a set of data points and construct a lower dimensional linear subspace that best explains the variation of these data points from their mean. This method (also known as the Karhunen–Loéve transform) is a classical technique from statistical pattern recognition (Duda and Hart, 1973, Oja, 1983, Fukunaga, 1990).

Assume we have a set of n feature vectors $\boldsymbol{x}_i$ $(i = 1, \ldots, n)$ in $\mathbb{R}^d$. The mean of this set of feature vectors is $\boldsymbol{\mu}$ (you should think of the mean as the center of gravity in this case), and their covariance is Σ (you can think of the variance as a matrix of second moments). We use the mean as an origin and study the offsets from the mean $(\boldsymbol{x}_i - \boldsymbol{\mu})$.

Our features are linear combinations of the original features; this means it is natural to consider the projection of these offsets onto various different directions. A unit vector $\boldsymbol{v}$ represents a direction in the original feature space; we can interpret this direction as a new feature $v(\boldsymbol{x})$. The value of u on the ith data point is given by $v(\boldsymbol{x}_i) = \boldsymbol{v}^T(\boldsymbol{x}_i - \boldsymbol{\mu})$. A good feature captures as much of the variance of the original dataset as possible. Notice that v has zero mean; then the variance of v is

$$\begin{aligned}
\text{var}(v) &= \frac{1}{n-1}\sum_{i=1}^{n} v(\boldsymbol{x}_i)v(\boldsymbol{x}_i)^T \\
&= \frac{1}{n}\sum_{i=1}^{n-1} \boldsymbol{v}^T(\boldsymbol{x}_i - \boldsymbol{\mu})(\boldsymbol{v}^T(\boldsymbol{x}_i - \boldsymbol{\mu}))^T \\
&= \boldsymbol{v}^T\left\{\sum_{i=1}^{n-1}(\boldsymbol{x}_i - \boldsymbol{\mu})(\boldsymbol{x}_i - \boldsymbol{\mu})^T\right\}\boldsymbol{v} \\
&= \boldsymbol{v}^T\Sigma\boldsymbol{v}.
\end{aligned}$$

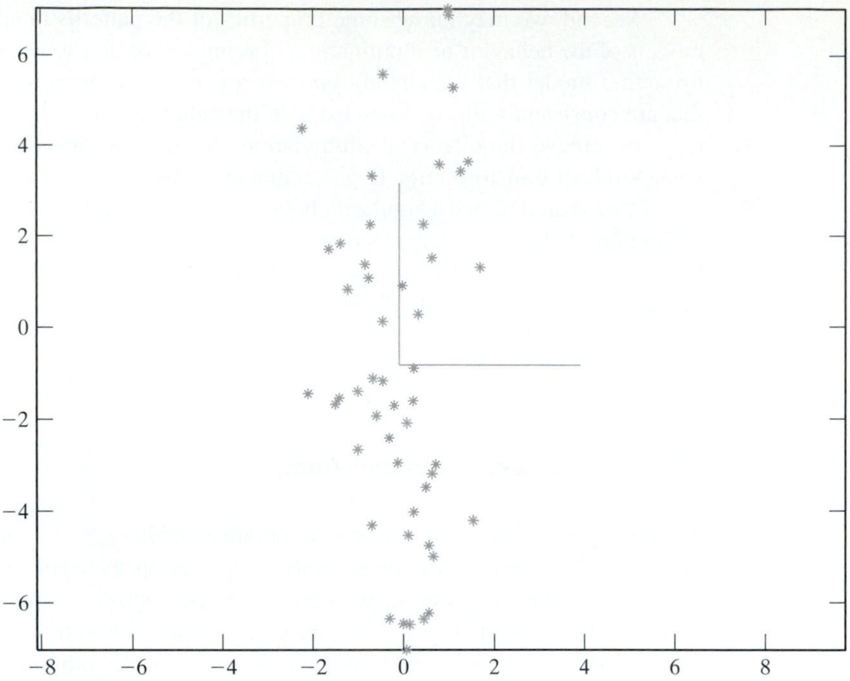

Figure 22.6 A dataset that is well represented by a principal component analysis. The axes represent the directions obtained using PCA; the vertical axis is the first principal component, and is the direction in which the variance is highest.

Now we should like to maximize $v^T \Sigma v$ subject to the constraint that $v^T v = 1$. This is an eigenvalue problem; the eigenvector of Σ corresponding to the largest eigenvalue is the solution. Now if we were to project the data onto a space *perpendicular* to this eigenvector, we would obtain a collection of $d - 1$ dimensional vectors. The highest variance feature for this collection would be the eigenvector of Σ with second largest eigenvalue, and so on.

This means that the eigenvectors of Σ—which we write as $v_1, v_2, \ldots, v_d$, where the order is given by the size of the eigenvalue and v_1 has the largest eigenvalue—give a set of features with the following properties:

- They are independent (because the eigenvectors are orthogonal).
- Projection onto the basis $\{v_1, \ldots, v_k\}$ gives the k-dimensional set of linear features that preserves the most variance.

You should notice that, depending on the data source, principal components can give a good or a bad representation of a data set (see Figures 22.6 and 22.7, and Figure 22.9).

22.3.2 Identifying Individuals with Principal Components Analysis

People are extremely good at remembering and recognizing a large number of faces, and mimicking this ability in an automated computer system has a wide range of applications, including human computer interaction and security. Kanade (1973) developed the first fully automated system for face recognition, and many other approaches have since been proposed to address various instances of this problem (e.g., fixed head orientation, fixed expression etc.; see Chel-

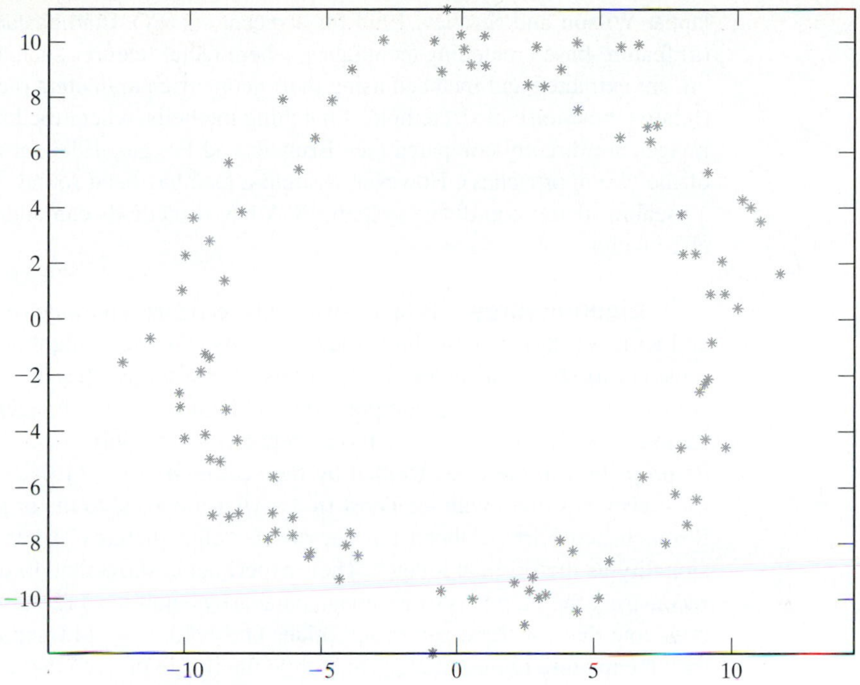

Figure 22.7 Not every dataset is well represented by PCA. The principal components of this dataset are relatively unstable, because the variance in each direction is the same for the source. This means that we may well report significantly different principal components for different datasets from this source. This is a secondary issue—the main difficulty is that projecting the dataset onto some axis suppresses the main feature, its circular structure.

Algorithm 22.5: Principal components analysis identifies a collection of linear features that are independent, and capture as much variance as possible from a dataset.

Assume we have a set of n feature vectors $\boldsymbol{x}_i$ $(i = 1, \ldots, n)$ in $\mathbb{R}^d$. Write

$$\boldsymbol{\mu} = \frac{1}{n}\sum_i \boldsymbol{x}_i$$

$$\Sigma = \frac{1}{n-1}\sum_i (\boldsymbol{x}_i - \boldsymbol{\mu})(\boldsymbol{x}_i - \boldsymbol{\mu})^T$$

The unit eigenvectors of Σ—which we write as $\boldsymbol{v}_1, \boldsymbol{v}_2, \ldots, \boldsymbol{v}_d$, where the order is given by the size of the eigenvalue and $\boldsymbol{v}_1$ has the largest eigenvalue—give a set of features with the following properties:

- They are independent.
- Projection onto the basis $\{\boldsymbol{v}_1, \ldots, \boldsymbol{v}_k\}$ gives the k-dimensional set of linear features that preserves the most variance.

lappa, Wilson and Sirohey, 1995 for a recent survey). Finding faces typically involves either: (a) feature-based matching techniques, where facial features such as the nose, lips, eyes, and so on, are extracted and matched using their geometric parameters (height, width) and relationship (relative position); or (b) template matching methods, where the brightness patterns of two face images are directly compared (see Brunelli and Poggio, 1993 for a discussion and comparison of the two approaches). However, assume a face has been found: Whose face is it? If we have a useable spatial coordinate system, PCA based methods can address this problem simply and effectively.

Eigenpictures People can quickly recognize enormous numbers of faces, and Sirovitch and Kirby (1987) suggested that the human visual system might only use a small number of parameters to store and index face pictures. Accordingly, they investigated the use of principal component analysis as a compression technique for face images. Indeed, PCA allows each sample $\boldsymbol{s}_i \in \mathbb{R}^d$ $(i = 1, \dots, n)$ to be represented by only $p \ll d$ numbers, the coordinates of its projection in the basis formed by the vectors $\boldsymbol{u}_j$ $(j = 1, \dots, p)$. Of course, these vectors must also be stored, with total cost $(n + d)p$ as opposed to the original nd storage requirement. Sirovich and Kirby dubbed the vectors $\boldsymbol{u}_i$ "eigenpictures" since they have the same dimensionality as the original images. Their experiments show that 40 eigenpictures are sufficient to reconstruct the members of an image database containing 115 128×128 faces with a 3% error rate.Note that for these parameter values ($p = 40$, $n = 115$ and $d = 128 \times 128$), the size of the eigenpicture representation (including the image projections) is less than half the size of the original database. Sirovich and Kirby also show that face images of persons who are not part of this database yield error rates inferior to 8%, even under adverse lighting conditions, which suggests an excellent extrapolating power.

Eigenfaces Although the idea of using eigenpictures for face recognition is implicit in their paper, Sirovich and Kirby did not propose an explicit recognition algorithm. This was done by Turk and Pentland (1991*a*), who presented a full recognition system based on a variant of the nearest neighbor classification scheme summarized at the beginning of this section. In the process, they renamed eigenpictures "eigenfaces."

The recognition algorithm is divided into the following steps:

Off-line:

1. Collect a set of pictures of m persons reflecting variations in expression, pose, and lighting.
2. Compute the eigenfaces $\boldsymbol{u}_i$ $(i = 1, \dots, p)$.
3. For each person in the database, calculate the corresponding representative vector $\boldsymbol{w}_j$ $(j = 1, \dots, m)$ in the subspace V_p spanned by the eigenfaces.

On-line:

4. Compute the projection $\boldsymbol{w}$ of any new image $\boldsymbol{t}$ onto V_p.
5. If the distance $d = |\boldsymbol{t} - \boldsymbol{w}|$ is greater than some preset threshold ε_1, classify the image as "non-face."
6. Otherwise, if the minimum distance $d_k = |\boldsymbol{w} - \boldsymbol{w}_k|$ between the projection of the new image and the known face representatives is smaller than some present threshold ε_2, classify the image as "person number k."
7. In the remaining case ($d < \varepsilon_1$ and $d_k \geq \varepsilon_2$), classify the image as "unknown person", and (optional) add the new image to the database and recompute the eigenfaces.

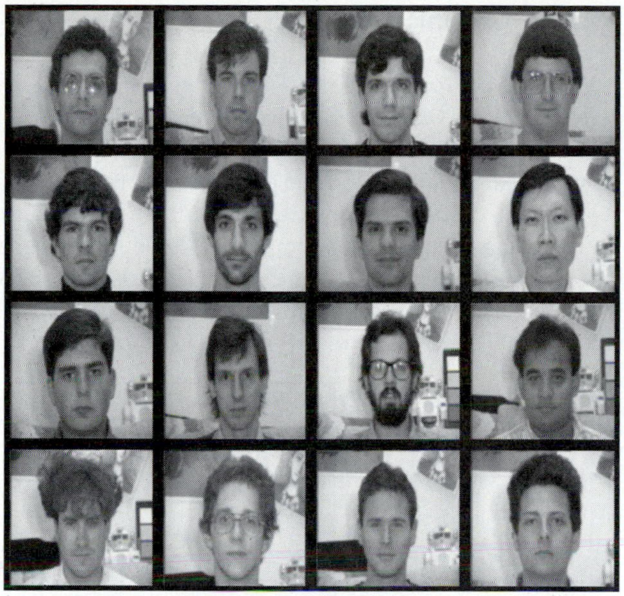

Figure 22.8 A subset of the image database used in the experiments of Turk and Pentland (1991*b*). Variations in lighting, orientation and scale are also included in the actual dataset but are not shown here. *Reprinted from "Face recognition using eigenfaces," by M. Turk and A. Pentland, Proc. Computer Vision and Pattern Recognition, 1991 © 1991, IEEE*

At this point, we should clarify how the representative vectors $\boldsymbol{w}_j$ are computed (step 3). Turk and Pentland (1991*a*) propose averaging the eigenface pattern vectors, (i.e., the projections of the images associated with class number j onto S_p). This is an alternative to the nearest-neighbor classification approach presented earlier.

In their experiments, Turk and Pentland use an image database of 2,500 128 × 128 images of 16 different subjects, corresponding to all combinations of three face orientations, three head scales, and three lighting conditions (Figure 22.8).

Table 22.1 gives quantitative recognition results. In the corresponding experiments, training sets are chosen among various groups of 16 images in the original database, making sure each person appears in each training set. All images in the database are then classified. Statistics are collected by measuring the mean variation between training and test conditions. Illumination, scale, and orientation are varied independently.

TABLE 22.1 Recognition results. The experimental conditions are set by changing the value of ε_1 (e.g., $\varepsilon_1 = +\infty$ to force classification). Lower (resp. higher) values of ε_1 yields more (resp. less) accurate recognition results, but higher (resp. lower) unknown classification results.

Experimental	Correct/Unknown Recognition Percentage		
Condition	Lighting	Orientation	Scale
Forced classification	96/0	85/0	64/0
Forced 100% accuracy	100/19	100/39	100/60
Forced 20% unknown rate	100/20	94/20	74/20

22.3.3 Canonical Variates

Principal component analysis yields a set of linear features of a particular dimension that best represents the variance in a high-dimensional dataset. There is no guarantee that this set of features is good for *classification*. For example, Figure 22.9 shows a dataset where the first principal component would yield a bad classifier, and the second principal component would yield quite a good one, despite not capturing the variance of the dataset.

Linear features that emphasize the distinction between classes are known as *canonical variates*. To construct canonical variates, assume that we have a set of data items $\boldsymbol{x}_i$, for $i \in \{1, \ldots, n\}$. We assume that there are p features (i.e., that the $\boldsymbol{x}_i$ are p-dimensional vectors). We

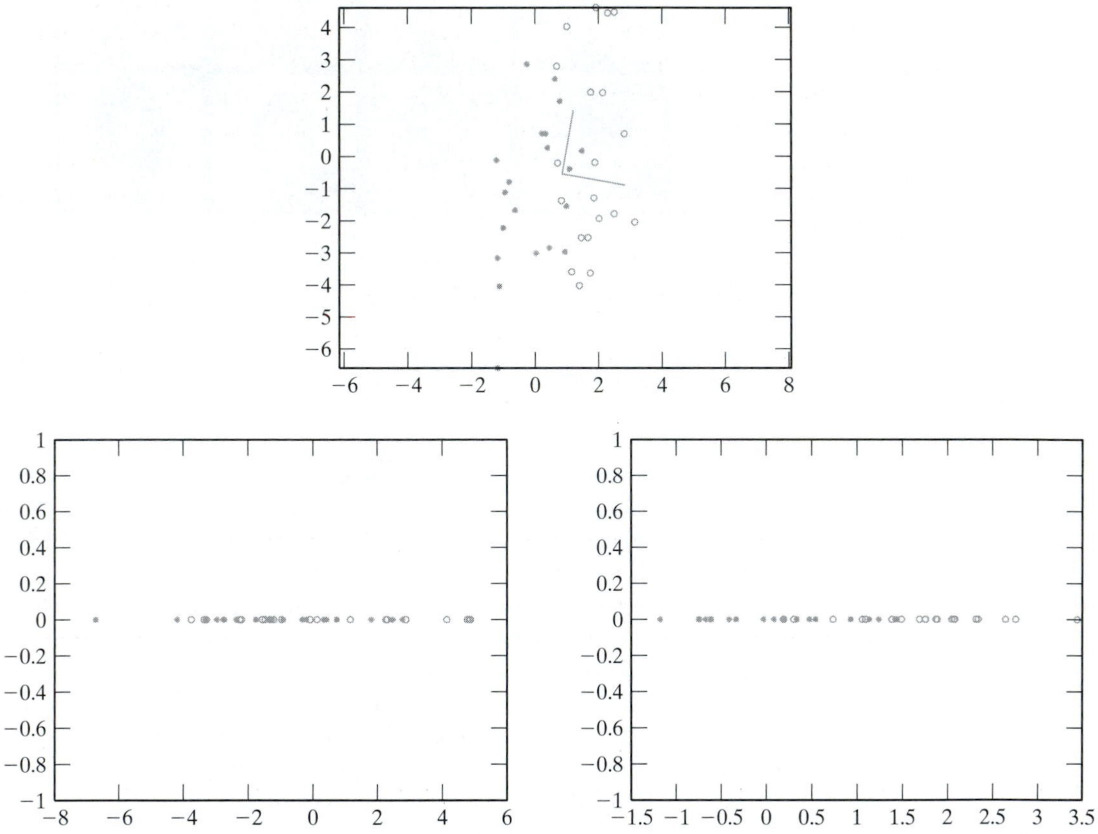

Figure 22.9 Principal component analysis doesn't take into account the fact that there may be more than one class of item in a dataset. This can lead to significant problems. For a classifier, we would like to obtain a set of features that reduces the number of features and makes the difference between classes most obvious. For the dataset on the **top**, one class is indicated by circles and the other by stars. PCA would suggest projection onto a vertical axis, which captures the variance in the dataset, but cannot be used to discriminate it as we can see from the axes obtained by PCA, which are overlaid on the dataset. The **bottom row** shows the projections onto those axes. On the **bottom left**, we show the projection onto the first principal component, which has higher variance but separates the classes poorly, and on the **bottom right**, we show the projection onto the second principal component, which has significantly lower variance (look at the axes) and gives better separation.

have g different classes, and the jth class has mean $\boldsymbol{\mu}_j$. Write $\overline{\boldsymbol{\mu}}$ for the mean of the class means, that is,

$$\overline{\boldsymbol{\mu}} = \frac{1}{g}\sum_{j=1}^{g}\boldsymbol{\mu}_j,$$

Write

$$\mathcal{B} = \frac{1}{g-1}\sum_{j=1}^{g}(\boldsymbol{\mu}_j - \overline{\boldsymbol{\mu}})(\boldsymbol{\mu}_j - \overline{\boldsymbol{\mu}})^T.$$

Note that $\mathcal{B}$ gives the variance of the class means. In the simplest case, we assume that each class has the same covariance Σ, and that this has full rank. We would like to obtain a set of axes where the clusters of data points belonging to a particular class group together tightly, whereas the distinct classes are widely separated. This involves finding a set of features that maximizes the ratio of the separation (variance) between the class means to the variance within each class. The separation between the class means is typically referred to as the *between-class variance*, and the variance within a class is typically referred to as the *within-class variance*.

Now we are interested in linear functions of the features, so we concentrate on

$$v(\boldsymbol{x}) = \boldsymbol{v}^T\boldsymbol{x}.$$

We should like to maximize the ratio of the between-class variances to the within-class variances for $\boldsymbol{v}_1$.

Using the same argument as for principal components, we can achieve this by choosing $\boldsymbol{v}$ to maximize

$$\frac{\boldsymbol{v}_1^T\mathcal{B}\boldsymbol{v}_1}{\boldsymbol{v}_1^T\Sigma\boldsymbol{v}_1}.$$

This problem is the same as maximizing $\boldsymbol{v}_1^T\mathcal{B}\boldsymbol{v}_1$ subject to the constraint that $\boldsymbol{v}_1^T\Sigma\boldsymbol{v}_1 = 1$. In turn, a solution has the property that

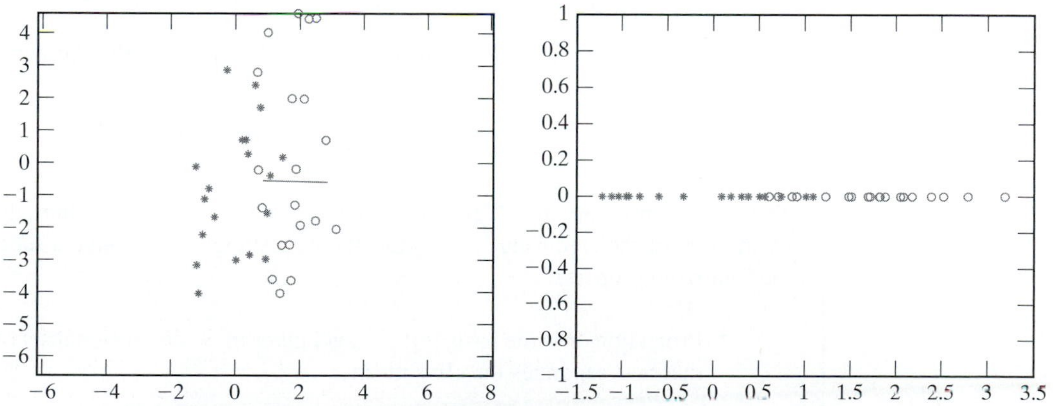

Figure 22.10 Canonical variates use the class of each data item as well as the features in estimating a good set of linear features. In particular, the approach constructs axes that separate different classes as well as possible. The dataset used in Figure 22.9 is shown on the **left**, with the axis given by the first canonical variate overlaid. On the **bottom right**, we show the projection onto that axis, where the classes are rather well separated.

$$\mathcal{B}v_1 + \lambda\Sigma v_1 = 0$$

for some constant λ. This is known as a *generalized eigenvalue problem*—if Σ has full rank, we can solve it by finding the eigenvector of $\Sigma^{-1}\mathcal{B}$ with largest eigenvalue (otherwise we use specialized routines within the relevant numerical software environment).

Now for each v_l, for $2 \leq l \leq p$, we should like to find features that extremize the criterion and are independent of the the previous v_l. These are provided by the other eigenvectors of $\Sigma^{-1}\mathcal{B}$. The eigenvalues give the variance along the features (which are independent). By choosing the $m < p$ eigenvectors with the largest eigenvalues, we obtain a set of features that reduces the dimension of the feature space while best preserving the separation between classes. This doesn't guarantee the best error rate for a classifier on a reduced number of features, but it offers a good place to start by reducing the number of features while respecting the category structure (Figure 22.11). Details and examples appear in McLachlan and Krishnan (1996), or in Ripley (1996).

Algorithm 22.6: *Canonical variates* identifies a collection of linear features that separating the classes as well as possible

Assume that we have a set of data items of g different classes. There are n_k items in each class, and a data item from the kth class is $x_{k,i}$, for $i \in \{1, \ldots, n_k\}$. The jth class has mean μ_j. We assume that there are p features (i.e., that the x_i are p-dimensional vectors).

Write $\overline{\mu}$ for the mean of the class means, that is,

$$\overline{\mu} = \frac{1}{g}\sum_{j=1}^{g}\mu_j,$$

Write

$$\mathcal{B} = \frac{1}{g-1}\sum_{j=1}^{g}(\mu_j - \overline{\mu})(\mu_j - \overline{\mu})^T.$$

Assume that each class has the same covariance Σ, which is either known or estimated as

$$\Sigma = \frac{1}{N-1}\sum_{c=1}^{g}\left\{\sum_{i=1}^{n_c}(x_{c,i} - \mu_c)(x_{c,i} - \mu_c)^T\right\}.$$

The unit eigenvectors of $\Sigma^{-1}\mathcal{B}$, which we write as $v_1, v_2, \ldots, v_d$, where the order is given by the size of the eigenvalue and v_1 has the largest eigenvalue, give a set of features with the following property:

- Projection onto the basis $\{v_1, \ldots, v_k\}$ gives the k-dimensional set of linear features that best separates the class means.

If the classes don't have the same covariance, it is still possible to construct canonical variates. In this case, we estimate a Σ as the covariance of all the offsets of each data item *from its own class mean* and proceed as before. Again, this is an approach without a guarantee of optimality, but one that can work quite well in practice.

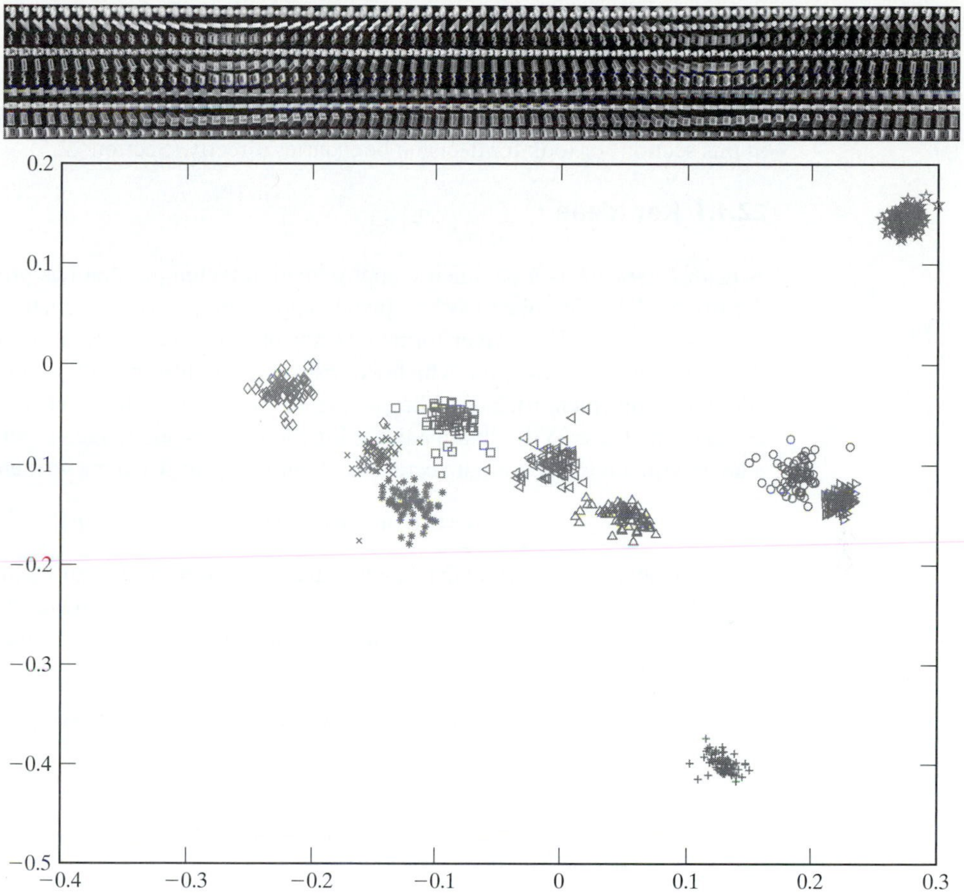

Figure 22.11 Canonical variates are effective for a variety of simple template matching problems. The figure on top shows views of 10 objects at a variety of poses on a black background (these images are smoothed and resampled versions of images in the well-known COIL database due to Nene and Nayar and available at `http://www.cs.columbia.edu/CAVE/research/softlib/coil-20.html`—a version with 20 objects—or `http://www.cs.columbia.edu/CAVE/research/softlib/coil-100.html`—a version with 100 objects) Identifying an object from one of these images is a relatively simple matter because objects appear on a constant background; they do not need to be segmented. We then used 60 of the images of each object to determine a set of canonical variates. The figure below shows the first two canonical variates for 71 images—the 60 training images and 11 others—of each object (different symbols correspond to different objects). Note that the clusters are tight and well separated; on these two canonical variates alone, we could probably get quite good classification.

22.4 NEURAL NETWORKS

It is commonly the case that neither simple parametric density models nor histogram models can be used. In this case, we must either use more sophisticated density models (an idea we explore in this section) or look for decision boundaries directly (Section 22.5).

22.4.1 Key Ideas

A *neural network* is a parametric approximation technique that has proved useful for building density models. Neural networks typically approximate a vector function $\boldsymbol{f}$ of some input $\boldsymbol{x}$ with a series of *layers*. Each layer forms a vector of outputs each of which is obtained by applying the same nonlinear function, which we write as ϕ, to different affine functions of the inputs. We adopt the convenient trick of adding an extra component to the inputs and fixing the value of this component at one so we obtain a linear function of this augmented input vector. This means that a layer with augmented input vector $\boldsymbol{u}$ and output vector $\boldsymbol{v}$ can be written as

$$\boldsymbol{v} = [\phi(\boldsymbol{w}_1 \cdot \boldsymbol{u}), \phi(\boldsymbol{w}_2 \cdot \boldsymbol{u}), \ldots \phi(\boldsymbol{w}_n \cdot \boldsymbol{u})],$$

where the $\boldsymbol{w}_i$ are parameters that can be adjusted to approve the approximation.

Typically, a neural net uses a sequence of layers to approximate a function. Each layer uses augmented input vectors. For example, if we are approximating a vector function $\boldsymbol{g}$ of a vector $\boldsymbol{x}$ with a two layer net, we obtain

$$\boldsymbol{g}(\boldsymbol{x}) \approx \boldsymbol{f}(\boldsymbol{x}) = \left[\phi(\boldsymbol{w}_{21} \cdot \boldsymbol{y}), \phi(\boldsymbol{w}_{22} \cdot \boldsymbol{y}), \ldots \phi(\boldsymbol{w}_{2n} \cdot \boldsymbol{y})\right],$$

where

$$\boldsymbol{y}(z) = [\phi(\boldsymbol{w}_{11} \cdot z), \phi(\boldsymbol{w}_{12} \cdot z), \ldots \phi(\boldsymbol{w}_{1m} \cdot z), 1]$$

and

$$z(\boldsymbol{x}) = \left[x_1, x_2, \ldots, x_p, 1\right].$$

Some of the elements of $\boldsymbol{w}_{1k}$ or $\boldsymbol{w}_{2k}$ could be clamped at zero; in this case, we are insisting that some elements of $\boldsymbol{y}$ do not affect $\boldsymbol{f}(\boldsymbol{x})$. If this is the case, the layer is referred to as a *partially connected layer*; otherwise it is known as a *fully connected layer*. Of course, layer two could be either fully or partially connected as well. The parameter n is fixed by the dimension of $\boldsymbol{f}$ and p is fixed by the dimension of $\boldsymbol{x}$, but there is no reason that m should be the same as either n or p. Typically, m is larger than either. A similar construction yields three layer networks (or networks with more layers, which are uncommon). Neural networks are often drawn with circles indicating variables and arrows indicating possibly nonzero connections; this gives a representation that exposes the basic structure of the approximation (Figure 22.12).

Choosing a Nonlinearity There are a variety of possibilities for ϕ. For example, we could use a *threshold function*, which has value one when the argument is positive and zero otherwise. It is quite easy to visualize the response of a layer of that uses a threshold function; each component of the layer changes from zero to one along a hyperplane. This means that the output vector takes different values in each cell of an arrangement of hyperplanes in the input space. Networks that use layers of this form are hard to train, because the threshold function is not differentiable.

It is more common to use a ϕ that changes smoothly (but rather quickly) from zero to one, often called a *sigmoid function* or *squashing function*. The *logistic function* is one popular example. This is a function of the form

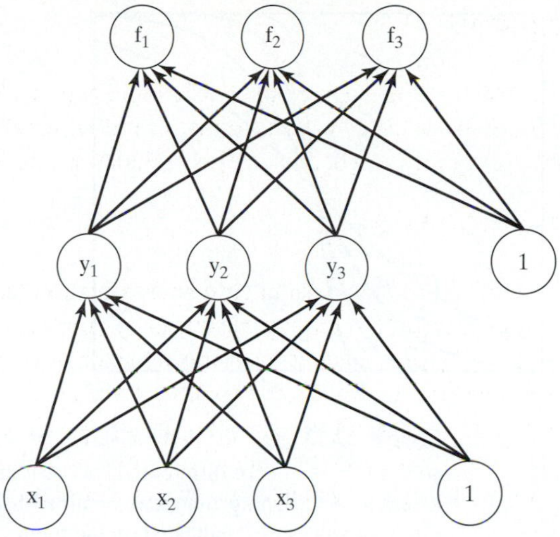

Figure 22.12 Neural networks are often illustrated by diagrams of the form shown here. Each circle represents a variable, and the circles are typically labeled with the variable. The "layers" are obvious in such a drawing. This network is the two-layer network given in the text; the arrows indicate that the coefficient coupling the two variables in the affine function could be nonzero. This network is fully connected because all arrows are present. It is possible to have arrows skip layers.

$$\phi(x; \nu) = \frac{e^{x/\nu}}{1 + e^{x/\nu}},$$

where ν controls how sharply the function changes at $x = 0$. It isn't crucial that the horizontal assymptotes are zero and one. Another popular squashing function is

$$\phi(x; \nu, A) = A \tanh(\nu x),$$

which has horizontal assymptotes at A and $-A$. Figure 22.13 illustrates these nonlinearities.

Producing a Classifier Using a Neural Net To produce a neural net that approximates some function $\boldsymbol{g}(\boldsymbol{x})$, we collect a series of examples $\boldsymbol{x}^e$. We construct a network that has one output for each dimension of $\boldsymbol{g}$. Write this network as $\boldsymbol{n}(\boldsymbol{x}; \boldsymbol{p})$, where $\boldsymbol{p}$ is a parameter vector that contains all the $\boldsymbol{w}_{ij}$. We supply a desired output vector $\boldsymbol{o}^e$ for this input; typically, $\boldsymbol{o}^e = \boldsymbol{g}(\boldsymbol{x}^e)$. We now obtain $\hat{\boldsymbol{p}}$ that minimizes

$$Error(\boldsymbol{p}) = \left(\frac{1}{2}\right) \sum_e |\boldsymbol{n}(\boldsymbol{x}^e; \boldsymbol{p}) - \boldsymbol{o}^e|^2$$

using appropriate optimization software (the half simplifies a little notation later on, but is of no real consequence).

The most significant case occurs when $\boldsymbol{g}(\boldsymbol{x})$ is intended to approximate the posterior on classes given the data. We do not know this posterior, and so cannot supply its value to the training procedure. Instead, we require that our network has one output for each class. Given an example $\boldsymbol{x}^e$, we construct a desired output $\boldsymbol{o}^e$ as a vector that contains a one in the component corresponding to that example's class and a zero in each other component. We now train the net as before, and regard the output of the neural network as a model of the posterior probability. An

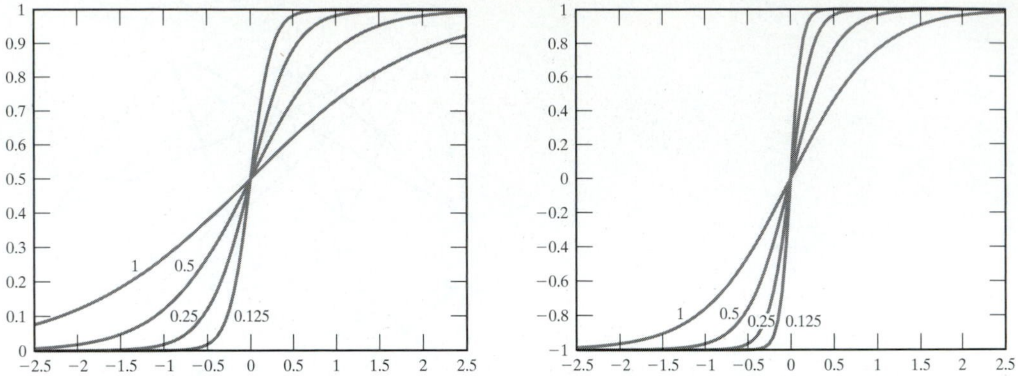

Figure 22.13 On the **left**, a series of squashing functions obtained using $\phi(x; \nu) = \frac{e^{x/\nu}}{1+e^{x/\nu}}$, for different values of ν indicated on the figure. On the **right**, a series of squashing functions obtained using $\phi(x; \nu, A) = A \tanh(x/\nu)$ for different values of ν indicated on the figure. Generally, for x close to the center of the range, the squashing function is linear; for x small or large, it is strongly nonlinear.

input $\boldsymbol{x}$ is then classified by forming $\boldsymbol{n}(\boldsymbol{x}; \hat{\boldsymbol{p}})$, and then choosing the class that corresponds to the largest component of this vector.

22.4.2 Minimizing the Error

Recall that we are training a net by minimizing the sum over the examples of the difference between the desired output and the actual output, that is, by minimizing

$$Error(\boldsymbol{p}) = \left(\frac{1}{2}\right) \sum_e |\boldsymbol{n}(\boldsymbol{x}^e; \boldsymbol{p}) - \boldsymbol{o}^e|^2$$

as a function of the parameters $\boldsymbol{p}$. There are a variety of strategies for obtaining $\hat{\boldsymbol{p}}$, the set of parameters that minimize this error. One is gradient descent; from some initial point $\boldsymbol{p}_i$, we compute a new point $\boldsymbol{p}_{i+1}$ by

$$\boldsymbol{p}_{i+1} = \boldsymbol{p}_i - \epsilon(\nabla\ Error),$$

where ϵ is some small constant.

Stochastic Gradient Descent Write the error for example e as $Error(\boldsymbol{p}; \boldsymbol{x}^e)$ so the total error is $Error(\boldsymbol{p}) = \sum_e Error(\boldsymbol{p}; \boldsymbol{x}^e)$. Now if we use gradient descent, we are updating parameters using the algorithm

$$\boldsymbol{p}_{i+1} = \boldsymbol{p}_i - \epsilon \nabla\ Error$$

(where the gradient is with respect to $\boldsymbol{p}$ and is evaluated at $\boldsymbol{p}_i$). This works because, if ϵ is sufficiently small, we have

$$\begin{aligned} Error(\boldsymbol{p}_{i+1}) &= Error(\boldsymbol{p}_i - \epsilon \nabla\ Error) \\ &\approx Error(\boldsymbol{p}_i) - \epsilon(\nabla\ Error \cdot \nabla\ Error) \\ &\leq Error(\boldsymbol{p}_i), \end{aligned}$$

with equality only at an extremum. This creates a problem: Evaluating the error and its gradient is going to involve a sum over all examples, which may be a large number. We should like to avoid this sum; it turns out that it is possible to do so by selecting an example at random, computing the gradient *for that example alone*, and updating the parameters using that gradient. In this process, known as *stochastic gradient descent*, we update the parameters using the algorithm

$$\boldsymbol{p}_{i+1} = \boldsymbol{p}_i - \epsilon \nabla \, Error(\boldsymbol{p}; \boldsymbol{x}^e)$$

(where the gradient is with respect to $\boldsymbol{p}$, is evaluated at $\boldsymbol{p}_i$, and we choose the example uniformly at random, making a different choice at each step). In this case, the error doesn't necessarily go down for each particular choice, but the *expected* value of the error *does* go down for a sufficiently small value of ϵ. In particular, we have

$$\begin{aligned}
\mathrm{E}(Error(\boldsymbol{p}_{i+1})) &= \mathrm{E}(Error(\boldsymbol{p}_i - \epsilon \nabla \, Error(\boldsymbol{p}; \boldsymbol{x}^e)) \\
&\approx \mathrm{E}(Error(\boldsymbol{p}_i) - \epsilon(\nabla \, Error \cdot \nabla \, Error(\boldsymbol{p}; \boldsymbol{x}^e)) \\
&= Error(\boldsymbol{p}_i) - \epsilon \frac{1}{n} \sum_e (\nabla \, Error \cdot \nabla \, Error(\boldsymbol{p}; \boldsymbol{x}^e)) \\
&= Error(\boldsymbol{p}_i) - \epsilon \left(\nabla \, Error \cdot \left(\frac{1}{n} \sum_e \nabla \, Error(\boldsymbol{p}; \boldsymbol{x}^e) \right) \right) \\
&= Error(\boldsymbol{p}_i) - \frac{\epsilon}{n}(\nabla \, Error \cdot \nabla \, Error) \\
&< Error(\boldsymbol{p}_i) \text{ if } |\nabla E| > 0.
\end{aligned}$$

By taking sufficient steps down a gradient computed using only one example (selected uniformly and at random each time we take a step), we can in fact minimize the function. This is because the expected value goes down for each step unless we're at the minimum. The gradient can be computed in a number of ways; one efficient trick is *backpropagation*, described in Section 22.7.

Algorithm 22.7: *Stochastic gradient descent* minimizes the error of a neural net approximation using backpropagation to compute the derivatives

Choose $\boldsymbol{p}_o$ (randomly)
Use backpropagation (Algorithm 22.9) to compute
    $\nabla \, Error(\boldsymbol{x}^e; \, \boldsymbol{p}_o)$
  $\boldsymbol{p}_n = \boldsymbol{p}_o - \epsilon \nabla \, Error(\boldsymbol{x}^e; \, \boldsymbol{p}_o)$
Until $|Error(\boldsymbol{p}_n) - Error(\boldsymbol{p}_o)|$ is small
    or $|\boldsymbol{p}_o - \boldsymbol{p}_n|$ is small

    $\boldsymbol{p}_o = \boldsymbol{p}_n$
    Choose an example $(\boldsymbol{x}^e, \boldsymbol{o}^e)$ uniformly and
      at random from the training set
    Use backpropagation (Algorithm 22.9) to compute
      $\nabla \, Error(\boldsymbol{x}^e; \, \boldsymbol{p}_o)$
    $\boldsymbol{p}_n = \boldsymbol{p}_o - \epsilon \nabla \, Error(\boldsymbol{x}^e; \, \boldsymbol{p}_o)$
end

22.4.3 When to Stop Training

Typically, gradient descent is not continued until an exact minimum is found. Surprisingly, this is a source of robustness. The easiest way to understand this is to consider the shape of the error function around the minimum. If the error function changes sharply at the minimum, then the performance of the network is quite sensitive to the choice of parameters. This suggests that the network generalizes badly. You can see this by assuming that the training examples are one half of a larger set; if we had trained the net on the other half we'd have obtained slightly different set of parameters. This means that the net with our current parameters will perform badly on this other half, because the error changes sharply with a small change in the parameters.

Now if the error function doesn't change sharply at the minimum, there is no particular point in expending effort to be at the minimum value as long as we are reasonably close—we know that this minimum error value won't be attained on a training set. It is common practice to continue with stochastic gradient descent until (a) each example has been visited on average rather more than once, and (b) the decrease in the value of the function goes below some threshold.

A more difficult question is how many layers to use and how many units to use in each layer. This question, which is one of model selection, tends to be resolved by experiment. We refer interested readers to Ripley (1996) and Haykin (1999).

22.4.4 Finding Faces Using Neural Networks

Face finding is an application that illustrates the usefulness of classifiers. In frontal views at a fairly coarse scale, all faces look basically the same. There are bright regions on the forehead, the cheeks, and the nose, and dark regions around the eyes, the eyebrows, the base of the nose, and the mouth. This suggests approaching face finding as a search over all image windows of a fixed size for windows that look like a face. Larger or smaller faces can be found by searching coarser or finer scale images.

Because a face illuminated from the left looks different than a face illuminated from the right, the image windows must be corrected for illumination. Generally, illumination effects look enough like a linear ramp (one side is bright, the other side is dark, and there is a smooth transition between them) that we can simply fit a linear ramp to the intensity values and subtract that from the image window. Another way to do this would be to log-transform the image and then subtract a linear ramp fitted to the logs. This has the advantage that (using a rather rough model) illumination effects are additive in the log transform. There doesn't appear to be any evidence in the literature that the log transform makes much difference in practice. Another approach is to histogram equalize the window to ensure that its histogram is the same as that of a set of reference images (histogram equalisation is described in Figure 22.14).

Once the windows have been corrected for illumination, we need to determine whether there is a face present. The orientation isn't known, and so we must either determine it or produce a classifier that is insensitive to orientation. Rowley, Baluja and Kanade (1998*b*) produced a face finder that finds faces very successfully by firstly estimating the orientation of the window, using one neural net, and then reorienting the window so that it is frontal, and passing the frontal window onto another neural net (see Figure 22.15; the paper is a development of Rowley, Baluja and Kanade, 1996 and, 1998*a*). The orientation finder has 36 output units, each coding for a 10° range of orientations; the window is reoriented to the orientation given by the largest output. Examples of the output of this system are given in Figure 22.16.

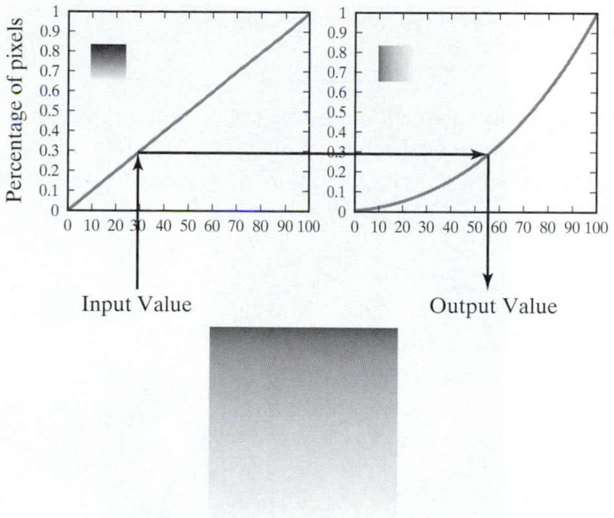

Figure 22.14 Histogram equalization uses cumulative histograms to map the gray levels of one image so that it has the same histogram as another image. The figure at the **top** shows two cumulative histograms with the relevant images inset in the graphs. To transform the left image so that it has the same histogram as the right image, we take a value from the left image, read off the percentage from the cumulative histogram of that image, and obtain a new value for that gray level from the inverse cumulative histogram of the right image. The image on the left is a linear ramp (it looks nonlinear because the relationship between brightness and lightness is not linear); the image on the right is a cube root ramp. The result—the linear ramp, with gray levels remapped so that it has the same histogram as the cube root ramp—is shown on the **bottom** row.

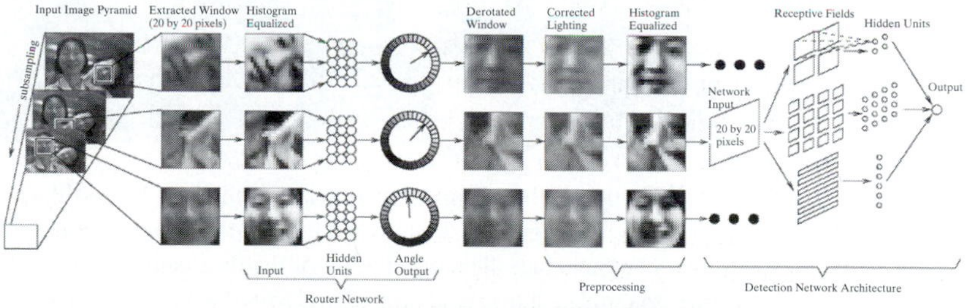

Figure 22.15 The architecture of Rowley, Baluja, and Kanade's system for finding faces. Image windows of a fixed size are corrected to a standard illumination using histogram equalization; they are then passed to a neural net that estimates the orientation of the window. The windows are reoriented and passed to a second net that determines whether a face is present. *Reprinted from "Rotation invariant neural-network based face detection," by H.A. Rowley, S. Baluja and T. Kanade, Proc. Computer Vision and Pattern Recognition, 1998,* © *1998, IEEE*

Figure 22.16 Typical responses for the Rowley, Baluja, and Kanade system for face finding; a mask icon is superimposed on each window that is determined to contain a face. The orientation of the face is indicated by the configuration of the eye holes in the mask. *Reprinted from "Rotation invariant neural-network based face detection," by H.A. Rowley, S. Baluja and T. Kanade, Proc. Computer Vision and Pattern Recognition, 1998, © 1998, IEEE*

22.4.5 Convolutional Neural Nets

Neural networks are not confined to the architecture sketched before; there is a wide variety of alternatives (a good start is to look at Bishop, 1995 or at Haykin, 1999). One architecture that has proved useful in vision applications is the *convolutional neural network*. The motivating idea here is that it appears to be useful to represent image regions with filter outputs. Furthermore, we can obtain a compositional representation we apply filters to a representation itself obtained using filter outputs. For example, assume that we are looking for handwritten characters; the

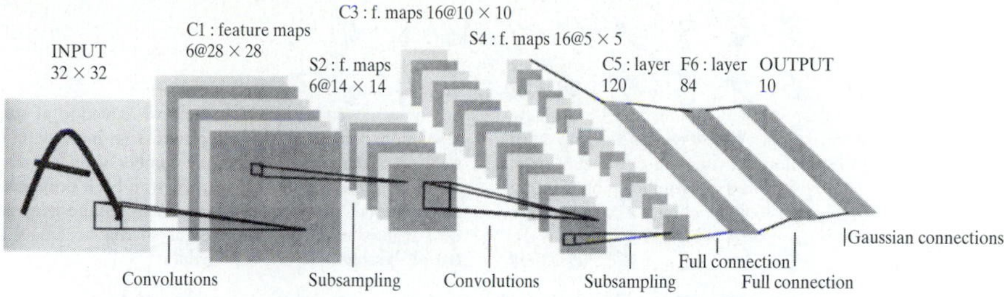

Figure 22.17 The architecture of LeNet 5, a convolutional neural net used for recognizing handwritten characters. The layers marked C are convolutional layers; those marked S are subsampling layers. The general form of the classifier uses an increasing number of features at increasingly coarse scales to represent the image window. Finally, the window is passed to a fully connected neural net, which produces a rectified output that is classified by looking at its distance from a set of canonical templates for characters. *Figure from "Gradient-Based Learning Applied to Document Recognition," by Y. Lecun et al. Proc. IEEE, 1998 © 1998, IEEE*

response of oriented bar filters is likely to be useful here. If we obtain a map of the oriented bars in the image, we can apply another filter to this map, and the output of this filter indicates spatial relations between the bars.

These observations suggest using a system of filters to build up a set of relations between primitives, and then using a conventional neural network to classify on the resulting representation. There is no particular reason to specify the filters in advance; instead, we could learn them, too.

Lecun, Bottou, Bengio and Haffner (1998) built a number of classifiers for handwritten digits using a convolutional neural network (Lecun et al., 1998). The basic architecture is given in Figure 22.17. The classifier is applied to a 32×32 image window. The first stage, C1 in the figure, consists of six feature maps. The feature maps are obtained by convolving the image with a 5×5 filter kernel, adding a constant, and applying a sigmoid function. Each map uses a different kernel and constant, and these parameters are learned.

Because the exact position of a feature should not be important, the resolution of the feature maps is reduced, leading to a new set of six feature maps—S2 in the figure. These maps are subsampled versions of the previous layer; this subsampling is achieved by averaging 2×2 neighborhoods, multiplying by a parameter, adding a parameter, and passing the result through a sigmoid function. The multiplicative and additive parameters are learned. A series of pairs of layers of this form follows, with the number of feature maps increasing as the resolution decreases. Finally, there is a layer with 84 outputs; each of these outputs is supplied by a unit that takes every element of the previous layer as an input.

This network is used to recognize hand printed characters. The outputs are seen as a 7×12 image of a character that has been rectified from its hand-printed version, and can now be compared with a canonical pattern for that character. The network can rectify distorted characters successfully. The input character is given the class of the character whose canonical pattern is closest to the rectified version. The resulting network has a test error rate of 0.95% (Figure 22.18).

22.5 THE SUPPORT VECTOR MACHINE

From the perspective of the vision community, classifiers are not an end in themselves, but a means. Thus, when a technique that is simple, reliable, and effective becomes available, it tends to

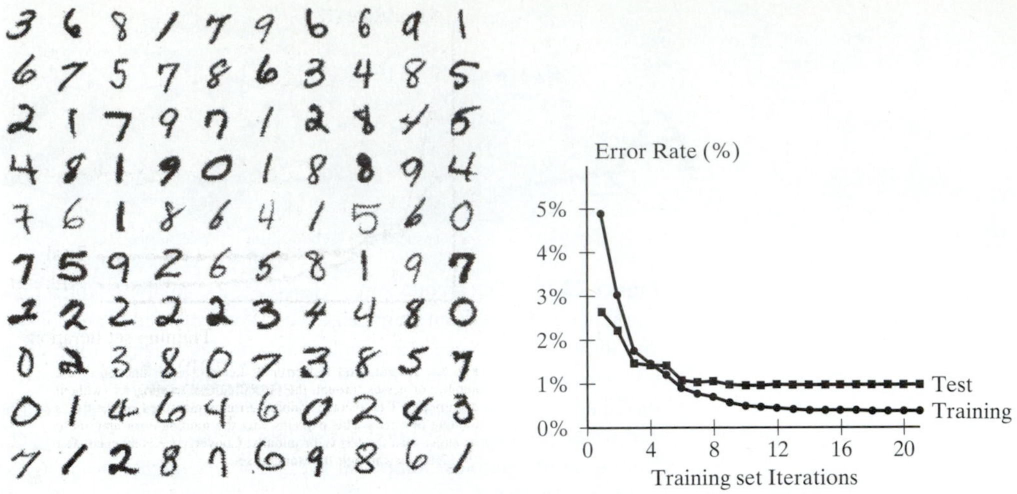

Figure 22.18 On the **left**, a small subset of the MNIST database of handwritten characters used to train and test LeNet 5. Note the fairly wide variation in the appearance of each character. On the **right**, the error rate of LeNet 5 on a training set and on a test set, plotted as a function of the number of gradient descent passes through the entire training set of 60,000 examples (i.e., if the horizontal axis reads six, the training has taken 360, 000 gradient descent steps). Note that at some point the training error goes down but the test error doesn't; this phenomenon occurs because the system's performance is optimized on the training data. A substantial difference would indicate overfitting. *Figure from "Gradient-Based Learning Applied to Document Recognition," by Y. Lecun et al. Proc. IEEE, 1998 © 1998, IEEE*

be adopted quite widely. The *support vector machine* is such a technique. This should be the first classifier you think of when you wish to build a classifier from examples (unless the examples come from a known distribution, which hardly ever happens). We give a basic introduction to the ideas, and show some examples where the technique has proved useful.

Assume we have a set of N points $\boldsymbol{x}_i$ that belong to two classes, which we indicate by 1 and -1. These points come with their class labels, which we write as y_i; thus, our dataset can be written as

$$\{(\boldsymbol{x}_1, y_1), \ldots, (\boldsymbol{x}_N, y_N)\}.$$

We should like to determine a rule that predicts the sign of y for any point $\boldsymbol{x}$; this rule is our classifier.

At this point, we distinguish between two cases: Either the data is linearly separable, or it isn't. The linearly separable case is much easier, and we deal with it first.

22.5.1 Support Vector Machines for Linearly Separable Datasets

In a linearly separable dataset, there is some choice of $\boldsymbol{w}$ and b (which represent a hyperplane) such that

$$y_i\,(\boldsymbol{w} \cdot \boldsymbol{x}_i + b) > 0$$

for every example point (notice the devious use of the sign of y_i). There is one of these expressions for each data point, and the set of expressions represents a set of constraints on the choice of $\boldsymbol{w}$ and b. These constraints express the constraint that all examples with a negative y_i should be on one side of the hyperplane and all with a positive y_i should be on the other side.

In fact, because the set of examples is finite, there is a family of separating hyperplanes. Each of these hyperplanes must separate the convex hull of one set of examples from the convex hull of the other set of examples. The most conservative choice of hyperplane is the one that is furthest from both hulls. This is obtained by joining the closest points on the two hulls, and constructing a hyperplane perpendicular to this line and through its midpoint. This hyperplane is as far as possible from each set, in the sense that it maximizes the minimum distance from example points to the hyperplane (Figure 22.19).

Now we can choose the scale of $\boldsymbol{w}$ and b because scaling the two together by a positive number doesn't affect the validity of the constraints $y_i(\boldsymbol{w} \cdot \boldsymbol{x}_i + b) > 0$. This means that we can choose $\boldsymbol{w}$ and b such that for every data point we have

$$y_i\,(\boldsymbol{w} \cdot \boldsymbol{x}_i + b) \geq 1$$

and such that equality is achieved on at least one point on each side of the hyperplane. Now assume that $\boldsymbol{x}_k$ achieves equality and $y_k = 1$, and $\boldsymbol{x}_l$ achieves equality and $y_l = -1$. This means that $\boldsymbol{x}_k$ is on one side of the hyperplane and $\boldsymbol{x}_l$ is on the other. Furthermore, the distance from $\boldsymbol{x}_l$

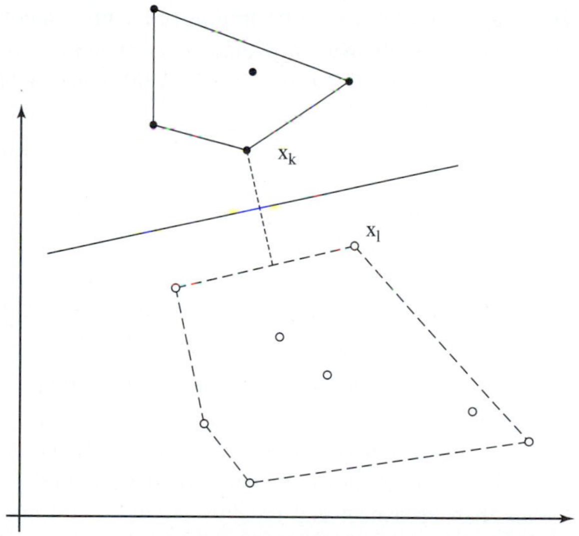

Figure 22.19 The hyperplane constructed by a support vector classifier for a plane dataset. The filled circles are data points corresponding to one class, and the empty circles are data points corresponding to the other. We have drawn in the convex hull of each dataset. The most conservative choice of hyperplane is one that maximizes the minimum distance from each hull to the hyperplane. A hyperplane with this property is obtained by constructing the shortest line segment between the hulls and then obtaining a hyperplane perpendicular to this line segment and through its midpoint. Only a subset of the data determines the hyperplane. Of particular interest are points on each convex hull that are associated with a minimum distance between the hulls. We use these points to find the hyperplane in the text.

to the hyperplane is minimal (among the points on the same side as $\boldsymbol{x}_l$) as is the distance from $\boldsymbol{x}_k$ to the hyperplane. Notice that there might be several points with these properties.

This means that $\boldsymbol{w} \cdot (\boldsymbol{x}_1 - \boldsymbol{x}_2) = 2$, so that

$$
\begin{aligned}
dist(\boldsymbol{x}_k, \text{hyperplane}) + dist(\boldsymbol{x}_l, \text{hyperplane}) &= \left(\frac{\boldsymbol{w}}{|\boldsymbol{w}|} \cdot \boldsymbol{x}_k + \frac{b}{|\boldsymbol{w}|}\right) - \left(\frac{\boldsymbol{w}}{|\boldsymbol{w}|} \cdot \boldsymbol{x}_1 + \frac{b}{|\boldsymbol{w}|}\right) \\
&= \frac{\boldsymbol{w}}{|\boldsymbol{w}|} \cdot (\boldsymbol{x}_1 - \boldsymbol{x}_2) = \frac{2}{|\boldsymbol{w}|}.
\end{aligned}
$$

This means that maximizing the distance is the same as *minimizing* $(1/2)\boldsymbol{w} \cdot \boldsymbol{w}$. We now have the constrained minimization problem:

$$
\begin{aligned}
&\text{minimize } (1/2)\boldsymbol{w} \cdot \boldsymbol{w} \\
&\text{subject to } y_i\,(\boldsymbol{w} \cdot \boldsymbol{x}_i + b) \geq 1,
\end{aligned}
$$

where there is one constraint for each data point.

Solving for the Support Vector Machine We can solve this problem by introducing Lagrange multipliers α_i to obtain the Lagrangian

$$
(1/2)\boldsymbol{w} \cdot \boldsymbol{w} - \sum_1^N \alpha_i \left(y_i\,(\boldsymbol{w} \cdot \boldsymbol{x}_1 + b) - 1\right).
$$

This Lagrangian needs to be minimized with respect to $\boldsymbol{w}$ and $\boldsymbol{b}$ and maximized with respect to α_i—these are the Karush-Kuhn-Tucker conditions, described in optimization textbooks (e.g., see (Gill, Murray and Wright 1981)). A little manipulation leads to the requirements that

$$
\sum_1^N \alpha_i y_i = 0
$$

and

$$
\boldsymbol{w} = \sum_1^N \alpha_i y_i \boldsymbol{x}_i.
$$

This second expression is why the device is known as a *support vector machine*. Generally, the hyperplane is determined by a relatively small number of example points, and the position of other examples is irrelevant (see Figure 22.19; everything inside the convex hull of each set of examples is irrelevant to choosing the hyperplane, and most of the hull vertices are, too). This means that we expect that most α_i are zero, and the data points corresponding to nonzero α_i, which are the ones that determine the hyperplane, are known as the *support vectors*.

Now by substituting these expressions into the original problem and manipulating, we obtain the *dual problem* given by

$$
\begin{aligned}
&\text{maximize } \sum_i^N \alpha_i - \frac{1}{2} \sum_{i,j=1}^N \alpha_i (y_i y_j \boldsymbol{x}_i \cdot \boldsymbol{x}_j) \alpha_j \\
&\text{subject to } \alpha_i \geq 0 \\
&\text{and } \sum_{i=1}^N \alpha_i y_i = 0
\end{aligned}
$$

You should notice that the criterion is a quadratic form in the Lagrange multipliers. This problem is a standard numerical problem known as *quadratic programming*. One can use standard packages quite successfully for this problem, but it does have special features—while there may be a large number of variables, most are zero at a solution point—which can be exploited (Smola et al., 2000).

Algorithm 22.8: Finding an SVM for a Linearly Separable Problem

Notation: We have a training set of N examples $\{(\boldsymbol{x}_1, y_1), \ldots, (\boldsymbol{x}_N, y_N)\}$ where y_i is either 1 or -1.

Solving for the SVM: Set up and solve the dual optimization problem:

$$\text{maximize} \sum_{i}^{N} \alpha_i - \frac{1}{2} \sum_{i,j=1}^{N} \alpha_i (y_i y_j \boldsymbol{x}_i \cdot \boldsymbol{x}_j) \alpha_j$$

$$\text{subject to } \alpha_i \geq 0$$

$$\text{and} \sum_{i=1}^{N} \alpha_i y_i = 0.$$

Now $\boldsymbol{w} = \sum_1^N \alpha_i y_i \boldsymbol{x}_i$ and for any example point $\boldsymbol{x}_i$ where α_i is nonzero, we have that $y_i(\boldsymbol{w} \cdot \boldsymbol{x}_i + b) = 1$, which yields the value of b.

Classifying a point: Any new data point is classified by

$$\begin{aligned} f(\boldsymbol{x}) &= \text{sign}\,(\boldsymbol{w} \cdot \boldsymbol{x} + b) \\ &= \text{sign}\left(\left(\sum_1^N \alpha_i y_i \boldsymbol{x} \cdot \boldsymbol{x}_i\right) + b\right) \\ &= \text{sign}\left(\sum_1^N (\alpha_i y_i \boldsymbol{x} \cdot \boldsymbol{x}_i + b)\right). \end{aligned}$$

22.5.2 Finding Pedestrians Using Support Vector Machines

At a fairly coarse scale, pedestrians have a characteristic, lollipoplike appearance—a wide torso on narrower legs. This suggests that they can be found using a support vector machine. The general strategy is the same as for the face-finding example in Section 22.4.4: Each image window of a fixed size is presented to a classifier, which determines whether the window contains a pedestrian. The number of pixels in the window may be large, and we know that many pixels may be irrelevant. In the case of faces, we could deal with this by cropping the image to an oval shape that would contain the face. This is harder to do with pedestrians because their outline is of a rather variable shape.

We need to identify features that can help determine whether a window contains a pedestrian. It is natural to try to obtain a set of features from a set of examples. A variety of feature selection algorithms might be appropriate here (all of them are variants of search). Oren, Papageorgiou, Sinha, Osuna and Poggio (1997) chose to look at local features—*wavelet coefficients*,

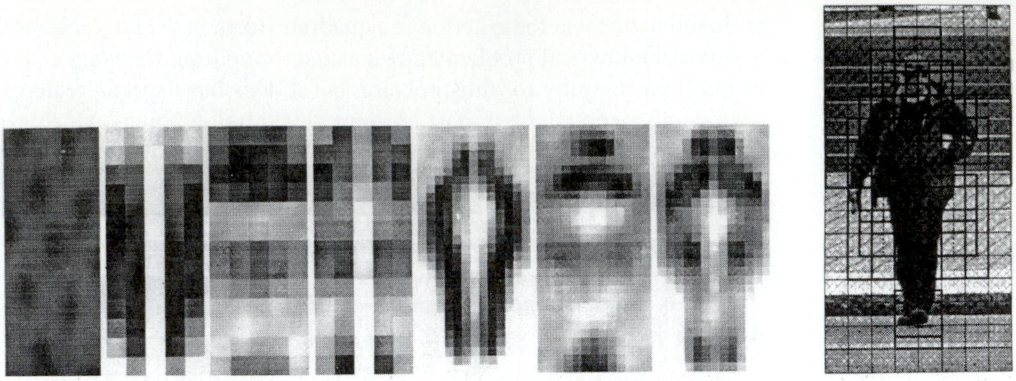

Figure 22.20 On the **left**, averages over the training set of different wavelet coefficients at different positions in the image. Coefficients that are above the (spatial) average value are shown dark, and those that are below are shown light. We expect that noise has the average value, meaning that coefficients that are very light or very dark contain information that could identify pedestrians. On the **right**, a grid showing the support domain for the features computed. Notice that this follows the boundary of the pedestrian fairly closely. *Reprinted from, "A general framework for object detection," by by C. Papageorgiou, M. Oren and T. Poggio, Proc. Int. Conf. Computer Vision, 1998, © 1998, IEEE*

which are the response of specially selected filters with local support—and to use an averaging approach. In particular, they argued that the background in a picture of a pedestrian looks like noise, images that don't contain pedestrians look like noise, and the average noise response of their filters is known. This means that attractive features are ones whose average over many images of pedestrians is different from their noise response. If we average the response of a particular filter in a particular position over a large number of images, and the average is similar to a noise response, that filter in that position is not particularly informative (Figure 22.20).

Now that features have been chosen, training follows the lines of Section 22.5. The approach is effective (Figures 22.21 and 22.22). Bootstrapping (Section 22.4) appears to improve performance significantly.

22.6 NOTES

What classifier to use where is a topic one on which no orthodoxy is yet established. Instead, one tries to use methods that seem likely to work on the problem in hand. For the sake of brevity, we have omitted a vast number of useful techniques; there are several very useful books covering this area (Bishop, 1995, Hastie, Tibshirani and Friedman, 2001, Haykin, 1999, McLachlan and Krishnan, 1996, Ripley, 1996, Vapnik, 1996 and 1998 are good places to start).

Choosing a decision boundary is strictly easier than fitting a posterior model. However, with a decision boundary, there is no reliable indication of the extent to which an example belongs to one or another class, as there is with a posterior model. Furthermore, fitting a decision boundary requires that we know the classes to which the example objects should be allocated. It is by no means obvious that one can construct an unambiguous class hierarchy for the objects we encounter in recognition problems. Both approaches can require large numbers of examples to build useful classifiers. Typically, the stronger the model applied, the fewer examples required to build a classifier.

Figure 22.21 Examples of pedestrians detected using the method of Papageorgiou, Oren, and Poggio. While not all pedestrians are found, there is a fairly high detection rate. The ROC is in Figure 22.22. *Reprinted from, "A general framework for object detection," by by C. Papageorgiou, M. Oren and T. Poggio, Proc. Int. Conf. Computer Vision, 1998, © 1998, IEEE*

It is difficult to build classifiers that are really successful when objects have a large number of degrees of freedom without a detailed model of this variation, and classifiers tend to be difficult to use if the number of features can vary from example to example. In both cases, some form of structural model appears to be necessary. However, estimating, representing, and manipulating probability densities in the high-dimensional spaces that occur in vision problems is practically impossible unless very strong assumptions are applied. Furthermore, it is easy to build probability

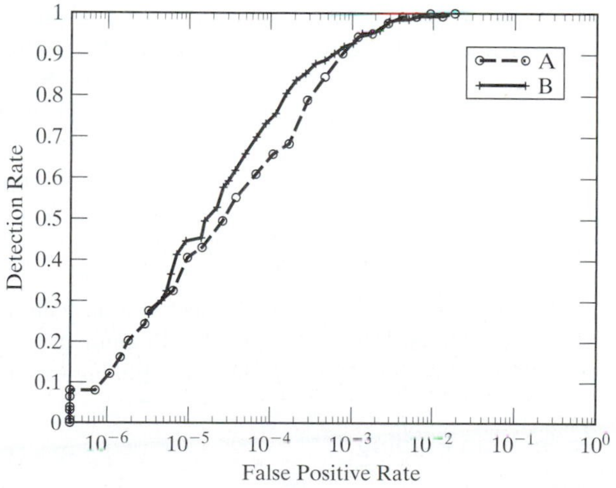

Figure 22.22 The receiver operating curve for the pedestrian detection system of Papageorgiou, Oren, and Poggio. *Reprinted from, "A general framework for object detection," by by C. Papageorgiou, M. Oren and T. Poggio, Proc. Int. Conf. Computer Vision, 1998, © 1998, IEEE*

models for which inference is again practically impossible. It isn't yet known how to build models that are easy to handle of the scale required for vision problems.

This subject is currently at the cutting edge of research in vision and learning. It's hard to know how to choose a method for a given problem, and opportunism seems to be the best approach at present. The examples in this chapter and the next illustrate a range of approaches that have been taken—some are very successful—but don't yet represent a clear theory.

An alternative approach to training one grand classifier is to train multiple classifiers and combine their outputs. This strategy is usually known as *boosting*. Boosting is most useful for classifiers with quite simple decision boundaries; these are usually easy to train, but have quite poor performance. Typically, we train a classifier and then determine the examples in the training set that it gets wrong. These examples are then emphasized—either by weighting errors on them more heavily or inserting copies into the training set—and a new classifier is trained. We now find out what the new classifier gets wrong and emphasize these examples and train again; this proces continues through many iterations. Now the outputs of all the classifiers are combined using a set of weights.

Skin Detection

There are a number of skin-detection papers; we selected one that fit with our didactic needs. Uses include finding naked people (Fleck, Forsyth and Bregler, 1996, Forsyth and Fleck, 1999) and finding and tracking faces and hands (Park, Seo, An and Chung, 2000, Yoo and Oh, 1999). There are a variety of techniques; for a start, look at the comparison of approaches in (Brand and Mason, 2000). The performance of face finders seems to be improved by using skin detectors tuned one way or another; it seems natural to learn the skin detector at the same time as one learns the face finder, although we are not aware of work that does this.

Face Finding

There is substantial interest in building face finders using template-matching techniques. The general strategy is to present image regions, possibly corrected for illumination variations, to classifiers of one form or another. We have taken liberties with history in our presentation; Schneiderman and Kanade's work appeared after Rowley, Baluja and Kanade's work, which appeared at the same time as important work by Sung and Poggio (1998). We do not discuss this work for didactic reasons only. Osuna, Freund and Girosi (1997) produced a face finder that successfully used a support vector machine, and this class of method is now widely used.

Face recognition (whose face is this) is also important, and it isn't clear that recognition and finding should be divorced. Important papers include Brunelli and Poggio (1993 and 1992). Significant technologies include recognition from very small numbers of examples (one would be great) (Beymer and Poggio, 1995); recognition from different views (Beymer, 1994); and managing variations induced by changes in illumination (Adini, Moses and Ullman, 1997, Adini, Moses and Ullman, 1994, Georghiades, Belhumeur and Kriegman, 2000, Jacobs, Belhumeur and Basri, 1998, Georghiades, Kriegman and Belhumeur, 1998).

Pedestrian Finding

The observation that pedestrians can be found using a template matcher is useful. As we have seen, the template looks rather like a lollipop. Although pedestrians may disappear if they raise their arms (because they no longer look like lollipops), they typically spend a fair amount of time with their arms at their sides, meaning that a reasonable count of the number of pedestrians may be available. Furthermore, pedestrian movements are characteristic (Papageorgiou, Oren and Poggio, 1998; Papageorgiou and Poggio, 1999 and 2000).

PROBLEMS

22.1. Assume that we are dealing with measurements $\boldsymbol{x}$ in some feature space S. There is an open set D where any element is classified as class one, and any element in the interior of $S - D$ is classified as class two.

(a) Show that

$$\begin{aligned} R(s) &= Pr\{1 \to 2 \mid \text{using } s\}\, L(1 \to 2) + Pr\{2 \to 1 \mid \text{using } s\}\, L(2 \to 1) \\ &= \int_{S-D} p(1 \mid \boldsymbol{x})\, d\boldsymbol{x} L(1 \to 2) + \int_{D} p(2 \mid \boldsymbol{x})\, d\boldsymbol{x} L(2 \to 1). \end{aligned}$$

(b) Why are we ignoring the boundary of D (which is the same as the boundary of $S - D$) in computing the total risk?

22.2. In Section 22.2, we said that if each class-conditional density had the same covariance, the classifier of Algorithm 22.2 boiled down to comparing two expressions that are linear in $\boldsymbol{x}$.

(a) Show that this is true.

(b) Show that if there are only two classes, we need only test the sign of a linear expression in $\boldsymbol{x}$.

22.3. In Section 22.3.1, we set up a feature u, where the value of u on the ith data point is given by $u_i = \boldsymbol{v} \cdot (\boldsymbol{x}_i - \boldsymbol{\mu})$. Show that u has zero mean.

22.4. In Section 22.3.1, we set up a series of features u, where the value of u on the ith data point is given by $u_i = \boldsymbol{v} \cdot (\boldsymbol{x}_i - \boldsymbol{\mu})$. We then said that the $\boldsymbol{v}$ would be eigenvectors of Σ, the covariance matrix of the data items. Show that the different features are independent using the fact that the eigenvectors of a symmetric matrix are orthogonal.

22.5. In Section 22.2.1, we said that the ROC was invariant to choice of prior. Prove this.

Programming Assignments

22.6. Build a program that marks likely skin pixels on an image; you should compare at least two different kinds of classifier for this purpose. It is worth doing this carefully because many people have found skin filters useful.

22.7. Build one of the many face finders described in the text.

22.7 APPENDIX I: BACKPROPAGATION

The difficulty in training neural networks using stochastic gradient descent is that ∇ *Error* could be quite hard to compute. There is an effective strategy for computing ∇ *Error* called *backpropagation*. This approach exploits the layered structure of the neural network as a function of a function of a function, etc. to obtain the derivative.

Now recall the two layer neural net, which we wrote as

$$\boldsymbol{f}(\boldsymbol{x}) = \left[\phi(\boldsymbol{w}_{21} \cdot \boldsymbol{y}), \phi(\boldsymbol{w}_{22} \cdot \boldsymbol{y}), \ldots \phi(\boldsymbol{w}_{2n} \cdot \boldsymbol{y})\right],$$

where

$$\boldsymbol{y}(z) = [\phi(\boldsymbol{w}_{11} \cdot z), \phi(\boldsymbol{w}_{12} \cdot z), \ldots \phi(\boldsymbol{w}_{1m} \cdot z), 1]$$

and

$$z(\boldsymbol{x}) = \left[x_1, x_2, \ldots, x_p, 1\right].$$

We would like to compute

$$\frac{\partial\, Error}{\partial w_{kl,m}},$$

where $w_{kl,m}$ is the mth component of $\mathbf{w}_{kl}$. Let us deal with the coefficients of the output layer first, so that we are interested in $w_{2l,m}$ and get

$$\begin{aligned}
\frac{\partial\, Error}{\partial w_{2l,m}} &= \sum_k \frac{\partial\, Error}{\partial f_k}\frac{\partial f_k}{\partial w_{2l,m}} \\
&= \frac{\partial\, Error}{\partial f_l}\frac{\partial f_l}{\partial w_{2l,m}} \\
&= \sum_e \left\{ (f_l(\mathbf{x}^e) - o_l^e)\frac{\partial f_l}{\partial w_{2l,m}} \right\} \\
&= \sum_e \left\{ (f_l(\mathbf{x}^e) - o_l^e)\phi'_{2l}(y_m(\mathbf{x}^e)) \right\} \\
&= \sum_e \left\{ \delta_{2l}^e (y_m(\mathbf{x}^e)) \right\}.
\end{aligned}$$

Here, we use the notation

$$\phi'_{2l} = \frac{\partial \phi}{\partial u},$$

where the derivative is evaluated at $u = \mathbf{w}_{21} \cdot \mathbf{y}$, and we write

$$\delta_{2l}^e = (f_l(\mathbf{x}^e) - o_l^e)\phi'_{2l}.$$

Notice that evaluating this derivative involves terms in the input of the layer—the terms $y_m(\mathbf{x}^e)$—and in its output—the terms δ_{2l}^e.

Now consider the coefficients of the second layer. We are interested in $w_{1l,m}$, and we get

$$\begin{aligned}
\frac{\partial\, Error}{\partial w_{1l,m}} &= \sum_k \left\{ \frac{\partial\, Error}{\partial f_k}\frac{\partial f_k}{\partial w_{1l,m}} \right\} = \sum_{i,j} \left\{ \frac{\partial\, Error}{\partial f_i}\frac{\partial f_i}{\partial y_j}\frac{\partial y_j}{\partial w_{1l,m}} \right\} \\
&= \left\{ \sum_k \frac{\partial\, Error}{\partial f_k}\frac{\partial f_k}{\partial y_l} \right\} \frac{\partial y_l}{\partial w_{1l,m}} \\
&= \sum_e \left\{ \sum_k \left\{ (f_k(\mathbf{x}^e) - o_k^e)\frac{\partial f_k}{\partial y_l} \right\} \frac{\partial y_l}{\partial w_{1l,m}} \right\} \\
&= \sum_e \left\{ \sum_k \left\{ (f_k(\mathbf{x}^e) - o_k^e)\phi'_{2k} w_{2k,l} \right\} \frac{\partial y_l}{\partial w_{1l,m}} \right\} \\
&= \sum_e \left\{ \sum_k \left\{ (f_k(\mathbf{x}^e) - o_k^e)\phi'_{2k} w_{2k,l} \right\} \phi'_{1l} z_m \right\} \\
&= \sum_e \left\{ \sum_k \left\{ \delta_{2k}^e w_{2k,l} \right\} \phi'_{1l} z_m \right\}.
\end{aligned}$$

In this expression,

$$\phi'_{2k} = \frac{\partial \phi}{\partial u},$$

evaluated at $u = \boldsymbol{w}_{2k} \cdot \boldsymbol{y}$, and

$$\phi'_{1l} = \frac{\partial \phi}{\partial u},$$

evaluated at $u = \boldsymbol{w}_{1l} \cdot z$. Now if we write

$$\delta^e_{1l} = \sum_k \left\{ \delta^e_{2k} w_{2k,l} \right\} \phi'_{1l},$$

we get

$$\frac{\partial E}{\partial w_{1l,m}} = \sum_e \delta^e_{1l} z_m(\boldsymbol{x}^e).$$

Again, this sum involves a term obtained computing the previous derivative, terms in the derivatives within the layer, and terms in the input. You should convince yourself that, if we had a third layer, the derivative of the error with respect to parameters within this third layer would have a similar form—a function of terms in the derivative of the second layer, terms in the derivatives within the third layer, and terms in the input (all this comes from aggressive application of the chain rule). This suggests a two-pass algorithm:

1. Evaluate the net's output on each example. This is usually referred to as a *forward pass*.
2. Evaluate the derivatives using the intermediate terms. This is usually referred to as a *backward pass*.

This process yields the derivatives of the total error with respect to the parameters. We can obtain another simplification: We adopted stochastic gradient descent to avoid having to sum the value of the error and of its gradient over all examples. Because computing a gradient is linear, to compute the gradient of the error on one example alone, we simply drop the sum at the front of our expressions for the gradient. The whole is given in Algorithm 22.7.

Algorithm 22.9: Backpropagation to compute the derivative of the fitting error of a two-layer neural net on a single example with respect to its parameters

Notation:
Write the two-layer neural net as

$$\begin{aligned} \boldsymbol{f}(\boldsymbol{x}; \boldsymbol{p}) &= \left[\phi(\boldsymbol{w}_{21} \cdot \boldsymbol{y}), \phi(\boldsymbol{w}_{22} \cdot \boldsymbol{y}), \ldots \phi(\boldsymbol{w}_{2n} \cdot \boldsymbol{y})\right] \\ \boldsymbol{y}(z) &= [\phi(\boldsymbol{w}_{11} \cdot z), \phi(\boldsymbol{w}_{12} \cdot z), \ldots \phi(\boldsymbol{w}_{1m} \cdot z), 1] \\ z(\boldsymbol{x}) &= \left[x_1, x_2, \ldots, x_p, 1\right] \end{aligned}$$

($\boldsymbol{p}$ is a vector containing all parameters). Write the error on a single example as

$$\begin{aligned} Error^e &= Error(\boldsymbol{p}; \boldsymbol{x}^e) \\ &= \left(\tfrac{1}{2}\right) |\boldsymbol{f}(\boldsymbol{x}^e; \boldsymbol{p}) - \boldsymbol{o}^e|^2. \end{aligned}$$

We would like to compute

$$\frac{\partial\, Error^e}{\partial w_{kl,m}},$$

where $w_{kl,m}$ is the mth component of $\boldsymbol{w}_{kl}$.

Forward pass: Compute $\boldsymbol{f}(\boldsymbol{x}^e; \boldsymbol{p})$ saving all intermediate variables.

Backward pass: Compute

$$\delta_{2l}^e = (f_l(\boldsymbol{x}^e) - o_l^e)\phi'_{2l}$$

$$\phi'_{2l} = \frac{\partial \phi}{\partial u} \text{ evaluated at } u = \boldsymbol{w}_{21} \cdot \boldsymbol{y}$$

$$\frac{\partial\, Error^e}{\partial w_{2l,m}} = \sum_e \left\{ \delta_{2l}^e (y_m(\boldsymbol{x}^e)) \right\}.$$

Now compute

$$\delta_{1l}^e = \sum_k \left\{ \delta_{2k}^e w_{2k,l} \right\} \phi'_{1l}$$

$$\phi'_{1l} = \frac{\partial \phi}{\partial u} \text{ evaluated at } u = \boldsymbol{w}_{11} \cdot \boldsymbol{z}$$

$$\frac{\partial E^e}{\partial w_{1l,m}} = \delta_{1l}^e z_m(\boldsymbol{x}^e).$$

22.8 APPENDIX II: SUPPORT VECTOR MACHINES FOR DATASETS THAT ARE NOT LINEARLY SEPARABLE

In many cases, a separating hyperplane does not exist. To allow for this case, we introduce a set of *slack variables*, $\xi_i \geq 0$, which represent the amount by which the constraint is violated. We can now write our new constraints as

$$y_i\, (\boldsymbol{w} \cdot \boldsymbol{x}_1 + b) \geq 1 - \xi_i,$$

and we modify the objective function to take account of the extent of the constraint violations to get the problem

$$\begin{aligned} &\text{minimize } \frac{1}{2}\boldsymbol{w} \cdot \boldsymbol{w} + C \sum_{i=1}^{N} \xi_i \\ &\text{subject to } y_i\, (\boldsymbol{w} \cdot \boldsymbol{x}_1 + b) \geq 1 - \xi_i \\ &\text{and } \xi_i \geq 0. \end{aligned}$$

Here C gives the significance of the constraint violations with respect to the distance between the points and the hyperplane. The dual problem becomes

$$\text{maximize} \sum_i^N \alpha_i - \frac{1}{2}\sum_{i,j=1}^N \alpha_i (y_i y_j \boldsymbol{x}_i \cdot \boldsymbol{x}_j)\alpha_j$$

$$\text{subject to } C \geq \alpha_i \geq 0$$

$$\text{and} \sum_{i=1}^N \alpha_i y_i = 0.$$

Again, we have

$$\boldsymbol{w} = \sum_1^N \alpha_i y_i \boldsymbol{x}_i,$$

but recovering b from the solution to the dual problem is slightly more interesting. For each example where $C > \alpha_i > 0$ (note that these are strict inequalities, unlike the constraints), the slack variable ξ_i will be zero. This means that

$$\sum_{j=1}^N y_j \alpha_j \boldsymbol{x}_i \cdot \boldsymbol{x}_j + b = y_i$$

for these values of i. This expression yields b. Again, the optimization problem is a quadratic programming problem, although there is no guarantee that many points will have $\alpha_i = 0$.

22.9 APPENDIX III: USING SUPPORT VECTOR MACHINES WITH NON-LINEAR KERNELS

For many datasets, it is unlikely that a hyperplane will yield a good classifier. Instead, we want a decision boundary with a more complex geometry. One way to achieve this is to map the feature vector into some new space and look for a hyperplane in that new space. For example, if we had a plane dataset that we were convinced could be separated by plane conics, we might apply the map

$$(x, y) \rightarrow (x^2, xy, y^2, x, y)$$

to the dataset. A classifier boundary that is a hyperplane in this new feature space is a conic in the original feature space. In this form, this idea is not particularly useful because we might need to map the data into a high-dimensional space (e.g., assume that we know the classifier boundary has degree two, and the data is 10 dimensional—we would need to map the data into a 65 dimensional space).

Write the map as $\boldsymbol{x}' = \phi(\boldsymbol{x})$. Write out the optimization problem for the new points $\boldsymbol{x}'_i$; you will notice that the only form in which $\boldsymbol{x}'_i$ appears is in the terms

$$\boldsymbol{x}'_i \cdot \boldsymbol{x}'_j,$$

which we could write as $\phi(\boldsymbol{x}_i) \cdot \phi(\boldsymbol{x}_j)$. Apart from always being positive, this term doesn't give us much information about ϕ. In particular, the map doesn't appear explicitly in the optimization problem. If we did solve the optimization problem, the final classifier would be

$$f(\boldsymbol{x}) = \text{sign}\left(\sum_1^N (\alpha_i y_i \boldsymbol{x}' \cdot \boldsymbol{x}'_i + b)\right) = \text{sign}\left(\sum_1^N (\alpha_i y_i \phi(\boldsymbol{x}) \cdot \phi(\boldsymbol{x}_i) + b)\right).$$

Assume that we have a function $k(\boldsymbol{x}, \boldsymbol{y})$ that is positive for all pairs of $\boldsymbol{x}$, $\boldsymbol{y}$. It can be shown that, under various technical conditions of no interest to us, there is some ϕ such that $k(\boldsymbol{x}, \boldsymbol{y}) = \phi(\boldsymbol{x}) \cdot \phi(\boldsymbol{y})$. All this allows us to adopt a clever trick—instead of constructing ϕ explicitly, we obtain some appropriate $k(\boldsymbol{x}, \boldsymbol{y})$ and use it in place of ϕ. In particular, the dual optimization problem becomes

$$\text{maximize} \sum_{i}^{N} \alpha_i - \frac{1}{2} \sum_{i,j=1}^{N} \alpha_i (y_i y_j k(\boldsymbol{x}_i, \boldsymbol{x}_j)) \alpha_j$$

$$\text{subject to } \alpha_i \geq 0$$

$$\text{and} \sum_{i=1}^{N} \alpha_i y_i = 0,$$

and the classifier becomes

$$f(\boldsymbol{x}) = \text{sign}\left(\sum_{1}^{N} (\alpha_i y_i k(\boldsymbol{x}, \boldsymbol{x}_i) + b) \right).$$

Of course, these equations assume that the dataset are separable in the new feature space represented by k. This may not be the case, in which case the problem becomes

$$\text{maximize} \sum_{i}^{N} \alpha_i - \frac{1}{2} \sum_{i,j=1}^{N} \alpha_i (y_i y_j k(\boldsymbol{x}_i, \boldsymbol{x}_j)) \alpha_j$$

$$\text{subject to } C \geq \alpha_i \geq 0$$

$$\text{and} \sum_{i=1}^{N} \alpha_i y_i = 0,$$

and the classifier becomes

$$f(\boldsymbol{x}) = \text{sign}\left(\sum_{1}^{N} (\alpha_i y_i k(\boldsymbol{x}, \boldsymbol{x}_i) + b) \right).$$

There are a variety of possible choices for $k(\boldsymbol{x}, \boldsymbol{y})$. The main issue is that it must be positive for all values of $\boldsymbol{x}$ and $\boldsymbol{y}$. Some typical choices are shown in Table 22.2. There doesn't appear to be any principled method for choosing between kernels; one tries different forms and uses the one that gives the best error rate measured using cross-validation.

TABLE 22.2 Some support vector kernels

Kernel form	Qualitative properties of ϕ represented by this kernel
$(\boldsymbol{x} \cdot \boldsymbol{y})^d$	ϕ is all monomials of degree d
$(\boldsymbol{x} \cdot \boldsymbol{y} + c)^d$	ϕ is all monomials of degree d or below
$\tanh(a\boldsymbol{x} \cdot \boldsymbol{y} + b)$	
$\exp\left(-\frac{(\boldsymbol{x}-\boldsymbol{y})^T(\boldsymbol{x}-\boldsymbol{y})}{2\sigma^2}\right)$	

23 Recognition by Relations Between Templates

An object may have internal degrees of freedom, which mean that its appearance is highly variable (e.g., people can move arms and legs, fish deform to swim, snakes wriggle, and so on). This phenomenon can make template matching extremely difficult because one may require either a classifier with a flexible boundary (and a lot of examples) or many different templates.

Many of these objects have small components that have a fairly orderly appearance. We could try to match these components as templates, and then determine what objects are present by looking for suggestive relationships between the templates that have been found. For example, instead of finding a face by looking for a single complete face template, we could find one by looking for eyes, nose, and a mouth that all lie in an appropriate configuration.

This approach has several possible advantages. First, it may be easier to learn an eye template than it is to learn a face template because the structure could be simpler. Second, it may be possible to obtain and use relatively simple probability models. This is because there may be some independence properties that can be exploited. Third, we may be able to match a large number of objects with a relatively small number of templates. Animal faces are a good example of this phenomenon—pretty much all animals with recognizable faces have eyes, nose, and a mouth, but with slightly different spatial layouts. Finally, it means that the simple individual templates can be used to construct complex objects. For example, people can move their arms and legs around, and it appears to be much more difficult to learn a single, explicit template for finding whole people than to obtain individual templates for bits of people and a probability model that describes their degrees of freedom.

This topic is not yet well enough understood for there to be a standard approach. However, the main issue—how does one encode a set of relationships between templates in a form that is easily managed?—is quite clear. In this chapter, we explore a series of different approaches to this problem. First, we could allow each template to vote for the objects that it could represent and then count the votes in some way (Section 23.1). We could place more weight on the specifics of the spatial relations by building some explicit probability model. This could come from the likelihood; in essence, we need a probability distribution function that has a high value when the components are configured like the object and a low value otherwise. Finding objects then becomes a matter of searching for templates that can be plugged into the probability model to get a high value (Section 23.2). Pruning the search requires care; we show one approach in Section 23.3. The difficulty with this approach is that, even when pruned, the search could be expensive. A particular class of probability model allows efficient search. We introduce these models in Section 23.4, and describe two applications in Sections 23.5 and 23.6.

23.1 FINDING OBJECTS BY VOTING ON RELATIONS BETWEEN TEMPLATES

Simple object models can result in quite effective recognition. The simplest model is to think of an object as a collection of image *patches*—small image neighborhoods of characteristic appearance—of several different types, forming an image *pattern*. To tell what pattern is present in an image, we find each patch that is present and allow it to vote for every pattern in which it appears. The pattern in the image is the one with the most votes. Although this strategy is simple, it is quite effective. We sketch methods for finding patches and then describe a series of increasingly sophisticated versions of the strategy.

23.1.1 Describing Image Patches

Small image patches can have a quite characteristic appearance, usually when they have many nonzero derivatives (e.g., at corners). We describe a system due to Schmid and Mohr (1997*a,b*) that takes advantage of this property. They find image corners—often called *interest points* (see Schmid, Mohr and Bauckhage, 2000). They then estimate a set of derivatives of the image grey-level at those corners, and evaluate a set of functions of the image derivatives that are invariant to rotation, translation, some scaling, and illumination changes. These features are called *invariant local jets*. Describing these features in detail would take us out of our way. The value of this approach is that because the combinations are invariant, we expect that they will take the same value on different views of an object.

We now assume that the image patches fall into a number of classes. We can obtain representatives for each class by having multiple pictures of each object—typically, corresponding patches will be of the same class, but probably have somewhat different invariant local jets as a result of image noise. We can determine an appropriate set of classes either by classifying the patches by hand or by clustering example patches (a somewhat better method!). We need to tell when two sets of invariant local jets represent the same class of image patch. Schmid and Mohr test the Mahalanobis distance between the feature vectors of a patch to be tested and an example patch; if it is below some threshold, the patch being tested is the same as the example.

Notice that this is a classifier—it is allocating patches to classes represented by the examples or deciding not to classify them—and that the patches are templates. We could build a template matcher with any of the techniques of chapter 22 without using features that are invariant to rotation. To do this, we would use a training set that contained rotated and scaled versions of each example patch, under varying illuminant conditions, so the classifier could learn that rotation, scaling, and illuminant changes don't affect the identity of the patch. The advantage of using invariant features is that the classifier doesn't need to learn this invariance from the training set.

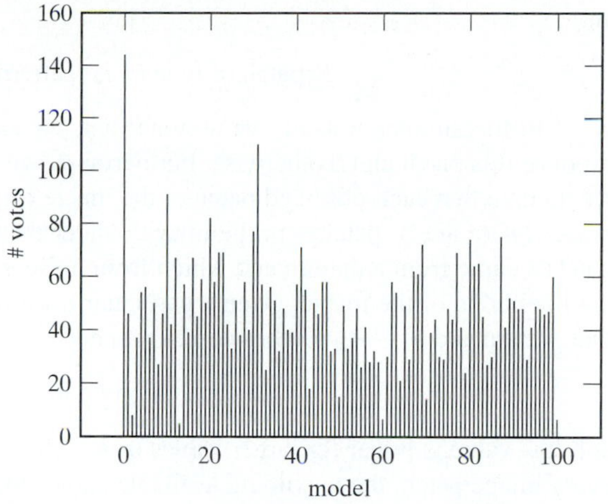

Figure 23.1 The graph shows the number of votes for each pattern recorded for a particular image under the simple voting scheme. Notice that, although the correct match (model # 0) receives the maximum number of votes, three other candidates receives more than half as many votes. *Reprinted from "Local gray-value invariants for image retrieval," by C. Schmid and R. Mohr, IEEE Trans. Pattern Analysis and Machine Intelligence, 1997 © 1997, IEEE.*

23.1.2 Voting and a Simple Generative Model

For a given image, we find the interest points and classify the image patch at each image point. Now, what pattern lies in the image? We can answer this question by constructing a correspondence between image patches and patterns. Assume that there are N_i patches in the image. Furthermore, we assume that there is either a single pattern from our collection in the image or there is no pattern there. An individual patch could have come either from whatever pattern is present, or from noise. However, patterns typically do not contain every class of patch. This means that asserting that a particular pattern is present is the same as asserting that some of the image patches came from noise (because only one pattern can be present, and these image patches belong to classes that are not in the current pattern).

We now have a (simple) generative model for an image. When a pattern is present, it produces patches of some classes, but not others. By elaborating this model, we obtain a series of algorithms for matching patterns to images.

The simplest version of this model is obtained by assuming that a pattern produces all patches of the classes that it can produce and then require that as few patches as possible come from noise. This assumption boils down to voting. We take each image patch and record a vote for every pattern that is capable of producing that class of patch. The pattern with the most votes, wins, and we say that this pattern is present. This strategy can be effective, but has some problems (Figure 23.1).

23.1.3 Probabilistic Models for Voting

We can interpret our simple voting process in terms of a probabilistic model. This is worth doing because it casts some light on the strengths and weaknesses of the approach. Our generative model can be made probabilistic by assuming that the patches are produced independently and at random, assuming that the object is present. Let us write

$$P\{\text{patch of type } i \text{ appears in image} \mid j\text{th pattern is present}\} = p_{ij}$$

and

$$P\{\text{patch of type } i \mid \text{no pattern is present}\} = p_{ix}.$$

In the simplest model, we assume that, for each pattern j, $p_{ij} = \mu$ if the pattern can produce this patch and 0 otherwise. Furthermore, we assume that $p_{ix} = \lambda < \mu$ for all i. Finally, we assume that each observed patch in the image can come from either a single pattern or from noise. There are n_i patches in the image. Under these assumptions, we need only know which patches came from a pattern and which from noise to compute a likelihood value. In particular, the likelihood of the image, given a particular pattern, and assuming that n_p patches came from that pattern and $n_i - n_p$ patches come from noise, is

$$P(\text{interpretation} \mid \text{pattern}) = \lambda^{n_p} \mu^{(n_i - n_p)},$$

and this value is larger for larger values of n_p. However because not every pattern can produce every image patch, the maximum available choice of n_p is *dependent on the pattern we choose.* Our voting method is equivalent to choosing the pattern with the maximum possible likelihood under this (simple) generative model.

This suggests the source of some difficulties: If the pattern is unlikely, we should take that (prior) information into account. Furthermore, noise may be able to produce some patches more easily than others—ignoring this fact can confuse the vote. Finally, some patches may be more likely given an object than others. For example, corners appear much more often on a checkerboard pattern than they do on a zebra stripe pattern.

Elaborating the Generative Model Dealing with these issues in the framework of our current model—that the patches occur independently given that the pattern is present—is relatively simple. Assume that there are N different types of patch. We now assume that each different type of patch is generated with a different probability by different patterns and by noise. Now assume that there are n_{il} instances of the lth type of patch in the image. Furthermore, n_k of these are generated by the pattern, and the rest are generated by noise.

The likelihood function for the jth pattern is

$$P\left(\begin{array}{c|c} n_1 \text{ of type 1 from pattern,} & \\ \ldots, & \\ n_N \text{ patches of type } N \text{ from pattern} & j\text{th pattern} \\ \text{and } n_{i1} - n_1 \text{ of type 1 from noise,} & \\ \ldots, & \\ n_{iN} - n_N \text{ from noise} & \end{array}\right).$$

Now because the patches arise independently given the pattern and the noise is independent of the pattern, this likelihood is

$$P(\text{patches from pattern} \mid j\text{th pattern}) P(\text{patches from noise}).$$

The first term is

$$P(\text{type } 1 \mid j\text{th pattern})^{n_1} P(\text{type } 2 \mid j\text{th pattern})^{n_2} \cdots P(\text{type } N \mid j\text{th pattern})^{n_N} P(\text{noise})$$

which is evaluated as

$$p_{1j}^{n_1} p_{2j}^{n_2} \cdots p_{Nj}^{n_N}.$$

We now assume that patches that arise from noise do so independently of one another. This means we can write the noise term as

$$P(\text{type } 1 \mid \text{noise})^{(n_{i1} - n_1)} \cdots P(\text{type } N \mid \text{noise})^{(n_{iN} - n_N)},$$

which is evaluated as

$$p_{1x}^{(n_{i1}-n_1)} \cdots p_{Nx}^{(n_{iN}-n_N)}.$$

This means that the likelihood can be written out as

$$p_{1j}^{n_1} p_{2j}^{n_2} \cdots p_{Nj}^{n_N} p_{1x}^{(n_{i1}-n_1)} \cdots p_{Nx}^{(n_{iN}-n_N)}.$$

There are two cases for each type of patch k; if $p_{kj} > p_{kx}$, then this is maximized by $n_k = n_{ik}$; otherwise it is maximized by $n_k = 0$. We write π_j for the prior probability that the image contains the jth pattern and π_0 for the prior probability it contains no object. This means that for each pattern type j, the maximal value of the posterior will look like

$$\left(\prod_m p_{mj}^{n_{im}} \prod_l p_{lx}^{n_{il}}\right) \pi_j,$$

where m runs through features for which $p_{mj} > p_{mx}$ and l runs through features for which $p_{lj} < p_{lx}$. We could form this value for each object and choose the one with the highest posterior value. Notice that this *is* a relational model, although we're not actually computing geometric features linking the templates. This is because the patches are linked by the conditional probability that they occur, given a pattern, *which is different from pattern to pattern*. You should compare this model with the face detector of Section 22.2.2; the probabilistic models are identical in spirit.

23.1.4 Voting on Relations

We can use geometric relations to improve on the simple voting strategy fairly easily. A patch should match to an object only if there are nearby patches that also match to the object and are in an appropriate configuration. The term "appropriate configuration" can be a source of difficulties, but for the moment assume that we are matching objects up to *plane* rotation, translation, and scale (this is a reasonable assumption for such things as frontal faces).

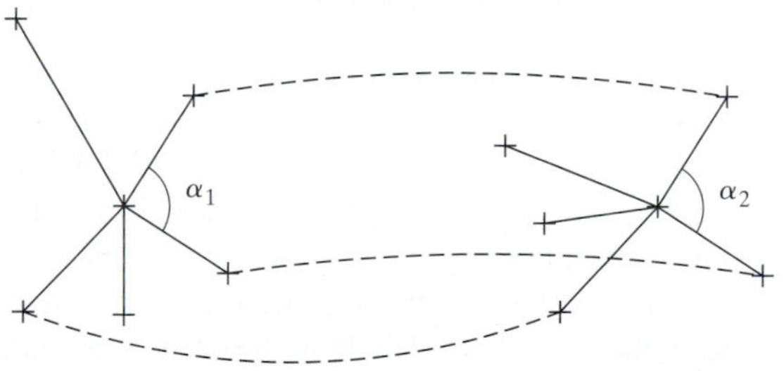

Figure 23.2 Instead of voting for patches that match, we can vote for collections of patches. In this approach, we register a vote only if a patch matches to a particular object and a percentage of its neighbors match too, and the angles between triples of matching patches have appropriate values as the figure illustrates. This strategy significantly reduces the number of random matches, as Figure 23.3 indicates. *Reprinted from "Local grayvalue invariants for image retrieval," by C. Schmid and R. Mohr, IEEE Trans. Pattern Analysis and Machine Intelligence, 1997 © 1997, IEEE.*

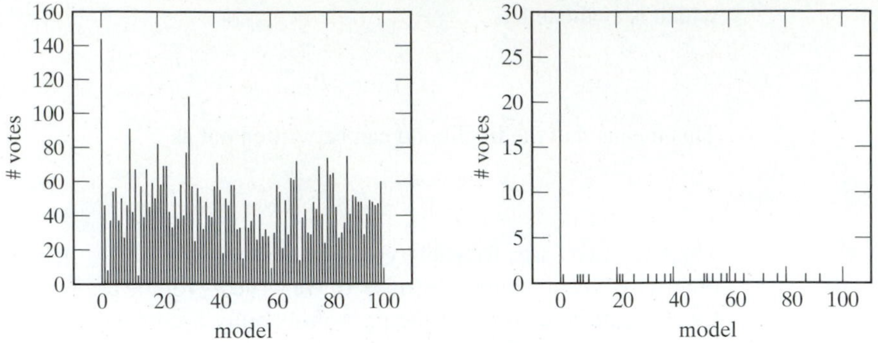

Figure 23.3 In practice, the use of semilocal constraints greatly increases the discriminating power of the original voting scheme. The figure on the **left** shows the number of votes recorded for each model under the original voting scheme, for a particular image. Although the correct match (model # 0) receives the maximum number of votes, three other candidates receive more than half as many votes. When semilocal constraints are added (**right**), the correct match stands out much more clearly. *Reprinted from "Local grayvalue invariants for image retrieval," by C. Schmid and R. Mohr, IEEE Trans. Pattern Analysis and Machine Intelligence, 1997 © 1997, IEEE.*

Now assume that we have a patch that matches to some object. We now take the p nearest patches and check that more than 50% of them match to the same objects and that the angles between the triples of matching patches are the same as the corresponding angles on the object—these are referred to as *semilocal constraints* (see Figures 23.2 and 23.3). If these two tests are passed, we register a vote for that object from the patch. You should compare this strategy (which is due to Schmid and Mohr, 1997*a,b*) quite closely with geometric hashing (Section 18.4.2). It is harder to construct a probabilistic interpretation for this approach—the emission probabilities of the patches now depend on *where* they are in the pattern, as well as identity of the pattern that generated them.

23.1.5 Voting and 3D Objects

Although we described Schmid and Mohr's approach in terms of 2D patterns, it can be extended fairly easily to 3D object recognition. We do this by regarding each of a series of views of the object as a different 2D pattern. Given enough views, this works because the small changes in the 2D pattern caused by a slight change in viewing angle will be compensated for by the available error range in the process that matches invariant local jets and angles.

This strategy for turning 3D recognition into 2D matching applies quite generally, but comes with difficulties. The main problem is the large number of models that result, which can make the voting procedure difficult. It isn't known what the minimum number of views required for matching in a scheme like this is. This was the approach used to obtain Figure 25.9 (in Chapter 25).

23.2 RELATIONAL REASONING USING PROBABILISTIC MODELS AND SEARCH

The previous section explored methods that assumed that templates were conditionally independent given the pattern. This assumption is a nonsense for most objects (although it can work extremely well in practice) because there are usually quite strong relations between features. For example, there are seldom more than two eyes, one nose, and one mouth in a face; the distance between the eyes is roughly the same as that from the bridge of the nose to the mouth; the line joining the eyes is roughly perpendicular to the line from bridge of nose to mouth; and so on. We incorporated some of these constraints into a voting strategy in Section 23.1.4, but didn't really indicate any principled framework within which they can be exploited.

This remains a difficult problem. We need to build models that represent what is significant and allow for efficient inference. We can't present current orthodoxy on this subject because it doesn't exist; instead, we describe the main issues and approaches.

23.2.1 Correspondence and Search

In this section, we explore the core issues in using a probabilistic model for matching. The general approach is as follows. We obtain an interpretation for the image and compute the value of the posterior probability of the interpretation given the image. We accept interpretations with a sufficiently large value of the posterior probability.

This account hides a myriad of practical problems. The most basic is that we do not know which image information comes from objects and which comes from noise. Generally, we evade this difficulty and turn our problem into a correspondence problem by matching templates to the image and then reasoning about the relations between templates. For example, if we wish to find faces, we apply a series of different detectors—say eye, nose, and mouth detectors—to the image and then look for suggestive configurations.

Correspondence This brings us to the second important problem, which is correspondence. We can't evaluate a posterior (or joint) probability density if we don't know the values of each variable. This means that we must, in essence, engage in a process that hypothesizes that one response is the left eye, another the right eye, a third the nose, and a fourth the mouth and *then* evaluates the posterior. It is clearly important to manage this search carefully so that we do not have to look at all possible correspondences.

The techniques of chapter 18 transfer relatively easily; we need to apply a cloak of probability, but the basic line of reasoning remains. The most basic fact of object recognition—for rigid objects, a small number of correspondences can be used to generate a large number of correspondences—translates easily into the language of probability easily, where it reads: for rigid objects, correspondences are not independent.

In general, our probability model evaluates the joint probability density for some set of variables. We call a set of values for these variables an *assembly* (other terms are a *group* or an *hypothesis*). An assembly consists of a set of detector outputs—each may respond with position, position and orientation, or even more—and a label for each output. These labels give the correspondences, and the labeling is important: if, say, the eye-detector does not differentiate between a left eye and a right eye, we will have to label the eye responses left and right, but there is no point in labeling a mouth detector response as an eye.

It is not possible to form and test all assemblies because there are usually far too many. For example, assume we have eye detectors, which don't differentiate between left and right eyes, nose detectors, and mouth detectors, and a face consists of two eyes, a nose, and a mouth. If there are N_e responses from eye detectors, N_n responses from nose detectors, and N_m responses from mouth detectors, we would have $O(N_e^2 N_n N_m)$ assemblies to look at. If you read chapter 18 carefully, you should be convinced this is a wild overestimate—the whole point of the chapter is that quite small numbers of correspondences predict other correspondences.

Incremental Assembly and Search This suggests thinking of the matching process in the following way. We assemble collections of detector responses, labeling them as necessary to establish their role in the final assembly. The assembly process is incremental—we expand small assemblies to make big ones. We take each working assembly and then determine whether it can be pruned, accepted *as is*, or should be expanded. Any particular assembly consists of a set of pairs, each of which is a detector response and a label. Some available labels may not have a detector response associated with them. Generally, a correspondence search has a working collection of hypotheses, which may be quite large. The search involves taking some correspondence hypothesis, attaching some new pairs, and then accepting the result as an object, removing it from consideration entirely, or returning it to the pool of working hypotheses.

There are three important components in a correspondence search:

- **When to accept an hypothesis:** If a correspondence hypothesis is sufficiently good, it may be possible to stop expanding it.
- **What to do next:** In a correspondence search, we have to determine which correspondence hypothesis to work on next. Generally, we wish to work on an hypothesis that is likely to succeed. Usually, it is easier to determine what *not* to do next.
- **When to prune an hypothesis:** If there is no set of detector responses that can be associated with any subset of the empty labels such that the resulting hypothesis would pass the classifier criterion, there is no point in expanding that search.

Setting up the Search Problem We continue to work with the face model to illustrate the ideas. Assume that the left eye detector responds at $\boldsymbol{x}_1$, the right eye detector responds at $\boldsymbol{x}_2$, the mouth detector responds at $\boldsymbol{x}_3$, and the nose detector responds at $\boldsymbol{x}_4$. We assume that the face is at $\boldsymbol{F}$, and that all other detector responses are due to noise. Furthermore, we assume that we are determining whether there is either a single face or none in the image. This involves comparing the value of

$$P(\text{one face at } \boldsymbol{F} \mid \boldsymbol{X}_{\text{le}} = \boldsymbol{x}_1, \boldsymbol{X}_{\text{re}} = \boldsymbol{x}_2, \boldsymbol{X}_{\text{m}} = \boldsymbol{x}_3, \boldsymbol{X}_{\text{n}} = \boldsymbol{x}_4, \text{all other responses})$$

with

$$P(\text{no face} \mid \boldsymbol{X}_{\text{le}} = \boldsymbol{x}_1, \boldsymbol{X}_{\text{re}} = \boldsymbol{x}_2, \boldsymbol{X}_{\text{m}} = \boldsymbol{x}_3, \boldsymbol{X}_{\text{n}} = \boldsymbol{x}_4, \text{all other responses}).$$

Stopping a Search: Detection Assume that noise responses are independent of the presence of a face (which is fairly plausible). We have

$$P(\text{one face at } \boldsymbol{F} \mid \boldsymbol{X}_{\text{le}} = \boldsymbol{x}_1, \boldsymbol{X}_{\text{re}} = \boldsymbol{x}_2, \boldsymbol{X}_{\text{m}} = \boldsymbol{x}_3, \boldsymbol{X}_{\text{n}} = \boldsymbol{x}_4, \text{all other responses})$$

is equal to

$$P(\text{one face at } \boldsymbol{F} \mid \boldsymbol{x}_1, \boldsymbol{x}_2, \boldsymbol{x}_3, \boldsymbol{x}_4) P(\text{all other responses})$$

which, in turn, is proportional to

$$P(\boldsymbol{x}_1, \boldsymbol{x}_2, \boldsymbol{x}_3, \boldsymbol{x}_4 \mid \text{one face at } \boldsymbol{F}) P(\text{all other responses}) P(\text{one face at } \boldsymbol{F})$$

(where we have suppressed some notation). We can classify particular groups of detector responses as coming from a face or from noise by comparing the posterior that this configuration comes from a face with the posterior that it comes from noise. In particular, we compare

$$P(\boldsymbol{x}_1, \boldsymbol{x}_2, \boldsymbol{x}_3, \boldsymbol{x}_4 \mid \text{one face at } \boldsymbol{F})$$

with

$$(P(\text{noise responses}) P(\text{no face}) / P(\text{one face at } \boldsymbol{F})) \text{ (term in relative loss)}.$$

Recall that there are only two options—face or no face—and check the chapter on probability on the website if the remark seems puzzling. Thus, the likelihood is the main term of interest. This tells us whether a complete assembly represents a face, but an incomplete assembly could represent a face, too. We can score configurations that lack features, too. This involves determining the posterior that a face is present given only some features and comparing that with the posterior that the features arose from noise.

When a group of features satisfies the classification criterion (that the posterior that a face is present exceeds the posterior that it is not), we can certainly stop searching. It may not be necessary to observe all possible features to determine that a face is present. If a configuration is strongly suggestive of a face and unlikely to have arisen from noise, then we may wish to assert that a face is present and stop searching at that point.

We illustrate with an example. Assume we wish to determine whether a right eye, a mouth, and a nose represent a face. To evaluate the joint, we need to evaluate a series of noise terms and the term

$$P(\boldsymbol{X}_{\text{le}} = \text{missing}, \boldsymbol{X}_{\text{re}} = \boldsymbol{x}_2, \boldsymbol{X}_{\text{m}} = \boldsymbol{x}_3, \boldsymbol{X}_{\text{n}} = \boldsymbol{x}_4 \mid \text{one face at } \boldsymbol{F})$$

using our model. This requires a model to explain how a feature went missing; the simplest is to assume that the detector did not respond (a variety of others are possible), and that this failure to respond is independent of the other feature detector responses. This yields

$$P(\text{missing}, \boldsymbol{x}_2, \boldsymbol{x}_3, \boldsymbol{x}_4 \mid \text{face})$$

is equal to

$$\int P(\text{le does not respond} \mid \boldsymbol{X}_1) P(\boldsymbol{X}_1, \boldsymbol{x}_2, \boldsymbol{x}_3, \boldsymbol{x}_4 \mid \text{face})\, d\boldsymbol{X}_1$$

(again, suppressing some notation). Now if the probability of detector failure is independent of feature position (which is the usual case in vision applications), we have

$$P(\text{missing}, \boldsymbol{x}_2, \boldsymbol{x}_3, \boldsymbol{x}_4 \mid \text{face})$$

is equal to

$$P(\text{le does not respond}) \int P(\boldsymbol{X}_1, \boldsymbol{x}_2, \boldsymbol{x}_3, \boldsymbol{x}_4 \mid \text{face}) d\boldsymbol{X}_1.$$

This *does not mean* that we can prune a search if a configuration doesn't represent a face when it has a missing feature. This is because there might be some position for the missing feature that did represent a face; as Section 23.3 shows, to prune we need a bound. If we do decide that an object is present without matching all components, this is because either (a) missing components didn't get detected, but the configuration of the components found is so distinctive it doesn't matter; or (b) we haven't yet found the other components, but the configuration of the other detectors is so distinctive it doesn't matter. Notice that in Case (b) we may still want to look for possible responses for the other detectors.

23.2.2 Example: Finding Faces

Perona and colleagues have built a series of face finders that use various forms of probability model (Burl and Perona, 1996, Leung, Burl and Perona, 1995, Weber, Einhaeuser, Welling and Perona, 2000). Each follows broadly the line of incremental search among assemblies, identifying assemblies that should be expanded. Features are obtained from the outputs of various filters, as in Figure 23.4. We sketch the system of Leung, Burl and Perona (1995).

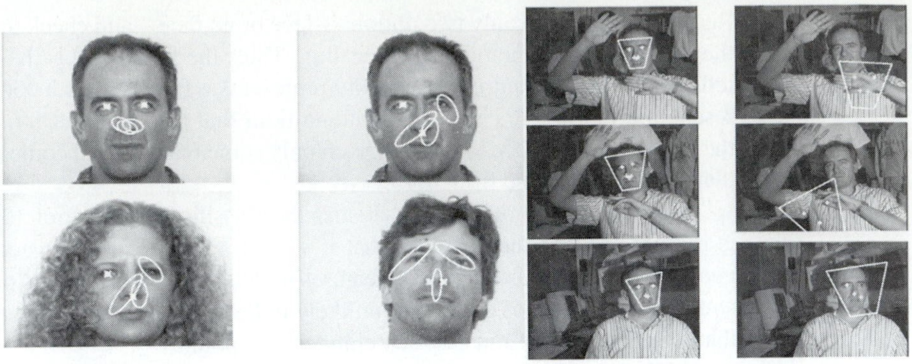

Figure 23.4 Perona et al.'s face detector looks for local patterns, which are classified on their appearance using filter responses. Appropriate arrangements of these patterns are faces. Relations between face points are represented by the interelement distances, and the class-conditional density for these distances is gaussian. In turn, this means that, once some feature responses have been found, the position of others can be predicted. The figure on the **left** illustrates predictions for different feature points. The features are points around the eyes, nose, and mouth, and the variation in interpoint distances comes from individual variations in facial structure. The figure on the **right** illustrates the behavior of the overall face detector, the column on the left shows the best face in the image, and the column in the right shows the second best, which typically has a posterior value that is smaller by an order of magnitude. *Reprinted from, "Finding faces in cluttered scenes using random labeled graph matching," by Leung, T., Burl, M., and Perona, P., Proc. Int. Conf. on Computer Vision, 1995 © 1995, IEEE.*

An assembly is represented by a vector of distances between features. This vector is assumed to have a Gaussian probability distribution, conditioned on the presence of a face. Missing elements can be dealt with by marginalizing out the relevant elements of this distribution. An attractive feature of this model is that, once one has a sufficiently large assembly, the position of new features can be predicted by (a) identifying distances that would lead to sufficiently large values of the conditional probability and (b) predicting a range of feature positions from these distances (Figure 23.4).

23.3 USING CLASSIFIERS TO PRUNE SEARCH

Assume that we have an assembly that consists of a right eye, a mouth, and a nose. If we can determine from the probability model that there is no possible position for the left eye that would result in an assembly that is acceptable as a face, there is no point in trying to grow our assembly. Recall that if we had a position for the left eye, we would test to see whether the likelihood was greater than some function of the number of noise responses, the priors, and the relative losses. In particular, there is some fixed value that the likelihood must exceed before we can assert that a face is present. If we can assert that *there is no value of* $\boldsymbol{x}_1$ that would result in a likelihood that is over threshold, the search can be pruned.

Notice that this line of reasoning extends. If we have a left eye response and a right eye response, there is no value of the nose and mouth responses that would result in a posterior that is over threshold, we can prune this search—and so not attempt to add either a nose or a mouth to this assembly. It is often quite tricky to supply the necessary bounds, however. One way to do this is to use a classifier.

You can think of a classifier as representing a (crude) probability model for the posterior. This model has a zero value when the posterior is below threshold and a non-zero value when it is above. The great advantage of this model is that it is quite easy to prune; we can reject any assembly if there was no element that could be added to yield something that would lie in the nonzero region.

23.3.1 Identifying Acceptable Assemblies Using Projected Classifiers

There is no point in growing a working assembly unless there is some conceivable way in which it could become an acceptable final assembly. Assume that we have a classifier that can determine whether a final assembly is acceptable. This classifier can be trained on a large sequence of examples.

We can use this final classifier to predict tests that determine whether small groups should be expanded. Essentially, we discard groups for which there is no prospect that any conceivable new set of elements would make the group acceptable. In turn, this boils down to projecting the decision boundaries of the classifier onto a set of different factors. For example, assume that we have a two eye detector responses: Should they form a pair or not? We can look at only features that come from those eyes because we don't know what other elements lie in the assembly, and therefore can't use them to compute features. In particular, the issue here is this: Is there some set of elements (i.e., a nose and a mouth) such that, if they were attached to this group, the resulting group would pass the classifier? This question is answered by a classifier whose decision boundary is obtained by projecting the decision boundary of the face assembly onto the space spanned by the features computed from our two eyes (Figure 23.5).

By projecting the classifier onto components that can handle small groups, we can prune the collection of groups that must be searched to find a large collection of segments that passes the main classifier. Using this strategy in practice requires a certain amount of care. It is important to have classifiers that project well—the decision boundaries must project to decision boundaries

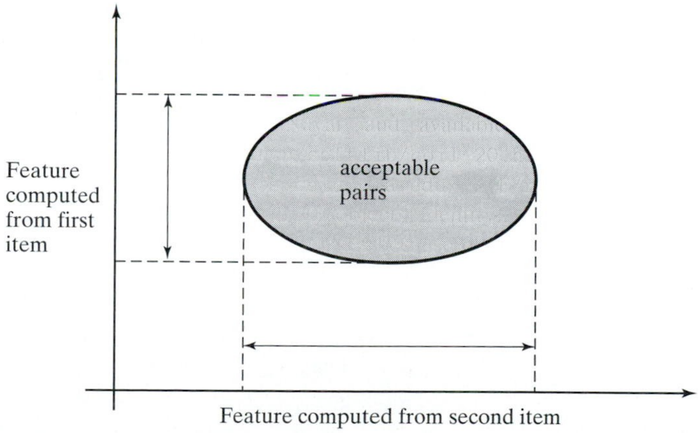

Figure 23.5 A classifier that accepts large groups can be used to predict tests that identify potentially useful small groups. Assume we want to group two items together: This grouping is acceptable only if a point, consisting of one feature computed from each item, lies in the shaded set. There are some items that could not appear in these groups because the feature computed from the items lies out of the acceptable range, so there does not exist a second item that could make the pair acceptable. It is possible to obtain classifiers that identify the acceptable range by projecting the main classifier onto its factors, as the picture indicates.

that can be represented reasonably easily. This can be dealt with—the details are rather beyond the scope of this account—and the resulting classifiers can be used to find people and horses relatively efficiently in quite simple cases (Figure 23.6 and Figure 23.7; Forsyth and Fleck, 1997, and Ioffe and Forsyth, 1998).

23.3.2 Example: Finding People and Horses Using Spatial Relations

Assume we wish to find people in images; a natural approach is to find possible body segments and reason about their configuration. This can be done in some specialized cases. One that may become important is if the people concerned are not wearing clothing (which would allow people to search for or avoid images associated with strong opinions). In this case, body segments can be found by looking for skin (as in Section 22.2.1) and then constructing extended image regions with roughly parallel sides (as in Section 24.1) that contain skin color. The resulting extended image regions tend to represent many of the body segments in the image.

Now we can search for people by searching for assemblies of these image segments. For example, assume that we see only frontal views; then we can look for assemblies of nine image segments containing a left upper arm, a left lower arm, and so on, together with a torso. It isn't practical to take all sets of nine segments from the image, so we use the pruning methodology described previously. This yields a system that can find people in simple images fairly reliably (Ioffe and Forsyth, 1998).

This methodology extends to finding horses as well. We identify all image pixels that could be hide (this means they lie in the right range of colors and have little texture locally). We then form extended regions of hide (as for people) and regard these as potential body segments. We use a crude model of a horse as a four-segment group (a body, a neck, and two legs); the order in which the tests are performed is illustrated in Figure 23.6. The crude model is used because (a) it is robust to changes in aspect, and (b) it doesn't require accurate localization of all segments.

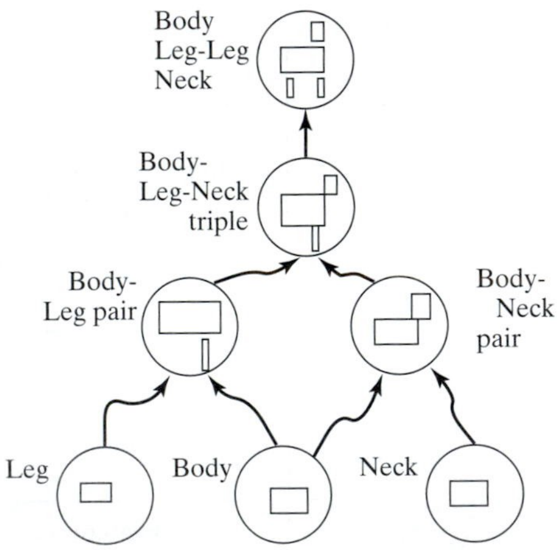

Figure 23.6 The model used for a horse involves two legs, a body, and a neck. Segments are first tested to see whether they are acceptable members of one of these classes. We then test pairs of body and leg segments, and pairs of body and neck segments. We then take body-leg pairs and body-neck pairs *which share the same body* and test to see if they form a valid body-leg-neck triple. Pairs of triples that *share the same body and the same neck* are tested to form quadruples. *Reprinted from "Body Plans," by D.A. Forsyth and M.M. Fleck, Proc. Computer Vision and Pattern Recognition, 1997, © 1997, IEEE.*

Figure 23.7 Horses can be found by building a classifier of the type described in the text. This example illustrates the effectiveness of the approach. The system finds image regions that have the color and texture of hide and look like cylinders (as in Section 24.1). A horse is defined as a collection of four cylinders—a body, a neck, and two legs—and a classifier that accepts this assembly is learned from examples. This classifier is projected onto factors, as described in the text, and these projected versions are used to build assemblies in an arbitrarily selected order. The result is a system that, although not spectacularly accurate, is useful. The figure shows all images that the classifier thinks contains horses, obtained using a control set of 1,086 nonhorse images and a test set of 100 horse images. The test images recovered contain horses in a wide range of aspects; one control image contains an animal that might reasonably pass for a horse. This figure appears also as Figure 25.10 (it's reproduced here for convenience). *Reprinted from "Body Plans," by D.A. Forsyth and M.M. Fleck, Proc. Computer Vision and Pattern Recognition, 1997, © 1997, IEEE.*

This method of finding objects has been useful in some applications, as Figures 23.6 and 23.7 suggest, but there are significant open questions. First, it isn't obvious what sequence of intermediate groups and tests is best. It is possible to think about this problem by asking which projections of the decision boundary lead to small projected volumes (and so, hopefully, to relatively few passing groups). This looks like a difficult problem because it deals with combinatorial properties. Second, it isn't obvious how we deal with multiple objects efficiently. If a single template occurs only on one object, then when we observe the template, we immediately have an object hypothesis. Things are much more difficult when templates could have come from many objects. For example, consider recognizing people *and* horses: A single segment could come from either, as could many pairs. How do we arrange the sequence of tests to minimize the work required to decide what we are dealing with? Finally, the use of a classifier generates some annoying problems with false negatives.

23.4 TECHNIQUE: HIDDEN MARKOV MODELS

Up to this point, adopting a probability model hasn't changed much of significance in our discussion of recognition. While we verify by evaluating a joint probability model, we are still engaged in a correspondence search, very like those of chapter 18. Our method of pruning is analogous to the interpretation tree of that chapter. However, more is possible: Some probability models have a structure that make it possible to get an exactly optimal correspondence efficiently.

A program that reads American Sign Language from a video sequence of someone signing must infer a state, internal to the user, for each sign. The program infers state from measurements of hand position that are unlikely to be accurate, but depends, hopefully quite strongly, on the state. The signs change state in a random (but quite orderly) fashion. In particular, some sequences of states occur seldom (e.g., a sequence of letter signs for the sequence "wkwk" is extremely unlikely). This means that both the measurements *and* the relative probabilities of different sequences of signs can be used to determine what actually happened.

The elements of this kind of problem are as follows:

- there is a sequence of random variables (in our example, the signs), each of which is conditionally independent of all others given its predecessor;
- each random variable generates a measurement (the measurements of hand position) whose probability distribution depends on the state.

Similar elements are to be found in such examples as interpreting the movement of dancers or of martial artists. There is an extremely useful formal model, known as a *hidden Markov model*, corresponding to these elements.

The sequence of random variables does not have to be temporal. Instead, we could order the variables by spatial relations, too. Consider an arm: The configuration of the lower arm is (roughly) independent of the configuration of the rest of the body given the configuration of the upper arm; the configuration of the upper arm is (roughly!) independent of the rest of the body given the configuration of the torso; and so on. This gives us a sequence of random variables with the conditional independence properties described before. Now we don't really know the configuration of the lower arm—we have only some image measurements that have some probabilistic relationship with the configuration of the lower arm (as before).

23.4.1 Formal Matters

A sequence of random variables X_n is said to be a *Markov chain* if

$$P(X_n = \boldsymbol{a} \mid X_{n-1} = \boldsymbol{b}, X_{n-2} = \boldsymbol{c}, \ldots, X_0 = \boldsymbol{x}) = P(X_n = \boldsymbol{a} \mid X_{n-1} = \boldsymbol{b})$$

and a *homogeneous Markov chain* if this probability does not depend on n. Markov chains can be thought of as sequences with little memory; the new state depends on the previous state, but not on the whole history. It turns out that this property is surprisingly useful in modeling because many physical variables appear to have it, and because it enables a variety of simple inference algorithms. There are slightly different notations for Markov chains on discrete and continuous state spaces; we shall discuss only the discrete case.

Assume that we have a discrete state space. It doesn't really matter what dimension the space is, although finite spaces are somewhat easier to imagine. Write the elements of the space as s_i and assume that there are k elements. Assume that we have a sequence of random variables taking values in that state space that forms a homogeneous Markov chain. Now we write

$$P(X_n = s_j \mid X_{n-1} = s_i) = p_{ij},$$

and because the chain is independent of n, so is p_{ij}. We can write a matrix $\mathcal{P}$ with i, jth element p_{ij} which describes the behavior of the chain. This matrix is called the *state transition matrix*. Assume that X_0 has probability distribution $P(X_0 = s_i) = \pi_i$, and we write $\boldsymbol{\pi}$ as a vector with ith element π_i. This means that

$$\begin{aligned} P(X_1 = s_j) &= \sum_{i=1}^{k} P(X_1 = s_j \mid X_0 = s_i) P(X_0 = s_i) \\ &= \sum_{i=1}^{k} P(X_1 = s_j \mid X_0 = s_i) \pi_i \\ &= \sum_{i=1}^{k} p_{ij} \pi_i, \end{aligned}$$

and so the probability distribution for the state of X_1 is given by $\mathcal{P}^T \boldsymbol{\pi}$. By a similar argument, the probability distribution for the state of X_n is given by $(\mathcal{P}^T)^n \boldsymbol{\pi}$. For all Markov chains, there is at least one distribution $\boldsymbol{\pi}^s$ such that $\boldsymbol{\pi}^s = \mathcal{P}^T \boldsymbol{\pi}^s$. This is known as the *stationary distribution* of the chain. Markov chains allow quite simple and informative pictures. We can draw a weighted, directed graph with a node for each state and the weight on each edge indicating the probability of a state transition (Figure 23.8).

If we observe the random variable X_n, then inference is easy—we know what state the chain is in. This is a poor observation model, however. A much better model is to say that, for each element of the sequence, we observe *another* random variable whose probability distribution depends on the state of the chain. That is, we observe some Y_n, where the probability distribution is some $P(Y_n \mid X_n = s_i) = q_i(Y_n)$. We can arrange these elements into a matrix $\mathcal{Q}$. Specifying a hidden Markov model requires providing the state transition process, the relationship between state and the probability distribution on Y_n and the initial distribution on states. This means that the model is given by $(\mathcal{P}, \mathcal{Q}, \boldsymbol{\pi})$. We assume that the state space has k elements.

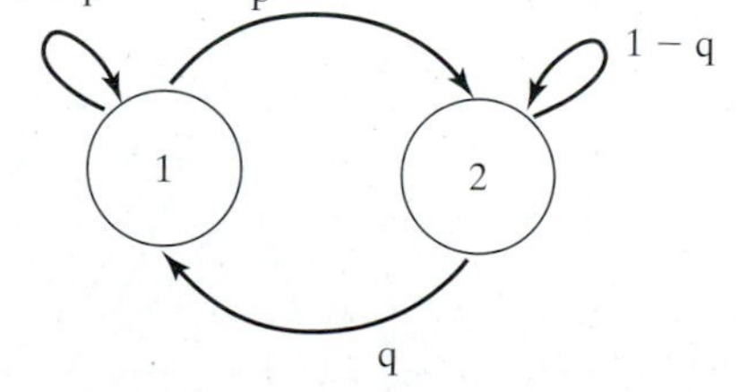

Figure 23.8 A simple, two-state Markov chain. In this chain, the probability of going from state one to state two is p; from state one to state one is $1 - p$; and so on. We could describe this chain with the state transition matrix. Its stationary distribution is $(q/(p+q), p/(p+q))$. This makes sense; for example, if p is small and q is close to one, the chain will spend nearly all its time in state one. Notice that, if p and q are both very small, the chain will stay in one state for a long time and then flip to the other state, where it will stay for a long time.

23.4.2 Computing with Hidden Markov Models

We assume that we are dealing with a hidden Markov model on a discrete state space—this simplifies computation considerably usually at no particular cost. There are two important problems:

- **Inference:** We need to determine what underlying set of states gave rise to our observations. This makes it possible to, for example, infer what the dancer is doing or the signer is saying.
- **Fitting:** We need to choose a hidden Markov model that represents a sequence of past observations well.

Each has an efficient, standard solution.

The Trellis Model Assume that we have a series of N measurements $\boldsymbol{Y}_i$ that we believe to be the output of a hidden Markov model. We can set up these measurements in a structure called a *trellis*. This is a weighted, directed graph consisting of N copies of the state space, which we arrange in columns. There is a column corresponding to each measurement. We weight the node representing state $\boldsymbol{X}_i$ in the column corresponding to $\boldsymbol{Y}_j$ with $\log q_i(\boldsymbol{Y}_j)$.

We join up the elements from column to column as follows. Consider the column corresponding the $\boldsymbol{Y}_j$; we join the element in this column representing state $\boldsymbol{X}_k$ to the element in the column corresponding to $\boldsymbol{Y}_{j+1}$ representing state $\boldsymbol{X}_l$ if p_{kl} is nonzero. This arc represents the fact that there is a possible transition between these states. This arc is weighted with $\log p_{kl}$. Figure 23.9 shows a trellis constructed from an HMM.

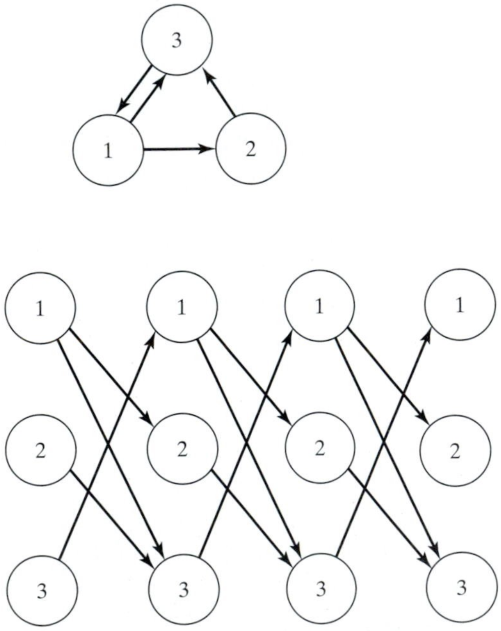

Figure 23.9 At the top, a simple state transition model. Below, the trellis corresponding to that model. Notice that each path through the trellis corresponds to a legal sequence of states for a sequence of four measurements. We weight the arcs with the log of the transition probabilities, and the nodes with the log of the emission probabilities; weights are not shown here to reduce the complexity of the drawing.

The trellis has the following interesting property: Each (directed) path through the trellis represents a legal sequence of states. Since each node of the trellis is weighted with the log of the emission probability and each arc is weighted with the log of the transition probability, the likelihood of sequence of states can be obtained by identifying the path corresponding to this sequence and summing the weights (of arcs and nodes) along the path. This yields an extremely effective algorithm for finding the maximum likelihood path known as *dynamic programming* or the *Viterbi algorithm*.

We start at the *final* column of the trellis. We know the log-likelihood of a one-state path, ending at each node, as this is just the weight of that node. Now consider a two-state path, which will start at the second last column of the trellis. We can easily obtain the best path leaving each node in this column. Consider a node: We know the weight of each arc leaving the node and the weight of the node at the far end of the arc, so we can choose the path segment with the largest value of the sum—this arc is the best we can do leaving that node. For each node, we add the weight at the node to the value of the best path segment leaving that node (i.e., the arc weight plus the weight of the node at the far end). This sum is the best value obtainable on reaching that node, which we call the *node value*.

Since we know the best value obtainable on reaching each node in the second-last column, we can figure out the best value obtainable on reaching each node in the third-last column. At

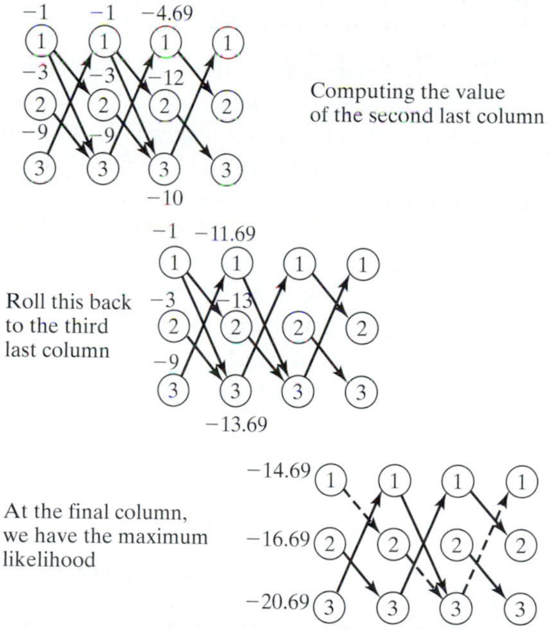

Figure 23.10 It is a simple matter to find the best path through the trellis of Figure 23.9 (or any other trellis). We assume that each 1 node has log-probability −1, each 2 node has log-probability −3 and each 3 node has log probability −9. We also assume that the probabilities of leaving a node are uniform (check our numbers). Now the value of each node in the second-last column is the value of the node plus the best value to be obtained by leaving that node. This is easily computed. The algorithm involves computing the value of each node in the second-last column, then of each node in the third last column; and so on, as described in the text. Once we get to the start of the trellis, the largest weight is the maximum of the log-likelihood; since we erased all but the best path segments, we have the best path, too (indicated by a dashed line).

each node in the third-last column, we have a choice of arcs, each reaching a node whose value we know. We choose the arc with the largest value of (arc weight plus node value), add this value to the weight at the starting node in the third last column, and this yields the value of the starting node. We can repeat this process until we have a value for each of the nodes in the first column; the largest value is the maximum likelihood.

We can also get the path with the maximum likelihood value. When we compute the value of a node, we erase all but the best arc leaving that node. Once we reach the first column, we simply follow the path from the node with the best value. Figure 23.10 illustrates this extremely simple and powerful algorithm.

In the following sections, we describe dynamic programming more formally. We can find the best path either by searching into the future, or by looking into the past. When we set up the trellis, we searched the future. For the formal description, we shall look into the past.

Inference and Dynamic Programming For inference, we have a series of observations $\{Y_0, Y_1, \ldots, Y_n\}$ and we would like to obtain the sequence of $n+1$ states $\boldsymbol{S} = \{S_0, S_1, \ldots, S_n\}$ that maximizes

$$P(\boldsymbol{S} \mid \{Y_0, Y_1, \ldots, Y_n\}, (\mathcal{P}, \mathcal{Q}, \pi)),$$

which is the same as maximizing the joint distribution

$$P(\boldsymbol{S}, \{Y_0, Y_1, \ldots, Y_n\} \mid (\mathcal{P}, \mathcal{Q}, \pi)).$$

There is a standard algorithm for this purpose, the *Viterbi algorithm*. We seek an $n+1$ element path through the states (from S_0 to S_n). There are k^{n+1} such paths because we could choose from each state for every element of the path (assuming that there are no zeros in $\mathcal{P}$—in most cases, there are certainly $O(k^{n+1})$ paths). We can't look at every path, but in fact we don't have to. The approach is as follows: Assume that, for each possible state s_l, we know the value of the joint for the best n step path that ends in $S_{n-1} = s_l$; then the path that maximizes the joint for an $n+1$ step path must consist of one of these paths combined with another step. All we have to do is find the missing step.

We can approach finding the path with the maximum value of the joint as an induction problem. Assume that, for each value j of S_{t-1}, we know the value of the joint for the best path that ends in $S_{t-1} = j$, which we write as

$$\delta_{t-1}(j) = \max_{S_0, S_1, \ldots, S_{n-2}} P(\{S_0, S_1, \ldots, S_{t-1} = j\}, \{Y_0, Y_1, \ldots, Y_{t-1}\} \mid (\mathcal{P}, \mathcal{Q}, \pi)).$$

Now we have

$$\delta_t(j) = \left(\max_i \delta_{t-1}(i) P_{ij}\right) q_j(Y_t).$$

We need not only the maximum *value*, but also the path that gave rise to this value. We define another variable

$$\psi_t(j) = \arg\max \left(\delta_{t-1}(i) P_{ij}\right)$$

(i.e., the best path that ends in $S_t = j$). This gives us an inductive algorithm for getting the best path.

The reasoning is as follows: I know the best path to each state for the $t-1$th measurement; for each state for the tth measurement, I can look backward and choose the best state for the $t-1$th measurement; but I know the best path from there, so I have the best path to each state for the tth measurement. I also know the best path to each of the available states at the first step. We have put everything together in Algorithm 23.1.

Algorithm 23.1: The Viterbi algorithm yields the path through an HMM that maximizes the joint and the value of the joint at this path. Here δ and ψ are convenient bookkeeping variables (as in the text); p^* is the maximum value of the joint; and q_t^* is the tth state in the optimal path.

1. **Initialization:**

$$\delta_0(j) = \pi_j b_j(Y_0) \quad 1 \leq j \leq N$$
$$\psi_0(j) = 0$$

2. **Recursion:**

$$\delta_t(j) = \left(\max_i \delta_{t-1}(i) P_{ij}\right) q_j(Y_t)$$
$$\psi_t(j) = \arg\max \left(\delta_{t-1}(i) P_{ij}\right)$$

3. **Termination:**

$$p^* = \max_i (\delta_n(i))$$
$$q_n^* = \arg\max_i (\delta_n(i))$$

4. **Path backtracking:**

$$q_t^* = \psi_{t+1}(q_{t+1}^*)$$

Fitting an HMM with EM We have a dataset $\boldsymbol{Y}$ for which we believe a hidden Markov model is an appropriate model, but which hidden Markov model should we use? We wish to choose a model that best represents a set of data. To do this, we use a version of the Expectation-Maximization algorithm of chapter 16. In this algorithm, we assume that we have an HMM, $(\mathcal{P}, \mathcal{Q}, \pi)$; we now want to use this model and our dataset to estimate a new set of values for these parameters. We now estimate $(\overline{\mathcal{P}}, \overline{\mathcal{Q}}, \overline{\pi})$ using a procedure given next. There are two possibilities (a fact which we won't prove). Either $P(\boldsymbol{Y} \mid (\overline{\mathcal{P}}, \overline{\mathcal{Q}}, \overline{\pi})) > P(\boldsymbol{Y} \mid (\mathcal{P}, \mathcal{Q}, \pi))$ or $(\overline{\mathcal{P}}, \overline{\mathcal{Q}}, \overline{\pi}) = (\mathcal{P}, \mathcal{Q}, \pi)$.

The updated values of the model parameters have the form:

$$\overline{\pi_i} = \text{expected frequency of being in state } s_i \text{ at time 1}$$
$$\overline{p_{ij}} = \frac{\text{expected number of transitions from } s_i \text{ to } s_j}{\text{expected number of transitions from state } s_i}$$
$$\overline{q_j(k)} = \frac{\text{expected number of times in } s_j \text{ and observing } Y = y_k}{\text{expected number of times in state } s_j}.$$

We need to evaluate these expressions. In particular, we need to determine

$$P(X_t = s_i, X_{t+1} = s_j \mid \boldsymbol{Y}, (\mathcal{P}, \mathcal{Q}, \pi)),$$

which we write as $\xi_t(i, j)$. If we know $\xi_t(i, j)$, we have

$$\text{expected number of transitions from } s_i \text{ to } s_j = \sum_{t=0}^{n} \xi_t(i, j);$$

$$\text{expected number of times in } s_i = \text{expected number of transitions from } s_i$$

$$= \sum_{t=0}^{n} \sum_{j=1}^{k} \xi_t(i, j);$$

$$\text{expected frequency of being in } s_i \text{ at time } 0 = \sum_{j=1}^{k} \xi_0(i, j);$$

$$\text{expected number of times in } s_i \text{ and observing } (Y = y_k) = \sum_{t=0}^{n} \sum_{j=1}^{k} \xi_t(i, j)\delta(Y_t, y_k),$$

where $\delta(u, v)$ is one if its arguments are equal and zero otherwise.

To evaluate $\xi_t(i, j)$, we need two intermediate variables: a *forward variable* and a *backward variable*. The forward variable is $\alpha_t(j) = P(Y_0, Y_1, \ldots, Y_t, X_t = s_j \mid (\mathcal{P}, \mathcal{Q}, \pi))$. The backward variable is $\beta_t(j) = P(\{Y_{t+1}, Y_{t+2}, \ldots, Y_n\} \mid X_t = s_j, (\mathcal{P}, \mathcal{Q}, \pi))$.

If we assume that we know the values of these variables, we have

$$\begin{aligned}
\xi_t(i, j) &= P(X_t = s_i, X_{t+1} = s_j \mid \mathbf{Y}, (\mathcal{P}, \mathcal{Q}, \pi)) \\
&= \frac{P(\mathbf{Y}, X_t = s_i, X_{t+1} = s_j \mid (\mathcal{P}, \mathcal{Q}, \pi))}{P(\mathbf{Y} \mid (\mathcal{P}, \mathcal{Q}, \pi))} \\
&= \frac{\left\{ \begin{array}{c} P(Y_0, Y_1, \ldots, Y_t, X_t = s_i \mid (\mathcal{P}, \mathcal{Q}, \pi)) \\ \times P(Y_{t+1} \mid X_{t+1} = s_j, (\mathcal{P}, \mathcal{Q}, \pi)) \\ \times P(X_{t+1} = s_j \mid X_t = s_i, (\mathcal{P}, \mathcal{Q}, \pi)) \\ \times P(Y_{t+2}, \ldots, Y_N \mid X_{t+1} = s_j, (\mathcal{P}, \mathcal{Q}, \pi)) \end{array} \right\}}{P(\mathbf{Y} \mid (\mathcal{P}, \mathcal{Q}, \pi))} \\
&= \frac{\alpha_t(i) p_{ij} q_j(Y_{t+1}) \beta_{t+1}(j)}{P(\mathbf{Y} \mid (\mathcal{P}, \mathcal{Q}, \pi))} \\
&= \frac{\alpha_t(i) p_{ij} q_j(Y_{t+1}) \beta_{t+1}(j)}{\sum_{i=1}^{N} \sum_{j=1}^{N} \alpha_t(i) p_{ij} q_j(Y_{t+1}) \beta_{t+1}(j)}.
\end{aligned}$$

Both the forward and backward variables can be evaluated by induction. We get $\alpha_t(j)$, by observing that

$$\begin{aligned}
\alpha_0(j) &= P(Y_0, X_0 = s_j \mid (\mathcal{P}, \mathcal{Q}, \pi)) \\
&= \pi_j q_j(Y_0); \\
\alpha_{t+1}(j) &= P(Y_0, Y_1, \ldots, Y_{t+1}, X_{t+1} = s_j \mid (\mathcal{P}, \mathcal{Q}, \pi)) \\
&= P(Y_0, Y_1, \ldots, Y_t, X_{t+1} = s_j \mid (\mathcal{P}, \mathcal{Q}, \pi)) P(Y_{t+1} \mid X_{t+1} = s_j) \\
&= \sum_{l=1}^{k} \big[P(Y_0, Y_1, \ldots, Y_t, X_t = s_l, X_{t+1} = s_j \mid (\mathcal{P}, \mathcal{Q}, \pi)) \\
&\quad \times P(Y_{t+1} \mid X_{t+1} = s_j) \big]
\end{aligned}$$

$$= \left[\sum_{l=1}^{k} P(Y_0, Y_1, \ldots, Y_t, X_t = s_l \mid (\mathcal{P}, \mathcal{Q}, \boldsymbol{\pi})) P(X_{t+1} = s_j \mid X_t = s_l) \right]$$
$$\times P(Y_{t+1} \mid X_{t+1} = s_j)$$
$$= \left[\sum_{l=1}^{k} \alpha_t(l) p_{lj} \right] q_j(Y_{t+1}) \qquad 1 \le t \le n-1.$$

The backward variable can also be obtained by induction as

$$\beta_n(j) = P(\text{no further states} \mid X_n = s_j, (\mathcal{P}, \mathcal{Q}, \boldsymbol{\pi}))$$
$$= 1;$$
$$\beta_t(j) = P(\{Y_{t+1}, Y_{t+2}, \ldots, Y_n\} \mid X_t = s_j, (\mathcal{P}, \mathcal{Q}, \boldsymbol{\pi}))$$
$$= \sum_{l=1}^{k} \left[P(\{Y_{t+1}, Y_{t+2}, \ldots, Y_n\}, X_t = s_l \mid X_{t+1} = s_j, (\mathcal{P}, \mathcal{Q}, \boldsymbol{\pi})) \right]$$
$$= \left[\sum_{l=1}^{k} P(X_t = s_l, Y_{t+1} \mid X_{t+1} = s_j) \right] P(\{Y_{t+2}, \ldots, Y_n\} \mid X_{t+1} = s_j, (\mathcal{P}, \mathcal{Q}, \boldsymbol{\pi}))$$
$$= \left[\sum_{l=1}^{k} p_{jl} q_l(Y_{t+1}) \right] \beta_{t+1}(j) \qquad 1 \le t \le k-1.$$

As a result, we have a simple fitting algorithm collected in Algorithm 23.2.

Algorithm 23.2: Fitting Hidden Markov Models to a data sequence Y is achieved by a version of EM. We assume a model $(\mathcal{P}, \mathcal{Q}, \boldsymbol{\pi})_i$ and then compute the coefficients of a new model; this iteration is guaranteed to converge to a local maximum of $P(Y \mid (\mathcal{P}, \mathcal{Q}, \boldsymbol{\pi}))$.

Until $(\mathcal{P}, \mathcal{Q}, \boldsymbol{\pi})_{i+1}$ is the same as $(\mathcal{P}, \mathcal{Q}, \boldsymbol{\pi})_i$
  compute the forward variables α and β
  using the procedures of algorithms 23.4 and 23.5

  compute $\xi_t(i, j) = \frac{\alpha_t(i) p_{ij} q_j(Y_{t+1}) \beta_{t+1}(j)}{\sum_{i=1}^{N} \sum_{j=1}^{N} \alpha_t(i) p_{ij} q_j(Y_{t+1}) \beta_{t+1}(j)}$
  compute the updated parameters using the procedures of Algorithm 23.3

  These values are the elements of $(\mathcal{P}, \mathcal{Q}, \boldsymbol{\pi})_{i+1}$
end

23.4.3 Varieties of HMMs

We have not spoken about the topology of the graph underlying our model. This could be a *complete graph* (one where every node is connected to every other in both directions), but does not have to be. One disadvantage of using complete graphs is that there are a large number of parameters to estimate. In applications, a number of variants have proved useful (Figure 23.11).

Algorithm 23.3: Computing the new values of parameters for the hidden Markov model fitting process

$$\overline{\pi_i} = \text{expected frequency of being in state } s_i \text{ at time } 0$$
$$= \sum_{j=1}^{k} \xi_0(i, j)$$

$$\overline{p_{ij}} = \frac{\text{expected number of transitions from } s_i \text{ to } s_j}{\text{expected number of transitions from state } s_i}$$
$$= \frac{\sum_{t=0}^{n} \xi_t(i, j)}{\sum_{t=0}^{n} \sum_{j=1}^{k} \xi_t(i, j)}$$

$$\overline{q_i(k)} = \frac{\text{expected number of times in } s_i \text{ and observing } Y = y_k}{\text{expected number of times in state } s_i}$$
$$= \frac{\sum_{t=0}^{n} \sum_{j=1}^{k} \xi_t(i, j)\delta(Y_t, y_k)}{\sum_{t=0}^{n} \sum_{j=1}^{k} \xi_t(i, j)}$$

here $\delta(u, v)$ is one if its arguments are equal and zero otherwise.

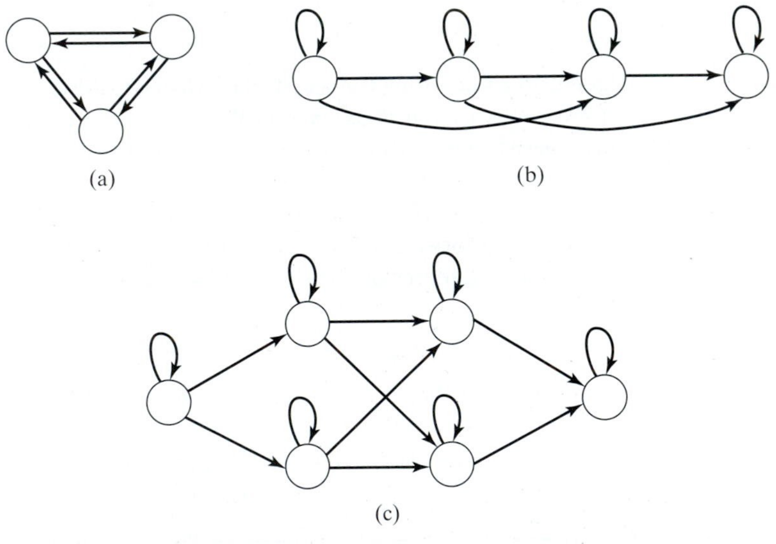

Figure 23.11 A variety of HMM topologies have proven useful in practice. At the **top**, we show an ergodic or fully connected model. In the **middle**, we show a four state left-right model. This model has been constrained so that $p_{ij} = 0$ for $j > i + 2$ as well. At the **bottom**, we show a six state parallel path left-right model. This model has, in essence, two parallel left-right models with the option of switching between them.

A *left-right model* or *Bakis model* has the property that, from state i, the model can move only to states $j > i$. This means that for $j < i$, $p_{ij} = 0$. Furthermore, $\pi_1 = 1$ and for all $i \neq 1$, $p_i = 0$. This means that the state transition matrix $\mathcal{P}$ is upper triangular. Furthermore, an N state left-right model will stay in the Nth state once it has been reached. You should notice that this model captures a notion of the order in which events can occur, which could be convenient if we were using HMMs to encode various kinds of motion information (or speech information, where the model originates).

However, a general left-right model suggests that large numbers of events may be skipped (the state may advance freely). This isn't really consistent with a reasonable model of motion because although we may miss a measurement or two or a state or two, it is unlikely that we will miss a large collection of states. It is common to use the additional constraint that for $j > i + \delta$, $p_{ij} = 0$. Here δ tends to be a small number (two is often used).

Surprisingly, constraining the topology of the model does not create any problems with the algorithms of the previous sections. You should verify that the estimation algorithm preserves zeros, meaning that if we start it with model of a particular topology—which has zeros in particular spots of the state transition matrix—it produces a new estimate of the state transition matrix that also has zeros in those spots.

Algorithm 23.4: Computing the forward variable for the hidden Markov model fitting process

$$\alpha_0(j) = \pi_j q_j(Y_0)$$

$$\alpha_{t+1}(j) = \left[\sum_{l=1}^{k} \alpha_t(l) p_{lj}\right] q_j(Y_{t+1}) \qquad 0 \leq t \leq n-1$$

Algorithm 23.5: Computing the backward variable for the hidden Markov model fitting process

$$\beta_n(j) = 1$$

$$\beta_t(j) = \left[\sum_{l=1}^{k} p_{jl} q_l(Y_{t+1})\right] \beta_{t+1}(j) \qquad 0 \leq t \leq n-1$$

23.5 APPLICATION: HIDDEN MARKOV MODELS AND SIGN LANGUAGE UNDERSTANDING

Sign language is language rendered using a system of gestures, rather than by manipulating the vocal tract. For most people, learning to understand sign language takes an effort; it would be attractive to have a device that could be pointed at someone who was signing and would generate speech. This is a problem that has some strong analogies with speech recognition, an area now quite highly developed.

Hidden Markov models have been highly successful in speech understanding applications. The hidden states describe the vocal system; the observations are various acoustic measurements. Typically, there is a hidden Markov model associated with each word. These models are attached together using a *language model*, which specifies the probability of a word occurring given some other word has already occurred. The resulting object is another—possibly big—HMM. The sentence represented by a set of acoustic measurements is then obtained by an inference algorithm applied to the language model. This technique has worked well for the speech community, and it is natural to try to suborn the model to deal with sign language.

Human gestures are rather like human sounds—there is typically a sequence of events, and we have measurements that result from these events, but do not determine them. Although there is no guarantee that the rather stiff conditional independence assumptions underlying hidden Markov models are actually true for human gestures, the model is certainly worth trying because there is no such guarantee for speech models, and HMMs work in that application.

We describe systems that use HMMs to interpret sign language, but you should realize that this approach also works for formal systems of gestures. For example, if I wish to build a system that opens a window when I move my hand one way and closes if I move my hand another way, I could regard that set of gestures as a rather restricted sign language.

There are a number of systems that use HMMs to interpret sign language. Generally, word models are left-right models, but constrained not to skip too many states (i.e., $p_{ij} = 0$ for $j > i+\Delta$ and $|\Delta|$ fairly small). Typically, there are a small number of states; each state has a self-loop, meaning that the model can stay in that state for a number of ticks; and there are transitions that skip some states, meaning that it is possible to move through the hidden states rather fast. Only two issues must now be resolved to build a system. First, the word models need to be connected together with some form of language model; second, one must decide what to measure.

23.5.1 Language Models: Sentences from Words

We wish to recognize expressions in sign language, rather than isolated words, so we need to specify how words appear. Language models are a specification of how word HMMs should be strung together to give an HMM that represents full sentences. The simplest language model has words that occur independent of the previous word. The language model can be expressed by a graph, as in Figure 23.12. In this model, the state moves to a start state, emits a word, then either stops or loops back from where another word can be emitted. It is now necessary to find an extremal path through this larger graph; the Viterbi algorithm still applies, there is just no observation corresponding to the shaded states.

More sophisticated language models are desirable—a language model for English that had words drawn independently at random would generate quite long strings of "and" 's and "the" 's—but do generate computational problems. The natural first step is to have a bigram language model, where the probability that a word is produced depends on the word that was

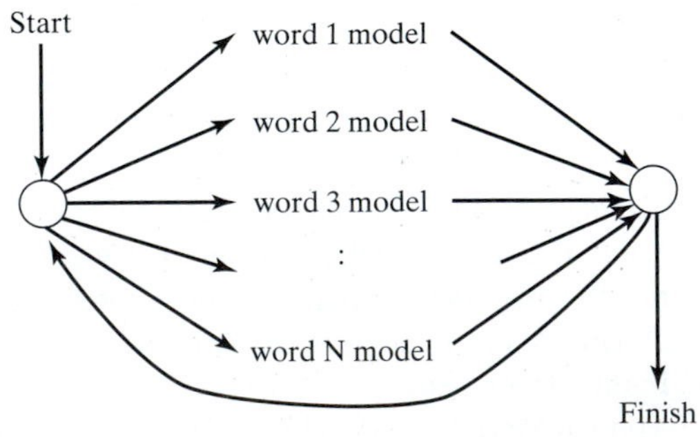

Figure 23.12 If we take an HMM for each word in the vocabulary and string these models together with independent emission probabilities, we obtain a language model. This is of the simplest kind, with independent words, and no constraint on the length of a sentence; although it has little syntax, it is still an HMM and allows extremely simple inference.

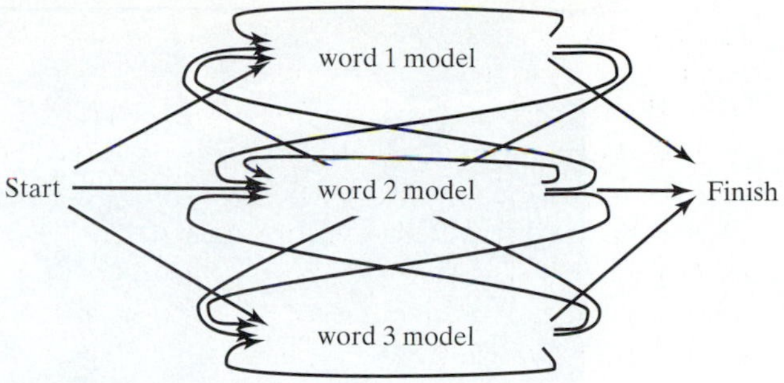

Figure 23.13 In a bigram language model, the probability that a word is emitted depends on the word that was emitted previously. Such models can be laid out as HMMs, again by stringing the word models together. Notice that the topology is now considerably more complicated; for simplicity, we have allowed only three words.

produced previously. This generates a language model of the form of Figure 23.13. More sophisticated language models involve trigrams—that is, the probability that a word is emitted depends on the last two words—or more detailed context from the preceding sentence. The difficulty with this approach is that, for a reasonably sized vocabulary, the number of states the Viterbi algorithm has to search can be prodigious. There are a variety of techniques for pruning this search, which are beyond our scope—the book by Jelinek (1999) gives one approach in chapter 5, or see Manning and Schütze (1999).

Features and Typical Levels of Performance There are a few current programs for sign language recognition. Starner has written several programs using a variety of features (see Starner, Weaver and Pentland, 1998). The simplest approach requires a user to wear a yellow glove on the right hand and a red glove on the left hand. An alternative approach to wearing gloves, which is something of a nuisance, is to segment hands using skin color (always assuming there is nothing else skin-colored in the vicinity). Pixels of interesting colors (yellow and red, or skin) are identified in the image. Once an appropriate pixel has been found, its eight neighbors are checked to determine whether their color is acceptable, too; this process continues to obtain a blob of pixels.

Blobs admit a variety of features. The center of gravity of the blob gives two features, and the change in center of gravity from the previous frame yields another two. The area of the blob is another feature. The orientation and size of the blob can be measured by forming a matrix of second moments,

$$\begin{pmatrix} \int x^2\,dx\,dy & \frac{1}{2}\int xy\,dx\,dy \\ \frac{1}{2}\int xy\,dx\,dy & \int y^2\,dx\,dy \end{pmatrix}.$$

The ratio of eigenvalues of this matrix gives an indication of the eccentricity of the blob, the largest eigenvalue is an estimate of the size along the principal direction, and the orientation of the eigenvector corresponding to this eigenvalue gives the orientation of the blob.

Starner's system works on a vocabulary of 40 words representing a total of four parts of speech. The topology of the HMMs for the words is given, and the parameters are estimated using the EM algorithm of chapter 16. For both isolated word-recognition tasks and for recognition using a language model that has five word sentences (words always appearing in the order

Figure 23.14 Starner and colleagues have built a sign language recognition system that uses HMMs to yield word models. This figure shows the view of a signer presented to the system; it is obtained from a camera on the desktop. *Reprinted from "Real time American sign language recognition using desk and wearable computer based video," by T. Starner, et al. Proc. Int. Symp. on Computer Vision, 1995, © 1995, IEEE.*

`pronoun verb noun adjective pronoun`), the system displays a word accuracy on the order of 90%. Values are slightly larger or smaller depending on the features and the task.

Vogler and Metaxas (1998 and 1999) have built a system that uses estimates of arm position recovered either from a physical sensor mounted on the body or from a system of three cameras that measures arm position fairly accurately (Vogler and Metaxas, 1998, 1999). For a vocabulary of 53 words and an independent word language model, they report a word recognition accuracy on the order of 90%.

The Future All the results in the literature are for small vocabularies and simple language models. Nonetheless, HMMs seem a promising method for recognizing sign language. It isn't clear that simple features are sufficient to recognize complex signs. One possible direction for progress is to use features that give a better estimate of what the fingers are doing.

Good language models—and appropriate inference algorithms—are at the core of the success of modern speech recognition systems. These models are typically obtained by measuring the frequency with which a word occurs in some context (e.g., trigram frequencies). Building these models requires tremendous quantities of data because we need an accurate estimate of the relative frequencies of quite infrequent events. For example, as Jelinek (1999) points out it typically takes a body of 640,000 words to assemble a set of 15,000 different words. This means that some words are used often, and that measuring word frequencies on a small body of data is extremely dangerous. Both the speech recognition community and the natural language community have a lot of experience with these difficulties and a variety of sophisticated tricks for dealing with them (e.g., Jelinek, 1999 or Manning and Schütze, 1999). Future research in sign language recognition should involve learning these tricks and transferring them to the vision domain.

23.6 APPLICATION: FINDING PEOPLE WITH HIDDEN MARKOV MODELS

It is fairly easy to see how a hidden Markov model is a reasonable choice for recognizing sign language—signs occur in a constrained random order and generate noisy measurements. There are other applications that are not as obvious. The important property of a hidden Markov model is not the temporal sequence; it is conditional independence—that X_{i+1} is independent of the past, given X_i.

This sort of independence occurs in a variety of cases. By far the most important is finding people. We assume that people appear in images as dolls, consisting of nine body segments (upper and lower left and right arms and legs, respectively, and a torso) each of which is rectangular. In particular, we assume that the left lower arm is independent of all other segments given the left upper arm, that the left upper arm is independent of all segments given the torso, and extend these assumptions in the obvious way to include the right arm and the legs, too. This gives us a hidden Markov model. We can write the model out to emphasize this point. We write X_{lua} for the configuration of the left upper arm, and so on, and have

$$\begin{aligned} &P(X_t, X_{lua}, X_{lla}, X_{rua}, X_{rla}, X_{lul}, X_{lll}, X_{rul}, X_{rll}) \\ &\quad = P(X_t)P(X_{lua} \mid X_t)P(X_{lla} \mid X_{lua})P(X_{rua} \mid X_t)P(X_{rla} \mid X_{rua})P(X_{lul} \mid X_t) \\ &\qquad \times P(X_{lll} \mid X_{lul})P(X_{rul} \mid X_t)P(X_{rll} \mid X_{rul}) \end{aligned}$$

which we can draw as a tree indicating the dependencies (Figure 23.15).

Now assume that we observe some image measurements relating to segment configuration. For the moment, we assume that each body segment can occupy one of a finite collection of discrete states—we write the event that a particular limb is in a particular state as, say, $X_{lua} = \boldsymbol{x}_{lua}$. This event results in a measurement whose conditional probability can be written as $P(M = \boldsymbol{m} \mid X_{lua} = \boldsymbol{x}_{lua})$ etc. In particular, we have a system of segments in the image, some of which correspond to limb segments and some of which result from noise. We assume that there are N_s segments in the image. Furthermore, we assume that the probability that a segment arises from noise is independent of anything we can measure from the segment. Now for each correspondence of image segments to limb segments, we can evaluate a likelihood function. We need to write out the correspondence; let us write $\{i_1, \ldots, i_9\}$ as the event that image segment i_1 is the torso, i_2 is the left upper arm, through to i_9 is the right lower leg, and that all others are noise. We write $\boldsymbol{m}_{i_k}$ for the image measurements associated with the i_kth image segment.

Assume for the moment that we expect all body segments to be present at every stage. We now wish to determine which choice of image segments represents the body segments of a person who is present and what the configuration of that person is. We have a log-likelihood that looks like this:

$$\begin{aligned} &\log P(\{i_1, \ldots, i_9\} \mid X_t = \boldsymbol{x}_t, \ldots, X_{rll} = \boldsymbol{x}_{rll}) \\ &\quad = \log P(\boldsymbol{m}_{i_1} \mid \boldsymbol{x}_t) + \log P(\boldsymbol{m}_{i_2} \mid \boldsymbol{x}_{lua}) \cdots + \log P(\boldsymbol{m}_{i_9} \mid \boldsymbol{x}_{rll}) \\ &\qquad + (N_s - 9)P(\text{image segment from noise}). \end{aligned}$$

Now if we write $P(X_{lua} = \boldsymbol{x}_{lua} \mid X_t = \boldsymbol{x}_t)$ as $P(\boldsymbol{x}_{lua} \mid \boldsymbol{x}_t)$, the the log of the joint probability is

$$\begin{aligned} &\log P(\{i_1, \ldots, i_9\}, X_t = \boldsymbol{x}_t, \ldots, X_{rll} = \boldsymbol{x}_{rll}) \\ &\quad = \log P(\{i_1, \ldots, i_9\} \mid X_t = \boldsymbol{x}_t, \ldots, X_{rll} = \boldsymbol{x}_{rll}) \\ &\qquad + \log P(\boldsymbol{x}_t, \boldsymbol{x}_{lua}, \boldsymbol{x}_{lla}, \boldsymbol{x}_{rua}, \boldsymbol{x}_{rla}, \boldsymbol{x}_{lul}, \boldsymbol{x}_{lll}, \boldsymbol{x}_{rul}, \boldsymbol{x}_{rll}) \end{aligned}$$

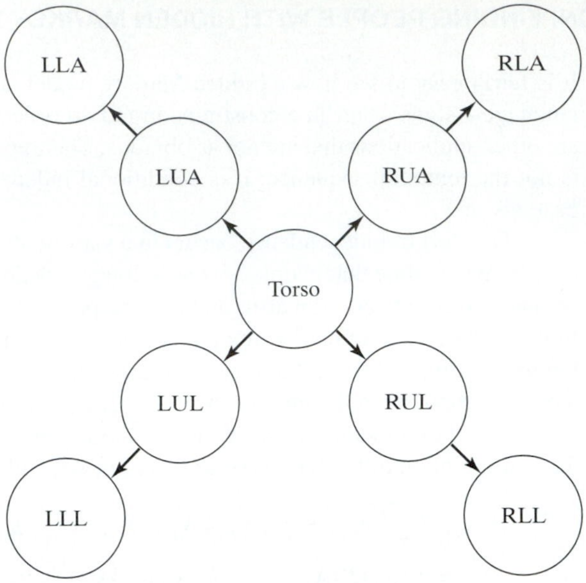

Figure 23.15 One can think of a human as forming a hidden Markov model. The tree in the figure illustrates the structure of one possible model: The torso generates various structures—arms and legs—whose properties are conditionally independent given the torso. The lower leg is conditionally independent of the rest of the model given the upper leg, and so on. We can encode these independence properties by drawing a node for each variable and a directed arc from a variable to another if the second depends directly on the first. If this drawing (which is a directed graph) is a tree, then we have a hidden Markov model. Notice that the semantics of this drawing are somewhat different from those of the drawing of Figure 23.8; that drawing showed the possible state transitions and their probabilities, whereas this shows the variable dependencies.

$$
\begin{aligned}
&= \log P(\{i_1, \ldots, i_9\} \mid X_t = \boldsymbol{x}_t, \ldots, X_{rll} = \boldsymbol{x}_{rll}) + \log P(\boldsymbol{x}_t) P(\boldsymbol{x}_{lua} \mid \boldsymbol{x}_t) \\
&\quad + \log P(\boldsymbol{x}_{lla} \mid \boldsymbol{x}_{lua}) + \log P(\boldsymbol{x}_{rua} \mid \boldsymbol{x}_t) \\
&\quad + \log P(\boldsymbol{x}_{rla} \mid \boldsymbol{x}_{rua}) + \log P(\boldsymbol{x}_{lul} \mid \boldsymbol{x}_t) \\
&\quad + \log P(\boldsymbol{x}_{lll} \mid \boldsymbol{x}_{lul}) + \log P(\boldsymbol{x}_{rul} \mid \boldsymbol{x}_t) \\
&\quad + \log P(\boldsymbol{x}_{rll} \mid \boldsymbol{x}_{rul}) \\
&= \log P(\boldsymbol{m}_{i_1} \mid \boldsymbol{x}_t) + \log P(\boldsymbol{m}_{i_2} \mid \boldsymbol{x}_{lua}) \cdots \\
&\quad + \log P(\boldsymbol{m}_{i_9} \mid \boldsymbol{x}_{rll}) + (N_s - 9) P(\text{image segment from noise}) \\
&\quad + \log P(\boldsymbol{x}_t) + \log P(\boldsymbol{x}_{lua} \mid \boldsymbol{x}_t) + \log P(\boldsymbol{x}_{lla} \mid \boldsymbol{x}_{lua}) \\
&\quad + \log P(\boldsymbol{x}_{rua} \mid \boldsymbol{x}_t) + \log P(\boldsymbol{x}_{rla} \mid \boldsymbol{x}_{rua}) \\
&\quad + \log P(\boldsymbol{x}_{lul} \mid \boldsymbol{x}_t) + \log P(\boldsymbol{x}_{lll} \mid \boldsymbol{x}_{lul}) \\
&\quad + \log P(\boldsymbol{x}_{rul} \mid \boldsymbol{x}_t) + \log P(\boldsymbol{x}_{rll} \mid \boldsymbol{x}_{rul}).
\end{aligned}
$$

Now all this fits into the lattice structure, which is the core of dynamic programming, rather easily. For each body segment, we establish a column of nodes, one for each pair of the form (image segment, body segment state). Attached to each of these nodes is a term of the form

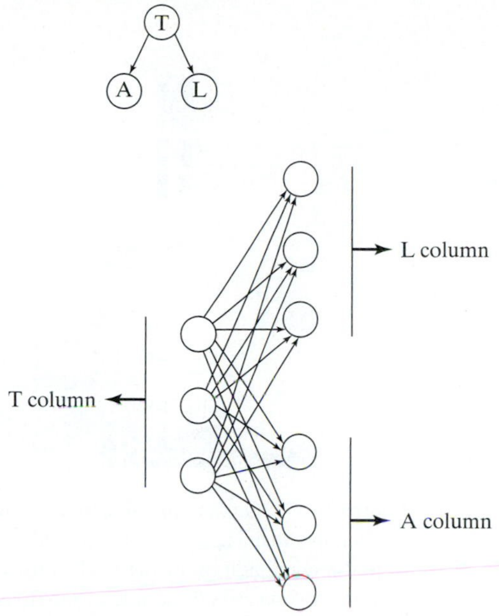

Figure 23.16 The figure shows a trellis derived from a simple tree based around a simplified human model. You can read leg for L, and so on. The elements of a column correspond to different possible correspondences among image segments, body segments, and body segment configuration variables. For example, a node might represent the fact that image segment two corresponds to a torso segment at a particular position. The fact that nodes in the column marked "T" have two children each—instead of the single child in our previous example—creates no problem. For each node in this column, we can determine the best available option and its value. This is obtained by computing the best available "A" option (and value) and the best available "L" option (and its value). In turn, this gives us the value of the "T" node. This observation means that we can use dynamic programming for any tree model.

$\log P(\boldsymbol{m}_{i_1} \mid \boldsymbol{x}_t)$, whose value we know because we know both the image segment and the body segment state represented by the node. There is a directed arc from each element of the column representing a body segment in the tree to each element in the columns representing its children. These arcs are labeled with the logs of the appropriate transition probabilities—for example $P(\boldsymbol{x}_{lua} \mid \boldsymbol{x}_t)$ (see Figure 23.16). We wish to find the directed path with the largest sum of node and arc values along the path.

The model's structure means that some nodes along the path have more than one child; this doesn't matter. The element of dynamic programming is that it is possible to process the lattice so that, at each node, we know the value of the best choice available at that node. This element is present here, too. We transfer values back up directed arcs as before; when a node has more than one child, we sum the best choices available over the children. Figure 23.16 illustrates the process for a simplified lattice.

Models of this form can be used to find people in images in a fairly straightforward manner. First, we need some model linking image observations to the configuration of the limb segments—i.e. a likelihood model. Felzenszwalb and Huttenlocher (2000) assume that segments have known color patterns—typically, a mixture of skin color and blue—and then compare the actual image color with these patterns. This leads to a fairly satisfactory matcher (Figure 23.17) with the proviso that the person's clothing be known in advance.

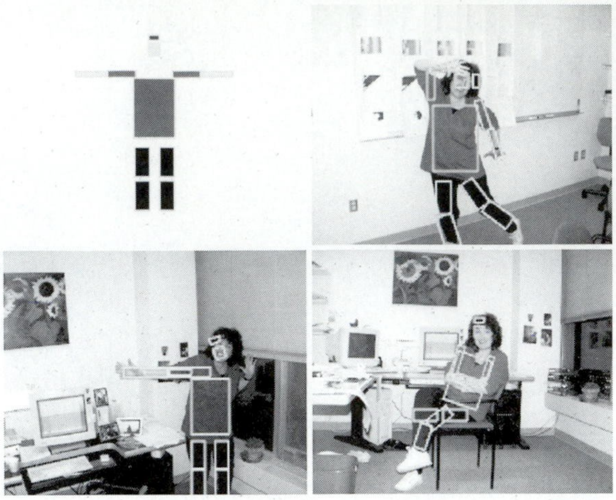

Figure 23.17 On the **top left**, a tree structured model of a person. Each segment is colored with the image color expected within this segment. The model attempts to find a configuration of these 11 body segments (9 limb segments, face, and hair) that (a) matches these colors, and (b) is configured like a person. This can be done with dynamic programming as described in the text. The other three frames show matches obtained using the method. *Reprinted from "Efficient Matching of Pictorial Structures," by P. Felzenszwalb and D.P. Huttenlocher, Proc. Computer Vision and Pattern Recognition 2000, © 2000, IEEE.*

23.7 NOTES

This topic is one on which substantial research is being conducted as we write; by the time anyone reads it, the chapter may not represent the cutting edge of research. We have tried to identify the intellectual strands we think prove most significant, which explains the emphasis on inference as search. The primary—and stunning—attraction of an HMM is that inference is simple (or, equivalently, the correspondence search is easily structured).

Hidden Markov Models

Dynamic programming has been used to find people in various ways. In Ioffe and Forsyth (1999), it is used to recommend hypotheses to a process that accepts or rejects them based on whether segments are shared or not—the conditional independence assumption prevents one compelling the model to ensure that body segments do not overlap. Song, Goncalves, di Bernardo and Perona (1999, 2000*b*, and 2000*a*) formulate finding people using motion information as a correspondence search through a probabilistic model and use dynamic programming to solve the search.

HMMs have some serious difficulties as devices for visual inference. The first is that their discriminative performance on vision problems hasn't been all that good to date. The second is that the property that inference is easy comes from the strong structural constraints on the model—these conditional independence constraints can make it difficult to impose quite natural constraints. For example, the model used for finding people does not admit the constraint that different parts of the image should correspond to different body segments (because that would involve adding an edge to the tree of Figure 23.17, which would result in a model that was not a tree). The difficulty with models that have more complex conditional independence relations is that inference can be difficult. An important research topic in object recognition involves finding models that (a) are quite a good representation of the world, in the sense that they can be used to achieve efficient discrimination; (b) admit quite simple inference algorithms; and (c) can be composed easily to yield new models.

This question of composition is important and hasn't had much attention yet. Basically, if one wishes to recognize large numbers of different types of object, it is probably difficult to do so by thinking of this collection as flat—in the sense that one (simultaneously or in sequence) attempts to match to each model independently. However, if each object gave rise to a different set of features, we would be forced to do this (by finding the features corresponding to the first object, the second, etc.). A more attractive model involves having features that, in general form, apply to many different types of object, with the distinctions between these types emerging from (a) the relations between the features, and (b) the detailed appearance of the features. How this process works remains extremely vague.

24

Geometric Templates from Spatial Relations

Up to this point, templates have been small blocks of pixels with a characteristic appearance. More complicated templates can be extremely useful: For example, spheres have roughly circular outlines in most practical perspective cameras. This suggests that, if one is looking for spheres, putting together edge points that form a circle is a useful thing to do. This is a form of template obtained by looking at the spatial relations between image components. It suggests the following way of thinking about matching:

- determine relations that constrain the appearance of the objects of interest,
- find image structures that satisfy those relations, and
- match only these structures.

Because most objects have a complicated geometric structure, this approach isn't helpful for whole objects (because the relations might be extremely complicated). Instead, we are forced to assume that objects are made up of parts, each of which has a simple structure. We then proceed to find parts, as before and to link those parts, together into an object.

The line of reasoning is attractive, because it connects object representation to segmentation, because it suggests how we might recognise lots of objects—assume they're made of parts, differently assembled, find the parts and then reason about assembly—and because there is a thread of inferential reasoning running through the argument.

24.1 SIMPLE RELATIONS BETWEEN OBJECT AND IMAGE

The simplest case of an object with a constrained outline is a polyhedron—its outline consists of a set of chains of line segments. In most applications, we are dealing with polyhedra whose faces are relatively large compared with an image pixel, so we expect these line segments to be many edge points long. In turn, this means there is no point in looking for polyhedra in images without lines, or (more usefully) that, if one wants to find polyhedra, forming chains of line segments is a good idea. There are further constraints on the outline of a polyhedron as long as *internal boundaries*—components of the outline on which there is more than one visible face incident—can be identified (Sugihara, 1986, Huffman, 1977, Clowes, 1971, Rothwell, Forsyth, Zisserman and Mundy, 1993). Typically, it is hard to identify internal boundaries reliably enough to exploit these constraints.

24.1.1 Relations for Curved Surfaces

Recall that the outline of a surface is formed by slicing a cone of rays with the image plane. The cone of rays consists of rays tangent to the surface that pass through the focal point of the camera—for a perspective camera—or are parallel—for an affine camera. Call this cone the *viewing cone*. If the affine camera is orthographic, which is by far the most common case, then the slice is taken perpendicular to the rays. The viewing cone is usually easier to analyze than the outline.

Cones and Cylinders A *cone* is a surface obtained by sweeping a family of rays through a point—the *vertex* of the cone—along a plane curve called the *generator*. Notice that this definition is more general than that of a *right circular cone*, a phrase many people incorrectly simplify to "cone"; right circular cones have a rotational symmetry (Figure 24.1). A cone consists of scaled and translated copies of its generator. Choose a coordinate frame where the generator can be written as $(x(t), y(t), 1)$. Then the cone can be written as

$$(x(t)s, y(t)s, s),$$

and the vertex occurs at $(0, 0, 0)$. A *cylinder* is a special case of a cone, where the vertex is at infinity—this means that, in an appropriate coordinate frame, a cylinder can be written as

$$(x(t), y(t), s).$$

A *right circular cylinder* is a cylinder whose generator is a circle.

The viewing cone for a cone is a family of planes, all of which pass through the focal point and the vertex of the cone. This means that our **grouping constraint** is:

> the outline of a cone consists of a set of lines passing through a vertex (Figure 24.2).

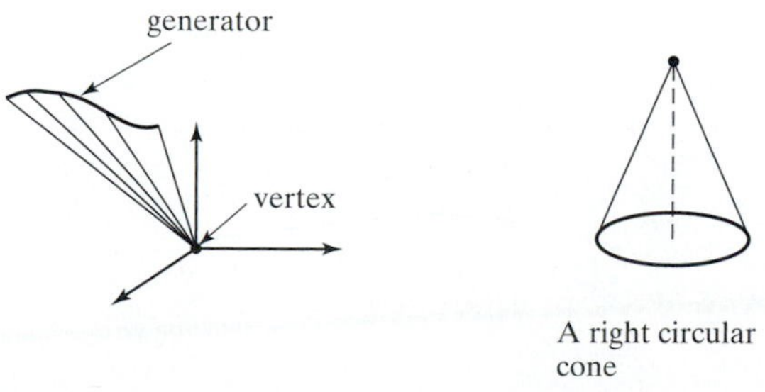

Figure 24.1 Cones are surfaces obtained by sweeping rays through a vertex along a generator. A right circular cone is a special cone, where the generator is a circle and the line joining the vertex with the center of the circle is normal to the circle's plane.

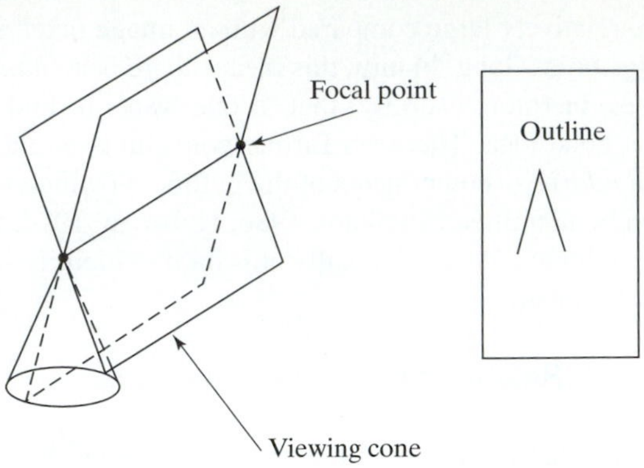

Figure 24.2 The viewing cone of a cone is a set of planes tangent to the cone and passing through both the vertex of the cone and the focal point. The outline is obtained by slicing the viewing cone with a plane and is a set of lines through a single point. These observations apply to a cylinder, too; the only difference between a cylinder and a cone is that a cylinder's vertex is at infinity.

All this should be obvious (the exercises ask for a proof), but is surprisingly useful. Because the only difference between a cylinder and a cone is that the vertex of a cylinder is infinitely far away, this grouping constraint applies to a cylinder as well. Notice that the *image vertex* for a cylinder is *not* necessarily at infinity unless the image is obtained in an orthographic view.

Canal Surfaces Take a sphere of fixed radius and sweep its center along a curve. Now form the envelope of the resulting family of spheres. This envelope is a surface, known as a *canal surface*; the curve is usually referred to as a *generating curve*. Canal surfaces typically look like twisted pipes (Figure 24.9), but can be singular. An easy example is obtained by taking a sphere of fixed radius and sweeping its center along a circle. If the sphere's radius is smaller than the

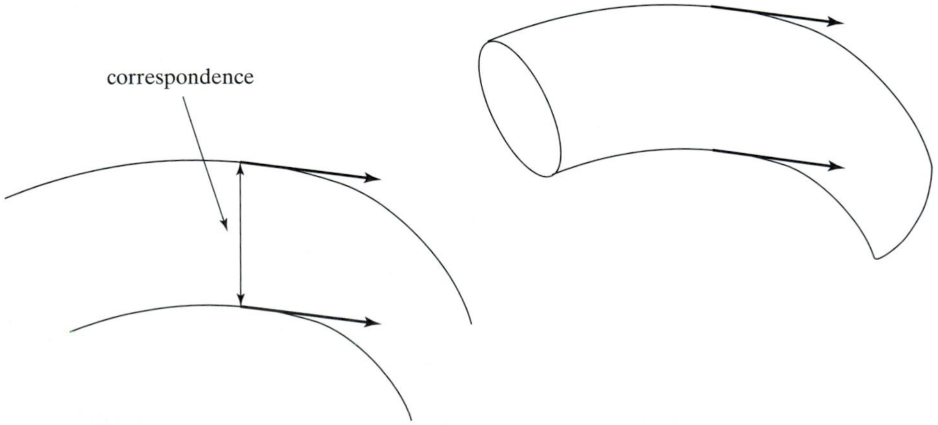

Figure 24.3 A pair of curves is parallel if there is a smooth correspondence between them such that the tangents are parallel (**left**). The outline of a canal surface in an orthographic view consists of a set of parallel symmetric curves (**right**).

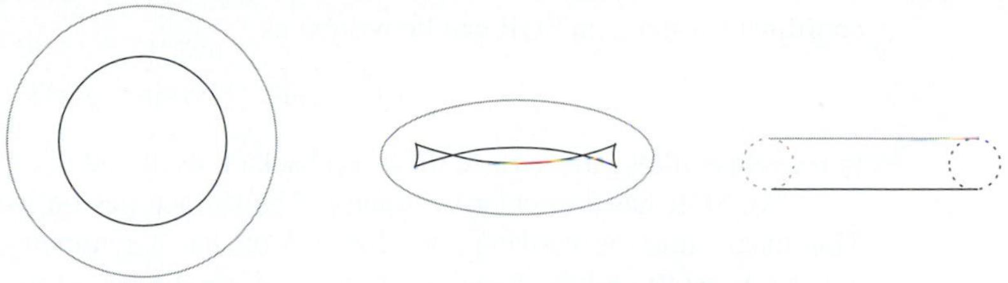

Figure 24.4 Outlines under orthographic projection of different views of a canal surface, in this case a torus. Notice that, although we can set up a 1-1 correspondence between the two sides such that every tangent on one side is parallel to the corresponding tangent on the other side, this doesn't mean that the curves are translates of one another (the case in the **center**). We have rendered one side of the correspondence in gray, the other in black to simplify interpreting the figure; in the figure on the **right**, the corresponding curves naturally form two groups, which we have indicated using dashes. Notice that we have not removed invisible segments of the outline; in practice, these would disrupt the process of constructing correspondences.

radius of the circle, we get a torus, but if it is larger, we get a surface that self-intersects along two circles.

A canal surface is locally a right circular cylinder (again, this is the same as being locally a right circular cone, but with the vertex infinitely far away). Now assume that we have orthographic projection. Under orthographic projection, the outline of a cylinder is two parallel lines; this means that, locally, the outline of a canal surface is two parallel lines, too. "Locally" means that the tangents to the outline of the canal surface are parallel at corresponding points. This doesn't necessarily mean that the curves are translated versions of one another (Figure 24.4), but it does supply the following **grouping cue:**

> the outline of a canal surface is going to consist of two curves, and one is parallel to the other (see Figure 24.3).

Grouping can proceed as follows: We first find fragments of outline that are parallel and then try and assemble these fragments into an outline. For the first step, we find an edge point and all others where the edge runs in the same direction; we now step along the first edge point's direction, keeping any of the possible matches that remain parallel; we keep doing this until we run out of points that stay parallel. Some care is required to ensure that we do not run faster along one side than along the other. For the second step, we find fragments whose ends are close, and (roughly) pointed in the same direction. These can be grouped together under the hypothesis that their separation had to do with the edge finder dropping out or with a change in outline visibility.

If we are dealing with a perspective view, we could assume that the depth range over the object is small compared with the distance to the object, meaning that the perspective view can be approximated by scaled orthography, and (since nothing we did depended on scale) there's nothing more to do. Alternatively, there are constraints on the inflections of the outline, which are true under perspective, but provide only weak grouping cues for surfaces with plane generating curves (Zisserman, Mundy, Forsyth, Liu, Pillow, Rothwell and Utcke, 1995*b*).

Surfaces of Revolution A *surface of revolution*—or *SOR* for short—is a surface obtained by sweeping a circle along a straight axis, perpendicular to the circle's plane and passing through the circle's center; the circle can grow or shrink as it is swept. This means that, in some coordinate system, an SOR can be written as

$$(f(s)\cos t,\ f(s)\sin t,\ g(s))$$

(a form that allows the surface to sweep back on itself and to be singular, etc.).

An SOR has a circular symmetry: The surface lies on itself it it is rotated about its axis. This means that the viewing cone for an SOR has a symmetry. Imagine the plane through the axis of the SOR and the focal point; the cone must have a flip symmetry about this plane. This *does not* mean that the outline of an SOR has a mirror symmetry because the outline is obtained by slicing the viewing cone with the image plane. If the image plane is not at right angles to the plane of symmetry then the outline will not have a mirror symmetry (Figure 24.5). Instead, we would have a form of symmetry in the image, sometimes known as a *conjugate symmetry*. In practical cameras, the angle between the image plane and the plane of symmetry is fairly close

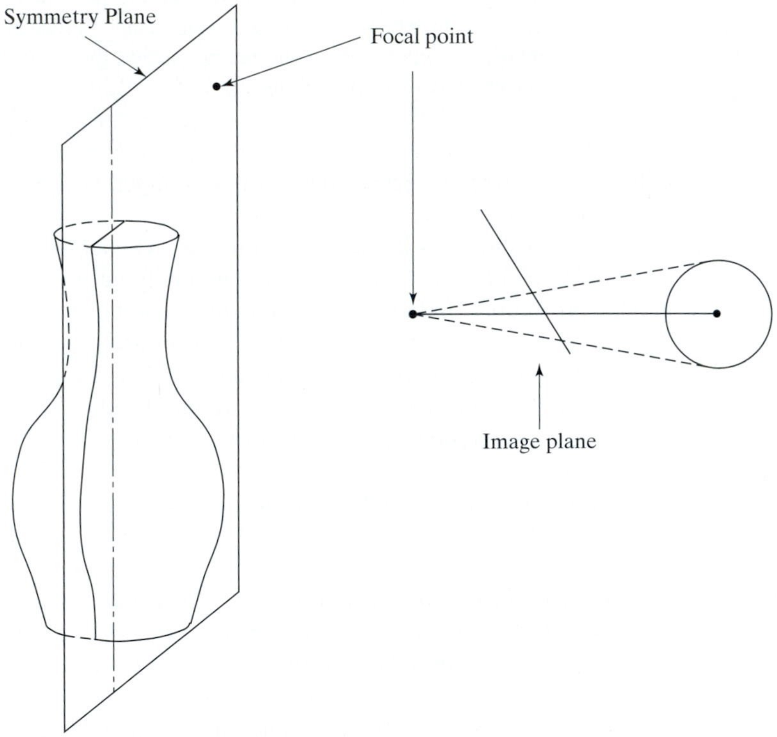

Figure 24.5 A surface of revolution and a focal point together give a plane of symmetry that passes through the axis of the SOR and the focal point (**left**). The contour generator must have a mirror symmetry in this plane. This doesn't mean that the outline has an exact mirror symmetry because the outline is obtained by slicing the viewing cone—which isn't shown for simplicity—by a plane *that may not be at right angles to the plane of symmetry*. On the **right**, we show a view from above; the viewing cone is dashed and is sliced by an image plane to form the outline. In this figure, the image plane is some way from being at right angles to the plane of symmetry, meaning that, in the image plane, the object is some way from the camera center so that the outline of an SOR does not have an exact mirror symmetry, but instead a conjugate symmetry. Typically, cameras have relatively small fields of view, meaning that extreme angles like this do not appear much in practice. For the vast majority of practical cameras, this conjugate symmetry is indistinguishable from a mirror symmetry.

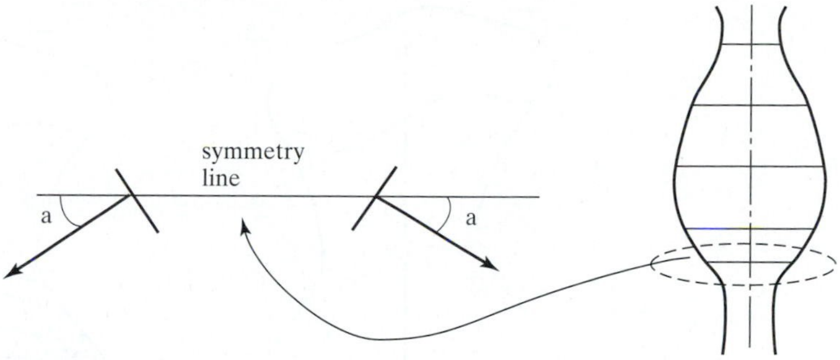

Figure 24.6 A local symmetry is a pair of contour points where the tangents to the contours are at roughly the same angle to the line joining the contours (the symmetry line). Such symmetries are often seen in the outlines of surfaces of revolution.

to a right angle, or the object would lie out of the field of view. Lens systems are also quite often adjusted to reduce the distortions that perspective can create in the far field (Fleck 1995). All this means that the distortion is small for practical cameras, and for all intents and purposes the outline of an SOR can be regarded as having a mirror symmetry.

All this means that corresponding points on either side of the outline of an SOR can be identified by the **grouping constraint** that

> the outline of an SOR has a (near) mirror symmetry.

This can be restated in an extremely useful local form:

> two points on image curves where the tangent is at about the same angle to the line joining the points (Figure 24.6) could be on opposite sides of a symmetry, and so could lie on an SOR.

We call this configuration a *local symmetry*, and the line joining the points is the *symmetry line*. We could find the outline of a surface of revolution by looking for local symmetries whose midpoints lie on a straight line roughly perpendicular to their symmetry lines. The main difficulty with this strategy is that most images contain an awful lot of symmetries, and there may be many groups of symmetries that satisfy this test.

Straight Homogeneous Generalized Cylinders A *straight homogeneous generalised cylinder* (*SHGC*) is a generalization of an SOR that is obtained by sweeping a plane generating curve along a line at right angles to the curve's plane and shrinking or growing it it as one sweeps (Figure 24.7). With an appropriate choice of coordinate system, the generating curve is $(x(t), y(t), 0)$, and the surface can be written as

$$(x(t)f(s), y(t)f(s), g(s)).$$

Again, this form means the surface can turn back on itself. SHGCs do not have convenient symmetries, but they do share some other properties with SORs. An SOR is *locally* a right circular cone, and an SHGC is *locally* a cone. To see this, fix a value of $s = s_0$ and consider the strip of surface from s_0 to $s_0 + \epsilon$. If the strip is small enough, then $f(s)$ can be approximated by $us + v$, and $g(s)$ can be approximated by $cs+d$, for u, v, c, and d some constants (this is what derivatives are all about!). The d can be disposed of by translation and the c by reparametrization, so this strip is a cone. This means that all the tangent vectors in the s direction at $s = s_0$ form a cone.

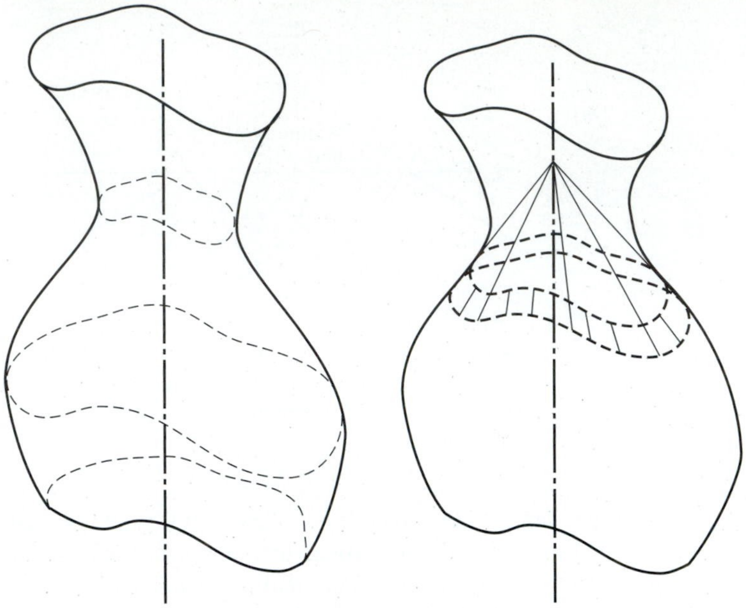

Figure 24.7 An SHGC is obtained by sweeping a plane generator along a line perpendicular to its plane and growing or shrinking it as it sweeps (**left**). This gives a surface that is *locally* a cone, in the sense that there is a cone tangent to the surface along any generator. In the **right** hand figure, we have cut a narrow strip of the surface between two generators and produced its tangents. These tangents meet at a vertex on the axis, meaning that each such strip of an SHGC is a strip of cone (i.e., it is *locally* a cone).

The vertex is easily determined in our coordinate system; when $us + v = 0$, then, whatever the value of t, the tangent vectors all pass through $(0, 0, g(s))$. This means that the vertex of each tangent cone lies on the z-axis in our coordinate frame.

Although we obtained this result in coordinates, this was merely for convenience. Because the result is about incidence properties of tangents, which aren't affected by change of coordinates, it applies to any SHGC in any coordinate frame. So for any SHGC, in any coordinate frame, there is a well-defined axis, which is the line along which the vertices of the tangent cones fall.

Because the surface is locally a cone, the outline is (locally) the outline of a cone. This means that, if we produce the tangents at points on the outline corresponding to some value of s, they intersect in a single point. As the value of s is changed, this point—the projected vertex of the tangent cone—moves along a straight line. So we have the following **grouping constraint:**

> There is some correspondence between components of the outline such that tangents at corresponding sides, when produced, will meet along a straight line.

This is a segmentation criterion because not all sets of curves satisfy it. It is somewhat tricky to use in this form, however.

PRCGCs Another case that leads to a clear grouping cue is a *planar right constant cross-section generalized cylinder* (PRCGC or tube). This is a surface obtained by sweeping a fixed plane curve perpendicular to some axis. If the plane curve is a circle, then the surface is a canal surface, but we can generalize to more complex cross-section curves.

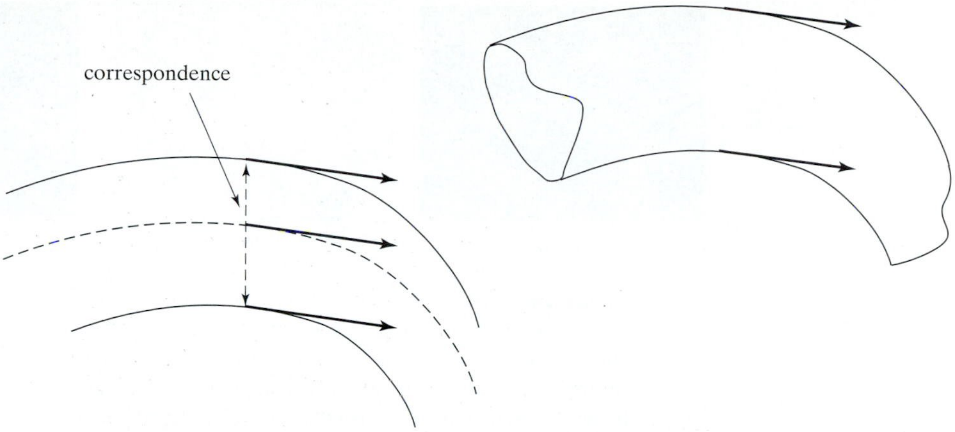

Figure 24.8 A pair of curves is parallel symmetric if there is a smooth correspondence between them such that the tangents are parallel (**left**). The outline of a PRCGC in an orthographic view consists of a set of parallel symmetric curves (**right**). We have shown one end of the surface to give a stronger sense that the picture is an outline.

As in the case of a canal surface, a PRCGC is locally a cylinder, but this may not be a right circular cylindar. All this means that the viewing cone must, locally, be the viewing cone of a cylinder, and so in an *orthographic* view the viewing cone is a set of sheets that are parallel and parallel to the axis.

As a result, in an orthographic view, we have the **grouping constraint** that

> the *outline* of a PRCGC consists of a set of parallel curves (i.e., that there is a smooth correspondence between points on the curves such that the tangents at corresponding points are parallel; Figure 24.8).

Sets of curves with this property are known as *parallel symmetric* curves. We have already discussed grouping a set of parallel symmetric curves.

24.1.2 Class-Based Grouping

The constraints described apply to whole classes of objects, which means that, when we group, we need not know *what* object we are grouping. Since different kinds of object produce somewhat different image constraints, we must use different strategies to group together outline components that come from different types of object—so-called *class-based grouping*. This isn't a problem. In fact, it's an example of an attractive feature. This is because, quite early in the grouping process, information appears that focuses attention on particular image curves and particular object classes. For example, if there aren't any lines, there isn't much point in trying to group polyhedra; if there aren't any mirror symmetries, there isn't much point in trying to group surfaces of revolution. Similarly, the process of grouping canal surfaces looks only at pairs of edge points with parallel tangents, and so may ignore a fair quantity of edge information; as it extends these pairs into curve segments, it is able to discard more spurious information.

Thus, we are searching a small set of object classes for grouping strategies, each of which produces a different set of outline points. These sets must, in turn, be searched to identify the objects present. This means that, by searching in two stages, we have potentially reduced the total amount of searching we need to do—if there are no lines, there is no need to look at the

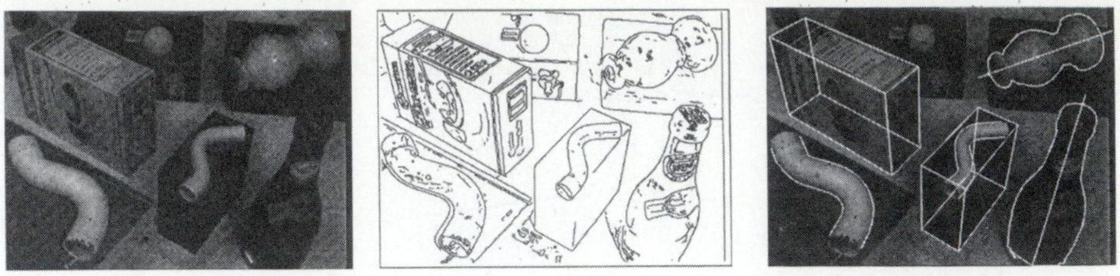

Figure 24.9 An example of a class-based grouping system; on the **left**, an input image whose edges are shown in the **center**; and on the **right**, groups of curves that appear to represent the outlines of objects of a known *type*. Edges for the image have dropouts, occlusion points, and so on. These can be repaired by processes that are specialized to each object type (i.e., the polyhedral grouper fixes problems with line segments; the SOR grouper fixes up pairs of near mirror symmetric curves using the mirror symmetry to fill in missing detail; etc.). Because each class uses a somewhat different grouper, the outlines are classified at the end of the process; however, this classification depends only on the type of the object, not its identity. More detailed examples of the process appear in Figure 24.10. *Reprinted from "Class-based grouping in perspective images," by A. Zisserman et al. Proc. Int. Conf. Computer Vision, 1995 © 1995, IEEE.*

polyhedral models. Although the example is small, the principle is powerful—it would be nice to extract small pieces of information early in the process of recognition that significantly reduce the number of models that must be looked at and (ideally) the processes that can be applied to the rest of the image. Figures 24.9 and 24.10 illustrate the behavior of one such grouper, built by Zisserman, Mundy, Forsyth, Liu, Pillow, Rothwell and Utcke (1995*b*).

24.2 PRIMITIVES, TEMPLATES, AND GEOMETRIC INFERENCE

The examples in the previous section illustrate the attraction of the line of reasoning, but are restricted to quite simple shapes. There are more interesting shape examples, but with somewhat less tight relations between surface and image. Reasoning about these cases has led to several well-known object recognition systems.

24.2.1 Generalized Cylinders as Volumetric Primitives

Volumetric primitives are interesting because (as we saw before) they offer a kind of geometric template: One collects together a chunk of image evidence because it takes a typical, constrained form suggesting the presence of a primitive and then looks at relations between these collections. Ideally, the primitive is expressed in a way that suppresses irrelevant surface detail (e.g., the bumps and ridges formed on the top of my fingers by the underlying veins and tendons don't actually prevent the fingers being an assembly of cylinders). However, the primitives we discussed earlier were pretty limited.

The *generalized cylinders* (*GC*s) proposed by Binford in a famous (but, paradoxically, unpublished) 1971 paper are the most popular form of volumetric primitives to date. They were originally defined in terms of "generalized translational invariance". Roughly speaking, this means that a GC is the solid swept by a one-dimensional set of cross-sections that smoothly

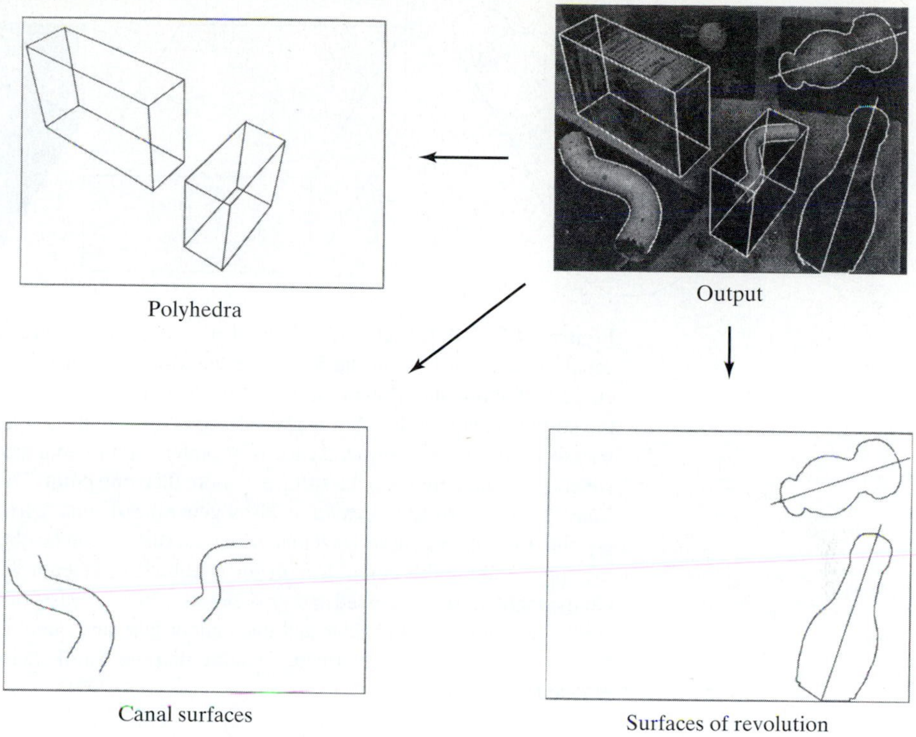

Figure 24.10 The edges for the image of Figure 24.9 have dropouts, occlusion points, and so on. These edges are routed to different groupers. The polyhedron grouper constructs groups and repairs edges by finding lines and grouping them into chains. The canal surface grouper constructs groups and repairs edges by finding locally parallel edge points and grouping them into parallel curves. The SOR grouper constructs groups and repairs edges by finding pairs of curves with mirror symmetries. These repairs are possible because the constraints mean that edge points observed at one point can do duty for dropouts in the presence of a grouping hypothesis. *Reprinted from "Class-based grouping in perspective images," by A. Zisserman et al. Proc. Int. Conf. Computer Vision, 1995 © 1995, IEEE.*

deform into one another. This definition captures the idea that many interesting objects, at an appropriate level of detail, are swept—pipes, tubes, and cylinders are easy examples, but one could think of a car as (roughly!) a rectangle swept along its drive train and made bigger and smaller (other arrangements are necessary for the wheels). One way to firm up our definition of a generalized cylinder is to expand our definition of a canal surface to allow the sphere being swept to increase or decrease in radius (Figure 24.11; left). Not all surfaces are *generalized canal surfaces*, however. For such surfaces, if we construct the locus of centers of maximal inscribed spheres, we expect to get a curve; but for most surfaces, this transform yields a surface (e.g., Figure 24.11).

An alternative way to firm up our definition would be to construct a surface by sweeping a plane curve along a space curve and shrinking, growing, and tilting it as we swept. This is what is usually thought of as the definition of a *generalized cylinder* (Figure 24.12; left). This is obviously a large class of surfaces—we have free choice of cross-section, scaling rule, generating

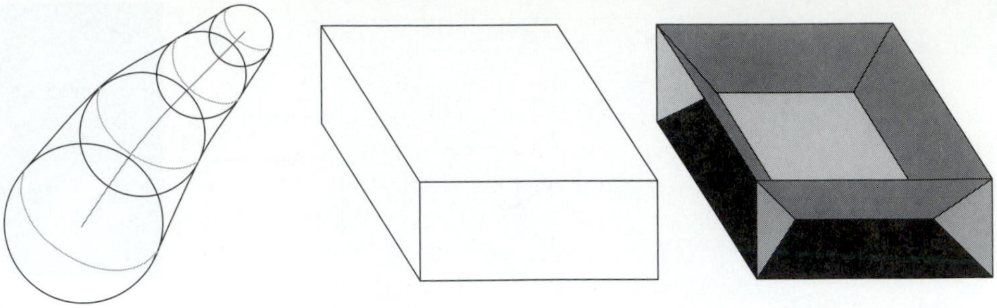

Figure 24.11 Potential realizations of a generalized cylinder. A generalized canal surface appears on the **left**; here we sweep a sphere along some space curve, shrinking and growing it as we go. Note that the generating balls touch the surface along circles. Not every surface is a generalized canal surface. One way to see this is to construct the locus of centers of maximal included spheres—spheres that are tangent to the surface at more than one point. This locus is sometimes called the surface's *skeleton*. For a generalized canal surface, the skeleton is going to contain a curve, such that the whole surface can be obtained by sweeping spheres along this curve. Not all surfaces have a skeleton with this property. On the **right**, a parallelepiped and its skeleton, which consists of polygonal faces. With some thought, it is clear that one cannot take any curve out of this set of faces with the property described, meaning that the parallelepiped is not a generalized canal surface.

curve, angle of tilt, and so on—and it is not known whether there is any surface that is not a generalized cylinder by this definition. Furthermore, it is fairly clear that, at least for some generalized cylinders, multiple choices of cross-section work—a sphere is a good example. Many different variant definitions are possible depending on the shape chosen for the generators (discs, say, or arbitrary plane regions) and on the class of deformations allowed (a simple scaling as in the two-dimensional case or some other type of smooth deformation). In either case, it is not obvious how to compute the cross-section, scaling rule, and generating curve from the surface, let alone image data.

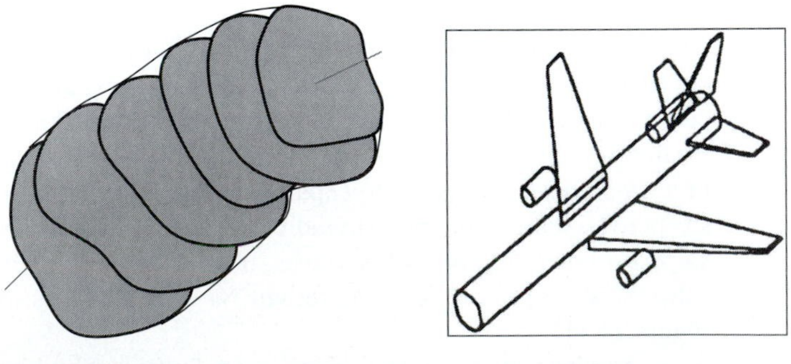

Figure 24.12 Generalized cylinders: **left**, the original definition: planar cross-sections deforming into one another; **right**, the ACRONYM model of an L-1011 model. This instance of a generic wide-body aircraft model is assembled out of cylindrical and conical primitives with circular and polygonal cross-sections.

24.2.2 Ribbons

The relationship between a generalized cylinder and its image is obscure (not least because no one has found it helpful to be precise about what a generalized cylinder is). Thus, we don't have access to the kind of precise geometric reasoning that illuminated the last section. However, if we accept the view of a generalized cylinder as a surface built up of swept, deforming cross-sections, we can guess that, in the image, it is going to look like a region built up by sweeping a deforming segment along a curve. This suggests looking for image regions that look as if they have been swept. Such regions are known as *ribbons*.

A ribbon is the envelope of a family of instances of a geometric figure—the *generator* of the ribbon—which is swept along some trajectory—the *spine* of the ribbon—and shrunk or expanded along the way (Figure 24.13). Two important cases are usually distinguished. A *Brooks ribbon* is obtained when the generator is a line segment, and a *Blum ribbon* is obtained when the generator is a circle (Figure 24.14).

We now assume that this swept form is related to the swept form of the surface we are studying. We can concentrate on three problems:

- **Interpretation:** Given a plane region, is it possible to recover a useful representation of this region in terms of ribbons?
- **Segmentation:** Can we identify the image evidence that forms ribbons, and does doing so help us identify objects?
- **Matching:** Can we identify objects using ribbon-based representations?

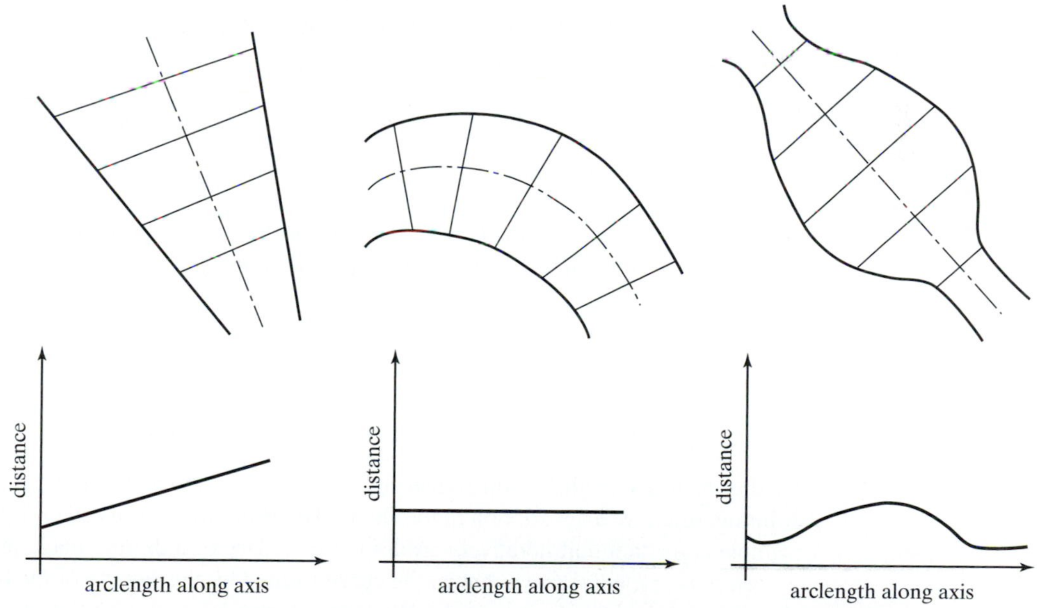

Figure 24.13 Interesting objects often look like ribbons in images; a reasonable model for a ribbon is obtained by taking a curve—the axis—and walking out equal distances along the normal on either side. The particular distance varies along the axis. For each of these example ribbons, we show the axis and some cross-sections to indicate the structure. Below each of these example ribbons, we show a plot of the distance perpendicular to the axis with arc length along the axis.

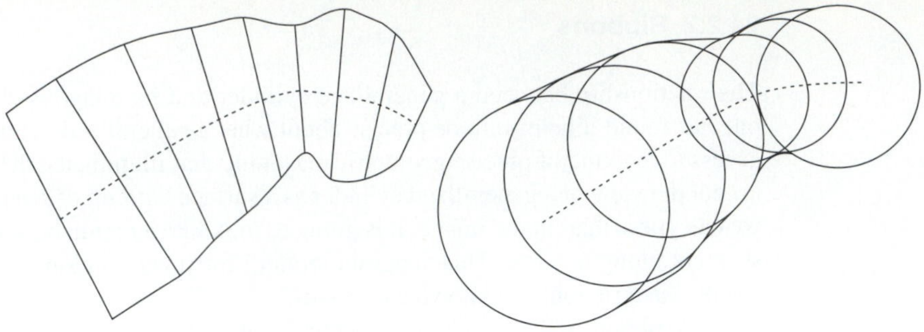

Figure 24.14 A Brooks ribbon (left) and a Blum ribbon (right). In each case, the central curve is the spine of the ribbon, defined as the trajectory of the center of its generator. Notice that, for a Brooks ribbon, the spine can be indeterminate by our definition; it is usual to mark the trajectory of the midpoint of the generating line segment.

Notice that these problems are interrelated; if any set of image evidence yields a ribbon, then being a ribbon is not much help as a segmentation cue.

Interpretation Most work has concentrated on the interpretation problem of determining a useful representation of a given plane region in terms of ribbons. This is probably because it seems to be easier and because it seems to offer more insight into object representation. Quite simple results are possible for Blum ribbons. In particular, we define the *Blum transform*, often known as the *grassfire transform*. For a given two-dimensional region, we construct the family of inscribed disks that are tangent to the boundary in at least two points. The locus of the centers of these discs is the Blum transform of the shape. If we count tangency with multiplicity, then the Blum transform extends into corners properly.

Now assume that we have some plane curve, C. Then we have

$$C \in \text{Blum Transform}(\text{Blum Ribbon}(C, \text{shrinking rule}))$$

i.e., if we sweep a circle along C with any rule for shrinking (or growing), we get a region whose Blum transform contains C. Furthermore, there is some special shrinking rule R such that

$$\text{Blum Ribbon}(\text{Blum Transform}(\text{Blum Ribbon}(C, \text{shrinking rule})), R)$$

is the same as

$$\text{Blum Ribbon}(C, \text{shrinking rule}).$$

All this means that the Blum transform—or some components of it—taken together with a shrinking rule give a description of the shape. The Blum transform can easily be implemented by a simple algorithm that iteratively erodes the boundary of a digital image region and constructs its *skeleton* (or *medial axis*), that is, the curve formed by the centers of the bitangent discs. This property yields another view of the Blum transform: If we regard the region as being a piece of grassland and start fires on the boundary, the Blum transform is the set of points where the firefronts meet (and, hopefully, cancel). This is the origin of the nomenclature "grassfire transform". The bifurcations of the skeleton, (i.e., the multiple points where several smooth branches meet), define a shape decomposition into parts that may, or may not, be intuitively appealing (Figures 24.15 and 24.20). The appeal of Blum ribbons in object recognition is unfortunately balanced by the fact that the Blum transform is susceptible to boundary noise. For example, an

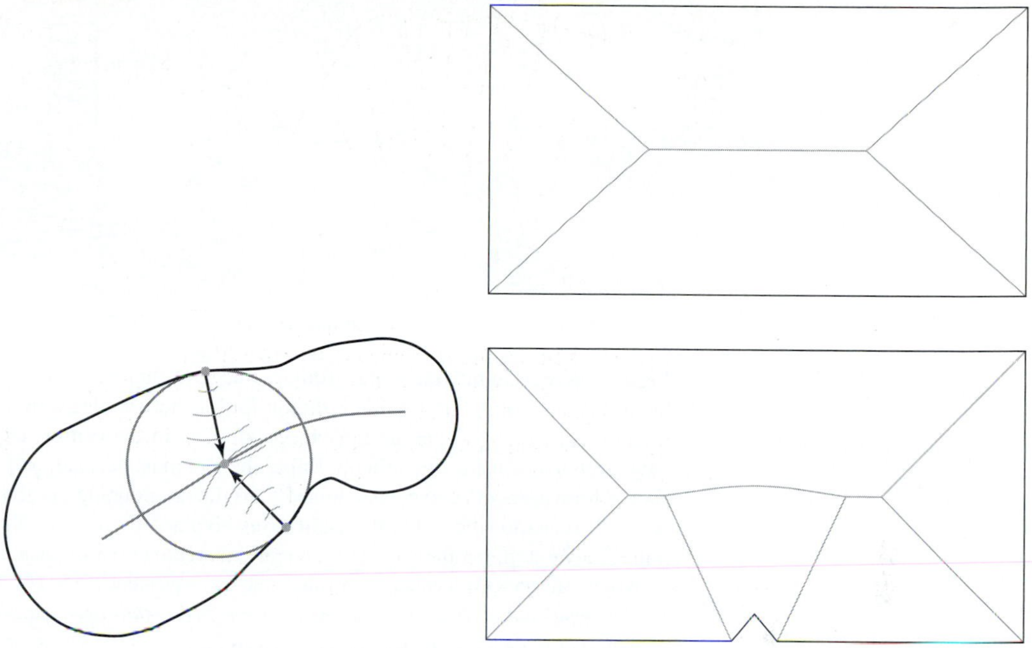

(a) (b)

Figure 24.15 The Blum transform: (a) the skeleton of a planar region is the locus of the discs inscribed in this region and bitangent to its boundary; (b) the skeletons of a perfect rectangle and its noisy version. Note that for most points inside a planar region, the minimum distance to the region boundary is reached at a unique point. At a skeleton point, in contrast, the minimum is reached at two points simultaneously: the two places where the corresponding disc touches the boundary. Thus, the skeleton can be thought of as the locus of points where a wavefront issued from the boundary collides with itself. In discrete images, this yields a skeletonization algorithm that iteratively erodes a shape by removing its boundary points layer by layer until only skeleton points are left. See the exercises for details.

arbitrarily small notch on the boundary of a rectangle dramatically alters the skeleton structure (Figure 24.15; right). Likewise, the bifurcations of the skeleton are also rather unstable under perturbations of the boundary.

Can a similar Brooks transform be defined rigorously? This is not quite as clear, although an attempt is described in the exercises, and we focus on the Blum transform in the rest of this section; Brooks ribbons and their three-dimensional relatives, generalized cylinders, are discussed again in the next section.

Segmentation Segmentation is an interesting and somewhat difficult problem. When the form of the ribbons is heavily constrained—as in the examples we gave before—there tend to be fewer ribbons available in the image, and these are usually relatively easy to find. When the form is less strongly constrained, a variety of heuristics need to be used to focus on the relevant ribbons. We describe a system due to Mohan and Nevatia (1989 and 1992).

The first step is to try and find joins between edge fragments; we want joins that fix drop outs—where the curves join up (roughly) smoothly and are (roughly) close—or joins where two curves appear to form a corner—where the endpoints are (roughly) close. We now look for curves that have sections that are close to parallel, rather as in the grouping of canal surfaces described

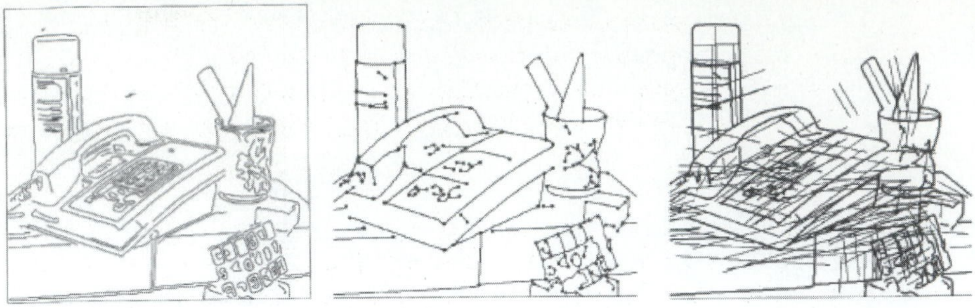

Edges Joined curves Symmetry axes

Figure 24.16 On the **left**, edges from an image of familiar objects; the edges break at points, and object outlines do not form coherent edge curves. Furthermore, there are edges that are not object outlines. In the **center**, edge curves have been joined using two criteria. Either the gap must be small and the edges should form a smooth curve when joined (curvilinear grouping), or the gap must be small (cotermination). On the **right**, axes obtained by marking the midway points between all parallel curves. There are a large number of such axes, most of which are unhelpful; many are ruled out by consistency criteria. *Reprinted from "Segmentation and description based on perceptual organisation," by R. Mohan and R. Nevatia, Proc. Computer Vision and Pattern Recognition, 1989, © 1989, IEEE.*

Contradiction Best symmetry axes Surface patches

Figure 24.17 On the **left**, an illustration of one consistency criterion; two axes cannot share the same segment of curve (because otherwise the view would be highly nongeneric). We must now identify a set of axes that is best in terms of continuity, and so on, discounted by any consistency violations. This is an optimization problem; various methods can be applied (Mohan and Nevatia used a technique known as relaxation). In the **center**, the extremal set of axes is shown. These can be linked to their boundary curves, yielding a hypothetical set of surfaces; further processing, related to occlusion reasoning, yields the set of hypothetical surfaces on the **right**. *Reprinted from "Segmentation and description based on perceptual organisation," by R. Mohan and R. Nevatia, Proc. Computer Vision and Pattern Recognition, 1989, © 1989, IEEE.*

before. This results in a collection of hypothesized generating curves; typically, this collection is much too large to be helpful (Figure 24.16).

There are interpretations that are inconsistent with one another. For example, as Figure 24.17 indicates, two generating curves can't share an outline as that would imply a nongeneric view. Ideally, there is an endcap available for each ribbon, and this constraint should help determine which generating curve survives when there is a contradiction. As another example, we typically wish to ignore markings on objects and so can suppress ribbons that lie wholly inside other ribbons (this has consequences for occlusion reasoning, however). We can now set up an optimization problem: We score ribbons for the smoothness of their joins, and so on, and then discount this score for constraint violations. We must now find the set of ribbons that extremises this score (the authors do this using a technique known as relaxation; there are a variety of possibilities). As Figure 24.17 indicates, the approach can be quite successful. The fact that the relation between 3D and 2D has been expressed somewhat informally is not an impediment. The optimization framework has the advantage that we could, in principle, discount hypotheses that lead to implausible objects; this may lead to a difficult optimization problem, however.

Matching Symmetries do seem to yield representations that can be useful in practice—for example, it is possible to decide, with quite limited accuracy, whether a picture has unclad people in it or not using a representation based on symmetries. Such systems were described in greater detail in Section 23.3.2; we recapitulate that description here to see the material in the light of primitives.

The reasoning is as follows: People are made up of (roughly!) cylindrical segments. These cylinders form long, straight ribbons in an image. Some configurations of ribbons suggest a person is present, whereas others do not. So we find people by first finding ribbons and then finding assemblies of ribbons that suggest people. The same strategy, *mutatis mutandis*, was used for horses and is illustrated in Figure 24.18.

The line of reasoning is attractive: We first check relatively small sets of image components to tell whether they are part of a ribbon; then assemble the survivors, which now form ribbons, into a larger region that looks like a person. This means that we are using a partial model of what (some) objects look like to help assemble together the evidence we need to tell whether an object is present. In particular, at each step we invoke a little model evidence to help us winnow down a range of possibilities. You should notice that, despite the fact that we have no real probabilistic model here, we seem to be doing something rather like inference.

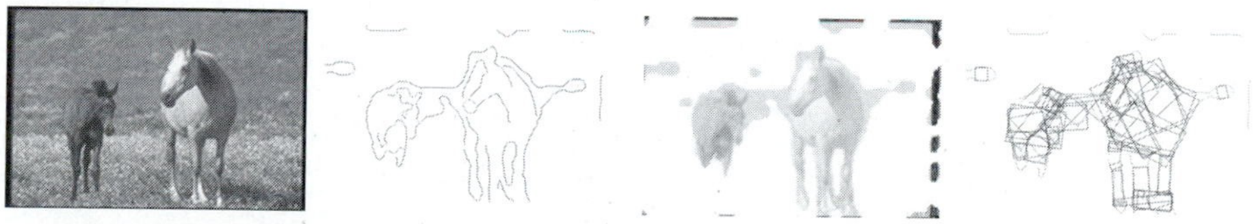

Figure 24.18 An example of the kind of representation that can be obtained from straight segments with near parallel sides. On the **left**, a color image of two horses; on the **center left**, pixels with the right color and texture to be hide have been retained and the others masked off; on the **center right**, the edges of this set of pixels; and on the **far right**, all straight segments with near parallel sides obtained using the mechanisms described before. These segments have been displayed using the abstraction maintained by the program (i.e., the sides of the abstract segment are shown superimposed on the edges that yield the segments). Notice that the components do not exactly correspond to body segments. Although it looks unpromising, this representation can be used to find the horses (Section 25.3.3) despite that it captures image information quite poorly.

24.2.3 What Can One Represent with Ribbons?

Now assume we are presented with object silhouettes (and so can dodge the segmentation problem); what can we represent with ribbons? The FORMS 2D recognition system, developed by Zhu and Yuille (1996), suggests that we can represent, at least, people and animals. Objects are represented by decomposing their silhouettes into ribbons. The difficulty is in avoiding ambiguous decompositions; Zhu and Yuille avoid this problem by finding circles that are multitangent to the silhouette, have a large radius, and cover little of the background. These circles, found by an extremization process, provide seed points from which a representation is grown.

A recursive algorithm now traces out the incident branches of the skeleton starting at the seed points and then looking for bifurcations in the skeleton to boundary distance function. These bifurcations could indicate that the skeleton is splitting—perhaps two fingers join a palm—and a new branch of the skeleton must be started for each component. Once the skeleton has been found, each branch becomes an object part; the discs associated with its endpoints are included in the part description (Figure 24.19).

As shown by Figure 24.19, this approach gives intuitively satisfying results for side views of vertebrate animals like mammals and fish. Once the parts have been found, they are turned into two classes of primitives: *worms* associated with segments of the spine that lie between two junctions, or terminate but are long relative to the radius of the discs attached to their endpoints, and *circular sectors* associated with the remaining (terminal) segments (see the fish decomposition in Figure 24.19). Each primitive is represented by a five-dimensional feature vector recording its length (for worms) or angular extent (for circular sectors) plus four deformation parameters rep-

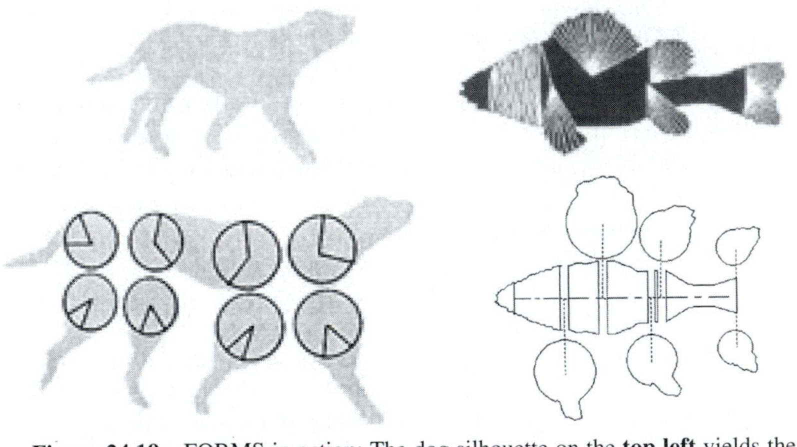

Figure 24.19 FORMS in action: The dog silhouette on the **top left** yields the decomposition on the **bottom left**. This is obtained by fitting maximal circles that are multitangent to the outline. These circles are, putatively, the points where the skeleton branches and so can be used to break the silhouette into parts. The silhouette of a fish on the **top right** is decomposed in the same way to yield the parts indicated below it. Notice that two parts can lie on top of one another. Furthermore, the top (invisible) boundary of the body parts of the fish are inferred from the (visible) belly using a symmetry assumption. *Reprinted from "FORMS: a flexible object recognition and modelling system," by S-C. Zhu and A. Yuille, Proc. Int. Conf. Computer Vision, 1995, © 1995, IEEE.*

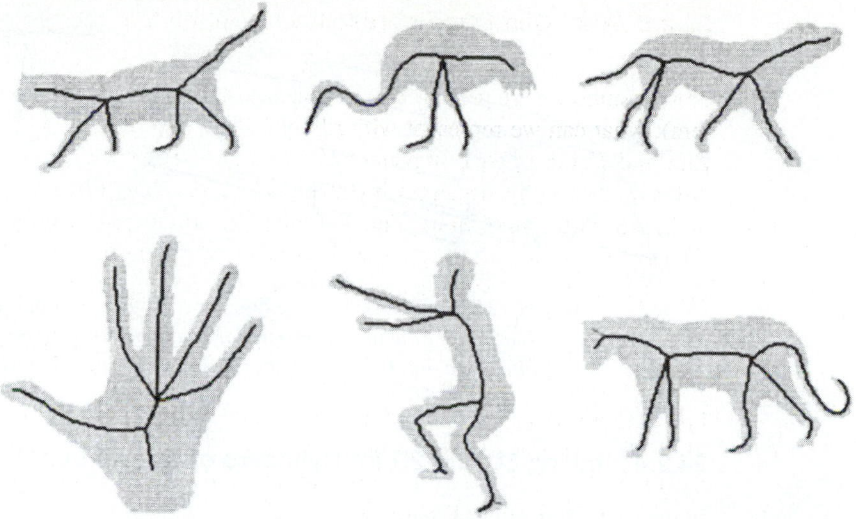

Figure 24.20 Once FORMS has determined the bifurcations of the skeleton, as in Figure 24.19, it fills in the skeleton. These figures show skeletons for a number of silhouettes. The skeletons look reasonable in the sense that they reflect our intuition that animals have heads, bodies, legs, and tails; hands have fingers; and so on. *Reprinted from "FORMS: a flexible object recognition and modelling system," by S-C. Zhu and A. Yuille, Proc. Int. Conf. Computer Vision, 1995, © 1995, IEEE.*

resenting the ribbon width or radius function and calculated using principal component analysis on the model database.

Matching is done by a graph-matching scheme using a measure of similarity among parts, hash tables, and a voting scheme for efficient retrieval of promising model parts and object categories. The similarity of a model part m and an observed primitive o is simply defined as $s(m, o) = \exp(-|\boldsymbol{m} - \boldsymbol{o}|^2)$, where $\boldsymbol{m}$ and $\boldsymbol{o}$ are the five-dimensional feature vectors associated with the two primitives.

Each instance of an object is represented by a graph whose nodes are skeleton junctions and terminations, and edges are the primitive parts between them. Each object category may be described by several of these *skeleton graphs* corresponding to its observed instances. The skeleton graphs associated with the most promising models are selected by similarity-based hashing and voting and then matched to the observed data using a graph-matching procedure that prunes away matches whose cost exceeds some threshold. This procedure is made robust via *adaptive matching*, which allows the skeleton graphs of the models to change during matching under the action of skeleton operators that adjust for the errors that may occur during the bottom-up skeleton extraction step (e.g., missing branch due to occlusion, extra branch due to clutter, splitting of a junction into two due to boundary noise, or conversely merging of two junctions into a single one). Each application of one of these operators incurs a cost, and we seek matches that have a minimum cost that is below some threshold.

Figure 24.21 shows a recognition example. The approach worked quite well on the database of 35 instances and 17 categories used in the FORMS experiments, suggesting that if we could get presegmented views of objects we might be able to represent a decent number of objects with this sort of ribbon.

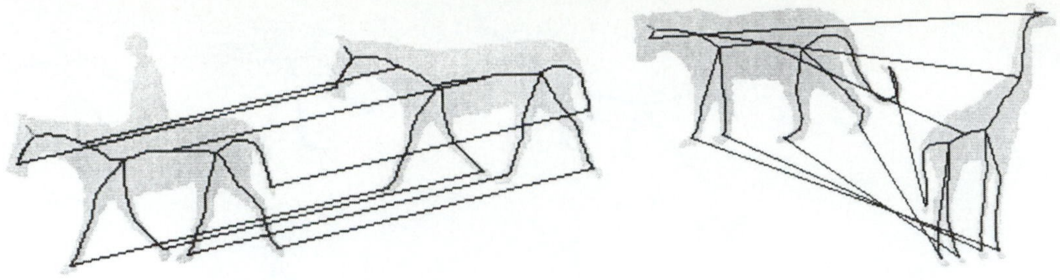

Figure 24.21 FORMS in action: recognition examples. *Reprinted from "FORMS: a flexible object recognition and modelling system," by S-C. Zhu and A. Yuille, Proc. Int. Conf. Computer Vision, 1995, © 1995, IEEE.*

24.2.4 Linking 3D and 2D for Cylinders of Known Length

Assume that we have a single view of a human figure in an unknown camera—do we know the configuration of the person in 3D? The answer is yes (slightly qualified). This is because the prior structure of a person strongly constrains the relationship between their 3D configuration and the 2D configuration, as Taylor (2000) shows.

We model a person as being an assembly of cylinders. Now we can assume that the lengths of these cylinders are known at least relative to one another. This is the key step, and it is an assumption that people with arms long with respect to their torso are uncommon (or that it doesn't matter if we misestimate their configuration). We can obtain the relative lengths of these cylinders from various anthropometric publications; one source is the Anthropology Research Project (1978).

Now assume that we have a projection of a cylinder of known length, l, in a scaled orthographic camera with known scale factor, s. We assume that we can identify the ends of the cylinder so that we can think of the cylinder as an arbitrarily thin line segment (note that, except at extreme angles, for cylinders long with respect to their radius that the ends don't make much contribution). Assuming these approximations, the length of the projected cylinder is then

$$sl \cos\theta,$$

where θ is the angle between the image plane and the cylinder's axis. Now this means that we know where the cylinder is in 3D up to two cases (it points toward or away from the camera). In turn, if we had an open chain of n such cylinders, then we would know the configuration of the chain up to 2^n cases—the cylinders join up, and each has a two-fold ambiguity. The human figure is usually modeled with 9 cylinders and a sphere (the neck and head could be fused into a cylinder, but that's not often done), and so this ambiguity is tolerable particularly since some configurations may be kinematically impermissible and so can be pruned.

Of course, we usually don't have s. This means that we cannot use the equation, but it is quite easy to use a derived version of the equation. In particular, assume that we have two cylinders of length l_1 and l_2; then the ratio of their lengths is l_1/l_2, and the ratio of their image lengths is

$$r_{12} = \left(\frac{l_1}{l_2}\right)\left(\frac{\cos\theta_1}{\cos\theta_2}\right).$$

Here θ_1 and θ_2 are the angles between the segments and the image plane. Now assume we have a chain of segments of known lengths. We can read the ratios of the angles off the image, meaning that if we supply one angle—say θ_1—we know all others. Furthermore, there are constraints on θ_1. We know the value of every r_{ij} because we can measure it. Now we must have

$$\cos\theta_1 = r_{1k}\cos\theta_k\left(\frac{l_k}{l_1}\right)$$
$$\leq r_{lk}\left(\frac{l_k}{l_1}\right),$$

a set of constraints that could be helpful. We have now shown that, for any value of $\cos\theta_1$ satisfying these constraints, we can reconstruct the 3D configuration of a set of cylinders of known length ratios up to a discrete ambiguity.

The reconstruction must depend on the choice of θ_1, but this choice is constrained not only as before, but also by the kinematics of the human figure. Taylor's work suggests that a good choice is the largest value of θ_1 that the data allow (Figures 24.22 and 24.23).

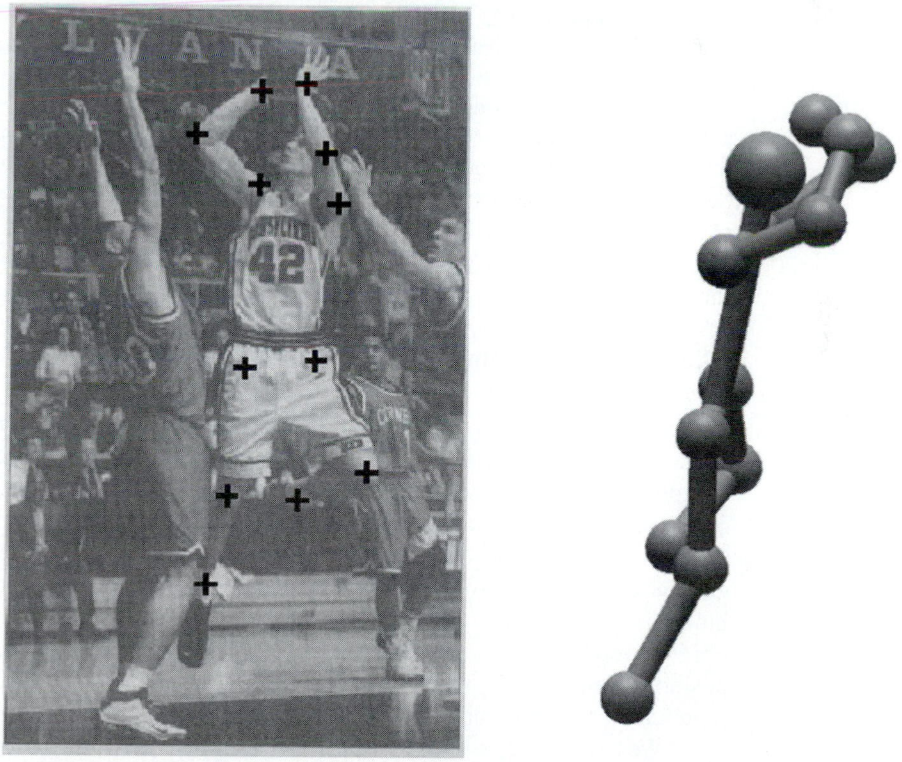

Figure 24.22 On the **left**, a human figure with the endpoints of body segments marked, by hand, with crosses. Because the relative lengths of these segments are known, an assumption of a scaled orthographic view allows a 3D reconstruction of the figure up to a small set of ambiguities. On the **right**, a lateral view of one such reconstruction. Some examples of the ambiguities are shown in Figure 24.23. *Reprinted from "Reconstruction of Articulated Objects from Point Correspondences in a Single Uncalibrated Image," by C.J. Taylor, Proc. Computer Vision and Pattern Recognition, 2000, © 2000, IEEE.*

Figure 24.23 On the **top**, a view of a crouching basketball player. Body segment endpoints are identified by hand; the ambiguity in reconstruction consists of a series of discrete choices and one continuous parameter, which represents the angle between one segment and the image plane. On the **bottom**, different reconstructions implied by different choices of this parameter are shown. In essence, choosing a larger value of this angle produces a longer reconstructed body in a more extreme crouch. *Reprinted from "Reconstruction of Articulated Objects from Point Correspondences in a Single Uncalibrated Image," by C.J. Taylor, Proc. Computer Vision and Pattern Recognition, 2000, © 2000, IEEE.*

24.2.5 Linking 3D and Image Data Using Explicit Geometric Reasoning

We have a (fairly vague) definition of a generalized cylinder and a correspondingly vague notion of the relation between the geometry of a generalized cylinder and corresponding image information. In fact, it is possible to work with these notions to build systems that infer the presence of assemblies of simple cylinders from image data.

Let us model objects as hierarchical assemblies of cylinders or of cones, both with either circular or polygonal cross-section. Each class of object is now represented by a distinct combinatorial structure (i.e., different types of cylinder, assembled in different numbers), and each object is represented by different parameter values (e.g. length of the axis, cross-section radius, position of one cylinder with respect to another, etc.). A further categorical structure can now be represented by sets of inequalities on these parameters.

We have already studied the outlines generated by each such object part. Some components of the combinatorial structure of an object class are apparent in the image because parts can go missing in projection, but not be acquired. If we have a camera model, then we may also be able to reason about inequalities defining object categories.

This is the approach taken by the ACRONYM system developed by Brooks, Greiner and Binford (1979 and 1981*a*). ACRONYM recognizes objects by a succession of prediction, description and interpretation steps. These steps are driven by a geometric reasoning system that can make deductions about constraints existing between spatial relationships in 3D (between object parts) and in 2D (between image components).

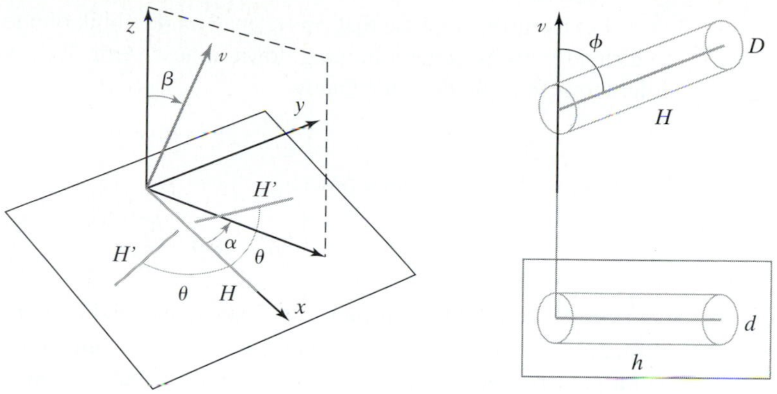

Figure 24.24 The parameters defining the projection of a simple airplane model: (left) the model, with the fuselage and wings depicted by their spines to avoid clutter; (right) orthographic projection of a cylinder.

Example: Relations Between a Fuselage and Wings For example, consider a simplistic airplane model that consists of a central cylinder (fuselage) with height H and diameter D and two symmetrical cylinders (wings) with height H' and diameter D' at angle θ from the fuselage. This plane is observed by a weak perspective camera, with a viewing direction $\boldsymbol{v}$ defined by its spherical coordinates (α, β) in a coordinate system whose x axis is aligned with the spine of the fuselage and whose y axis is in the plane that contains the axes of the fuselage and the two wings (Figure 24.24; left).

As shown in Figure 24.24(right), a cylinder with height H and diameter D seen from a direction $\boldsymbol{v}$ at an angle ϕ from its axis projects orthographically onto a Brooks ribbon with length $h = H \sin\phi$ and diameter $d = D$ (we ignore the elliptical ends of ribbons here). These values become $h = \mu H \sin\phi$ and $d = \mu D$ under weak perspective projection with magnification μ. Thus, the three cylinders representing the fuselage and the two wings project onto ribbons with spine lengths h, h'_l, and h'_r and diameters d, d'_l, and d'_r, and

$$\begin{cases} h = \mu H\sqrt{1-\sin^2\beta\cos^2\alpha}, \\ h'_l = \mu H'\sqrt{1-\sin^2\beta\cos^2(\theta+\alpha)}, \\ h'_r = \mu H'\sqrt{1-\sin^2\beta\cos^2(\theta-\alpha)}, \end{cases} \qquad \begin{cases} d = \mu D, \\ d'_l = \mu D', \\ d'_r = \mu D'. \end{cases} \tag{24.1}$$

In turn, we obtain immediately the following viewpoint-invariant image constraints:

$$d'_l = d'_r \quad \text{and} \quad \frac{d'_l}{d} = \frac{d'_r}{d} = \frac{D'}{D}. \tag{24.2}$$

Let us define the angles θ_l and θ_r between the axis of the ribbon associated with the fuselage and the spines of the left and right wings. It is easy to show (see Exercises) that

$$\begin{cases} \tan\theta_l = \dfrac{\cos\beta\sin\theta}{\cos\theta - \sin^2\beta\cos(\theta+\alpha)\cos\alpha}, \\ \tan\theta_r = \dfrac{\cos\beta\sin\theta}{\cos\theta - \sin^2\beta\cos(\theta-\alpha)\cos\alpha}. \end{cases} \tag{24.3}$$

Note that these equations imply that a planar bilaterally symmetric figure does not project onto another bilaterally symmetric figure (i.e., in general, $\tan\theta_l \neq \tan\theta_r$).

Let us further assume that β is a small angle (think of a reconnaissance aircraft flying over an airfield with its camera looking down almost vertically). Neglecting the second and higher order terms in β yields immediately

$$\theta_l = \theta_r = \theta. \tag{24.4}$$

Likewise, we have, to second order,

$$h'_l = h'_r \quad \text{and} \quad \frac{h'_l}{h} = \frac{h'_r}{h} = \frac{H'}{H}. \tag{24.5}$$

In ACRONYM, invariants, quasi invariants, and bounds on sizes and angles are all used during prediction. They are derived from object and camera models using a set of heuristic rules and a powerful geometric reasoning system capable of manipulating and simplifying complex trigonometric and algebraic expressions, as well as deriving lower and upper bounds on these expressions.

Structuring Geometric Deduction There are two ways to use Equations (24.1) to (24.3). First, given some camera constraint, we can narrow the collection of ribbons that are relevant to our problem. Second, given the parameters of detected ribbons and a correspondence hypothesis, we can obtain a camera constraint. Clearly, we need to take errors in the image description process into account. ACRONYM does this by propagating bounds using methods that are now obsolete and do not need to be discussed. More interesting is the process, which you should compare with the correspondence reasoning of chapter 18.

Assume that a set of ribbons that could correspond to an object has been identified. This, in turn, yields a camera constraint, which should narrow down the available interpretations for other ribbons. This means, in turn, that a small set of correspondences can be parlayed into a large set of recognition hypotheses. We now have what is known as a consistent labeling problem: We should like the largest set of recognition hypotheses that is consistent with the geometric constraints that apply.

ACRONYM attacks this problem expanding correspondence hypotheses. It constructs a prediction graph representing possible object instances and an observation graph, representing image data and then it attempts to match these graphs (details are shown in Algorithm 24.1). Individual ribbon correspondences are first hypothesized and pruned using the associated back

Algorithm 24.1: The ACRONYM recognition algorithm

1. **Prediction**: Construct a prediction graph whose nodes are predicted image ribbons with associated parameter ranges and whose arcs link adjacent ribbons.
2. **Description**: Construct a similar observation graph whose nodes are image ribbons with parameters in ranges compatible with the prediction graph.
3. **Interpretation**:
   - **3.1.** Construct an interpretation graph whose nodes are potential matches between predicted and observed ribbons, using back constraints associated with the prediction graph to ensure parameter compatibility;
   - **3.2.** Use constraint propagation to identify connected components of the interpretation graph that are consistent with the associated back constraints;
   - **3.3.** Use combinatorial search to identify maximal sets of consistent connected components.

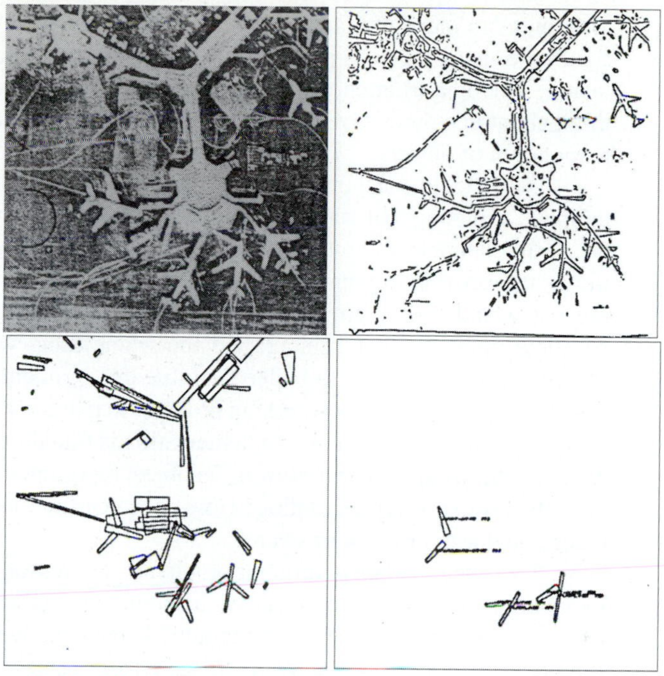

Figure 24.25 The interpretation of an overhead picture of an airfield by ACRONYM. The input to the system consists of the image (**top left**), of a generic model of a wide-bodied passenger aircraft, and of the class specifications to L-1011s instances (Figure 24.12), plus a model of the imaging situation in the form of a calibrated aerial camera whose height is constrained to be between 1,000 and 12,000 meters. On the **top right**, the result of edge detection on the input image, and the **bottom left** shows the ribbons found by ACRONYM under the supervision of its prediction module. These ribbons are about the right size to be aircraft parts (body, wings, tail) viewed by this camera from this height. The recognized airplanes are shown in **bottom right**. Clearly, several planes have been missed. The most likely reasons for this are that (a) components of the ribbons were missed during segmentation, or (b) the bounds propagated by the geometric reasoning system have expanded to such an extent that they no longer usefully constrain correspondence hypotheses. *Reprinted from "Model-based three-dimensional interpretations of two-dimensional images," by R. A. Brooks, IEEE Trans. PAMI, 1983, © 1983, IEEE.*

constraints. Pairs of ribbons are then matched, using again the back constraints associated with the corresponding prediction graph arcs to test consistency. Matching then expands to triples, and so on, global consistency being maintained at each step via constraint propagation. At the end of this process, consistent connected components of the *interpretation graph* have been identified, and they correspond to candidate object models (e.g., individual planes in an airfield). A final global consistency check is done by searching for maximal sets of connected components whose back constraints can all be satisfied (e.g., all the planes in the same airfield should yield consistent viewing parameters). Figure 24.25 illustrates this strategy in action.

24.3 AFTERWORD: OBJECT RECOGNITION

Computer vision has come a long way since its start in the 1960s. Some of this advance has to do with the astonishing drop in price of fast computers and imaging systems. A lot more has to do with an increased understanding of the component problems. The result is that there are, right now, many practical problems that can usefully be solved using techniques from computer vision. This is a triumph. However, there are core problems that are unsolved and difficult to think about in a productive fashion. These core problems have to do with the representation and recognition of objects.

How should one think about recognition in general? The answer probably involves repeated template matching, but what the templates, what the relations, what the object hierarchy should be is still completely obscure. In this view, the basic recognition processes would start by using image information to prune the number of object categories likely to be present; the categories present would supply grouping routines; these would collect together more information, and so allow more detailed pruning; and the whole thing would continue until objects had been identified. It is important to understand that this is an entirely hypothetical construct. Although the idea appeals, there is no working program that one can point to that suggests it might actually be successful. This is because implementation is fraught with practical difficulties: For example, what are the grouping cues? How is the object representation to be organized? How do we attach new objects to this representation? How do we ensure that the pruning process starts in the right place and doesn't run out of steam?

These are extremely difficult problems; it is reasonable that progress should be slow. Remember, it isn't particularly clear what should be regarded as an object—is a face an object, or is it a composite of eyes, nose and mouth? Is mud an object? Is sin an object? It isn't clear what should be recognized. We probably do want to recognize running, but may not want to think of it as an object; it might be quite hard to recognize sin (many people find this difficult); regardless of whether a face is an object or not, it's a good idea to recognize faces. We don't know what distinctions should be primarily visual and what should come later. For example, although tigers and leopards have strong similarities at some semantic level, they look very different; equally, while there is no zoological similarity between a small dolphin and a large penguin, they can be visually quite similar. We do not currently have a satisfactory intellectual framework for discussing these issues.

24.3.1 The Facts on the Ground

People can name many thousands of different kinds of object. This facility is not affected by superficial changes in individual objects (e.g., disrupting the spot pattern on a cheetah, or changing the upholstery on, or the design of, a chair). Furthermore, people need to see only few examples of a new object to get it and be able to recognize other instances of this object at some later date.

It would be useful to have computer programs that, even partially, shared these skills. People probably possess them because they have practical value (knowing what to eat, who owes you food, when to fight, when to flee, what is going to eat you, etc.). The key matter seems to be one of building object representations that behave well when there are large numbers of different objects to be recognized.

Hierarchies are the reflex response induced in computer scientists to problems of scale, and it is quite widely believed that the key to representing and recognizing large numbers of objects is to organize the objects in some form of hierarchy. Generally, this hierarchy is seen as one where high-level distinctions are drawn early, and distinctions of detail are achieved later. Typically, these distinctions are distinctions of appearance, meaning that objects that look similar but are, in fact, very different (e.g. a small dolphin and a large penguin) might be distinguished only very late in the recognition process, and objects that look different, but are similar (e.g. an eel and a fish) might be distinguished early in the recognition process.

The ideal object recognition system would

- recognize many different objects;
  [This is much more difficult than it sounds: To recognize large numbers of objects, we need to know how to organize them into a data structure that is easily searched given image data. In particular, we need to know what measurements can be used to distinguish between objects as opposed to distinguishing between instances (one cat may be tabby, the other gray; they are both cats).]
- recognize objects seen against many different backgrounds;
  [Again, this appears to be difficult. Ideally, an appropriate object representation would help by organizing the image into segments that might have come from an object category (without reference to a particular instance) and those that could not.]
- recognize objects at an appropriate level of abstraction.
  [Humans do not need to have seen a particular chair before they know it is a chair. Ideally, our programs would be able to recognize both leopards and cheetahs as spotted cats before drawing a distinction. Just precisely what is an appropriate level of abstraction is mysterious; at least part of the issue is tied up in the question of recognizing many different objects.]

Current recognition strategies typically perform rather poorly when measured against these requirements, as we see. This is not because they are bad: The problem is just difficult.

24.3.2 Current Approaches to Object Recognition

Pose-consistency Approaches use geometric mechanisms to identify a sufficient number of matches between image and model features. They include alignment techniques and affine and projective invariants, as discussed in chapter 18. In the former case, matching proceeds as a tree search whose potentially exponential cost is kept under control by exploiting that few matches are sufficient to completely determine the object pose and predict the image position of any further correspondences. In the latter case, small groups of points are used to directly compute a feature vector independent of the viewpoint that, in turn, can be used to index a hash table storing all models. An advantage of this approach is that indexing can be done in sublinear time.

Template Matchers record a description of all images of each object. As discussed in Chapter 22, they have been used successfully in tasks such as face identification and three-dimensional object recognition. Their main virtue is that, unlike purely geometric approaches, they exploit the great discriminatory power of image brightness/color information. However, they normally require a separate segmentation step that separates the objects of interest from the image background, and they are potentially sensitive to changes in illumination.

Relational Matchers describe objects in terms of relations between templates. Typically, one looks for rather stylized image patches and then reasons about relations between them (as in chapter 23). There are two difficulties here: First, as that chapter indicated, some relational models are easy to match, but some can be difficult to match; second, current methods handle local patches (like eye corners) and simple objects (like faces) well, but it remains hard to see how one would build matchers that find, say, animals based on relations between image patches.

Aspect Graphs explicitly record the qualitative changes in oject appearance due to viewpoint variation. Recognition techniques based on aspect graphs (discussed in Chapter 20) lie somewhere between appearance-based and structural methods since they actually describe the appearance of an object by the evolution of its structure as a function of viewpoint. Since similar objects (hopefully) have a similar appearance, they may have similar aspect graphs, and the understanding of image structure may serve as a guide for image segmentation. In practice, however, *exact* aspect graphs have not fullfilled their promise partly because the reliable extraction of contour features such as terminations and T-junctions from real images is extremely hard, and partly because even relative simple objects may have extremely complicated aspect graphs.

24.3.3 Limitations

The methods we have described are substantially limited. The primary issue is scale—managing lots of different objects on lots of different backgrounds, but the limitations appear to fit into three (rough) classes:

Segmentation Relational matchers can easily be overwhelmed by large numbers of candidates; similarly, template matchers perform poorly if unknown regions of the patterns to be matched contain irrelevant information. It is probably important that the basic currency of recognition be easily segmented, in the sense that it is easy to tell which image pieces belong together without knowing what object we are looking at.

Categories and Abstraction It is not known how to find objects at the right abstraction level. Although we might not expect to recognize animals at the level of mammals, we should to be able to find spotted mammals, before we worry about whether we are dealing with a leopard or a cheetah. It is important to understand that very little is known about *what* is an appropriate level of abstraction: How do you describe (visually) a farm? The methods we have described so far are incapable of categorization. Notice that, rather than being a deep cognitive mystery, categorization may be a phenomenon that has quite practical origins—it may be much easier to recognize large numbers of objects by drawing rough distinctions first, and fine ones later. How one should categorize is still a mystery.

Generality Successful recognition strategies should apply to many types or object categories without significant tuning. Ideally, because appropriate approaches would be organized according to basic rules of object structure, it should be easy to learn satisfactory models of new objects from a small number of examples. The methods described can usually generalize, but not in particularly satisfactory ways.

24.4 NOTES

Thinking about classes of surface-forming geometric templates goes in and out of fashion (at time of writing, it is resolutely unfashionable, one well-informed insider contending that only one believer remains and that believer somewhat heretical). Nonetheless, we discuss it here because the overall line of thought is probably right. It is hard to believe that the answer in the book of vision is generalized cylinders; it is easy to believe that the answer contains lines of the form "pay close attention to relations between 3D structure and image structure, and particularly close attention to any cue that allows you to group together 2D image components." Many areas of vision have already benefited from a light fertilization of ideas from probabilistic inference; it is surprising to us that this area hasn't. One difficulty is that, as most ideas in this area are formulated with relative imprecision, a certain amount of swashbuckling is required. Another may be that it is uncertain which thread to pull. Is the key idea that constraints on surface structure produce constraints in the image, or is it that if you can't find constraints by which to segment an object, then you can't recognize it?

The first computer programs capable of some form of three-dimensional object recognition date back to Roberts (1965), and models of the recognition process in both people and computers are discussed in Biederman (1987), Bülthoff, Tarr, Blanz and Zabinski (1995), Marr (1982), Palmer (1999), Rosch (1988), Tarr, Hayward, Gauthier and Williams (1995), and Ullman (1996), for example. The object models used by the human mind in recognition tasks remain elusive, however, and the debate between proponents of view-based theories (e.g., Bülthoff et al., 1995, or Tarr et al., 1995) and primitive-based representations (e.g., Marr, 1982, or Biederman, 1987) continues.

The skeleton (or medial axis) was introduced by Blum (1967). Skeletons in digital images are studied in mathematical morphology (Serra, 1982). Comparisons of various types of ribbons, including Blum and Brooks ribbons, but also smooth local symmetries (Brady and Asada, 1984) and skewed symmetries (Kanade, 1981) can be found in Rosenfeld (1986), or Ponce (1990). The FORMS system was developed by Zhu and Yuille (1996). See also Siddiqi, Shokoufandeh, Dickinson and Zucker (1999*b*) for related work.

Generalized cylinders were introduced by Tom Binford (1971). They are also known as generalized cones (Marr and Nishihara, 1978, Brooks, 1981*a*). Most of the early attempts at extracting GC descriptions from images focused on range data (e.g., Agin, 1972). Among those, the work of Nevatia and Binford (1977) is particularly noteworthy since it does implement a version of generalized translational invariance: their algorithm tries all possible cross-section orientations of objects such as dolls, horses, and snakes, then selects subsets of the cross-section candidates with smoothly varying parameters. The ACRONYM system was developed by Brooks and Binford (1979, 1981*a*, 1981*b*). SHGCs were introduced by Shafer and Kanade (1983, 1985*a*) as part of a general taxonomy of generalized cylinders. As noted earlier, imiting the class of GCs under consideration makes it possible to predict viewpoint-independent properties of their projections. For example, Nalwa (1987) proved that the silhouette of a solid of revolution observed under orthographic projection is bilaterally symmetric, and Ponce et al. (1989) showed that, under both orthographic and perspective projection, the tangents to the silhouette of an SHGC at points corresponding to the same cross-section intersect on the image of the SHGC's axis. Such analytical predictions provide a rigorous basis for finding individual GC instances in images or recognizing GC instances based on projective invariants. Impressive results have been achieved (see Zerroug and Medioni, 1995 for example). Preliminary efforts toward the definition of a three-dimensional *generalized cylinder transform* analogous to the Blum transform in the plane are described in Ponce, Cepeda, Pae and Sullivan (1999) (see also the exercises below). The medial axis transform (Blum transform, grassfire transform) was rather frowned on until recently because of its instability. It has recently been rehabilitated, in two forms: in one, one studies shock-graphs which represent its properties (Kimia, Tannenbaum and Zucker, 1990, 1995, Giblin and Kimia, 1999, Siddiqi, Kimia, Tannenbaum and Zucker, 1999*a*, *b*); in another, one subjects the boundary to a series of small deformations and then looks at the "average" of the resulting transforms (Zhu, 1999).

A discussion of shading primitives appears in Haddon and Forsyth (1998*a*). Other primitives that have been used with some success in recognition tasks include superquadrics (e.g., Pentland, 1986). Initial attempts at exploring the role of function in object recognition are described in Stark and Bowyer (1996).

PROBLEMS

24.1. Defining a Brooks transform: Consider a 2D shape bounded by a curve Γ defined by $\boldsymbol{x} : I \rightarrow \mathbb{R}^2$ and parameterized by arc length. The line segment joining any two points $\boldsymbol{x}(s_1)$ and $\boldsymbol{x}(s_2)$ on Γ defines a cross-section of the shape, with length $l(s_1, s_2) = |\boldsymbol{x}(s_1) - \boldsymbol{x}(s_2)|$. We can thus reduce the problem of studying the set of cross-sections of the shape to the study of the topography of the surface S associated with the height function $h : I^2 \to \mathbb{R}^+$ defined by $h(s_1, s_2) = \frac{1}{2}l(s_1, s_2)^2$. In this context, the ribbon associated with Γ can be defined (Ponce et al. 1999) as the set of cross-sections whose end-points correspond to valleys of S, (i.e., according to Haralick, 1983 or Haralick, Watson and Laffey, 1983, the set of pairs (s_1, s_2) where the gradient ∇h of h is an eigenvector of the Hessian $\nabla^2 h$, and where the eigenvalue associated with the other eigenvector of the Hessian is positive).

If $\boldsymbol{t}_i$ and $\boldsymbol{n}_i$ denote, respectively, the unit tangent and normals in $\boldsymbol{x}_i$ $(i = 1, 2)$, and θ_i and κ_i denote, respectively, the angle between the vectors u and t_i and the curvature in $\boldsymbol{x}_i$, show that the ribbon associated with Γ is the set of cross-sections of this shape whose endpoints satisfy

$$(cos^2\theta_1 - \cos^2\theta_2)cos(\theta_1 - \theta_2) + l\cos\theta_1\cos\theta_2(\kappa_1\sin\theta_1 + \kappa_2\sin\theta_2) = 0.$$

24.2. Generalized cylinders: The definition of a valley given in the previous exercise is valid for height surfaces defined over n-dimensional domains and valleys form curves in any dimension. Briefly explain how to extend the definition of ribbons given in that exercise to a new definition for generalized cylinders. Are difficulties not encountered in the two-dimensional case to be expected?

24.3. Skewed symmetries: A skewed symmetry is a Brooks ribbon with a straight axis and generators at a fixed angle θ from the axis. Skewed symmetries play an important role in line-drawing analysis because it can be shown that a bilaterally symmetric planar figure projects onto a skewed symmetry under orthographic projection (Kanade, 1981). Show that two contour points P_1 and P_2 forming a skewed symmetry verify the equation

$$\frac{\kappa_2}{\kappa_1} = -\left[\frac{\sin\alpha_2}{\sin\alpha_1}\right]^3,$$

where κ_i denotes the curvature of the skewed symmetry's boundary in P_i $(i = 1, 2)$, and α_i denotes the angle between the line joining the two points and the normal to this boundary (Ponce, 1990).

Hint: Construct a parametric representation of the skewed symmetry.

Programming Assignments

24.4. Write an erosion-based skeletonization program. The program should iteratively process a binary image until it does not change anymore. Each iteration is divided into eight steps. In the first one, pixels from the input image whose 3×3 neighborhood matches the left pattern below (where "*" means that the corresponding pixel value does not matter) are assigned a value of zero in an auxiliary image; all other pixels in that picture are assigned their original value from the input image.

0	0	0
*	1	*
1	1	1

0	0	*
0	1	1
*	1	1

The auxiliary picture is then copied into the input image, and the process is repeated with the right pattern. The remaining steps of each iteration are similar and use the six patterns obtained by consecutive 90-degree rotations of the original ones. The output of the program is the 4-connected skeleton of the original region (Serra, 1982).

24.5. Implement the FORMS approach to skeleton detection.

24.6. Implement the Brooks transform.

24.7. Write a program for finding skewed symmetries. You can implement either (a) a naive $O(n^2)$ algorithm comparing all pairs of contour points, or (b) the $O(kn)$ projection algorithm proposed by Nevatia and Binford (1977). The latter method can be summarized as follows: Discretize the possible orientations of local ribbon axes; for each of these k directions, project all contour points into buckets and verify the local skewed symmetry condition for points within the same bucket only; finally, group the resulting ribbon pairs into ribbons.

PART VII
Applications

25

Application: Finding in Digital Libraries

Large collections of digital pictures seem to spring up quite easily. Some collections of pictures are being digitized in the hope of better conservation, easier distribution, and better access. Others are intrinsically digital; examples include individual collections of family photographs (which can be big and digital); the Web, which is a big, disorganized collection; and home videos (again, some collections are big and many are now digital).

Tools for interacting with collections of documents or of data are now quite sophisticated. Typically, one can search a collection using various kinds of text matching, one can cluster collections of text, and one can use *data mining* techniques. Data mining involves using statistical fitting procedures to look for trends not previously known (this useful pastime used to be known as "exploratory data analysis," a less exciting name, and is sometimes called "data dredging" by those who disapprove). Generally, a significant component of the value of a collection comes from the presence of such tools. To see why this might be, imagine visiting a large secondhand book shop that has its books sorted by, say, the color of the dust-jacket; although the collection may be large, it's hard to imagine that you'd use the shop unless you were desparate.

It is currently difficult to organize or search image collections in a satisfactory fashion, meaning they are somewhat analogous to a poorly organized bookshop. The difficulty lies in building appropriate representations of the image information. It is no help to annotate each picture by hand because preparing a good text description of an image is difficult. Furthermore, some collections are enormous (tens of millions of pictures; Enser, 1995). Indexing a large collection by hand involves a substantial volume of work. Furthermore, there is the prospect of having to reindex sections of the collection; for example, if a news event makes a previously unknown person famous, it would be nice to know if the collection contained pictures of that person. Finally, it is often hard to know what a picture is about.

Despite all these difficulties, any technology that helps manage collections of pictures has a tremendous range of practical applications. One important tool is search—find me a picture matching these criteria—but this is by no manner of means the only need. We might wish to organize the pictures in a way that supports browsing so that pictures with similar content are near to one another. We might wish to search for trends or to have tools that identify important changes.

Typical applications include the following

- **Planning and government:** There is a lot of satellite imagery of the earth, that can be used to inform important political debates. For example, how far does urban sprawl extend? What acreage is under crops? How large will the maize crop be? How much rainforest is left?, etc. (e.g., Smith, 1996).
- **Military intelligence:** Satellite imagery can contain important military information. Typical queries involve finding militarily interesting changes—for example, is there a concentration of force? How much damage was caused by the last bombing raid? What happened today? and so on—occuring at particular places on the earth (e.g., Mundy and Vrobel, 1994).
- **Stock photo and stock footage:** Commercial libraries, which often have extremely large and diverse collections, survive by selling the rights to use particular images (e.g., Enser, 1993, 1995, and 1993).
- **Access to museums:** Museums are increasingly creating Web views of their collections, typically at restricted resolutions, to entice viewers into visiting the museum (e.g., Holt and Hartwick, 1994*a*, Holt and Hartwick, 1994*b*, and Psarrou, Konstantinou, Morse and O'Reilly, 1997). Ideally, viewers should be able to browse the collection to get a sense of what is at the museum.
- **Trademark and copyright enforcement:** As electronic commerce grows, so does the opportunity for automatic searches to find violations of trademark or of copyright (e.g., Eakins, Boardman and Graham, 1998, Jain and Vailaya, 1998, Kato, Shimogaki, Mizutori and Fujimura, 1988, and Kato and Fujimura, 1990). For example, at time of writing, the owner of rights to a picture could register it with an organization called BayTSP, who would then search for stolen copies of the picture on the Web.
- **Indexing the web:** Indexing web pages appears to be a profitable activity. Users may also wish to have tools that allow them to avoid offensive images or advertising. A number of tools have been built to support searches for images on the web using techniques described later (e.g., Cascia, Sethi and Sclaroff, 1998, Chang, Smith, Beigi and Benitez, 1997*b*, or Smith and Chang, 1997).
- **Medical information systems:** Recovering medical images similar to a given query example might give more information on which to base a diagnosis or to conduct epidemiological studies (e.g., Kofakis and Orphanoudakis, 1991 or Wong, 1998). Furthermore, one might be able to cluster medical images in ways that suggest interesting and novel hypotheses to experts.
- **Image data mining:** The attraction of data mining is that one can go on fishing expeditions using large data sets. Sometimes a data mining method will suggest a genuinely useful or novel hypothesis that can be verified by domain experts. Many image collections could support a similar activity. For example, there is a large collection of digitised images of Buddhist art, which is collected together with geolocation data (where was the object found?) and various expert comments. If we could recover representations from the images, we could look for, say, trends in the depiction of the human figure across both space and time.

The core issue for all of these applications is the nature of the underlying representation of the images. Once we have decided on a representation, it is relatively easy to search (by finding images with a representation like this), organize (by putting images with similar representations near to one another), or search for trends (by looking for relationships between components of representations). One (possibly desirable) representation would be a complete description of all the objects present in an image. There is little prospect of generating descriptions like this using computer programs in the foreseeable future. However, quite crude representations seem to have been helpful to date. In this chapter, we first review some general material on information retrieval. We then show a series of current methods for organizing and searching image collections using computer vision tools.

25.1 BACKGROUND: ORGANIZING COLLECTIONS OF INFORMATION

Information retrieval is the study of systems that recover items from collections using various kinds of information. The topic is interesting to us because information retrieval researchers have become adept at performance analysis, which is often quite difficult.

25.1.1 How Well Does the System Work?

Typically, the performance of information retrieval systems is described in terms of *recall* —the percentage of relevant items actually recovered—and *precision* —the percentage of recovered items actually relevant. The word relevant is the difficulty here—to make these measurements, we need to know what items are relevant to a query. This is a question on which competent human informants can differ.

What Is Good Depends on the Application Typically, as one changes the configuration of a system to make the recall go up, the precision goes down. It is tempting to believe that good systems should have high recall and high precision, but this is not the case. Instead, what is required for a system to be good depends on the application.

Patent searches: Patents are invalidated by the existence of prior art, material that predates the patent and contains similar ideas. This means that it is valuable to be able to search for such material, either to check that it doesn't exist (when applying for a patent) or to find it (and so overturn an inconvenient patent). It can be expensive to patent an idea, and it is often catastrophic for the patent owner (and lucrative for the challenger) when a patent is overturned. This means that it is usually much cheaper to pay someone to wade through irrelevant material than it is to miss relevant material. This means that high recall is essential, even at the cost of low precision.

Web and E-mail filtering: There is a quite widespread demand for Web and e-mail filtering services. For example, US companies are often anxious that a substantial volume of internal email containing sexually explicit pictures may create legal liabilities to do with harassment. One way such services might be delivered would be to have a program that searched e-mail traffic for problem pictures. A manager would expect to get a warning and to be shown the pictures that the program thinks are problematic. At time of writing, there are several vendors of such programs; it isn't yet clear whether this is a lucrative application, so they may have gone out of business when you read this. In an application like this, high recall is not important, although it wouldn't present a problem. If the program has only 10% recall, it will still be difficult to get more than a small number of pictures past it. High precision is important because of the "boy who cried wolf" effect. People tend to ignore systems that generate large numbers of false alarm, and so would not use—and, more important, not pay for upgrades and maintenance on—a system that had low precision.

Looking for a news item: There are various services that provide stock photographs or video footage to news organizations. These collections tend to have many photographs of celebrities—one would expect a good stock photo service to have many thousands of photographs of Nelson Mandela, or Princess Di, and so on. This means that a high-recall search can be a serious nuisance—no picture editor really wants to wade through thousands of pictures. Typically, staff at stock photo organizations use their expertise and interviews with customers to provide only a small subset of relevant pictures.

Assessing Systems It is usually quite difficult to assess a system properly. It is common to plot precision at various different levels of recall (obtained by varying match thresholds) and then average these plots over typical queries. One can weight recall and precision to reflect their relative importance in a particular application and compute an average utility score, too. Good experiments are quite hard to do; what is worse, bad experiments are quite easy to do. This is because it is often quite hard to tell what is relevant (i.e., what should have been recovered by a query—what pictures should the query term "queen" return?). It is even harder to tell how many relevant items appear in a large collection (imagine counting all the pictures relevant to the term "queen" in a collection of ten million). It's obviously a bad idea to use the system being tested to determine what pictures are relevant (this could lead to an inaccurate claim of 100% recall), which means we have to find some other way to count the relevant items in the collection. If we count carelessly—in particular, if we *undercount*—then our estimate of the system's recall is higher.

25.1.2 What Do Users Want?

The most comprehensive study of the behavior of users of image collections is Enser's work on the then Hulton–Deutsch collection (Armitage and Enser, 1997, Enser, 1993, Enser, 1995) (the collection has been acquired by a new owner since these papers were written and is now known as the Hulton–Getty collection). This is a collection of prints, negatives, slides, and the like used mainly by media professionals. Enser studied the request forms on which client requests are logged; he classified requests into four semantic categories depending on whether a unique instance of an object class is required and whether that instance is refined. Significant points include the fact that the specialized indexing language used gives only a "blunt pointer to regions of the Hulton collections" (Enser, 1993, p. 35) and the broad and abstract semantics used to describe images. For example, users requested images of hangovers, physicists, and the smoking of kippers. All these concepts are well beyond the reach of current image analysis techniques. As a result, there are few cases where one can obtain a tool that directly addresses a need. For the foreseeable future, the main constraint on the design of tools for finding images is our quite limited understanding of vision.

However, useful tools can be built even with a limited understanding of vision (this extremely important point seems to be quite widely missed). It is hard to measure success. Enser suggests that the most reliable measure of the success of Hulton–Getty's indexing system is that the organization is profitable. This test is a bit difficult to apply in practice, but there are a number of products available. IBM has produced a product for image search—QBIC (Query By Image Content)—which has appeared in mass market advertising and appears to be successful. Similarly, Virage—a company whose main product is an image search engine—appears to be thriving (the company is described at (Virage n.d.); a description of their technology appears in Hampapur, Gupta, Horowitz and Shu, 1997).

25.1.3 Searching for Pictures

Rather roughly, there are three ways to represent an image: at the iconic level, where one is interested in exact pixel values; at the compositional level, where one is interested in the overall appearance of the image; or at the level of object semantics, where one is interested in the things depicted in the image.

Iconic Matching In iconic matching, we are seeking images that look as much like an example picture—which we might draw or supply—as possible. The ideal match would have exactly the same pixel value in each location. The best known system of this form is due to Jacobs, Finkelstein and Salesin (1995), who apply a series of filters to images at different scales, and then compares the filter responses for the example image and other images. Scale can be used to structure this comparison; if coarse scale responses match poorly, then there may be no reason to check fine-scale responses.

Iconic matchers are helpful only if the user knows what the picture being sought looks like. This situation doesn't arise all that often for naive users (users who don't know the collection well), but there are some important applications where iconic matching is helpful. One is in copyright protection. It is easy to steal digital images—that is, use them without paying a fee to the copyright owner. With an appropriate iconic matcher, it is also quite easy to find thieves.

The process works something like this. The copyright owner registers the image with an organization that specializes in finding thieves and pays a fee. This organization then uses some form of spider to search the Web, downloading images as it goes. These images are compared to the collection of registered images, using an iconic matcher; any hit generates a stiff letter from a lawyer and the prospect of fines and copyright fees. Although some extra work needs to be done to ensure that cropped and rotated images match, and that the matching process is efficient, matching based on filter responses is quite sufficient for this application.

There are other possible applications for iconic matching. For example, it is apparently the case that the vast majority of child pornography in current circulation is relatively old, dating to the 1970s or before. Furthermore, agencies investigating a charge of child pornography need to determine whether the material involved is new material—in which case, there is more to do than just prosecute for possession or distribution because the material documents what may be ongoing abuse. This need can be met by matching to a reference library of known material. Furthermore, this matching process can be used to connect investigations by connecting prosecutors in different jurisdictions prosecuting cases involving different defendants but the same material. At time of writing, law enforcement agencies apparently do not use a reference collection in this way, but such use is the subject of quite extensive discussions. We won't discuss iconic matching further in this chapter as our main interest is in organizing collections.

Matching Using the Whole Image In some applications, the structure of the whole image is important. In these applications, we think of the image as an arrangement of colored pixels, rather than a picture of objects. This abstraction is often called **appearance**. The distinction between appearance and object semantics is somewhat empty—how do we know what's in an image except by its appearance?—but the approach is important in practice because it is quite easy to represent appearance automatically. This is because we don't need to segment the image.

Appearance is particularly helpful when the *composition* of the image is important. For example, one could search for stock photos using a combination of appearance cues and keywords, and require the user to exclude images with the right composition but the wrong semantics. The central technical problem in using appearance to represent images is defining a useful notion of image similarity; Section 25.2 illustrates a variety of different strategies.

Object Level Semantics Enser's study suggests that people using stock photo collections are searching for images with particular semantics (e.g., smoking kippers). It is difficult to cope with such queries using appearance tools. Furthermore, it is almost never possible to construct object recognition programs that can deal with semantic queries. However, we can build representations that try to respect object level semantics. Typically, this involves segmenting images and then building representations around the segments. It is not known how to build finding tools that can handle high-level semantic queries, nor how to build a user interface for a general finding tool. Nonetheless, current technology can produce quite useful tools for various special cases (Section 25.3).

25.1.4 Structuring and Browsing

Searching for images raises some difficult problems. For example, assume you had a perfect object recognition system; how would you describe the picture you wanted? However, search is often not as important as it seems. Typically, it is hard to frame a helpful search unless one has some kind of model of what lies in a collection. For example, think about how you behave in a new shop; you first attempt to determine the kinds of things it stocks and then look for things that you expect to be able to find. You wouldn't ask an assistant in a bookshop for a motorcar. This suggests that browsing is an important interaction, although it is likely to be helpful only if the collection is appropriately organized. Ideally, browsing tools should

- display images that are similar—this could mean that they look similar, have similar appearance, lie close to one another in the collection, or contain similar content, etc.—in a way that makes their similarity apparent;
- display a representation of clusters of images that makes it easy for a user to get a sense of what is in the collection (e.g., similar clusters might be close together and big clusters might be big, etc.);
- provide some form of interaction that makes it possible to see the collection at different levels of detail (perhaps one wants to see only the elements of a particular cluster or to see a summary of all the images near a set of clusters) and to move through the collection in different directions.

Browsing and search tools naturally complement one another. A user could first browse the collection and then frame a search. Having searched, the user might then choose to look at items "near" to any hits returned by the search tool.

Building useful browsing tools also requires an effective notion of image similarity. Constructing a good user interface for such systems is difficult (as the following examples indicate); desirable features include a clear and simple query specification process, and a clear presentation of the internal representation used by the program so that failures are not excessively puzzling. Typically, users are expected to offer an example image or to fill in a form-based interface to search for the first image, and then can move around the collection by clicking on samples offered by the browsing tool.

25.2 SUMMARY REPRESENTATIONS OF THE WHOLE PICTURE

Images are often highly stylized, particularly when the intent of the artist is to emphasize a particular object or a mood. This means that the overall layout of an image can be a guide to what it depicts, so that useful query mechanisms can be built by looking for images that look similar to a sample image, a sketched sample, or textual specification of appearance. The success of such methods rests on the sense in which images look similar. It is important to convey to the user the sense in which images look similar because otherwise mildly annoying errors can become extremely puzzling. A good notion of similarity is also important for efficient browsing because a user interface that can tell how different images are can lay out a display of images to suggest the overall structure of the section of the collection being displayed.

25.2.1 Histograms and Correlograms

A popular measurement of similarity compares counts of the number of pixels in particular color categories. For example, a sunset scene and a pastoral scene would be different by this measure because the sunset scene contains many red, orange, and yellow pixels and the pastoral scene has a preponderance of green (grass), blue (sky), and perhaps white (cloud) pixels (e.g., Figure 25.1). Furthermore, sunset scenes tend to be similar; most have many red, orange, and yellow pixels and few others.

A *color histogram* is a record of the number of pixels in an image or a region that fall into particular quantization buckets in some color space (RGB is popular, for reasons we cannot explain). If the color histogram for an image of an object fits into the histogram for an image, then it is possible that that object is present in the image—if the illumination is not expected to vary all that much. This test can be quite sensitive to viewing direction and scale because the relative number of pixels of a given color can change sharply. Nonetheless, it has the advantage of being quick and easy, and applies to things like clothing that may have bright colors but little or no recognisable shape.

Colour histogram matching has been extremely popular; it dates back at least to the work of Swain and Ballard (1991), and has been used in a number of systems used in practice (Flickner,

Figure 25.1 Results from a query to the Calphotos collection at U.C. Berkeley that sought pastoral scenes. The query was composed by searching for images that contain many green and light blue pixels. As the results suggest, such color histogram queries can be quite effective.

Sawhney, Niblack and Ashley, 1995, Holt and Hartwick, 1994*b*, Ogle and Stonebraker, 1995). The usefulness of color histograms is slightly surprising, given how much image information the representation discards. For example, Chappelle *et al.* at ATT have shown that images from the Corel collection—a collection of 60,000 images quite commonly used in vision research which used to be available in three series from the Corel corporation, 1600 Carling Avenue Ottawa, Ontario, Canada K1Z 8R7—can be classified by their category in the collection using color histogram information alone (Chapelle, Haffner and Vapnik, 1999).

There is no record in a color histogram of *where* colored pixels are with respect to one another. Thus, for example, pictures of the French and UK flags are extremely similar according to a color histogram measure. Each has red, blue, and white pixels in about the same number; it is the spatial layout of the pixels that differs. One problem that can result is that pictures taken from slightly different viewing positions look substantially different by a color histogram measure. This effect can be alleviated by considering the probability that a pixel of some color lies within a particular pixel of another color (which can be measured by counting the number of pixels at various distances). For small movements of the camera, these probabilities are largely unchanged, so that similarity between these *color correlograms* yields a measure of similarity between images. Requiring that color correlograms be similar provides another measure of image similarity. The computational details have been worked out by Zabih and colleagues (Huang, Kumar, Mitra, Zhu and Zabih, 1997, Huang and Zabih, 1998).

25.2.2 Textures and Textures of Textures

Colour histograms contain no information about the layout of color pixels. An explicit record of layout is the next step. For example, a snowy mountain image will have bluer regions on top, whiter regions in the middle, and a bluer region at the bottom (the lake at the foot of the mountain), whereas a waterfall image will have a darker region on the left and right and a lighter vertical stripe in the center. These layout templates were introduced by Lipson, Grimson and Sinha (1997); they can be learned for a range of images, and appear to provide a significant improvement over a color histogram.

Looking at image texture is a natural next step because texture is the difference between, say, a field of flowers (many small orange blobs) and a single flower (one big orange blob), or between a dalmation and a zebra. Most people know texture when they see it, although the concept is either difficult or impossible to define. Typically, textures are thought of as spatial arrangements of small patterns (e.g., a tartan is an arrangement of small squares and lines, and the texture of a grassy field is an arrangement of thin bars).

The usual strategy for finding these subpatterns is to apply a linear filter to the image (see chapter 7 or Section 7.6), where the kernel of the filter looks similar to the pattern element. From filter theory, we have that strong responses from these filters suggest the presence of the particular pattern. Several different filters can be applied, and the statistics of the responses in different places then yield a decomposition of the picture into spotty regions, barred regions, and the like (Ma and Manjunath, 1997*b*, Malik and Perona, 1990).

A histogram of filter responses is a first possible description of texture. For example, one might query for images with few small yellow blobs. This mechanism is used quite successfully in the Calphotos collection at Berkeley (`http://elib.cs.berkeley.edu/photos`; there are many thousands of images of California natural resources, flowers, and wildlife).

Texture histograms have some problems with camera motion; as the camera approaches the scene, details get bigger in the image. A strategy for minimizing the impact of this effect is to define a family of allowable transformations on the image—(e.g., scaling the image by a factor in some range). We now apply each of these transformations and measure the similarity between two images as the smallest difference that can be obtained using a transformation. For

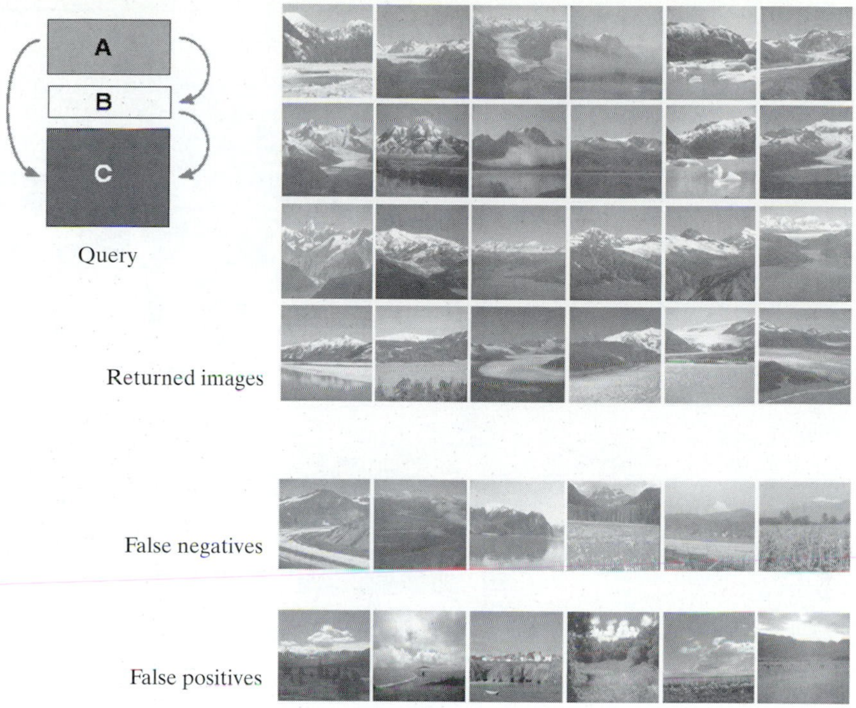

Figure 25.2 Spatial layout of colored regions is a natural guide to the content of many types of image. The figure on the **top left** shows a layout of colored regions that suggests a scene showing snowy mountains; **top right**, the figures recovered by this criterion that actually do show snowy mountains; **center**, views of mountains that were in the collection but not recovered; and **bottom**, images that meet the criterion but do not actually show a view of a snowy mountain. *Figures from "Configuration based scene classification and image indexing," by P. Lipson, E. Grimson and P. Sinha. Proc. IEEE Conf. on Computer Vision and Pattern Recognition, 1997 © 1997, IEEE.*

example, we could scale one image by each legal factor and look for the smallest difference between color and texture histograms. This *earth-mover's distance*, due to Rubner, Tomasi and Guibas (1998) allows a wide variety of transformations. Furthermore, it has been coupled with a process for laying out images that makes the distance between images in the display reflect the dissimilarity between images in the collection. This approach allows for rapid and intuitive browsing (Figure 25.3).

The spatial layout of textures is a powerful cue. For example, in aerial images, housing developments have a fairly characteristic texture, and the layout of this texture gives cues to the region sought. In the Netra system, built by Ma and Manjunath at U.C. Santa Barbara, textures are classified into stylized families (yielding a texture thesaurus), which are used to segment large aerial images; this approach exploits the fact that, although there is a large family of possible textures, only some texture distinctions are significant. Users can then use example regions to query a collection for similar views; for example, obtaining aerial pictures of a particular region at a different time or date to keep track of such matters as the progress of development, traffic patterns, or vegetation growth (Figure 25.4; Ma and Manjunath, 1997*b* and 1998, Manjunath and Ma, 1996*a* and 1996*b,c*).

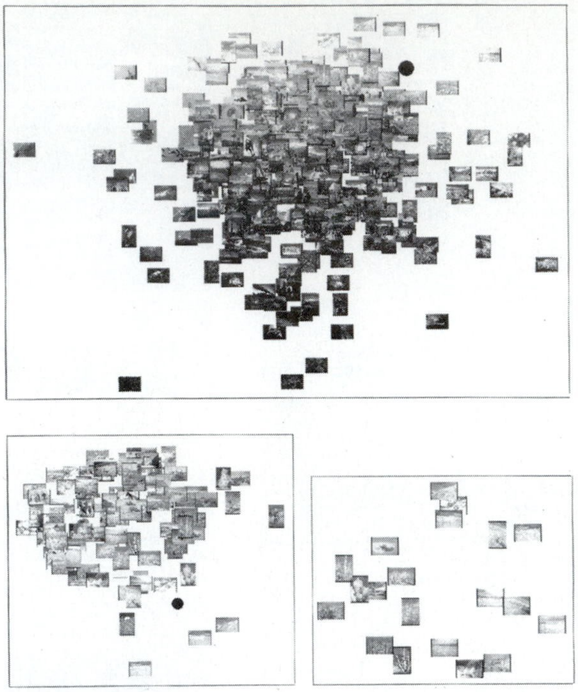

Figure 25.3 Images laid out according to their similarity using the earth mover's distance (EMD). The EMD can be computed fast so that displays like this—where distances between images on the display reflect the EMDs between them as faithfully as possible—can be created on-line. The figure on the **top** shows a large number of pictures returned from a query. This display suggests the overall collection at a glance, and a mouse click in the neighborhood of pictures that look similar to what the user is looking for tells the retrieval system where to search next (the black circle in the image on the top shows where the user clicked; this leads to the display on the **bottom right**, and in turn to that on the **bottom left**). With this technology, users browse and navigate in an image database just as they would browse through a department store. Because of the large number of images displayed, and their spatially intuitive layout, users quickly form a mental model of what is in the database and rapidly learn where to find the pictures they need. *Reprinted from "A metric for distributions with applications to image databases" by Y. Rubner, C. Tomasi, and L.J. Guibas, Proc. 1998 IEEE Int. Conf. Computer Vision, © 1998, IEEE.*

Regions of texture responses form patterns, too. For example, if an image shows a pedestrian in a spotted shirt there will be many strong responses from spot-detecting filters; the region of strong responses looks roughly like a large bar. A group of pedestrians in spotted shirts will look like a family of bars, which is itself a texture. These observations suggest applying texture finding filters to the outputs of texture finding filters—perhaps recurring several times—and using measures of similarity of these responses as a measure of image similarity. This approach—due to de Bonet and Viola (1997) and to Tieu and Viola (2000)—involves a large number of features, so it is impractical to ask users to fill in a form. One way to proceed is to have the user specify some example images that illustrate the type of pictures being sought. We then choose a small random subset of the collection to serve as negative examples (this works because a picture chosen at random is almost certainly not something the user is looking for). We now use an efficient

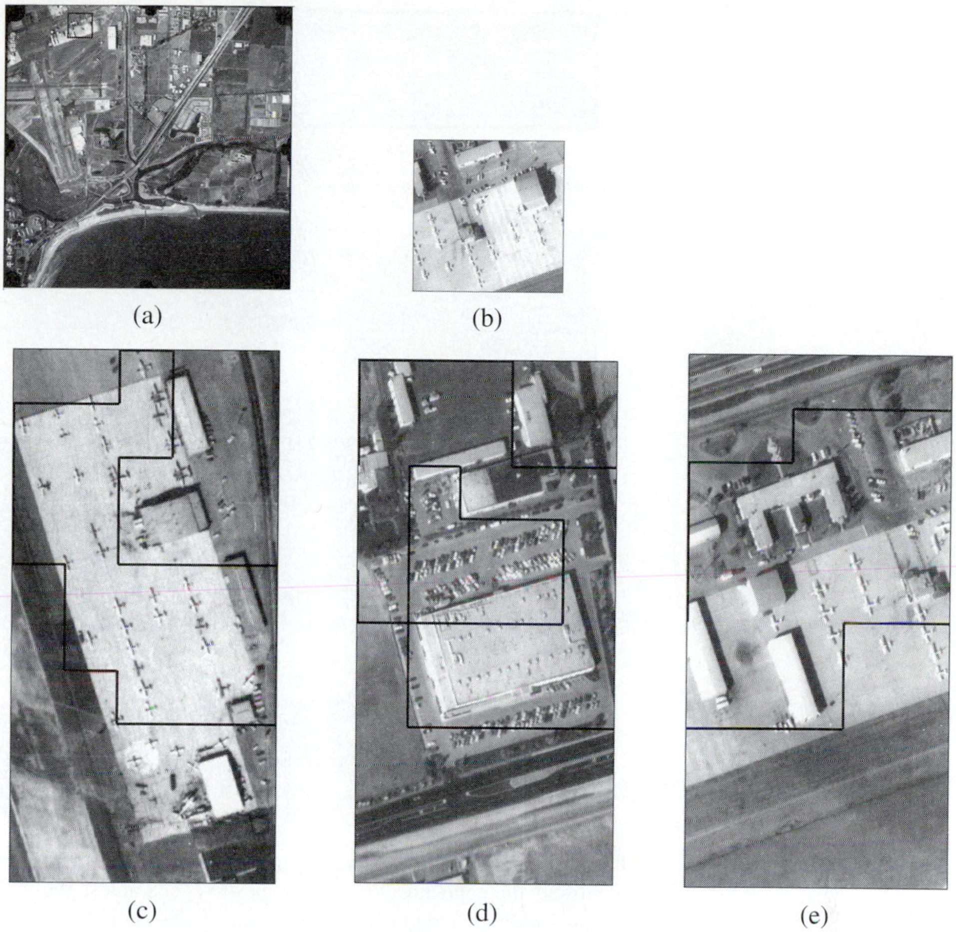

Figure 25.4 A texture-based search in an aerial image. (a) shows the down-sampled version of the aerial photograph from which the query is derived. (b) shows a full-resolution detail of the region used for the query. The region contains aircraft, cars and buildings. (c)-(e) show the ordered three best results of the query. Once again, the results come from three different aerial photographs. This time, the second and third results are from the same year (1972) as the query photograph but the first match is from a different year (1966). More details appear in (Ma and Manjunath, 1997*a*). Figure by kind permission of B.S. Manjunath.

mechanism to build a classifier (which classifies images as relevant or irrelevant) using this set of positive and negative examples (the details would take us somewhat out of our way; chapter 22 gives some more information on classifiers). Figure 25.5 gives one example of the method in use.

25.3 REPRESENTATIONS OF PARTS OF THE PICTURE

The tools described in this section try to estimate object-level semantics more or less directly. Typically, such systems first segment the image and focus on some of the image segments.

Structure in a collection is helpful in finding semantics because it can be used to guide the choice of particular search mechanisms. Photobook—due to Pentland, Picard and Sclaroff (1996)—is a system that provides three main search categories: Shape Photobook searches for

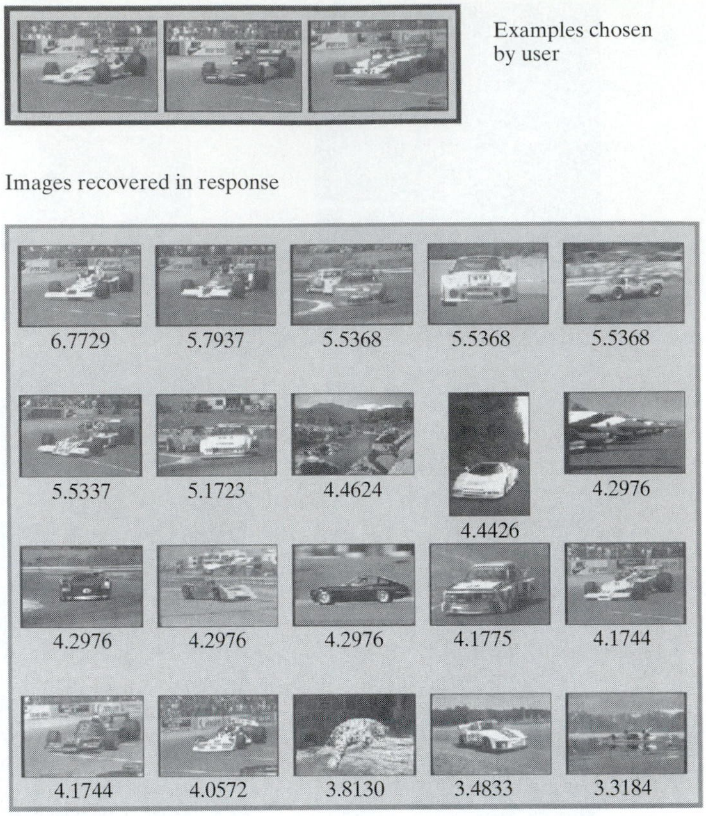

Figure 25.5 Querying using the texture of textures approach. The user has identified three pictures of cars as positive examples; these would respond strongly to large horizontal bar filters among others. The system then chooses a collection of images at random (to represent irrelevant images); this data is then used to build a classifier to identify other images that are likely to be relevant. This query results in a number of returned images containing several images of cars. *Reprinted from "Boosting image retrieval" by K. Tieu and P. Viola, Proc. 2000 IEEE Computer Vision and Pattern Recognition © 2000, IEEE.*

isolated objects (e.g., tools or fishes) using contour shape measured as elastic deformations of a contour; appearance Photobook can find faces using a small number of principal components; and texture Photobook uses a texture representation to find textured swatches of material.

25.3.1 Segmentation

Humans decompose images into pieces corresponding to the objects we are interested in, and classification is one way to achieve this *segmentation*. Segmentation is a crucial idea because it means that irrelevant information can be discarded in comparing images. For example, if we are searching for an image of a tiger, it should not matter whether the background is snow or grass; the tiger is the issue. However, if the whole image is used to generate measures of similarity, a tiger on grass is different from a tiger on snow. These observations suggest segmenting an image into regions of pixels that belong together in an appropriate sense and then allowing the user to search on the properties of particular regions. The most natural sense in which pixels belong together is that they come from a single object. Currently, it is almost never possible to use this criterion because we don't know how to tell when this is the case. However, objects usually result in image regions of coherent color and texture so that pixels that belong to the same region have a good prospect of belonging to an object.

VisualSEEK—due to Smith and Chang (1996)—automatically breaks images into regions of coherent color, and allows users to query on the spatial layout and extent of colored regions. Thus, a query for a sunset image might specify an orange background with a yellow blob lying on that background.

Blobworld is a system built by Belongie, Carson, Greenspan and Malik (1998) that represents images in terms of a collection of regions of coherent color and texture (Belongie et al., 1998). The representation is displayed to the user, with region color and texture displayed inside elliptical blobs, which represent the shape of the image regions. The shape of these regions is represented crudely because details of the region boundaries are not cogent. A user can query the system by specifying which blobs in an example image are important and what spatial relations should hold (Figure 25.6). These queries can incorporate text information, too; as Figure 25.7 and 25.8 indicate, images and text complement one another. Section 25.3.4 shows some of the uses that can be made of this complementary nature.

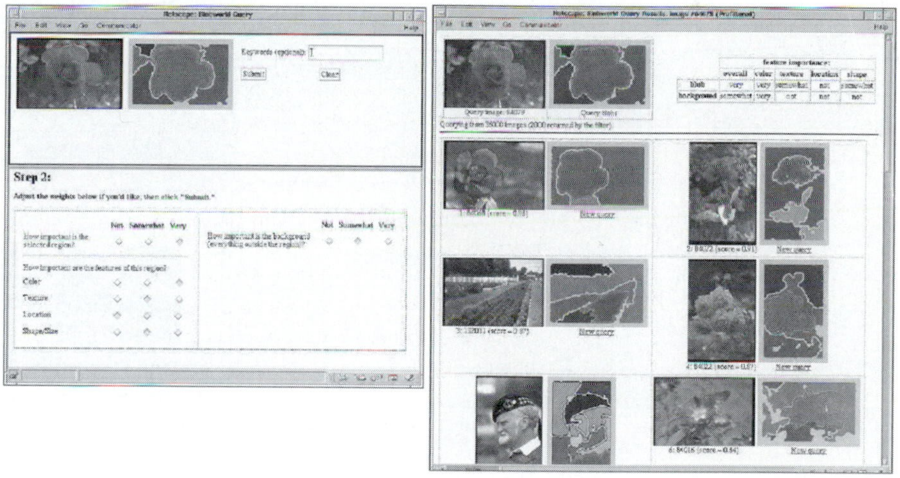

Figure 25.6 A Blobworld query for images of roses. Users of image databases generally want to find images containing particular objects, not images with particular global statistics. The Blobworld representation facilitates such queries by representing each image as a collection of regions (or blobs), which correspond to objects or parts of objects. The image is segmented into regions automatically, and each region's color, texture, and shape characteristics are encoded. The user constructs a query by selecting regions of interest (on the **left**). Blobworld recovers images and scores matches on similarity, producing the result on the **right**. The Blobworld version of each retrieved image is shown, with matching regions highlighted; displaying the system's internal representation in this way makes the query results more understandable and aids the user in creating and refining queries. Experiments show that queries for distinctive objects such as tigers and cheetahs have much higher precision using the Blobworld system than using a similar system based only on global color and texture descriptions. Blobworld is described in greater detail in (Belongie et al., 1998, Carson et al., 1999); a demonstration version can be found at `http://elib.cs.berkeley.edu`.

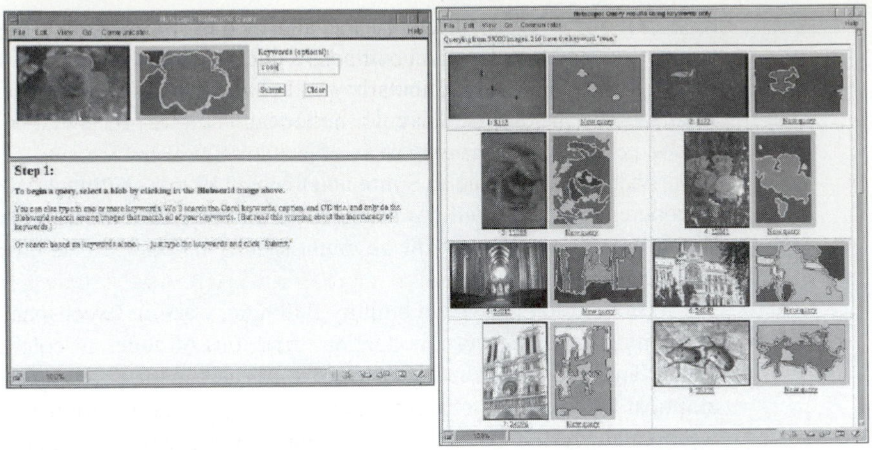

Figure 25.7 Blobworld also allows simple text queries. On the **left**, a query for images to which the word rose is attached. The figure on the **right** shows the top of the set of images recovered. Notice the presence of roseate spoonbills and rose beetles among the pictures of roses; words tend to be ambiguous.

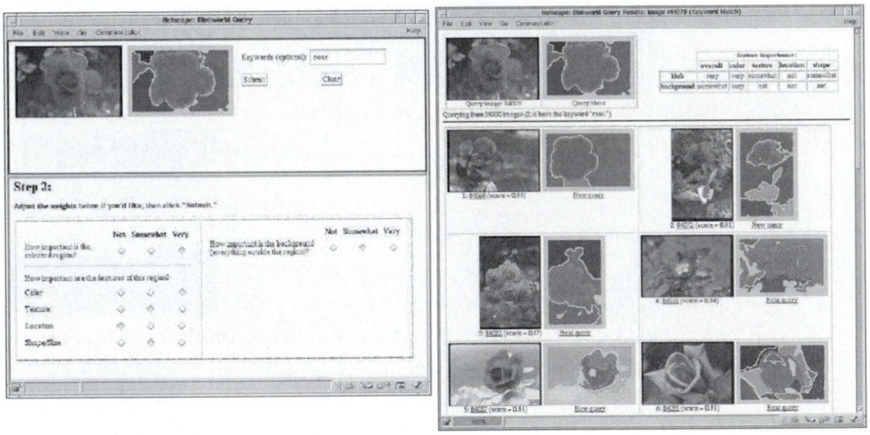

Figure 25.8 Images and words are complementary to a remarkable extent. In Figure 25.6, there are images that contained red blobs that did not happen to be roses; similarly, in Figure 25.7, the word "rose" is ambiguous. On the **left**, a query for images containing a big red blob and have the word rose is attached. The figure on the **right** shows the top of the set of images recovered. Most are pictures of flowers.

25.3.2 Template Matching

Some objects have quite characteristic appearance for a wide range of viewing directions and conditions. Template matching is an object recognition strategy that finds objects by matching image patches with example templates. We discuss these mechanisms in detail in chapter 22, with considerable emphasis on face finding (which we summarize here for convenience). A natural application of template matching is to construct whole-image templates that correspond to particular semantic categories (Figure 25.13 and Chang, Chen and Sundaram, 1998*a*). These templates can be constructed off-line, and used to simplify querying by allowing a user to use an existing template, rather than compose a query.

Face finding is a particularly good case for template matching. Frontal views of faces are extremely similar, particularly when the face is viewed at low resolution—the main features are then a dark bar at the mouth, dark blobs where the eyes are, and lighter patches at the forehead, nose, and mouth. This means that faces can be found, independent of the identity of the person, by looking for this pattern. Typical face finding systems extract small image windows of a fixed size, prune these windows to be oval, correct for lighting across the window, and then use a learned classifier to tell whether a face is present in the window (Rowley, Baluja and Kanade, 1998*a*, Rowley et al., 1998*b*, Sung and Poggio, 1998). This process works for both large and small faces because windows are extracted from images at a variety of resolutions (windows from low-resolution images yield large faces, and those from high-resolution images yield small faces). Because the pattern changes when the face is tilted to the side, this tilt must be estimated and corrected for; this is done using a mechanism learned from data (Rowley, Baluja and Kanade, 1998*c*). Knowing where the faces are is extremely useful because many natural queries refer to the people present in an image or a video.

25.3.3 Shape and Correspondence

If object appearance can vary, template matching becomes more difficult as one is forced to adopt many more templates. There is a good template matching system for finding pedestrians, which appears to work because pedestrians tend to be seen at low-resolution with their arms at their sides (Oren et al., 1997) However, building a template matching system to find people is intractable because clothing and configuration can vary too widely. The general strategy for dealing with this difficulty is to look for smaller templates—perhaps corresponding to parts—and then look for legal configurations.

One version of this technique involves finding interest points—points where combinations of measurements of intensity and its derivatives take on unusual values (e.g., at corners). As Schmid and Mohr (1997*a,b*) have shown, the spatial arrangement of these points is quite distinctive in many cases. For example (as Figure 25.9 illustrates), relations between interest points yield quite effective matching for even three-dimensional objects. This matching can be extended to obtain an image-image transformation, which can be used to register the images. Registration yields further evidence to support the match, and can be used to compare two images at specific points. For example, one might compare traffic flows at different times of day by applying this approach to aerial pictures.

This form of correspondence reasoning extends to matching image components with object parts at a more abstract level. People and many animals can be thought of as assemblies of cylinders (corresponding to body segments). A natural finding representation uses grouping stages assemble image components that could correspond to appropriate body segments or other components.

Forsyth and Fleck (1999) used this representation to identify pictures containing people wearing little or no clothing. This is an interesting example: First, it is much easier than finding clothed people because skin displays little variation in color and texture in images, whereas the appearance of clothing varies widely; second, many people are interested in avoiding or finding images based on whether they contain unclad people. This program has been tested on an usually large and diverse set of images; on a test collection of 565 images known to contain lightly clad people and 4289 control images with widely varying content, one tuning of the program marked 241 test images and 182 control images. A second example used a representation whose combinatorial structure—the order in which tests were applied—was built by hand, but where the tests were learned from data. This program identified pictures containing horses and is described

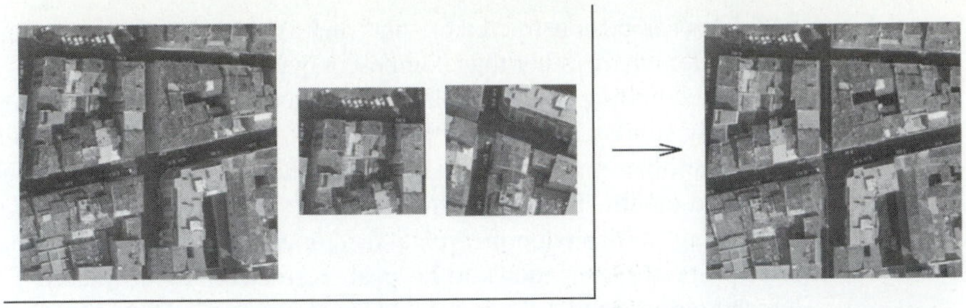

(a)

(b)

Figure 25.9 Recognition results/ (a) image matching in aerial photo interpretation: the image on the **right** is correctly retrieved using any of the images on the **left**; (b) three-dimensional object recognition: a toy dinosaur is correctly recognized in both images despite a large amount of background clutter. *Reprinted from "Local grayvalue invariants for image retrieval," by C. Schmid and R. Mohr, IEEE Trans. Pattern Analysis and Machine Intelligence, 1997 © 1997, IEEE.*

in greater detail in Forsyth and Fleck (1997). Tests used 100 images containing horses and 1,086 control images with widely varying content; for a typical configuration, the program marks 11 images of horses and 4 control images (Figure 25.10).

25.3.4 Clustering and Organizing Collections

An alternative strategy to searching for particular segments (or groups of segments with suggestive relations) is to cluster images. Typically, we would like to form clusters of images that are similar; this should include similarity in semantics as well as similarity in visual appearance.

It is a remarkable fact that, although text and images are separately ambiguous, jointly they tend not to be. The writers of text descriptions of images tend to leave out what is visually obvious (the color of flowers, etc.) and to mention properties that are difficult to infer using computer vision (the species of the flower, say). This suggests that one should attempt to form clusters of images using both image information and the text associated with the images. Using text involves us in a series of issues that are somewhat out of our remit. First, words are ambiguous ("bank" as in money, or on which the wild thyme grows?). Second, words often come in sentences or,

Figure 25.10 Images of horses recovered using a body plan representation from a test collection consisting of 100 images containing horses and 1,086 control images with widely varying content. Note that the method is relatively insensitive to aspect, but can be fooled by brown, horse-shaped regions. More details appear in (Forsyth and Fleck 1997). *Reprinted from "Body Plans," by D.A. Forsyth and M.M. Fleck, Proc. Computer Vision and Pattern Recognition, 1997, © 1997, IEEE.*

worse, paragraphs, and we need to decide which words to ignore and which to use. Third, we need to know how to manage association between words and picture elements in models.

Barnard and colleagues have scratched the surface of this potentially interesting topic (Barnard and Forsyth, 2001, Barnard, Duygulu and Forsyth, 2001). They have demonstrated their work on two collections: a set of images released by the Corel corporation, each of which comes with a small set of keywords, and a collection of pictures of art belonging to the Fine Arts Museum of San Francisco, each of which comes with a free text annotation written by volunteers (who did not have computerized analysis in mind when they wrote the annotations). The free text is reduced to a collection of nouns, verbs, adjectives, and adverbs by a part-of-speech

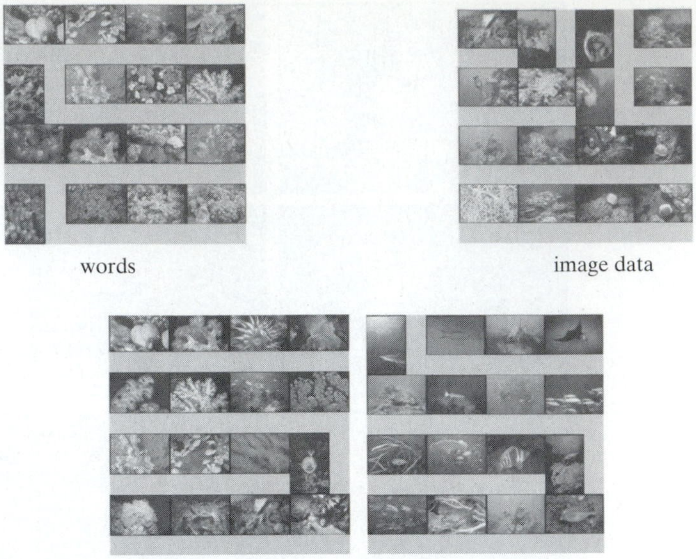

Figure 25.11 The **top right** figure shows a collection of images from a cluster, where the images were clustered using words alone. You should notice a general ocean theme, but the images look quite different; there are divers on a background of blue sea, coral, and so on. On the **top left**, a collection of images from a cluster, where images were clustered using image segment features alone. Now the images look similar, but are not semantically coherent; some are pictures of coral, and others are pictures of flowers. On the **bottom**, two clusters obtained using both text and image segment features. Generally, the pictures share a theme *and* look similar; these are both desirable properties. *Reprinted from "Learning the semantics of words and pictures," by Proc. IEEE Int. Conf. Computer Vision, 2001, © 2001, IEEE.*

tagger (Brill, 1992). A sense is selected for words using a voting strategy that compares possible senses for each word with possible senses for nearby words on the assumption that nearby words have similar senses. Finally, the images and words are clustered by fitting a generative model.

The generative model is, in essence, a mixture model. Each component of the mixture emits words and blobs—image regions, whose features encode color, texture, and shape rather roughly—with probabilities that are conditionally independent given the component and vary from component to component. This model can be fitted to the image data using an EM algorithm; once it is fitted, an image belongs to the mixture component that is most likely to have produced it. There is no reason to believe that this is the best way to proceed, but it does produce rather good clusters. Figure 25.11 shows some clusters obtained for the corel dataset comparing clustering based on image data with clustering based on text data and clustering based on both cues.

Good clusters can be used in a number of ways. Clearly, they should yield a browsing mechanism, although that demonstrated by Barnard *et al.* is primitive. We have more, however; by fitting a generative model, we have constructed a joint probability distribution for image features and words. This means that we can (a) search for pictures using words (which Barnard *et al.* call "auto-illustrate") and (b) search for words using pictures ("auto-annotate"—but you should notice that this is rather like object recognition). Either can be startlingly successful with appropriate clusters. Figure 25.12 shows an auto illustration result.

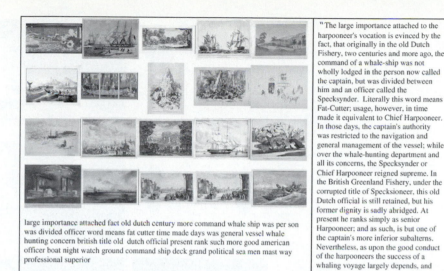

Figure 25.12 On the **left**, a block of text from Moby Dick. This text is processed to obtain nouns, verbs, adjectives and adverbs and the terms are disambiguated by a voting process. The resulting text is used as a query to Barnard *et al.*'s joint probability model, where the search returns images that have high joint probability with the collection of words. On the **right**, the images returned by this query. The query appears to be successful (among other things, there's a picture of a whaleboat with sailors in it harpooning a whale). *Reprinted from "Clustering Art," by K. Barnard, P. Duygulu and D.A. Forsyth, Proc. Computer Vision and Pattern Recognition, 2001, © 2001, IEEE.*

25.4 VIDEO

Although video represents a richer source of information than still images, the issues remain largely the same. Videos are typically segmented into *shots*—short sequences that contain similar content—and techniques of the form described applied within shots. We described shot boundary detection briefly in Section 14.3.2.

The motion of individual pixels in a video is often called *optic flow* and is measured by attempting to find pixels in the next frame that correspond to a pixel in this (correspondence being measured by similarity in color, intensity, and texture). In principle, there is an optic flow vector at each pixel, forming a *motion field*. In practice, it is extremely hard to measure optic flow reliably at featureless pixels because they could correspond to pretty much anything. For example, consider the optic flow of an egg rotating on its axis; there is little information about what the pixels inside the boundary of the egg are doing because each looks like the other.

Motion fields can be extremely complex; however, particularly if there are no moving objects in the frame, it is possible to classify motion fields corresponding to the camera shot used. For example, a pan shot leads to strong lateral motion, and a zoom leads to a radial motion field. This classification is usually obtained by comparing the measured motion field with a parametric family (e.g., Sawhney and Ayer, 1996, Smith and Kanade, 1997).

Complex motion sequences are difficult to query without segmentation, because much of the motion may be irrelevant to the query. For example, in a soccer match, the motion of many players may not be significant. In Chang's system VideoQ, motion sequences are segmented moving blobs and then queried on the color and motion of a particular blob (Figure 25.13

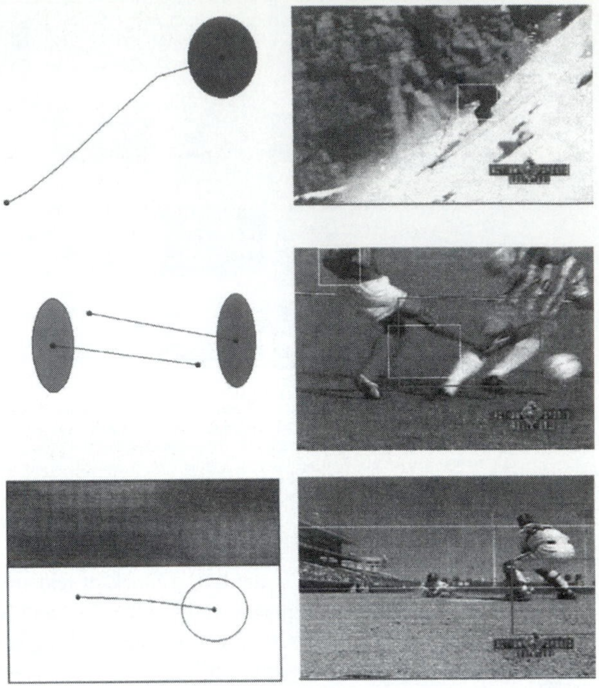

Figure 25.13 Video can be represented by moving blobs; sequences can then be queried by specifying blob properties and motion properties desired. The **left column** shows queries for various types of moving blob sketched in the user interface for Chang's VideoQ system. The **right column** shows frames from two sequences returned. *Reprinted from "A Fully Automated Content-Based Video Search Engine Supporting Spatiotemporal Queries," by S-F. Chang et al., IEEE Transactions on Circuits and Systems for Video Technology, 1998, © 1998, IEEE.*

and Chang, Chen, Meng, Sundaram and Zhong, 1997*a*, Chang, Chen, Meng, Sundaram and Zhong, 1998*b*).

The Informedia project has studied preparing detailed skims of video sequences. In this case, a segment of video is broken into shots, shots are annotated with the camera motion in shot, with the presence of faces, with the presence of text in shot, with keywords from the transcript and with audio level (Figure 25.14). This information yields a compact representation—the "skim"—which gives the main content of the video sequence (details in Smith and Kanade, 1997, Wactlar, Kanade, Smith and Stevens, 1996, Smith and Christel, 1995, and Smith and Hauptmann, 1995).

25.5 NOTES

This area is easily the most interesting application of computer vision. It is quite an old area (important early papers include Chang and Yang, 1983, Kato, Shimogaki, Mizutori and Fujimura, 1988, Kofakis and Orphanoudakis, 1991), but has undergone a recent period of popularity, probably because it is now possible to solve some problems that were simply unmanageable 30 years ago (for this problem, having more disk, faster computers, etc. can make a huge difference). We have tried to give pointers into the (huge) literature in the text, but are conscious that the relatively large interest in the area means that things go out of date quickly.

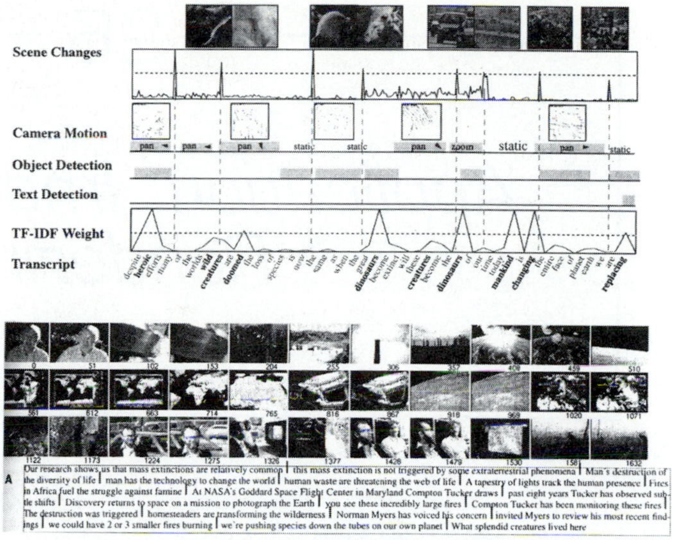

Figure 25.14 On the **top**, characterizing a video sequence to create a skim. The video is segmented into scenes. Camera motions are detected along with significant objects (faces and text). Bars indicate frames with positive results. Word relevance is evaluated in the transcript by comparing term frequency with inverse document frequency (essentially, favouring terms that are common in a document, but uncommon overall). All this data is used to extract a shortened set of frames and words, given on the **bottom**, which represents the original video sequence, but is substantially shorter. By using text, motion, face, and shot information, it is possible to obtain skims that are a significant improvement on (say) subsampling frames because long sequences with relatively little new in them are more aggressively compressed. *Reprinted from "Video skimming and characterization through the combination of image and language understanding techniques," by M. Smith and T. Kanade, Proc. Computer Vision and Pattern Recognition, 1997 © 1997 IEEE.*

For applications where the colors, textures and layout of the image are all strongly correlated with the kind of content desired, a number of usable tools exist to find images based on content. Because color, texture and layout are, at best, a rough guide to image content, puzzling search results are pretty much guaranteed. There is not yet a clear theory of how to build interfaces that minimize the impact of this effect. The most widely adopted strategy is to allow quite fluid browsing.

Successful performance at serious applications requires techniques that get some notion of semantics from an image. In particular, to build tools that work well, we need to engage with deep and poorly understood problems in object recognition. As we have seen, object recognition seems to require segmenting images into coherent pieces and reasoning about both the pieces (as in face finding) and the relationships between those pieces. This rather vague view of recognition can be exploited to produce segmented representations that allow searches for objects independent of their backgrounds. Furthermore, some special cases of object recognition can be handled explicitly. It is not known how to build a system that could search for a wide variety of objects; building a user interface for such a system would present substantial problems, too.

26

Application: Image-Based Rendering

The entertainment industry touches hundreds of millions of people every day, and synthetic pictures of real scenes, often mixed with actual film footage, are now common place in computer games, sports broadcasting, TV advertising, and feature films. Creating these images is what *image-based rendering*—defined here as the synthesis of new views of a scene from prerecorded pictures—is all about, and it does require the recovery of quantitative (although not necessarily three-dimensional) shape information from images. This chapter presents a number of representative approaches to image-based rendering, dividing them, rather arbitrarily, into (a) techniques that first recover a three-dimensional scene model from a sequence of pictures, then render it with classical computer graphics tools (naturally, these approaches are often related to stereo and motion analysis); (b) methods that do not attempt to recover the camera or scene parameters, but construct instead an explicit representation of the set of all possible pictures of the observed scene, then use the image position of a small number of tie points to specify a new view of the scene and *transfer* all the other points into the new image, in the photogrammetric sense already mentioned in chapter 10; and (c) approaches that model images by a two-dimensional set of light rays (or more precisely by the value of the radiance along these rays) and the set of all pictures of a scene by a four-dimensional set of rays, the *light field* (Figure 26.1).

26.1 CONSTRUCTING 3D MODELS FROM IMAGE SEQUENCES

This section addresses the problem of building and rendering a three-dimensional object model from a sequence of pictures. It is, of course, possible to construct such a model by fusing registered depth maps acquired by range scanners as described in chapter 21, but we focus here on the case where the input images are digitized photographs or film clips of a rigid or dynamic scene.

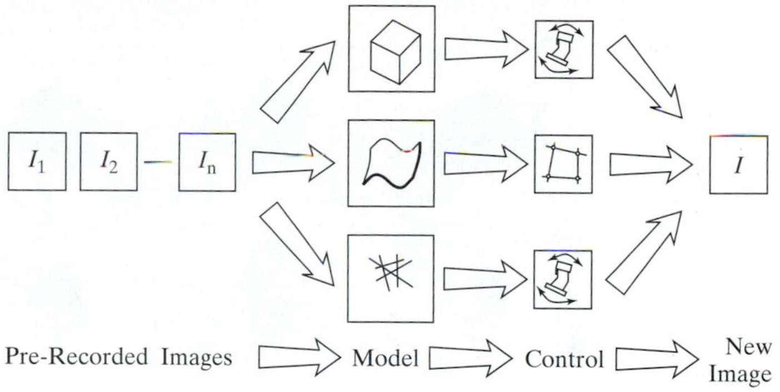

Figure 26.1 Approaches to image-based rendering. From top to bottom: three-dimensional model construction from image sequences, transfer-based image synthesis, the light field. From left to right, the image-based rendering pipeline: A scene model (that may not be three-dimensional) is constructed from sample images, and used to render new images of the scene. The rendering engine may be controlled by a joystick (or equivalently by the specification of camera parameters) or, in the case of transfer-based techniques, by setting the image position of a small number of tie points.

26.1.1 Scene Modeling from Registered Images

Volumetric Reconstruction Let us assume that an object has been delineated (perhaps interactively) in a collection of photographs registered in the same global coordinate system. It is impossible to uniquely recover the object shape from the image contours since, as observed in chapter 19, the concave portions of its surface never show up on the image contour. Still, we should be able to construct a reasonable approximation of the surface from a large enough set of pictures. There are two main global constraints imposed on a solid shape by its image contours: (a) it lies in the volume defined by the intersection of the viewing cones attached to each image, and (b) the cones are tangent to its surface (there are other local constraints; e.g., as shown in chapter 19, convex [resp. concave] parts of the contour are the projections of convex [resp. saddle-shaped] parts of the surface). Baumgart exploited the first of these constraints in his 1974 PhD thesis to construct polyhedral models of various objects by intersecting the polyhedral cones associated with polygonal approximations of their silhouettes. His ideas have inspired a number of approaches to object modeling from silhouettes, including the technique presented in the rest of this section (Sullivan and Ponce, 1998) that also incorporates the tangency constraint associated with the viewing cones. As in Baumgart's system, a polyhedral approximation of the observed object is first constructed by intersecting the visual cones associated with a few photographs (Figure 26.2). The vertices of this polyhedron are then used as the control points of a smooth *spline surface*, which is deformed until it is tangent to the visual rays. We focus here on the construction and deformation of this surface.

Spline Construction. A *spline curve* is a piecewise polynomial parametric curve that satisfies certain smoothness conditions. For example, it may be C^k (i.e., differentiable with continuous derivatives of order up to k), with k usually taken to be 1 or 2, or G^k (i.e., not necessarily differentiable everywhere, but with continuous tangents in the G^1 case and continuous curvatures in the G^2 case). Spline curves are usually constructed by stitching together *Bézier arcs*. A Bézier curve of degree n is a polynomial parametric curve $P : [0, 1] \rightarrow \mathbb{E}^3$ defined as the barycentric

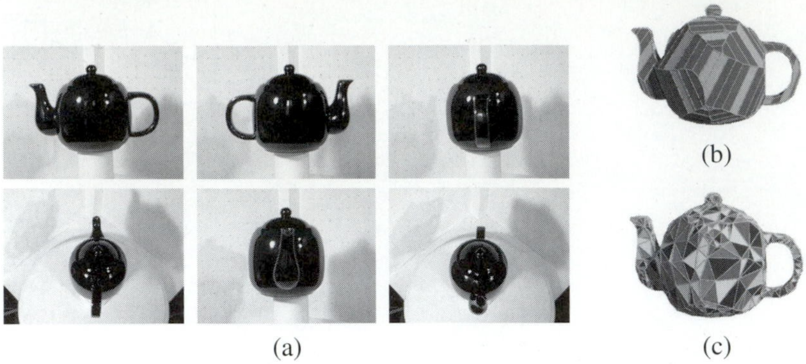

Figure 26.2 Constructing object models by intersecting (polyhedral) viewing cones: (a) six photographs of a teapot, (b) the raw intersection of the corresponding viewing cones, (c) the triangulation obtained by splitting each face into triangles and simplifying the resulting mesh. *Reprinted from "Automatic Model Construction, Pose Estimation, and Object Recognition from Photographs Using Triangular Splines," by S. Sullivan and J. Ponce, IEEE Transactions on Pattern Analysis and Machine Intelligence, 20(10):1091–1096, (1998). © 1998 IEEE.*

combination

$$P(t) = \sum_{i=0}^{n} b_i^{(n)}(t) P_i$$

of $n + 1$ *control points* $P_0, \dots, P_n$, where the weights $b_i^{(n)}(t) \stackrel{\text{def}}{=} \binom{n}{i} t^i (1-t)^{n-i}$ are called the *Bernstein polynomials* of degree n.[1] A Bézier curve interpolates its first and last control points, but not the other ones (Figure 26.3a). As shown in the exercises, the tangents at its endpoints are along the first and last line segments of the *control polygon* formed by the control points.

The definition of Bézier arcs and spline curves naturally extends to surfaces: A triangular Bézier patch of degree n is a parametric surface $P : [0, 1] \times [0, 1] \to \mathbb{E}^3$ defined as the barycentric combination

$$P(u, v) = \sum_{i+j+k=n} b_{ijk}^{(n)}(u, v, 1-u-v) P_{ijk}$$

of a triangular array of control points P_{ijk}, where the homogeneous polynomials $b_{ijk}^{(n)}(u, v, w) \stackrel{\text{def}}{=} \frac{n!}{i!j!k!} u^i v^j w^k$ are the *trivariate Bernstein polynomials* of degree n. In the rest of this section, we use *quartic* Bézier patches ($n = 4$), each defined by 15 control points (Figure 26.3b). Their boundaries are the quartic Bézier curves $P(u, 0)$, $P(0, v)$, and $P(u, 1-u)$. By definition, a G^1 *triangular spline* is a network of triangular Bézier patches that share the same tangent plane along their common boundaries. A necessary (but not sufficient) condition for G^1 continuity is that all control points surrounding a common vertex be coplanar. We first construct these points, then place the remaining control points to ensure that the resulting spline is indeed G^1 continuous. As discussed in Loop (1994), a set of coplanar points $Q_1, \dots, Q_p$ can be created as a barycentric

[1]This is indeed a barycentric combination (as defined in chapter 12) since the Bernstein polynomials are easily shown to always add to 1. In particular, Bézier curves are affine constructs—a desirable property since it allows the definition of these curves purely in terms of their control points and independently of the choice of any external coordinate system.

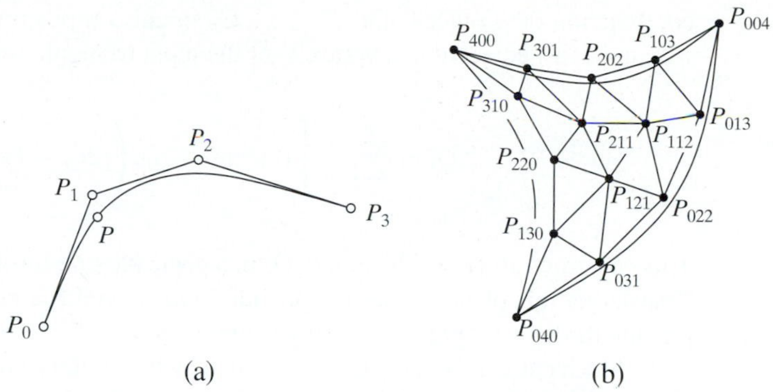

Figure 26.3 Bézier curves and patches: (a) a cubic Bézier curve and its control polygon; (b) a quartic triangular Bézier patch and its control mesh. *Tensor-product* Bézier patches can also be defined using a rectangular array of control points (Farin 1993). Triangular patches are, however, more appropriate for modeling free-form *closed* surfaces.

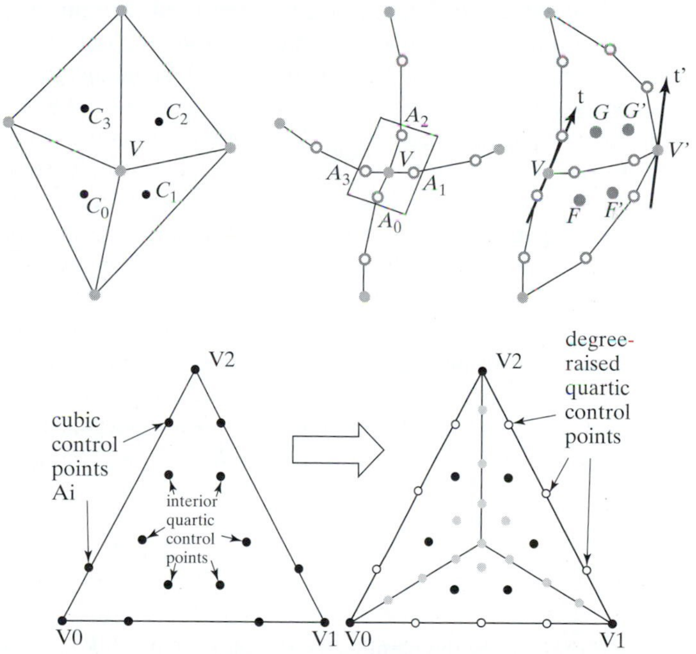

Figure 26.4 Construction of a triangular spline over a triangular polyhedral mesh. Top, from left to right: The cubic boundary control points, the boundary curves surrounding a mesh vertex, and the construction of internal control points from tangent specification. Bottom: Splitting a patch three ways to enforce G^1 continuity: The white points are the control points obtained by raising the degree of the control curves, and the gray points are the remaining control points, computed to ensure G^1 continuity. After Sullivan and Ponce (1998).

combination of p other points $C_1, \ldots, C_p$ in general position (in our case, the centroids of the p triangles T_j adjacent to a vertex V of the input triangulation, Figure 26.4, top left) as

$$Q_i = \sum_{j=1}^{p} \frac{1}{p} \left\{ 1 + \cos\frac{\pi}{p} \cos\left([2(j-i)-1]\frac{\pi}{p} \right) \right\} C_j.$$

This construction places the points Q_i in a plane passing through the centroid O of the points C_i. Translating this plane so that O coincides with V yields a new set of points A_i lying in a plane passing through V (Figure 26.4, top center).

Since cubic Bézier curves are defined by four points, we can interpret two adjacent vertices V and V' and the points A_i and A'_i associated with the corresponding edge as the control points of a cubic curve. This yields a set of cubic arcs that interpolate the vertices of the control mesh and form the boundaries of triangular patches. Once these curves have been constructed, the control points on both sides of a boundary can be chosen to satisfy interpatch G^1 continuity. In this construction, the cross-boundary tangent field linearly interpolates the tangents at the two endpoints of the boundary curve. At the endpoint V, the tangent $\boldsymbol{t}$ across the curve that contains the point A_i is taken to be parallel to the line joining A_{i-1} to A_{i+1}. The tangent $\boldsymbol{t}'$ is obtained by a similar construction. The interior control points F, F', G, and G' (Figure 26.4, top right) are constructed by solving the set of linear equations associated with this geometric condition (Chiyokura, 1983). However, there are not enough degrees of freedom in a quartic patch to allow the simultaneous setting of the interior points for all three boundaries. Thus, each patch must be split three ways, using, for example, the method of Shirman and Sequin (1987) to ensure continuity among the new patches: Performing *degree elevation* on the boundary curves replaces them by quartic Bézier curves with the same shape (see Exercises). Three quartic triangular patches can then be constructed from the boundaries as shown in Figure 26.4, bottom. The result is a set of three quartic patches for each mesh face, which are G^1 continuous across all boundaries.

Spline Deformation. We have given a method for constructing a G^1-continuous triangular spline approximation of a surface from a triangulation such as the one shown in Figure 26.2(b). Let us now show how to deform this spline to ensure that it is tangent to the viewing cones associated with the input photographs. The shape of the spline surface S is determined by the position of its control vertices $V_1, \ldots, V_p$. We denote by V_{jk} ($k = 1, 2, 3$) the coordinates of the point V_j ($j = 1, \ldots, p$) in some reference Euclidean coordinate system, and use these $3p$ coefficients as shape parameters. Given a set of rays $R_1, \ldots, R_q$, we minimize the energy function

$$\frac{1}{q}\sum_{i=1}^{q} d^2(R_i, S) + \lambda \sum_{i=1}^{r} \iint \left[|P_{uu}|^2 + 2|P_{uv}|^2 + |P_{vv}|^2 \right] du\, dv$$

with respect to the parameters V_{jk} of S. Here, $d(R, S)$ denotes the distance between the ray R and the surface S, the integral is a *thin-plate* spline energy term used to enforce smoothness in areas of sparse data, and λ is a constant weight introduced to balance the distance and smoothness terms. The variables u and v in this integral are the patch parameters, and the summation is done over the r patches that form the spline surface. The signed distance between a ray and a surface patch can be computed using Newton's method. For rays that do not intersect the surface, we define $d(R, S) = \min\{|\overrightarrow{QP}|, Q \in R, P \in S\}$, and compute the distance by minimizing $|\overrightarrow{QP}|^2$. For those rays that intersect the surface, we follow Brunie, Lavallée, and Szeliski (1992) and measure the distance to the *farthest* point from the ray that lies on the surface in the direction of the surface normal at the corresponding occluding contour point. In both cases, Newton iterations are initialized from a sampling of the surface S. During surface fitting, the spline is deformed

Figure 26.5 Shaded and texture-mapped models of a teapot, gargoyle and dinosaur. The teapot was constructed from six registered photographs; the gargoyle and dinosaur models were each built from nine images. *Reprinted from "Automatic Model Construction, Pose Estimation, and Object Recognition from Photographs Using Triangular Splines," by S. Sullivan and J. Ponce, IEEE Transactions on Pattern Analysis and Machine Intelligence, 20(10):1091–1096, (1998). © 1998 IEEE.*

to minimize the mean-squared ray-surface distance using a simple gradient descent technique. Although each distance is computed numerically, its derivatives with respect to the surface parameters V_{jk} are easily computed by differentiating the constraints satisfied by the surface and ray points where the distance is reached.

The three object models shown in Figure 26.5 have been constructed using the method described in this section. This technique does not require establishing any correspondence across the input pictures, but its scope is (currently) limited to static scenes. In contrast, the approach presented next is based on multicamera stereopsis, and, as such, requires correspondences, but it handles dynamic scenes as well as static ones.

Virtualized Reality Kanade and his colleagues (1997) have proposed the concept of *Virtualized Reality* as a new visual medium for manipulating and rendering prerecorded and synthetic images of real scenes captured in a controlled environment. The first physical implementation of this concept at Carnegie-Mellon University consisted of a geodesic dome equipped with 10 synchronized video cameras hooked to consumer-grade VCRs. As of this writing, the latest implementation is a "3D Room", where a volume of $20 \times 20 \times 9$ cubic feet is observed by 49 color cameras connected to a PC cluster and registered in the same world coordinate system, with the capability of digitizing in real-time the synchronized video streams of all cameras. Three-dimensional scene models are acquired by fusing dense depth maps acquired via multiple-camera stereo (see Okutami and Kanade, 1993, chapter 11). One such map is acquired by each camera and a small number of its neighbors (between three and six). Every range image is then

Figure 26.6 Multicamera stereo. From left to right: the range map associated with a cluster of cameras; a texture-mapped image of the corresponding mesh, observed from a different viewpoint; note the dark areas associated with depth discontinuities in the map; a texture-mapped image constructed from two adjacent camera clusters; note that the gaps have been filled. *Reprinted from "Virtualized Reality: Constructing Virtual Worlds From Real Scenes," by T. Kanade, P.W. Rander and J.P. Narayanan, IEEE Multimedia, 4(1):34–47, (1997). © 1997 IEEE.*

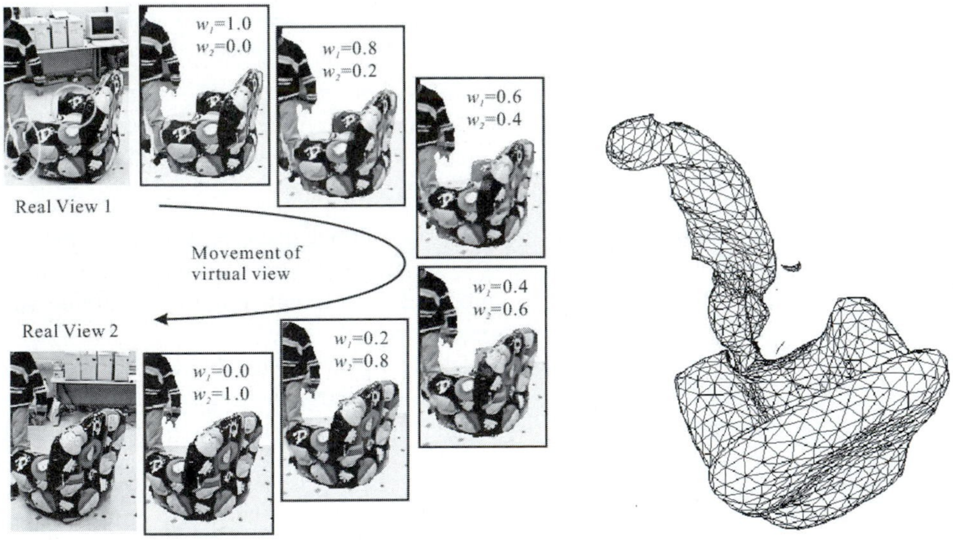

(a) (b)

Figure 26.7 Virtualized Reality: (a) a sequence of synthetic images; note that occlusion in the two elliptical regions of the first view is handled correctly; (b) the corresponding mesh model. *Reprinted from "Appearance-Based Virtual View Generation of Temporally-Varying Events from Multi-Camera Images in the 3D Room," by H. Saito, S. Baba, M. Kimura, S. Vedula, and T. Kanade, Tech. Rep. CMU-CS-99-127, School of Computer Science, Carnegie-Mellon University, (1999).*

converted to a surface mesh that can be rendered using classical computer graphics techniques such as texture mapping. As shown by Figure 26.6, images of a scene constructed from a single depth map may exhibit gaps. These gaps can be filled by rendering in the same image the meshes corresponding to adjacent cameras.

It is also possible to directly merge the surface meshes associated with different cameras into a single surface model. This task is challenging since: (a) multiple, conflicting measurements of the same surface patches are available in areas where the fields of view of several cameras overlap, and (b) certain scene patches are not observed by any camera. Both problems can be solved using the volumetric technique for range image fusion proposed by Curless and Levoy (1996) and introduced in chapter 21. Once a global surface model has been constructed, it can of course be texture mapped as before. Synthetic animations can also be obtained by interpolating two arbitrary views in the input sequence. First, the surface model is used to establish correspondences between these two views: The optical ray passing through any point in the first image is intersected with the mesh and the intersection point is reprojected in the second image, yielding the desired match.[2] Once the correspondences are known, new views are constructed by linearly interpolating both the positions and colors of matching points. As discussed in Saito et al. (1999), this simple algorithm only provides an approximation of true perspective imaging, and additional logic has to be added in practice to handle points that are visible in the first image but not in the second one. Nevertheless, it can be used to generate realistic animations of dynamic scenes with changing occlusion patterns, as demonstrated by Figure 26.7.

26.1.2 Scene Modeling from Unregistered Images

This section addresses again the problem of acquiring and rendering three-dimensional object models from a set of images, but this time the positions of the cameras observing the scene are not known a priori and must be recovered from image information using methods related to those presented in chapters 12 and 13. The techniques presented in this section are, however, explicitly geared toward computer graphics applications.

The Façade System The *Façade* system for modeling and rendering architectural scenes from digitized photographs was developed at UC Berkeley by Debevec, Taylor, and Malik (1996). This system takes advantage of the relatively simple overall geometry of many buildings to simplify the estimation of scene structure and camera motion, and it uses the simple but powerful idea of *model-based stereopsis*, to be described in a minute, to add detail to rough building outlines. Figure 26.8 shows an example.

Façade models are constrained hierarchies of parametric primitives such as boxes, prisms, and solids of revolution. These primitives are defined by a small number of coefficients (e.g., the height, width, and breadth of a box) and related to each other by rigid transformations. Any of the parameters defining a model is either a constant or variable, and constraints can be specified between the various unknowns (e.g., two blocks may be constrained to have the same height). Model hierarchies are defined interactively with a graphical user interface, and the main computational task of the Façade system is to use image information to assign definite values to the unknown model parameters. The overall system is divided into three main components: The first one, or *photogrammetric module*, recasts structure and motion estimation as the nonlinear optimization problem of minimizing the discrepency between line segments selected by hand in the photographs and the projections of the corresponding parts of the parametric model (see Exercises for details). As shown in Debevec et al. (1996), this process involves relatively few

[2]Classical narrow-baseline methods like correlation would be ineffective in this context since the two views may be far from each other. A similar method is used in the Façade system described later in this chapter to establish correspondences between widely separated images when the rough shape of the observed surface is known.

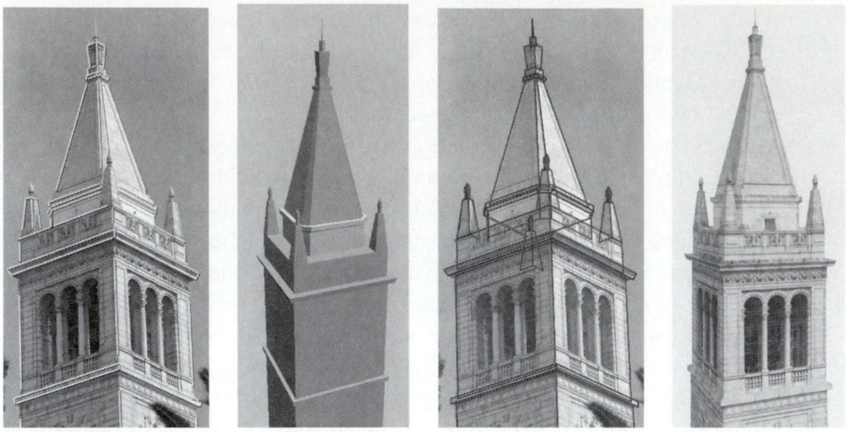

Figure 26.8 Façade model of the Berkeley Campanile. From left to right: A photograph of the Campanile, with selected edges overlaid; the 3D model recovered by photogrammetric modeling; reprojection of the model into the photograph; a texture-mapped view of the model. *Reprinted from "Modeling and Rendering Architecture from Photographs: A Hybrid Geometry- and Image-Based Approach," by P. Debevec, C.J. Taylor, and J. Malik, Proc. SIGGRAPH, (1996). © 1996 ACM, Inc. Included here by permission.*

variables, namely the positions and orientations of the cameras used to photograph a building and the parameters of the building model, and when the orientation of some of the model edges is fixed relative to the world coordinate system, an initial estimate for these parameters is easily found using linear least squares.

The second main component of Façade is the *view-dependent texture-mapping module* that renders an architectural scene by mapping different photographs onto its geometric model according to the user's viewpoint. Conceptually, the cameras are replaced by slide projectors that project the original images onto the model. Of course, each camera only sees a portion of a building, and several photographs must be used to render a complete model. In general, parts of a building are observed by several cameras, so the renderer must not only pick, but also appropriately merge, the pictures relevant to the synthesis of a virtual view. The solution adopted in Façade is to assign to each pixel in a new image a weighted average of the values predicted from the overlapping input pictures, with weights inversely proportional to the angle between the corresponding light rays in the input and virtual views.

The last component of Façade is the *model-based stereopsis module*, which uses stereo pairs to add fine geometric detail to the relatively rough scene description constructed by the photogrammetric modeling module. The main difficulty in using stereo vision in this setting is the wide separation of the cameras, which prevents the straightforward use of correlation-based matching techniques. The solution adopted in Façade is to exploit a priori shape information to map the stereo images into the same reference frame (Figure 26.9, top). Specifically, given *key* and *offset* pictures, the offset image can be projected onto the scene model before being rendered from the key camera's viewpoint, yielding a *warped offset* picture similar to the key image (Figure 26.9, bottom). In turn, this allows the use of correlation to establish correspondences between these two images, and thus between the key and offset images as well. Once the matches between these two pictures have been established, stereo reconstruction reduces to the usual triangulation process.

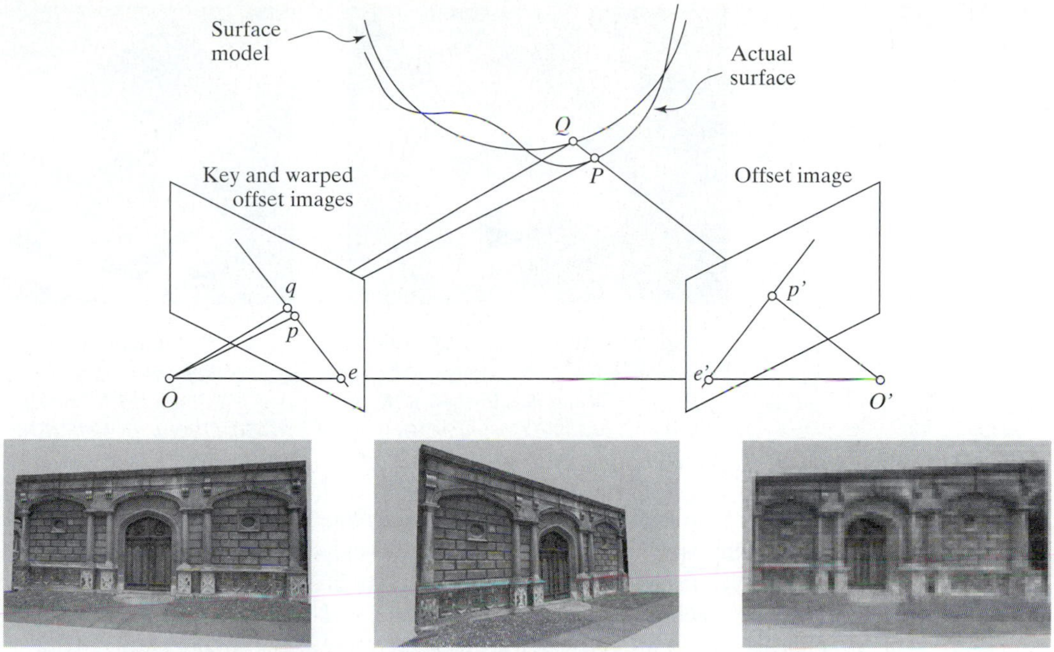

Figure 26.9 Model-based stereopsis. Top: Synthesis of a warped offset image. The point p' in the offset image is mapped onto the point Q of the surface model, then reprojected onto the point q of the warped offset image. The actual surface point P observed by both cameras projects onto the point p of the key image. Note that the point q must lie on the epipolar line ep, which facilitates the search for matches as in the conventional stereo case. Note also that the disparity between p and q along the epipolar line measures the discrepancy between the modeled and actual surfaces. After Debevec et al. (1996, Figure 15). Bottom, from left to right: A key image, an offset image, and the corresponding warped offset image. *Reprinted from "Modeling and Rendering Architecture from Photographs: A Hybrid Geometry- and Image-Based Approach," by P. Debevec, C.J. Taylor, and J. Malik, Proc. SIGGRAPH, (1996). © 1996 ACM, Inc. Included here by permission.*

26.2 TRANSFER-BASED APPROACHES TO IMAGE-BASED RENDERING

This section explores a completely different approach to image-based rendering. In this framework, an explicit three-dimensional scene reconstruction is never performed. Instead, new images are created directly from a (possibly small) set of views among which point correspondences have been established by feature tracking or conventional stereo matching. This approach is related to the classical transfer problem from photogrammetry already mentioned in chapter 10. Given the image positions of a number of *tie points* in a set of reference images and in a new image, and given the image positions of a ground point in the reference images, predict the position of that point in the new image.

Transfer-based techniques for image-based rendering were introduced in the projective setting by Laveau and Faugeras (1994), who proposed to first estimate the pairwise epipolar geometry between reference views, then reproject the scene points into a virtual image, specified by the projections of the new optical center in two reference pictures (i.e., the epipoles) and the position of four tie points in the new view. By definition, the epipolar geometry constrains the

Figure 26.10 Augmented reality experiment. The (affine) world coordinate system is defined by corners of the black polygons. *Reprinted from "Calibration-Free Augmented Reality," by K. Kutulakos and J. Vallino, IEEE Transactions on Visualization and Computer Graphics, 4(1):1–20, (1998). © 1998 IEEE.*

possible reprojections of points in the reference images. In the new view, the projection of the scene point is at the intersection of the two epipolar lines associated with the point and two reference pictures. Once the feature points have been reprojected, realistic pictures are synthesized using ray tracing and texture mapping. As noted by Laveau and Faugeras, however, since the Euclidean constraints associated with calibrated cameras are not enforced, the rendered images are separated from correct pictures by arbitrary planar projective transformations unless additional scene constraints are taken into account. The rest of this section explores two affine variants of the transfer-based approach that circumvent this difficulty. Both techniques construct a parameterization of the set of all images of a rigid scene: In the first case (Section 26.2.1), the affine structure of the space of affine images is used to render synthetic objects in an augmented reality system. Because the tie points in this case are always geometrically valid image features (e.g., the corners of calibration polygons; see Figure 26.10), the synthesized images are automatically Euclidean ones. In the second instance (Section 26.2.2), the metric constraints associated with calibrated cameras are explicitly taken into account in the image space parameterization, guaranteeing once again the synthesis of correct Euclidean images.

Let us note again a particularity of transfer-based approaches to image-based rendering already mentioned in the introduction: Because no three-dimensional model is ever constructed, a joystick cannot be used to control the synthesis of an animation. Instead, the position of tie points must be specified interactively by a user. This is not a problem in an augmented reality context, but whether this is a viable user interface for virtual reality applications remains to be shown.

26.2.1 Affine View Synthesis

Here we address the problem of synthesizing new (affine) images of a scene from old ones *without* setting an explicit three-dimensional Euclidean coordinate system. Recall from chapter 12 that if we denote the coordinate vector of a scene point P in some world coordinate system by $\boldsymbol{P} = (x, y, z)^T$ and denote by $\boldsymbol{p} = (u, v)^T$ the coordinate vector of the projection p of P onto the image plane, the affine camera model of Eq. (2.19) can be written as

$$\boldsymbol{p} = \mathcal{A}\boldsymbol{P} + \boldsymbol{b}, \quad \text{where} \quad \mathcal{A} = \begin{pmatrix} \boldsymbol{a}_1^T \\ \boldsymbol{a}_2^T \end{pmatrix}, \tag{26.1}$$

$\boldsymbol{b}$ is the position of the projection into the image of the object coordinate system's origin, and $\boldsymbol{a}_1$ and $\boldsymbol{a}_2$ are vectors in $\mathbb{R}^3$.

Let us consider four (noncoplanar) scene points, say P_0, P_1, P_2, and P_3. We can choose (without loss of generality) these points as an affine reference frame so their coordinate vectors are

$$\boldsymbol{P}_0 = \begin{pmatrix} 0 \\ 0 \\ 0 \end{pmatrix}, \quad \boldsymbol{P}_1 = \begin{pmatrix} 1 \\ 0 \\ 0 \end{pmatrix}, \quad \boldsymbol{P}_2 = \begin{pmatrix} 0 \\ 1 \\ 0 \end{pmatrix}, \quad \boldsymbol{P}_3 = \begin{pmatrix} 0 \\ 0 \\ 1 \end{pmatrix}.$$

The points P_i ($i = 1, 2, 3$) are *not* in general at a unit distance from P_0, nor are the vectors $\overrightarrow{P_0P_i}$ and $\overrightarrow{P_0P_j}$ orthogonal to each other when $i \neq j$. This is irrelevant since we work in an affine setting. Since the 3×3 matrix with columns $\boldsymbol{P}_1$, $\boldsymbol{P}_2$, and $\boldsymbol{P}_3$ is the identity, Eq. (26.1) can be rewritten as

$$\boldsymbol{p} = \mathcal{A}\boldsymbol{P} + \boldsymbol{b} = \begin{pmatrix} \boldsymbol{a}_1^T \\ \boldsymbol{a}_2^T \end{pmatrix} [\boldsymbol{P}_1 | \boldsymbol{P}_2 | \boldsymbol{P}_3] \begin{pmatrix} x \\ y \\ z \end{pmatrix} + \boldsymbol{b}.$$

Finally, since we have chosen P_0 as the origin of the world coordinate system, we have $\boldsymbol{b} = \boldsymbol{p}_0$ and we obtain

$$\boldsymbol{p} = (1 - x - y - z)\boldsymbol{p}_0 + x\boldsymbol{p}_1 + y\boldsymbol{p}_2 + z\boldsymbol{p}_3. \tag{26.2}$$

This result is related to the affine structure of affine images as discussed in chapter 12. In the context of image-based rendering, it follows from Eq. (26.2) that x, y, and z can be computed from $m \geq 2$ images of the points P_0, P_1, P_2, P_3, and P through linear least squares. Once these values are known, new images can be synthesized by specifying the image positions of the points p_0, p_1, p_2, p_3 and using Eq. (26.2) to compute all the other point positions (Kutulakos and Vallino, 1998). In addition, since the affine representation of the scene is truly three-dimensional, the relative depth of scene points can be computed and used to eliminate hidden surfaces in the z-buffer part of the graphics pipeline. It should be noted that specifying arbitrary positions for the points p_0, p_1, p_2, p_3 generally produces affinely deformed pictures. This is not a problem in augmented reality applications, where graphical and physical objects co-exist in the image. In this case, the anchor points p_0, p_1, p_2, p_3 can be chosen among true image points, guaranteed to be in the correct Euclidean position. Figure 26.10 shows an example where synthetic objects have been overlaid on real images.

When longer image sequences are available, a variant of this approach that takes into account all scene points in a uniform manner can be obtained as follows. Suppose we observe a fixed set of points $P_0, \dots, P_{n-1}$ with coordinate vectors $\boldsymbol{P}_i$ ($i = 0, \dots, n-1$) and let $\boldsymbol{p}_i$ denote the coordinate vectors of the corresponding image points. Writing Eq. (26.1) for all the scene points yields

$$\begin{pmatrix} \boldsymbol{p}_0 \\ \dots \\ \boldsymbol{p}_{n-1} \end{pmatrix} = \begin{pmatrix} \boldsymbol{P}_0^T & \boldsymbol{0}^T & 1 & 0 \\ \boldsymbol{0}^T & \boldsymbol{P}_0^T & 0 & 1 \\ \dots & \dots & \dots & \dots \\ \boldsymbol{P}_{n-1}^T & \boldsymbol{0}^T & 1 & 0 \\ \boldsymbol{0}^T & \boldsymbol{P}_{n-1}^T & 0 & 1 \end{pmatrix} \begin{pmatrix} \boldsymbol{a}_1 \\ \boldsymbol{a}_2 \\ \boldsymbol{b} \end{pmatrix}.$$

In other words, the set of all affine images of n *fixed* points is an eight-dimensional vector space V embedded in $\mathbb{R}^{2n}$ and parameterized by the vectors $\boldsymbol{a}_1$, $\boldsymbol{a}_2$, and $\boldsymbol{b}$.[3] Given $m \geq 8$ views

[3]This does not contradict the result established in chapter 12, which states that the set of m *fixed* views of an *arbitrary* collection of points is a three-dimensional affine subspace of $\mathbb{R}^{2m}$.

of $n \geq 4$ points, a basis for this vector space can be identified by performing the singular value decomposition of the $2n \times m$ matrix

$$\begin{pmatrix} \boldsymbol{p}_0^{(1)} & \dots & \boldsymbol{p}_0^{(m)} \\ \dots & \dots & \dots \\ \boldsymbol{p}_{n-1}^{(1)} & \dots & \boldsymbol{p}_{n-1}^{(m)} \end{pmatrix},$$

where $\boldsymbol{p}_i^{(j)}$ denotes the position of the image point number i in frame number j.[4] Once a basis for V has been constructed, new images can be constructed by assigning arbitrary values to $\boldsymbol{a}_1$, $\boldsymbol{a}_2$ and $\boldsymbol{b}$. For interactive image synthesis purposes, a more intuitive control of the imaging geometry can be obtained by specifying as before the position of four image points, solving for the corresponding values of $\boldsymbol{a}_1$, $\boldsymbol{a}_2$, and $\boldsymbol{b}$, and computing the remaining image positions.

26.2.2 Euclidean View Synthesis

As discussed earlier, a drawback of the method presented in the previous section is that specifying arbitrary positions for the points p_0, p_1, p_2, p_3 generally yields affinely deformed pictures. This can be avoided by taking into account from the start the Euclidean constraints associated with calibrated cameras. We saw in chapter 12 that a weak-perspective camera is an affine camera satisfying the two quadratic constraints

$$\boldsymbol{a}_1 \cdot \boldsymbol{a}_2 = 0 \quad \text{and} \quad |\boldsymbol{a}_1|^2 = |\boldsymbol{a}_2|^2.$$

The previous section showed that the affine images of a fixed scene form an eight-dimensional vector space V. Now if we restrict our attention to weak-perspective cameras, the set of images becomes the six-dimensional subspace defined by these two polynomial constraints. Similar constraints apply to paraperspective and true perspective projection, and they also define a six-dimensional *variety* (i.e., a subspace defined by polynomial equations) in each case.

Let us suppose that we observe three points P_0, P_1, P_2 whose images are not collinear. We can choose (without loss of generality) a *Euclidean* coordinate system such that the coordinate vectors of the four points in this system are

$$\boldsymbol{P}_0 = \begin{pmatrix} 0 \\ 0 \\ 0 \end{pmatrix}, \quad \boldsymbol{P}_1 = \begin{pmatrix} 1 \\ 0 \\ 0 \end{pmatrix}, \quad \boldsymbol{P}_2 = \begin{pmatrix} p \\ q \\ 0 \end{pmatrix},$$

where p and q are nonzero, but (a priori) unknown. Let us denote as before by p_i the projection of the point P_i ($i = 0, 1, 2$). Since P_0 is the origin of the world coordinate system, we have $\boldsymbol{b} = \boldsymbol{p}_0$. We are also free to pick p_0 as the origin of the image coordinate system (this amounts to submitting all image points to a known translation), so Eq. (26.1) simplifies into

$$\boldsymbol{p} = \mathcal{A}\boldsymbol{P} = \begin{pmatrix} \boldsymbol{a}_1^T \boldsymbol{P} \\ \boldsymbol{a}_2^T \boldsymbol{P} \end{pmatrix}. \tag{26.3}$$

Now applying Eq. (26.3) to P_1, P_2, and P yields

$$\boldsymbol{u} \stackrel{\text{def}}{=} \begin{pmatrix} u_1 \\ u_2 \\ u \end{pmatrix} = \mathcal{P}\boldsymbol{a}_1 \quad \text{and} \quad \boldsymbol{v} \stackrel{\text{def}}{=} \begin{pmatrix} v_1 \\ v_2 \\ v \end{pmatrix} = \mathcal{P}\boldsymbol{a}_2, \tag{26.4}$$

[4]Requiring at least eight images may seem like overkill since the affine structure of a scene can be recovered from two pictures as shown in chapter 12. Indeed, as shown in the exercises, a basis for V can in fact be constructed from two images of at least four points.

where

$$\mathcal{P} \stackrel{\text{def}}{=} \begin{pmatrix} \boldsymbol{P}_1^T \\ \boldsymbol{P}_2^T \\ \boldsymbol{P}^T \end{pmatrix} = \begin{pmatrix} 1 & 0 & 0 \\ p & q & 0 \\ x & y & z \end{pmatrix}.$$

In turn, this implies that

$$\boldsymbol{a}_1 = \mathcal{Q}\boldsymbol{u} \quad \text{and} \quad \boldsymbol{a}_2 = \mathcal{Q}\boldsymbol{v}, \tag{26.5}$$

where

$$\mathcal{Q} \stackrel{\text{def}}{=} \mathcal{P}^{-1} = \begin{pmatrix} 1 & 0 & 0 \\ \lambda & \mu & 0 \\ \alpha/z & \beta/z & 1/z \end{pmatrix} \quad \text{and} \quad \begin{cases} \lambda = -p/q, \\ \mu = 1/q, \\ \alpha = -(x + \lambda y), \\ \beta = -\mu y. \end{cases}$$

Using Eq. (26.5) and letting $\mathcal{R} \stackrel{\text{def}}{=} z^2 \mathcal{Q}^T \mathcal{Q}$, the weak-perspective constraints of Eq. (12.10) can be rewritten as

$$\begin{cases} \boldsymbol{u}^T \mathcal{R} \boldsymbol{u} - \boldsymbol{v}^T \mathcal{R} \boldsymbol{v} = 0, \\ \boldsymbol{u}^T \mathcal{R} \boldsymbol{v} = 0, \end{cases} \tag{26.6}$$

with

$$\mathcal{R} = \begin{pmatrix} \xi_1 & \xi_2 & \alpha \\ \xi_2 & \xi_3 & \beta \\ \alpha & \beta & 1 \end{pmatrix} \quad \text{and} \quad \begin{cases} \xi_1 = (1 + \lambda^2) z^2 + \alpha^2, \\ \xi_2 = \lambda \mu z^2 + \alpha \beta, \\ \xi_3 = \mu^2 z^2 + \beta^2. \end{cases}$$

Equation (26.6) defines a pair of linear constraints on the coefficients ξ_i $(i = 1, 2, 3)$, α, and β. These can be rewritten as

$$\begin{pmatrix} \boldsymbol{d}_1^T \\ \boldsymbol{d}_2^T \end{pmatrix} \boldsymbol{\xi} = 0, \tag{26.7}$$

where

$$\boldsymbol{d}_1 \stackrel{\text{def}}{=} \begin{pmatrix} u_1^2 - v_1^2 \\ 2(u_1 u_2 - v_1 v_2) \\ u_2^2 - v_2^2 \\ 2(u_1 u - v_1 v) \\ 2(u_2 u - v_2 v) \\ u^2 - v^2 \end{pmatrix}, \quad \boldsymbol{d}_2 \stackrel{\text{def}}{=} \begin{pmatrix} u_1 v_1 \\ u_1 v_2 + u_2 v_1 \\ u_2 v_2 \\ u_1 v + u v_1 \\ u_2 v + u v_2 \\ u v \end{pmatrix}, \quad \text{and} \quad \boldsymbol{\xi} \stackrel{\text{def}}{=} \begin{pmatrix} \xi_1 \\ \xi_2 \\ \xi_3 \\ \alpha \\ \beta \\ 1 \end{pmatrix}.$$

When the four points P_0, P_1, P_2, and P are rigidly attached to each other, the five structure coefficients ξ_1, ξ_2, ξ_3, α, and β are fixed. For a rigid scene formed by n points, choosing three of the points as a reference triangle and writing Eq. (26.7) for the remaining ones yields a set of $2n - 6$ quadratic equations in $2n$ unknowns, which define a parameterization of the set of all weak-perspective images of the scenes. This is the *parameterized image variety (PIV)* of Genc and Ponce (2001).

Note again that the weak-perspective constraints of Eq. (26.7) are linear in the five structure coefficients. Thus, given a collection of images and point correspondences, these coefficients can be estimated through linear least squares. Once the vector $\boldsymbol{\xi}$ has been estimated, arbitrary image

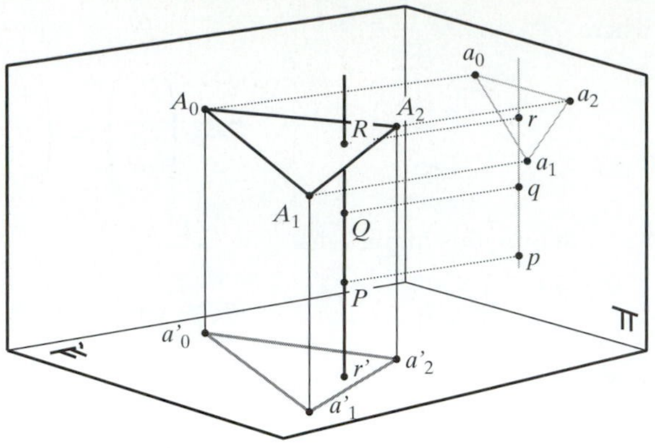

Figure 26.11 Z-buffering. *Reprinted from "Parameterized Image Varieties: A Novel Approach to the Analysis and Synthesis of Image Sequences," by Y. Genc and J. Ponce, Proc. International Conference on Computer Vision, (1998). © 1998 IEEE.*

positions can be assigned to the three reference points. Equation (26.7) yields, for each feature point, two quadratic constraints on the two unknowns u and v. Although this system should *a priori* admit four solutions, it admits, as shown in the exercises, exactly two real solutions. In fact, given n point correspondences and the image positions of the three tie points, it can also be shown (Genc and Ponce, 2001) that the pictures of the remaining $n-3$ points can be determined in closed form up to a two-fold ambiguity. Once the positions of all feature points have been determined, the scene can be rendered by triangulating these points and texture-mapping the triangles. Interestingly, hidden-surface removal can also be performed via traditional z-buffer techniques, although no explicit three-dimensional reconstruction is performed: The idea is to assign relative depth values to the vertices of the triangulation, and it is closely related to the method used in the affine structure-from-motion theorem from chapter 12. Let Π denote the image plane of one of our input images, and Π' the image plane of our synthetic image. To render correctly two points P and Q that project onto the same point r' in the synthetic image, we must compare their depths (Figure 26.11).

Let R denote the intersection of the viewing ray joining P to Q with the plane spanned by the reference points A_0, A_1, and A_2, and let p, q, r denote the projections of P, Q, and R into the reference image. Suppose for the time being that P and Q are two of the points tracked in the input image; it follows that the positions of p and q are known. The position of r is easily computed by remarking that its coordinates in the affine basis of Π formed by the projections a_0, a_1, a_2 of the reference points are the same as the coordinates of R in the affine basis formed by the points A_0, A_1, A_2 in their own plane, and thus are also the same as the coordinates of r' in the affine basis of Π' formed by the projections a'_0, a'_1, a'_2 of the reference points. The ratio of the depths of P and Q relative to the plane Π is simply the ratio $\overline{pr}/\overline{qr}$. Not that deciding which point is actually visible requires orienting the line supporting the points p, q, r, which is simply the epipolar line associated with the point r'. A coherent orientation should be chosen for all epipolar lines (this is easy since they are all parallel to each other). Note that this does not require explicitly computing the epipolar geometry: Given a first point p', one can orient the line pr and then use the same orientation for all other point correspondences. The orientations chosen should also be consistent over successive frames, but this is not a problem since the direction of the epipolar lines changes slowly from one frame to the next,

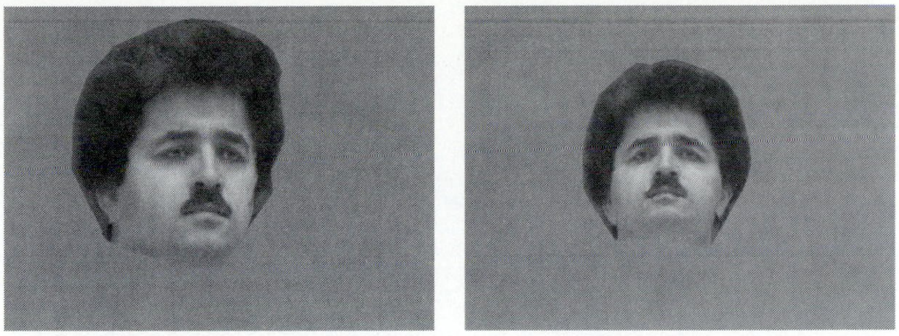

Figure 26.12 Two images of a face synthesized using parameterized image varieties. *Courtesy of Yakup Genc.*

and one can simply choose the new orientation so that it makes an acute angle with the previous one. Examples of synthetic pictures constructed using this method are shown in Figure 26.12.

26.3 THE LIGHT FIELD

This section discusses a different approach to image-based rendering, whose only similarity with the techniques discussed in the previous section is that, like them, it does not require the construction of any implicit or explicit 3D model of a scene. Let us consider, for example, a panoramic camera that optically records the radiance along rays passing through a single point and covering a full hemisphere (see, e.g., Peri and Nayar, 1997; Figure 26.13, left). It is possible to create

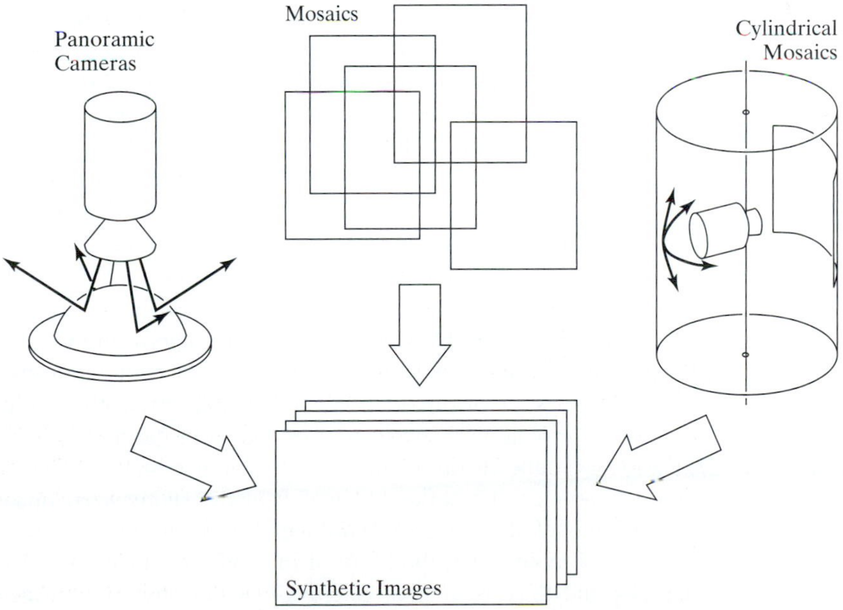

Figure 26.13 Constructing synthetic views of a scene from a fixed viewpoint.

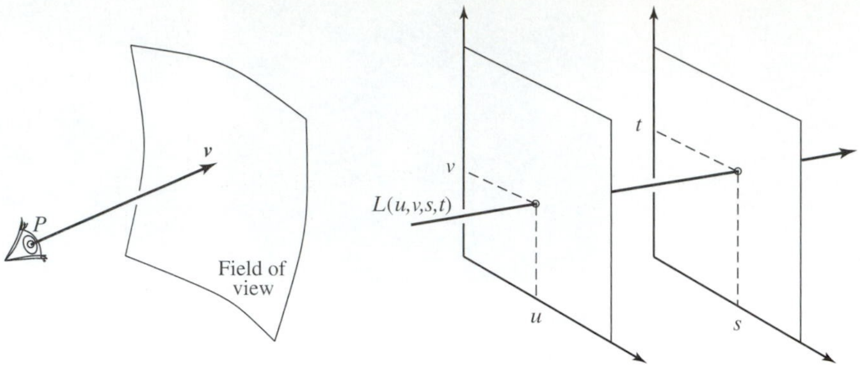

Figure 26.14 The plenoptic function and the light field. Left: the plenoptic function can be parameterized by the position P of the observer and the viewing direction $\mathbf{v}$. Right: the light field can be parameterized by the four parameters u, v, s, t defining a light slab. In practice, several light slabs are necessary to model a whole object and obtain full spherical coverage.

any image observed by a virtual camera whose pinhole is located at this point by mapping the original image rays onto virtual ones. This allows a user to arbitrarily pan and tilt the virtual camera and interactively explore his or her visual environment. Similar effects can be obtained by stitching together close-by images taken by a hand-held camcorder into a mosaic (see, e.g., Shum and Szeliski, 1998; Figure 26.13, middle), or by combining the pictures taken by a camera panning (and possibly tilting) about its optical center into a cylindrical mosaic (see, e.g., Chen, 1995; Figure 26.13, right).

These techniques have the drawback of limiting the viewer motions to pure rotations about the optical center of the camera. A more powerful approach can be devised by considering the *plenoptic function* (Adelson and Bergen, 1991) that associates with each point in space the (wavelength-dependent) radiant energy along a ray passing through this point at a given time (Figure 26.14, left). The *light field* (Levoy and Hanrahan, 1996) is a snapshot of the plenoptic function for light traveling in vacuum in the absence of obstacles. This relaxes the dependence of the radiance on time and on the position of the point of interest along the corresponding ray (since radiance is constant along straight lines in a nonabsorbing medium) and yields a representation of the plenoptic function by the radiance along the four-dimensional set of light rays. These rays can be parameterized in many different ways (e.g., using the Plücker coordinates introduced in chapter 3), but a convenient parameterization in the context of image-based rendering is the *light slab*, where each ray is specified by the coordinates of its intersections with two arbitrary planes (Figure 26.14, right).

The light slab is the basis for a two-stage approach to image-based rendering. During the learning stage, many views of a scene are used to create a discrete version of the slab that can be thought of as a four-dimensional lookup table. At synthesis time, a virtual camera is defined, and the corresponding view is interpolated from the lookup table. The quality of the synthesized images depends on the number of reference images. The closer the virtual view is to the reference images, the better the quality of the synthesized image. Note that constructing the light slab model of the light field does not require establishing correspondences between images. It should be noted that, unlike most methods for image-based rendering that rely on texture mapping and thus assume (implicitly) that the observed surfaces are Lambertian, light-field techniques can be used to render (under a fixed illumination) pictures of objects with *arbitrary* BRDFs.

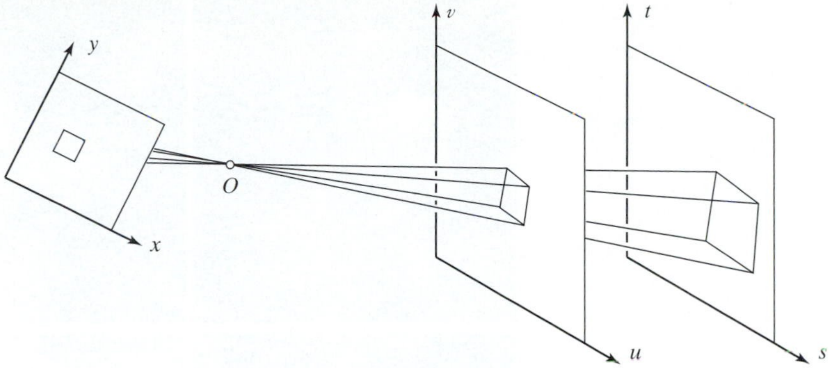

Figure 26.15 The acquisition of a light slab from images and the synthesis of new images from a light slab can be modeled via projective transformations between the (x, y) image plane and the (u, v) and (s, t) planes defining the slab.

In practice, a sample of the light field is acquired by taking a large number of images and mapping pixel coordinates onto slab coordinates. Figure 26.15 illustrates the general case: The mapping between any pixel in the (x, y) image plane and the corresponding areas of the (u, v) and (s, t) plane defining a light slab is a planar projective transformation. Hardware- or software-based texture mapping can thus be used to populate the light field on a four-dimensional rectangular grid. In the experiments described in Levoy and Hanrahan (1996), light slabs are acquired in the simple setting of a camera mounted on a planar gantry and equipped with a pan-tilt head so it can rotate about its optical center and always point toward the center of the object of interest. In this context, all calculations can be simplified by taking the (u, v) plane to be the plane in which the camera's optical center is constrained to remain.

At rendering time, the projective mapping between the (virtual) image plane and the two planes defining the light slab can once again be used to efficiently synthesize new images. Figure 26.16 shows sample pictures generated using the light-field approach. The top three image pairs were generated using synthetic pictures of various objects to populate the light field. The last pair of images was constructed by using the planar gantry mentioned earlier to acquire 2048 256×256 images of a toy lion, grouped into four slabs each consisting of 32×16 images.

An important issue is the size of the light slab representation: The raw input images of the lion take 402MB of disk space. There is, of course, much redundancy in these pictures, as in the case of successive frames in a motion sequence. A simple but effective two-level approach to image (de)compression is proposed in Levoy and Hanrahan (1996): The light slab is first decomposed into four-dimensional tiles of color values. These tiles are encoded using *vector quantization* (Gersho and Gray, 1992), a lossy compression technique where the 48-dimensional vectors representing the RGB values at the 16 corners of the original tiles are replaced by a relatively small set of reproduction vectors, called *codewords*, that best approximate in the mean-squared-error sense the input vectors. The light slab is thus represented by a set of indexes in the *codebook* formed by all codewords. In the case of the lion, the codebook is relatively small (0.8MB) and the size of the set of indexes is 16.8MB. The second compression stage consists of applying the *gzip* implementation of *entropy coding* (Ziv and Lempel, 1977) to the codebook and the indexes. The final size of the representation is only 3.4MB, corresponding to a compression rate of 118:1. At rendering time, entropy decoding is performed as the file is loaded in main memory. Dequantization is performed on demand during display, and it allows interactive refresh rates.

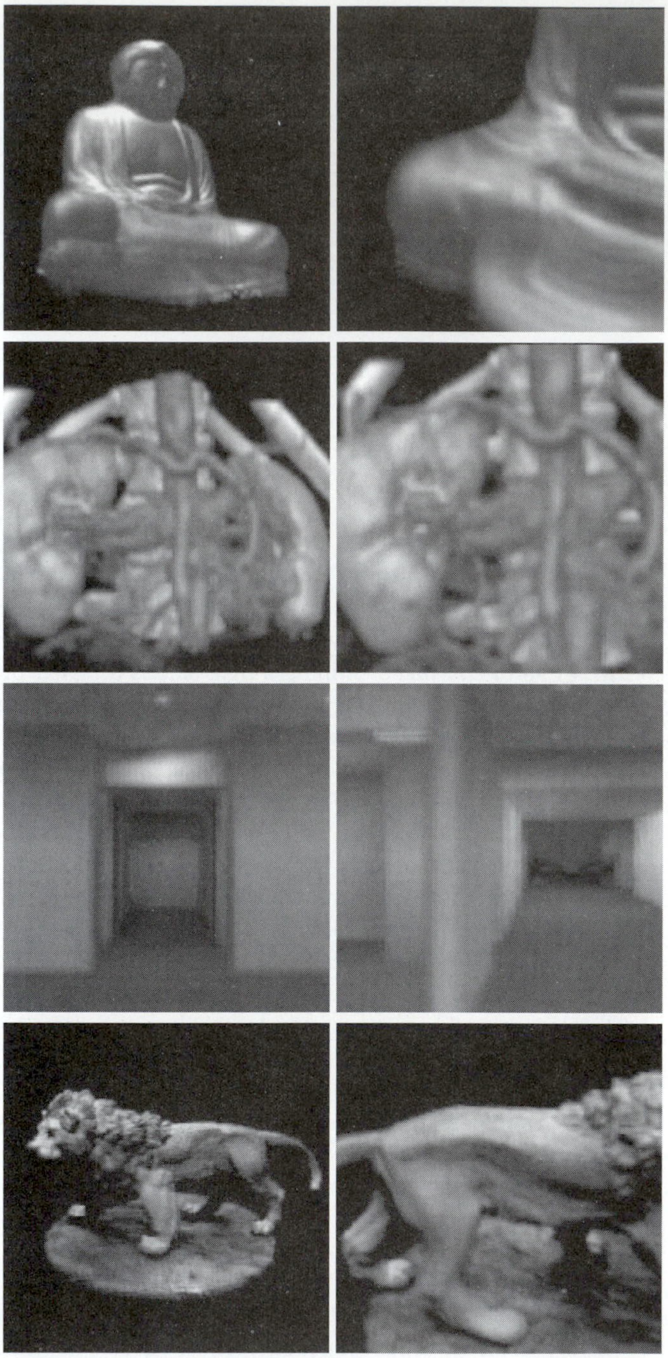

Figure 26.16 Images synthesized with the light field approach. *Reprinted from "Light Field Rendering," by M. Levoy and P. Hanrahan, Proc. SIGGRAPH, (1996). © 1996 ACM, Inc. Included here by permission.*

26.4 NOTES

Image-based rendering is a quickly expanding field. To close this chapter, let us just mention a few alternatives to the approaches already mentioned in the previous sections. The intersection of all of the cones that graze the surface of a solid forms its *visual hull* (Laurentini, 1995). A solid is always contained in its visual hull, which, in turn, is contained in the solid's convex hull. The volumetric approach to object modeling from registered silhouettes presented in Section 26.1 is aimed at constructing an approximation of the visual hull from a finite set of photographs. Variants use polyhedra or octrees (Martin and Aggarwal, 1983, Connolly and Stenstrom, 1989, Srivastava and Ahuja, 1990) to represent the cones and their intersection, and include a commercial system, Sphinx3D (Niem and Buschmann, 1994), for automatically constructing polyhedral models from images. See also Kutulakos and Seitz (1999) for a related approach, called *space carving*, where empty voxels are iteratively removed using brightness or color coherence constraints. The tangency constraint has been used in various approaches for reconstructing a surface from a continuous sequence of outlines under known or unknown camera motions (Arbogast and Mohr, 1991, Cipolla and Blake, 1992, Vaillant and Faugeras, 1992, Cipolla et al., 1995, Boyer and Berger, 1996, Cross et al., 1999, Joshi et al., 1999). Variants of the view interpolation method discussed in Section 26.1 include Williams and Chen (1993) and Seitz and Dyer (1995, 1996). Transfer-based approaches to image-based rendering include, besides those discussed in Section 26.2, Havaldar et al. (1996) and Avidan and Shashua (1997). As briefly mentioned in Section 26.3, a number of techniques have been developed for interactively exploring a user's visual environment from a fixed viewpoint. These include a commercial system, *QuickTime VR*, developed at Apple by Chen (1995), and algorithms that reconstruct pinhole perspective images from panoramic pictures acquired by special-purpose cameras (see, e.g., Peri and Nayar, 1997). Similar effects can be obtained in a less controlled setting by stitching together close-by images taken by a hand-held camcorder into a mosaic (see, e.g., Irani et al., 1996, Shum and Szeliski, 1998). For images of distant terrains or cameras rotating about their optical center, the mosaic can be constructed by registering successive pictures via planar homographies. In this context, estimating the *optical flow* (i.e., the vector field of apparent image velocities at every image point, a notion that has, admittedly, largely been ignored in this book), may also prove important for fine registration and *deghosting* (Shum and Szeliski, 1998). Variants of the light field approach discussed in Section 26.3 include McMillan and Bishop (1995) and Gortler et al. (1996). An excellent introduction to Bézier arcs and patches and spline curves and surfaces can be found in Farin (1993).

PROBLEMS

26.1. Given $n+1$ point $P_0, \dots, P_n$, we recursively define the parametric curve $P_i^k(t)$ by $P_i^0(t) = P_i$ and

$$P_i^k(t) = (1-t)P_i^{k-1}(t) + tP_{i+1}^{k-1}(t) \quad \text{for} \quad k = 1, \dots, n \quad \text{and} \quad i = 0, \dots, n-k.$$

We show in this exercise that $P_0^n(t)$ is the Bézier curve of degree n associated with the $n+1$ points $P_0, \dots, P_n$. This construction of a Bézier curve is called the *de Casteljeau algorithm*.

(a) Show that Bernstein polynomials satisfy the recursion

$$b_i^{(n)}(t) = (1-t)b_i^{(n-1)}(t) + tb_{i-1}^{(n-1)}(t)$$

with $b_0^{(0)}(t) = 1$ and, by convention, $b_j^{(n)}(t) = 0$ when $j < 0$ or $j > n$.

(b) Use induction to show that

$$P_i^k(t) = \sum_{j=0}^{k} b_j^{(k)}(t) P_{i+j} \quad \text{for} \quad k = 0, \dots, n \quad \text{and} \quad i = 0, \dots, n-k.$$

26.2. Consider a Bézier curve of degree n defined by $n+1$ control points $P_0, \dots, P_n$. We address here the problem of constructing the $n+2$ control points $Q_0, \dots, Q_{n+1}$ of a Bézier curve of degree $n+1$ with the same shape. This process is called *degree elevation*. Show that $Q_0 = P_0$ and

$$Q_j = \frac{j}{n+1} P_{j-1} + \left(1 - \frac{j}{n+1}\right) P_j \quad \text{for} \quad j = 1, \dots, n+1.$$

Hint: Write that the same point is defined by the barycentric combinations associated with the two curves, and equate the polynomial coefficients on both sides of the equation.

26.3. Show that the tangent to the Bézier curve $P(t)$ defined by the $n+1$ control points $P_0, \dots, P_n$ is

$$P'(t) = n \sum_{j=0}^{n-1} b_j^{(n-1)}(t)(P_{j+1} - P_j).$$

Conclude that the tangents at the endpoints of a Bézier arc are along the first and last line segments of its control polygon.

26.4. Show that the construction of the points Q_i in Section 26.1.1 places these points in a plane that passes through the centroid O of the points C_i.

26.5. Façade's photogrammetric module. We saw in the exercises of Chapter 3 that the mapping between a line δ with Plücker coordinate vector $\mathbf{\Delta}$ and its image δ with homogeneous coordinates $\mathbf{\Delta}$ can be represented by $\rho\boldsymbol{\delta} = \tilde{\mathcal{M}}\mathbf{\Delta}$. Here, $\mathbf{\Delta}$ is a function of the model parameters, and $\tilde{\mathcal{M}}$ depends on the corresponding camera position and orientation.

(a) Assuming that the line δ has been matched with an image edge e of length l, a convenient measure of the discrepancy between predicted and observed data is obtained by multiplying by l the mean squared distance separating the points of e from δ. Defining $d(t)$ as the signed distance between the edge point $p = (1-t)p_0 + tp_1$ and the line δ, show that

$$E = \int_0^1 d^2(t)\, dt = \frac{1}{3}\left(d(0)^2 + d(0)d(1) + d(1)^2\right),$$

where d_0 and d_1 denote the (signed) distances between the endpoints of e and δ.

(b) If $\boldsymbol{p}_0$ and $\boldsymbol{p}_1$ denote the homogeneous coordinate vectors of these points, show that

$$d_0 = \frac{1}{|[\tilde{\mathcal{M}}\mathbf{\Delta}]_2|} \boldsymbol{p}_0^T \tilde{\mathcal{M}}\mathbf{\Delta} \quad \text{and} \quad d_1 = \frac{1}{|[\tilde{\mathcal{M}}\mathbf{\Delta}]_2|} \boldsymbol{p}_1^T \tilde{\mathcal{M}}\mathbf{\Delta},$$

where $[\boldsymbol{a}]_2$ denotes the vector formed by the first two coordinates of the vector $\boldsymbol{a}$ in $\mathbb{R}^3$.

(c) Formulate the recovery of the camera and model parameters as a non-linear least-squares problem.

26.6. Show that a basis for the eight-dimensional vector space V formed by all affine images of a fixed set of points $P_0, \dots, P_{n-1}$ can be constructed from at least two images of these points when $n \geq 4$.

Hint: Use the matrix

$$\begin{pmatrix} u_0^{(1)} & v_0^{(1)} & \dots & u_0^{(m)} & v_0^{(m)} \\ \dots & \dots & \dots & & \\ u_{n-1}^{(1)} & v_{n-1}^{(1)} & \dots & u_{n-1}^{(m)} & v_{n-1}^{(m)} \end{pmatrix},$$

where $(u_i^{(j)}, v_i^{(j)})$ are the coordinates of the projection of the point P_i into image number j.

26.7. Show that the set of all projective images of a fixed scenes is an eleven-dimensional variety.

26.8. Show that the set of all perspective images of a fixed scene (for a camera with constant intrinsic parameters) is a six-dimensional variety.

26.9. In this exercise, we show that Eq. (26.7) only admits two solutions.

(a) Show that Eq. (26.6) can be rewritten as

$$\begin{cases} X^2 - Y^2 + e_1 - e_2 = 0, \\ 2XY + e = 0, \end{cases} \tag{26.8}$$

where

$$\begin{cases} X = u + \alpha u_1 + \beta u_2, \\ Y = v + \alpha v_1 + \beta v_2, \end{cases}$$

and e, e_1, and e_2 are coefficients depending on u_1, v_1, u_2, v_2 and the structure parameters.

(b) Show that the solutions of Eq. (26.8) are given by

$$\begin{cases} X' = \sqrt[4]{(e_1 - e_2)^2 + e^2} \cos\left(\frac{1}{2} \arctan(e, e_1 - e_2)\right), \\ Y' = \sqrt[4]{(e_1 - e_2)^2 + e^2} \sin\left(\frac{1}{2} \arctan(e, e_1 - e_2)\right), \end{cases}$$

and $(X'', Y'') = (-X', -Y')$.

Hint: Use a change of variables to rewrite Eq. (26.8) as a system of trigonometric equations.

Bibliography

Adelson, E. and Bergen, J. (1991), The plenoptic function and the elements of early vision, *in* M. Landy and J. Movshon, eds, 'Computational Models of Visual Processing', MIT Press.

Adelson, E. and Weiss, Y. (1996), A unified mixture framework for motion segmentation: Incorporating spatial coherence and estimating the number of models, *in* 'IEEE Conference on Computer Vision and Pattern Recognition', 1996, pp. 321–326.

Adini, Y., Moses, Y. and Ullman, S. (1994), Face recognition: The problem of compensating for changes in illumination direction, *in* 'European Conference on Computer Vision', pp. A:286–296.

Adini, Y., Moses, Y. and Ullman, S. (1997), 'Face recognition: The problem of compensating for changes in illumination direction', *IEEE Trans. Pattern Analysis and Machine Intelligence* **19**(7), 721–732.

Agin, G. (1972), Representation and description of curved objects, PhD thesis, Stanford University.

Agin, G. (1981), Fitting ellipses and general second-order curves, Technical report, CMU Robotics Institute.

Aho, A., Hopcroft, J. and Ullman, J. (1974), *The Design and Analysis of Computer Algorithms*, Addison-Wesley.

Ahuja, N. and Abbott, A. (1993), 'Active stereo: Integrating disparity, vergence, focus, aperture, and calibration for surface estimation', *IEEE Trans. Patt. Anal. Mach. Intell.* **15**(10), 1007–1029.

Aikens, R., Agard, D. and Sedat, J. (1989), 'Solid-state imagers for microscopy', *Methods Cell Biol.* **29**, 291–313.

Aloimonos, Y. (1986), Detection of surface orientation from texture. I. the case of planes., *in* 'IEEE Conf. on Computer Vision and Pattern Recognition', pp. 584–593.

Aloimonos, Y. (1990), 'Perspective approximations', *Image and Vision Computing* **8**(3), 177–192.

Aloimonos, Y., Weiss, I. and Bandyopadhyay, A. (1987), 'Active vision', *Int. J. of Comp. Vision* **1**(4), 333–356.

Amelio, G., Tompsett, M. and Smith, G. (1970), 'Experimental verification of the charge coupled device concept', *Bell Syst. Tech. J.* **49**, 593–600.

Amenta, N., Bern, M. and Kamvysselis, M. (1998), A new Voronoi-based surface reconstruction algorithm, *in* 'SIGGRAPH-98', pp. 415–421.

Amir, A. and Lindenbaum, M. (1996), Quantitative analysis of grouping processes, *in* 'European Conference on Computer Vision', 1996, pp. I:371–384.

Anderson, B. and Nayakama, K. (1994), 'Toward a general theory of stereopsis—binocular matching, occluding contours, and fusion', *Psych. Review* **101**(3), 414–445.

Anthropology Research Project, (1978), *Anthropometric Source Book*, Webb Associates. NASA reference publication 1024, 3 Vols.

Arbogast, E. and Mohr, R. (1991), '3D structure inference from image sequences', *Journal of Pattern Recognition and Artificial Intelligence*.

Arend, L. and Reeves, A. (1986), 'Simultaneous colour constancy', *J. Opt. Soc. America - A* **3**, 1743–1751.

Armitage, L. and Enser, P. (1997), 'Analysis of user need in image archives', *Journal of Information Science* **23**(4), 287–299.

Arnol'd, V. (1984), *Catastrophe Theory*, Springer-Verlag, Heidelberg.

Arnon, D., Collins, G. and McCallum, S. (1984), 'Cylindrical algebraic decomposition I and II', *SIAM J. Comput.* **13**(4), 865–889.

Avidan, S. and Shashua, A. (1997), Novel view synthesis in tensor space, *in* 'Proc. IEEE Conf. Comp. Vision Patt. Recog.', pp. 1034–1040.

Ayache, N. (1995*a,b*), 'Medical computer vision, virtual-reality and robotics', *Image and Vision Computing* **13**(4), 295–313.

Ayache, N. and Faugeras, O. (1986), 'Hyper: a new approach for the recognition and positioning of two-dimensional objects', *IEEE Trans. Patt. Anal. Mach. Intell.* **8**(1), 44–54.

Ayache, N. and Faugeras, O. (1987), Building, registrating, and fusing noisy visual maps, *in* 'Proc. Int. Conf. Comp. Vision', pp. 73–82.

Ayache, N. and Faverjon, B. (1997), 'Efficient registration of stereo images by matching graph descriptions of edge segments', *Int. J. of Comp. Vision* pp. 101–137.

Ayache, N. and Lustman, F. (1987), Fast and reliable passive trinocular stereovision, *in* 'Proc. Int. Conf. Comp. Vision', pp. 422–427.

Bajcsy, R. (1988), 'Active perception', *Proceedings of the IEEE* **76**(8), 996–1005.

Bajcsy, R. and Solina, F. (1987), Three-dimensional object representation revisited, *in* 'Proc. Int. Conf. Comp. Vision', pp. 231–240.

Baker, H. and Binford, T. (1981), Depth from edge- and intensity-based stereo, *in* 'Proc. International Joint Conference on Artificial Intelligence', pp. 631–636.

Baker, S., Szeliski, R. and Anandan, P. (1998), A layered approach to stereo reconstruction, *in* 'IEEE Conference on Computer Vision and Pattern Recognition', 1998, pp. 434–441.

Barnard, K. (2000), Improvements to gamut mapping colour constancy algorithms, *in* 'European Conference on Computer Vision', 2000, pp. 390–402.

Barnard, K. and Forsyth, D. (2001), Learning the semantics of words and pictures, *in* 'Int. Conf. on Computer Vision', pp. 408–15.

Barnard, K. and Funt, B. (2002), 'Camera characterization for color research', *Color research and application*. Accepted for publication.

Barnard, K., Duygulu, P. and Forsyth, D. (2001), Clustering art, *in* 'IEEE Conf. on Computer Vision and Pattern Recognition'. To appear.

Barnard, K., Finlayson, G. and Funt, B. (1997), 'Color constancy for scenes with varying illumination', *Computer Vision and Image Understanding* **65**(2), 311–321.

Barrett, E., Brill, M., Haag, N. and Payton, P. (1992), Some invariant linear methods in photogrammetry and model-matching, *in* 'IEEE Conference on Computer Vision and Pattern Recognition', 1992, pp. 122–128.

Barrett, E., Payton, P., Haag, N. and Brill, M. (1991), 'General methods for determining projective invariants in imagery', *CVGIP: Image Understanding* **53**(1), 46–65.

Basri, R. (1996), 'Parraperspective = affine', *Int. J. of Comp. Vision* **19**(2), 169–179.

Beardsley, P., Zisserman, A. and Murray, D. (1997), 'Sequential updating of projective and affine structure from motion', *Int. J. of Comp. Vision* **23**(3), 235–259.

Beckmann, P. and Spizzichino, A. (1987), *Scattering of Electromagnetic Waves from Rough Surfaces*, Artech House.

Belhumeur, P. and Kriegman, D. (1998), 'What is the set of images of an object under all possible illumination conditions', *International Journal of Computer Vision* **28**(3), 245–260.

Belhumeur, P., Kriegman, D. and Yuille, A. (1999), 'The bas-relief ambiguity', *International Journal of Computer Vision* **35**(1), 33–44.

Belongie, S., Carson, C., Greenspan, H. and Malik, J. (1998), Color- and texture-based image segmentation using the expectation-maximization algorithm and its application to content-based image retrieval, *in* 'Proceedings, Sixth International Conference on Computer Vision', 1998, pp. 675–682.

Berger, M. (1987), *Geometry*, Springer-Verlag.

Bergholm, F. (1987), 'Edge focusing', *IEEE Trans. Pattern Analysis and Machine Intelligence* **9**(6), 726–741.

Bertero, M., Poggio, T. and Torre, V. (1988), 'Ill-posed problems in early vision', *Proceedings of IEEE* **76**(8), 869–889.

Bertsekas, D. (1995), *Nonlinear Programming*, Athena Scientific.

Berzins, V. (1984), 'Accuracy of laplacian edge detectors', *CVGIP: Image Understanding* **27**(2), 195–210.

Besl, P. (1989), 'Active optical range imaging sensors', *Machine vision and applications* **1**, 127–152.

Besl, P. and Jain, R. (1988), 'Segmentation through variable-order surface fitting', *IEEE Trans. Patt. Anal. Mach. Intell.* **10**(2), 167–192.

Besl, P. and McKay, N. (1992), 'A method for registration of 3D shapes', *IEEE Trans. Patt. Anal. Mach. Intell.* **14**(2), 239–256.

Beymer, D. (1994), Face recognition under varying pose, *in* 'IEEE Conference on Computer Vision and Pattern Recognition', 1994, pp. 756–761.

Beymer, D. and Poggio, T. (1995), Face recognition from one example view, *in* 'Proceedings, Fifth International Conference on Computer Vision', 1995, pp. 500–507.

Beymer, D., McLauchlan, P., Coifman, B. and Malik, J. (1997), A real time computer vision system for measuring traffic parameters, *in* 'IEEE Conference on Computer Vision and Pattern Recognition', 1997, pp. 495–501.

Biederman, I. (1987), 'Recognition-by-components: A theory of human image understanding', *Psych. Review* **94**(2), 115–147.

Binford, T. (1971), Visual perception by computer, *in* 'Proc. IEEE Conference on Systems and Control'.

Binford, T. (1984), Stereo vision: complexity and constraints, *in* 'Int. Symp. on Robotics Research', MIT Press, pp. 475–487.

Bishop, C. (1995), *Neural Networks for Pattern Recognition*, Oxford University Press.

Black, M. and Anandan, P. (1996), 'The robust estimation of multiple motions: Parametric and piecewise-smooth flow-fields', *Computer Vision and Image Understanding* **63**(1), 75–104.

Blackman, S. and Popoli, R. (1999), *Design and Analysis of Modern Tracking Systems*, Artech House.

Blake, A. (1985), 'Boundary conditions for lightness computation in mondrian world', *CVGIP: Image Understanding* **32**(3), 314–327.

Blake, A. and Marinos, C. (1990), 'Shape from texture: estimation, isotropy and moments', *Artificial Intelligence* **45**(3), 323–380.

Blake, A. and Zisserman, A. (1987), *Visual Reconstruction*, MIT Press.

Blum, H. (1967), A transformation for extracting new descriptors of shape, *in* W. Wathen-Dunn, ed., 'Models for perception of speech and visual form', MIT Press.

Boissonnat, J.-D. (1984), 'Geometric structures for three-dimensional shape representation', *ACM Transaction on Computer Graphics* **3**(4), 266–286.

Boissonnat, J.-D. and Germain, F. (1981), A new approach to the problem of acquiring randomly-oriented workpieces from a bin, *in* 'Proc. International Joint Conference on Artificial Intelligence'.

Bookstein, F. (1979), 'Fitting conic sections to scattered data', *Computer Graphics Image Processing* **9**(1), 56–71.

Boult, T. and Brown, L. (1991), Factorization-based segmentation of motions, *in* 'IEEE Workshop on Visual Motion', pp. 179–186.

Bowyer, K., Kranenburg, C. and Dougherty, S. (1999), Edge detector evaluation using empirical roc curves, *in* 'IEEE Conference on Computer Vision and Pattern Recognition', 1999, pp. I:354–359.

Boyer, E. (1996), Object Models from Contour Sequences, *in* 'Proceedings of Fourth European Conference on Computer Vision, Cambridge, (England)', pp. 109–118. Lecture Notes in Computer Science, volume 1065.

Boyer, E. and Berger, M.-O. (1996), '3D surface reconstruction using occluding contours', *Int. J. of Comp. Vision*. To appear.

Boyle, W. and Smith, G. (1970), 'Charge coupled semiconductor devices', *Bell Syst. Tech. J.* **49**, 587–593.

Bracewell, R. (1995), *Two-Dimensional Imaging*, Prentice Hall.

Bracewell, R. (2000), *The Fourier Transform and Its Applications, 3ed*, McGraw-Hill.

Brady, J. and Asada, H. (1984), 'Smoothed local symmetries and their implementation', *International Journal of Robotics Research*.

Brady, J., Ponce, J., Yuille, A. and Asada, H. (1985*a*), 'Describing surfaces', *Computer Vision, Graphics and Image Processing* **32**(1), 1–28.

Brady, J., Ponce, J., Yuille, A. and Asada, H. (1985*b*), Describing surfaces, *in* H. Hanafusa and H. Inoue, eds, 'Proceedings of the 2nd International Symposium on Robotics Research', MIT Press, pp. 5–16.

Brainard, D. and Wandell, B. (1986), 'Analysis of the retinex theory of color vision', *Journal of the Optical Society of America* **3**, 1651–1661.

Brand, J. and Mason, J. (2000), A comparative assessment of three approaches to pixel-level human skin-detection, *in* 'Proceedings, International Conference on Pattern Recognition', pp. Vol I: 1056–1059.

Brelstaff, G. and Blake, A. (1987), 'Computing lightness', *Pattern Recognition Letters* **5**, 129–138.

Brelstaff, G. and Blake, A. (1988), Detecting specular reflection using lambertian constraints, *in* 'Proceedings, Second International Conference on Computer Vision', 1988, pp. 297–302.

Brill, E. (1992), A simple rule-based part of speech tagger, *in* 'Proc. Third Conference on Applied Natural Language Processing'.

Brooks, R. (1981*a*), 'Symbolic reasoning among 3-D models and 2-D images', *Artificial Intelligence Journal* **17**(1-3), 285–348.

Brooks, R. (1981*b*), Symbolic Reasoning among 3-D Models and 2-D Images, PhD thesis, Stanford University Computer Science Dept.

Brooks, R., Greiner, R. and Binford, T. (1979), Model-based three-dimensional interpretation of two-dimensional images, *in* 'Proc. International Joint Conference on Artificial Intelligence', Tokyo, Japan, pp. 105–113.

Brostow, G. and Essa, I. (1999), Motion based decompositing of video, *in* 'Proceedings, Seventh International Conference on Computer Vision', 1999, pp. 8–13.

Bruce, J., Giblin, P. and Tari, F. (1996*a*), 'Parabolic curves of evolving surfaces', *Int. J. of Comp. Vision* **17**(3), 291–306.

Bruce, J., Giblin, P. and Tari, F. (1996*b*), 'Ridges, crests and sub-parabolic lines of evolving surfaces', *Int. J. of Comp. Vision* **18**(3), 195–210.

Brunelli, R. and Poggio, T. (1992), Face recognition through geometrical features, *in* Sandini (1992), pp. 792–800.

Brunelli, R. and Poggio, T. (1993), 'Face recognition: Features versus templates', *IEEE Trans. Pattern Analysis and Machine Intelligence* **15**(10), 1042–1052.

Brunie, L., Lavallée, S. and Szeliski, R. (1992), Using force fields derived from 3D distance maps for inferring the attitude of a 3D rigid object, *in* G. Sandini, ed., 'Proc. European Conf. Comp. Vision', Vol. 588 of *Lecture Notes in Computer Science*, Springer-Verlag, pp. 670–675.

Brunnström, K., Ekhlund, J.-O. and Uhlin, T. (1996), 'Active fixation for scene exploration', *Int. J. of Comp. Vision* **17**(2), 137–162.

Bubna, K. and Stewart, C. (2000), 'Model selection techniques and merging rules for range data segmentation algorithms', *Computer Vision and Image Understanding* **80**(2), 215–245.

Buchsbaum, G. (1980), 'A spatial processor model for object colour perception', *J. Franklin Inst.* **310**, 1–26.

Bülthoff, H., Tarr, M., Blanz, V. and Zabinski, M. (1995), 'To what extent do unique parts influence recognition across changes in viewpoint?', *Perception* **24**, 3.

Burl, M. and Perona, P. (1996), Recognition of planar object classes, *in* 'IEEE Conference on Computer Vision and Pattern Recognition', 1996, pp. 223–230.

Cabrera, J. and Meer, P. (1996), 'Unbiased estimation of ellipses by bootstrapping', *IEEE Trans. Pattern Analysis and Machine Intelligence* **18**(7), 752–756.

Callahan, J. and Weiss, R. (1985), A model for describing surface shape, *in* 'Proc. IEEE Conf. Comp. Vision Patt. Recog.', pp. 240–245.

Canny, J. (1986), 'A computational approach to edge detection', *IEEE Trans. Pattern Analysis and Machine Intelligence* **8**(6), 679–698.

Canny, J. (1988), *The Complexity of Robot Motion Planning*, MIT Press.

Carmichael, O., Huber, D. and Hebert, M. (1999), Large data sets and confusing scenes in 3-D surface matching and recognition, *in* 'Second International Conference on 3-D Digital Imaging and Modeling (3DIM'99)', pp. 358–367.

Carson, C., Thomas, M., Belongie, S., Hellerstein, J. and Malik, J. (1999), Blobworld: A system for region-based image indexing and retrieval, *in* 'Third Int. Conf. on Visual Information Systems', Lecture Notes in Computer Science 1614, Springer-Verlag, pp. 509–516.

Cascia, M. L., Sethi, S. and Sclaroff, S. (1998), Combining textual and visual cues for content based image retrieval on the web, *in* 'IEEE Workshop on Content Based Access of Image and Video Libraries', pp. 24–28.

Castore, G. (1984), Solid Modeling, Aspect Graphs, and Robot Vision, *in* Pickett and Boyse, eds, 'Solid modeling by computer', Plenum Press, pp. 277–292.

Chakravarty, I. (1982), The use of characteristic views as a basis for recognition of three-dimensional objects, Image Processing Laboratory IPL-TR-034, Rensselaer Polytechnic Institute.

Chang, S. and Yang, C. (1983), 'Picture information measures for similarity retrieval', *CVGIP: Image Understanding* **23**, 366–375.

Chang, S.-F., Chen, W. and Sundaram, H. (1998*a*), Semantic visual templates—linking visual features to semantics, *in* 'IEEE Int. Conf. Image Processing', pp. 531–535.

Chang, S.-F., Chen, W., Meng, H., Sundaram, H. and Zhong, D. (1997*a*), Videoq—an automatic content-based video search system using visual cues, *in* 'ACM Multimedia Conference'.

Chang, S.-F., Chen, W., Meng, H., Sundaram, H. and Zhong, D. (1998*b*), 'A fully automated content based video search engine supporting spatiotemporal queries', *IEEE Trans. Circuits and Systems for Video Technology* **8**(8), 602–615.

Chang, S.-F., Smith, J., Beigi, M. and Benitez, A. (1997*b*), 'Visual information retrieval from large distributed online repositories', *Comm. ACM* **40**(12), 63–71.

Chapelle, O., Haffner, P. and Vapnik, V. (1999), 'Support vector machines for histogram-based image classification', *IEEE Neural Networks* **10**(5), 1055–1064.

Chasles, M. (1855), 'Question no. 296', *Nouv. Ann. Math.*

Chellappa, R. and Jain, A. (1993), *Markov Random Fields: Theory and Applications*, Academic Press.

Chellappa, R., Wilson, C. and Sirohey, S. (1995), 'Human and machine recognition of faces: a survey', *Proceedings IEEE* **83**(5), 705–740.

Chen, S. (1995), Quicktime VR: An image-based approach to virtual environment navigation, *in* 'SIGGRAPH', pp. 29–38.

Chen, S. and Freeman, H. (1991), On the characteristic views of quadric-surfaced solids, *in* 'IEEE Workshop on Directions in Automated CAD-Based Vision', pp. 34–43.

Chin, R. and Dyer, C. (1986), 'Model-based recognition in robot vision', *ACM Computing Surveys* **18**(1), 67–108.

Chiyokura, B. and Kimura, F. (1983), 'Design of solids with free-form surfaces', *Computer Graphics* **17**(3), 289–298.

Cho, K., Meer, P. and Cabrera, J. (1997), 'Performance assessment through bootstrap', *IEEE Trans. Pattern Analysis and Machine Intelligence* **19**(11), 1185–1198.

Cho, K., Meer, P. and Cabrera, J. (1998), 'Performance assessment through bootstrap', *IEEE Trans. Pattern Analysis and Machine Intelligence* **20**(1), 94.

Christy, S. and Horaud, R. (1996), 'Euclidean shape and motion from multiple perspective views by affine iterations', *IEEE Trans. Patt. Anal. Mach. Intell.* **18**(11), 1098–1104.

Chua, C. and Jarvis, R. (1996), 'Point signatures: a new representation for 3D object recognition', *Int. J. of Comp. Vision* **25**(1), 63–85.

Chui, C. (1991), *Kalman Filtering: With Real-Time Applications*, Springer-Verlag.

Cipolla, R. and Blake, A. (1992), 'Surface shape from the deformation of the apparent contour', *Int. J. of Comp. Vision* **9**(2), 83–112.

Cipolla, R., Astrom, K. and Giblin, P. (1995), Motion from the frontier of curved surfaces, *in* 'Proc. Int. Conf. Comp. Vision', pp. 269–275.

Clarkson, K. (1988), 'A randomized algorithm for closest-point queries', *SIAM J. Computing* **17**, 830–847.

Clerc, M. and Mallat, S. (1999), Shape from texture through deformations, *in* 'Int. Conf. on Computer Vision', pp. 405–410.

Clowes, M. (1971), 'On seeing things', *Artificial Intelligence* **2**(1), 79–116.

Cohen, J. (1964), 'Dependency of the spectral reflectance curves of the munsell color chips', *Psychon. Sci.* **1**, 369–370.

Cohen, M. and Wallace, J. (1993), *Radiosity and realistic image synthesis*, Academic Press.

Collins, G. (1971), 'The calculation of multivariate polynomial resultants', *Journal of the ACM* **18**(4), 515–522.

Collins, G. (1975), *Quantifier Elimination for Real Closed Fields by Cylindrical Algebraic Decomposition*, Vol. 33 of *Lecture Notes in Computer Science*, Springer-Verlag, New York.

Connolly, C. and Stenstrom, J. (1989), 3D scene reconstruction from multiple intensity images, *in* 'Proc. IEEE Workshop on Interpretation of 3D Scenes', pp. 124–130.

Cook, R. and Torrance, K. (1987), A reflectance model for computer graphics, *in* 'ARPA Image Understanding Workshop', pp. 1–19.

Costeira, J. and Kanade, T. (1998), 'A multi-body factorization method for motion analysis', *Int. J. of Comp. Vision* **29**(3), 159–180.

Cover, T. and Thomas, J. (1991), *Elements of Information Theory*, Wiley-Interscience.

Cox, I., Zhong, Y. and Rao, S. (1996), Ratio regions: A technique for image segmentation, *in* 'Proceedings, International Conference on Pattern Recognition', pp. 557–564.

Coxeter, H. (1974), *Projective Geometry*, Springer-Verlag. Second edition.

Craig, J. (1989), *Introduction to Robotics: Mechanics and Control*, Addison-Wesley. Second edition.

Cross, G. and Jain, A. (1983), 'Markov random field texture models', *IEEE Trans. Pattern Analysis and Machine Intelligence* **5**(1), 25–39.

Cross, G., Fitzgibbon, A. and Zisserman, A. (1999), Parallax geometry of smooth surfaces in multiple views, *in* 'Proc. Int. Conf. Comp. Vision', pp. 323–329.

Csurka, G. and Faugeras, O. (1998), 'Computing 3-Dimensional project invariants from a pair of images using the grassmann-cayley algebra', *Image and Vision Computing* **16**(1), 3–12.

Csurka, G. and Faugeras, O. (1999), 'Algebraic and geometric tools to compute projective and permutation invariants', *IEEE Trans. Pattern Analysis and Machine Intelligence* **21**(1), 58–65.

Curless, B. and Levoy, M. (1996), A volumetric method for building complex models from range images, *in* 'SIGGRAPH'.

Darrell, T. and Simoncelli, E. (1993), 'Nulling' filters and the separation of transparent motions, *in* 'IEEE Conference on Computer Vision and Pattern Recognition', 1993, pp. 738–739.

Davis, L. (1975), 'A survey of edge detection techniques', *Computer Graphics Image Processing* **4**(3), 248–270.

de Bonet, J. and Viola, P. (1997), Rosetta: An image database retrieval system, *in* 'Proc. DARPA IU Workshop', pp. 655–660.

Debevec, P., Taylor, C. and Malik, J. (1996), Modeling and rendering architecture from photographs: a hybrid geometry- and image-based approach, *in* 'SIGGRAPH', pp. 11–20.

Dellaert, F., Seitz, S., Thorpe, C. and Thrun, S. (2000), Structure from motion without correspondence, *in* 'IEEE Conference on Computer Vision and Pattern Recognition', 2000, pp. II:557–564.

Demazure, M. (2000), *Bifurcations and Catastrophes*, Springer-Verlag. Translated by D. Chillingworth.

Dempster, A., Laird, N. and Rubin, D. (1977), 'Maximum likelihood from incomplete data via the EM algorithm', *Journal of the Royal Statistical Society* **39 (Series B)**, 1–38.

Deriche, R. (1987), 'Using Canny's criteria to derive a recursively implemented optimal edge detector', *International Journal of Computer Vision* **1**(2), 167–187.

Deriche, R. (1990), 'Fast algorithms for low-level vision', *IEEE Trans. Pattern Analysis and Machine Intelligence* **12**(1), 78–87.

Devernay, F. and Faugeras, O. (1994), Computing differential properties of 3D shapes from stereopsis without 3D models, *in* 'Proc. IEEE Conf. Comp. Vision Patt. Recog.', pp. 208–213.

Devroye, L., Gyorfi, L. and Lugosi, G. (1996), *A Probabilistic Theory of Pattern Recognition*, Springer Verlag.

Devy, M., Garric, V. and Orteu, J. (1997), Camera calibration from multiple views of a 2D object using a lgobal non-linear minimization method, *in* 'IEEE/RSJ International Conference on Intelligent Robots and Systems', pp. 1583–1589.

do Carmo, M. (1976), *Differential Geometry of Curves and Surfaces*, Prentice-Hall.

Dougherty, S. and Bowyer, K. (1998), Objective evaluation of edge detectors using a formally defined framework, *in* 'Workshop on Empirical Evaluation Methods in Computer Vision', p. xx.

Dove, H. (1841), 'Über stereoskopie', *Annals Phys. Series 2* **110**, 494–498.

Drew, M. and Funt, B. (1990), Calculating surface reflectance using a single-bounce model of mutual reflection, *in* 'Proceedings, Third International Conference on Computer Vision', 1990, pp. 393–399.

Driscoll, W. and Vaughan, W., eds (1978), *Handbook of Optics*, McGraw-Hill.

Duda, R. and Hart, P. (1973), *Pattern Classification and Scene Analysis*, Wiley.

Duncan, J. and Ayache, N. (2000), 'Medical image analysis: Progress over two decades and the challenges ahead', *IEEE Trans. Pattern Analysis and Machine Intelligence* **22**(1), 85–106.

D'Zmura, M. and Lennie, P. (1986), 'Mechanisms of colour constancy', *J. Opt. Soc. America - A* **3**, 1662–1672.

Eakins, J., Boardman, J. and Graham, M. (1998), 'Similarity retrieval of trademark images', *IEEE Multimedia* **5**(2), 53–63.

Ebert, D. S., Musgrave, F. K., Peachey, D., Worley, S. and Perlin, K., eds (1998), *Texturing and Modeling*, Morgan Kaufmann.

Efros, A. and Leung, T. (1999), Texture synthesis by non-parametric sampling, *in* 'Proceedings, Seventh International Conference on Computer Vision', 1999, pp. 1033–1038.

Eggert, D. and Bowyer, K. (1989), Computing the orthographic projection aspect graph of solids of revolution, *in* 'Proc. IEEE Workshop on Interpretation of 3D Scenes', pp. 102–108.

Eggert, D. and Bowyer, K. (1991), Perspective projection aspect graphs of solids of revolution: An implementation, *in* 'IEEE Workshop on Directions in Automated CAD-Based Vision', pp. 44–53.

Eggert, D., Bowyer, K., Dyer, C., Christensen, H. and Goldgof, D. (1993), 'The scale space aspect graph', *IEEE Trans. Patt. Anal. Mach. Intell.* **15**(11), 1114–1130.

Elder, J. and Zucker, S. (1998), 'Local scale control for edge detection and blur estimation', *IEEE Trans. Pattern Analysis and Machine Intelligence* **20**(7), 699–716.

Enser, P. (1993), 'Query analysis in a visual information retrieval context', *J. Document and Text Management* **1**(1), 25–52.

Enser, P. (1995), 'Pictorial information retrieval', *Journal of Documentation* **51**(2), 126–170.

Ettinger, G. (1988), Large hierarchical object recognition using libraries of parameterized model sub-parts, *in* 'IEEE Conference on Computer Vision and Pattern Recognition', 1988, pp. 32–41.

Fairchild, M. (1998), *Color Appearance Models*, Addison-Wesley.

Fan, T., Médioni, G. and Nevatia, R. (1987), 'Segmented descriptions of 3D surfaces', *IEEE Transactions on Robotics and Automation* **3**(6), 527–538.

Farin, G. (1993), *Curves and Surfaces for Computer Aided Geometric Design*, Academic Press, San Diego, CA.

Faugeras, O. (1992), What can be seen in three dimensions with an uncalibrated stereo rig?, *in* G. Sandini, ed., 'Proc. European Conf. Comp. Vision', Vol. 588 of *Lecture Notes in Computer Science*, Springer-Verlag, pp. 563–578.

Faugeras, O. (1993), *Three-Dimensional Computer Vision*, MIT Press.

Faugeras, O. (1995), 'Stratification of 3D vision: projective, affine and metric representations', *J. Opt. Soc. Am. A* **12**(3), 465–484.

Faugeras, O. and Hebert, M. (1986), 'The representation, recognition, and locating of 3-D objects', *International Journal of Robotics Research* **5**(3), 27–52.

Faugeras, O. and Maybank, S. (1990), 'Motion from point matches: multiplicity of solutions', *Int. J. of Comp. Vision* **4**(3), 225–246.

Faugeras, O. and Mourrain, B. (1995), On the geometry and algebra of the point and line correspondences between *n* images, Technical Report 2665, INRIA Sophia-Antipolis.

Faugeras, O. and Papadopoulo, T. (1997), Gaussman-Caylay algebra for modeling systems of cameras and the algebraic equations of the manifold of trifocal tensors, Technical Report 3225, INRIA Sophia-Antipolis.

Faugeras, O., Hebert, M., Pauchon, E. and Ponce, J. (1984), Object representation, identification, and positioning from range data, *in* 'Robotics Research: The First International Symposium', MIT Press, pp. 425–446.

Faugeras, O., Luong, Q.-T. and Papadopoulo, T. (2001), *The Geometry of Multiple Images*, MIT Press.

Felzenszwalb, P. and Huttenlocher, D. (2000), Efficient matching of pictorial structures, *in* 'IEEE Conference on Computer Vision and Pattern Recognition', 2000, pp. II:66–73.

Feng, X. and Perona, P. (1998), Scene segmentation from 3D motion, *in* 'IEEE Conference on Computer Vision and Pattern Recognition', 1998, pp. 225–231.

Ferryman, J., Maybank, S. and Worrall, A. (2000), 'Visual surveillance for moving vehicles', *International Journal of Computer Vision* **37**(2), 187–197.

Finlayson, G. and Hordley, S. (1999), 'Selection for gamut mapping colour constancy', *Image and Vision Computing* **17**(8), 597–604.

Finlayson, G. and Hordley, S. (2000), 'Improving gamut mapping color constancy', *IEEE Trans. Image Processing* **9**(10), 1774–1783.

Finlayson, G., Chatterjee, S. and Funt, B. (1996), Color angular indexing, *in* 'European Conference on Computer Vision', 1996, pp. II:16–27.

Finlayson, G., Drew, M. and Funt, B. (1994*a*), 'Color constancy: Generalized diagonal transforms suffice', *Journal of the Optical Society of America* **11**(11), 3011–3019.

Finlayson, G., Drew, M. and Funt, B. (1994*b*), 'Spectral sharpening: Sensor transformations for improved color constancy', *Journal of the Optical Society of America* **11**(5), 1553–1563.

Fischler, M. and Bolles, R. (1981), 'Random sample consensus: A paradigm for model fitting with applications to image analysis and automated cartography', *Communications of the ACM* **24**(6), 381–395.

Fitzgibbon, A. and Zisserman, A. (1998), Automatic 3D model acquisition and generation of new images from video sequences, *in* 'European Signal Processing Conference', pp. 311–326.

Fleck, M. (1992*a*), 'Multiple widths yield reliable finite differences', *IEEE Trans. Pattern Analysis and Machine Intelligence* **14**(4), 412–429.

Fleck, M. (1992*b*), 'Some defects in finite-difference edge finders', *IEEE Trans. Pattern Analysis and Machine Intelligence* **14**(3), 337–345.

Fleck, M. (1995), Cs-tr 95-01: Perspective projection: the wrong imaging model, Technical report, University of Iowa.

Fleck, M., Forsyth, D. and Bregler, C. (1996), Finding naked people, *in* 'European Conference on Computer Vision', 1996, pp. II:593–602.

Flickner, M., Sawhney, H., Niblack, W. and Ashley, J. (1995), 'Query by image and video content: the QBIC system', *Computer* **28**(9), 23–32.

Flock, H. (1984), 'Illumination: inferred or observed?', *Perception and Psychophysics*.

Foley, J., van Dam, A., Feiner, S. and Hughes, J. (1990), *Computer Graphics: Principle and Practice*, Addison-Wesley. Second edition.

Forney, G. (1973), 'The Viterbi algorithm', *Proceedings of the IEEE*.

Forsyth, D. (1990), 'A novel approach to color constancy', *International Journal of Computer Vision* **5**(1), 5–36.

Forsyth, D. (1996), 'Recognizing algebraic surfaces from their outlines', *International Journal of Computer Vision* **18**(1), 21–40.

Forsyth, D. (1999), Sampling, resampling and colour constancy, *in* 'IEEE Conference on Computer Vision and Pattern Recognition', 1999, pp. I:300–305.

Forsyth, D. and Fleck, M. (1997), Body plans, *in* 'IEEE Conference on Computer Vision and Pattern Recognition', 1997, pp. 678–683.

Forsyth, D. and Fleck, M. (1999), 'Automatic detection of human nudes', *International Journal of Computer Vision* **32**(1), 63–77.

Forsyth, D. and Zisserman, A. (1989), Mutual illumination, *in* 'IEEE Conference on Computer Vision and Pattern Recognition', 1989, pp. 466–473.

Forsyth, D. and Zisserman, A. (1990), 'Shape from shading in the light of mutual illumination', *Image and Vision Computing* **8**, 42–29.

Forsyth, D. and Zisserman, A. (1991), 'Reflections on shading', *IEEE Trans. Pattern Analysis and Machine Intelligence* **13**(7), 671–679.

Forsyth, D., Mundy, J., Zisserman, A. and Rothwell, C. (1992), Recognizing rotationally symmetric surfaces from their outlines, *in* Sandini (1992), pp. 639–647.

Forsyth, D., Mundy, J., Zisserman, A. and Rothwell, C. (1994), Using global consistency to recognise euclidean objects with an uncalibrated camera, *in* 'IEEE Conference on Computer Vision and Pattern Recognition', 1994, pp. 502–507.

Forsyth, D., Mundy, J., Zisserman, A., Coelho, C., Heller, A. and Rothwell, C. (1991), 'Invariant descriptors for 3-D object recognition and pose', *IEEE Trans. Pattern Analysis and Machine Intelligence* **13**(10), 971–991.

Freeman, W. and Adelson, E. (1991), 'The design and use of steerable filters', *IEEE Trans. Pattern Analysis and Machine Intelligence* **13**(9), 891–906.

Freeman, W. and Brainard, D. (1997), 'Bayesian color constancy', *Journal of the Optical Society of America* **14**(7), 1393–1411.

Freeman, W., Anderson, D. and et al., P. B. (1998), 'Computer vision for interactive computer graphics', *Computer Graphics and Applications* pp. 42–53.

Freeman, W., Pasztor, E. and Carmichael, O. (2000), 'Learning low-level vision', *International Journal of Computer Vision* **40**(1), 25–47.

Friedman, J., Bentley, J. and Finkel, R. (1977), 'An algorithm for finding best matches in logarithmic expected time', *ACM Trans. on Math. Software*.

Frisby, J. (1980), *Seeing: Illusion, Brain and Mind*, Oxford University Press.

Fu, K. and Mui, J. (1981), 'A survey of image segmentation', *Pattern Recognition* **13**(1), 3–16.

Fukunaga, K. (1990), *Introduction to Statistical Pattern Recognition*, Academic Press. 2nd edition.

Funt, B. and Drew, M. (1988), Color constancy computation in near-mondrian scenes using a finite dimensional linear model, *in* 'IEEE Conference on Computer Vision and Pattern Recognition', 1988, pp. 544–549.

Funt, B. and Drew, M. (1993), 'Color space analysis of mutual illumination', *IEEE Trans. Pattern Analysis and Machine Intelligence* **15**(12), 1319–1326.

Funt, B. and Finlayson, G. (1995), 'Color constant color indexing', *IEEE Trans. Pattern Analysis and Machine Intelligence* **17**(5), 522–529.

Funt, B., Barnard, K. and Martin, L. (1998), Is machine colour constancy good enough?, *in* 'European Conference on Computer Vision', 1998, pp. 445–459.

Funt, B., Drew, M. and Ho, J. (1991), 'Color constancy from mutual reflection', *International Journal of Computer Vision* **6**(1), 5–24.

Garcia-Bermejo, J., Diaz Pernas, F. and Coronado, J. (1996), An approach for determining bidirectional reflectance parameters from range and brightness data, *in* 'Proceedings, International Conference on Image Processing', p. 16A2.

Garding, J. (1992), Shape from texture for smooth curved surfaces, *in* 'European Conference on Computer Vision', pp. 630–638.

Garding, J. (1995), Surface orientation and curvature from differential texture distortion, *in* 'Int. Conf. on Computer Vision', pp. 733–739.

Gaston, P. and Lozano-Pérez, T. (1984), 'Tactile recognition and localization using object models: The case of polyhedra in the plane', *IEEE Trans. Patt. Anal. Mach. Intell.*

Gear, C. (1998), 'Multibody grouping in moving objects', *Int. J. of Comp. Vision* **29**(2), 133–150.

Genc, Y. and Ponce, J. (1998), Parameterized image varieties: A novel approach to the analysis and synthesis of image sequences, *in* 'Proc. Int. Conf. Comp. Vision', pp. 11–16.

Genc, Y. and Ponce, J. (2001), 'Image-based rendering using parameterized image varieties', *Int. J. of Comp. Vision* **41**(3), 143–170.

Gennery, D. (1980), Modelling the environment of an exploring vehicle by means of stereo vision, PhD thesis, Stanford University.

Georghiades, A., Belhumeur, P. and Kriegman, D. (2000), From few to many: Generative models for recognition under variable pose and illumination, *in* 'International Conference on Automatic Face and Gesture Recognition', 1900, pp. 277–284.

Georghiades, A., Kriegman, D. and Belhumeur, P. (1998), Illumination cones for recognition under variable lighting: Faces, *in* 'IEEE Conference on Computer Vision and Pattern Recognition', 1998, pp. 52–59.

Gerig, G., Pun, T. and Ratib, O. (1994), 'Image analysis and computer vision in medicine', *Computerized Medical Imaging and Graphics* **18**(2), 85–96.

Gersho, A. and Gray, R. (1992), *Vector quantization and signal compression*, Kluwer Academic Publishers.

Gershon, R. (1987), The Use of Color in Computational Vision, PhD thesis, University of Rochester.

Gershon, R., Jepson, A. and Tsotsos, J. (1986), 'Ambient illumination and the determination of material changes', *J. Opt. Soc. America* **A-3**(10), 1700–1707.

Giblin, P. and Kimia, B. (1999), On the local form and transitions of symmetry sets, medial axes, and shocks, *in* 'Proceedings, Seventh International Conference on Computer Vision', 1999, pp. 385–391.

Gigus, Z. and Malik, J. (1990), 'Computing the aspect graph for line drawings of polyhedral objects', *IEEE Trans. Patt. Anal. Mach. Intell.* **12**(2), 113–122.

Gigus, Z., Canny, J. and Seidel, R. (1991), 'Efficiently computing and representing aspect graphs of polyhedral objects', *IEEE Trans. Patt. Anal. Mach. Intell.*

Gilchrist, A., Kossyfidis, C., Bonato, F., Agostini, T., Cataliotti, J., Li, X., Spehar, B., Annan, V. and Economou, E. (1999), 'An anchoring theory of lightness perception', *Psychological Review* **106**(4), 795–834.

Gill, P., Murray, W. and Wright, M. (1981), *Practical Optimization*, Academic Press.

Gordon, I. (1997), *Theories of Visual Perception*, John Wiley and Son.

Gortler, S., Grzeszczuk, R., Szeliski, R. and Cohen, M. (1996), The lumigraph, *in* 'SIGGRAPH', pp. 43–54.

Greenspan, H., Belongie, S., Perona, P., Goodman, R., Rakshit, S. and Anderson, C. (1994), Overcomplete steerable pyramid filters and rotation invariance, *in* 'IEEE Conference on Computer Vision and Pattern Recognition', 1994, pp. 222–228.

Grimson, W. (1981*a*), 'A computer implementation of a theory of human stereo vision', *Philosophical Transactions of the Royal Society of London* pp. 217–253.

Grimson, W. (1981*b*), *From images to surfaces:A Computational Study of the Human Early Visual System*, MIT Press.

Grimson, W. (1992), 'The cost of choosing the wrong model in object recognition by constrained search', *International Journal of Computer Vision* **7**(3), 195–210.

Grimson, W. and Huttenlocher, D. (1990*a*), On the sensitivity of geometric hashing, *in* 'Proceedings, Third International Conference on Computer Vision', 1990, pp. 334–338.

Grimson, W. and Huttenlocher, D. (1990*b*), 'On the sensitivity of the Hough transform for object recognition', *IEEE Trans. Pattern Analysis and Machine Intelligence* **12**(3), 255–274.

Grimson, W. and Huttenlocher, D. (1991), 'On the verification of hypothesized matches in model-based recognition', *IEEE Trans. Pattern Analysis and Machine Intelligence* **13**(12), 1201–1213.

Grimson, W. and Lozano-Perez, T. (1984), 'Model-based recognition and localization from sparse range or tactile data', *International Journal of Robotics Research* **3**(3), 3–35.

Grimson, W. and Lozano-Pérez, T. (1987), 'Localizing overlapping parts by searching the interpretation tree', *IEEE Trans. Patt. Anal. Mach. Intell.* **9**(4), 469–482.

Grimson, W., Huttenlocher, D. and Alter, T. (1992), Recognizing 3D objects from 2D images: An error analysis, *in* 'IEEE Conference on Computer Vision and Pattern Recognition', 1992, pp. 316–321.

Grimson, W., Huttenlocher, D. and Jacobs, D. (1994), 'A study of affine matching with bounded sensor error', *International Journal of Computer Vision* **13**(1), 7–32.

Grimson, W., Lozano-Perez, T. and Huttenlocher, D. (1990), *Object Recognition by Computer: The Role of Geometric Constraints*, MIT Press.

Gross, A. and Boult, T. (1988), Error of fit measures for recovering parametric solids, *in* 'Proc. Int. Conf. Comp. Vision', pp. 690–694.

Haddon, J. and Forsyth, D. (1998*a*), Shading primitives: Finding folds and shallow grooves, *in* 'Proceedings, Sixth International Conference on Computer Vision', 1998, pp. 236–241.

Haddon, J. and Forsyth, D. (1998*b*), Shape representations from shading primitives, *in* 'European Conference on Computer Vision', 1998, pp. 415–431.

Hamilton, W. (1844), 'On a new species of imaginary quantities connected with a theory of quaternions', *Transactions of the Royal Irish Academy* **2**, 424–434.

Hampapur, A., Gupta, A., Horowitz, B. and Shu, C.-F. (1997), Virage video engine, *in* 'Storage and Retrieval for Image and Video Databases V—Proceedings of the SPIE', Vol. 3022, pp. 188–198.

Haralick, R. (1983), 'Ridges and valleys in digital images', *Computer Vision, Graphics and Image Processing* **22**, 28–38.

Haralick, R. and Shapiro, L. (1985), 'Image segmentation techniques', *CVGIP: Image Understanding* **29**(1), 100–132.

Haralick, R. and Shapiro, L. (1992), *Computer and robot vision*, Addison Wesley.

Haralick, R., Watson, L. and Laffey, T. (1983), 'The topographic primal sketch', *International Journal of Robotics Research* **2**, 50–72.

Hardin, C. and Maffi, L. (1997), *Color Categories in thought and language*, Cambridge University Press.

Harris, C. and Stephens, M. (1988), A combined corner and edge detector, *in* 'Alvey Conference', pp. 147–152.

Hartley, R. (1994*a*), An algorithm for self calibration from several views, *in* 'Proc. IEEE Conf. Comp. Vision Patt. Recog.', pp. 908–912.

Hartley, R. (1994*b*), 'Projective reconstruction and invariants from multiple images', *IEEE Trans. Patt. Anal. Mach. Intell.* **16**(10), 1036–1041.

Hartley, R. (1995), In defence of the 8-point algorithm, *in* 'Proc. Int. Conf. Comp. Vision', pp. 1064–1070.

Hartley, R. (1997), 'Lines and points in three views and the trifocal tensor', *Int. J. of Comp. Vision* **22**(2), 125–140.

Hartley, R. (1998), Computation of the quadrifocal tensor, *in* 'Proc. European Conf. Comp. Vision', pp. 20–35.

Hartley, R. and Sturm, P. (1997), 'Triangulation', *Computer Vision and Image Understanding* **68**(2), 146–157.

Hartley, R. and Zisserman, A. (2000), *Multiple view geometry in computer vision*, Cambridge University Press.

Hartley, R., Gupta, R. and Chang, T. (1992), Stereo from uncalibrated cameras, *in* 'Proc. IEEE Conf. Comp. Vision Patt. Recog.', pp. 761–764.

Hartley, R., Wang, C., Kitchen, L. and Rosenfeld, A. (1982), 'Segmentation of FLIR images: A comparative study', *IEEE Trans. Systems, Man and Cybernetics* **12**(4), 553–566.

Hastie, T., Tibshirani, R. and Friedman, J. (2001), *The Elements of Statistical Learning: Data Mining, Inference and Prediction*, Springer Verlag.

Havaldar, P., Lee, M. and Medioni, G. (1996), View synthesis from unregistered 2D images, *in* 'Graphics Interface'96', pp. 61–69.

Haykin, S. (1999), *Neural Networks: A Comprehensive Introduction*, Prentice-Hall.

Healey, G. and Kondepudy, R. (1994), 'Radiometric CCD camera calibration and noise estimation', *IEEE Trans. Patt. Anal. Mach. Intell.* **16**(3), 267–276.

Heath, M. (2002), *Scientific Computing: An Introductory Survey*, McGraw-Hill. Second edition.

Heath, M., Sarkar, S., Sanocki, T. and Bowyer, K. (1997), 'A robust visual method for assessing the relative performance of edge detection algorithms', *IEEE Trans. Pattern Analysis and Machine Intelligence* **19**(12), 1338–1359.

Hebert, M. (2000), Active and passive range sensing for robotics, *in* 'IEEE Int. Conf. on Robotics and Automation'.

Hebert, M. and Kanade, T. (1985), The 3D profile method for object recognition, *in* 'Proc. IEEE Conf. Comp. Vision Patt. Recog.', pp. 458–463.

Hecht, E. (1987), *Optics*, Addison-Wesley.

Hel-Or, Y. and Teo, P. (1996), Canonical decomposition of steerable functions, *in* 'IEEE Conference on Computer Vision and Pattern Recognition', 1996, pp. 809–816.

Helmholtz, H. (1909), *Physiological Optics*, Dover. 1962 edition of the English translation of the 1909 German original, first published by the Optical Society of America in 1924.

Helson, H. (1934), 'Some factors and implications of colour constancy', *J. Opt. Soc. America* **48**, 555–567.

Helson, H. (1938*a*), Fundamental problems in color vision, i, *in* 'Journal of Experimental Psychology', Vol. 23.

Helson, H. (1938*b*), Fundamental problems in color vision, ii, *in* 'Journal of Experimental Psychology', Vol. 26.

Herskovits, A. and Binford, T. (1970), On boundary detection, Technical report, MIT AI Lab.

Hesse, O. (1863), 'Die cubische Gleichung, von welcher die Lösung des Problems der Homographie von M. Chasles abhängt', *J. Reine Angew. Math.* **62**, 188–192.

Heyden, A. (1995), Geometry and algebra of multiple projective transformations, PhD thesis, Lund University, Sweden.

Heyden, A. (1998), A common framework for multiple view tensors, *in* 'Proc. European Conf. Comp. Vision', pp. 3–19.

Heyden, A. and Åström, K. (1996), Euclidean reconstruction from constant intrinsic parameters, *in* 'International Conference on Pattern Recognition', pp. 339–343.

Heyden, A. and Åström, K. (1998), Minimal conditions on intrinsic parameters for Euclidean reconstruction, *in* 'Asian Conference on Computer Vision'.

Heyden, A. and Åström, K. (1999), Flexible calibration: minimal cases for auto-calibration, *in* 'Proc. Int. Conf. Comp. Vision', pp. 350–355.

Hilbert, D. and Cohn-Vossen, S. (1952), *Geometry and the Imagination*, Chelsea.

Holst, G. (1998), *CCD Arrays, Cameras and Displays*, SPIE Press.

Holt, B. and Hartwick, L. (1994*a*), 'Quick, who painted fish?': searching a picture database with the QBIC project at UC Davis', *Information Services and Use* **14**(2), 79–90.

Holt, B. and Hartwick, L. (1994*b*), Retrieving art images by image content: the UC Davis QBIC project, *in* 'ASLIB Proceedings', Vol. 46, pp. 243–8.

Horn, B. (1970), Shape from shading: A method for obtaining the shape of a smooth opaque object from one view, Technical report, MIT AI Lab.

Horn, B. (1971), The Binford-Horn line finder, Technical report, MIT AI Lab.

Horn, B. (1974), 'Determining lightness from an image', *Computer Graphics Image Processing* **3**(1), 277–299.

Horn, B. (1975), Obtaining shape from shading information, *in* 'The Psychology of Computer Vision', McGraw-Hill, pp. 115–155.

Horn, B. (1977), 'Understanding image intensities', *Artificial Intelligence* **8**(2), 201–231.

Horn, B. (1984), 'Extended Gaussian images', *Proceedings of the IEEE* **72**(12), 1671–1686.

Horn, B. (1986), *Robot Vision*, MIT Press, Cambridge, MA.

Horn, B. (1987), 'Closed-form solution of absolute orientation using unit quaternions', *J. Opt. Soc. Am. A* **4**(4), 629–642.

Horn, B. (1990), 'Height and gradient from shading', *International Journal of Computer Vision* **5**(1), 37–76.

Horn, B. and Brooks, M. (1989), *Shape from Shading*, MIT Press.

Horn, B. and Schunck, B. (1981), 'Determining optical flow', *Artificial Intelligence* **17**, 185–203.

Horn, B., Woodham, R. and Silver, W. (1978), Determining shape and reflectance using multiple images, Technical report, MIT AI Lab.

Hsu, S., Anandan, P. and Peleg, S. (1994), Accurate computation of optical flow by using layered motion representations, *in* 'Proceedings, International Conference on Pattern Recognition', pp. A:743–746.

Huang, J. and Zabih, R. (1998), Combining color and spatial information for content-based image retrieval, *in* 'European Conference on Digital Libraries'. Web version at http://www.cs.cornell.edu/html/rdz/Papers/ECDL2/spatial.htm.

Huang, J., Kumar, S., Mitra, M., Zhu, W.-J. and Zabih, R. (1997), Image indexing using color correlograms, *in* 'IEEE Conf. on Computer Vision and Pattern Recognition', pp. 762–768.

Huang, T. (1981), *Image Sequence Analysis*, Springer-Verlag.

Huang, T. and Faugeras, O. (1989), 'Some properties of the E-matrix in two-view motion estimation', *IEEE Trans. Patt. Anal. Mach. Intell.* **11**(12), 1310–1312.

Huber, P. (1981), *Robust Statistics*, Wiley.

Hueckel, M. (1971), 'An operator which locates edges in digitized pictures', *Journal of the ACM* **18**(1), 113–125.

Huffman, D. (1977), 'Realizable configurations of lines in pictures of polyhedra', *Machine Intelligence* **8**, 493–509.

Huttenlocher, D. and Ullman, S. (1987), Object recognition using alignment, *in* 'Proc. Int. Conf. Comp. Vision', pp. 102–111.

Huttenlocher, D. and Ullman, S. (1990), 'Recognizing solid objects by alignment with an image', *International Journal of Computer Vision* **5**(2), 195–212.

Huttenlocher, D. and Wayner, P. (1992), 'Finding convex edge groupings in an image', *International Journal of Computer Vision* **8**(1), 7–27.

Ikeuchi, K. (1987*a*), 'Determining a depth map using a dual photometric stereo', *International Journal of Robotics Research*.

Ikeuchi, K. (1987*b*), Precompiling a geometrical model into an interpretation tree for object recognition in bin-picking tasks, *in* 'Proc. DARPA Image Understanding Workshop', pp. 321–339.

Ikeuchi, K. and Kanade, T. (1988), 'Automatic generation of object recognition programs', *Proceedings of the IEEE* **76**(8), 1016–35.

Ioffe, S. and Forsyth, D. (1998), Learning to find pictures of people, *in* 'NIPS'.

Ioffe, S. and Forsyth, D. (1999), Finding people by sampling, *in* 'Proceedings, Seventh International Conference on Computer Vision', 1999, pp. 1092–1097.

Irani, M., Anandan, P., Bergen, J., Kumar, R. and Hsu, S. (1996), 'Mosaic representations of video sequences and their applications', *Signal Processing: Image Communication*.

Jacobs, C., Finkelstein, A. and Salesin, D. (1995), Fast multiresolution image querying, *in* 'Proc SIGGRAPH-95', pp. 277–285.

Jacobs, D., Belhumeur, P. and Basri, R. (1998), Comparing images under variable illumination, *in* 'IEEE Conference on Computer Vision and Pattern Recognition', 1998, pp. 610–617.

Jacobsen, A. and Gilchrist, A. (1988), 'The ratio principle holds over a million-to-one range of illumination', *Perception and Psychophysics* **43**, 1–6.

Jain, A. and Vailaya, A. (1998), 'Shape-based retrieval: a case study with trademark image databases', *Pattern Recognition* **31**(9), 1369–1390.

Janesick, J., Elliott, T., Collins, S., Blouke, M. and Freeman, J. (1987), 'Scientific charge-coupled devices', *Optical Engineering* **26**, 692–714.

Jarvis, R. (1983), 'A perspective on range finding techniques in computer vision', *IEEE Trans. Patt. Anal. Mach. Intell.* **5**(2), 122–139.

Jelinek, F. (1999), *Statistical Methods for Speech Recognition (Language, Speech and Communication)*, MIT Press.

Jepson, A. and Black, M. (1993), Mixture models for optical flow computation, *in* 'IEEE Conference on Computer Vision and Pattern Recognition', 1993, pp. 760–761.

Johnson, A. and Hebert, M. (1998), 'Surface matching for object recognition in complex three-dimensional scenes', *Image and Vision Computing* **16**, 635–651.

Johnson, A. and Hebert, M. (1999), 'Using spin images for efficient object recognition in cluttered 3D scenes', *IEEE Trans. Patt. Anal. Mach. Intell.* **21**(5), 433–449.

Jones, M. and Rehg, J. (1999), Statistical color models with application to skin detection, *in* 'IEEE Conference on Computer Vision and Pattern Recognition', 1999, pp. I:274–280.

Joshi, T., Ahuja, N. and Ponce, J. (1999), 'Structure and motion estimation from dynamic silhouettes under perspective projection', *Int. J. of Comp. Vision* **31**(1), 31–50.

Judd, D. (1940), 'Hue, saturation and lightness of surface colors with chromatic illumination', *Journal of the Optical Society of America* **30**(1), 2–32.

Julesz, B. (1960), 'Binocular depth perception of computer-generated patterns', *The BellSystem Technical Journal* **39**(5), 1125–1162.

Julesz, B. (1971), *Foundations of Cyclopean Perception*, The University of Chicago Press.

Julez, B. (1959), 'A method of coding tv signals based on edge detection', *Bell System Tech. J.* **38**(4), 1001–1020.

Kakarala, R. and Hero, A. (1992), 'On achievable accuracy in edge localization', *IEEE Trans. Pattern Analysis and Machine Intelligence* **14**(7), 777–781.

Kanade, T. (1973), Picture processing by computer complex and recognition of human faces, Technical report, Kyoto University, Dept. of Information Science.

Kanade, T. (1981), 'Recovery of the three-dimensional shape of an object from a single view', *Artificial Intelligence Journal* **17**, 409–460.

Kanade, T., Rander, P. and Narayanan, J. (1997), 'Virtualized reality: Constructing virtual worlds from real scenes', *IEEE Multimedia* **4**(1), 34–47.

Kanatani, K. (1998), 'Geometric information criterion for model selection', *International Journal of Computer Vision* **26**(3), 171–189.

Kanizsa, G. (1976), 'Subjective contours', *Scientific American*.

Kanizsa, G. (1979), *Organization in Vision: Essays on Gestalt Perception*, Praeger.

Karasaridis, A. and Simoncelli, E. (1996), A filter design technique for steerable pyramid image transforms, *in* 'International Conference on Acoustics, Speech and Signal Processing', pp. 2387–90.

Kato, T. and Fujimura, K. (1990), 'Trademark: Multimedia image database system with intelligent human interface', *Systems and Computers in Japan* **21**(11), 33–45.

Kato, T., Shimogaki, H., Mizutori, T. and Fujimura, K. (1988), Trademark: Multimedia database with abstracted representation on knowledge base, *in* 'Proc. Second Int Symp on Interoperable Information Systems', pp. 245–252.

Kelly, R., McConnell, P. and Mildenberger, S. (1977), 'The Gestalt photomapping system', *Photogrammetric Engineering and Remote Sensing* **43**(11), 1407–1417.

Keren, D., Cooper, D. and Subrahmonia, J. (1994), 'Describing complicated objects by implicit polynomials', *IEEE Trans. Patt. Anal. Mach. Intell.* **16**(1), 38–53.

Kergosien, Y. (1981), 'La famille des projections orthogonales d'une surface et ses singularités', *C.R. Acad. Sc. Paris* **292**, 929–932.

Kimia, B., Tannenbaum, A. and Zucker, S. (1990), Toward a computational theory of shape: An overview, *in* 'European Conference on Computer Vision', 1990, pp. 402–407.

Kimia, B., Tannenbaum, A. and Zucker, S. (1995), 'Shapes, shocks, and deformations i: The components of 2-dimensional shape and the reaction-diffusion space', *International Journal of Computer Vision* **15**(3), 189–224.

Kinoshita, K. and Lindenbaum, M. (2000), Camera model selection based on geometric AIC, *in* 'IEEE Conference on Computer Vision and Pattern Recognition', 2000, pp. II:514–519.

Klinker, G., Shafer, S. and Kanade, T. (1987*a,b*), Using a color reflection model to separate highlights from object color, *in* 'Proceedings, First International Conference on Computer Vision', pp. 145–150.

Klinker, G., Shafer, S. and Kanade, T. (1990), 'A physical approach to color image understanding', *International Journal of Computer Vision* **4**(1. January 1990), 7–38.

Koenderink, J. (1984), 'What does the occluding contour tell us about solid shape?', *Perception* **13**, 321–330.

Koenderink, J. (1986), An internal representation for solid shape based on the topological properties of the apparent contour, *in* W. Richards and S. Ullman, eds, 'Image Understanding: 1985-86', Ablex Publishing Corp., chapter 9, pp. 257–285.

Koenderink, J. (1990), *Solid Shape*, MIT Press.

Koenderink, J. and Van Doorn, A. (1976*a*), 'Geometry of binocular vision and a model for stereopsis', *Biological Cybernetics* **21**, 29–35.

Koenderink, J. and Van Doorn, A. (1976*b*), 'The singularities of the visual mapping', *Biological Cybernetics* **24**, 51–59.

Koenderink, J. and Van Doorn, A. (1979), 'The internal representation of solid shape with respect to vision', *Biological Cybernetics* **32**, 211–216.

Koenderink, J. and van Doorn, A. (1980), 'Photometric invariants related to solid shape', *Optica Acta* **27**(7), 981–996.

Koenderink, J. and van Doorn, A. (1983), 'Geometrical modes as a general method to treat diffuse interreflections in radiometry', *Journal of the Optical Society of America* **73**(6), 843–850.

Koenderink, J. and van Doorn, A. (1986), 'Dynamic shape', *Biological Cybernetics* **53**, 383–396.

Koenderink, J. and Van Doorn, A. (1990), 'Affine structure from motion', *J. Opt. Soc. Am. A* **8**, 377–385.

Koenderink, J. and Van Doorn, A. (1997), 'The generic bilinear calibration-estimation problem', *Int. J. of Comp. Vision* **23**(3), 217–234.

Koenderink, J., van Doorn, A., Dana, K. and Nayar, S. (1999), 'Bidirectional reflection distribution function of thoroughly pitted surfaces', *International Journal of Computer Vision* **31**(2/3), 129–144.

Kofakis, P. and Orphanoudakis, S. (1991), Graphical tools and retrieval strategies for medical image databases, *in* 'Proceedings of the International Symposium on Computer Assisted Radiology', Springer-Verlag, pp. 519–524.

Koffka, K. (1935), *Principles of Gestalt Psychology*, Harcourt Brace.

Kriegman, D. and Belhumeur, P. (1998), What shadows reveal about object structure, *in* 'European Conference on Computer Vision', 1998, pp. 399–414.

Kriegman, D. and Ponce, J. (1990*a*), 'Computing exact aspect graphs of curved objects: solids of revolution', *Int. J. of Comp. Vision* **5**(2), 119–135.

Kriegman, D. and Ponce, J. (1990*b*), 'On recognizing and positioning curved 3-D objects from image contours', *IEEE Trans. Pattern Analysis and Machine Intelligence* **12**(12), 1127–1137.

Krinov, E. (1947), Spectral reflectance properties of natural formations, Technical report, National Research Council of Canada, Technical Translation: TT-439.

Kruppa, E. (1913), 'Zur ermittung eines objektes aus zwei perspektiven mit innerer orientierung', *Sitz.-Ber. Akad. Wiss., Wien, Math. Naturw. Kl., Abt. Ila.* **122**, 1939–1948.

Kube, P. and Perona, P. (1996), 'Scale-space properties of quadratic feature-detectors', *IEEE Trans. Pattern Analysis and Machine Intelligence* **18**(10), 987–999.

Kutulakos, K. and Seitz, S. (1999), A theory of shape by space carving, *in* 'Proc. Int. Conf. Comp. Vision', pp. 307–314.

Kutulakos, K. and Vallino, J. (1998), 'Calibration-free augmented reality', *IEEE Transactions on Visualization and Computer Graphics* **4**(1), 1–20.

Lamb, T. and Bourriau, J., eds (1995), *Colour Art and Science*, Cambridge University Press.

Lamdan, Y., Schwartz, J. and Wolfson, H. (1990), 'Affine invariant model-based object recognition', *IEEE Trans. Robotics and Automation* **6**, 578–589.

Land, E. (1959*a*), 'Color vision and the natural image: Part i', *Proceedings National Academy Science USA* **45**(1), 115–129.

Land, E. (1959*b*), 'Color vision and the natural image: Part ii', *Proceedings National Academy Science USA* **45**(4), 636–644.

Land, E. (1959*c*), 'Experiments in color vision', *Scientific American* **200**, 84–89.

Land, E. (1983), 'Color vision and the natural image', *Proceedings National Academy Science USA* **80**, 5163–5169.

Land, E. and McCann, J. (1971), 'Lightness and retinex theory', *Journal of the Optical Society of America* **61**(1), 1–11.

Laurentini, A. (1995), 'How far 3D shapes can be understood from 2D silhouettes', *IEEE Trans. Patt. Anal. Mach. Intell.* **17**(2), 188–194.

Lavallée, S. (1996), Registration for computer integrated surgery: Methodology and state of the art, *in* R. Taylor, S. Lavallée, G. Burdea and R. Mosges, eds, 'Computer Integrated Surgery', MIT Press.

Laveau, S. and Faugeras, O. (1994), 3D scene representation as a collection of images and fundamental matrices, Technical Report 2205, INRIA Sophia-Antipolis.

Lecun, Y., Bottou, L., Bengio, Y. and Haffner, P. (1998), 'Gradient-based learning applied to document recognition', *Proceedings of the IEEE* **86**(11), 2278–2324.

Lee, H. (1986), 'Method for computing the scene-illuminant chromaticity from specular highlights', *J. Opt. Soc. Am.-A* **3**, 1694–1699.

Lee, S. and Bajcsy, R. (1992*a*), Detection of specularity using colour and multiple views, *in* Sandini (1992), pp. 99–114.

Lee, S. and Bajcsy, R. (1992*b*), 'Detection of specularity using colour and multiple views', *Image and Vision Computing* **10**, 643–653.

Leung, T. and Malik, J. (1997), On perpendicular texture or: Why do we see more flowers in the distance?, *in* 'IEEE Conference on Computer Vision and Pattern Recognition', 1997, pp. 807–813.

Leung, T., Burl, M. and Perona, P. (1995), Finding faces in cluttered scenes using labelled random graph matching, *in* 'Proceedings, Fifth International Conference on Computer Vision', 1995, pp. 637–644.

Levoy, M. and Hanrahan, P. (1996), Light field rendering, *in* 'SIGGRAPH', pp. 31–42.

Lim, H. and Binford, T. (1988), Curved surface reconstruction using stereo correspondence, *in* 'Proc. DARPA Image Understanding Workshop', pp. 809–819.

Lipson, P., Grimson, W. L. and Sinha, P. (1997), Configuration based scene classification and image indexing, *in* 'IEEE Conf. on Computer Vision and Pattern Recognition', pp. 1007–1013.

Ljung, L. (1995), System identification, *in* W. S. Levine, ed., 'The Control Handbook', CRC Press, in cooperation with IEEE Press.

Longuet-Higgins, H. (1981), 'A computer algorithm for reconstructing a scene from two projections', *Nature* **293**, 133–135.

Loop, C. (1994), 'Smooth spline surfaces over irregular meshes', *Computer Graphics* pp. 303–310.

Lorensen, W. and Cline, H. (1987), 'Marching cubes: a high resolution 3D surface construction algorithm', *Computer Graphics* **21**, 163–169.

Lowe, D. (1985), *Perceptual Organization and Visual Recognition*, Kluwer.

Luenberger, D. (1984), *Linear and Nonlinear Programming*, Addison-Wesley. Second edition.

Luong, Q.-T. (1992), Matrice fondamentale et calibration visuelle sur l'environnement: vers une plus grande autonomie des systèmes robotiques, PhD thesis, University of Paris XI, Orsay, France.

Luong, Q.-T. and Faugeras, O. (1996), 'The fundamental matrix: theory, algorithms, and stability analysis', *Int. J. of Comp. Vision* **17**(1), 43–76.

Luong, Q.-T., Deriche, R., Faugeras, O. and Papadopoulo, T. (1993), On determining the fundamental matrix: analysis of different methods and experimental results, Technical Report 1894, INRIA Sophia-Antipolis.

Lynch, D. and Livingston, W. (2001), *Color and Light in Nature*, Cambridge University Press.

Lyvers, E. and Mitchell, O. (1988), 'Precision edge contrast and orientation estimation', *IEEE Trans. Pattern Analysis and Machine Intelligence* **10**(6), 927–937.

Ma, W. and Manjunath, B. (1995), A comparison of wavelet features for texture annotation, *in* 'Proceedings, International Conference on Image Processing', 1995, pp. II: 256–259.

Ma, W. and Manjunath, B. (1996), Texture features and learning similarity, *in* 'IEEE Conference on Computer Vision and Pattern Recognition', 1996, pp. 425–430.

Ma, W. and Manjunath, B. (1997*a*), Netra: a toolbox for navigating large image databases, *in* 'IEEE Int. Conf. Image Processing', pp. 568–571.

Ma, W. and Manjunath, B. (1998), 'A texture thesaurus for browsing large aerial photographs', *Journal of the American Society for Information Science (special issue on AI Techniques for Emerging Information Systems Applications)* **49**(7), 633–648.

Ma, W. Y. and Manjunath, B. S. (1997*b*), Edgeflow: a framework for boundary detection and image segmentation, *in* 'IEEE Conf. on Computer Vision and Pattern Recognition', pp. 744–749.

MacAdam, D. (1942), 'Visual sensitivities to small color differences in daylight', *Journal of the Optical Society of America* **32**, 247.

Macaulay, F. (1916), *The Algebraic Theory of Modular Systems*, Cambridge University Press.

Mahamud, S. and Hebert, M. (2000), Iterative projective reconstruction from multiple views, *in* 'Proc. IEEE Conf. Comp. Vision Patt. Recog.', pp. II–430–437.

Mahamud, S., Hebert, M., Omori, Y. and Ponce, J. (2001), Provably-convergent iterative methods for projective structure from motion, *in* 'Proc. IEEE Conf. Comp. Vision Patt. Recog.', pp. 1018–1025.

Maintz, J. and Viergever, M. (1998), 'A survey of medical image registration', *Medical Image Analysis* **2**(1), 1–16.

Malik, J. and Perona, P. (1989), A computational model of texture segmentation, *in* 'IEEE Conference on Computer Vision and Pattern Recognition', 1989, pp. 326–332.

Malik, J. and Perona, P. (1990), 'Preattentive texture discrimination with early visual mechanisms', *J. Opt. Soc. America* **7A**(5), 923–932.

Malik, J. and Rosenholtz, R. (1997), 'Computing local surface orientation and shape from texture for curved surfaces', *Int. J. Computer Vision* pp. 149–168.

Maloney, L. (1984), Computational Approaches to Color Vision, PhD thesis, Stanford University.

Maloney, L. (1986), 'Evaluation of linear models of surface spectral reflectance with small numbers of parameters', *Journal of the Optical Society of America* **3**(10), 1673–1683.

Maloney, L. and Wandell, B. (1986), 'Color constancy: A method for recovering surface spectral reflectance', *Journal of the Optical Society of America* **3**, 29–33.

Manjunath, B. and Chellappa, R. (1991), 'Unsupervised texture segmentation using markov random field models', *IEEE Trans. Pattern Analysis and Machine Intelligence* **13**(5), 478–482.

Manjunath, B. and Ma, W. (1996*a*), Browsing large satellite and aerial photographs, *in* 'IEEE Int. Conf. Image Processing'.

Manjunath, B. and Ma, W. (1996*b,c*), 'Texture features for browsing and retrieval of image data', *IEEE Trans. Pattern Analysis and Machine Intelligence* **18**(8), 837–842.

Manning, C. and Schütze, H. (1999), *Foundations of Statistical Natural Language Processing*, MIT Press.

Manocha, D. (1992), Algebraic and Numeric Techniques for Modeling and Robotics, PhD thesis, Computer Science Division, Univ. of California, Berkeley.

Marimont, D. and Wandell, B. (1992), 'Linear models of surface and illuminant spectra', *J. Opt. Soc. Am.-A* **9**, 1905–1913.

Marr, D. (1977), 'Analysis of occluding contour', *Proc. Royal Society, London* **B-197**, 441–475.

Marr, D. (1982), *Vision*, Freeman.

Marr, D. and Hildreth, E. (1980), 'Theory of edge detection', *Proceedings of Royal Society of London* **B-207**, 187–217.

Marr, D. and Nishihara, K. (1978), 'Representation and recognition of the spatial organization of three-dimensional shapes', *Proc. Royal Society, London* **B-200**, 269–294.

Marr, D. and Poggio, T. (1976), 'Cooperative computation of stereo disparity', *Science* **194**, 283–287.

Marr, D. and Poggio, T. (1979), 'A computational theory of human stereo vision', *Proceedings of the Royal Society of London* **B 204**, 301–328.

Martin, W. and Aggarwal, J. (1983), 'Volumetric description of objects from multiple views', *IEEE Trans. Patt. Anal. Mach. Intell.* **5**(2), 150–158.

Maxwell, B. and Shafer, S. (2000), 'Segmentation and interpretation of multicolored objects with highlights', *Computer Vision and Image Understanding* **77**(1), 1–24.

Maybank, S. and Faugeras, O. (1992), 'A theory of self-calibration of a moving camera', *Int. J. of Comp. Vision* **8**(2), 123–151.

Maybank, S. and Sturm, P. (1999), MDL, collineations and the fundamental matrix, *in* 'British Machine Vision Conference'.

McCann, J., McKee, S. and Taylor (1976), 'Quantitative studies in retinex theory', *Vision Research* **16**, 445–458.

McInerney, T. and Terzopolous, D. (1996), 'Deformable models in medical image analysis: a survey', *Medical Image Analysis* **1**(2), 91–108.

McKee, S., Levi, D. and Brown, S. (1990), 'The imprecision of stereopsis', *Vision Research* **30**(11), 1763–1779.

McLachlan, G. and Krishnan, T. (1996), *The EM Algorithm and Extensions*, John Wiley & Sons.

McMillan, L. and Bishop, G. (1995), Plenoptic modeling: an image-based rendering approach, *in* 'SIGGRAPH', pp. 39–46.

Medioni, G. and Nevatia, R. (1984), 'Matching images using linear features', *IEEE Trans. Patt. Anal. Mach. Intell.* **6**(6), 675–685.

Meer, P., Mintz, D., Kim, D. and Rosenfeld, A. (1991), 'Robust regression methods for computer vision: A review', *International Journal of Computer Vision* **6**(1), 59–70.

Milenkovic, V. and Kanade, T. (1985), Trinocular vision using photometric and edge orientation constraints, *in* 'Proc. DARPA Image Understanding Workshop', pp. 163–175.

Minnaert, M. (1993), *Light and Color in the Outdoors*, Springer Verlag. Translator: L. Seymour.

Mohan, R. and Nevatia, R. (1989), Segmentation and description based on perceptual organization, *in* 'IEEE Conference on Computer Vision and Pattern Recognition', 1989, pp. 333–341.

Mohan, R. and Nevatia, R. (1992), 'Perceptual organization for scene segmentation and description', *IEEE Trans. Pattern Analysis and Machine Intelligence* **14**(6), 616–635.

Mohr, R., Morin, L. and Grosso, E. (1992), Relative positioning with uncalibrated cameras, *in* J. Mundy and A. Zisserman, eds, 'Geometric Invariance in Computer Vision', MIT Press, pp. 440–460.

Mollon, J. (1995), Seeing colour, *in* T. Lamb and J. Bourriau, eds, 'Colour Art and Science', Cambridge University Press.

Montel, P., ed. (1972), *Toute la photographie*, Librairie Larousse and Publications Montel.

Morgan, A. (1987), *Solving Polynomial Systems using Continuation for Engineering and Scientific Problems*, Prentice Hall.

Morita, T. and Kanade, T. (1997), 'A sequential factorization method for recovering shape and motion from image sequences', *IEEE Trans. Patt. Anal. Mach. Intell.*

Mukawa, N. (1990), Estimation of shape, reflection coefficients and illuminant direction from image sequences, *in* 'Proceedings, Third International Conference on Computer Vision', 1990, pp. 507–512.

Mumford, D. and Shah, J. (1985), Boundary detection by minimizing functionals, *in* 'IEEE Conference on Computer Vision and Pattern Recognition', IEEE Press, pp. 22–26.

Mumford, D. and Shah, J. (1988), 'Optimal approximations by piecewise smooth functions and variational problems', *Communications on Pure and Applied Mathematics* **XLII**(5), 577–685.

Mundy, J. and Vrobel, P. (1994), The role of IU technology in RADIUS phase ii, *in* 'Proc. Image Understanding Workshop', pp. 251–264.

Mundy, J. and Zisserman, A. (1992), *Geometric Invariance in Computer Vision*, MIT Press.

Mundy, J., Zisserman, A. and Forsyth, D. (1993), *Applications of Invariance in Computer Vision*, Springer-Verlag.

Nalwa, V. (1987), Line-drawing interpretation: bilateral symmetry, *in* 'Proc. DARPA Image Understanding Workshop', pp. 956–967.

Nalwa, V. (1988), 'Line-drawing interpretation: A mathematical framework', *Int. J. of Comp. Vision* **2**, 103–124.

Nathans, J., Piantanida, T., Eddy, R., Shows, T. and Hogness, D. (1986*a*), 'Molecular genetics of inherited variation in human color vision', *Science* **232**, 203–210.

Nathans, J., Thomas, D. and Hogness, D. (1986*b*), 'Molecular genetics of human color vision: The genes encoding blue, green, and red pigments', *Science* **232**, 193–203.

Navy, U. (1969), *Basic Optics and Optical Instruments*, Dover. Prepared by the Bureau of Naval Personnel.

Nayar, S. (1997), Catadioptric omnidirectional camera, *in* 'Proc. IEEE Conf. Comp. Vision Patt. Recog.', pp. 482–488.

Nayar, S. and Oren, M. (1993), Diffuse reflectance from rough surfaces, *in* 'IEEE Conference on Computer Vision and Pattern Recognition', 1993, pp. 763–764.

Nayar, S. and Oren, M. (1995), 'Visual appearance of matte surfaces', *Science* **267**(5201), 1153–1156.

Nayar, S., Ikeuchi, K. and Kanade, T. (1990), 'Determining shape and reflectance of hybrid surfaces by photometric sampling', *IEEE Trans. Robotics and Automation* **6**(4), 418–431.

Nayar, S., Ikeuchi, K. and Kanade, T. (1991*a*), 'Shape from interreflections', *International Journal of Computer Vision* **6**(3), 173–195.

Nayar, S., Ikeuchi, K. and Kanade, T. (1991*b*), 'Surface reflection: Physical and geometrical perspectives', *IEEE Trans. Pattern Analysis and Machine Intelligence* **13**(7), 611–634.

Nevatia, R. (1986), Image segmentation, *in* K. Fu and T. Young, eds, 'Handbook of Pattern Recognition and Image Processing', Academic Press, pp. 215–231.

Nevatia, R. and Binford, T. (1977), 'Description and recognition of complex curved objects', *Artificial Intelligence Journal* **8**, 77–98.

Nielsen, M., Johansen, P., Olsen, O. F. and Weickert, J., eds (1999), *Scale-Space Theory in Computer Vision*, Vol. 1682, Springer Verlag LNCS.

Niem, W. and Buschmann, R. (1994), Automatic modelling of 3D natural objects from multiple views, *in* 'European Workshop on Combined Real and Synthetic Image Processing for Broadcast and Video Production'.

Nitzan, D. (1988), 'Three-dimensional vision structure for robot applications', *IEEE Trans. Patt. Anal. Mach. Intell.* **10**(3), 291–309.

Noble, A., Wilson, D. and Ponce, J. (1997), 'On Computing Aspect Graphs of Smooth Shapes from Volumetric Data', *Computer Vision and Image Understanding: special issue on Mathematical Methods in Biomedical Image Analysis* **66**(2), 179–192.

Ogle, V. and Stonebraker, M. (1995), 'Chabot: retrieval from a relational database of images', *Computer* **28**, 40–48.

Ohlander, R., Price, K. and Reddy, R. (1978), 'Picture segmentation by a recursive region splitting method', *Computer Graphics Image Processing* **8**, 313–333.

Ohta, Y. and Kanade, T. (1985), 'Stereo by intra- and inter-scanline search', *IEEE Trans. Patt. Anal. Mach. Intell.* **7**(2), 139–154.

Ohta, Y., Maenobu, K. and Sakai, T. (1981), Obtaining surface orientation from texels under perspective projection, *in* 'Proc. International Joint Conference on Artificial Intelligence', pp. 746–751.

Oja, E. (1983), *Subspace methods of pattern recognition*, Research Study Press.

Okutami, M. and Kanade, T. (1993), 'A multiple-baseline stereo system', *IEEE Trans. Patt. Anal. Mach. Intell.* **15**(4), 353–363.

Olson, C. (1998), Variable-scale smoothing and edge detection guided by stereoscopy, *in* 'IEEE Conference on Computer Vision and Pattern Recognition', 1998, pp. 80–85.

Oren, M. and Nayar, S. (1995), 'Generalization of the lambertian model and implications for machine vision', *International Journal of Computer Vision* **14**(3), 227–251.

Oren, M., Papageorgiou, C., Sinha, P., Osuna, E. and Poggio, T. (1997), Pedestrian detection using wavelet templates, *in* 'IEEE Conference on Computer Vision and Pattern Recognition', 1997, pp. 193–199.

Osuna, E., Freund, R. and Girosi, F. (1997), Training support vector machines: An application to face detection, *in* 'IEEE Conference on Computer Vision and Pattern Recognition', 1997, pp. 130–136.

Pae, S. and Ponce, J. (1999), Toward a scale-space aspect graph: Solids of revolution, *in* 'Proc. IEEE Conf. Comp. Vision Patt. Recog.', Vol. II, pp. 196–201.

Pae, S. and Ponce, J. (2001), 'On computing structural changes in evolving surfaces and their appearance', *Int. J. of Comp. Vision* **43**(2), 113–131.

Pal, N. and Pal, S. (1993), 'A review on image segmentation techniques', *Pattern Recognition* **26**(9), 1277–1294.

Palmer, S. (1999), *Vision Science : Photons to Phenomenology*, MIT Press.

Papageorgiou, C. and Poggio, T. (1999), A pattern classification approach to dynamical object detection, *in* 'Proceedings, Seventh International Conference on Computer Vision', 1999, pp. 1223–1228.

Papageorgiou, C. and Poggio, T. (2000), 'A trainable system for object detection', *International Journal of Computer Vision* **38**(1), 15–33.

Papageorgiou, C., Oren, M. and Poggio, T. (1998), A general framework for object detection, *in* 'Proceedings, Sixth International Conference on Computer Vision', 1998, pp. 555–562.

Park, J., Seo, J., An, D. and Chung, S. (2000), Detection of human faces using skin color and eyes, *in* 'Multimedia and Exposition', pp. 133–136.

Pentland, A. (1986), 'Perceptual organization and the representation of natural form', *Artificial Intelligence Journal* **28**, 293–331.

Pentland, A., Picard, R. and Sclaroff, S. (1996), 'Photobook: content-based manipulation of image databases', *Int. J. Computer Vision* **18**(3), 233–254.

Peri, V. and Nayar, S. (1997), Generation of perspective and panoramic video from omnidirectional video, *in* 'Proc. DARPA Image Understanding Workshop'.

Perlin, K. (1985), An image synthesizer, *in* 'SIGGRAPH-85', pp. 287–296.

Perona, P. (1992), Steerable-scalable kernels for edge detection and junction analysis, *in* Sandini (1992), pp. 3–18.

Perona, P. (1995), 'Deformable kernels for early vision', *IEEE Trans. Pattern Analysis and Machine Intelligence* **17**(5), 488–499.

Perona, P. and Freeman, W. (1998), A factorization approach to grouping, *in* 'European Conference on Computer Vision', 1998, pp. 655–670.

Perona, P. and Malik, J. (1990*a,b*), Detecting and localizing edges composed of steps, peaks and roofs, *in* 'Proceedings, Third International Conference on Computer Vision', 1990, pp. 52–57.

Perona, P. and Malik, J. (1990*c*), 'Scale-space and edge detection using anisotropic diffusion', *IEEE Trans. Patt. Anal. Mach. Intell.* **12**(7), 629–639.

Petitjean, S. (1995), Géométrie énumérative et contacts de variétés linéaires: application aux graphes d'aspects d'objets courbes, PhD thesis, Institut National Polytechnique de Lorraine.

Petitjean, S., Ponce, J. and Kriegman, D. (1992), 'Computing exact aspect graphs of curved objects: Algebraic surfaces', *Int. J. of Comp. Vision* **9**(3), 231–255.

Phillips, P. and Vardi, Y. (1996), 'Efficient illumination normalization of facial images', *Pattern Recognition Letters* **17**(8), 921–927.

Plantinga, H. and Dyer, C. (1990), 'Visibility, occlusion, and the aspect graph', *Int. J. of Comp. Vision* **5**(2), 137–160.

Platonova, O. (1981), 'Singularities of the mutual disposition of a surface and a line', *Russian Mathematical Surveys* **36**(1), 248–249.

Poelman, C. and Kanade, T. (1997), 'A paraperspective factorization method for shape and motion recovery', *IEEE Trans. Patt. Anal. Mach. Intell.* **19**(3), 206–218.

Poggio, T., Torre, V. and Koch, C. (1985), 'Computational vision and regularization theory', *Nature* **317**, 314–319.

Pollard, S., Mayhew, J. and Frisby, J. (1970), 'A stereo correspondence algorithm using a disparity gradient limit', *Perception* **14**, 449–470.

Pollefeys, M. (1999), Self-calibration and metric 3D reconstruction from uncalibrated image sequences, PhD thesis, Katholieke Universiteit Leuven.

Pollefeys, M., Koch, R. and Van Gool, L. (1999), 'Self-calibration and metric reconstruction in spite of varying and unknown internal camera parameters', *Int. J. of Comp. Vision* **32**(1), 7–26.

Ponce, J. (1990), 'On characterizing ribbons and finding skewed symmetries', *Computer Vision, Graphics and Image Processing* **52**, 328–340.

Ponce, J. (2000), Metric upgrade of a projective reconstruction under the rectangular pixel assumption, *in* 'Second Workshop on Structure from Multiple Images of Large Scale Environments', pp. 18–27. Preprints.

Ponce, J. and Brady, J. (1987), Toward a surface primal sketch, *in* T. Kanade, ed., 'Three-dimensional machine vision', Kluwer Publishers, pp. 195–240.

Ponce, J., Cass, T. and Marimont, D. (1993), Relative stereo and motion reconstruction, Technical Report UIUC-BI-AI-RCV-93-07, Beckman Institute, University of Illinois.

Ponce, J., Cepeda, M., Pae, S. and Sullivan, S. (1999), Shape models and object recognition, *in* D. Forsyth, J. Mundy, V. Gesu and R. Cipolla, eds, 'Shape, Contour and Grouping in Computer Vision', Vol. 1681 of *Lecture Notes in Computer Science*, Springer-Verlag.

Ponce, J., Chelberg, D. and Mann, W. (1989), 'Invariant properties of straight homogeneous generalized cylinders and their contours', *IEEE Trans. Patt. Anal. Mach. Intell.* **11**(9), 951–966.

Porrill, J. (1990), 'Fitting ellipses and predicting confidence envelopes using a bias corrected kalman filter', *Image and Vision Computing* **8**(1), 37–41.

Pritchett, P. and Zisserman, A. (1998), Wide baseline stereo matching, *in* 'Proc. Int. Conf. Comp. Vision', pp. 754–760.

Psarrou, A., Konstantinou, V., Morse, P. and O'Reilly, P. (1997), Content based search in mediaeval manuscripts, *in* 'TENCON-97 - Proc. IEEE Region 10 Conf. Speech and Image Technologies for Computing and Telecommunications', pp. 187–190.

Raja, Y., McKenna, S. and Gong, S. (1998), Colour model selection and adaptation in dynamic scenes, *in* 'European Conference on Computer Vision', 1998, pp. 460–474.

Ranade, S. and Prewitt, J. (1980), A comparison of some segmentation algorithms for cytology, *in* 'Proceedings, International Conference on Pattern Recognition', pp. 561–564.

Rieger, J. (1987), 'On the classification of views of piecewise-smooth objects', *Image and Vision Computing* **5**, 91–97.

Rieger, J. (1990), 'The geometry of view space of opaque objects bounded by smooth surfaces', *Artificial Intelligence Journal* **44**(1-2), 1–40.

Rieger, J. (1992), 'Global bifurcations sets and stable projections of non-singular algebraic surfaces', *Int. J. of Comp. Vision* **7**(3), 171–194.

Ripley, B. (1996), *Pattern Recognition and Neural Networks*, Cambridge University Press.

Riseman, E. and Arbib, M. (1977), 'Computational techniques in the visual segmentation of static scenes', *Computer Graphics Image Processing* **6**(3), 221–276.

Rissanen, J. (1983), 'A universal prior for integers and estimation by minimum description length', *Annals of Statistics* **11**, 416–431.

Rissanen, J. (1987), 'Stochastic complexity (with discussion)', *J. Roy. Stat. Soc. Series B* **49**, 223–239.

Robert, L. and Faugeras, O. (1991), Curve-based stereo: figural continuity and curvature, *in* 'Proc. IEEE Conf. Comp. Vision Patt. Recog.', pp. 57–62.

Roberts, L. (1965), Machine perception of 3-D solids, *in* J. T. *et al.*, ed., 'Optical and Electro-Optical Information Processing', MIT Press, pp. 159–197.

Rosch, E. (1988), Principles of categorisation, *in* A. Collins and E. Smith, eds, 'Readings in Cognitive Science: A Perspective from Psychology and Artificial Intelligence', Morgan Kauffman, pp. 312–322.

Rosenfeld, A. (1986), 'Axial representations of shape', *Computer Vision, Graphics and Image Processing* **33**, 156–173.

Rosenholtz, R. and Malik, J. (1997), 'Surface orientation from texture: isotropy or homogeneity (or both)?', *Vision Research* **37**(16), 2283–2293.

Rothwell, C. (1995), *Object Recognition through Invariant Indexing*, Oxford University Press.

Rothwell, C., Forsyth, D., Zisserman, A. and Mundy, J. (1993), Extracting projective structure from single perspective views of 3d point sets, *in* 'Proceedings, Fourth International Conference on Computer Vision', pp. 573–582.

Rothwell, C., Zisserman, A., Forsyth, D. and Mundy, J. (1995), 'Planar object recognition using projective shape representation', *International Journal of Computer Vision* **16**(1), 57–99.

Rothwell, C., Zisserman, A., Marinos, C., Forsyth, D. and Mundy, J. (1992), 'Relative motion and pose from arbitrary plane curves', *Image and Vision Computing* **10**, 250–262.

Rousseeuw, P. (1987), *Robust Regression and Outlier Detection*, Wiley.

Rowley, H., Baluja, S. and Kanade, T. (1996), Neural network-based face detection, *in* 'IEEE Conference on Computer Vision and Pattern Recognition', 1996, pp. 203–208.

Rowley, H., Baluja, S. and Kanade, T. (1998*a*), 'Neural network-based face detection', *IEEE Trans. Pattern Analysis and Machine Intelligence* **20**(1), 23–38.

Rowley, H., Baluja, S. and Kanade, T. (1998*b*), Rotation invariant neural network-based face detection, *in* 'IEEE Conference on Computer Vision and Pattern Recognition', 1998, pp. 38–44.

Rowley, H., Baluja, S. and Kanade, T. (1998*c*), Rotation invariant neural network-based face detection, *in* 'IEEE Conf. on Computer Vision and Pattern Recognition', pp. 38–44.

Rubner, Y., Tomasi, C. and Guibas, L. J. (1998), A metric for distributions with applications to image databases, *in* 'Int. Conf. on Computer Vision', pp. 59–66.

Russ, J. (1995), *The Image Processing Handbook*, CRC Press. Second edition.

Sabin, M. (1994), *Acta Numerica 1994*, Cambridge University Press, chapter Numerical Geometry of Surfaces.

Saito, H., Baba, S., Kimura, M., Vedula, S. and Kanade, T. (1999), Appearance-based virtual view generation of temporally-varying events from multi-camera images in the 3D room, Technical Report CMU-CS-99-127, Carnegie-Mellon University.

Samuel, P. (1988), *Projective Geometry*, Springer-Verlag. English translation of "Géométrie Projective", Presses Universitaires de France, 1986.

Sandini, G., ed. (1992), *European Conference on Computer Vision*, Vol. 588 of *Lecture Notes in Computer Science*, Springer-Verlag.

Sarachik, K. and Grimson, W. (1993), Gaussian error models for object recognition, *in* 'IEEE Conference on Computer Vision and Pattern Recognition', 1993, pp. 400–406.

Sarkar, S. and Boyer, K. (1993), 'Integration, inference, and management of spatial information using Bayesian networks: Perceptual organization', *IEEE Trans. Pattern Analysis and Machine Intelligence* **15**(3), 256–274.

Sarkar, S. and Boyer, K. (1994), *Computing Perceptual Organization in Computer Vision*, World Scientific.

Sarkar, S. and Boyer, K. (1998), 'Quantitative measures of change based on feature organization: Eigenvalues and eigenvectors', *Computer Vision and Image Understanding* **71**(1), 110–136.

Saund, E. and Moran, T. (1995), Perceptual organization in an interactive sketch editing application, *in* 'Proceedings, Fifth International Conference on Computer Vision', 1995, pp. 597–604.

Sawhney, H. and Ayer, S. (1996), 'Compact representations of videos through dominant and multiple motion estimation', *IEEE T. Pattern Analysis and Machine Intelligence* **18**(8), 814–830.

Schmid, C. and Mohr, R. (1997*a,b*), 'Local grayvalue invariants for image retrieval', *IEEE Trans. Patt. Anal. Mach. Intell.* **19**(5), 530–535.

Schmid, C., Mohr, R. and Bauckhage, C. (2000), 'Evaluation of interest point detectors', *International Journal of Computer Vision* **37**(2), 151–172.

Schneiderman, H. and Kanade, T. (1998), Probabilistic formulation for object recognition, *in* 'IEEE Conference on Computer Vision and Pattern Recognition', 1998, pp. 45–51.

Seitz, S. and Dyer, C. (1995), Physically-valid view synthesis by image interpolation, *in* 'Workshop on Representations of Visual Scenes'.

Seitz, S. and Dyer, C. (1996), Toward image-based scene representation using view morphing, Technical Report 1298, University of Wisconsin.

Serra, J. (1982), *Image Analysis and Mathematical Morphology*, Academic Press.

Shade, J., Gortler, S., Li-wei, H. and Szeliski, R. (1998), Layered depth images, *in* 'SIGGRAPH 98', pp. 231–242.

Shafer, S. (1985*a*), *Shadows and Silhouettes in Computer Vision*, Kluwer Academic Publishers.

Shafer, S. (1985*b*), 'Using color to separate reflection components', *Color Res. App.* **10**(4), 210–218.

Shafer, S. and Kanade, T. (1983), The theory of straight homogeneous generalized cylinders and a taxonomy of generalized cylinders, Technical Report CMU-CS-83-105, Carnegie-Mellon University.

Shashua, A. (1993), Projective depth: a geometric invariant for 3D reconstruction from two perspective/orthographic views and for visual recognition, *in* 'Proc. Int. Conf. Comp. Vision', pp. 583–590.

Shashua, A. (1995), 'Algebraic functions for recognition', *IEEE Trans. Patt. Anal. Mach. Intell.* **17**(8), 779–789.

Shi, J. and Malik, J. (1997), Normalized cuts and image segmentation, *in* 'IEEE Conference on Computer Vision and Pattern Recognition', 1997, pp. 731–737.

Shi, J. and Malik, J. (1998*a*), Motion segmentation and tracking using normalized cuts, *in* 'Proceedings, Sixth International Conference on Computer Vision', 1998, pp. 1154–1160.

Shi, J. and Malik, J. (1998*b*), Self-inducing relational distance and its application to image segmentation, *in* 'European Conference on Computer Vision', 1998, pp. 528–543.

Shi, J. and Malik, J. (2000), 'Normalized cuts and image segmentation', *IEEE Trans. Pattern Analysis and Machine Intelligence* **22**(8), 888–905.

Shimshoni, I. and Ponce, J. (1997), 'Finite-resolution aspect graphs of polyhedral objects', *IEEE Trans. Patt. Anal. Mach. Intell.* **19**(4), 315–327.

Shin, M., Goldgof, D. and Bowyer, K. (1998), An objective comparison methodology of edge detection algorithms using a structure from motion task, *in* 'IEEE Conference on Computer Vision and Pattern Recognition', 1998, pp. 190–195.

Shin, M., Goldgof, D. and Bowyer, K. (1999), Comparison of edge detectors using an object recognition task, *in* 'IEEE Conference on Computer Vision and Pattern Recognition', 1999, pp. I:360–365.

Shirai, Y. (1972), 'Recognition of polyhedrons with a range finder', *Pattern Recognition* **4**, 243–250.

Shirman, L. and Sequin, C. (1987), 'Local surface interpolation with Bezier patches', *CAGD* **4**, 279–295.

Shum, H. and Szeliski, R. (1998), Construction and refinement of panoramic mosaics with global and local alignment, *in* 'Proc. Int. Conf. Comp. Vision', pp. 953–958.

Shum, H., Ikeuchi, K. and Reddy, R. (1995), 'Principal component analysis with missing data and its application to polyhedral object modeling', *IEEE Trans. Patt. Anal. Mach. Intell.* **17**(9), 854–867.

Siddiqi, K., Kimia, B., Tannenbaum, A. and Zucker, S. (1999*a*), 'Shapes, shocks and wiggles', *Image and Vision Computing* **17**(5/6), 365–373.

Siddiqi, K., Shokoufandeh, A., Dickinson, S. J. and Zucker, S. W. (1999*b*), 'Shock graphs and shape matching', *Int. J. of Comp. Vision* **35**(1), 13–32.

Sillion, F. (1994), *Radiosity and Global Illumination*, Morgan-Kauffman.

Simon, D., Hebert, M. and Kanade, T. (1994), Real-time 3D pose estimation using a high-speed range sensor, *in* 'IEEE Int. Conf. on Robotics and Automation', pp. 2235–2241.

Simoncelli, E. and Farid, H. (1995), Steerable wedge filters, *in* 'Proceedings, Fifth International Conference on Computer Vision', 1995, pp. 189–194.

Simoncelli, E. and Freeman, W. (1995), The steerable pyramid: A flexible architecture for multi-scale derivative computation, *in* 'Proceedings, International Conference on Image Processing', 1995, pp. 444–447.

Sirovitch, L. and Kirby, M. (1987), 'Low-dimensional procedure for the characterization of human faces', *J. Opt. Soc. Am. A* **2**, 586–591.

Slama, C., Theurer, C. and Henriksen, S., eds (1980), *Manual of photogrammetry*, American Society of Photogrammetry. Fourth edition.

Smith, J. and Chang, S.-F. (1996), Visualseek: A fully automated content-based image query system, *in* 'ACM Multimedia Conference'.

Smith, J. and Chang, S.-F. (1997), 'Visually searching the web for content', *IEEE Multimedia* **4**(3), 12–20.

Smith, M. and Christel, M. (1995), Automating the creation of a digital video library, *in* 'ACM Multimedia'.

Smith, M. and Hauptmann, A. (1995), Text, speech and vision for video segmentation: The informedia project, *in* 'AAAI Fall 1995 Symposium on Computational Models for Integrating Language and Vision'.

Smith, M. and Kanade, T. (1997), Video skimming for quick browsing based on audio and image characterization, *in* 'IEEE Conf. on Computer Vision and Pattern Recognition'.

Smith, T. (1996), 'A digital library for geographically referenced materials', *Computer* **29**(5), 54–60.

et al., A. S., ed. (2000), *Advances in Large Margin Classifiers*, MIT Press.

Snapper, E. and Troyer, R. (1989), *Metric Affine Geometry*, Dover Publications Inc. Reprinted from Academic Press, 1971.

Snyder, D., Hammoud, A. and White, R. (1993), 'Image recovery from data acquired with a charge-coupled-device camera', *J. Opt. Soc. Am. A* **10**(5), 1014–1023.

Song, Y., Feng, X. and Perona, P. (2000*a*), Towards detection of human motion, *in* 'IEEE Conference on Computer Vision and Pattern Recognition', 2000, pp. I:810–817.

Song, Y., Goncalves, L. and Perona, P. (2000*b*), Monocular percepton of biological motion: Clutter and partial occlusion, *in* 'European Conference on Computer Vision', 2000, pp. 719–733.

Song, Y., Goncalves, L., di Bernardo, E. and Perona, P. (1999), Monocular perception of biological motion: Detection and labeling, *in* 'Proceedings, Seventh International Conference on Computer Vision', 1999, pp. 805–813.

Spacek, L. (1986), 'Edge detection and motion detection', *Image and Vision Computing* **4**(1), 43–56.

Speis, A. and Healey, G. (1996), 'Feature-extraction for texture-discrimination via random-field models with random spatial interaction', *IEEE Trans. Image Processing* **5**(4), 635–645.

Spetsakis, M. and Aloimonos, Y. (1990), 'Structure from motion using line correspondences', *Int. J. of Comp. Vision* **4**(3), 171–183.

Srivastava, S. and Ahuja, N. (1990), 'Octree generation from object silhouettes in perspective views', *Computer Vision, Graphics and Image Processing* **49**(1), 68–84.

Stark, L. and Bowyer, K. (1996), *Generic Object Recognition Using Form and Function*, World Scientific Publishing.

Starner, T., Weaver, J. and Pentland, A. (1998), 'Real-time american sign language recognition using desk and wearable computer based video.', *IEEE T. Pattern Analysis and Machine Intelligence* **20**(12), 1371–1375.

Stein, F. and Medioni, G. (1992), 'Structural indexing: efficient 3D object recognition', *IEEE Trans. Patt. Anal. Mach. Intell.*

Stewart, C. (1999), 'Robust parameter estimation in computer vision', *SIAM-Review* **41**(3), 513–537.

Stewman, J. and Bowyer, K. (1987), Aspect graphs for planar-face convex objects, *in* 'Proc. IEEE Workshop on Computer Vision', pp. 123–130.

Stewman, J. and Bowyer, K. (1988), Creating the perspective projection aspect graph of polyhedral objects, *in* 'Proc. Int. Conf. Comp. Vision', pp. 495–500.

Strang, G. (1980), *Linear Algebra and its Applications*, Academic Press, Inc. Second edition.

Struik, D. (1988), *Lectures on Classical Differential Geometry*, Dover. Reprint of the second edition (1961) of the work first published by Addison-Wesley in 1950.

Sturm, P. and Triggs, B. (1996), A factorization-based algorithm for multi-image projective structure and motion, *in* 'Proc. European Conf. Comp. Vision', pp. 709–720.

Sugihara, K. (1986), *Machine Interpretation of Line Drawings*, MIT Press.

Sullivan, G., Baker, K., Worrall, A., Attwood, C. and Remagnino, P. (1997), 'Model-based vehicle detection and classification using orthographic approximations', *Image and Vision Computing* **15**(8), 649–654.

Sullivan, S. and Ponce, J. (1998), 'Automatic model construction, pose estimation, and object recognition from photographs using triangular splines', *IEEE Trans. Patt. Anal. Mach. Intell.* **20**(10), 1091–1096.

Sullivan, S., Sandford, L. and Ponce, J. (1994*a,b*), 'Using geometric distance fits for 3D object modelling and recognition', *IEEE Trans. Patt. Anal. Mach. Intell.* **16**(12), 1183–1196.

Sung, K.-K. and Poggio, T. (1998), 'Example-based learning for view-based human face detection', *IEEE T. Pattern Analysis and Machine Intelligence* **20**, 39–51.

Swain, M. and Ballard, D. (1991), 'Color indexing', *Int. J. Computer Vision* **7**(1), 11–32.

Szeliski, R., Avidan, S. and Anandan, P. (2000), Layer extraction from multiple images containing reflections and transparency, *in* 'IEEE Conference on Computer Vision and Pattern Recognition', 2000, pp. I:246–253.

Tagare, H. and de Figueiredo, R. (1991), 'A theory of photometric stereo for a class of diffuse non-Lambertian surfaces', *IEEE Trans. Pattern Analysis and Machine Intelligence* **13**(2), 133–152.

Tagare, H. and de Figueiredo, R. (1992), 'Simultaneous estimation of shape and reflectance map from photometric stereo', *CVGIP: Image Understanding* **55**(3), 275–286.

Tagare, H. and de Figueiredo, R. (1993), 'A framework for the construction of reflectance maps for machine vision', *CVGIP: Image Understanding* **57**(3), 265–282.

Tao, H., Sawhney, H. and Kumar, R. (2000), Dynamic layer representation with applications to tracking, *in* 'IEEE Conference on Computer Vision and Pattern Recognition', 2000, pp. II:134–141.

Tarr, M., Hayward, W., Gauthier, I. and Williams, P. (1995), 'Is object recognition mediated by viewpoint invariant parts or viewpoint dependent features', *Perception* **24**, 4.

Taubin, G., Cukierman, F., Sullivan, S., Ponce, J. and Kriegman, D. (1994*a,b*), 'Parameterized families of polynomials for bounded algebraic and surface curve fitting', *IEEE Trans. Patt. Anal. Mach. Intell.* **16**(3), 287–303.

Taylor, C. (2000), Reconstruction of articulated objects from point correspondences in a single uncalibrated image, *in* 'IEEE Conf. on Computer Vision and Pattern Recognition', pp. 677–684.

ter Haar Romeny, B. (1994), Geometry-driven diffusion in computer vision, *in* 'Geometry Driven Diffusion in Computer Vision', Kluwer Academic Press.

ter Haar Romeny, B., Florack, L. M., Koenderink, J. J. and Viergever, M. A., eds (1997), *Scale-Space Theory in Computer Vision*, Vol. 1252, Springer Verlag LNCS.

Terzopoulos, D. (1984), Multiresolution Computation of Visible-Surface Representations, PhD thesis, Massachusetts Institute of Technology.

Thom, R. (1972), *Structural Stability and Morphogenesis*, Benjamin.

Thompson, D. and Mundy, J. (1987), Three dimensional model matching from an unconstrained viewpoint, *in* 'International Conference on Robotics and Automation', pp. 208–220.

Thompson, M., Eller, R., Radlinski, W. and Speert, J., eds (1966), *Manual of Photogrammetry*, American Society of Photogrammetry. Third edition.

Tieu, K. and Viola, P. (2000), Boosting image retrieval, *in* 'IEEE Conference on Computer Vision and Pattern Recognition', 2000, pp. I:228–235.

Todd, J. (1946), *Projective and Analytical Geometry*, Pitman Publishing Corporation.

Tomasi, C. and Kanade, T. (1991), Factoring image sequences into shape and motion, *in* 'IEEE Workshop on Visual Motion', pp. 21–28.

Tomasi, C. and Kanade, T. (1992), 'Shape and motion from image streams under orthography: a factorization method', *Int. J. of Comp. Vision* **9**(2), 137–154.

Torr, P. (1997), An assessment of information criteria for motion model selection, *in* 'IEEE Conference on Computer Vision and Pattern Recognition', 1997, pp. 47–52.

Torr, P. (1999), Model selection for two view geometry: a review, *in* D. Forsyth, J. Mundy, V. diGesu and R. Cipolla, eds, 'Shape, Contour and Grouping in Computer Vision', Springer-Verlag, pp. 277–301.

Torr, P. and Murray, D. (1997), 'The development and comparison of robust methods for estimating the fundamental matrix', *International Journal of Computer Vision* **24**(3), 271–300.

Torr, P. and Zisserman, A. (1998), Concerning Bayesian motion segmentation, model averaging, matching and the trifocal tensor, *in* 'European Conference on Computer Vision', 1998, pp. 511–527.

Torr, P., Fitzgibbon, A. and Zisserman, A. (1999*a*), 'The problem of degeneracy in structure and motion recovery from uncalibrated image sequences', *International Journal of Computer Vision* **32**(1), 27–44.

Torr, P., Szeliski, R. and Anandan, P. (1999*b*), An integrated Bayesian approach to layer extraction from image sequences, *in* 'Proceedings, Seventh International Conference on Computer Vision', 1999, pp. 983–990.

Torrance, K. and Sparrow, E. (1967), 'Theory for off-specular reflection from roughened surfaces', *Journal of the Optical Society of America* **57**, 1105–1114.

Torre, V. and Poggio, T. (1986), 'On edge detection', *IEEE Trans. Pattern Analysis and Machine Intelligence* **8**(2), 147–163.

Triesman, A. (1982), 'Perceptual grouping and attention in visual search for features and objects', *Journal of Experimental Psychology: Human Perception and Performance* **8**(2), 194–214.

Triggs, B. (1995), Matching constraints and the joint image, *in* 'Proc. Int. Conf. Comp. Vision', pp. 338–343.

Triggs, B., McLauchlan, P., Hartley, R. and Fitzgibbon, A. (2000), Bundle adjustment—a modern synthesis, *in* B. Triggs, A. Zisserman and R. Szeliski, eds, 'Vision Algorithms: Theory and Practice', Springer-Verlag, pp. 298–372. Lecture Notes in Computer Science 1883.

Triggs, W. (1997), Auto-calibration and the absolute quadric, *in* 'Proc. IEEE Conf. Comp. Vision Patt. Recog.', pp. 609–614.

Trussell, H., Allebach, J., Fairchild, M., Funt, B. and Wong, P. (1997), 'Special issue: Digital color imaging', *IEEE Trans. Image Processing* **6**(7), 897–900.

Tsai, R. (1987*a*), 'A versatile camera calibration technique for high-accuracy 3D machine vision metrology using off-the-shelf TV cameras', *IEEE Journal of Robotics and Automation* **RA-3**(4), 323–344.

Tsai, R. (1987*b*), 'A versatile camera calibration technique for high-accuracy 3D machine vision metrology using off-the-shelf TV cameras', *Journal of Robotics and Automation* **RA-3**(4), 323–344.

Tsai, R. and Huang, T. (1984), 'Uniqueness and estimation of 3D motion parameters of rigid bodies with curved surfaces', *IEEE Trans. Patt. Anal. Mach. Intell.* **6**, 13–27.

Turk, G. and Levoy, M. (1994), Zippered polygon meshes from range images, *in* 'SIGGRAPH', pp. 311–318.

Turk, M. and Pentland, A. (1991*a*), 'Face recognition using eigenfaces', *J. of Cognitive Neuroscience*.

Turk, M. and Pentland, A. (1991*b*), Face recognition using eigenfaces, *in* 'IEEE Conf. on Computer Vision and Pattern Recognition', pp. 586–591.

Tyson, J. (1990), 'Progress in low-light-level charge-coupled device imaging in astronomy', *J. Opt. Soc. Am. A* **7**, 1231–1236.

Ullman, S. (1979), *The Interpretation of Visual Motion*, MIT Press.

Ullman, S. (1996), *High-Level Vision: Object Recognition and Visual Cognition*, MIT Press.

Ullman, S. and Basri, R. (1991), 'Recognition by linear combination of models', *IEEE Trans. Patt. Anal. Mach. Intell.* **13**(10), 992–1006.

Vaillant, R. and Faugeras, O. (1992), 'Using extremal boundaries for 3D object modeling', *IEEE Trans. Patt. Anal. Mach. Intell.* **14**(2), 157–173.

Vapnik, V. (1996), *The Nature of Statistical Learning Theory*, Springer Verlag.

Vapnik, V. N. (1998), *Statistical Learning Theory*, John Wiley & Sons.

Vasconcelos, N. and Lippman, A. (1997), Empirical Bayesian EM based motion segmentation, *in* 'IEEE Conference on Computer Vision and Pattern Recognition', 1997, pp. 527–532.

Viéville, T. and Faugeras, O. (1995), Motion analysis with a camera with unknown, and possibly varying intrinsic parameters, *in* 'Proc. Int. Conf. Comp. Vision', pp. 750–756.

Virage (n.d.), 'Virage home page at `http://www.virage.com/`'.

Vogler, C. and Metaxas, D. (1998), ASL recognition based on a coupling between HMMs and 3D motion analysis, *in* 'Proceedings, Sixth International Conference on Computer Vision', 1998, pp. 363–369.

Vogler, C. and Metaxas, D. (1999), Parallel hidden markov models for American sign language recognition, *in* 'Proceedings, Seventh International Conference on Computer Vision', 1999, pp. 116–122.

Vora, P., Farell, J., Tietz, J. and Brainard, D. (1997), Digital color cameras 1: Response models, Technical Report HPL-97-53, Hewlett-Packard Laboratory.

Wactlar, H., Kanade, T., Smith, M. and Stevens, S. (1996), 'Intelligent access to digital video: The informedia project', *IEEE Computer*.

Wallace, C. and Freeman, P. (1987), 'Estimation and inference by compact encoding (with discussion)', *J. Roy. Stat. Soc. Series B* **49**, 240–265.

Wandell, B. (1987), 'The synthesis and analysis of color images', *IEEE Trans. Pattern Analysis and Machine Intelligence* **9**(1), 2–13.

Wandell, B. (1995), *Foundations of Vision*, Sinauer Associates, Inc.

Wang, J. and Adelson, E. (1994), 'Representing moving images with layers', *IEEE Trans. Image Processing* **3**(5), 625–638.

Wang, R. and Freeman, H. (1990), Object recognition based on characteristic views, *in* 'International Conference on Pattern Recognition', pp. 8–12.

Watts, N. (1987), Calculating the principal views of a polyhedron, CS Tech. Report 234, Rochester University.

Weber, M., Einhaeuser, W., Welling, M. and Perona, P. (2000), Viewpoint-invariant learning and detection of human heads, *in* 'International Conference on Automatic Face and Gesture Recognition', 1900, pp. 20–27.

Weinshall, D. (1993), Model-based invariants for 3-D vision, *in* 'IEEE Conference on Computer Vision and Pattern Recognition', 1993, pp. 695–696.

Weinshall, D. and Tomasi, C. (1995), 'Linear and incremental acquisition of invariant shape models from image sequences', *IEEE Trans. Patt. Anal. Mach. Intell.*

Weiss, Y. (1997), Smoothness in layers: Motion segmentation using nonparametric mixture estimation, *in* 'IEEE Conference on Computer Vision and Pattern Recognition', 1997, pp. 520–526.

Weiss, Y. (1999), Segmentation using eigenvectors: A unifying view, *in* 'Proceedings, Seventh International Conference on Computer Vision', 1999, pp. 975–982.

Wells, W., Grimson, W., Kikinis, R. and Jolesz, F. (1996), 'Adaptive segmentation of MRI data', *IEEE Transactions on Medical Imaging* **15**(4), 429–442.

Weng, J., Huang, T. and Ahuja, N. (1992), 'Motion and structure from line correspondences: closed form solution, uniqueness, and optimization', *IEEE Trans. Patt. Anal. Mach. Intell.* **14**(3), 318–336.

West, J., Fitzpatrick, M., Wang, M., Dawant, B., Maurer, C.R., J., Kessler, R., Maciunas, R., Barillot, C., Lemoine, D., Collignon, A., Maes, F., Suetens, P., Vandermeulen, D., van den Elsen, P., Napel, S., Sumanaweera, T., Harkness, B., Hemler, P., Hill, D., Hawkes, D., Studholme, C., Antoine Maintz, J., Viergever, M., Malandain, G., Pennec, X., Noz, M., Maguire, G.Q., J., Pollack, M., Pelizzari, C., Robb, R., Hanson, D. and Woods, R. (1997), 'Comparison and evaluation of retrospective intermodality registration techniques', *J. Computer Assisted Tomography* **21**(4), 554–566.

West, M. and Harrison, J. (1997), *Bayesian Forecasting and Dynamic Models*, Springer Verlag.

Wheatstone, C. (1838), 'On some remarkable, and hitherto unobserved, phenomena of binocular vision', *Philosophical Transactions of the Royal Society (London)* **128**, 371–394.

Wheeler, M. and Ikeuchi, K. (1995), 'Probabilistic hypothesis generation and robust localization for object recognition', *IEEE Trans. Patt. Anal. Mach. Intell.* **17**(3), 252–265.

Whitney, H. (1955), 'On singularities of mappings of Euclidean spaces. I. Mappings of the plane into the plane', *Annals of Mathematics* **62**(3), 374–410.

Wilkinson, J. and Reinsch, C. (1971), *Linear Algebra - Vol. II of Handbook for Automatic Computation*, Springer-Verlag. Chapter I.10 by G.H. Golub and C. Reinsch.

Williams, L. and Chen, E. (1993), 'View interpolation for image synthesis', *SIGGRAPH*.

Williamson, S. and Cummins, H. (1983), *Light and Color in Nature and Art*, John Wiley & Sons.

Witkin, A. (1981), 'Recovering surface shape and orientation from texture', *Artificial Intelligence* **17**, 17–45.

Witkin, A. (1983), Scale-space filtering, *in* 'International Joint Conference on Artificial Intelligence', pp. 1019–1022.

Wolff, L., Nayar, S. and Oren, M. (1998), 'Improved diffuse reflection models for computer vision', *International Journal of Computer Vision* **30**(1), 55–71.

Wolfson, H. (1990), Model-based object recognition by geometric hashing, *in* 'European Conference on Computer Vision', 1990, pp. 526–536.

Wolfson, H. and Lamdan, Y. (1988), Geometric hashing: A general and efficient model-based recognition scheme, *in* 'Proceedings, Second International Conference on Computer Vision', 1988, pp. 238–249.

Wong, S. (1998), CBIR in medicine: still a long way to go, *in* 'IEEE Workshop on Content Based Access of Image and Video Libraries', p. 114.

Woodham, R. (1979), Analyzing curved surfaces using reflectance map techniques, *in* 'Artificial Intelligence: An MIT Perspective', MIT Press, pp. 161–182.

Woodham, R. (1980), 'Photometric method for determining surface orientation from multiple images', *Optical Engineering* **19**(1), 139–144.

Woodham, R. (1989), Determining surface curvature with photometric stereo, *in* 'International Conference on Robotics and Automation', pp. 36–42.

Woodham, R. (1994), 'Gradient and curvature from the photometric-stereo method, including local confidence estimation', *Journal of the Optical Society of America* **11**(11), 3050–3068.

Wu, Z. and Leahy, R. (1993), 'An optimal graph theoretic approach to data clustering: Theory and its application to image segmentation', *IEEE Trans. Pattern Analysis and Machine Intelligence* **15**(11), 1101–1113.

Wyszecki, G. and Stiles, W. (1982), *Color Science: Concepts and Methods, Quantitative Data and Formulas*, Wiley.

Yachida, M., Kitamura, Y. and Kimachi, M. (1986), Trinocular vision: new approach for correspondence problem, *in* 'Proceedings IAPR International Conference on Pattern Recognition', pp. 1041–1044.

Yasnoff, W., Mui, W. and Bacus, J. (1977), 'Error measures in scene segmentation', *Pattern Recognition* **9**(4), 217–231.

Yoo, T. and Oh, I. (1999), 'A fast algorithm for tracking human faces based on chromatic histograms', *Pattern Recognition Letters* **20**(10), 967–978.

Zerroug, M. and Medioni, G. (1995), The challenge of generic object recongnition, *in* M. Hebert, J. Ponce, T. Boult and A. Gross, eds, 'Object Representation for Computer Vision', number 994 *in* 'Lecture Notes in Computer Science', Springer-Verlag, pp. 217–232.

Zhang, Y. (1996), 'A survey on evaluation methods for image segmentation', *Pattern Recognition* **29**(8), 1335–1346.

Zhang, Z. (1994), 'Iterative point matching for registration of free-form curves and surfaces', *Int. J. of Comp. Vision* **13**(2), pages 119–152.

Zhang, Z., Deriche, R., Faugeras, O. and Luong, Q.-T. (1995), 'A robust technique for matching two uncalibrated images through the recovery of the unknown epipolar geometry', *Artificial Intelligence Journal* **78**, 87–119.

Zhu, S. (1999), 'Stochastic jump-diffusion process for computing medial axes in markov random fields', *IEEE Trans. Pattern Analysis and Machine Intelligence* **21**(11), 1158–1169.

Zhu, S. and Yuille, A. (1996), 'FORMS: a flexible object recognition and modeling system', *Int. J. of Comp. Vision* **20**(3), 187–212.

Zhu, S., Wu, Y. and Mumford, D. (1998), 'Filters, random-fields and maximum-entropy (frame): Towards a unified theory for texture modeling', *International Journal of Computer Vision* **27**(2), 107–126.

Zisserman, A., Forsyth, D., Mundy, J., Rothwell, C., Liu, J. and Pillow, N. (1995*a*), '3d object recognition using invariance', *Artificial Intelligence* **78**(1-2), 239–288.

Zisserman, A., Mundy, J., Forsyth, D., Liu, J., Pillow, N., Rothwell, C. and Utcke, S. (1995*b*), Class-based grouping in perspective images, *in* 'Proceedings, Fifth International Conference on Computer Vision', 1995, pp. 183–188.

Ziv, J. and Lempel, A. (1977), 'A universal algorithm for sequential data compression', *IEEE Transactions on Information Theory* **IT-23**, 337–343.

Index